CESMM4
Carbon & Price Book
2013

Edited by
Mott MacDonald

CESMM4 Carbon & Price Book 2013

Published by ICE Publishing, One Great George Street, Westminster, London SW1P 3AA, UK.

Published 2013

A CIP catalogue record for this book is available from the British Library
ISBN: 978-0-7277-5812-5
ISSN: 02687690

Printed and bound in Great Britain by CPI Group (UK) Ltd, Croydon, CR0 4YY

Foreword

Formally known as the CESMM3 Price Database, The Institution of Civil Engineers' CESMM4 Carbon & Price Book is now in its eighth edition. Building upon its standard in the provision of civil engineering cost information, in this latest edition the name has changed to reflect the latest Civil Engineering Standard Method of Measurement (CESMM4).

The CESMM Price Database evolved from a document known as the Wessex Database for Civil Engineering. The originator of this publication was Richard McGill, who was a partner of EC Harris at the time. Richard not only developed the Wessex Database of Civil Engineering in 1985, but also led the team that produced the first edition of the CESMM3 Price Database in 1992.

It has now been thoroughly revised and updated based the latest version of the standard method of measurement (CESMM4). The data for this new publication has been compiled using the unparalleled industry knowledge and expertise of Mott MacDonald, one of the UK's foremost whole life cost and carbon advisers. In particular their dedicated Economic Research Unit (www.economic-research-unit.com) that produces high quality cost and carbon data for many price books and databases for all sectors of the construction industry.

In 2010, James Fiske, project director of Mott MacDonald, was responsible for introducing the carbon values, and for the first time establishing a link between cost and carbon measurement in the civil engineering industry. In this latest edition, James has facilitated the inclusion of the wealth of carbon data held by the BRE to provide this information to the reader. The research collaboration between the Institution of Civil Engineering, Mott MacDonald and BRE, sets a new standard in the availability of cost and carbon data for the civil engineering industry.

The CESMM4 Carbon & Price Book should be on the desk of every contractor, specifyer, estimator, engineer, quantity surveyor and project manager actively involved in civil engineering work. It is the only definitive guide to publish estimates of cost and carbon emissions, allowing reference to both to enable an optimum design solution.

The information will undoubtedly be used for appraisal purposes at various stages in civil engineering contracts and will be an invaluable cost and carbon management tool. However, whilst every effort has been made to ensure the accuracy of the data, it must be borne in mind that resource availability, location, procurement nature and size of project etc. will all have a bearing on the cost and carbon values recorded for any individual project. The CESMM4 data may only be regarded as 'typical' under 'normal conditions' and are current the first quarter of 2013. Prices exclude Value Added Tax and contractors' overhead and profit.

The data provided here should be substituted with actual data where available, functionality that is available through the on-line version of the database available from www.capit-online.com.

ICE Publishing welcomes comments from users on changes or amendments which would further enhance the usefulness of the publication. Please address any correspondence to:

ICE Publishing, One Great George Street, Westminster, London SW1P 3AA, UK.
Telephone (0)20 7987 2026
email info@icepublishing.com Internet http://www.icepublishing.com

Acknowledgements

The publishers wish to acknowledge the invaluable assistance given by the following individuals in Mott MacDonald:

James Fiske – Project Director
Nang Vo Kham Murng - Senior Editor
Karl Horton – Project Manager
Robert Walker – Estimating Manager
Peter Taylor – Estimating Support
John Tarpay - Estimating Support
David Green – Project Director (Tunnels)
Masoud Taskindoust – Assistant Engineer (Tunnels)
Aaron Wright – Cost Research
Sam Wenham – Cost Research
Jingxin Shen – Web Developer
Rachael Croft – Typesetter
Emma Will – IMS and Commercial Support

James Fiske would also like to give acknowledgement and appreciation for the efforts of the BRE team who were responsible for the quantification of the carbon data. In particular the following individuals:

David Richardson – Group Director – Building Technology Group
Tim Barrow-Williams – Business Leader – Centre for Sustainable Products
Emma Franklin - Carbon Data Analyst
Julia Barnard – Carbon Data Analyst

James would also like to give special recognition of the major contribution of Nang Vo Kham Murng, the Senior Editor, for pulling this project together across multidisciplinary streams within Mott MacDonald.

All queries should be made to the Senior Editor, Nang Vo Kham Murng, via email at eru@mottmac.com.

The publishers would like to thank the CESMM Review Committee and the many organisations, manufacturers, contractors, suppliers and associations, including those who have given their kind permission for the reproduction and publication of copyright material. In particular, special thanks to the following for their continued and valued assistance in the production of the data within this publication.

CPM Group Ltd
Mells Road
Frome
Somerset BAll 3PD
Tel (0117) 9812791
sales@cpm-group.co.uk

Ennis Prismo Traffic Products
Unit 48 First Avenue
Westfield Trading Centre
Midsomer Norton
Radstock
Bath, BA3 4BS
Tel (01761) 414824
ptp@ennisprismo.com

FP McCann Ltd
Doseley
Telford
Shropshire TF4 3BX
Tel (01952) 630300

Hepworth Building Products
Edlington Lane
Edlington
Doncaster
South Yorkshire
DN12 1BY
United Kingdom
Tel (0170) 985 6300

HSS Hire Service Group Ltd
U3 14 wates Way
Mitcham
Surrey
CR4 4HR
Tel (0208) 685 9500

Jacksons Fencing
Ashford
Kent
TN25 6BN
Tel (0123) 3750393
Fax (0123) 3750403
sales@jacksons-fencing.co.uk

Jewson Ltd
Merchant House
Binley Business Park
Coventry CV3 2TT
Tel (0247) 643 8400

Tarmac Limited
Millfields Road
Ettingshall
Wolverhampton
West Midlands
WV4 6JP
Tel (0190) 235 3522

CONTENTS

Introduction

This is the eight edition of the CESMM Price Database. Completely redesigned and updated for 2013, this new edition still following the same structure and presentation as earlier editions, the bulk of the book consisting of pricing information covering all aspects of civil engineering activity structured in accordance with the requirements of CESMM4 and broken down into the 26 classes of CESMM4, with an additional section, Class ZZ, which covers alteration works.

This exciting new edition of CESMM Price Database sets a standard in the construction industry. This is the second update of the first civil engineering price book to now include an estimate of Carbon emissions for all activities of work – as well as retaining the accurate pricing that has made this price book a definitive guide for civil engineering professions.

The price information is complemented by sections covering approximate estimating, plant hire rates and outputs, economic forecasts, the Working Rule Agreement, professional, government and trade bodies and, finally, a section of technical information useful for those preparing costs and quantities for civil engineering projects.

It is generally accepted in the construction industry that unit costs for civil engineering works are more profoundly influenced by such factors as scale, nature and characteristics of the project than the similar unit costs for building works. The reasons for this are many and varied but the greater opportunity for the contractor to exercise his expertise and ingenuity in arriving at the optimum solution to a particular problem, combined with the wide variety of available plant with its different outputs and applications, must be major factors.

Consequently the build up of the civil engineering unit rate is of particular significance and it is for this reason that the book provides detailed information on how the rates have been derived in order to enable users to make adjustments to allow for the various factors affecting a particular project.

The information contained in this book will have many applications including providing valuable data for use in the preparation of budget estimates at the various stages of civil engineering projects from inception to completion, supplementing the information already in Contractors' estimating departments and forming a basis for the negotiation of new rates within existing contracts. However, rates, prices and discounts can fluctuate dramatically over short periods of time and between various regions of the country. The costs shown are at first quarter 2013. Prices exclude Value Added Tax, general items, contractor's profit and overheads (see Section 1, Unit Pricing).

Whilst it has been our intention throughout the preparation of this book to be factual and accurate, neither the Editing Authors nor the Publishers can in any way accept liability for any loss, direct or otherwise, of any kind arising out of the use or application of the contents of this book.

CapIT: Online Carbon & Cost Estimator

CapIT is a powerful online estimating tool and the only such system to offer comprehensive and detailed embodied CO_2e data for civil engineering works. Fully customisable, it enables users to build up cost and carbon estimates using the data found in this book, but also their own items, carbon values, unit rates, and plant or material costs.

The cost and carbon data is updated quarterly and users can select specific regional settings as well as dates in the past or future for ever greater accuracy in their estimates. (For a full listing of the many benefits of CapIT: Online Carbon & Cost Estimator, readers are encouraged to visit: www.capit-online.com).

As a purchaser of the book, you are entitled to a 12-month period of access to CapIT. Please refer to the back page for instructions on activating your user account.

SECTION 1:
UNIT PRICING

SECTION 1 - UNIT PRICING

Introduction to Unit Pricing Section

Prices used have been calculated as at the first quarter of 2013.

The price shown is the net cost to the contractor for the activity and does not include preliminary and general items, head office overheads and profit, nor VAT.

Coherence has been maintained throughout the Unit Pricing section by basing the construction activities on a typical civil engineering contract, with pricing levels applicable to a UK average location. Access to the site is free from restraint. The total value of the construction works being £8 to £10m.

Our unit pricing assumes a general average level of conditions and circumstances. You should modify the appropriate parts of a unit price in line with your own knowledge and experience of the factors which may affect them. Prices and rates can fluctuate considerably, so any application other than a broad, approximate estimate requires up-to-date quotations for the constituent parts of a unit price.

Unless shown otherwise a percentage addition has been made to the supply cost of the materials to cover unloading, storing and rehandling as well as wastage. A small general purpose gang has been allowed for in the general items which will assist with unloading. Where unloading and stacking are a significant part of the activity, such as pipe stringing in a pipe laying contract, the activity has been separately priced and included in the total cost of the activity. The percentage addition to reinforcing steel covers stools, spacers, binding wire as well as wastage.

It is common practice in Civil Engineering estimating to build up a price for an activity on a weekly gang cost of labour and plant and this has been adopted here, the weekly cost of the gang being divided by 39 to arrive at an hourly cost.

Many contractors these days sub-let non-specialist as well as specialist activities according to their current commitments; for instance, there has been a move by contractors to favour sub-letting major elements of a contract to obtain low competitive sub-contract quotes so enabling them to make low tender submissions.

The hire rates used for the plant can be improved upon if a firm quote is called for. The rates include fuel and consumables and driver where appropriate. Fuel consumptions have been averaged out for each broad sector of machine. Particular attention has to be paid to the cost of fuel which can fluctuate widely.

All gang hours have been calculated to 3 places of decimals but, due to printing requirements, we can only show the quantity to 2 places of decimals. Thus, where an hourly quantity is less than one hundredth part of an hour, although the calculation has been correctly inserted into the net price, it will appear in the labour gang hours or plant gang hours column at the 'rounded-up' value of 0.01 (but in the calculation of the NET UNIT PRICE the 'rounding-up' has been ignored).

If the reader wishes to determine the actual quantity of hours, this can be derived by dividing the relevant gang cost given at the front of the section into the net labour or plant price of the item.

Guidance Notes to Use of Pricing Section

At the beginning of each pricing section the cost of the labour and plant gangs respectively has been shown, with the details of the gang make up; also shown is the cost of the material reduced to the appropriate unit used in the material content of the unit price.

The hours shown in the 'labour gang hours' and 'plant gang hours' columns have to be multiplied by the appropriate hourly cost to arrive at the 'net labour price' and the 'net plant price' which, in turn have to be added to the 'net materials price' to give the total 'net unit price'.

PRICING EXAMPLE
A typical unit price build-up is shown here, followed by a detailed analysis to explain how the price is built-up.

CLASS G - CONCRETE ANCILLARIES

G5   REINFORCEMENT

		Unit	Labour Gang	Plant Gang	Labour Net	Plant Net	Material Net	Unit Net
			Hours	Hours	Price £	Price £	Price £	PRICE £
G5.2	**Deformed high yield steel bars, BS 4449; bent and cut to length**							
G5.2.4	Nominal size: 12mm							
G5.2.4.01	Generally	t	11.82	11.82	563.58	136.76	666.79	**1,367.13**

1)   LABOUR GANGS, PLANT GANGS AND MATERIAL BUILD-UPS

Labour Cost Analysis

Labour gangs
Steel fixing labour gang

Craftsman WRA	2 x £17.33	= £34.66
Labourer (general operative)	1 x £13.02	= £13.02

Total hourly cost of gang = **£47.68**

Plant Cost Analysis
Plant gangs
Steel fixings plant gang

Excavators Cable - NCK 305A 0.67m³	0.25 x £46.29	= £11.57

Total hourly cost of gang = **£11.57**

Materials Cost Analysis

Reinforcement
NOTE: The waste factor of 5% is inclusive of offloading, tying wire, chairs and spacers.
Grade 460 deformed type 2 high tensile reinforcing bars, BS 4449

	Supply Price £	Waste Factor %	Unloading Labour £	Plant £	TOTAL UNIT COST £	Unit
Cut, bent and bundled: 12mm	654.09	5	-	-	686.80	t

2)   PRICE BUILD-UP SUMMARY

Net Labour Price:
Labour gang hours x Total hourly cost of gang = 11.82 x £47.68 — 563.58
Net Plant Price:
Plant gang hours x Total hourly cost of gang = 11.82 x £11.57 — 136.76
Net Material Price:
12mm reinforcement bar — 686.80
Net Unit Price — £1,387.13 t

The reader may adjust any of the variables in the resource build-ups to suit his own requirements e.g. size of labour gang, plant type, material costs or wastage on materials.

Coding System

The codes used to identify items refer to the coding structure of the Civil Engineering Standard Method of Measurement, 4th Edition, 2012. Each item and each group of items has been allocated a unique code.

The codes used in the pricing example for reinforcement on page 99, are defined below:

CODE	CESMM, 4th Edition, 2012 Reference

G   Class G: Concrete Ancillaries

5.2.4.02
524 - This is an item build-up from the divisions of the CESMM

t .02 - Is a secondary code, giving the item a unique identity (i.e. as opposed to .01, .03, etc.)

5 - The first numerical digit refers to the First Division within Class G: Concrete Ancillaries. In this case;
5 - Reinforcement

2 - The second numerical digit refers to the Second Division within Class G appertaining to reinforcement;
2 - Deformed high yield steel bars to BS 4449

4 - The third numerical digit refers to the Third Division within Class G appertaining to Deformed high yield steel bars to BS 4449; 4 - Nominal size 12mm

t - Contrary to CESMM3 section 4 paragraph 4.7 two numerical digits have been used to clarify secondary coding.

Units Used in Pricing Section

The following units have been used in the pricing sections:

Kg - kilogramme	m - linear metre	Nr - number
t - tonne	m^2 - square metre	h - hour
ha - hectare	m^3 - cubic metre	l - litre

Introduction to Carbon Dioxide Equivalents Emissions

Climate change is considered to be one of the greatest environmental threats facing the world today. Manmade climate change is caused predominantly by the release of carbon dioxide (CO_2), a greenhouse gas. Other significant greenhouse gases linked to manmade climate change include nitrous oxides and methane. The global warming potential (GWP) of greenhouse gases are stated relative to the GWP of CO_2. As such, the unit 'carbon dioxide equivalent' (CO_2e or sometimes CO_2eq.) can be used to report on the sum total GWP (of more than one greenhouse gas) released as a result of a particular activity.

Various human activities cause the release of greenhouse gases, ranging from the burning of fossil fuels for energy generation and transport to agriculture and land use change. The construction industry consumes large amounts of materials and it is estimated that the emissions associated with their production are responsible for around 13% of annual CO_2 emissions from human activity. Greenhouse gas emissions that have occurred as a result of the production of a particular material are often referred to as 'embodied carbon' – that is, the CO_2 or CO_2e (if available) embodied in the material. Embodied carbon is estimated using a technique called life cycle assessment (LCA).

LCA studies that are conducted using a peer reviewed and publically available methodology, which is in accordance with international standards, are widely held to be of the highest quality. Even so, there is sufficient scope in the standards for different methodologies to produce different results for the same product. Therefore, it is advisable to use data produced according to a single methodology. BRE worked with many industry sectors to produce the data used in this publication, which is based on a single, cross-sector, industry-agreed method and represents typical UK emissions.

Study Boundaries

In order to correctly interpret the results of an LCA study it is important to understand some of the key assumptions in the model. In particular, the boundary of the LCA (what is included in the study) can have a significant effect on the results generated.

The most common boundary types encountered are:

1) Cradle-to-Gate
2) Cradle-to-Site
3) Cradle-to-Grave
4) Cradle-to-Cradle

Cradle-to-gate includes all emissions up to the point at which the product under assessment is finished – at the factory gate. Cradle-to-site then includes emissions associated with transport to site. Cradle-to-grave includes all the cradle-to-site emissions plus the future emissions associated with installation, maintenance, replacement, disposal, and recycling/reuse that occur during a specified life cycle scenario of the product.

While a single scenario can influence the results considerably it cannot reflect every possible life cycle. As such, rather than providing cradle-to-grave values, only cradle-to-gate values are provided in this publication. It is strongly recommended that before comparing options, users of the carbon data apply their project-specific knowledge to understand the influence of durability and disposal options (reuse, recycling, energy recovery and/or landfill).

Labour

The CO2e impact of labour is not included within the calculations as these impacts would occur regardless of if undertaking construction activity. Allowances for hand tools are made within the plant totals. Please see the 'Transportation' section with regard to quantification of labour transportation.

Materials

The CO_2e rates for materials have been calculated by BRE according its Environmental Profiles methodology. The peer reviewed methodology has been prepared to conform to the relevant ISO standards - ISO 21930, ISO 14025, and standards relating to Life Cycle Assessment in general, ISO 14040 and 14044. For more information on the Environmental Profiles methodology, visit: http://www.bre.co.uk/greenguide/podpage.jsp?id=2126 .

With any material the embodied CO_2e rate will differ between manufacturers and depends on factors such as their power sources, manufacturing process, transport distances etc. Although manufacturer specific data exists in the form of Environmental Product Declarations (EPD) it is not yet comprehensive enough to cover all of the options available. To overcome this issue the rates given are generic, based on UK industry average production data provided to BRE by trade associations.

The use of recycled materials has an impact on the embodied CO_2e for materials. The BRE Environmental Profiles methodology makes allowance for typical levels of recycled content and also contains allowances for production waste.

Timber and natural fibre materials absorb CO_2 as they grow. The carbon contained in such materials is referred to as 'sequestered carbon' because it is locked into the material until such time as it is released through decay or combustion. The amount of sequestered carbon in timber and natural fibres is highly significant and so the emissions from disposal are important to consider. This publication therefore lists two values for such materials: one that includes sequestered (in *italic* font) carbon and one that excludes it (in **bold** font). Users of the data are encouraged to consider the effect that the inclusion or exclusion of sequestered carbon has on the overall carbon assessment and also to consider that for carbon sequestration to offer benefits the materials must be grown and harvested sustainably.

Please see the 'Transportation' section with regard to quantification of material transportation.

Plant

The embodied CO_2e of the production and maintenance of plant is not included within the guide figures. However, CO_2e rates for fuels are provided and can be used along with fuel consumption data to estimate construction stage emissions. The plant selected for the guide uses what the author considers the more traditional sources. Reductions will be recorded if alternate sources of renewable energy or biofuels are used.

As with the estimates of cost, elements of non-productive standing time have been included, but additional allowance would need to be made for project specifics such as ground conditions and weather.

Please see the 'Transportation' section with regard to quantification of the transportation of plant.

Transportation

The CO_2e impact for transporting labour, materials and plant resources has not been included within the unit rate data. This is due to the levels of disparity recorded on previous projects studied.

It is important that quantification of transport is included when you are considering procurement options for locations with variations in transportation distance as this will have a significant impact on the final result, especially when quantifying international procurement options.

To assist in estimating emissions from the transport of labour, the following table has been produced indicating an estimate of CO_2e, per kilometre travelled, per person:

Method	CO_2e emissions (kgCO$_2$e/km)
Aeroplane: Long Haul International	0.109
Aeroplane: Short Haul International	0.0952
Average Diesel Car	0.187
Average Local Bus	0.112
Average Petrol Car	0.2019
Average Petrol Hybrid Car	0.2168
Average Petrol Motorcycles	0.119
Average Van Up To 3.5 Tonne	0.2472
Bus - Coach	0.0287
CNG Van Up To 3.5 Tonne	0.2389
Diesel Van Up To 3.5 Tonne	0.249
Domestic Flight (Average)	0.1669
Ferry	0.1161
International Rail	0.0151
Light Rail And Tram	0.0675
LPG Van Up To 3.5 Tonne	0.4233
Mini: Diesel	0.1056
Mini: Petrol	0.1487
MPV: Diesel	0.1968
MPV: Petrol	0.2217
National Rail	0.0582
Petrol Van Up To 3.5 Tonne	0.3436
SUV: Diesel	0.1681
SUV: Petrol	0.2538
Taxi - Black Cab	0.1571
Taxi - Regular Taxi	0.1476
Underground	0.0719

To quantify the transport of materials and plant additionally the following table has been included to allow estimates of the kilograms of CO_2e produced per tonne of weight transported per kilometre travelled:

Method	CO_2e emissions (kgCO$_2$e/t/km)
Road	0.1067
Rail	0.0369
Water	0.0154

Both tables above have been provided to allow estimates only, actual data will vary depending on age and state of repair of vehicle, weather conditions and numbers of people being transported.

CLASS A:
GENERAL ITEMS

CLASS A – GENERAL ITEMS

Build Up of Labour Rates

The following is an abridged extract from the Working Rule Agreement for the Building and Civil Engineering Industry:

Rule WR.3 The normal working hours in the industry shall generally be:

Monday to Thursday	8 hrs per day
Friday	7 hrs
Total for week	39 hrs

Rule WR.4 Overtime shall be calculated as follows:
- Monday to Friday
For the first four hours after the completion of the normal working hours of the day time and a half; thereafter at the rate of double time until the starting time the following day.
- Saturday
Time and a half, until the completion of the first four hours, and thereafter at double time.
- Sunday
At the rate of double time, until the starting time on Monday morning.

Overtime shall be calculated on the normal hourly rate. Additional payments for intermittent skill or responsibility or adverse conditions and bonus shall not be included when calculating overtime payments.

In practice the working week is commonly five days of 9.5 hours in the summer and five days of 8.5 hours in the winter. The following calculation is used to determine the actual cost of productive hours.

Summertime Production hours

5 days @ 9.5 hours =	47.5 x 35.0wks =	1,662.50 hours	
Non-Productive hours			
4 days @ 0.75 hours =	3.00		
1 day @ 1.25 hours =	1.25		
	4.25 x 35 wks =	148.75 hours	**1,811.25 hours**

Wintertime Production hours

5 days @ 8.5 hours =	42.5 x 11.2wks =	476.00 hours	
Non-Productive hours			
4 days @ 0.25 hours =	1.00		
1 day @ 0.75 hours =	0.75		
	1.75 x 11.2wks =	19.60 hours	
			495.60 hours

Total number of paid hours =   **2,306.85 hours**

Total number of production hours = (1662.50 + 476.00) =	2,138.50	
Allow 5% wet time = (say)	106.93	
		2,031.60 hours

Basic wage rates: on 7 January 2013

General Operative =	£313.70 for 39 hour week: £8.03 per hour
Craft Rate =	£416.13 for 39 hour week: £10.67 per hour
Average paid hours/week =	2,306.85 divided by 46.2 = 49.93
Average production hours/week =	2,031.60 divided by 46.2 = 43.97

GENERAL ITEMS - A

Build Up of Labour Rates continued...

		General Operative	Skill Rate 4	Skill Rate 3	Skill Rate 2	Skill Rate 1
Total paid hours 49.93 @	£	8.03	8.65	9.16	9.79	10.17
per week	£	400.95	431.91	457.38	488.83	507.81
Overtime payment at premium rate: (49.93 -39) = 10.93 hrs		43.88	47.27	50.06	53.50	55.58
Service allowance @ 5 hours per week		-	-	-	-	-
Gross weekly wage (A)	£	**444.84**	**479.18**	**507.43**	**542.33**	**563.39**
Employers liability and public liability insurance at 2.5% of labour content (A)		11.12	11.98	12.69	13.56	14.08
Construction industry training board levy @ 0.50% of (A)		2.22	2.40	2.54	2.71	2.82
Daily travel allowance (say 5 days x 25km one way per day)		10.35	10.35	10.35	10.35	10.35
Annual holiday allowance		41.88	45.12	47.78	51.06	53.05
Public holidays with pay		15.41	16.60	17.57	18.78	19.51
National Insurance contributions based on 12.8% of gross weekly wage above earnings threshold (ET) of £110.00 per week		41.52	46.26	50.15	54.97	57.88
B&CE EasyBuild Stakeholder pension contribution (say)		5.00	5.00	5.00	5.00	5.00
Total weekly cost	£	572.34	616.88	653.51	698.77	726.07
Total Lablour cost per Hour	£	**13.02**	**14.03**	**14.86**	**15.89**	**16.51**

		Craftsman	Plant Operator SR 4	Plant Operator SR 3	Plant Operator SR 2	Plant Operator SR 1
Total paid hours 49.93 @	£	10.67	8.65	9.16	9.79	10.17
per week	£	532.77	431.91	457.38	488.83	507.81
Overtime payment at premium rate: (49.93 -39) = 10.93 hrs		58.31	47.27	50.06	53.50	55.58
Service allowance @ 5 hours per week		-	43.25	45.80	48.95	50.85
Gross weekly wage (A)	£	**591.08**	**522.43**	**553.23**	**591.28**	**614.24**
Employers liability and public liability insurance at 2.5% of labour content (A)		14.78	13.06	13.83	14.78	15.36
Construction industry training board levy @ 0.50% of (A)		2.96	2.61	2.77	2.96	3.07
Daily travel allowance (say 5 days x 25km one way per day)		10.35	10.35	10.35	10.35	10.35
Annual holiday allowance		55.65	49.19	52.09	55.67	57.83
Public holidays with pay		20.47	18.09	19.16	20.48	21.27
National Insurance contributions based on 12.8% of gross weekly wage above earnings threshold (ET) of £110.00 per week		61.70	52.22	56.47	61.73	64.89
B&CE EasyBuild Stakeholder pension contribution (say)		5.00	5.00	5.00	5.00	5.00
Total weekly cost	£	761.99	672.96	712.91	762.25	792.01
Total Labour cost per Hour	£	**17.33**	**15.30**	**16.21**	**17.33**	**18.01**

GENERAL ITEMS - A

Build Up of Labour Rates continued...

NOTES: There has been an allowance included for daily fare and travel allowances as required under WR.5 on the assumption that sites will be away from major conurbations, entailing the provision of transport or travel allowances to operatives (which will not always be the case) and that craftsmen, in particular, will in some cases require subsistence payments, as required under WR.15. These allowances should be reviewed when considering the labour costs for specific sites.

Other 'Labour cost per hour' rates have been similarly calculated and applied in this book to the following classes of operatives:

Banksman	£14.03	Ganger	£17.64	Pipelayer (standard)	£14.86
Pipelayer (large pipes)	£16.51	Driller	£14.86	Shot Firer	£14.86
Fitters & Welders	£17.33	Timbermen	£14.86	Piling Hand	£14.03
Painter	£17.33	Painter (chargehand)	£18.63	Brush Hand	£13.02
Spray Painter	£17.33	Asphalt Layer	£14.86	Plasterer	£17.33
Tiler	£17.33	Bricklayer (chargehand)	£18.63	Waller	£17.33
Miner*	£39.89	Carpenter (chargehand)	£18.63		

* Not based on WRA

PRICING EXAMPLE

Generally

The figures used in this section are based on a typical civil engineering project of value £8.50m, located on the fringe of a large conurbation with reasonable access to the resources necessary for the project.

Principal quantities

Clear and level site	300,000m^2
Excavation	130,000m^3
Concrete	17,000m^3
Formwork	28,000m^2
Reinforcement	1,700t

Tender build-up

General items	1,562,900
Profit and H.O overheads	510,000
Labour (including Labour only Subcontractors)	1,487,500
Materials	2,092,640
Plant and fuel	629,000
Direct Sub-contractors	920,100
Nominated Sub-contractors	1,211,160
Engineers contingencies	86,700
Contract value	£8,500,000

Cost analysis of General Items

The total cost of the General Items represents 24.32% of the net cost of the works calculated as follows:

Total tender cost	8,500,000
Less Profit and H.O. Overheads (say 6.0%)	510,000
Net tender cost	7,990,000
Less General Items	1,566,652
	6,423,348

Total cost of General Items as a percentage of net cost of the works
$$\frac{1,566,652}{6,423,348} \times 100 = 24.39\%$$

The following table shows a cost analysis of the General Items with reference to the General Items section of the Civil Engineering Standard Method of Measurement.

CESMM Class 'A' Code	CESMM Class 'A' First division Section	Amount £	% of General Items	% of net Cost of Works
1	Contractual Requirements	111,136	7.09%	1.73%
2	Specified Requirements	224,396	14.32%	3.49%
3	Method-Related Charges	1,231,121	78.58%	19.17%
4-	Provisional Sums	excl	-	-
5	Nominated Sub-contracts (incl work on site)	excl	-	-
6	Nominated Sub-contracts (excl work on site)	excl	-	-
1-6	All Sections	1,566,652	100.00%	24.39%

GENERAL ITEMS - A

EXAMPLE continued/ ....

NOTES: Performance Bond
It is usual for a local authority to call for a performance bond from the Contractor. This is normally a percentage of the tender sum (0.25%) and will vary according to the nature of the work and the standing of the Contractor. An estimator will be required to obtain firm quotations from an insurance company or bank. The percentage sum will be levied on an annual basis for the actual duration of the contract, that is until the issue of a completion certificate and thereafter at a reduced percentage from the period of maintenance. It is important to have the details fixed before the final tender figure is submitted.

Insurance of the Works
The premium will depend on the nature of the work and the insurance experience of the Contractor. Normally this will form part of an insurance package which will include public liability, fire and theft and possibly professional indemnity (Employers liability is dealt with in the labour rate build-up).

Provisional Sums and Nominated Sub-contracts
Provisional Sums and Nominated Sub-contracts have been excluded from the General Items.

A1   Contractual Requirements

A1.1	**Performance Bond**			
	Not normally requested from the Contractor for this type of Contract			
A1.2	**Insurance**			
	Allow 1.00% of total contract value			
			£	
	Contract Value	=	8,500,000	
	Inflation Allowance 2.0%	=	170,000	
			8,670,000	
	Additional costs including professional fees 8%	=	693,600	
			9,363,600	
	1.00% of £9,363,600			93,636
A1.3	**Parent company guarantee**			
	Allow lump sum			17,500
	TOTAL FOR CONTRACTUAL REQUIREMENTS			**£111,136**

A2   Specified Requirements

A2.1   Accommodation for the Contract Administrators

A2.1.1   Offices
A2.1.1.01 Establish and remove

Transport, erect and dismantle		595

A2.1.1.02 Maintain and operate

40m² @ £3.25 per m² x 78wks	10,140	
Heat, light and clean: £87 per week x 78wks	6,786	
		16,926

A2.1.2   Laboratories
A2.1.2.01 Establish and remove

Transport, erect and dismantle		595

A2.1.2.02 Maintain and operate

20m² @ £3.25 per m² x 78wks	5,070	
Heat, light and clean: Included with offices		5,070

A2.1.3   Cabins
A2.1.3.01 Establish and remove Ablution Block

Connect to cesspit or town sewer	1,480

A2.1.3.02 Maintain and operate Ablution Block

£107 per wk x 78wks	8,346

A2.1.3.03 Establish and remove Inspectors Hut

Mobile: on/off charges	245

A2.1.3.04 Maintain and operate Inspectors Hut

8m² @ £3.25 per wk x 78 wks	2,028

A2.2   Services for the Contract Administrators

A2.2.1   Transport vehicles
A2.2.1.01 Maintain and operate

2000cc car - £ 216 per wk x 78wks	16,848	
Landrover - £405.00 per wk x 78wks	31,590	
Petrol - 120 litres per wk @ £0.90 per litre x 78 wks	12,168	
		60,606

A2.2.2   Telephones
A2.2.2.01 Establish and remove

Installation: BT charges	330

A2.2.2.02 Maintain and operate

Calls and rental: £107 per wk x 78wks	8,346
Carried forward	£104,567

A2 Specified Requirements continued/...

	Brought forward	£104,567

A2.3 Equipment for use by the **Contract Administrators**

A2.3.1 Office equipment
A2.3.1.01	Establish and remove	400
A2.3.1.02	Maintain and operate	

2 PC's, 4 desks, 4 tables, 1 conference table, 6 chairs, 2 cabinets:
£110 per wk x 78wks   8,580

A2.3.2 Laboratory equipment
A2.3.2.01	Establish and remove	400
A2.3.2.02	Maintain and operate	

£90 per wk x 78 wks   7,020

A2.3.3 Surveying equipment
233.1	Establish and remove	400
233.2	Maintain and operate	

£90 per wk x 78wks   7,020

A2.4 Attendance upon the Contract Administrators

A2.4.1 Drivers
A2.4.1.01 Establish and remove

Recruitment - included   -

A2.4.1.02 Maintain and operate

£12.56 per hr x 39 x 78wks   38,208

A2.4.2 Chainmen
A2.4.2.01 Establish and remove

Recruitment - included   -

A2.4.2.02 Maintain and operate

£12.56 per hr x 39 x 78wks   38,208

A2.4.3 Laboratory assistants
A2.4.3.01 Establish and remove

Recruitment - included   -

A2.4.3.01 Maintain and operate

£12.56 per hr x 39 x 40wks   19,594

A2.5 Testing of materials
Included   -

A2.6 Testing of the Works
Included   -

TOTAL FOR SPECIFIED REQUIREMENTS   **£224,396**

A3 Method-Related Charges

A3.1 Accommodation and buildings

A3.1.1 Offices

A3.1.1.01 Establish and remove

Transport	1,650	
Erect and dismantle	600	
Telephone installation: BT charges	600	
Furniture, PC's, etc.	6,600	9,450
		(Fixed)

A3.1.1.02 Maintain and operate

Mobile offices: 10 staff x 8m2 each $= 80m^2$
Conference room $= 20m^2$
$100m^2$

100m2 @ £2.00 per m2 x 60wks	12,000	
Heat, light and clean: £140 per wk x 60wks	8,400	
Telephone calls and rental: £120 per wk x 60wks	7,200	
Stationery £125 per wk x 60wks	7,500	35,100
		(Time Related)

A3.1.3 Cabins

A3.1.3.01 Establish and remove Ablution Blocks

Connect to cesspit or town sewer		1,500
		(Time Related)

A3.1.3.02 Maintain and operate Ablution Blocks

£100 per wk x 78wks	7,800	7,800
		(Time Related)

A3.1.3.03 Supply, maintain and operate Toilet Blocks

lnr @ £100 per wk x 60wks	6,000	
Maintain: 0.5 man @ £12.56 per man per hr x 39 x 60wks	14,695	20,695
		(Time Related)

A3.1.3.04 Supply, maintain and operate Drying Room

£33 per wk x 60wks		1,980
		(Time Related)

A3.1.4 Stores

A3.1.4.01 Establish and remove

Erect and dismantle		600
		(Fixed)

A3.1.4.02 Maintain and operate

£30 per wk x 60wks		1,800
		(Time Related)

A3.1.5 Canteens and Messrooms

A3.1.5.01 Maintain and operate

4nr mobiles @ £25 each per wk x 60wks	6,000	
Heat, light and clean: £28 per wk x 60wks	1,680	7,680
		(Time Related)

A3.2 Services

A3.2.1 Electricity

A3.2.1.01 Establish and remove

Electricity to compound and batching plant:

Cable and switches	3,000	3,000
(Time Related)		

A3.2.2 Water

A3.2.2.01 Establish and remove

Water main to compound and batching plant:

50mm dia. UPVC - 600m @ £13.33 including fittings	7,998	7,998
		(Fixed)
	Carried forward	£97,603

A3 Method-Related Charges continued/ ...

Brought forward   97,603

A3.3.5 Concrete Mixing
A3.3.5.01 Establish and remove
Transport, erect and dismantle: concrete batching plant,
aggregate bins, cement silos and concrete pump.

Site preparation: level area, spread hardcore, lay concrete:
Plant:
Dresser 1004 - 2 days @ £28.68 per hr   459

Labour:
4 labourers - 2 days @ £12.56 per hr   804
Carpenter -1 day @ £17.93 per hr   143

Material:
Hardcore - 400m2 x 0.25m deep @ £23.90 per m^3 2,390
Concrete - 50m3 @ £87.88 per m^3 4,394
                                                                            8,190

Transport batcher and erect:
Plant:
2nr flat top lorries - 1 day @ £42.63 each per hr   682
Inr crane - 3 days @ £42.28 per hr   1,015

Labour:
4 labourers - 5 days @ £12.56 per hr   2,010
Supervision   340
Timber for bins - £1,350 less residual value   675
Dismantle and reinstate   1,390
                                                                            6,112

                                                                                    14,302
                                                                                   (fixed)
                                         Carried forward   £111,905

A3   Method-Related Charges continued/ ...

Brought forward   111,905

A3.5   Temporary Works

A3.5.3   Access roads
A3.5.3.01 Establish and remove access roads
        1000 x 6m wide = 6000m²
        Construct 100m run per day

        Plant:
        Dresser 1004 - 10 days @ £28.68 per hr   2,294
        Vibratory roller - 5 days @ £2.04 per hr   82
        JCB 3CX - 5 days @ £27.01 per hr   1,080

        Labour:
        2 labourers -10 days @ £12.56 per hr   2,010

        Materials:
        Hardcore or ballast - 6000m2 x 0.20m deep = 1200m³ @
        £23.90 per m³   28,680
        Lotrak - 6000m² @ £0.93 per m²   5,580   39,726
                                                                                 (Fixed)

A3.5.3.02 Maintain access roads
        Dresser 1004 - 0.5 days per wk: 60 wks x 4hrs @ £28.68 per hr   6,883

                                                                              (Time Related)

A3.6   Temporary Works

A3.6.6   Hardstandings
A3.6.6.01 Establish and remove administration and storage compound hardstanding
        Allow 5 days for the operation

        Plant:
        Dresser 1004 - 3 days @ £28.68 per hr   688
        Vibratory roller - 3 days @ £2.04 per hr   49
        JCB 3CX - 2 days @ £27.01 per hr   432

        Labour:
        Tradesman - 5 days @ £17.93 per hr   717
        2 labourers - 5 days @ £12.56 per hr   1,005

        Materials:
        Sub-base - spread, level 1000m² x 0.20m deep = 200m³ @
        £24.03 per m³   4,806
        Concrete - 600m2 x 0.15m deep = 90m³ @ £87.88 per m³   7,909
        Drainage - septic tank or connection into main sewer   2,000   17,606
                                                                                 (Fixed)
                                                    Carried forward   £176,120

A3 Method-Related Charges continued/ ...

Brought forward   176,120

A3.7	Supervision and Labour

Regional/Area Manager - Incl. in H.O. Overheads
Regional/Area Q.S. - Incl. in H.O. Overheads
Planning Engineer - Incl. in H.O. Overheads
Trade Foremen - Incl. in rates
Staff subsistence - Incl. in weekly charges
Overtime - Incl. in weekly charges

NOTE: * weekly rate includes allowance for Company car

A3.7.1 Supervision

Agent/Senior Engineer - 60wks @ £1,776 per wk*	106,560	
Section Engineer: Tanks - 60wks @ £1,247 per wk*	74,820	
Section Engineer: Wharf - 40wks @ £1,247 per wk*	49,880	
Section Engineer: Excavating Roads, etc. 60wks @ £1,247 per wk	74,820	
Assistant Engineer - 50wks @ £975 per wk	48,750	
Technicians - 60wks @ £585 per wk	35,100	
Senior Quantity Surveyor - 78wks @ £1,550 per wk x 50%*	60,450	
Assistant Quantity Surveyor - 50 wks @ £1,260 per wk*	63,000	
General Foreman -78wks @ £1,535 per wk (Landrover incl.)	119,730	
Sectional Foreman: Concrete - 50wks @ £1,404 per wk	70,200	
Sectional Foreman: Excavating Roads etc. - 60wks @ £1,536 per wk (Landrover incl.)	92,160	795,470
		(Time Related)

A3.7.2 Administration

Office Manager - 60wks @ £773 per wk	46,380	
Wages Clerk - 60wks @ £514 per wk x 50%	15,420	
Timekeeper - 60wks @ £500 per wk	30,000	
Storekeeper - 10wks @ £585 per wk	5,850	
Typist/Receptionist/Telephonist - 60wks @ £463 per wk	27,780	
Office cleaners/tea - 60wks @ £104 per wk (part time)	6,240	
Security man - 60wks @ £803 per wk	48,180	179,850
		(Time Related)

A3.7.3 Labour teams
Offloading materials and maintaining services gang

Plant:		
Tractor and trailer - 40wks @ £1,125 per wk	45,000	
NCK 305 - 10wks @ £1,524 per wk	15,240	
Labour:		
1.5 labourers - 40wks @ £486 per wk	19,440	79,680
		(Time Related)

TOTAL FOR METHOD-RELATED CHARGES	**£1,231,121**

SUMMARY OF GENERAL ITEMS

CONTRACTUAL REQUIREMENTS	£111,136	
SPECIFIED REQUIREMENTS	£224,396	
METHOD-RELATED CHARGES	£1,231,121	
TOTAL GENERAL ITEMS	**£1,566,652**	

CLASS B:
GROUND INVESTIGATION

Calculations used throughout Class B - Ground Investigation

Labour

			Qty		Rate		Total
L B0001ICE	**Cable Percussion Borehole Labour Gang**						
	L A9000ICE	Craftsman WRA	1	x	17.33	=	£17.33
	L A9002ICE	Labourer (General Operative)	1	x	13.02	=	£13.02
		Total hourly cost of gang				=	**£30.35**
L A0120ICE	**General Earthworks Labour Gang**						
	L A9016ICE	Ganger	1	x	17.64	=	£17.64
	L A9002ICE	Labourer (General Operative)	1	x	13.02	=	£13.02
	L A9015ICE	Banksman	1	x	14.03	=	£14.03
		Total hourly cost of gang				=	**£44.69**
L B0004ICE	**Rotary Bored Borehole Labour Gang**						
	L A9000ICE	Craftsman WRA	1	x	17.33	=	£17.33
	L A9002ICE	Labourer (General Operative)	2	x	13.02	=	£26.03
		Total hourly cost of gang				=	**£43.36**
L A0331ICE	**Trial Hole Labour Gang**						
	L A9016ICE	Ganger	1	x	17.64	=	£17.64
	L A9014ICE	Labourer (Skill Rate 4)	1	x	14.03	=	£14.03
	L A9002ICE	Labourer (General Operative)	2	x	13.02	=	£26.03
	L A9024ICE	Plant Operator (Class 3)	1	x	16.21	=	£16.21
		Total hourly cost of gang				=	**£73.91**

Plant

			Qty		Rate		Total
P A1333ICE	**Trial holes excavation Plant Gang**						
	P A0594ICE	Hydraulic Excavator - JCB 3CX Sitemaster	1	x	28.91	=	£28.91
	P A0550ICE	Dumper - 3.0t 4WD	1	x	25.44	=	£25.44
	P A0454ICE	Compressor - 180 cfm	1	x	7.66	=	£7.66
	P A0673ICE	Hydraulic Breaker (Independant) - Two Tool Towable Breaker	1	x	2.94	=	£2.94
	P A0461ICE	Hose & Breaker & 3 Steels	1	x	0.38	=	£0.38
	P A0853ICE	Vibrating Plate Diesel 24kN	1	x	2.53	=	£2.53
		Total hourly cost of gang				=	**£67.86**
P B0001ICE	**Cable Percussion Borehole Plant Gang**						
	P A1064ICE	Cable Percussion Rig	1	x	8.63	=	£8.63
	P A1023ICE	Landrover 4WD	1	x	16.16	=	£16.16
	P A0442ICE	Bowsers (250 gallon water / fuel)	1	x	0.94	=	£0.94
		Total hourly cost of gang				=	**£25.73**
P B0003ICE	**Rotary Bored Borehole Plant Gang**						
	P A1074ICE	Rotary Borehole Rig	1	x	13.17	=	£13.17
	P A1023ICE	Landrover 4WD	1	x	16.16	=	£16.16
	P A0442ICE	Bowsers (250 gallon water / fuel)	1	x	0.94	=	£0.94
	P A0442ICE	Bowsers (250 gallon water / fuel)	1	x	0.94	=	£0.94
	P A0458ICE	Compressor - 480 cfm	1	x	20.17	=	£20.17
	P A1069ICE	Grouting Rig (Anchorage Items)	1	x	10.53	=	£10.53
		Total hourly cost of gang				=	**£61.91**

Ground Investigation

Note(s): NOTE: The rates in this section are based on historical specialist sub-contractor prices and are provided as guide prices only.

B1 Trial Pits And Trenches

Note(s): NOTE: The prices under Trial Pits and Trenches assume that the pits and trenches are inspected and backfilled the same day.

		Unit	Labour Hours	Labour Net £	Plant Net £	Materials Net £	Unit Net £	CO$_2$e Kg
B1.1	**Number in material other than rock**							
B1.1.1	maximum depth: not exceeding 1 m							
B1.1.1.01	plan area at bottom of pit 2 x 1m	Nr	0.22	16.26	14.93	-	31.19	6.142
B1.1.1.02	plan area at bottom of pit 2 x 1m; excavated by hand	Nr	1.00	73.91	1.52	-	75.43	0.370
B1.1.2	maximum depth: 1 - 2 m							
B1.1.2.01	plan area at bottom of pit 2 x 1m	Nr	0.43	31.78	29.18	-	60.96	12.006
B1.1.3	maximum depth: 2 - 3 m							
B1.1.3.01	plan area at bottom of pit 2 x 1m	Nr	0.46	34.00	31.21	-	65.21	12.843
B1.1.4	maximum depth: 3-5 m							
B1.1.4.01	plan area at bottom of pit 2 x 1m	Nr	0.52	38.43	35.28	-	73.71	14.518
B1.2	**Number in material which includes rock**							
B1.2.1	maximum depth: not exceeding 1 m							
B1.2.1.01	plan area at bottom of pit 2 x 1m	Nr	0.24	17.74	16.28	-	34.02	6.701
B1.2.2	maximum depth: 1 - 2 m							
B1.2.2.01	plan area at bottom of pit 2 x 1m	Nr	0.45	33.26	30.53	-	63.79	12.564
B1.2.3	maximum depth: 2 - 3 m							
B1.2.3.01	plan area at bottom of pit 2 x 1m	Nr	0.49	36.22	33.25	-	69.47	13.681
B1.2.4	maximum depth: 3-5 m							
B1.2.4.01	plan area at bottom of pit 2 x 1m	Nr	0.58	42.87	39.35	-	82.22	16.194
B1.3	**Depth in material other than rock**							
B1.3.0	Plan area at bottom of pit 2 x 1m							
B1.3.0.01	Maximum depth not exceeding 1m	m	0.24	17.74	16.28	-	34.02	6.701
B1.3.0.02	Maximum depth not exceeding 1m (excavated by hand)	m	1.00	73.91	1.52	-	75.43	0.370
B1.3.0.03	Maximum depth 1 - 2m	m	0.50	36.95	33.92	-	70.87	13.960
B1.3.0.04	Maximum depth 2 - 3m	m	0.56	41.39	38.00	-	79.39	15.635
B1.3.0.05	Maximum depth 3 - 5m	m	0.64	47.30	43.42	-	90.72	17.869
B1.4	**Depth in rock**							
B1.4.0	Plan area at bottom of pit 2 x 1m							
B1.4.0.01	Maximum depth not exceeding 1m	m	0.32	23.65	21.71	-	45.36	8.934
B1.4.0.02	Maximum depth 1 - 2m	m	0.59	43.61	40.03	-	83.64	16.473
B1.4.0.03	Maximum depth 2 - 3m	m	0.67	49.30	45.26	-	94.56	18.623
B1.4.0.04	Maximum depth 3 - 5m	m	0.79	58.39	53.60	-	111.99	22.057
B1.5	**Depth supported**							
B1.5.0	Plan area at bottom of pit 2 x 1m							
B1.5.0.01	Maximum depth not exceeding 1m; in soft ground	m	0.43	19.17	-	8.78	27.95	12.354	-5.561
B1.5.0.02	Maximum depth 1 - 2m; in soft ground	m	0.51	22.65	-	12.87	35.52	17.545	-7.905

B1 Trial Pits And Trenches continued...

	Unit	Labour Hours	Labour Net £	Plant Net £	Materials Net £	Unit Net £	CO_2e Kg	
B1.5	**Depth supported**							
B1.5.0	Plan area at bottom of pit 2 x 1m							
B1.5.0.03	Maximum depth 2 - 3m; in soft ground	m	0.62	27.88	-	21.06	48.94	27.962	-12.599
B1.5.0.04	Maximum depth 3 - 5m; in soft ground	m	0.90	40.08	-	32.78	72.86	41.666	-18.774
B1.5.0.05	Maximum depth not exceeding 1m; in hard ground	m	0.10	4.33	-	0.87	5.20	2.889	-1.302
B1.5.0.06	Maximum depth 1 - 2m; in hard ground	m	0.12	5.23	-	1.15	6.38	3.841	-1.730
B1.5.0.07	Maximum depth 2 - 3m; in hard ground	m	0.16	6.97	-	4.10	11.07	13.632	-6.142
B1.5.0.08	Maximum depth 3 - 5m; in hard ground	m	0.39	17.43	-	6.03	23.46	20.068	-9.042
B1.6	**Depth backfilled, material stated**							
B1.6.0	Materials arising from excavation							
B1.6.0.01	In soft ground	m	0.65	9.12	0.38	-	9.50	0.092
B1.6.0.02	In hard ground	m	1.65	83.03	4.82	-	87.85	30.689

B2 Light Cable Percussion Boreholes

	Unit	Labour Hours	Labour Net £	Plant Net £	Materials Net £	Unit Net £	CO_2e Kg
B2.1	**Number**						
B2.1.0	including transport, erection, dismantling						
B2.1.0.01	diameter at base 150mm	Nr	1.00	30.34	33.01	-	63.35	27.422
B2.3	**Depth**						
B2.3.2	In holes of maximum depth: 5 - 10m						
B2.3.2.01	In soft ground; borehole at base; 150mm diameter	m	0.50	15.17	12.87	-	28.04	9.647
B2.3.3	In holes of maximum depth: 10 - 20 m						
B2.3.3.01	In soft ground; borehole at base; 150mm diameter	m	0.70	21.24	18.01	-	39.25	13.505
B2.3.4	in holes of maximum depth: 20 - 30 m						
B2.3.4.01	In soft ground; borehole at base; 150mm diameter	m	0.80	24.27	20.58	-	44.85	15.434

B3 Rotary Drilled Boreholes

	Unit	Labour Hours	Labour Net £	Plant Net £	Materials Net £	Unit Net £	CO_2e Kg
B3.1	**Number**						
B3.1.0	including transport, erection, dismantling						
B3.1.0.01	diameter at base 150mm	Nr	1.00	43.36	69.20	-	112.56	52.981
B3.3	**Depth; without core recovery**						
B3.3.1	In holes of maximum depth: not exceeding 5m						
B3.3.1.01	150mm diameter	m	0.40	17.34	24.77	-	42.11	17.941
B3.3.2	In holes of maximum depth: 5 - 10 m						
B3.3.2.01	150mm diameter	m	0.40	17.34	24.77	-	42.11	17.941
B3.3.3	In holes of maximum depth: 10 - 20 m						
B3.3.3.01	150mm diameter	m	0.50	21.68	30.96	-	52.64	22.426

B3 Rotary Drilled Boreholes continued...

	Unit	Labour Hours	Labour Net £	Plant Net £	Materials Net £	Unit Net £	CO_2e Kg
B3.4	**Depth; with core recovery**						
B3.4.1	In holes of maximum depth: not exceeding 5m						
B3.4.1.01	Maximum 76mm diameter	m	1.15	49.86	71.21	-	121.07	51.580
B3.4.2	In holes of maximum depth: 5 - 10 m						
B3.4.2.01	Maximum 76mm diameter	m	1.15	49.86	71.21	-	121.07	51.580
B3.4.3	In holes of maximum depth: 10 - 20 m						
B3.4.3.01	Maximum 76mm diameter	m	1.25	54.20	77.40	-	131.60	56.065
B3.5	**Depth cased**						
B3.5.0	Additional cost						
B3.5.0.01	Additional cost of casting in drift deposits	m	0.15	6.50	9.29	-	15.79	6.728
B3.5.0.02	Additional cost of casing rock	m	0.15	6.50	9.29	-	15.79	6.728
B3.7	**Core boxes, length of core stated**						
B3.7.0	3m capacity						
B3.7.0.01	to be retained until 1 month after submission of report	Nr	0.20	8.67	12.38	-	21.05	8.970
B3.7.0.02	to be retained by client	Nr	0.40	17.34	24.77	-	42.11	17.941

B4 Samples

	Unit	Labour Hours	Labour Net £	Plant Net £	Materials Net £	Unit Net £	CO_2e Kg
B4.2	**From boreholes**						
B4.2.1	Open tube						
B4.2.1.01	Undisturbed sample; 102 x 450mm long	Nr	-	-	-	-	22.18	-
B4.2.2	Distributed						
B4.2.2.01	Bulk disturbed samples	Nr	-	-	-	-	4.88	-
B4.2.3	Groundwater						
B4.2.3.01	Jar samples of soil or ground water	Nr	-	-	-	-	4.20	-

B5 Site Tests And Observations

	Unit	Labour Hours	Labour Net £	Plant Net £	Materials Net £	Unit Net £	CO_2e Kg
B5.1	**Site tests and observations**						
B5.1.3	Standard penetration						
B5.1.3.01	maximum 75 nr. blows	Nr	-	-	-	-	22.18	-

B7 Laboratory Tests

	Unit	Labour Hours	Labour Net £	Plant Net £	Materials Net £	Unit Net £	CO_2e Kg
B7.1	**Classification**						
B7.1.1	Moisture content						
B7.1.1.01	Moisture content	Nr	-	-	-	-	5.77	-
B7.1.3	Specific gravity						
B7.1.3.01	Bulk density	Nr	-	-	-	-	18.63	-
B7.1.4	Particle size analysis by sieve						
B7.1.4.01	P.S.D. sieve analysis	Nr	-	-	-	-	48.27	-
B7.1.5	Particle size analysis by pipette or hydrometer						
B7.1.5.01	P.S.D. sedimentation analysis	Nr	-	-	-	-	67.44	-

B7 Laboratory Tests continued...

	Unit	Labour Hours	Labour Net £	Plant Net £	Materials Net £	Unit Net £	CO_2e Kg
B7.2	**Chemical content**						
B7.2.1 Organic matter							
B7.2.1.01 Organic content	Nr	-	-	-	-	26.62	-
B7.2.2 Sulphate							
B7.2.2.01 Quantitative estimation of sulphate content of water or soil	Nr	-	-	-	-	17.75	-
B7.2.3 pH value							
B7.2.3.01 pH determination of water or soil	Nr	-	-	-	-	10.65	-
B7.3	**Compaction**						
B7.3.1 Standard compaction							
B7.3.1.01 Density/water content relationship	Nr	-	-	-	-	106.49	-
B7.3.2 Heavy compaction							
B7.3.2.01 Density/water content relationship	Nr	-	-	-	-	115.36	-
B7.4	**Consolidation test**						
B7.4.1 Oedometer cell							
B7.4.1.01 Oedometer cell	Nr	-	-	-	-	124.23	-
B7.5	**Permeability**						
B7.5.1 Constant head							
B7.5.1.01 granular soils	Nr	-	-	-	-	133.11	-
B7.5.1.02 cohesive soils	Nr	-	-	-	-	177.48	-
B7.6	**Soil strength**						
B7.6.1 Quick undrained triaxial							
B7.6.1.01 Set of 3 nr, 38mm diameter specimens	Nr	-	-	-	-	79.86	-
B7.6.1.02 102mm diameter specimen	Nr	-	-	-	-	79.86	-
B7.6.2 Consolidated undrained triaxial, with pore water pressure measurement							
B7.6.2.01 Multi-stage; 102mm diameter specimen	Nr	-	-	-	-	79.86	-
B7.6.2.02 Set of 3 nr, 38mm diameter samples: maximum 3 days	Nr	-	-	-	-	244.03	-
B7.6.2.03 Set of 3 nr, 38mm diameter samples: exceeding 3 days	Nr	-	-	-	-	244.03	-
B7.6.3 Consolidated drained triaxial, with volume change measurement							
B7.6.3.01 Set of 3 nr, 38mm diameter samples: maximum 4 days	Nr	-	-	-	-	266.21	-
B7.6.3.02 Set of 3 nr, 38mm diameter samples: exceeding 4 days	Nr	-	-	-	-	266.21	-
B7.6.4 Shearbox: peak only							
B7.6.4.01 Quick undrained; 60 x 60mm	Nr	-	-	-	-	106.49	-
B7.6.4.02 Quick undrained; 100 x 100mm	Nr	-	-	-	-	133.11	-
B7.6.5 Shearbox: peak and residual							
B7.6.5.01 Consolidated drained: maximum 3 days	Nr	-	-	-	-	244.03	-
B7.6.6 Shearbox: residual only							
B7.6.6.01 Consolidated drained with measurement of residual strength: maximum 5 days	Nr	-	£	£	£	266.21	-

B7 Laboratory Tests continued...

	Unit	Labour Hours	Labour Net £	Plant Net £	Materials Net £	Unit Net £	CO_2e Kg
B7.6	**Soil strength**						
B7.6.6 Shearbox: residual only							
B7.6.6.02 Consolidated drained with measurement of residual strength: exceeding 5 days	Nr	-	-	-	-	266.21	-
B7.6.8 California bearing ratio							
B7.6.8.01 Remoulded at natural m/c and density	Nr	-	-	-	-	70.99	-
B7.6.8.02 Remoulded at specified m/c and density	Nr	-	-	-	-	70.99	-
B7.6.8.03 On undisturbed sample	Nr	-	-	-	-	62.12	-
B7.7	**Rock strength**						
B7.7.1 Unconfined compressive strength of core samples; 100mm diameter samples: maximum 3 days							
B7.7.1.01 100mm diameter samples: maximum 3 days	Nr	-	-	-	-	75.43	-
B7.7.1.02 100mm diameter samples: exceeding 3 days	Nr	-	-	-	-	75.43	-
B7.7.2 Consolidated drained triaxial, with volume change measurement							
B7.7.2.01 Set of 3nr, 38mm diameter samples: maximum 3 days	Nr	-	-	-	-	244.03	-
B7.7.2.02 Set of 3nr, 38mm diameter samples: exceeding 3 days	Nr	-	-	-	-	244.03	-
B7.7.3 Brazilian: 100mm diameter samples							
B7.7.3.01 100mm diameter samples	Nr	-	-	-	-	399.32	-
B7.7.5 Point load							
B7.7.5.01 Minimum dimensions of the samples 60mm	Nr	-	-	-	-	14.20	-

B8 Professional Services

Note(s): NOTE: The following are rates for the various grades of technical staff who should be appropriately employed in the preparation of the report. Chargeable time shall include visiting sites to assess the soil, geological and topographical conditions, examination of samples and cores on site and in the laboratory, preparation of testing schedules, logs, sections, and analysis and writing of the report. Charges shall also be made in connection with site or office meetings investigated at the Client's / Engineer's request.

	Unit	Labour Hours	Labour Net £	Plant Net £	Materials Net £	Unit Net £	CO_2e Kg
B8.1	**Technician**						
B8.1.0 Technician							
B8.1.0.01 Technical assistant	Hr	1.00	31.06	-	-	31.06	-
B8.2	**Technician engineer**						
B8.2.0 Technician engineer							
B8.2.0.01 Assistant engineer	Hr	1.00	55.12	-	-	55.12	-
B8.3	**Engineer or geologist**						
B8.3.1 Graduate							
B8.3.1.01 Engineer or geologist	Hr	1.00	47.62	-	-	47.62	-
B8.3.2 Chartered							
B8.3.2.01 Senior engineer or geologist	Hr	1.00	68.07	-	-	68.07	-

B8 Professional Services continued...

	Unit	Labour Hours	Labour Net £	Plant Net £	Materials Net £	Unit Net £	CO_2e Kg
B8.3 **Engineer or geologist**							
B8.3.3 Principal or consultant							
B8.3.3.01 Chief engineer or geologist	Hr	1.00	88.52	-	-	88.52	-
B8.4 **Visits to the site**							
B8.4.0 Engineering grades							
B8.4.0.01 Assistant engineer	Nr	1.00	55.12	-	-	55.12	-
B8.4.0.02 Senior or chief engineer	Nr	1.00	68.07	-	-	68.07	-
B8.5 **Overnight stays in connection with visits to site**							
B8.5.0 Engineer							
B8.5.0.01 Assistant engineer	Nr	1.50	82.68	-	-	82.68	-
B8.5.0.02 Senior or chief engineer	Nr	1.50	102.10	-	-	102.10	-

CLASS C:
GEOTECHNICAL AND OTHER SPECIALIST PROCESSES

Calculations used throughout Class B - Ground Investigation

Labour

			Qty		Rate		Total
L A0150ICE	**Provision of concrete Labour Gang**						
	L A9003ICE	Labourer (Skill Rate 3)	1	x	14.86	=	£14.86
	L A9002ICE	Labourer (General Operative)	1	x	13.02	=	£13.02
	L A9015ICE	Banksman	0.5	x	14.03	=	£7.01
		Total hourly cost of gang				=	**£34.89**
L C0002ICE	**Bored Piling Labour Gang**						
	L A9016ICE	Ganger	2	x	17.64	=	£35.28
	L A9002ICE	Labourer (General Operative)	2	x	13.02	=	£26.03
	L A9000ICE	Craftsman WRA	2	x	17.33	=	£34.66
		Total hourly cost of gang				=	**£95.97**
L C0003ICE	**Diaphragam Wall Concreting Labour Gang**						
	L A9003ICE	Labourer (Skill Rate 3)	2	x	14.86	=	£29.72
	L A9002ICE	Labourer (General Operative)	2	x	13.02	=	£26.03
	L A9015ICE	Banksman	1	x	14.03	=	£14.03
		Total hourly cost of gang				=	**£69.78**
L C0004ICE	**Diaphragm Wall - Guide Wall Construction Labour Gang**						
	L A9016ICE	Ganger	2	x	17.64	=	£35.28
	L A9002ICE	Labourer (General Operative)	2	x	13.02	=	£26.03
	L A9015ICE	Banksman	2	x	14.03	=	£28.06
	L A9003ICE	Labourer (Skill Rate 3)	2	x	14.86	=	£29.72
	L A9002ICE	Labourer (General Operative)	2	x	13.02	=	£26.03
	L A9015ICE	Banksman	1	x	14.03	=	£14.03
	L A9000ICE	Craftsman WRA	2	x	17.33	=	£34.66
	L A9002ICE	Labourer (General Operative)	1	x	13.02	=	£13.02
		Total hourly cost of gang				=	**£206.83**
L C0005ICE	**Diaphragm Wall Excavation Labour Gang**						
	L A9016ICE	Ganger	2	x	17.64	=	£35.28
	L A9002ICE	Labourer (General Operative)	2	x	13.02	=	£26.03
	L A9015ICE	Banksman	2	x	14.03	=	£28.06
		Total hourly cost of gang				=	**£89.37**
L C0006ICE	**Diaphragm Wall Joints Labour Gang**						
	L A9000ICE	Craftsman WRA	2	x	17.33	=	£34.66
	L A9002ICE	Labourer (General Operative)	1	x	13.02	=	£13.02
		Total hourly cost of gang				=	**£47.68**
L C0007ICE	**Diaphragm Wall Steelfixing Labour Gang**						
	L A9000ICE	Craftsman WRA	2	x	17.33	=	£34.66
	L A9002ICE	Labourer (General Operative)	1	x	13.02	=	£13.02
		Total hourly cost of gang				=	**£47.68**

L C0008ICE   **Ground Anchorage Labour Gang**

L A9016ICE	Ganger	1	x	17.64	=	£17.64
L A9002ICE	Labourer (General Operative)	1	x	13.02	=	£13.02
L A9000ICE	Craftsman WRA	1	x	17.33	=	£17.33
	Total hourly cost of gang				=	**£47.99**

L C0009ICE   **Ground Anchorage Labour Gang (Mobilise / Demobilise / Move)**

L A9016ICE	Ganger	10	x	17.64	=	£176.40
L A9002ICE	Labourer (General Operative)	10	x	13.02	=	£130.15
L A9000ICE	Craftsman WRA	10	x	17.33	=	£173.28
	Total hourly cost of gang				=	**£479.83**

Plant

P C0002ICE   **Bored Piling (Medium Rig) Plant Gang**

P A1063ICE	Bored Piling Rig Medium	1	x	165.10	=	£165.10
P A0737ICE	Concrete Mixer - Schwing BP1000R Concrete Pumps	1	x	46.33	=	£46.33
	Total hourly cost of gang				=	**£211.43**

P C0003ICE   **Bored Piling (Medium) Plant Gang (Mobilise / Demobilise / Move)**

P A1063ICE	Bored Piling Rig Medium	8	x	165.10	=	£1,320.80
P A0737ICE	Concrete Mixer - Schwing BP1000R Concrete Pumps	8	x	46.33	=	£370.64
P A1044ICE	Articulated Lorry	8	x	68.51	=	£548.08
P A1045ICE	Low Loader Trailer	8	x	13.96	=	£111.68
P A1063ICE	Bored Piling Rig Medium	10	x	165.10	=	£1,651.00
P A0737ICE	Concrete Mixer - Schwing BP1000R Concrete Pumps	10	x	46.33	=	£463.30
	Total hourly cost of gang				=	**£4,465.50**

P C0006ICE   **Diaphragm Wall Hydrofraise Plant Gang**

P A1070ICE	Hydrofraise Set Up	1	x	530.67	=	£530.67
P A1039ICE	Bentonite Plant	1	x	115.52	=	£115.52
	Total hourly cost of gang				=	**£646.19**

P C0008ICE   **Ground Anchorage (Mobilise / Demobilise / Move) Plant Gang**

P A1068ICE	Ground Anchor Drilling Rig	10	x	119.16	=	£1,191.60
P A1069ICE	Grouting Rig (Anchorage Items)	10	x	10.53	=	£105.30
P A0458ICE	Compressor - 480 cfm	10	x	20.17	=	£201.70
P A0737ICE	Concrete Mixer - Schwing BP1000R Concrete Pumps	10	x	46.33	=	£463.30
P A0442ICE	Bowsers (250 gallon water / fuel)	10	x	0.94	=	£9.40
P A1044ICE	Articulated Lorry	8	x	68.51	=	£548.08
P A1045ICE	Low Loader Trailer	8	x	13.96	=	£111.68
	Total hourly cost of gang				=	**£2,631.06**

P C0009ICE   **Ground Anchorage Plant Gang**

P A1068ICE	Ground Anchor Drilling Rig	1	x	119.16	=	£119.16
P A1069ICE	Grouting Rig (Anchorage Items)	1	x	10.53	=	£10.53
P A0458ICE	Compressor - 480 cfm	1	x	20.17	=	£20.17
P A0737ICE	Concrete Mixer - Schwing BP1000R Concrete Pumps	1	x	46.33	=	£46.33
P A0442ICE	Bowsers (250 gallon water / fuel)	1	x	0.94	=	£0.94

Total hourly cost of gang   =   **£197.13**

P C0010ICE   **Grout Hole Plant Gang (Drilling rock or other artificial hard material)**

P A1060ICE	Anchor Drilling Rig (Rock)	1	x	85.26	=	£85.26
P A0458ICE	Compressor - 480 cfm	1	x	20.17	=	£20.17
P A0442ICE	Bowsers (250 gallon water / fuel)	1	x	0.94	=	£0.94

Total hourly cost of gang   =   **£106.37**

P C0011ICE   **Grout Hole Plant Gang (Drilling Soils)**

P A1061ICE	Anchor Drilling Rig (Soil)	1	x	78.91	=	£78.91
P A0458ICE	Compressor - 480 cfm	1	x	20.17	=	£20.17
P A0442ICE	Bowsers (250 gallon water / fuel)	1	x	0.94	=	£0.94

Total hourly cost of gang   =   **£100.02**

P C0012ICE   **Grout Hole Plant Gang (Grouting rock or other artificial hard material)**

P A1060ICE	Anchor Drilling Rig (Rock)	1	x	85.26	=	£85.26
P A0458ICE	Compressor - 480 cfm	1	x	20.17	=	£20.17
P A0442ICE	Bowsers (250 gallon water / fuel)	1	x	0.94	=	£0.94
P A1069ICE	Grouting Rig (Anchorage Items)	1	x	10.53	=	£10.53

Total hourly cost of gang   =   **£116.90**

P C0013ICE   **Grout Hole Plant Gang (Mobilise / Demobilise / Move)**

P A1061ICE	Anchor Drilling Rig (Soil)	12	x	78.91	=	£946.92
P A0458ICE	Compressor - 480 cfm	12	x	20.17	=	£242.04
P A0442ICE	Bowsers (250 gallon water / fuel)	12	x	0.94	=	£11.28
P A0737ICE	Concrete Mixer - Schwing BP1000R Concrete Pumps	12	x	46.33	=	£555.96
P A1071ICE	Pile Jetting Water Pump	12	x	522.14	=	£6,265.68
P A1044ICE	Articulated Lorry	10	x	68.51	=	£685.10
P A1045ICE	Low Loader Trailer	10	x	13.96	=	£139.60

Total hourly cost of gang   =   **£8,846.58**

Geotechnical and other Specialist Processes

Note(s): NOTE: The rates in this section are based on historical specialist sub-contractor prices and are provided as guide prices only.

C1 Drilling For Grout Holes Through Materials Other Than Rock Or Artificial Hard

Note(s): NOTE: An allowance for establishing on site all plant, equipment and staff and removing on completion is included in the rates for drilling, assuming a total of 50nr

		Unit	Labour Hours	Labour Net £	Plant Net £	Materials Net £	Unit Net £	CO_2e Kg
C1.1	**Vertically downwards**							
C1.1.1	In holes of depth not exceeding 5m							
C1.1.1.01	Generally	m	0.18	8.64	18.00	-	26.64	48.846
C1.1.2	In holes of depth; 5 - 10m							
C1.1.2.01	Generally	m	0.20	9.60	20.00	-	29.60	54.273
C1.1.3	In holes of depth; 10 - 20m							
C1.1.3.01	Generally	m	0.22	10.56	22.00	-	32.56	59.700
C1.1.4	In holes of depth; 20 - 30m							
C1.1.4.01	Generally	m	0.25	11.99	25.00	-	36.99	67.841
C1.2	**Downwards at an angle 0 - 45 degrees to the vertical**							
C1.2.1	In holes of depth not exceeding 5m							
C1.2.1.01	Generally	m	0.18	8.64	18.00	-	26.64	48.846
C1.2.2	In holes of depth; 5 - 10m							
C1.2.2.01	Generally	m	0.20	9.60	20.00	-	29.60	54.273
C1.2.3	In holes of depth; 10 - 20m							
C1.2.3.01	Generally	m	0.22	10.56	22.00	-	32.56	59.700
C1.2.4	In holes of depth; 20 - 30m							
C1.2.4.01	Generally	m	0.25	11.99	25.00	-	36.99	67.841
C1.3	**Horizontally or downwards at an angle less than 45 degrees to the horizontal**							
C1.3.1	In holes of depth not exceeding 5m							
C1.3.1.01	Generally	m	0.18	8.64	18.00	-	26.64	48.846
C1.3.2	In holes of depth; 5 - 10m							
C1.3.2.01	Generally	m	0.20	9.60	20.00	-	29.60	54.273
C1.3.3	In holes of depth; 10 - 20m							
C1.3.3.01	Generally	m	0.22	10.56	22.00	-	32.56	59.700
C1.3.4	In holes of depth; 20 - 30m							
C1.3.4.01	Generally	m	0.25	11.99	25.00	-	36.99	67.841
C1.4	**Upwards at an angle of 0 - 45 degrees to the horizontal**							
C1.4.1	In holes of depth not exceeding 5m							
C1.4.1.01	Generally	m	0.30	14.39	30.01	-	44.40	81.410
C1.4.2	In holes of depth; 5 - 10m							
C1.4.2.01	Generally	m	0.35	16.79	35.01	-	51.80	94.978
C1.4.3	In holes of depth; 10 - 20m							
C1.4.3.01	Generally	m	0.40	19.19	40.01	-	59.20	108.546
C1.4.4	In holes of depth; 20 - 30m							
C1.4.4.01	Generally	m	0.50	23.99	50.01	-	74.00	135.683

C1 Drilling For Grout Holes Through Materials Other Than Rock Or Artificial Hard Material continued...

	Unit	Labour Hours	Labour Net £	Plant Net £	Materials Net £	Unit Net £	CO_2e Kg
C1.5	**Upwards at an angle less than 45 degrees to the vertical**						
C1.5.1	In holes of depth not exceeding 5m						
C1.5.1.01	Generally	m	0.30	14.39	30.01	-	44.40	81.410
C1.5.2	In holes of depth; 5 - 10m						
C1.5.2.01	Generally	m	0.35	16.79	35.01	-	51.80	94.978
C1.5.3	In holes of depth; 10 - 20m						
C1.5.3.01	Generally	m	0.40	19.19	40.01	-	59.20	108.546
C1.5.4	In holes of depth; 20 - 30m						
C1.5.4.01	Generally	m	0.50	23.99	50.01	-	74.00	135.683

C2 Drilling for grout holes through rock or artificial hard material

Note(s): NOTE: An allowance for establishing on site all plant, equipment and staff and removing on completion is included in the rates for drilling, assuming a total of 50nr

	Unit	Labour Hours	Labour Net £	Plant Net £	Materials Net £	Unit Net £	CO_2e Kg
C2.1	**Vertically downwards**						
C2.1.1	In holes of depth not exceeding 5m						
C2.1.1.01	Generally	m	0.16	7.68	17.02	-	24.70	43.418
C2.1.2	In holes of depth; 5 - 10m						
C2.1.2.01	Generally	m	0.20	9.60	21.27	-	30.87	54.273
C2.1.3	In holes of depth; 10 - 20m						
C2.1.3.01	Generally	m	0.24	11.52	25.53	-	37.05	65.128
C2.1.4	In holes of depth; 20 - 30m						
C2.1.4.01	Generally	m	0.30	14.39	31.91	-	46.30	81.410
C2.2	**Downwards at an angle 0 - 45 degrees to the vertical**						
C2.2.1	In holes of depth not exceeding 5m						
C2.2.1.01	Generally	m	0.16	7.68	17.02	-	24.70	43.418
C2.2.2	In holes of depth; 5 - 10m						
C2.2.2.01	Generally	m	0.20	9.60	21.27	-	30.87	54.273
C2.2.3	In holes of depth; 10 - 20m						
C2.2.3.01	Generally	m	0.24	11.52	25.53	-	37.05	65.128
C2.2.4	In holes of depth; 20 - 30m						
C2.2.4.01	Generally	m	0.30	14.39	31.91	-	46.30	81.410
C2.3	**Horizontally or downwards at an angle less than 45 degrees to the horizontal**						
C2.3.1	In holes of depth not exceeding 5m						
C2.3.1.01	Generally	m	0.16	7.68	17.02	-	24.70	43.418
C2.3.2	In holes of depth; 5 - 10m						
C2.3.2.01	Generally	m	0.20	9.60	21.27	-	30.87	54.273
C2.3.3	In holes of depth; 10 - 20m						
C2.3.3.01	Generally	m	0.24	11.52	25.53	-	37.05	65.128
C2.3.4	In holes of depth; 20 - 30m						
C2.3.4.01	Generally	m	0.30	14.39	31.91	-	46.30	81.410

C2 Drilling for grout holes through rock or artificial hard material continued...

	Unit	Labour Hours	Labour Net £	Plant Net £	Materials Net £	Unit Net £	CO_2e Kg
C2.4	**Upwards at an angle of 0 - 45 degrees to the horizontal**						
C2.4.1 In holes of depth not exceeding 5m							
C2.4.1.01 Generally	m	0.27	12.95	28.72	-	41.67	73.269
C2.4.2 In holes of depth; 5 - 10m							
C2.4.2.01 Generally	m	0.30	14.39	31.91	-	46.30	81.410
C2.4.3 In holes of depth; 10 - 20m							
C2.4.3.01 Generally	m	0.35	16.79	37.23	-	54.02	94.978
C2.4.4 In holes of depth; 20 - 30m							
C2.4.4.01 Generally	m	0.40	19.19	42.55	-	61.74	108.546
C2.5	**Upwards at an angle less than 45 degrees to the vertical**						
C2.5.1 In holes of depth not exceeding 5m							
C2.5.1.01 Generally	m	0.38	18.23	40.42	-	58.65	103.119
C2.5.2 In holes of depth; 5 - 10m							
C2.5.2.01 Generally	m	0.45	21.59	47.87	-	69.46	122.114
C2.5.3 In holes of depth; 10 - 20m							
C2.5.3.01 Generally	m	0.51	24.47	54.25	-	78.72	138.396
C2.5.4 In holes of depth; 20 - 30m							
C2.5.4.01 Generally	m	0.60	28.79	63.82	-	92.61	162.819

C4 Grout Holes Materials And Injection

Note(s): NOTE: An allowance for establishing on site all plant, equipment and staff and removing on completion is included in the rates for 'number of injections', assuming 50nr injections.

	Unit	Labour Hours	Labour Net £	Plant Net £	Materials Net £	Unit Net £	CO_2e Kg
C4.1	**Number of holes**						
C4.1.0 Generally							
C4.1.0.01 Number of holes	Nr	0.75	35.98	87.68	-	123.66	203.524
C4.2	**Number of stages**						
C4.2.0 Generally							
C4.2.0.01 Number of stages (4nr)	Nr	0.25	11.99	29.23	-	41.22	67.841
C4.3	**Single water pressure tests**						
C4.3.0 Generally							
C4.3.0.01 Single water pressure test	Nr	0.25	11.99	9.10	-	21.09	31.003
C4.4	**Multiple water pressure test**						
C4.4.0 Generally							
C4.4.0.01 Multiple water pressure test	Nr	0.50	23.99	18.20	-	42.19	62.006
C4.5	**Materials**						
C4.5.1 Cement							
C4.5.1.01 Generally	t	-	-	-	156.68	156.68	159.238
C4.5.2 Pulverized fuel ash							
C4.5.2.01 Generally	t	-	-	-	216.63	216.63	729.833
C4.5.3 Sand							
C4.5.3.01 Generally	t	-	-	-	17.28	17.28	7.428

C4 Grout Holes Materials And Injection continued...

	Unit	Labour Hours	Labour Net £	Plant Net £	Materials Net £	Unit Net £	CO_2e Kg
C4.5	**Materials**						
C4.5.4	Pea Gravel						
C4.5.4.01	Generally	t	-	-	-	26.06	26.06	7.003
C4.5.5	Bentonite						
C4.5.5.01	Generally	t	-	-	-	263.79	263.79	18.149
C4.5.6	Chemicals						
C4.5.6.01	Colloidal Silica Chemical Grout	t	-	-	-	3,765.71	3,765.71	2,564.045
C4.5.6.02	Sodium Silicate Chemical Grout	t	-	-	-	2,124.55	2,124.55	2,564.045
C4.5.6.03	Acrylic Resin Chemical Grout	t	-	-	-	4,822.02	4,822.02	6,684.597
C4.5.6.04	Polyurethane Resin Chemical Grout	t	-	-	-	9,081.87	9,081.87	6,684.597
C4.6	**Injection**						
C4.6.1	Number of injections						
C4.6.1.01	Injection	Nr	0.02	14.87	89.49	-	104.36	66.711

C5 Diaphragm Walls

Note(s): NOTE: An allowance for establishing on site all plant, equipment and staff and removing on completion is included in the rates for excavation, assuming 1000m³ of excavation.

Diaphragm walls cannot comply with watertight structures codes and the degree of watertightness achievable must always be qualified.

	Unit	Labour Hours	Labour Net £	Plant Net £	Materials Net £	Unit Net £	CO_2e Kg
C5.1	**Excavation in material other than rock or artificial hard material**						
C5.1.1	Maximum depth; not exceeding 5m						
C5.1.1.01	Supply and disposal of bentonite slurry, walls 1m thick	m³	0.33	29.76	76.84	52.76	159.36	66.211
C5.1.2	Maximum depth; 5m - 10m						
C5.1.2.01	Supply and disposal of bentonite slurry, walls 1m thick	m³	0.33	29.76	76.84	52.76	159.36	66.211
C5.1.3	Maximum depth; 10m - 15m						
C5.1.3.01	Supply and disposal of bentonite slurry, walls 1m thick	m³	0.33	29.76	76.84	52.76	159.36	66.211
C5.4	**Concrete**						
C5.4.0	Grade 25, 20mm aggregate						
C5.4.0.01	Walls; 1m thick	m³	0.20	10.47	4.63	90.37	105.47	277.308
C5.5	**Plain round steel bar reinforcement**						
C5.5.3	nominal size: 10mm						
C5.5.3.01	Cut, bent and fixed in place	t	10.00	476.70	-	683.48	1,160.18	588.516
C5.5.5	nominal size: 16mm						
C5.5.5.01	Cut, bent and fixed in place	t	8.00	381.36	-	630.16	1,011.52	588.516
C5.5.6	nominal size: 20mm						
C5.5.6.01	Cut, bent and fixed in place	t	8.00	381.36	-	630.16	1,011.52	588.516
C5.5.7	nominal size: 25mm						
C5.5.7.01	Cut, bent and fixed in place	t	8.00	381.36	-	629.72	1,011.08	588.516

C5 Diaphragm Walls continued...

	Unit	Labour Hours	Labour Net £	Plant Net £	Materials Net £	Unit Net £	CO_2e Kg
C5.6	**Deformed high yield steel bar reinforcement**						
C5.6.3	nominal size: 10mm						
C5.6.3.01 Cut, bent and fixed in place	t	10.00	476.70	-	629.64	1,106.34	588.516
C5.6.5	nominal size: 16mm						
C5.6.5.01 Cut, bent and fixed in place	t	8.00	381.36	-	580.52	961.88	588.516
C5.6.6	nominal size: 20mm						
C5.6.6.01 Cut, bent and fixed in place	t	8.00	381.36	-	580.52	961.88	588.516
C5.6.7	nominal size: 25mm						
C5.6.7.01 Cut, bent and fixed in place	t	8.00	381.36	-	580.52	961.88	588.516
C5.7	**Waterproofed joints**						
C5.7.0	Generally						
C5.7.0.01 Waterproofed joints	Joint	2.00	90.41	3.80	15.85	110.06	13.312
C5.8	**Guided Walls**						
C5.8.0	Concrete grade 25; 20mm aggregate						
C5.8.0.01 Walls 1m wide x 1m deep, running each side of intended excavation	m	0.50	103.41	9.27	354.90	467.58	731.170

C6 Ground Reinforcement

Note(s): NOTES 1) Ground anchors are commonly used to tie back steel sheet piling or other forms of support to the sides of major excavations or waterside structures in order to leave the front face of the support free from obstructions. The tendons which can be steel cable or bar are anchored into the ground behind the support wall by either attaching to a 'deadman' anchor or drilling a bore into the ground inserting the tendon with a suitable end plate and grouting the hole. The free end of the tendon is then joined to the support wall and tensioned to a stress exceeding that of the load to be carried. Anchors can be temporary or permanent. 2) In the following examples ready fabricated anchorage assemblies between 5m and 10m long, complete with corrosion protection are inserted into the previuosly bored holes and then ground into position. 3) An allowance for establishing on site all plant, equipment and staff and removing on completion is included in the rates for 'number of ground anchorages', assuming a total of 50nr ground anchorages. 4) The rates for total length of tendons include drilling, temporary casing, supply of tendon and grouting in.

	Unit	Labour Hours	Labour Net £	Plant Net £	Materials Net £	Unit Net £	CO_2e Kg
C6.1	**Number in material other than rock or artificial hard material to a stated maximum depth**						
C6.1.1	Temporary						
C6.1.1.01 to a maximum depth of 15m	Nr	0.02	9.60	52.62	-	62.22	61.329
C6.1.2	Temporary with single corrosion protection						
C6.1.2.01 to a maximum depth of 15m	Nr	0.02	9.60	52.62	-	62.22	61.329
C6.1.3	Temporary with double corrosion protection						
C6.1.3.01 to a maximum depth of 15m	Nr	0.02	9.60	52.62	-	62.22	61.329
C6.1.4	Permanent						
C6.1.4.01 to a maximum depth of 15m	Nr	0.02	9.60	52.62	-	62.22	61.329
C6.1.5	Permanent with single corrosion protection						
C6.1.5.01 to a maximum depth of 15m	Nr	0.02	9.60	52.62	-	62.22	61.329
C6.1.6	Permanent with double corrosion protection						
C6.1.6.01 to a maximum depth of 15m	Nr	0.02	9.60	52.62	-	62.22	61.329

C6 Ground Reinforcement continued...

	Unit	Labour Hours	Labour Net £	Plant Net £	Materials Net £	Unit Net £	CO_2e Kg
C6.2	**Total length of tendons in material other than rock or artificial hard material**						
C6.2.1	Temporary						
C6.2.1.01 to a maximum depth of 15m	m	0.15	6.96	28.58	35.07	70.61	61.813
C6.2.2	Temporary with single corrosion protection						
C6.2.2.01 to a maximum depth of 15m	m	0.15	6.96	28.58	35.07	70.61	61.813
C6.2.3	Temporary with double corrosion protection						
C6.2.3.01 to a maximum depth of 15m	m	0.15	6.96	28.58	35.07	70.61	61.813
C6.2.4	Permanent						
C6.2.4.01 to a maximum depth of 15m	m	0.15	6.96	28.58	47.81	83.35	61.813
C6.2.5	Permanent with single corrosion protection						
C6.2.5.01 to a maximum depth of 15m	m	0.15	6.96	28.58	47.81	83.35	61.813
C6.2.6	Permanent with double corrosion protection						
C6.2.6.01 to a maximum depth of 15m	m	0.15	6.96	28.58	47.81	83.35	61.813
C6.3	**Number in material which includes rock or artificial hard material in a stated maximum depth**						
C6.3.1	Temporary						
C6.3.1.01 to a maximum depth of 15m	Nr	0.02	9.60	52.62	-	62.22	61.329
C6.3.2	Temporary with single corrosion protection						
C6.3.2.01 to a maximum depth of 15m	Nr	0.02	9.60	52.62	-	62.22	61.329
C6.3.3	Temporary with double corrosion protection						
C6.3.3.01 to a maximum depth of 15m	Nr	0.02	9.60	52.62	-	62.22	61.329
C6.3.4	Permanent						
C6.3.4.01 to a maximum depth of 15m	Nr	0.02	9.60	52.62	-	62.22	61.329
C6.3.5	Permanent with single corrosion protection						
C6.3.5.01 to a maximum depth of 15m	Nr	0.02	9.60	52.62	-	62.22	61.329
C6.3.6	Permanent with double corrosion protection						
C6.3.6.01 to a maximum depth of 15m	Nr	0.02	9.60	52.62	-	62.22	61.329
C6.4	**Total length of tendons in material which includes rock or artificial hard material**						
C6.4.1	Temporary						
C6.4.1.01 to a maximum depth of 15m	m	0.15	6.96	28.58	42.14	77.68	61.813
C6.4.2	Temporary with single corrosion protection						
C6.4.2.01 to a maximum depth of 15m	m	0.15	6.96	28.58	42.14	77.68	61.813
C6.4.3	Temporary with double corrosion protection						
C6.4.3.01 to a maximum depth of 15m	m	0.15	6.96	28.58	42.14	77.68	61.813
C6.4.4	Permanent						
C6.4.4.01 to a maximum depth of 15m	m	0.15	6.96	28.58	61.20	96.74	61.813
C6.4.5	Permanent with single corrosion protection						
C6.4.5.01 to a maximum depth of 15m	m	0.15	6.96	28.58	61.20	96.74	61.813
C6.4.6	Permanent with double corrosion protection						
C6.4.6.01 to a maximum depth of 15m	m	0.15	6.96	28.58	61.20	96.74	61.813

C7   Sand, band and wick drains

Note(s):   NOTE: An allowance for establishing on site all plant, equipment and staff and removing on completion is included in the rates for 'number of drains' assuming a total of 50nr sand drains.

		Unit	Labour Hours	Labour Net £	Plant Net £	Materials Net £	Unit Net £	CO_2e Kg
C7.1	**Number of drains**							
C7.1.3	Cross-sectional dimension: 200 - 300mm							
C7.1.3.01	Sand drains	Nr	0.02	11.89	89.31	-	101.20	82.044
C7.2	**Number of predrilled holes**							
C7.2.3	Cross-sectional dimension: 200 - 300mm							
C7.2.3.01	Sand drains	Nr	0.02	11.89	89.31	-	101.20	82.044
C7.4	**Depth of drains of maximum; not exceeding 10m**							
C7.4.2	Cross-sectional dimension: 100 - 200mm							
C7.4.2.01	Sand drains	m	0.02	1.63	3.59	0.29	5.51	3.715
C7.4.4	Cross-sectional dimension:300 - 400mm							
C7.4.4.01	Sand drains	m	0.02	2.02	4.44	1.21	7.67	4.954
C7.5	**Depth of drains of maximum; 10 - 15m**							
C7.5.2	Cross-sectional dimension: 100 - 200mm							
C7.5.2.01	Sand drains	m	0.02	1.63	3.59	0.29	5.51	3.715
C7.5.4	Cross-sectional dimension:300 - 400mm							
C7.5.4.01	Sand drains	m	0.03	2.40	5.29	1.21	8.90	5.798
C7.6	**Depth of drains of maximum; 15 - 20m**							
C7.6.2	Cross-sectional dimension: 100 - 200mm							
C7.6.2.01	Sand drains	m	0.03	3.17	6.98	0.29	10.44	7.093
C7.6.4	Cross-sectional dimension:300 - 400mm							
C7.6.4.01	Sand drains	m	0.05	4.80	10.57	1.21	16.58	11.076

CLASS D:
DEMOLITION AND SITE CLEARANCE

Calculations used throughout Class D - Demolition and Site Clearance

Labour

			Qty		Rate		Total
L A0330ICE	**Clearance Labour Gang**						
	L A9016ICE	Ganger	1	x	17.64	=	£17.64
	L A9002ICE	Labourer (General Operative)	2	x	13.02	=	£26.03
		Total hourly cost of gang				=	**£43.67**

Plant

			Qty		Rate		Total
P A1330ICE	**Clearance Plant Gang**						
	P A0945ICE	Crawler Tractor / Dozer - Dresser 1004 48kW	1	x	30.69	=	£30.69
	P A1002ICE	Tipping Waggon - 16t 6-Wheel (24t Gr)	1	x	51.34	=	£51.34
	P A0452ICE	Compressor - 2-Tool (Complete)	1	x	5.20	=	£5.20
	P A0404ICE	Rock Drill	3	x	0.95	=	£2.85
		Total hourly cost of gang				=	**£90.08**
P A1331ICE	**Demolition Plant Gang**						
	P A0945ICE	Crawler Tractor / Dozer - Dresser 1004 48kW	1	x	30.69	=	£30.69
	P A1002ICE	Tipping Waggon - 16t 6-Wheel (24t Gr)	1	x	51.34	=	£51.34
	P A0452ICE	Compressor - 2-Tool (Complete)	1	x	5.20	=	£5.20
	P A0404ICE	Rock Drill	3	x	0.95	=	£2.85
	P A1031ICE	Cutting and Burning Gear	1	x	3.77	=	£3.77
	P A0592ICE	Hydraulic Excavator - Cat 320 96kW	1	x	36.36	=	£36.36
		Total hourly cost of gang				=	**£130.21**

Demolition and Site Clearance

D1 Site Clearance

		Unit	Labour Hours	Labour Net £	Plant Net £	Materials Net £	Unit Net £	CO_2e Kg
D1.1	**General clearance**							
D1.1.0	Demolition and removal from site							
D1.1.0.01	Articles, objects and obstructions over the total area of the site	ha	8.00	349.36	720.64	-	1,070.00	439.360
D1.1.0.02	Wooded areas including hedges and small trees	ha	24.00	1,048.08	2,161.92	-	3,210.00	1,318.080

D2 Trees

		Unit	Labour Hours	Labour Net £	Plant Net £	Materials Net £	Unit Net £	CO_2e Kg
D2.1	**Girth: 500mm - 2m**							
D2.1.0	Clearance of trees (girth measured 1m above ground); excluding stumps							
D2.1.0.01	Girth: 500mm - 1m	Nr	0.39	17.03	35.13	-	52.16	21.419
D2.1.0.02	Girth: 1 - 2m	Nr	0.79	34.50	71.16	-	105.66	43.387
D2.2	**Girth: exceeding 2m**							
D2.2.0	Clearance of trees (girth measured 1m above ground); excluding stumps							
D2.2.0.01	Girth: 2 - 3m	Nr	3.00	131.01	270.24	-	401.25	164.760
D2.2.0.02	Girth: 3 - 4m	Nr	8.00	349.36	720.64	-	1,070.00	439.360
D2.2.0.03	Girth: exceeding 5m	Nr	12.50	545.88	1,126.00	-	1,671.88	686.500

D3 Stumps

		Unit	Labour Hours	Labour Net £	Plant Net £	Materials Net £	Unit Net £	CO_2e Kg
D3.1	**Diameter: less than 1m**							
D3.1.0	Generally							
D3.1.0.01	Clearance of stumps; Diameter: 150 - 500mm; Holes backfilled with topsoil from site	Nr	0.50	21.84	45.04	-	66.88	27.460
D3.1.0.02	Clearance of stumps; Diameter: 150 - 500mm; Holes backfilled with clean hardcore from site	Nr	0.60	26.20	54.05	-	80.25	32.952
D3.1.0.03	Clearance of stumps; Diameter: 500mm - 1m; Holes backfilled with topsoil from site	Nr	1.00	43.67	90.08	-	133.75	54.920
D3.1.0.04	Clearance of stumps; Diameter: 500mm - 1m; Holes backfilled with clean hardcore from site	Nr	1.20	52.40	108.10	-	160.50	65.904
D3.2	**Diameter: exceeding 1m**							
D3.2.0	Generally							
D3.2.0.01	Clearance of stumps; diameter: 1.5m; Holes backfilled with topsoil from site	Nr	1.25	54.59	112.60	-	167.19	68.650
D3.2.0.02	Clearance of stumps; diameter: 1.5m; Holes backfilled with clean hardcore from site	Nr	1.50	65.50	135.12	-	200.62	82.380

D4 Buildings

		Unit	Labour Hours	Labour Net £	Plant Net £	Materials Net £	Unit Net £	CO_2e Kg
D4.1	**Brickwork**							
D4.1.6	Volume: 1000 - 2500 m^3							
D4.1.6.01	Administration block; 2 storey height; 250 mm thick traditional cavity wall with timber flat roof and bitumen felt finish; block partitions; 2nr timber staircases (work below original surface excluded); overall size 30 × 12m × 6m high	Nr	27.32	1,193.06	2,460.99	-	3,654.05	1,500.414

D5 Other Structures

		Unit	Labour Hours	Labour Net £	Plant Net £	Materials Net £	Unit Net £	CO_2e Kg
D5.2	**Concrete**							
D5.2.6	Volume: 1000 - 2500 m^3							
D5.2.6.01	Settlement tanks (work below original surface); 500mm thick reinforced concrete; 20m diameter at original surface tapered to 5m diameter at base; 10m height; including 300mm thick reinforced concrete inlet chamber 3 × 3 × 4m deep containing 300mm bore cast Iron pipe and fitting together with 2Nr valves and penstocks	Nr	159.00	6,943.53	14,322.72	-	21,266.25	8,732.280
D5.2.6.02	Sludge Tanks (work above original surface); 400mm thick reinforced concrete; 15 × 15 × 8m high; sloping reinforced concrete base 300mm thick suspended above original surface (work below original surface excluded)	Nr	175.00	7,642.25	15,764.00	-	23,406.25	9,611.000
D5.5	**Timber**							
D5.5.1	Volume: not exceeding 50 m^3							
D5.5.1.01	Timber footbridge; above stream; 8 × 3m wide; timber rails 1.6m high	Nr	6.00	262.02	540.48	-	802.50	329.520

D6 Pipelines

		Unit	Labour Hours	Labour Net £	Plant Net £	Materials Net £	Unit Net £	CO_2e Kg
D6.1	**Nominal bore: 100 - 300mm**							
D6.1.0	Generally							
D6.1.0.01	Clay drains including concrete bed and surround 2m deep	m	0.08	3.41	10.16	-	13.57	5.548
D6.1.0.02	Concrete pipelines including concrete bed and surround 2m deep	m	0.09	3.71	11.07	-	14.78	6.045
D6.1.0.03	Cast iron pipeline on metal supports at 2m centres suspended 1m above final surface	m	0.08	3.41	10.16	-	13.57	5.548

D6 Pipelines continued...

	Unit	Labour Hours	Labour Net £	Plant Net £	Materials Net £	Unit Net £	CO_2e Kg
D6.2	**Nominal bore: 300 - 500mm**						
D6.2.0 Generally							
D6.2.0.01 Concrete pipelines including concrete bed and surround 2m deep	m	0.09	3.97	11.85	-	15.82	6.472
D6.2.0.02 Cast iron pipeline on metal supports at 2m centres suspended 1m above final surface	m	0.09	3.97	11.85	-	15.82	6.472
D6.3	**Nominal bore: exceeding 500mm**						
D6.3.0 Generally							
D6.3.0.01 Concrete pipelines including concrete bed and surround 2m deep; nominal bore: 600mm	m	0.10	4.15	12.37	-	16.52	6.757
D6.3.0.02 Cast iron pipeline on metal supports at 2m centres suspended 1m above final surface; nominal bore: 600mm	m	0.10	4.15	12.37	-	16.52	6.757
D6.3.0.03 Ductile iron pipeline on concrete supports at 4m centres suspended 1m above final surface; nominal bore: 600mm	m	0.10	4.15	12.37	-	16.52	6.757
D6.3.0.04 Ductile iron pipeline on concrete supports at 4m centres suspended 1m above final surface; nominal bore: 800mm	m	0.12	5.24	15.63	-	20.87	8.535

CLASS E:
EARTHWORKS

Calculations used throughout Class E - Earthworks

Labour

			Qty		Rate		Total
L A0120ICE	**General Earthworks Labour Gang**						
	L A9016ICE	Ganger	1	x	17.64	=	£17.64
	L A9002ICE	Labourer (General Operative)	1	x	13.02	=	£13.02
	L A9015ICE	Banksman	1	x	14.03	=	£14.03
		Total hourly cost of gang				**=**	**£44.69**
L A0122ICE	**Drill and blast Labour Gang**						
	L A9019ICE	Driller	1	x	14.86	=	£14.86
	L A9020ICE	Shot firer	0.5	x	14.86	=	£7.43
		Total hourly cost of gang				**=**	**£22.29**
L A0123ICE	**Landscape Labour Gang**						
	L A9003ICE	Labourer (Skill Rate 3)	1	x	14.86	=	£14.86
	L A9002ICE	Labourer (General Operative)	1	x	13.02	=	£13.02
		Total hourly cost of gang				**=**	**£27.88**
L A0320ICE	**Waterproofing Labour Gang**						
	L A9016ICE	Ganger	1	x	17.64	=	£17.64
	L A9002ICE	Labourer (General Operative)	3	x	13.02	=	£39.05
		Total hourly cost of gang				**=**	**£56.69**

Plant

			Qty		Rate		Total
P A1120ICE	**Excavate, haul and deposit (Motorway)**						
	P A0957ICE	Crawler Tractor / Dozer - Terex TS24 C Scraper (Wheeled) 18.4m³	2	x	103.02	=	£206.04
	P A0958ICE	Crawler Tractor / Dozer - Cat 633E Scraper	2	x	96.44	=	£192.88
	P A0937ICE	Crawler Tractor / Dozer - Cat 302kW Pusher	1	x	90.17	=	£90.17
	P A0959ICE	Crawler Tractor / Dozer - Tractor D8 Blade	0.5	x	68.05	=	£34.03
	P A0960ICE	Crawler Tractor / Dozer - Cat 16H Grader	0.5	x	87.86	=	£43.93
		Total hourly cost of gang				**=**	**£567.05**
P A1121ICE	**Compaction (Motorway) Plant Gang**						
	P A0959ICE	Crawler Tractor / Dozer - Tractor D8 Blade	0.5	x	68.05	=	£34.03
	P A0960ICE	Crawler Tractor / Dozer - Cat 16H Grader	0.5	x	87.86	=	£43.93
	P A0878ICE	Roller - Case Vibromax W651	1	x	7.62	=	£7.62
		Total hourly cost of gang				**=**	**£85.58**
P A1122ICE	**Ripping Plant Gang**						
	P A0953ICE	Crawler Tractor / Dozer - Fiatallis FD145 93kW	1	x	55.12	=	£55.12
	P A0937ICE	Crawler Tractor / Dozer - Cat 302kW Pusher	1	x	90.17	=	£90.17
		Total hourly cost of gang				**=**	**£145.29**
P A1123ICE	**Breaking Plant Gang**						
	P A0592ICE	Hydraulic Excavator - Cat 320 96kW	1	x	36.36	=	£36.36
		Total hourly cost of gang				**=**	**£36.36**

P A1124ICE   **General excavation Plant Gang**

P A0605ICE	Hydraulic Excavator - Cat 166kW	1	x	44.31	=	£44.31
P A0943ICE	Crawler Tractor / Dozer - Cat D6 LGP 160 Hp	0.5	x	52.02	=	£26.01

Total hourly cost of gang   =   **£70.32**

P A1125ICE   **Trimming Plant Gang / preparation Plant Gang / spread and level Plant Gang**

P A0960ICE	Crawler Tractor / Dozer - Cat 16H Grader	0.34	x	87.86	=	£29.87
P A0592ICE	Hydraulic Excavator - Cat 320 96kW	0.34	x	36.36	=	£12.36
P A0943ICE	Crawler Tractor / Dozer - Cat D6 LGP 160 Hp	0.34	x	52.02	=	£17.69

Total hourly cost of gang   =   **£59.92**

P A1126ICE   **General spoil haulage Plant Gang**

P A1011ICE	Dumper Truck - Volvo A25C 25t 6x6	3	x	60.30	=	£180.90

Total hourly cost of gang   =   **£180.90**

P A1127ICE   **General compaction Plant Gang**

P A0868ICE	Roller - Bomag 90 900mm	1	x	3.50	=	£3.50
P A0945ICE	Crawler Tractor / Dozer - Dresser 1004 48kW	1	x	30.69	=	£30.69

Total hourly cost of gang   =   **£34.19**

P A1128ICE   **Drill and blast Plant Gang**

P A0420ICE	Waggon Drill with Steel & bits	1	x	4.90	=	£4.90
P A0457ICE	Compressor - 375 cfm	1	x	16.00	=	£16.00
P A0421ICE	Exploder with Circuit Tester	1	x	2.64	=	£2.64

Total hourly cost of gang   =   **£23.54**

Earthworks

Note(s):   NOTE: The following rates are for a typical Motorway cutting.

E2   Excavation For Cuttings

	Unit	Labour Hours	Labour Net £	Plant Net £	Materials Net £	Unit Net £	CO_2e Kg
E2.1 **Topsoil**							
E2.1.0 Generally							
E2.1.0.01 Typical for a motorway cutting	m³	0.00	0.04	0.57	-	0.61	0.441
E2.2 **Material other than topsoil, rock or artificial hard material**							
E2.2.0 Generally							
E2.2.0.01 Typical for a motorway cutting	m³	0.00	0.04	0.57	-	0.61	0.441
E2.3 **Rock**							
E2.3.0 Generally							
E2.3.0.01 Typical for a motorway cutting; Rock (using D9 and towed ripper)	m³	0.01	0.36	5.55	-	5.91	4.189
E2.4 **Stated artificial hard material exposed at the Commencing Surface**							
E2.4.0 Generally							
E2.4.0.01 Typical for a motorway cutting; Concrete pavement; 200mm thick	m³	0.01	0.36	12.79	-	13.15	9.767
E2.4.0.02 Typical for a motorway cutting; Tarmacadam pavement; 200mm thick	m³	0.00	0.18	2.56	-	2.74	1.953
E2.5 **Stated artificial hard material not exposed at the Commencing Surface**							
E2.5.0 Generally							
E2.5.0.01 Typical for a motorway cutting; Plain concrete pavement (using Hymac with breaker)	m³	0.01	0.36	15.44	-	15.80	7.931
E2.5.0.02 Typical for a motorway cutting; Reinforced concrete pavement	m³	0.02	0.71	25.80	-	26.51	13.809

E3   Excavation For Foundations

	Unit	Labour Hours	Labour Net £	Plant Net £	Materials Net £	Unit Net £	CO_2e Kg
E3.1 **Topsoil**							
E3.1.1 Maximum depth: not exceeding 0.25m							
E3.1.1.01 Generally	m³	0.03	1.34	2.11	-	3.45	1.097
E3.2 **Material other than topsoil, rock or artificial hard material**							
E3.2.1 Maximum depth: not exceeding 0.25m							
E3.2.1.01 Generally	m³	0.03	1.43	2.25	-	3.68	1.170
E3.2.2 Maximum depth: 0.25 - 0.5m							
E3.2.2.01 Generally	m³	0.03	1.43	2.25	-	3.68	1.170
E3.2.3 Maximum depth: 0.5 - 1m							
E3.2.3.01 Generally	m³	0.03	1.43	2.25	-	3.68	1.170
E3.2.4 Maximum depth: 1 - 2m							
E3.2.4.01 Generally	m³	0.03	1.43	2.25	-	3.68	1.170

E3 Excavation For Foundations continued...

		Unit	Labour Hours	Labour Net £	Plant Net £	Materials Net £	Unit Net £	CO_2e Kg
E3.2	**Material other than topsoil, rock or artificial hard material**							
E3.2.5	Maximum depth: 2 - 5m							
E3.2.5.01	Generally	m³	0.04	1.74	2.74	-	4.48	1.426
E3.2.6	Maximum depth: 5 - 10m							
E3.2.6.01	Generally	m³	0.05	2.01	3.16	-	5.17	1.645
E3.2.7	Maximum depth: 10 - 15m							
E3.2.7.01	Generally	m³	0.05	2.23	3.52	-	5.75	1.828
E3.3	**Rock**							
E3.3.1	Maximum depth: not exceeding 0.25m							
E3.3.1.01	Rock, well broken, (excavation by machines - using drill and blast ripper gangs)	m³	0.14	4.69	6.57	-	11.26	4.085
E3.3.2	Maximum depth: 0.25 - 0.5m							
E3.3.2.01	Rock, well broken, (excavation by machines - using drill and blast ripper gangs)	m³	0.14	4.69	6.57	-	11.26	4.085
E3.3.3	Maximum depth: 0.5 - 1m							
E3.3.3.01	Rock, well broken, (excavation by machines - using drill and blast ripper gangs)	m³	0.15	5.02	7.04	-	12.06	4.377
E3.3.4	Maximum depth: 1 - 2m							
E3.3.4.01	Rock, well broken, (excavation by machines - using drill and blast ripper gangs)	m³	0.17	5.69	7.98	-	13.67	4.961
E3.3.5	Maximum depth: 2 - 5m							
E3.3.5.01	Rock, well broken, (excavation by machines - using drill and blast ripper gangs)	m³	0.21	7.03	9.86	-	16.89	6.128
E3.4	**Stated artificial hard material exposed at the Commencing Surface**							
E3.4.1	Maximum depth: not exceeding 0.25m							
E3.4.1.01	Plain Concrete Slab	m³	0.34	15.19	12.36	-	27.55	4.987
E3.4.1.02	Tarmacadam Pavement	m³	0.17	7.60	6.18	-	13.78	2.494
E3.5	**Stated artificial hard material not exposed at the Commencing Surface**							
E3.5.1	Maximum depth: not exceeding 0.25m							
E3.5.1.01	Plain concrete (using breaking gang)	m³	0.34	15.19	36.27	-	51.46	17.420
E3.5.1.02	Reinforced concrete	m³	0.51	22.79	51.13	-	73.92	24.810
E3.5.2	Maximum depth: 0.25 - 0.5m							
E3.5.2.01	Plain concrete (using breaking gang)	m³	0.34	15.19	36.27	-	51.46	17.420
E3.5.2.02	Reinforced concrete	m³	0.51	22.79	51.13	-	73.92	24.810
E3.5.3	Maximum depth: 0.5 - 1m							
E3.5.3.01	Plain concrete (using breaking gang)	m³	0.34	15.19	36.27	-	51.46	17.420
E3.5.3.02	Reinforced concrete	m³	0.51	22.79	51.13	-	73.92	24.810

E3 Excavation For Foundations continued...

	Unit	Labour Hours	Labour Net £	Plant Net £	Materials Net £	Unit Net £	CO_2e Kg
E3.5	**Stated artificial hard material not exposed at the Commencing Surface**						
E3.5.4	Maximum depth: 1 - 2m						
E3.5.4.01 Plain concrete (using breaking gang)	m³	0.39	17.43	41.61	-	59.04	19.982
E3.5.4.02 Reinforced concrete	m³	0.58	25.91	59.33	-	85.24	28.689

E4 General Excavation

Note(s): HARD ROCK
The following has been assumed for blasting of Hard Rock:-

1) Rock blasting in large open areas
2) The surface of the rock to be blasted has been cleaned of overburden and is suitable for access of wagon drills.
3) The rock is good hard rock not badly fissured.
4) Sufficient blasting agent has been used to ensure a high level of fragmentation, enabling straight forward excavation.

	Unit	Labour Hours	Labour Net £	Plant Net £	Materials Net £	Unit Net £	CO_2e Kg
E4.1	**Topsoil**						
E4.1.1	Maximum depth: not exceeding 0.25m						
E4.1.1.01 Generally	m³	0.02	0.89	1.41	-	2.30	0.731
E4.2	**Material other than topsoil, rock or artificial hard material**						
E4.2.1	Maximum depth: not exceeding 0.25m						
E4.2.1.01 Loaded into wagons	m³	0.02	1.03	1.62	-	2.65	0.841
E4.2.2	Maximum depth: 0.25 - 0.5m						
E4.2.2.01 Loaded into wagons	m³	0.02	1.03	1.62	-	2.65	0.841
E4.2.3	Maximum depth: 0.5 - 1m						
E4.2.3.01 Loaded into wagons	m³	0.02	1.03	1.62	-	2.65	0.841
E4.2.4	Maximum depth: 1 - 2m						
E4.2.4.01 Loaded into wagons	m³	0.03	1.30	2.04	-	3.34	1.060
E4.2.5	Maximum depth: 2 - 5m						
E4.2.5.01 Loaded into wagons	m³	0.03	1.30	2.04	-	3.34	1.060
E4.2.6	Maximum depth: 5 - 10m						
E4.2.6.01 Loaded into wagons	m³	0.04	1.56	2.46	-	4.02	1.280
E4.2.7	Maximum depth: 10 - 15m						
E4.2.7.01 Loaded into wagons	m³	0.04	1.65	2.60	-	4.25	1.353
E4.3	**Rock**						
E4.3.1	Maximum depth: not exceeding 0.25m						
E4.3.1.01 Rock; well broken (excavated by machine)	m³	0.08	3.49	5.48	-	8.97	2.852
E4.3.1.02 Rock; hard (including drilling and blasting and ripping prior to excavation)	m³	0.82	35.56	55.37	0.55	91.48	29.367
E4.3.1.03 For rail track; excavate ballast at formation level; Maximum depth: 150mm	m³	0.06	23.19	20.99	-	44.18	17.283
E4.3.1.04 For rail track; excavate ballast at formation level; unsuitable material; Maximum depth: 150mm	m³	0.06	21.35	19.32	-	40.67	15.911
E4.3.1.05 For rail track; excavate ballast at formation level; unsuitable material; Maximum depth: 250 mm	m³	0.14	25.66	76.97	-	102.63	52.128

E4 General Excavation continued...

		Unit	Labour Hours	Labour Net £	Plant Net £	Materials Net £	Unit Net £	CO_2e Kg
E4.3	**Rock**							
E4.3.2	Maximum depth: 0.25 - 0.5m							
E4.3.2.01	Rock; well broken (excavated by machine)	m³	0.09	3.80	5.98	-	9.78	3.108
E4.3.2.02	Rock; hard (including drilling and blasting and ripping prior to excavation)	m³	0.82	35.56	55.37	2.33	93.26	29.431
E4.3.2.03	For rail track; excavate ballast at formation level; Maximum depth: 300mm	m³	0.07	25.40	22.99	-	48.39	18.929
E4.3.3	Maximum depth: 0.5 - 1m							
E4.3.3.01	Rock; well broken (excavated by machine)	m³	0.09	3.80	5.98	-	9.78	3.108
E4.3.3.02	Rock; hard (including drilling and blasting and ripping prior to excavation)	m³	0.84	36.50	56.82	2.33	95.65	30.205
E4.3.4	Maximum depth: 1 - 2m							
E4.3.4.01	Rock; well broken (excavated by machine)	m³	0.09	4.11	6.47	-	10.58	3.364
E4.3.4.02	Rock; hard (including drilling and blasting and ripping prior to excavation)	m³	0.87	37.44	58.28	2.33	98.05	30.980
E4.3.5	Maximum depth: 2 - 5m							
E4.3.5.01	Rock; well broken (excavated by machine)	m³	0.11	4.83	7.59	-	12.42	3.949
E4.3.5.02	Rock; hard (including drilling and blasting and ripping prior to excavation)	m³	0.89	38.38	59.73	2.33	100.44	31.755
E4.4	**Stated artificial hard material exposed at the Commencing Surface**							
E4.4.1	Maximum depth: not exceeding 0.25m							
E4.4.1.01	Plain concrete slab (using breaking gang)	m³	0.30	13.40	10.91	-	24.31	4.401
E4.4.1.02	Tarmacadam pavement	m³	0.15	6.70	5.45	-	12.15	2.200
E4.5	**Stated artificial hard material not exposed at the Commencing Surface**							
E4.5.1	Maximum depth: not exceeding 0.25m							
E4.5.1.01	Plain concrete (using breaking gang)	m³	0.60	21.77	21.10	-	42.87	10.970
E4.5.1.02	Reinforced concrete	m³	0.92	33.38	32.35	-	65.73	16.820
E4.5.2	Maximum depth: 0.25 - 0.5m							
E4.5.2.01	Plain concrete (using breaking gang)	m³	0.60	21.77	21.10	-	42.87	10.970
E4.5.2.02	Reinforced concrete	m³	0.92	33.38	32.35	-	65.73	16.820
E4.5.3	Maximum depth: 0.5 - 1m							
E4.5.3.01	Plain concrete (using breaking gang)	m³	0.60	21.77	21.10	-	42.87	10.970
E4.5.3.02	Reinforced concrete	m³	0.92	33.38	32.35	-	65.73	16.820
E4.5.4	Maximum depth: 1 - 2m							
E4.5.4.01	Plain concrete (using breaking gang)	m³	0.78	28.30	27.42	-	55.72	14.261
E4.5.4.02	Reinforced concrete	m³	1.02	37.01	35.86	-	72.87	18.649

E5 Excavation Ancillaries

	Unit	Labour Hours	Labour Net £	Plant Net £	Materials Net £	Unit Net £	CO_2e Kg
E5.1	**Trimming of excavated surfaces**						
E5.1.1	Topsoil						
E5.1.1.01	Horizontal	m³	0.01	0.54	0.72	-	1.26	0.514
E5.1.1.02	Inclined at an angle of 10 - 45 degrees to the horizontal	m³	0.01	0.54	0.72	-	1.26	0.514
E5.1.2	Material other than topsoil, rock or artificial hard material						
E5.1.2.01	Horizontal	m³	0.01	0.54	0.72	-	1.26	0.514
E5.1.2.02	Inclined at an angle of 10 - 45 degrees to the horizontal	m³	0.01	0.54	0.72	-	1.26	0.514
E5.1.2.03	Inclined at an angle of 45 - 90 degrees to the horizontal	m³	0.02	0.67	0.90	-	1.57	0.642
E5.1.3	Rock						
E5.1.3.01	Horizontal	m³	0.23	10.19	13.66	-	23.85	9.762
E5.1.3.02	Inclined at an angle of 10 - 45 degrees to the horizontal	m³	0.23	10.19	13.66	-	23.85	9.762
E5.1.3.03	Inclined at an angle of 45 - 90 degrees to the horizontal	m³	0.33	14.66	19.65	-	34.31	14.043
E5.1.3.04	Vertical	m³	0.43	19.12	25.65	-	44.77	18.324
E5.2	**Preparation of excavated surfaces**						
E5.2.1	Topsoil						
E5.2.1.01	Horizontal	m³	0.02	1.07	1.44	-	2.51	1.028
E5.2.1.02	Inclined at an angle of 10 - 45 degrees to the horizontal	m³	0.02	1.07	1.44	-	2.51	1.028
E5.2.2	Material other than topsoil, rock or artificial hard material						
E5.2.2.01	Horizontal	m³	0.02	1.07	1.44	-	2.51	1.028
E5.2.2.02	Inclined at an angle of 10 - 45 degrees to the horizontal	m³	0.02	1.07	1.44	-	2.51	1.028
E5.2.2.03	Inclined at an angle of 45 - 90 degrees to the horizontal	m³	0.03	1.25	1.68	-	2.93	1.199
E5.2.3	Rock						
E5.2.3.01	Horizontal	m³	0.23	10.19	13.66	-	23.85	9.762
E5.2.3.02	Inclined at an angle of 10 - 45 degrees to the horizontal	m³	0.23	10.19	13.66	-	23.85	9.762
E5.2.3.03	Inclined at an angle of 45 - 90 degrees to the horizontal	m³	0.33	14.66	19.65	-	34.31	14.043
E5.2.3.04	Vertical	m³	0.43	19.12	25.65	-	44.77	18.324
E5.3	**Disposal of excavated material**						
E5.3.1	Topsoil						
E5.3.1.01	Remove from site (transporting to tip distance 5km)	m³	0.03	1.34	5.43	-	6.77	3.397
E5.3.1.02	Remove from site (transporting to tip distance 15km)	m³	0.08	3.35	13.57	-	16.92	8.492
E5.3.1.03	Stored on site for re-use; stockpiled at sides of excavation	m³	0.01	0.45	1.81	-	2.26	1.132
E5.3.1.04	Stored on site for re-use; stockpiled 100m from excavation	m³	0.02	0.67	2.71	-	3.38	1.698
E5.3.2	Material other than topsoil, rock or artificial hard material						
E5.3.2.01	Remove from site (transporting to tip distance 5km)	m³	0.03	1.34	5.43	-	6.77	3.397
E5.3.2.02	Remove from site (transporting to tip distance 15km)	m³	0.08	3.35	13.57	-	16.92	8.492

E5   Excavation Ancillaries continued...

	Unit	Labour Hours	Labour Net £	Plant Net £	Materials Net £	Unit Net £	CO_2e Kg
E5.3 **Disposal of excavated material**							
E5.3.2 Material other than topsoil, rock or artificial hard material							
E5.3.2.03 Stored on site for re-use; stockpiled at sides of excavation	m³	0.01	0.45	1.81	-	2.26	1.132
E5.3.2.04 Stored on site for re-use; stockpiled 100m from excavation	m³	0.02	0.67	2.71	-	3.38	1.698
E5.3.3 Rock							
E5.3.3.01 Remove from site (transporting to tip distance 5km)	m³	0.04	1.56	6.33	-	7.89	3.963
E5.3.3.02 Remove from site (transporting to tip distance 15km)	m³	0.08	3.57	14.47	-	18.04	9.059
E5.3.3.03 Stored on site for re-use; stockpiled at sides of excavation	m³	0.02	0.67	2.71	-	3.38	1.698
E5.3.3.04 Stored on site for re-use; stockpiled 100m from excavation	m³	0.02	0.89	3.62	-	4.51	2.265
E5.3.4 Stated artificial hard material							
E5.3.4.01 Remove from site (transporting to tip distance 5km)	m³	0.04	1.56	6.33	-	7.89	3.963
E5.3.4.02 Remove from site (transporting to tip distance 15km)	m³	0.08	3.57	14.47	-	18.04	9.059
E5.4 **Double handling of excavated material**							
E5.4.1 Topsoil							
E5.4.1.01 Removing from temporary stockpiles; distance to place of disposal (using general haulage gang) 30m	m³	0.03	1.47	3.43	-	4.90	1.973
E5.4.1.02 Removing from temporary stockpiles; distance to place of disposal 100m	m³	0.04	1.74	4.51	-	6.25	2.653
E5.4.1.03 Removing from temporary stockpiles; distance to place of disposal 500m	m³	0.05	2.19	6.32	-	8.51	3.785
E5.4.2 Material other than topsoil, rock or artificial hard material							
E5.4.2.01 Removing from temporary stockpiles; distance to place of disposal (using general haulage gang) 30m	m³	0.03	1.47	3.43	-	4.90	1.973
E5.4.2.02 Removing from temporary stockpiles; distance to place of disposal 100m	m³	0.04	1.74	4.51	-	6.25	2.653
E5.4.2.03 Removing from temporary stockpiles; distance to place of disposal 500m	m³	0.05	2.19	6.32	-	8.51	3.785
E5.4.3 Rock							
E5.4.3.01 Removing from temporary stockpiles; distance to place of disposal (using general haulage gang) 30m	m³	0.09	3.93	7.29	-	11.22	3.984
E5.4.3.02 Removing from temporary stockpiles; distance to place of disposal 100m	m³	0.09	4.20	8.38	-	12.58	4.664
E5.4.3.03 Removing from temporary stockpiles; distance to place of disposal 500m	m³	0.10	4.65	10.19	-	14.84	5.796

E5 Excavation Ancillaries continued...

	Unit	Labour Hours	Labour Net £	Plant Net £	Materials Net £	Unit Net £	CO_2e Kg
E5.4	**Double handling of excavated material**						
E5.4.4	Stated artificial hard material						
E5.4.4.01	Removing from temporary stockpiles; distance to place of disposal (using general haulage gang) 30m	m³	0.09	3.93	7.29	-	11.22	3.984
E5.4.4.02	Removing from temporary stockpiles; distance to place of disposal 100m	m³	0.09	4.20	8.38	-	12.58	4.664
E5.4.4.03	Removing from temporary stockpiles; distance to place of disposal 500m	m³	0.10	4.65	10.19	-	14.84	5.796

E6 Filling

Note(s): NOTE: Rates for filling include for spreading, levelling and compacting in layers where applicable (using spread and level and general haulage gangs)

		Unit	Labour Hours	Labour Net £	Plant Net £	Materials Net £	Unit Net £	CO_2e Kg
E6.1	**To structures**							
E6.1.1	Excavated topsoil							
E6.1.1.01	Generally	m³	0.01	0.49	0.66	-	1.15	0.471
E6.1.1.02	Generally; taken from temporary stockpile distance 100m	m³	0.01	0.49	3.42	-	3.91	2.119
E6.1.2	Imported topsoil							
E6.1.2.01	Generally	m³	0.01	0.49	0.66	19.05	20.20	1.079
E6.1.3	Non-selected excavated material other than topsoil or rock							
E6.1.3.01	using spread, level, general haulage and compaction gangs	m³	0.02	0.67	2.03	-	2.70	0.996
E6.1.3.02	including transporting directly from point of excavation distance 100m	m³	0.03	1.47	4.20	-	5.67	2.354
E6.1.3.03	taken from temporary stockpile distance 100m	m³	0.03	1.47	4.97	-	6.44	2.757
E6.1.4	Selected excavated material other than topsoil or rock							
E6.1.4.01	Generally	m³	0.02	0.67	2.03	-	2.70	0.996
E6.1.4.02	including transporting directly from point of excavation distance 100m	m³	0.03	1.47	4.20	-	5.67	2.354
E6.1.4.03	taken from temporary stockpile distance 100m	m³	0.03	1.47	4.97	-	6.44	2.757
E6.1.5	Imported natural material other than topsoil or rock							
E6.1.5.01	Imported subsoil	m³	0.01	0.49	2.03	24.38	26.90	1.968
E6.1.5.02	Imported granular material; DTp Spec. type 1	m³	0.01	0.49	2.03	24.72	27.24	11.592
E6.1.5.03	Imported granular material; DTp Spec. type 2	m³	0.01	0.49	2.03	26.46	28.98	12.200
E6.1.6	Excavated rock							
E6.1.6.01	Generally	m³	0.02	0.98	1.32	-	2.30	0.942
E6.1.6.02	including transporting directly from point of excavation distance 100m	m³	0.06	2.68	4.62	-	7.30	2.654
E6.1.6.03	taken from temporary stockpile distance 100m	m³	0.06	2.68	8.84	-	11.52	4.848

E6 Filling continued...

	Unit	Labour Hours	Labour Net £	Plant Net £	Materials Net £	Unit Net £	CO_2e Kg
E6.1 **To structures**							
E6.1.7 Imported rock							
E6.1.7.01 unscreened; maximum size 0.2m³	m³	0.02	0.76	2.96	25.88	29.60	11.960
E6.1.8 Imported artificial material							
E6.1.8.01 Imported broken brick hardcore	m³	0.01	0.49	2.03	19.94	22.46	10.847
E6.2 **Embankments**							
E6.2.1 Excavated topsoil							
E6.2.1.01 Generally	m³	0.00	0.04	0.57	-	0.61	0.441
E6.2.1.02 taken from temporary stockpile distance 500m	m³	0.00	0.09	1.22	-	1.31	0.951
E6.2.2 Imported topsoil							
E6.2.2.01 Generally	m³	0.00	0.04	0.09	19.05	19.18	0.676
E6.2.3 Non-selected excavated material other than topsoil or rock							
E6.2.3.01 Generally	m³	0.00	0.09	0.17	-	0.26	0.136
E6.2.3.02 including transporting directly from point of excavation distance 500m	m³	0.00	0.09	1.31	-	1.40	1.018
E6.2.3.03 taken from temporary stockpile distance 500m	m³	0.00	0.13	1.87	-	2.00	1.460
E6.2.4 Selected excavated material other than topsoil or rock							
E6.2.4.01 Generally	m³	0.00	0.09	0.17	-	0.26	0.136
E6.2.4.02 including transporting directly from point of excavation distance 500m	m³	0.00	0.09	1.31	-	1.40	1.018
E6.2.4.03 taken from temporary stockpile distance 500m	m³	0.00	0.13	1.87	-	2.00	1.460
E6.2.5 Imported natural material other than topsoil or rock							
E6.2.5.01 Imported subsoil	m³	0.00	0.09	0.29	15.24	15.62	0.829
E6.2.5.02 Imported granular material; DTp Spec. type 1	m³	0.00	0.09	0.29	24.72	25.10	10.818
E6.2.5.03 Imported granular material; DTp Spec. type 2	m³	0.00	0.09	0.29	26.46	26.84	11.426
E6.2.6 Excavated rock							
E6.2.6.01 Generally	m³	0.00	0.18	0.34	-	0.52	0.272
E6.2.6.02 including transporting directly from point of excavation distance 500m	m³	0.00	0.09	2.61	-	2.70	2.037
E6.2.6.03 taken from temporary stockpile distance 500m	m³	0.01	0.45	2.61	-	3.06	2.037
E6.2.7 Imported rock							
E6.2.7.01 unscreened; maximum size 0.2m³	m³	0.00	0.13	0.44	25.88	26.45	10.837
E6.3 **General**							
E6.3.1 Excavated topsoil							
E6.3.1.01 Generally	m³	0.01	0.49	0.66	-	1.15	0.471
E6.3.1.02 taken from temporary stockpile distance 100m (using spread and level gang and general haulage gang)	m³	0.01	0.49	3.42	-	3.91	2.119

E6 Filling continued...

	Unit	Labour Hours	Labour Net £	Plant Net £	Materials Net £	Unit Net £	CO_2e Kg
E6.3	**General**						
E6.3.2	Imported topsoil						
E6.3.2.01	Generally	m³	0.01	0.49	0.66	30.48	31.63	1.444
E6.3.3	Non-selected excavated material other than topsoil or rock						
E6.3.3.01	Generally	m³	0.03	1.47	1.79	-	3.26	0.824
E6.3.3.02	including transporting directly from point of excavation distance 100m	m³	0.03	1.47	3.78	-	5.25	2.070
E6.3.3.03	taken from temporary stockpile distance 100m	m³	0.03	1.47	4.55	-	6.02	2.472
E6.3.4	Selected excavated material other than topsoil or rock						
E6.3.4.01	Generally	m³	0.03	1.47	1.79	-	3.26	0.824
E6.3.4.02	including transporting directly from point of excavation distance 100m	m³	0.03	1.47	3.78	-	5.25	2.070
E6.3.4.03	taken from temporary stockpile distance 100m	m³	0.03	1.47	4.55	-	6.02	2.472
E6.3.5	Imported natural material other than topsoil or rock						
E6.3.5.01	Imported subsoil	m³	0.01	0.49	0.66	24.38	25.53	1.444
E6.3.5.02	Imported granular material; DTp Spec. type 1 (using spread, level and general compaction gangs)	m³	0.01	0.49	1.79	24.72	27.00	11.421
E6.3.5.03	Imported granular material; DTp Spec. type 2	m³	0.01	0.49	1.79	26.46	28.74	12.029
E6.3.6	Excavated rock						
E6.3.6.01	Generally	m³	0.03	1.47	3.17	-	4.64	1.809
E6.3.6.02	including transporting directly from point of excavation distance 100m	m³	0.03	1.47	5.16	-	6.63	3.055
E6.3.6.03	taken from temporary stockpile distance 100m	m³	0.03	1.52	7.55	-	9.07	4.298
E6.3.7	Imported rock						
E6.3.7.01	unscreened; maximum size 0.2m³	m³	0.02	0.76	2.66	25.88	29.30	11.746
E6.4	**To stated depth or thickness**						
E6.4.1	Excavated topsoil						
E6.4.1.01	100mm depth; Generally	m²	0.00	0.09	0.12	-	0.21	0.086
E6.4.1.02	100mm depth; taken from temporary stockpile distance 100m	m²	0.00	0.09	0.55	-	0.64	0.349
E6.4.1.03	150mm depth; Generally	m²	0.00	0.09	0.12	-	0.21	0.086
E6.4.1.04	150mm depth; taken from temporary stockpile distance 100m	m²	0.00	0.09	0.62	-	0.71	0.385
E6.4.1.05	200mm depth; Generally	m²	0.00	0.13	0.18	-	0.31	0.128
E6.4.1.06	200mm depth; taken from temporary stockpile distance 100m	m²	0.00	0.13	0.93	-	1.06	0.578
E6.4.1.07	250 mm depth; Generally	m²	0.00	0.13	0.18	-	0.31	0.128
E6.4.1.08	250 mm depth; taken from temporary stockpile distance 100m	m²	0.00	0.13	0.93	-	1.06	0.578
E6.4.1.09	300mm depth; Generally	m²	0.00	0.18	0.24	-	0.42	0.171

E6 **Filling continued...**

	Unit	Labour Hours	Labour Net £	Plant Net £	Materials Net £	Unit Net £	CO_2e Kg
E6.4	**To stated depth or thickness**						
E6.4.1	Excavated topsoil						
E6.4.1.10	300mm depth; taken from temporary stockpile distance 100m	m²	0.00	0.18	1.24	-	1.42	0.770
E6.4.1.11	500mm depth; Generally	m²	0.01	0.22	0.30	-	0.52	0.214
E6.4.1.12	500mm depth; taken from temporary stockpile distance 100m	m²	0.01	0.22	1.56	-	1.78	0.963
E6.4.2	Imported topsoil						
E6.4.2.01	100mm depth; Imported topsoil	m²	0.00	0.09	0.12	6.10	6.31	0.280
E6.4.2.02	150mm depth; Imported topsoil	m²	0.00	0.09	0.12	6.10	6.31	0.280
E6.4.2.03	200mm depth; Imported topsoil	m²	0.00	0.13	0.18	7.62	7.93	0.372
E6.4.2.04	250 mm depth; Imported topsoil	m²	0.00	0.13	0.18	8.53	8.84	0.401
E6.4.2.05	300mm depth; Imported topsoil	m²	0.00	0.18	0.24	10.29	10.71	0.500
E6.4.2.06	500mm depth; Imported topsoil	m²	0.01	0.22	0.30	15.24	15.76	0.700
E6.4.3	Non-selected excavated material other than topsoil or rock						
E6.4.3.01	100mm depth; Non-selected excavated material other than topsoil or rock	m²	0.01	0.27	0.32	-	0.59	0.150
E6.4.3.02	100mm depth; Non-selected excavated material other than topsoil or rock; including transporting directly from point of excavation distance 100m	m²	0.01	0.27	0.69	-	0.96	0.376
E6.4.3.03	100mm depth; Non-selected excavated material other than topsoil or rock; taken from temporary stockpile distance 100m	m²	0.01	0.27	0.83	-	1.10	0.449
E6.4.3.04	150mm depth; Non-selected excavated material other than topsoil or rock	m²	0.01	0.31	0.36	-	0.67	0.161
E6.4.3.05	150mm depth; Non-selected excavated material other than topsoil or rock; including transporting directly from point of excavation distance 100m	m²	0.01	0.31	0.72	-	1.03	0.387
E6.4.3.06	150mm depth; Non-selected excavated material other than topsoil or rock; taken from temporary stockpile distance 100m	m²	0.01	0.31	0.86	-	1.17	0.460
E6.4.3.07	200mm depth; Non-selected excavated material other than topsoil or rock	m²	0.01	0.36	0.45	-	0.81	0.214
E6.4.3.08	200mm depth; Non-selected excavated material other than topsoil or rock; including transporting directly from point of excavation distance 100m	m²	0.01	0.36	1.00	-	1.36	0.554
E6.4.3.09	200mm depth; Non-selected excavated material other than topsoil or rock; taken from temporary stockpile distance 100m	m²	0.01	0.36	1.21	-	1.57	0.664

E6 Filling continued...

	Unit	Labour Hours	Labour Net £	Plant Net £	Materials Net £	Unit Net £	CO_2e Kg
E6.4	**To stated depth or thickness**						
E6.4.3	Non-selected excavated material other than topsoil or rock						
E6.4.3.10	250 mm depth; Non-selected excavated material other than topsoil or rock	m^2	0.01	0.40	0.49	-	0.89	0.225
E6.4.3.11	250 mm depth; Non-selected excavated material other than topsoil or rock; including transporting directly from point of excavation distance 100m	m^2	0.01	0.40	0.72	-	1.12	0.468
E6.4.3.12	250 mm depth; Non-selected excavated material other than topsoil or rock; taken from temporary stockpile distance 100m	m^2	0.01	0.40	1.24	-	1.64	0.674
E6.4.3.13	300mm depth; Non-selected excavated material other than topsoil or rock	m^2	0.01	0.49	0.62	-	1.11	0.289
E6.4.3.14	300mm depth; Non-selected excavated material other than topsoil or rock; including transporting directly from point of excavation distance 100m	m^2	0.01	0.49	1.34	-	1.83	0.742
E6.4.3.15	300mm depth; Non-selected excavated material other than topsoil or rock; taken from temporary stockpile distance 100m	m^2	0.01	0.49	1.62	-	2.11	0.888
E6.4.3.16	500mm depth; Non-selected excavated material other than topsoil or rock	m^2	0.02	0.71	0.85	-	1.56	0.385
E6.4.3.17	500mm depth; Non-selected excavated material other than topsoil or rock; including transporting directly from point of excavation distance 100m	m^2	0.02	0.71	1.75	-	2.46	0.952
E6.4.3.18	500mm depth; Non-selected excavated material other than topsoil or rock; taken from temporary stockpile distance 100m	m^2	0.02	0.71	2.10	-	2.81	1.134
E6.4.4	Selected excavated material other than topsoil or rock						
E6.4.4.01	100mm depth; Selected excavated material other than topsoil or rock	m^2	0.01	0.27	0.32	-	0.59	0.150
E6.4.4.02	100mm depth; Selected excavated material other than topsoil or rock; including transporting directly from point of excavation distance 100m	m^2	0.01	0.27	0.69	-	0.96	0.376
E6.4.4.03	100mm depth; Selected excavated material other than topsoil or rock; taken from temporary stockpile distance 100m	m^2	0.01	0.27	0.83	-	1.10	0.449
E6.4.4.04	150mm depth; Selected excavated material other than topsoil or rock	m^2	0.01	0.31	0.36	-	0.67	0.161

E6 Filling continued...

	Unit	Labour Hours	Labour Net £	Plant Net £	Materials Net £	Unit Net £	CO_2e Kg
E6.4	**To stated depth or thickness**						
E6.4.4	Selected excavated material other than topsoil or rock						
E6.4.4.05	150mm depth; Selected excavated material other than topsoil or rock; including transporting directly from point of excavation distance 100m	m²	0.01	0.31	0.72	-	1.03	0.387
E6.4.4.06	150mm depth; Selected excavated material other than topsoil or rock; taken from temporary stockpile distance 100m	m²	0.01	0.31	0.86	-	1.17	0.460
E6.4.4.07	200mm depth; Selected excavated material other than topsoil or rock	m²	0.01	0.36	0.45	-	0.81	0.214
E6.4.4.08	200mm depth; Selected excavated material other than topsoil or rock; including transporting directly from point of excavation distance 100m	m²	0.01	0.36	1.00	-	1.36	0.554
E6.4.4.09	200mm depth; Selected excavated material other than topsoil or rock; taken from temporary stockpile distance 100m	m²	0.01	0.36	1.21	-	1.57	0.664
E6.4.4.10	250 mm depth; Selected excavated material other than topsoil or rock	m²	0.01	0.40	0.49	-	0.89	0.225
E6.4.4.11	250 mm depth; Selected excavated material other than topsoil or rock; including transporting directly from point of excavation distance 100m	m²	0.01	0.40	1.03	-	1.43	0.565
E6.4.4.12	250 mm depth; Selected excavated material other than topsoil or rock; taken from temporary stockpile distance 100m	m²	0.01	0.40	1.24	-	1.64	0.674
E6.4.4.13	300mm depth; Selected excavated material other than topsoil or rock	m²	0.01	0.49	0.62	-	1.11	0.289
E6.4.4.14	300mm depth; Selected excavated material other than topsoil or rock; including transporting directly from point of excavation distance 100m	m²	0.01	0.49	1.34	-	1.83	0.742
E6.4.4.15	300mm depth; Selected excavated material other than topsoil or rock; taken from temporary stockpile distance 100m	m²	0.01	0.49	1.62	-	2.11	0.888
E6.4.4.16	500mm depth; Selected excavated material other than topsoil or rock	m²	0.02	0.71	0.85	-	1.56	0.385
E6.4.4.17	500mm depth; Selected excavated material other than topsoil or rock; including transporting directly from point of excavation distance 100m	m²	0.02	0.71	1.75	-	2.46	0.952

E6 Filling continued...

	Unit	Labour Hours	Labour Net £	Plant Net £	Materials Net £	Unit Net £	CO_2e Kg
E6.4	**To stated depth or thickness**						
E6.4.4	Selected excavated material other than topsoil or rock						
E6.4.4.18	500mm depth; Selected excavated material other than topsoil or rock; taken from temporary stockpile distance 100m	m²	0.02	0.71	2.10	–	2.81	1.134
E6.4.5	Imported natural material other than topsoil or rock						
E6.4.5.01	100mm depth; Imported subsoil	m²	0.00	0.09	0.32	4.88	5.29	0.344
E6.4.5.02	100mm depth; Imported granular material; DTp Spec. type 1	m²	0.00	0.09	0.32	4.94	5.35	2.269
E6.4.5.03	100mm depth; Imported granular material; DTp Spec. type 2	m²	0.00	0.09	0.32	5.29	5.70	2.391
E6.4.5.04	150mm depth; Imported subsoil	m²	0.00	0.09	0.36	4.88	5.33	0.355
E6.4.5.05	150mm depth; Imported granular material; DTp Spec. type 1	m²	0.00	0.09	0.36	4.94	5.39	2.280
E6.4.5.06	150mm depth; Imported granular material; DTp Spec. type 2	m²	0.00	0.09	0.36	5.29	5.74	2.402
E6.4.5.07	200mm depth; Imported subsoil	m²	0.00	0.13	0.18	6.10	6.41	0.372
E6.4.5.08	200mm depth; Imported granular material; DTp Spec. type 1	m²	0.00	0.13	0.45	6.18	6.76	2.863
E6.4.5.09	200mm depth; Imported granular material; DTp Spec. type 2	m²	0.00	0.13	0.28	5.29	5.70	2.402
E6.4.5.10	250 mm depth; Imported subsoil	m²	0.00	0.13	0.49	6.86	7.48	0.498
E6.4.5.11	250 mm depth; Imported granular material; DTp Spec. type 1	m²	0.00	0.13	0.49	6.95	7.57	3.205
E6.4.5.12	250 mm depth; Imported granular material; DTp Spec. type 2	m²	0.00	0.13	0.49	7.44	8.06	3.376
E6.4.5.13	300mm depth; Imported subsoil	m²	0.00	0.18	0.62	8.23	9.03	0.617
E6.4.5.14	300mm depth; Imported granular material; DTp Spec. type 1	m²	0.00	0.18	0.62	8.34	9.14	3.865
E6.4.5.15	300mm depth; Imported granular material; DTp Spec. type 2	m²	0.00	0.18	0.62	8.93	9.73	4.071
E6.4.5.16	500mm depth; Imported subsoil	m²	0.02	0.71	0.85	12.19	13.75	0.872
E6.4.5.17	500mm depth; Imported granular material; DTp Spec. type 1	m²	0.02	0.71	0.85	12.36	13.92	5.684
E6.4.5.18	500mm depth; Imported granular material; DTp Spec. type 2	m²	0.02	0.71	0.85	13.23	14.79	5.988
E6.4.6	Excavated rock						
E6.4.6.01	100mm depth; Excavated rock	m²	0.00	0.13	0.56	–	0.69	0.321
E6.4.6.02	100mm depth; Excavated rock; including transporting directly from point of excavation distance 100m	m²	0.01	0.27	0.93	–	1.20	0.548
E6.4.6.03	100mm depth; Excavated rock; taken from temporary stockpile distance 100m	m²	0.01	0.27	1.35	–	1.62	0.767
E6.4.6.04	150mm depth; Excavated rock	m²	0.01	0.31	0.66	–	0.97	0.375
E6.4.6.05	150mm depth; Excavated rock; including transporting directly from point of excavation distance 100m	m²	0.01	0.31	1.02	–	1.33	0.601
E6.4.6.06	150mm depth; Excavated rock; taken from temporary stockpile distance 100m	m²	0.01	0.31	1.51	–	1.82	0.857

E6 Filling continued...

	Unit	Labour Hours	Labour Net £	Plant Net £	Materials Net £	Unit Net £	CO$_2$e Kg
E6.4	**To stated depth or thickness**						
E6.4.6	Excavated rock						
E6.4.6.07	200mm depth; Excavated rock	m²	0.01	0.36	0.81	-	1.17	0.471
E6.4.6.08	200mm depth; Excavated rock; including transporting directly from point of excavation distance 100m	m²	0.01	0.36	1.36	-	1.72	0.811
E6.4.6.09	200mm depth; Excavated rock; taken from temporary stockpile distance 100m	m²	0.01	0.36	1.99	-	2.35	1.140
E6.4.6.10	250 mm depth; Excavated rock	m²	0.01	0.40	0.91	-	1.31	0.525
E6.4.6.11	250 mm depth; Excavated rock; including transporting directly from point of excavation distance 100m	m²	0.01	0.40	1.45	-	1.85	0.864
E6.4.6.12	250 mm depth; Excavated rock; taken from temporary stockpile distance 100m	m²	0.01	0.40	2.15	-	2.55	1.230
E6.4.6.13	300mm depth; Excavated rock	m²	0.01	0.49	1.04	-	1.53	0.589
E6.4.6.14	300mm depth; Excavated rock; including transporting directly from point of excavation distance 100m	m²	0.01	0.49	1.76	-	2.25	1.042
E6.4.6.15	300mm depth; Excavated rock; taken from temporary stockpile distance 100m	m²	0.01	0.49	2.53	-	3.02	1.444
E6.4.6.16	500mm depth; Excavated rock	m²	0.02	0.71	1.57	-	2.28	0.899
E6.4.6.17	500mm depth; Excavated rock; including transporting directly from point of excavation distance 100m	m²	0.02	0.71	2.65	-	3.36	1.579
E6.4.6.18	500mm depth; Excavated rock; taken from temporary stockpile distance 100m	m²	0.02	0.76	3.85	-	4.61	2.200
E6.4.7	Imported rock						
E6.4.7.01	100mm depth; Imported rock	m²	0.00	0.13	0.56	3.86	4.55	1.890
E6.4.7.02	150mm depth; Imported rock	m²	0.01	0.31	0.66	5.80	6.77	2.728
E6.4.7.03	200mm depth; Imported rock	m²	0.01	0.36	0.81	7.73	8.90	3.608
E6.4.7.04	250 mm depth; Imported rock	m²	0.01	0.40	0.91	9.66	10.97	4.446
E6.4.7.05	300mm depth; Imported rock	m²	0.01	0.49	1.04	11.59	13.12	5.295
E6.4.7.06	500mm depth; Imported rock	m²	0.02	0.71	1.57	19.32	21.60	8.743

E7 Filling Ancillaries

	Unit	Labour Hours	Labour Net £	Plant Net £	Materials Net £	Unit Net £	CO$_2$e Kg
E7.1	**Trimming of filled surfaces**						
E7.1.1	Topsoil						
E7.1.1.01	Horizontal	m²	0.01	0.58	0.78	-	1.36	0.557
E7.1.1.02	Inclined at an angle of 10-45 degrees to the horizontal	m²	0.01	0.63	0.84	-	1.47	0.599
E7.1.1.03	Inclined at an angle of 45-90 degrees to the horizontal	m²	0.02	0.85	1.14	-	1.99	0.813
E7.1.2	Material other than topsoil, rock or artificial hard material						
E7.1.2.01	Horizontal	m²	0.01	0.58	0.78	-	1.36	0.557
E7.1.2.02	Inclined at an angle of 10-45 degrees to the horizontal	m²	0.01	0.63	0.84	-	1.47	0.599

E7 Filling Ancillaries continued...

	Unit	Labour Hours	Labour Net £	Plant Net £	Materials Net £	Unit Net £	CO_2e Kg
E7.1 **Trimming of filled surfaces**							
E7.1.2 Material other than topsoil, rock or artificial hard material							
E7.1.2.03 Inclined at an angle of 45-90 degrees to the horizontal	m²	0.02	0.85	1.14	-	1.99	0.813
E7.1.3 Rock							
E7.1.3.01 Horizontal	m²	0.23	10.28	13.78	-	24.06	9.847
E7.1.3.02 Inclined at an angle of 10-45 degrees to the horizontal	m²	0.25	11.30	15.16	-	26.46	10.832
E7.1.3.03 Inclined at an angle of 45-90 degrees to the horizontal	m²	0.37	16.67	22.35	-	39.02	15.970
E7.1.4 Stated artificial hard material							
E7.1.4.01 Plain concrete; Horizontal	m²	0.23	10.28	13.78	-	24.06	9.847
E7.1.4.02 Plain concrete; Inclined at an angle of 10-45 degrees to the horizontal	m²	0.25	11.30	15.16	-	26.46	10.832
E7.1.4.03 Plain concrete; Inclined at an angle of 45-90 degrees to the horizontal	m²	0.37	16.67	22.35	-	39.02	15.970
E7.2 **Preparation of filled surfaces**							
E7.2.1 Topsoil							
E7.2.1.01 Horizontal	m²	0.02	1.03	1.38	-	2.41	0.985
E7.2.1.02 Inclined at an angle of 10-45 degrees to the horizontal	m²	0.03	1.12	1.50	-	2.62	1.070
E7.2.1.03 Inclined at an angle of 45-90 degrees to the horizontal	m²	0.03	1.30	1.74	-	3.04	1.242
E7.2.2 Material other than topsoil, rock or artificial hard material							
E7.2.2.01 Horizontal	m²	0.02	1.03	1.38	-	2.41	0.985
E7.2.2.02 Inclined at an angle of 10-45 degrees to the horizontal	m²	0.03	1.12	1.50	-	2.62	1.070
E7.2.2.03 Inclined at an angle of 45-90 degrees to the horizontal	m²	0.03	1.30	1.74	-	3.04	1.242
E7.2.3 Rock							
E7.2.3.01 Horizontal	m²	0.14	6.26	8.39	-	14.65	5.994
E7.2.3.02 Inclined at an angle of 10-45 degrees to the horizontal	m²	0.15	6.79	9.11	-	15.90	6.508
E7.2.3.03 Inclined at an angle of 45-90 degrees to the horizontal	m²	0.20	8.89	11.92	-	20.81	8.520
E7.2.4 Stated artificial hard material							
E7.2.4.01 Plain concrete; Horizontal	m²	0.14	6.26	8.39	-	14.65	5.994
E7.2.4.02 Inclined at an angle of 10-45 degrees to the horizontal	m²	0.15	6.70	8.99	-	15.69	6.422
E7.2.4.03 Inclined at an angle of 45-90 degrees to the horizontal	m²	0.20	8.94	11.98	-	20.92	8.563
E7.3 **Geotextiles**							
E7.3.0 Typar 136 ground stabilising matting; 150mm laps							
E7.3.0.01 Laid upon a surface inclined at an angle to the horizontal; not exceeding 10 degrees	m²	0.01	0.40	-	2.67	3.07	1.040
E7.3.0.02 Laid upon a surface inclined at an angle to the horizontal; 10 - 45 degrees	m²	0.01	0.62	-	2.67	3.29	1.040

E7 Filling Ancillaries continued...

	Unit	Labour Hours	Labour Net £	Plant Net £	Materials Net £	Unit Net £	CO_2e Kg
E7.3	**Geotextiles**						
E7.3.0 Typar 136 ground stabilising matting; 150mm laps							
E7.3.0.03 Laid upon a surface inclined at an angle to the horizontal; 45 - 90 degrees	m²	0.02	0.85	-	2.67	3.52	1.040
E7.3.0.04 Vertical	m²	0.02	1.08	-	2.67	3.75	1.040

E8 Landscaping

Note(s): NOTE: Items for landscaping include fertilising, trimming and preparation of surfaces.

	Unit	Labour Hours	Labour Net £	Plant Net £	Materials Net £	Unit Net £	CO_2e Kg
E8.1	**Turfing**						
E8.1.0 Turfing with turf PC £2.50 per m²; laid with broken joints; rolling, watering, weeding, cutting, returfing until established							
E8.1.0.01 Laid upon a surface inclined at an angle to the horizontal; not exceeding 10 degrees	m²	0.12	3.46	-	2.50	5.96	-
E8.1.0.02 Laid upon a surface inclined at an angle to the horizontal; 10 - 45 degrees	m²	0.14	3.79	-	2.50	6.29	-
E8.1.0.03 Laid upon a surface inclined at an angle to the horizontal; 45 - 90 degrees	m²	0.24	6.58	-	2.50	9.08	-
E8.3	**Other grass seeding**						
E8.3.0 Seeding with grass seed PC £125 per 25 kg; sowing by hand at the rate of 0.05kg per m²; watering, weeding, raking, rolling, cutting, re-seeding until established							
E8.3.0.01 Laid upon a surface inclined at an angle to the horizontal; not exceeding 10 degrees	m²	0.00	0.11	-	2.50	2.61	-
E8.3.0.02 Laid upon a surface inclined at an angle to the horizontal; 10 - 45 degrees	m²	0.00	0.11	-	2.50	2.61	-
E8.3.0.03 Laid upon a surface inclined at an angle to the horizontal; 45 - 90 degrees	m²	0.01	0.14	-	2.50	2.64	-
E8.5	**Shrubs, stated species and size**						
E8.5.0 Providing and planting including excavating holes, backfilling etc							
E8.5.0.01 PC£2.90 Per Nr	Nr	0.13	3.62	-	2.90	6.52	-
E8.6	**Trees, stated species and size**						
E8.6.0 Trees including 50 x 50mm treated wrought softwood stakes and PVC ties							
E8.6.0.01 Medium; PC£25 Per Nr	Nr	0.18	5.02	-	25.00	30.02	-
E8.6.0.02 Large; PC£50 Per Nr	Nr	0.24	6.69	-	50.00	56.69	-
E8.6.0.03 Mature trees 10m high; PC£170 Per Nr; 3 Nr 100mm diameter treated hardwood stakes; galvanised twisted wire ties and PVC collars	Nr	1.54	42.94	-	170.00	212.94	-
E8.7	**Hedges, stated species, size and spacing**						
E8.7.1 Single row							
E8.7.1.01 Hedges, privet; PC£0.85 Per Nr; 300mm centres	m	0.07	1.95	-	2.55	4.50	-
E8.7.1.02 Hedges, privet; PC£0.85 Per Nr; 450mm centres	m	0.05	1.39	-	1.91	3.30	-

E8   Landscaping continued...

	Unit	Labour Hours	Labour Net £	Plant Net £	Materials Net £	Unit Net £	CO_2e Kg
E8.7	**Hedges, stated species, size and spacing**						
E8.7.1	Single row						
E8.7.1.03	Hedges, privet; PC£0.85 Per Nr; 600mm centres	m	0.04	1.12	-	1.27	**2.39**	-
E8.7.2	Double row						
E8.7.2.01	Hedges, beech; staggered PC£0.85 Per Nr; 300mm centres	m	0.14	3.90	-	5.10	**9.00**	-
E8.7.2.02	Hedges, beech; staggered PC£0.85 Per Nr; 450mm centres	m	0.09	2.51	-	3.82	**6.33**	-
E8.7.2.03	Hedges, beech; staggered PC£0.85 Per Nr; 600mm centres	m	0.07	1.95	-	1.27	**3.22**	-

CLASS F:
IN SITU CONCRETE

Calculations used throughout Class F - In Situ Concrete

Labour

			Qty		Rate		Total
L A0150ICE	**Provision of concrete Labour Gang**						
	L A9003ICE	Labourer (Skill Rate 3)	1	x	14.86	=	£14.86
	L A9002ICE	Labourer (General Operative)	1	x	13.02	=	£13.02
	L A9015ICE	Banksman	0.5	x	14.03	=	£7.01
		Total hourly cost of gang				=	**£34.89**
L A0155ICE	**Placing of concrete Labour Gang**						
	L A9003ICE	Labourer (Skill Rate 3)	1	x	14.86	=	£14.86
	L A9002ICE	Labourer (General Operative)	4	x	13.02	=	£52.06
	L A9015ICE	Banksman	0.25	x	14.03	=	£3.51
	L A9016ICE	Ganger	1	x	17.64	=	£17.64
	L A9000ICE	Craftsman WRA	0.5	x	17.33	=	£8.66
		Total hourly cost of gang				=	**£96.73**

Plant

			Qty		Rate		Total
P A1150ICE	**Provision of concrete Plant Gang**						
	P A0735ICE	Concrete Mixer - Liner Rolpanit	1	x	39.98	=	£39.98
	P A0473ICE	Cement Silo 50t	1	x	2.64	=	£2.64
	P A0737ICE	Concrete Mixer - Schwing BP1000R Concrete Pumps	1	x	46.33	=	£46.33
	P A0945ICE	Crawler Tractor / Dozer - Dresser 1004 48kW	1	x	30.69	=	£30.69
	P A0456ICE	Compressor - 250 cfm	1	x	11.57	=	£11.57
	P A0412ICE	Air Hose - 1 inch - 15m length	1	x	0.38	=	£0.38
		Total hourly cost of gang				=	**£131.59**
P A1155ICE	**Placing of concrete Plant Gang**						
	P A0405ICE	Air Vibrating Poker up to 75mm	3	x	1.33	=	£3.99
	P A0476ICE	Concrete Skip	2	x	0.75	=	£1.50
	P A0408ICE	Scabbler - Floor - 3 Headed	1	x	2.64	=	£2.64
	P A0409ICE	Scabbler - Floor - 5 Headed	1	x	3.77	=	£3.77
	P A0526ICE	Excavators Cable - NCK 305A 0.67m3	0.25	x	44.72	=	£11.18
	P A0457ICE	Compressor - 375 cfm	1	x	16.00	=	£16.00
		Total hourly cost of gang				=	**£39.08**

In Situ Concrete

Note(s):   NOTES:
Provision of Concrete
1) The prices are based on ready mixed concrete pumped from the concrete mixing plant to the point of placing.
2) The alternative means of transporting the concrete from the mixing plant by dumpers or agitator trucks would be covered by the prices shown given similar conditions.
3) The rates quoted are for medium workability. If high workability is necessary because of high reinforcing steel and / or duct content than an addition of 5.5% to the rates will be an adequate indication of the cost.

F1 Provision Of Concrete: Designed Concrete

		Unit	Labour Hours	Labour Net £	Plant Net £	Materials Net £	Unit Net £	CO_2e Kg
F1.1	**Strength C8/10**							
F1.1.1	Maximum aggregate size 10mm							
F1.1.1.01	Cement to BS EN 197-1	m³	-	-	-	87.31	87.31	246.147
F1.1.1.02	Sulphate resisting cement to BS EN 197-1	m³	-	-	-	97.27	97.27	198.280
F1.1.2	Maximum aggregate size 20mm							
F1.1.2.01	Cement to BS EN 197-1	m³	-	-	-	82.23	82.23	246.147
F1.1.2.02	Sulphate resisting cement to BS EN 197-1	m³	-	-	-	91.62	91.62	198.280
F1.1.3	Maximum aggregate size 40mm							
F1.1.3.01	Cement to BS EN 197-1	m³	-	-	-	86.04	86.04	246.147
F1.1.3.02	Sulphate resisting cement to BS EN 197-1	m³	-	-	-	95.83	95.83	198.280
F1.2	**Strength C12/15**							
F1.2.1	Maximum aggregate size 10mm							
F1.2.1.01	Cement to BS EN 197-1	m³	-	-	-	90.37	90.37	261.473
F1.2.1.02	Sulphate resisting cement to BS EN 197-1	m³	-	-	-	100.57	100.57	238.486
F1.2.2	Maximum aggregate size 20mm							
F1.2.2.01	Cement to BS EN 197-1	m³	-	-	-	85.10	85.10	261.473
F1.2.2.02	Sulphate resisting cement to BS EN 197-1	m³	-	-	-	94.98	94.98	238.486
F1.2.3	Maximum aggregate size 40mm							
F1.2.3.01	Cement to BS EN 197-1	m³	-	-	-	89.04	89.04	261.473
F1.2.3.02	Sulphate resisting cement to BS EN 197-1	m³	-	-	-	99.29	99.29	238.486
F1.3	**Strength C16/20**							
F1.3.1	Maximum aggregate size 10mm							
F1.3.1.01	Cement to BS EN 197-1	m³	-	-	-	93.42	93.42	276.799
F1.3.1.02	Sulphate resisting cement to BS EN 197-1	m³	-	-	-	104.19	104.19	278.693
F1.3.2	Maximum aggregate size 20mm							
F1.3.2.01	Cement to BS EN 197-1	m³	-	-	-	87.98	87.98	276.799
F1.3.2.02	Sulphate resisting cement to BS EN 197-1	m³	-	-	-	98.14	98.14	278.693
F1.3.3	Maximum aggregate size 40mm							
F1.3.3.01	Cement to BS EN 197-1	m³	-	-	-	92.05	92.05	276.799

F1   Provision Of Concrete: Designed Concrete continued...

	Unit	Labour Hours	Labour Net £	Plant Net £	Materials Net £	Unit Net £	CO_2e Kg
F1.3	**Strength C16/20**						
F1.3.3 Maximum aggregate size 40mm							
F1.3.3.02 Sulphate resisting cement to BS EN 197-1	m³	-	-	-	102.72	102.72	278.693
F1.4	**Strength C20/25**						
F1.4.1 Maximum aggregate size 10mm							
F1.4.1.01 Cement to BS EN 197-1	m³	-	-	-	95.96	95.96	276.799
F1.4.1.02 Sulphate resisting cement to BS EN 197-1	m³	-	-	-	107.10	107.10	278.693
F1.4.2 Maximum aggregate size 20mm							
F1.4.2.01 Cement to BS EN 197-1	m³	-	-	-	90.37	90.37	276.799
F1.4.2.02 Sulphate resisting cement to BS EN 197-1	m³	-	-	-	100.88	100.88	278.693
F1.4.3 Maximum aggregate size 40mm							
F1.4.3.01 Cement to BS EN 197-1	m³	-	-	-	94.56	94.56	276.799
F1.4.3.02 Sulphate resisting cement to BS EN 197-1	m³	-	-	-	105.59	105.59	278.693
F1.5	**Strength C25/30**						
F1.5.1 Maximum aggregate size 10mm							
F1.5.1.01 Cement to BS EN 197-1	m³	-	-	-	98.49	98.49	276.799
F1.5.1.02 Sulphate resisting cement to BS EN 197-1	m³	-	-	-	110.00	110.00	278.693
F1.5.2 Maximum aggregate size 20mm							
F1.5.2.01 Cement to BS EN 197-1	m³	-	-	-	92.75	92.75	276.799
F1.5.2.02 Sulphate resisting cement to BS EN 197-1	m³	-	-	-	103.62	103.62	278.693
F1.5.3 Maximum aggregate size 40mm							
F1.5.3.01 Cement to BS EN 197-1	m³	-	-	-	97.05	97.05	276.799
F1.5.3.02 Sulphate resisting cement to BS EN 197-1	m³	-	-	-	108.45	108.45	278.693
F1.8	**Strength C32/40**						
F1.8.1 Maximum aggregate size 10mm							
F1.8.1.01 Cement to BS EN 197-1	m³	-	-	-	103.80	103.80	276.799
F1.8.1.02 Sulphate resisting cement to BS EN 197-1	m³	-	-	-	116.02	116.02	278.693
F1.8.2 Maximum aggregate size 20mm							
F1.8.2.01 Cement to BS EN 197-1	m³	-	-	-	97.75	97.75	276.799
F1.8.2.02 Sulphate resisting cement to BS EN 197-1	m³	-	-	-	109.28	109.28	278.693

F3   Provision Of Concrete: Standardised Prescribed Concrete

	Unit	Labour Hours	Labour Net £	Plant Net £	Materials Net £	Unit Net £	CO_2e Kg
F3.1	**ST1**						
F3.1.2 Maximum aggregate size 20mm							
F3.1.2.01 Cement to BS EN 197-1	m³	-	-	-	83.38	83.38	246.147
F3.1.2.02 Sulphate resisting cement to BS EN 197-1	m³	-	-	-	92.89	92.89	198.280

F3 Provision Of Concrete: Standardised Prescribed Concrete continued...

		Unit	Labour Hours	Labour Net £	Plant Net £	Materials Net £	Unit Net £	CO_2e Kg
F3.1	**ST1**							
F3.1.3	Maximum aggregate size 40mm							
F3.1.3.01	Cement to BS EN 197-1	m^3	-	-	-	87.24	87.24	246.147
F3.1.3.02	Sulphate resisting cement to BS EN 197-1	m^3	-	-	-	97.21	97.21	198.280
F3.2	**ST2**							
F3.2.2	Maximum aggregate size 20mm							
F3.2.2.01	Cement to BS EN 197-1	m^3	-	-	-	87.98	87.98	261.473
F3.2.2.02	Sulphate resisting cement to BS EN 197-1	m^3	-	-	-	97.99	97.99	238.486
F3.2.3	Maximum aggregate size 40mm							
F3.2.3.01	Cement to BS EN 197-1	m^3	-	-	-	91.87	91.87	261.473
F3.2.3.02	Sulphate resisting cement to BS EN 197-1	m^3	-	-	-	102.24	102.24	238.486
F3.3	**ST3**							
F3.3.2	Maximum aggregate size 20mm							
F3.3.2.01	Cement to BS EN 197-1	m^3	-	-	-	89.70	89.70	261.473
F3.3.2.02	Sulphate resisting cement to BS EN 197-1	m^3	-	-	-	100.11	100.11	238.486
F3.3.3	Maximum aggregate size 40mm							
F3.3.3.01	Cement to BS EN 197-1	m^3	-	-	-	93.60	93.60	261.473
F3.3.3.02	Sulphate resisting cement to BS EN 197-1	m^3	-	-	-	104.37	104.37	238.486
F3.4	**ST4**							
F3.4.2	Maximum aggregate size 20mm							
F3.4.2.01	Cement to BS EN 197-1	m^3	-	-	-	93.15	93.15	276.799
F3.4.2.02	Sulphate resisting cement to BS EN 197-1	m^3	-	-	-	103.89	103.89	278.693
F3.4.3	Maximum aggregate size 40mm							
F3.4.3.01	Cement to BS EN 197-1	m^3	-	-	-	96.54	96.54	276.799
F3.4.3.02	Sulphate resisting cement to BS EN 197-1	m^3	-	-	-	107.70	107.70	278.693
F3.5	**ST5**							
F3.5.2	Maximum aggregate size 20mm							
F3.5.2.01	Cement to BS EN 197-1	m^3	-	-	-	97.18	97.18	276.799
F3.5.2.02	Sulphate resisting cement to BS EN 197-1	m^3	-	-	-	108.43	108.43	278.693
F3.5.3	Maximum aggregate size 40mm							
F3.5.3.01	Cement to BS EN 197-1	m^3	-	-	-	100.71	100.71	276.799
F3.5.3.02	Sulphate resisting cement to BS EN 197-1	m^3	-	-	-	112.41	112.41	278.693

F4 Provision Of Concrete: Prescribed Mix

		Unit	Labour Hours	Labour Net £	Plant Net £	Materials Net £	Unit Net £	CO_2e Kg
F4.0	**Generally**							
F4.0.1	Maximum aggregate size 10mm							
F4.0.1.01	Cement to BS EN 197-1; Strength C8/10	m^3	-	-	-	91.68	91.68	258.454

F4 Provision Of Concrete: Prescribed Mix continued...

	Unit	Labour Hours	Labour Net £	Plant Net £	Materials Net £	Unit Net £	CO_2e Kg

F4.0 Generally

F4.0.1 Maximum aggregate size 10mm

	Unit	Labour Hours	Labour Net £	Plant Net £	Materials Net £	Unit Net £	CO_2e Kg
F4.0.1.02 Cement to BS EN 197-1; Strength C12/15	m³	-	-	-	94.89	94.89	274.547
F4.0.1.03 Cement to BS EN 197-1; Strength C16/20	m³	-	-	-	98.09	98.09	274.547
F4.0.1.04 Cement to BS EN 197-1; Strength C20/25	m³	-	-	-	100.76	100.76	290.639
F4.0.1.05 Cement to BS EN 197-1; Strength C25/30	m³	-	-	-	103.41	103.41	290.639
F4.0.1.06 Cement to BS EN 197-1; Strength C32/40	m³	-	-	-	108.99	108.99	290.639
F4.0.1.07 Sulphate resisting cement to BS EN 197-1; Strength C8/10	m³	-	-	-	102.13	102.13	208.194
F4.0.1.08 Sulphate resisting cement to BS EN 197-1; Strength C12/15	m³	-	-	-	105.60	105.60	250.410
F4.0.1.09 Sulphate resisting cement to BS EN 197-1; Strength C16/20	m³	-	-	-	109.40	109.40	292.628
F4.0.1.10 Sulphate resisting cement to BS EN 197-1; Strength C20/25	m³	-	-	-	112.46	112.46	292.628
F4.0.1.11 Sulphate resisting cement to BS EN 197-1; Strength C25/30	m³	-	-	-	115.50	115.50	292.628
F4.0.1.12 Sulphate resisting cement to BS EN 197-1; Strength C32/40	m³	-	-	-	121.82	121.82	292.628

F4.0.2 Maximum aggregate size 20mm

	Unit	Labour Hours	Labour Net £	Plant Net £	Materials Net £	Unit Net £	CO_2e Kg
F4.0.2.01 Cement to BS EN 197-1; Strength C8/10	m³	-	-	-	82.23	82.23	246.147
F4.0.2.02 Cement to BS EN 197-1; Strength C12/15	m³	-	-	-	85.10	85.10	261.473
F4.0.2.03 Cement to BS EN 197-1; Strength C16/20	m³	-	-	-	87.98	87.98	261.473
F4.0.2.04 Cement to BS EN 197-1; Strength C20/25	m³	-	-	-	90.37	90.37	276.799
F4.0.2.05 Cement to BS EN 197-1; Strength C25/30	m³	-	-	-	92.75	92.75	276.799
F4.0.2.06 Cement to BS EN 197-1; Strength C32/40	m³	-	-	-	97.75	97.75	276.799
F4.0.2.07 Sulphate resisting cement to BS EN 197-1; Strength C8/10	m³	-	-	-	91.62	91.62	198.280
F4.0.2.08 Sulphate resisting cement to BS EN 197-1; Strength C12/15	m³	-	-	-	94.98	94.98	238.486
F4.0.2.09 Sulphate resisting cement to BS EN 197-1; Strength C16/20	m³	-	-	-	98.14	98.14	278.693
F4.0.2.10 Sulphate resisting cement to BS EN 197-1; Strength C20/25	m³	-	-	-	100.88	100.88	278.693
F4.0.2.11 Sulphate resisting cement to BS EN 197-1; Strength C25/30	m³	-	-	-	103.62	103.62	278.693
F4.0.2.12 Sulphate resisting cement to BS EN 197-1; Strength C32/40	m³	-	-	-	109.28	109.28	278.693

F4.0.3 Maximum aggregate size 40mm

	Unit	Labour Hours	Labour Net £	Plant Net £	Materials Net £	Unit Net £	CO_2e Kg
F4.0.3.01 Cement to BS EN 197-1; Strength C8/10	m³	-	-	-	86.04	86.04	246.147
F4.0.3.02 Cement to BS EN 197-1; Strength C12/15	m³	-	-	-	93.49	93.49	274.547
F4.0.3.03 Cement to BS EN 197-1; Strength C16/20	m³	-	-	-	96.65	96.65	290.639
F4.0.3.04 Cement to BS EN 197-1; Strength C20/25	m³	-	-	-	99.29	99.29	290.639

F4 Provision Of Concrete: Prescribed Mix continued...

	Unit	Labour Hours	Labour Net £	Plant Net £	Materials Net £	Unit Net £	CO_2e Kg
F4.0	**Generally**						
F4.0.3	Maximum aggregate size 40mm						
F4.0.3.05	Cement to BS EN 197-1; Strength C25/30	m³	-	-	-	101.90	101.90	290.639
F4.0.3.06	Cement to BS EN 197-1; Strength C32/40	m³	-	-	-	100.62	100.62	290.639
F4.0.3.07	Sulphate resisting cement to BS EN 197-1; Strength C8/10	m³	-	-	-	104.25	104.25	250.410
F4.0.3.08	Sulphate resisting cement to BS EN 197-1; Strength C12/15	m³	-	-	-	107.86	107.86	292.628
F4.0.3.09	Sulphate resisting cement to BS EN 197-1; Strength C16/20	m³	-	-	-	110.87	110.87	292.628
F4.0.3.10	Sulphate resisting cement to BS EN 197-1; Strength C20/25	m³	-	-	-	113.87	113.87	292.628

F6 Placing Of Concrete: Mass

	Unit	Labour Hours	Labour Net £	Plant Net £	Materials Net £	Unit Net £	CO_2e Kg
F6.1	**Blinding**						
F6.1.1	Thickness: not exceeding 150mm						
F6.1.1.01	Generally	m³	0.18	17.41	7.04	-	24.45	5.070
F6.1.2	Thickness: 150 - 300mm						
F6.1.2.01	Generally	m³	0.17	16.44	6.65	-	23.09	4.788
F6.1.3	Thickness: 300 - 500mm						
F6.1.3.01	Generally	m³	0.13	12.57	5.08	-	17.65	3.661
F6.1.4	Thickness: exceeding 500mm						
F6.1.4.01	Generally	m³	0.13	12.57	5.08	-	17.65	3.661
F6.2	**Bases, footings, pile caps and ground slabs**						
F6.2.1	Thickness: not exceeding 150mm						
F6.2.1.01	Generally	m³	0.19	18.38	7.43	-	25.81	5.351
F6.2.1.02	Placed against excavated surface	m³	0.20	19.35	7.82	-	27.17	5.633
F6.2.2	Thickness: 150 - 300mm						
F6.2.2.01	Generally	m³	0.18	17.41	7.04	-	24.45	5.070
F6.2.2.02	Placed against excavated surface	m³	0.19	18.38	7.43	-	25.81	5.351
F6.2.3	Thickness: 300 - 500mm						
F6.2.3.01	Generally	m³	0.14	13.54	5.47	-	19.01	3.943
F6.2.3.02	Placed against excavated surface	m³	0.15	14.51	5.87	-	20.38	4.225
F6.2.4	Thickness: exceeding 500mm						
F6.2.4.01	Generally	m³	0.14	13.54	5.47	-	19.01	3.943
F6.2.4.02	Placed against excavated surface	m³	0.15	14.51	5.87	-	20.38	4.225
F6.4	**Walls**						
F6.4.1	Thickness: not exceeding 150mm						
F6.4.1.01	Generally	m³	0.22	21.28	8.60	-	29.88	6.196
F6.4.2	Thickness: 150 - 300mm						
F6.4.2.01	Generally	m³	0.20	19.35	7.82	-	27.17	5.633
F6.4.3	Thickness: 300 - 500mm						
F6.4.3.01	Generally	m³	0.16	15.48	6.26	-	21.74	4.506

F6 Placing Of Concrete: Mass continued...

	Unit	Labour Hours	Labour Net £	Plant Net £	Materials Net £	Unit Net £	CO_2e Kg
F6.4	**Walls**						
F6.4.4	Thickness: exceeding 500mm						
F6.4.4.01 Generally	m^3	0.16	15.48	6.26	-	21.74	4.506
F6.8	**Other concrete forms**						
F6.8.4	Cross-sectional area: 0.25 - 1m^2						
F6.8.4.01 Plinths: 900 x 900 x 1200mm	m^3	0.40	38.69	15.64	-	54.33	11.266
F6.8.5	Cross-sectional area: exceeding 1m^2						
F6.8.5.01 Thrust Blocks: 1200 x 1200 x 1000mm	m^3	0.30	29.02	11.73	-	40.75	8.450
F6.8.5.02 Plinths: 1000 x 1000 x 1200mm	m^3	0.40	38.69	15.64	-	54.33	11.266

F7 Placing Of Concrete: Reinforced

	Unit	Labour Hours	Labour Net £	Plant Net £	Materials Net £	Unit Net £	CO_2e Kg
F7.2	**Bases, footings, pile caps and ground slabs**						
F7.2.1	Thickness: not exceeding 150mm						
F7.2.1.01 Generally	m^3	0.20	19.35	7.82	-	27.17	5.633
F7.2.2	Thickness: 150 - 300mm						
F7.2.2.01 Generally	m^3	0.19	18.38	7.43	-	25.81	5.351
F7.2.3	Thickness: 300 - 500mm						
F7.2.3.01 Generally	m^3	0.15	14.51	5.87	-	20.38	4.225
F7.2.4	Thickness: exceeding 500mm						
F7.2.4.01 Generally	m^3	0.15	14.51	5.87	-	20.38	4.225
F7.3	**Suspended slabs**						
F7.3.1	Thickness: not exceeding 150mm						
F7.3.1.01 Generally	m^3	0.24	23.22	9.38	-	32.60	6.760
F7.3.2	Thickness: 150 - 300mm						
F7.3.2.01 Generally	m^3	0.22	21.28	8.60	-	29.88	6.196
F7.3.3	Thickness: 300 - 500mm						
F7.3.3.01 Generally	m^3	0.20	19.35	7.82	-	27.17	5.633
F7.3.4	Thickness: exceeding 500mm						
F7.3.4.01 Generally	m^3	0.20	19.35	7.82	-	27.17	5.633
F7.4	**Walls**						
F7.4.1	Thickness: not exceeding 150mm						
F7.4.1.01 Generally	m^3	0.24	23.22	9.38	-	32.60	6.760
F7.4.2	Thickness: 150 - 300mm						
F7.4.2.01 Generally	m^3	0.24	23.22	9.38	-	32.60	6.760
F7.4.3	Thickness: 300 - 500mm						
F7.4.3.01 Generally	m^3	0.20	19.35	7.82	-	27.17	5.633
F7.4.4	Thickness: exceeding 500mm						
F7.4.4.01 Generally	m^3	0.20	19.35	7.82	-	27.17	5.633

F7 Placing Of Concrete: Reinforced continued...

		Unit	Labour Hours	Labour Net £	Plant Net £	Materials Net £	Unit Net £	CO_2e Kg
F7.5	**Columns and piers**							
F7.5.1	Cross-sectional area: not exceeding 0.03m^2							
F7.5.1.01	Generally	m^3	0.44	42.56	17.20	-	59.76	12.393
F7.5.2	Cross-sectional area: 0.03 - 0.1m^2							
F7.5.2.01	Generally	m^3	0.44	42.56	17.20	-	59.76	12.393
F7.5.3	Cross-sectional area: 0.1 - 0.25m^2							
F7.5.3.01	Generally	m^3	0.36	34.82	14.08	-	48.90	10.139
F7.5.4	Cross-sectional area: 0.25 - 1m^2							
F7.5.4.01	Generally	m^3	0.25	24.18	9.78	-	33.96	7.041
F7.5.5	Cross-sectional area: exceeding 1m^2							
F7.5.5.01	Generally	m^3	0.25	24.18	9.78	-	33.96	7.041
F7.6	**Beams**							
F7.6.1	Cross-sectional area: not exceeding 0.03m^2							
F7.6.1.01	Generally	m^3	0.44	42.56	17.20	-	59.76	12.393
F7.6.2	Cross-sectional area: 0.03 - 0.1m^2							
F7.6.2.01	Generally	m^3	0.44	42.56	17.20	-	59.76	12.393
F7.6.3	Cross-sectional area: 0.1 - 0.25m^2							
F7.6.3.01	Generally	m^3	0.36	34.82	14.08	-	48.90	10.139
F7.6.4	Cross-sectional area: 0.25 - 1m^2							
F7.6.4.01	Generally	m^3	0.25	24.18	9.78	-	33.96	7.041
F7.6.5	Cross-sectional area: exceeding 1m^2							
F7.6.5.01	Generally	m^3	0.25	24.18	9.78	-	33.96	7.041
F7.7	**Casing to metal sections**							
F7.7.1	Cross-sectional area: not exceeding 0.03m^2							
F7.7.1.01	Generally	m^3	0.50	48.37	19.55	-	67.92	14.083
F7.7.2	Cross-sectional area: 0.03 - 0.1m^2							
F7.7.2.01	Generally	m^3	0.50	48.37	19.55	-	67.92	14.083
F7.7.3	Cross-sectional area: 0.1 - 0.25m^2							
F7.7.3.01	Generally	m^3	0.45	43.53	17.60	-	61.13	12.674
F7.7.4	Cross-sectional area: 0.25 - 1m^2							
F7.7.4.01	Generally	m^3	0.40	38.69	15.64	-	54.33	11.266
F7.7.5	Cross-sectional area: exceeding 1m^2							
F7.7.5.01	Generally	m^3	0.40	38.69	15.64	-	54.33	11.266
F7.8	**Other concrete forms**							
F7.8.3	Cross-sectional area: 0.1 - 0.25m^2							
F7.8.3.01	Bollards; Diameter: 400mm	m^3	0.40	38.69	15.64	-	54.33	11.266
F7.8.4	Cross-sectional area: 0.25 - 1m^2							
F7.8.4.01	Box culverts: 1200 x 800mm wall thickness 250 mm	m^3	0.23	22.25	8.99	-	31.24	6.478
F7.8.5	Cross-sectional area: exceeding 1m^2							
F7.8.5.01	Box culverts: 1500 x 800mm wall thickness 250 mm	m^3	0.21	20.31	8.21	-	28.52	5.915

F8 Placing Of Concrete: Prestressed

		Unit	Labour Hours	Labour Net £	Plant Net £	Materials Net £	Unit Net £	CO_2e Kg
F8.3	**Suspended slabs**							
F8.3.1	Thickness: not exceeding 150mm							
F8.3.1.01	Generally	m³	0.29	28.05	11.34	-	39.39	8.168
F8.3.2	Thickness: 150 - 300mm							
F8.3.2.01	Generally	m³	0.29	28.05	11.34	-	39.39	8.168
F8.3.3	Thickness: 300 - 500mm							
F8.3.3.01	Generally	m³	0.20	19.35	7.82	-	27.17	5.633
F8.3.4	Thickness: exceeding 500mm							
F8.3.4.01	Generally	m³	0.20	19.35	7.82	-	27.17	5.633
F8.6	**Beams**							
F8.6.1	Cross-sectional area: not exceeding 0.03m²							
F8.6.1.01	Generally	m³	0.50	48.37	19.55	-	67.92	14.083
F8.6.2	Cross-sectional area: 0.03 - 0.1m²							
F8.6.2.01	Generally	m³	0.50	48.37	19.55	-	67.92	14.083
F8.6.3	Cross-sectional area: 0.1 - 0.25m²							
F8.6.3.01	Generally	m³	0.50	48.37	19.55	-	67.92	14.083
F8.6.4	Cross-sectional area: 0.25 - 1m²							
F8.6.4.01	Generally	m³	0.40	38.69	15.64	-	54.33	11.266
F8.6.5	Cross-sectional area: exceeding 1m²							
F8.6.5.01	Generally	m³	0.40	38.69	15.64	-	54.33	11.266

CLASS G:
CONCRETE ANCILLARIES

Calculations used throughout Class G - Concrete Ancillaries

Labour

			Qty		Rate		Total
L A0130ICE	**Formwork (make, fix and strike) Labour Gang**						
	L A9000ICE	Craftsman WRA	1	x	17.33	=	£17.33
	L A9003ICE	Labourer (Skill Rate 3)	1	x	14.86	=	£14.86
	L A9001ICE	Carpenter (charge hand)	1	x	18.63	=	£18.63
		Total hourly cost of gang				=	**£50.82**
L A0140ICE	**Steel fixing Labour Gang**						
	L A9000ICE	Craftsman WRA	2	x	17.33	=	£34.66
	L A9002ICE	Labourer (General Operative)	1	x	13.02	=	£13.02
		Total hourly cost of gang				=	**£47.68**
L A0145ICE	**Welding Labour Gang**						
	L A9003ICE	Labourer (Skill Rate 3)	1	x	14.86	=	£14.86
	L A9021ICE	Fitters and Welders	2	x	17.33	=	£34.66
		Total hourly cost of gang				=	**£49.52**
L A0152ICE	**Concrete finish Labour Gang**						
	L A9003ICE	Labourer (Skill Rate 3)	1	x	14.86	=	£14.86
	L A9014ICE	Labourer (Skill Rate 4)	1	x	14.03	=	£14.03
		Total hourly cost of gang				=	**£28.89**

Plant

			Qty		Rate		Total
P A1040ICE	**Formwork (make) Plant Gang**						
	P A0891ICE	Saw Bench - 24 inch Diesel / Electric	1	x	2.03	=	£2.03
	P A0703ICE	Kango Type Tool - Electric Power Woodauger	1	x	0.61	=	£0.61
	P A0704ICE	Kango Type Tool - Electric Nut Runner	1	x	0.52	=	£0.52
		Total hourly cost of gang				=	**£3.16**
P A1035ICE	**Formwork (fix and strike) Plant Gang**						
	P A0703ICE	Kango Type Tool - Electric Power Woodauger	1	x	0.61	=	£0.61
	P A0704ICE	Kango Type Tool - Electric Nut Runner	1	x	0.52	=	£0.52
	P A0507ICE	Cranes Crawler - NCK 305B - 20t	0.25	x	48.55	=	£12.14
		Total hourly cost of gang				=	**£13.27**
P A1140ICE	**Steel fixing Plant Gang**						
	P A0509ICE	Cranes Crawler - NCK 305C - 19t	0.25	x	46.29	=	£11.57
		Total hourly cost of gang				=	**£11.57**
P A1145ICE	**Welding Plant Gang**						
	P A0992ICE	Welding Set - 250 amp Diesel Electric Start Sil	1	x	3.76	=	£3.76
		Total hourly cost of gang				=	**£3.76**
P A1152ICE	**Concrete finish Plant Gang**						
	P A0452ICE	Compressor - 2-Tool (Complete)	1	x	5.20	=	£5.20
	P A0406ICE	Scabbler - Single Headed Hand	2	x	0.98	=	£1.96
	P A0408ICE	Scabbler - Floor - 3 Headed	1	x	2.64	=	£2.64
		Total hourly cost of gang				=	**£9.80**

Concrete Ancillaries

Note(s):   NOTES:
Formwork
1) Beams - For Isolated Beams, prices include for formwork to three faces. For Attached Beams, prices include for formwork to two faces.
2) Columns - For Isolated Columns, prices include for formwork to four faces. For Attached Columns, prices include for formwork to three faces.
3) Walls - For Battered and Vertical Walls, material prices have been enhanced to cover extra propping and strutting.
4) Six uses of formwork have been assumed in the formwork section and thus the labour, plant and material prices bear one sixth of the 'make' element.

G1　Formwork: Rough Finish

		Unit	Labour Hours	Labour Net £	Plant Net £	Materials Net £	Unit Net £	CO_2e Kg	
G1.1	**Plane horizontal**								
G1.1.1	Width not exceeding 0.1m.								
G1.1.1.01	Temporary	m	0.20	9.91	2.42	0.79	13.12	1.199	-0.371
G1.1.1.02	Left in	m	0.33	16.97	3.97	4.81	25.75	5.598	-2.246
G1.1.2	Width 0.1 - 0.2m.								
G1.1.2.01	Temporary	m	0.20	9.91	2.31	0.99	13.21	1.381	-0.460
G1.1.2.02	Left in	m	0.33	16.97	4.44	6.01	27.42	6.911	-2.807
G1.1.3	Width 0.2 - 0.4m.								
G1.1.3.01	Temporary	m²	0.42	21.34	4.97	4.09	30.40	5.008	-1.909
G1.1.3.02	Left in	m²	1.00	50.82	9.36	24.05	84.23	26.649	-11.228
G1.1.4	Width 0.4 - 1.22m.								
G1.1.4.01	Temporary	m²	0.42	21.34	4.97	4.09	30.40	5.008	-1.909
G1.1.4.02	Left in	m²	0.78	39.64	9.36	24.05	73.05	26.649	-11.228
G1.1.5	Width exceeding 1.22m.								
G1.1.5.01	Temporary	m²	0.42	21.34	4.97	4.09	30.40	5.008	-1.909
G1.1.5.02	Left in	m²	0.72	36.59	9.36	24.05	70.00	26.649	-11.228
G1.2	**Plane sloping**								
G1.2.1	Width not exceeding 0.1m.								
G1.2.1.01	Temporary	m	0.22	10.93	2.55	0.87	14.35	1.293	-0.404
G1.2.1.02	Left in	m	0.37	18.70	4.37	4.81	27.88	5.660	-2.246
G1.2.2	Width 0.1 - 0.2m.								
G1.2.2.01	Temporary	m	0.22	10.93	2.55	1.08	14.56	1.517	-0.505
G1.2.2.02	Left in	m	0.37	18.70	4.37	9.62	32.69	10.643	-4.491
G1.2.3	Width 0.2 - 0.4m.								
G1.2.3.01	Temporary	m²	0.91	46.25	5.10	4.33	55.68	5.276	-2.021
G1.2.3.02	Left in	m²	1.04	52.85	9.52	24.05	86.42	26.681	-11.228
G1.2.4	Width 0.4 - 1.22m.								
G1.2.4.01	Temporary	m²	0.69	35.07	5.10	4.35	44.52	5.301	-2.032
G1.2.4.02	Left in	m²	0.80	40.66	9.52	24.05	74.23	26.681	-11.228
G1.2.5	Width exceeding 1.22m.								
G1.2.5.01	Temporary	m²	0.43	21.85	5.10	4.35	31.30	5.301	-2.032
G1.2.5.02	Left in	m²	0.74	37.45	9.52	24.05	71.02	26.681	-11.228

G1 Formwork: Rough Finish continued...

	Unit	Labour Hours	Labour Net £	Plant Net £	Materials Net £	Unit Net £	CO_2e Kg	
G1.3	**Plane battered**							
G1.3.1	Width not exceeding 0.1m.							
G1.3.1.01	Temporary	m	0.23	11.89	2.78	0.94	15.61	1.403	*-0.438*
G1.3.1.02	Left in	m	0.40	20.23	4.72	5.63	30.58	6.561	*-2.627*
G1.3.2	Width 0.1 - 0.2m.							
G1.3.2.01	Temporary	m	0.23	11.89	2.78	1.18	15.85	1.652	*-0.550*
G1.3.2.02	Left in	m	0.40	20.23	4.72	7.07	32.02	8.056	*-3.301*
G1.3.3	Width 0.2 - 0.4m.							
G1.3.3.01	Temporary	m²	0.94	47.77	5.57	4.76	58.10	5.797	*-2.223*
G1.3.3.02	Left in	m²	1.59	80.80	9.46	24.05	114.31	26.382	*-11.228*
G1.3.4	Width 0.4 - 1.22m.							
G1.3.4.01	Temporary	m²	0.70	35.73	8.34	5.70	49.77	7.199	*-2.661*
G1.3.4.02	Left in	m²	1.20	60.73	14.17	24.05	98.95	27.114	*-11.228*
G1.3.5	Width exceeding 1.22m.							
G1.3.5.01	Temporary	m²	0.47	23.89	11.12	9.52	44.53	11.593	*-4.446*
G1.3.5.02	Left in	m²	0.80	40.66	18.91	24.05	83.62	27.849	*-11.228*
G1.4	**Plane vertical**							
G1.4.1	Width not exceeding 0.1m.							
G1.4.1.01	Temporary	m	0.25	12.86	3.00	0.94	16.80	1.438	*-0.438*
G1.4.1.02	Left in	m	0.40	20.28	3.12	5.63	29.03	6.354	*-2.627*
G1.4.2	Width 0.1 - 0.2m.							
G1.4.2.01	Temporary	m	0.25	12.86	3.00	1.18	17.04	1.687	*-0.550*
G1.4.2.02	Left in	m	0.40	20.28	3.12	7.07	30.47	7.849	*-3.301*
G1.4.3	Width 0.2 - 0.4m.							
G1.4.3.01	Temporary	m²	1.02	51.84	6.02	4.76	62.62	5.868	*-2.223*
G1.4.3.02	Left in	m²	0.86	43.91	6.94	28.57	79.42	30.989	*-13.339*
G1.4.4	Width 0.4 - 1.22m.							
G1.4.4.01	Temporary	m²	0.77	39.33	9.18	5.70	54.21	7.330	*-2.661*
G1.4.4.02	Left in	m²	1.73	87.92	10.84	34.20	132.96	37.711	*-15.966*
G1.4.5	Width exceeding 1.22m.							
G1.4.5.01	Temporary	m²	0.52	26.43	12.05	9.52	48.00	11.736	*-4.446*
G1.4.5.02	Left in	m²	0.86	43.71	14.30	57.14	115.15	62.188	*-26.678*
G1.5	**Curved to one radius in one plane**							
G1.5.1	Width not exceeding 0.1m.							
G1.5.1.01	1m radius; Temporary	m	0.31	15.55	3.61	0.94	20.10	1.531	*-0.438*
G1.5.1.02	1m radius; Left in	m	0.48	24.29	3.75	5.63	33.67	6.459	*-2.627*
G1.5.1.03	2m radius; Temporary	m	0.29	14.79	3.43	1.06	19.28	1.628	*-0.494*
G1.5.1.04	2m radius; Left in	m	0.46	23.28	3.58	5.63	32.49	6.432	*-2.627*
G1.5.2	Width 0.1 - 0.2m.							
G1.5.2.01	1m radius; Temporary	m	0.31	15.55	3.61	1.18	20.34	1.781	*-0.550*
G1.5.2.02	1m radius; Left in	m	0.48	24.29	3.75	7.07	35.11	7.954	*-3.301*
G1.5.2.03	2m radius; Temporary	m	0.29	14.79	3.43	1.35	19.57	1.927	*-0.629*
G1.5.2.04	2m radius; Left in	m	0.46	23.28	3.58	7.07	33.93	7.927	*-3.301*
G1.5.3	Width 0.2 - 0.4m.							
G1.5.3.01	1m radius; Temporary	m²	0.94	47.77	6.70	4.76	59.23	6.573	*-2.223*
G1.5.3.02	1m radius; Left in	m²	1.04	52.65	8.36	28.57	89.58	31.272	*-13.339*
G1.5.3.03	2m radius; Temporary	m²	1.17	59.46	6.92	5.46	71.84	6.730	*-2.549*

G1 Formwork: Rough Finish continued...

	Unit	Labour Hours	Labour Net £	Plant Net £	Materials Net £	Unit Net £	CO$_2$e Kg	
G1.5	**Curved to one radius in one plane**							
G1.5.3	Width 0.2 - 0.4m.							
G1.5.3.04	2m radius; Left in	m^2	1.98	100.62	7.97	28.57	137.16	31.196	*-13.339*
G1.5.4	Width 0.4 - 1.22m.							
G1.5.4.01	1m radius; Temporary	m^2	0.94	47.67	11.08	5.70	64.45	7.626	*-2.661*
G1.5.4.02	1m radius; Left in	m^2	1.98	100.62	13.13	34.20	147.95	38.172	*-15.966*
G1.5.4.03	2m radius; Temporary	m^2	0.89	45.23	10.52	6.54	62.29	8.410	*-3.054*
G1.5.4.04	2m radius; Left in	m^2	1.51	76.59	12.43	34.20	123.22	38.035	*-15.966*
G1.5.5	Width exceeding 1.22m.							
G1.5.5.01	1m radius; Temporary	m^2	0.62	31.51	14.54	9.52	55.57	12.125	*-4.446*
G1.5.5.02	1m radius; Left in	m^2	2.07	105.20	17.22	57.14	179.56	62.791	*-26.678*
G1.5.5.03	2m radius; Temporary	m^2	0.58	29.48	13.85	10.94	54.27	13.486	*-5.109*
G1.5.5.04	2m radius; Left in	m^2	0.99	50.31	16.44	57.14	123.89	62.637	*-26.678*
G1.6	**Other curved**							
G1.6.0	Curved to conical shaped surfaces; 1m minimum radius; 2m maximum radius							
G1.6.0.01	Temporary; Width: not exceeding 0.1m	m	0.41	20.84	4.11	1.08	26.03	1.986	*-0.505*
G1.6.0.02	Temporary; Width: 0.1 - 0.2m	m	0.51	25.92	4.11	1.35	31.38	2.260	*-0.629*
G1.6.0.03	Temporary; Width: 0.2 - 0.4m	m^2	1.02	51.84	9.04	5.44	66.32	7.534	*-2.538*
G1.6.0.04	Temporary; Width: 0.4 - 1.22m	m^2	1.55	78.77	12.65	6.54	97.96	9.441	*-3.054*
G1.6.0.05	Temporary; Width: Exceeding	m^2	1.02	51.84	16.68	10.94	79.46	14.848	*-5.109*
G1.6.0.06	Left in; Width: not exceeding 0.1m	m	0.80	40.66	4.76	5.63	51.05	6.833	*-2.627*
G1.6.0.07	Left in; Width: 0.1 - 0.2m	m	0.80	40.66	4.76	7.05	52.47	8.303	*-3.290*
G1.6.0.08	Left in; Width: 0.2 - 0.4m	m^2	3.55	180.41	13.06	28.55	222.02	32.324	*-13.328*
G1.6.0.09	Left in; Width: 0.4 - 1.22m	m^2	2.62	133.15	21.61	34.20	188.96	39.978	*-15.966*
G1.6.0.10	Left in; Width: Exceeding 1.22m	m^2	1.73	87.92	28.34	57.14	173.40	65.164	*-26.678*
G1.7	**For voids**							
G1.7.1	Small void: depth not exceeding 0.5m							
G1.7.1.01	Temporary; formwork left in; Cross-sectional area not exceeding 0.1m^2	Nr	0.95	48.28	7.62	31.51	87.41	34.166	*-14.709*
G1.7.1.02	Temporary; For voids; polystyrene void former; Cross-sectional area not exceeding 0.1m^2	Nr	0.70	35.57	-	4.10	39.67	4.194	
G1.7.2	Small void: depth 0.5 - 1m							
G1.7.2.01	Temporary; formwork left in; Cross-sectional area not exceeding 0.1m^2	Nr	1.29	65.56	10.40	62.77	138.73	67.133	*-29.305*
G1.7.2.02	Temporary; For voids; polystyrene void former; Cross-sectional area not exceeding 0.1m^2	Nr	0.70	35.57	-	8.19	43.76	8.388	
G1.7.3	Small void: depth 1 - 2m							
G1.7.3.01	Temporary; formwork left in; Cross-sectional area not exceeding 0.1m^2	Nr	1.72	87.41	13.82	126.74	227.97	134.071	*-59.172*

G1   Formwork: Rough Finish continued...

	Unit	Labour Hours	Labour Net £	Plant Net £	Materials Net £	Unit Net £	CO_2e Kg	
G1.7	**For voids**							
G1.7.3	Small void: depth 1 - 2m							
G1.7.3.02	Temporary; For voids; polystyrene void former; Cross-sectional area not exceeding 0.1m^2	Nr	1.00	50.82	-	16.39	67.21	16.775	
G1.7.4	Small void: depth stated exceeding 2m							
G1.7.4.01	Temporary; formwork left in; Cross-sectional area not exceeding 0.1m^2	Nr	2.58	131.12	20.61	171.24	322.97	181.463	-79.943
G1.7.4.02	Temporary; For voids; polystyrene void former; Cross-sectional area not exceeding 0.1m^2	Nr	1.00	50.82	-	24.58	75.40	25.163	
G1.7.5	Large void: depth not exceeding 0.5m							
G1.7.5.01	Temporary; formwork left in; Cross-sectional area not exceeding 0.1m2	Nr	1.29	65.56	10.40	62.77	138.73	67.111	-29.305
G1.7.5.02	Temporary; For voids; polystyrene void former; Cross-sectional area not exceeding 0.1m^2	Nr	0.70	35.57	-	20.48	56.05	20.969	
G1.7.6	Large void: depth 0.5 - 1m							
G1.7.6.01	Temporary; formwork left in; Cross-sectional area not exceeding 0.1m^2	Nr	1.72	87.41	13.82	126.74	227.97	134.071	-59.172
G1.7.6.02	Temporary; For voids; polystyrene void former; Cross-sectional area not exceeding 0.1m^2	Nr	1.00	50.82	-	40.97	91.79	41.938	
G1.7.7	Large void: depth 1 - 2m							
G1.7.7.01	Temporary; formwork left in; Cross-sectional area not exceeding 0.1m^2	Nr	2.58	131.12	20.61	171.24	322.97	181.463	-79.943
G1.7.7.02	Temporary; For voids; polystyrene void former; Cross-sectional area not exceeding 0.1m^2	Nr	1.00	50.82	-	81.94	132.76	83.876	
G1.7.8	Large void: depth stated exceeding 2m							
G1.7.8.01	Temporary; formwork left in; Cross-sectional area not exceeding 0.1m^2	Nr	4.80	243.94	45.40	238.10	527.44	255.134	-111.157
G1.7.8.02	Temporary; For voids; polystyrene void former; Cross-sectional area not exceeding 0.1m^2	Nr	1.50	76.23	-	122.91	199.14	125.814	
G1.8	**For concrete components of constant cross-section**							
G1.8.1	Beams							
G1.8.1.01	Isolated; 100 x 200mm	m	0.34	17.28	4.21	2.04	23.53	2.749	-0.954
G1.8.1.02	Isolated; 100 x 300mm	m	0.48	24.14	5.93	2.81	32.88	3.803	-1.314
G1.8.1.03	Isolated; 200 x 200mm	m	0.41	20.94	5.05	2.45	28.44	3.297	-1.145
G1.8.1.04	Isolated; 200 x 300mm	m	0.57	29.07	6.73	3.27	39.07	4.398	-1.527
G1.8.1.05	Isolated; 400 x 400mm	m	0.86	43.91	10.10	4.91	58.92	6.598	-2.291

G1 Formwork: Rough Finish continued...

		Unit	Labour Hours	Labour Net £	Plant Net £	Materials Net £	Unit Net £		CO_2e Kg
G1.8	**For concrete components of constant cross-section**								
G1.8.1	Beams								
G1.8.1.06	Isolated; 400 x 600mm	m	1.15	58.54	13.47	6.54	78.55	8.797	-3.054
G1.8.1.07	Isolated; 500 x 800mm	m	1.43	72.57	17.68	8.59	98.84	11.546	-4.008
G1.8.1.08	Attached; 100 x 200mm	m	0.34	17.28	4.21	1.64	23.13	2.325	-0.764
G1.8.1.09	Attached; 100 x 300mm	m	0.48	24.14	5.93	2.26	32.33	3.230	-1.055
G1.8.1.10	Attached; 200 x 200mm	m	0.41	20.94	5.05	1.97	27.96	2.799	-0.921
G1.8.1.11	Attached; 200 x 300mm	m	0.57	29.07	6.73	2.62	38.42	3.726	-1.224
G1.8.1.12	Attached; 400 x 400mm	m	0.86	43.91	10.10	3.92	57.93	5.576	-1.830
G1.8.1.13	Attached; 400 x 600mm	m	1.15	58.54	13.47	5.24	77.25	7.451	-2.448
G1.8.1.14	Attached; 500 x 800mm	m	1.43	72.57	17.68	6.88	97.13	9.777	-3.211
G1.8.2	Columns								
G1.8.2.01	Isolated; 100 x 200mm	m	0.40	20.12	4.90	2.45	27.47	3.283	-1.145
G1.8.2.02	Isolated; 100 x 300mm	m	0.53	26.99	6.53	3.27	36.79	4.371	-1.527
G1.8.2.03	Isolated; 200 x 200mm	m	0.53	26.99	6.53	3.27	36.79	4.371	-1.527
G1.8.2.04	Isolated; 200 x 300mm	m	0.55	27.95	6.81	4.09	38.85	5.264	-1.909
G1.8.2.05	Isolated; 400 x 400mm	m	0.77	39.13	10.44	6.54	56.11	8.338	-3.054
G1.8.2.06	Isolated; 400 x 600mm	m	0.88	44.67	10.52	8.18	63.37	10.092	-3.818
G1.8.2.07	Isolated; 500 x 800mm	m	1.15	58.44	13.71	10.63	82.78	13.124	-4.963
G1.8.2.08	Attached; 100 x 200mm	m	0.26	13.42	3.16	2.04	18.62	2.605	-0.954
G1.8.2.09	Attached; 100 x 300mm	m	0.35	17.99	4.26	2.86	25.11	3.620	-1.336
G1.8.2.10	Attached; 200 x 200mm	m	0.35	17.99	4.26	2.45	24.70	3.196	-1.145
G1.8.2.11	Attached; 200 x 300mm	m	0.44	22.41	5.22	3.27	30.90	4.194	-1.527
G1.8.2.12	Attached; 400 x 400mm	m	0.70	35.78	8.38	4.91	49.07	6.375	-2.291
G1.8.2.13	Attached; 400 x 600mm	m	0.88	44.67	10.52	6.54	61.73	8.398	-3.054
G1.8.2.14	Attached; 500 x 800mm	m	1.15	58.44	13.71	8.59	80.74	11.006	-4.008
G1.8.3	Walls								
G1.8.3.01	250 mm thick x 750mm high	m	0.76	38.62	9.18	6.13	53.93	7.757	-2.863
G1.8.3.02	250 mm thick x 1000mm high	m	1.12	56.92	13.65	8.18	78.75	10.544	-3.818
G1.8.3.03	400mm thick x 1000mm high	m	1.12	56.92	13.65	8.18	78.75	10.544	-3.818
G1.8.4	Other members								
G1.8.4.01	Box culverts; Internal dimensions 1 x 1m; wall thickness 250 mm	m	4.25	216.09	40.66	33.19	289.94	40.824	-15.495
G1.8.4.02	Box culverts; Internal dimensions 2 x 2m; wall thickness 250 mm	m	9.83	499.46	113.87	99.52	712.85	120.811	-46.461
G1.8.4.03	Box culverts; Internal dimensions 2 x 2m; wall thickness 400mm	m	10.34	525.58	81.41	104.09	711.08	124.972	-48.595
G1.8.5	Projections								
G1.8.5.01	Nibs; 50 x 50mm deep	m	0.36	18.30	4.47	2.41	25.18	3.160	-1.123
G1.8.5.02	Nibs; 100 x 100mm deep	m	0.36	18.30	4.47	4.81	27.58	5.652	-2.246
G1.8.6	Intrusions								
G1.8.6.01	Rebates 50 x 50mm deep	m	0.18	9.15	2.17	1.55	12.87	1.032	-1.769
G1.8.6.02	Rebates 100 x 100mm deep	m	0.18	9.15	2.17	5.16	16.48	2.681	-5.897

G2 Fair Finish

		Unit	Labour Hours	Labour Net £	Plant Net £	Materials Net £	Unit Net £		CO_2e Kg
G2.1	**Plane horizontal**								
G2.1.1	Width not exceeding 0.1m.								
G2.1.1.01	Generally	m	0.20	9.91	2.31	0.81	13.03	1.181	-0.371
G2.1.2	Width 0.1 - 0.2m.								
G2.1.2.01	Generally	m	0.20	9.91	2.31	1.01	13.23	1.381	-0.460
G2.1.3	Width 0.2 - 0.4m.								
G2.1.3.01	Generally	m²	0.91	46.25	4.97	4.17	55.39	5.008	-1.909

G2   Fair Finish continued...

	Unit	Labour Hours	Labour Net £	Plant Net £	Materials Net £	Unit Net £	CO$_2$e Kg	
G2.1	**Plane horizontal**							
G2.1.4	Width 0.4 - 1.22m.							
G2.1.4.01	Generally	m^2	0.69	35.07	4.97	4.17	44.21	5.008	-1.909
G2.1.5	Width exceeding 1.22m.							
G2.1.5.01	Generally	m^2	0.42	21.34	4.97	4.17	30.48	5.008	-1.909
G2.2	**Plane sloping**							
G2.2.1	Width not exceeding 0.1m.							
G2.2.1.01	Generally	m	0.22	10.93	1.81	0.88	13.62	1.278	-0.404
G2.2.2	Width 0.1 - 0.2m.							
G2.2.2.01	Generally	m	0.22	10.93	2.55	1.10	14.58	1.517	-0.505
G2.2.3	Width 0.2 - 0.4m.							
G2.2.3.01	Generally	m^2	0.69	35.07	5.10	4.44	44.61	5.301	-2.032
G2.2.4	Width 0.4 - 1.22m.							
G2.2.4.01	Generally	m^2	0.43	21.85	5.10	4.44	31.39	5.301	-2.032
G2.2.5	Width exceeding 1.22m.							
G2.2.5.01	Generally	m^2	0.47	23.89	5.10	4.44	33.43	5.301	-2.032
G2.3	**Plane battered**							
G2.3.1	Width not exceeding 0.1m.							
G2.3.1.01	Generally	m	0.23	11.89	2.78	0.96	15.63	1.403	-0.438
G2.3.2	Width 0.1 - 0.2m.							
G2.3.2.01	Generally	m	0.23	11.89	2.78	1.20	15.87	1.652	-0.550
G2.3.3	Width 0.2 - 0.4m.							
G2.3.3.01	Generally	m^2	0.94	47.77	5.57	4.86	58.20	5.797	-2.223
G2.3.4	Width 0.4 - 1.22m.							
G2.3.4.01	Generally	m^2	0.70	35.73	8.34	5.82	49.89	7.199	-2.661
G2.3.5	Width exceeding 1.22m.							
G2.3.5.01	Generally	m^2	0.47	23.89	11.12	9.72	44.73	11.593	-4.446
G2.4	**Plane vertical**							
G2.4.1	Width not exceeding 0.1m.							
G2.4.1.01	Generally	m	0.25	12.86	3.00	0.96	16.82	1.438	-0.438
G2.4.2	Width 0.1 - 0.2m.							
G2.4.2.01	Generally	m	0.25	12.86	2.96	1.18	17.00	1.656	-0.539
G2.4.3	Width 0.2 - 0.4m.							
G2.4.3.01	Generally	m^2	1.02	51.84	6.02	4.86	62.72	5.868	-2.223
G2.4.4	Width 0.4 - 1.22m.							
G2.4.4.01	Generally	m^2	0.77	39.33	9.18	5.82	54.33	7.330	-2.661
G2.4.5	Width exceeding 1.22m.							
G2.4.5.01	Generally	m^2	0.52	26.43	12.05	9.72	48.20	11.736	-4.446
G2.5	**Curved to one radius in one plane**							
G2.5.1	Width not exceeding 0.1m.							
G2.5.1.01	1m radius	m	0.31	15.55	3.61	0.96	20.12	1.531	-0.438

G2 Fair Finish continued...

	Unit	Labour Hours	Labour Net £	Plant Net £	Materials Net £	Unit Net £	CO_2e Kg	
G2.5	**Curved to one radius in one plane**							
G2.5.1	Width not exceeding 0.1m.							
G2.5.1.02	2m radius	m	0.29	14.79	3.43	1.08	19.30	1.628	-0.494
G2.5.2	Width 0.1 - 0.2m.							
G2.5.2.01	1m radius	m	0.31	15.55	3.61	1.20	20.36	1.781	-0.550
G2.5.2.02	2m radius	m	0.29	14.79	3.43	1.37	19.59	1.927	-0.629
G2.5.3	Width 0.2 - 0.4m.							
G2.5.3.01	1m radius	m²	1.23	62.51	7.27	4.86	74.64	6.061	-2.223
G2.5.3.02	2m radius	m²	0.58	29.68	6.92	5.57	42.17	6.730	-2.549
G2.5.4	Width 0.4 - 1.22m.							
G2.5.4.01	1m radius	m²	0.94	47.67	11.08	5.82	64.57	7.626	-2.661
G2.5.4.02	2m radius	m²	0.89	45.23	10.52	6.67	62.42	8.410	-3.054
G2.5.5	Width exceeding 1.22m.							
G2.5.5.01	1m radius	m²	0.62	31.51	14.54	9.72	55.77	12.125	-4.446
G2.5.5.02	2m radius	m²	1.17	59.36	13.85	11.17	84.38	13.486	-5.109
G2.6	**Other curved**							
G2.6.0	Curved to conical shaped surfaces; 1m minimum radius; 2m maximum radius							
G2.6.0.01	Width: not exceeding 0.1m	m	0.51	25.71	4.11	1.08	30.90	1.961	-0.494
G2.6.0.02	Width: 0.1 - 0.2m	m	0.51	25.71	4.11	1.37	31.19	2.260	-0.629
G2.6.0.03	Width: 0.2 - 0.4m	m²	2.03	103.16	8.30	5.57	117.03	7.403	-2.549
G2.6.0.04	Width: 0.4 - 1.22m	m²	1.55	78.77	12.65	6.67	98.09	9.441	-3.054
G2.6.0.05	Width: exceeding 1.22m	m²	1.02	51.84	16.68	11.17	79.69	14.848	-5.109
G2.8	**For concrete components of constant cross-section**							
G2.8.1	Beams							
G2.8.1.01	Isolated; 100 × 200mm	m	0.34	17.28	4.21	2.09	23.58	2.749	-0.954
G2.8.1.02	Isolated; 100 × 300mm	m	0.48	24.14	5.93	2.87	32.94	3.803	-1.314
G2.8.1.03	Isolated; 200 × 200mm	m	0.41	20.94	5.05	2.50	28.49	3.297	-1.145
G2.8.1.04	Isolated; 200 × 300mm	m	0.57	29.07	6.73	3.34	39.14	4.398	-1.527
G2.8.1.05	Isolated; 400 × 400mm	m	0.86	43.91	10.10	5.01	59.02	6.598	-2.291
G2.8.1.06	Isolated; 400 × 600mm	m	1.15	58.54	13.47	6.67	78.68	8.797	-3.054
G2.8.1.07	Isolated; 500 × 800mm	m	1.43	72.57	17.68	8.76	99.01	11.546	-4.008
G2.8.1.08	Attached; 100 × 200mm	m	0.34	17.28	4.21	1.67	23.16	2.325	-0.764
G2.8.1.09	Attached; 100 × 300mm	m	0.48	24.14	5.93	2.31	32.38	3.230	-1.055
G2.8.1.10	Attached; 200 × 200mm	m	0.41	20.94	5.05	2.01	28.00	2.799	-0.921
G2.8.1.11	Attached; 200 × 300mm	m	0.57	29.07	6.73	2.67	38.47	3.726	-1.224
G2.8.1.12	Attached; 400 × 400mm	m	0.86	43.91	10.10	4.00	58.01	5.576	-1.830
G2.8.1.13	Attached; 400 × 600mm	m	1.15	58.54	13.47	5.35	77.36	7.451	-2.448
G2.8.1.14	Attached; 500 × 800mm	m	1.43	72.57	17.68	7.02	97.27	9.777	-3.211
G2.8.2	Columns							
G2.8.2.01	Isolated; 100 × 200mm	m	0.40	20.12	4.92	2.50	27.54	3.289	-1.145
G2.8.2.02	Isolated; 100 × 300mm	m	0.53	26.99	4.35	3.34	34.68	4.305	-1.527
G2.8.2.03	Isolated; 200 × 200mm	m	0.53	26.99	6.53	3.34	36.86	4.371	-1.527
G2.8.2.04	Isolated; 200 × 300mm	m	0.55	27.95	6.81	4.17	38.93	5.264	-1.909
G2.8.2.05	Isolated; 400 × 400mm	m	0.77	39.13	10.44	6.67	56.24	8.338	-3.054
G2.8.2.06	Isolated; 400 × 600mm	m	0.88	44.67	10.52	8.34	63.53	10.092	-3.818
G2.8.2.07	Isolated; 500 × 800mm	m	1.15	58.44	13.71	10.85	83.00	13.124	-4.963
G2.8.2.08	Attached; 100 × 200mm	m	0.26	13.42	3.16	2.09	18.67	2.605	-0.954
G2.8.2.09	Attached; 100 × 300mm	m	0.35	17.99	4.26	2.92	25.17	3.620	-1.336
G2.8.2.10	Attached; 200 × 200mm	m	0.35	17.99	4.26	2.50	24.75	3.196	-1.145
G2.8.2.11	Attached; 200 × 300mm	m	0.44	22.41	5.22	3.34	30.97	4.194	-1.527
G2.8.2.12	Attached; 400 × 400mm	m	0.70	35.78	8.38	5.01	49.17	6.375	-2.291
G2.8.2.13	Attached; 400 × 600mm	m	0.88	44.67	10.54	6.67	61.88	8.400	-3.054

G2 Fair Finish continued...

	Unit	Labour Hours	Labour Net £	Plant Net £	Materials Net £	Unit Net £	CO_2e Kg	
G2.8	**For concrete components of constant cross-section**							
G2.8.2	Columns							
G2.8.2.14	Attached; 500 x 800mm	m	1.15	58.44	13.71	8.76	80.91	11.006	*-4.008*
G2.8.3	Walls							
G2.8.3.01	250 mm thick x 750mm high	m	0.76	38.62	9.18	6.26	54.06	7.757	*-2.863*
G2.8.3.02	250 mm thick x 1000mm high	m	1.12	56.92	13.65	8.34	78.91	10.544	*-3.818*
G2.8.3.03	400mm thick x 1000mm high	m	1.12	56.92	13.65	8.34	78.91	10.544	*-3.818*
G2.8.4	Other members							
G2.8.4.01	Box culverts; Internal dimensions 1 x 1m; wall thickness 250 mm	m	4.25	216.09	40.66	33.87	290.62	40.824	*-15.495*
G2.8.4.02	Box culverts; Internal dimensions 2 x 2m; wall thickness 250 mm	m	9.83	499.46	113.87	101.55	714.88	120.811	*-46.461*
G2.8.4.03	Box culverts; Internal dimensions 2 x 2m; wall thickness 400mm	m	10.34	525.58	116.88	106.23	748.69	126.077	*-48.606*
G2.8.5	Projections							
G2.8.5.01	Nibs; 50 x 50mm deep	m	0.36	18.30	4.47	2.45	25.22	3.160	*-1.123*
G2.8.5.02	Nibs; 100 x 100mm deep	m	0.36	18.30	4.47	4.91	27.68	5.652	*-2.246*
G2.8.6	Intrusions							
G2.8.6.01	Rebates 50 x 50mm deep	m	0.18	9.15	2.17	1.55	12.87	1.032	*-1.769*
G2.8.6.02	Rebates 100 x 100mm deep	m	0.18	9.15	2.17	8.20	19.52	8.646	*-3.750*

G3 Other Stated Finish

	Unit	Labour Hours	Labour Net £	Plant Net £	Materials Net £	Unit Net £	CO_2e Kg	
G3.1	**Plane horizontal**							
G3.1.1	Width not exceeding 0.1m.							
G3.1.1.01	Rough Board Finish	m	0.21	10.67	2.50	0.94	14.11	1.536	*-0.515*
G3.1.1.02	Extra Smooth Finish	m	0.21	10.67	2.50	1.06	14.23	1.216	*-0.371*
G3.1.2	Width 0.1 - 0.2m.							
G3.1.2.01	Rough Board Finish	m	0.21	10.67	2.50	1.17	14.34	1.813	*-0.639*
G3.1.2.02	Extra Smooth Finish	m	0.21	10.67	2.50	1.31	14.48	1.416	*-0.460*
G3.1.3	Width 0.2 - 0.4m.							
G3.1.3.01	Rough Board Finish	m²	0.91	46.25	5.00	4.86	56.11	6.671	*-2.651*
G3.1.3.02	Extra Smooth Finish	m²	0.91	46.25	5.00	5.44	56.69	5.024	*-1.909*
G3.1.4	Width 0.4 - 1.22m.							
G3.1.4.01	Rough Board Finish	m²	0.69	35.07	5.00	4.86	44.93	6.671	*-2.651*
G3.1.4.02	Extra Smooth Finish	m²	0.69	35.07	5.00	5.44	45.51	5.024	*-1.909*
G3.1.5	Width exceeding 1.22m.							
G3.1.5.01	Rough Board Finish	m²	0.43	21.85	5.00	4.86	31.71	6.671	*-2.651*
G3.1.5.02	Extra Smooth Finish	m²	0.43	21.85	5.00	5.44	32.29	5.024	*-1.909*
G3.2	**Plane sloping**							
G3.2.1	Width not exceeding 0.1m.							
G3.2.1.01	Rough Board Finish	m	0.23	11.74	2.73	1.03	15.50	1.671	*-0.561*
G3.2.1.02	Extra Smooth Finish	m	0.23	11.74	2.73	1.15	15.62	1.322	*-0.404*
G3.2.2	Width 0.1 - 0.2m.							
G3.2.2.01	Rough Board Finish	m	0.23	11.74	2.73	1.29	15.76	1.982	*-0.702*
G3.2.2.02	Extra Smooth Finish	m	0.23	11.74	2.73	1.44	15.91	1.546	*-0.505*
G3.2.3	Width 0.2 - 0.4m.							
G3.2.3.01	Rough Board Finish	m²	0.94	47.77	5.50	4.86	58.13	6.750	*-2.651*

G3 Other Stated Finish continued...

	Unit	Labour Hours	Labour Net £	Plant Net £	Materials Net £	Unit Net £	CO_2e Kg	
G3.2	**Plane sloping**							
G3.2.3	Width 0.2 - 0.4m.							
G3.2.3.02 Extra Smooth Finish	m²	0.94	47.77	5.50	5.79	59.06	5.377	-2.032
G3.2.4	Width 0.4 - 1.22m.							
G3.2.4.01 Rough Board Finish	m²	0.70	35.57	5.50	4.86	45.93	6.750	-2.651
G3.2.4.02 Extra Smooth Finish	m²	0.70	35.57	5.50	5.79	46.86	5.377	-2.032
G3.2.5	Width exceeding 1.22m.							
G3.2.5.01 Rough Board Finish	m²	0.47	24.04	5.50	4.86	34.40	6.750	-2.651
G3.2.5.02 Extra Smooth Finish	m²	0.47	24.04	5.50	5.79	35.33	5.377	-2.032
G3.3	**Plane battered**							
G3.3.1	Width not exceeding 0.1m.							
G3.3.1.01 Rough Board Finish	m	0.25	12.81	3.00	1.03	16.84	1.712	-0.561
G3.3.1.02 Extra Smooth Finish	m	0.25	12.81	3.00	1.25	17.06	1.438	-0.438
G3.3.2	Width 0.1 - 0.2m.							
G3.3.2.01 Rough Board Finish	m	0.25	12.81	3.00	1.29	17.10	2.023	-0.702
G3.3.2.02 Extra Smooth Finish	m	0.25	12.81	3.00	1.57	17.38	1.687	-0.550
G3.3.3	Width 0.2 - 0.4m.							
G3.3.3.01 Rough Board Finish	m²	0.52	26.22	6.86	5.66	38.74	7.904	-3.088
G3.3.3.02 Extra Smooth Finish	m²	0.52	26.22	6.86	6.34	39.42	5.985	-2.223
G3.3.4	Width 0.4 - 1.22m.							
G3.3.4.01 Rough Board Finish	m²	0.57	28.97	9.18	6.78	44.93	9.626	-3.696
G3.3.4.02 Extra Smooth Finish	m²	0.70	35.73	8.34	7.58	51.65	7.199	-2.661
G3.3.5	Width exceeding 1.22m.							
G3.3.5.01 Rough Board Finish	m²	1.03	52.34	9.21	11.33	72.88	15.170	-6.175
G3.3.5.02 Extra Smooth Finish	m²	1.03	52.34	9.21	12.67	74.22	11.333	-4.446
G3.4	**Plane vertical**							
G3.4.1	Width not exceeding 0.1m.							
G3.4.1.01 Rough Board Finish	m	0.29	14.48	3.00	1.12	18.60	1.815	-0.608
G3.4.1.02 Extra Smooth Finish	m	0.29	14.48	3.00	1.25	18.73	1.438	-0.438
G3.4.2	Width 0.1 - 0.2m.							
G3.4.2.01 Rough Board Finish	m	0.29	14.48	2.05	1.40	17.93	2.128	-0.764
G3.4.2.02 Extra Smooth Finish	m	0.29	14.48	2.05	1.57	18.10	1.653	-0.550
G3.4.3	Width 0.2 - 0.4m.							
G3.4.3.01 Rough Board Finish	m²	1.14	57.93	6.86	5.66	70.45	7.904	-3.088
G3.4.3.02 Extra Smooth Finish	m²	1.14	57.93	6.86	6.34	71.13	5.985	-2.223
G3.4.4	Width 0.4 - 1.22m.							
G3.4.4.01 Rough Board Finish	m²	0.94	47.77	11.08	6.78	65.63	9.922	-3.696
G3.4.4.02 Extra Smooth Finish	m²	0.94	47.77	11.08	7.58	66.43	7.626	-2.661
G3.4.5	Width exceeding 1.22m.							
G3.4.5.01 Rough Board Finish	m²	1.23	62.51	14.54	11.33	88.38	15.961	-6.175
G3.4.5.02 Extra Smooth Finish	m²	1.23	62.51	14.54	12.67	89.72	12.125	-4.446
G3.5	**Curved to one radius in one plane**							
G3.5.1	Width not exceeding 0.1m.							
G3.5.1.01 Rough Board Finish; 1m radius	m	0.34	17.38	3.60	1.12	22.10	1.907	-0.608
G3.5.1.02 Extra Smooth Finish; 1m radius	m	0.34	17.38	3.60	1.25	22.23	1.530	-0.438

G3 Other Stated Finish continued...

	Unit	Labour Hours	Labour Net £	Plant Net £	Materials Net £	Unit Net £	CO_2e Kg	

G3.5 Curved to one radius in one plane

G3.5.1 Width not exceeding 0.1m.

G3.5.1.03	Rough Board Finish; 2m radius	m	0.33	16.62	3.46	1.12	21.20	1.885	-0.608
G3.5.1.04	Extra Smooth Finish; 2m radius	m	0.33	16.62	3.46	1.25	21.33	1.508	-0.438

G3.5.2 Width 0.1 - 0.2m.

G3.5.2.01	Rough Board Finish; 1m radius	m	0.34	17.38	3.60	1.40	22.38	2.253	-0.764
G3.5.2.02	Extra Smooth Finish; 1m radius	m	0.34	17.38	3.60	1.57	22.55	1.779	-0.550
G3.5.2.03	Rough Board Finish; 2m radius	m	0.33	16.62	3.46	1.40	21.48	2.231	-0.764
G3.5.2.04	Extra Smooth Finish; 2m radius	m	0.33	16.62	3.46	1.79	21.87	1.931	-0.629

G3.5.3 Width 0.2 - 0.4m.

G3.5.3.01	Rough Board Finish; 1m radius	m²	1.37	69.62	8.23	5.66	83.51	8.114	-3.088
G3.5.3.02	Extra Smooth Finish; 1m radius	m²	1.37	69.62	8.23	6.34	84.19	6.196	-2.223
G3.5.3.03	Rough Board Finish; 2m radius	m²	1.31	66.57	7.90	5.66	80.13	8.068	-3.088
G3.5.3.04	Extra Smooth Finish; 2m radius	m²	1.31	66.57	7.90	6.34	80.81	6.150	-2.223

G3.5.4 Width 0.4 - 1.22m.

G3.5.4.01	Rough Board Finish; 1m radius	m²	0.68	34.56	12.05	6.78	53.39	10.071	-3.696
G3.5.4.02	Extra Smooth Finish; 1m radius	m²	0.94	47.67	11.08	7.58	66.33	7.626	-2.661
G3.5.4.03	Rough Board Finish; 2m radius	m²	0.65	33.03	8.77	6.78	48.58	9.580	-3.696
G3.5.4.04	Extra Smooth Finish; 2m radius	m²	0.89	45.23	10.52	8.70	64.45	8.410	-3.054

G3.5.5 Width exceeding 1.22m.

G3.5.5.01	Rough Board Finish; 1m radius	m²	1.37	69.52	16.43	11.33	97.28	16.227	-6.175
G3.5.5.02	Extra Smooth Finish; 1m radius	m²	1.37	69.52	16.43	12.67	98.62	12.391	-4.446
G3.5.5.03	Rough Board Finish; 2m radius	m²	1.31	66.63	15.75	11.33	93.71	16.122	-6.175
G3.5.5.04	Extra Smooth Finish; 2m radius	m²	1.31	66.63	15.75	14.56	96.94	13.756	-5.109

G3.6 Other curved

G3.6.0 Stated Finishes

G3.6.0.01	Rough Board Finish; Width: not exceeding 0.1m	m	0.59	29.98	4.19	1.56	35.73	2.502	-0.730
G3.6.0.02	Rough Board Finish; Width: 0.1 - 0.2m	m	0.59	29.98	4.19	1.95	36.12	2.900	-0.909
G3.6.0.03	Rough Board Finish; Width: 0.2 - 0.4m	m²	2.28	115.87	9.37	7.86	133.10	10.119	-3.672
G3.6.0.04	Rough Board Finish; Width: 0.4 - 1.22m	m²	1.71	86.90	14.05	9.40	110.35	12.700	-4.390
G3.6.0.05	Rough Board Finish; Width: exceeding 1.22m	m²	1.40	71.15	18.73	15.70	105.58	20.213	-7.332
G3.6.0.06	Extra Smooth Finish; Width: not exceeding 0.1m	m	0.59	29.98	4.19	2.08	36.25	2.502	-0.730
G3.6.0.07	Extra Smooth Finish; Width: 0.1 - 0.2m	m	0.59	29.98	4.19	2.59	36.76	2.900	-0.909
G3.6.0.08	Extra Smooth Finish; Width: 0.2 - 0.4m	m²	2.28	115.87	9.37	10.46	135.70	10.119	-3.672
G3.6.0.09	Extra Smooth Finish; Width: 0.4 - 1.22m	m²	1.71	86.90	14.05	12.51	113.46	12.700	-4.390
G3.6.0.10	Extra Smooth Finish; Width: exceeding 1.22m	m²	1.40	71.15	18.73	20.90	110.78	20.213	-7.332

G3.8 For concrete components of constant cross-section

G3.8.1 Beams

G3.8.1.01	Rough Board Finish; Isolated; 100 x 200mm	m	0.35	17.53	4.22	2.43	24.18	3.580	-1.325
G3.8.1.02	Extra Smooth Finish; Isolated; 100 x 200mm	m	0.35	17.53	4.22	2.72	24.47	2.757	-0.954

G3   Other Stated Finish continued...

	Unit	Labour Hours	Labour Net £	Plant Net £	Materials Net £	Unit Net £	CO_2e Kg	

G3.8	**For concrete components of constant cross-section**							
G3.8.1	Beams							
G3.8.1.03	Rough Board Finish; Isolated; 100 x 300mm	m	0.48	24.55	5.91	3.40	33.86	5.013	-1.856
G3.8.1.04	Extra Smooth Finish; Isolated; 100 x 300mm	m	0.48	24.55	5.91	3.81	34.27	3.860	-1.336
G3.8.1.05	Rough Board Finish; Isolated; 200 x 200mm	m	0.41	21.04	5.07	2.92	29.03	4.296	-1.591
G3.8.1.06	Extra Smooth Finish; Isolated; 200 x 200mm	m	0.41	21.04	5.07	3.26	29.37	3.308	-1.145
G3.8.1.07	Rough Board Finish; Isolated; 200 x 300mm	m	0.55	28.05	6.76	3.89	38.70	5.729	-2.121
G3.8.1.08	Extra Smooth Finish; Isolated; 200 x 300mm	m	0.55	28.05	6.76	4.35	39.16	4.411	-1.527
G3.8.1.09	Rough Board Finish; Isolated; 400 x 400mm	m	0.83	42.08	10.14	5.84	58.06	8.593	-3.181
G3.8.1.10	Extra Smooth Finish; Isolated; 400 x 400mm	m	0.83	42.08	10.14	6.53	58.75	6.617	-2.291
G3.8.1.11	Rough Board Finish; Isolated; 400 x 600mm	m	1.10	56.11	13.52	7.78	77.41	11.457	-4.242
G3.8.1.12	Extra Smooth Finish; Isolated; 400 x 600mm	m	1.10	56.11	13.52	8.70	78.33	8.822	-3.054
G3.8.1.13	Rough Board Finish; Isolated; 500 x 800mm	m	1.45	73.64	17.74	10.21	101.59	15.038	-5.567
G3.8.1.14	Extra Smooth Finish; Isolated; 500 x 800mm	m	1.45	73.64	17.74	11.42	102.80	11.579	-4.008
G3.8.1.15	Rough Board Finish; Attached; 100 x 200mm	m	0.35	17.53	2.79	1.95	22.27	2.941	-1.060
G3.8.1.16	Extra Smooth Finish; Attached; 100 x 200mm	m	0.35	17.53	2.79	2.18	22.50	2.282	-0.764
G3.8.1.17	Rough Board Finish; Attached; 100 x 300mm	m	0.48	24.55	5.91	2.60	33.06	4.044	-1.419
G3.8.1.18	Extra Smooth Finish; Attached; 100 x 300mm	m	0.48	24.55	5.91	3.01	33.47	3.237	-1.055
G3.8.1.19	Rough Board Finish; Attached; 200 x 200mm	m	0.41	21.04	5.07	2.35	28.46	3.604	-1.279
G3.8.1.20	Extra Smooth Finish; Attached; 200 x 200mm	m	0.41	21.04	5.07	2.62	28.73	2.810	-0.921
G3.8.1.21	Rough Board Finish; Attached; 200 x 300mm	m	0.55	28.05	6.76	3.12	37.93	4.794	-1.700
G3.8.1.22	Extra Smooth Finish; Attached; 200 x 300mm	m	0.57	29.07	6.73	3.49	39.29	3.726	-1.224
G3.8.1.23	Rough Board Finish; Attached; 400 x 400mm	m	0.83	42.08	10.14	4.66	56.88	7.174	-2.542
G3.8.1.24	Extra Smooth Finish; Attached; 400 x 400mm	m	0.86	43.91	10.10	5.22	59.23	5.576	-1.830
G3.8.1.25	Rough Board Finish; Attached; 400 x 600mm	m	1.10	56.11	13.52	6.24	75.87	9.589	-3.399
G3.8.1.26	Extra Smooth Finish; Attached; 400 x 600mm	m	1.15	58.54	13.47	6.98	78.99	7.451	-2.448
G3.8.1.27	Rough Board Finish; Attached; 500 x 800mm	m	1.45	73.64	17.74	8.18	99.56	12.581	-4.460
G3.8.1.28	Extra Smooth Finish; Attached; 500 x 800mm	m	1.45	73.64	17.74	9.15	100.53	9.810	-3.211
G3.8.2	Columns							
G3.8.2.01	Rough Board Finish; Isolated; 100 x 200mm	m	0.41	21.04	3.30	3.06	27.40	4.211	-1.669
G3.8.2.02	Extra Smooth Finish; Isolated; 100 x 200mm	m	0.41	21.04	3.30	3.42	27.76	3.175	-1.201

G3 Other Stated Finish continued...

	Unit	Labour Hours	Labour Net £	Plant Net £	Materials Net £	Unit Net £	CO_2e Kg	
G3.8	**For concrete components of constant cross-section**							
G3.8.2	Columns							
G3.8.2.03	Rough Board Finish; Isolated; 100 x 300mm	m	0.56	28.21	4.45	4.09	36.75	5.633	-2.230
G3.8.2.04	Extra Smooth Finish; Isolated; 100 x 300mm	m	0.56	28.21	4.45	4.58	37.24	4.247	-1.606
G3.8.2.05	Rough Board Finish; Isolated; 200 x 200mm	m	0.56	28.21	4.45	4.09	36.75	5.633	-2.230
G3.8.2.06	Extra Smooth Finish; Isolated; 200 x 200mm	m	0.56	28.21	4.45	4.58	37.24	4.247	-1.606
G3.8.2.07	Rough Board Finish; Isolated; 200 x 300mm	m	0.58	29.27	5.45	5.12	39.84	7.035	-2.791
G3.8.2.08	Extra Smooth Finish; Isolated; 200 x 300mm	m	0.58	29.27	5.45	5.73	40.45	5.301	-2.010
G3.8.2.09	Rough Board Finish; Isolated; 400 x 400mm	m	0.81	41.16	8.77	8.15	58.08	11.213	-4.444
G3.8.2.10	Extra Smooth Finish; Isolated; 400 x 400mm	m	0.81	41.16	8.77	9.12	59.05	8.452	-3.200
G3.8.2.11	Rough Board Finish; Isolated; 400 x 600mm	m	0.92	46.70	10.99	10.21	67.90	14.046	-5.567
G3.8.2.12	Extra Smooth Finish; Isolated; 400 x 600mm	m	0.92	46.70	10.99	11.42	69.11	10.587	-4.008
G3.8.2.13	Rough Board Finish; Isolated; 500 x 800mm	m	1.20	61.09	14.33	13.28	88.70	18.263	-7.236
G3.8.2.14	Extra Smooth Finish; Isolated; 500 x 800mm	m	1.20	61.09	14.33	14.85	90.27	13.768	-5.210
G3.8.2.15	Rough Board Finish; Attached; 100 x 200mm	m	0.28	14.03	3.30	25.46	42.79	31.305	-13.879
G3.8.2.16	Extra Smooth Finish; Attached; 100 x 200mm	m	0.28	14.03	3.30	28.48	45.81	22.682	-9.993
G3.8.2.17	Rough Board Finish; Attached; 100 x 300mm	m	0.37	18.80	4.45	3.58	26.83	5.010	-1.949
G3.8.2.18	Extra Smooth Finish; Attached; 100 x 300mm	m	0.37	18.80	4.45	4.00	27.25	3.799	-1.404
G3.8.2.19	Rough Board Finish; Attached; 200 x 200mm	m	0.37	18.80	4.45	3.06	26.31	4.387	-1.669
G3.8.2.20	Extra Smooth Finish; Attached; 200 x 200mm	m	0.37	18.80	4.45	3.42	26.67	3.350	-1.201
G3.8.2.21	Rough Board Finish; Attached; 200 x 300mm	m	0.46	23.43	5.45	4.09	32.97	5.789	-2.230
G3.8.2.22	Extra Smooth Finish; Attached; 200 x 300mm	m	0.46	23.43	5.45	4.58	33.46	4.404	-1.606
G3.8.2.23	Rough Board Finish; Attached; 400 x 400mm	m	0.74	37.40	8.77	6.12	52.29	8.757	-3.337
G3.8.2.24	Extra Smooth Finish; Attached; 400 x 400mm	m	0.74	37.40	8.77	6.85	53.02	6.683	-2.403
G3.8.2.25	Rough Board Finish; Attached; 400 x 600mm	m	0.92	46.70	10.99	8.18	65.87	11.589	-4.460
G3.8.2.26	Extra Smooth Finish; Attached; 400 x 600mm	m	0.92	46.70	10.99	9.15	66.84	8.819	-3.211
G3.8.2.27	Rough Board Finish; Attached; 500 x 800mm	m	1.20	61.09	14.33	10.73	86.15	15.183	-5.848
G3.8.2.28	Extra Smooth Finish; Attached; 500 x 800mm	m	1.20	61.09	14.33	12.00	87.42	11.550	-4.211
G3.8.3	Walls							
G3.8.3.01	Rough Board Finish; 250 mm thick x 750mm high	m	0.86	43.45	9.89	7.30	60.64	10.345	-3.976
G3.8.3.02	Extra Smooth Finish; 250 mm thick x 750mm high	m	0.86	43.45	9.89	8.16	61.50	7.875	-2.863

G3 Other Stated Finish continued...

	Unit	Labour Hours	Labour Net £	Plant Net £	Materials Net £	Unit Net £		CO$_2$e Kg

G3.8 For concrete components of constant cross-section

G3.8.3 Walls

G3.8.3.03 Rough Board Finish; 250 mm thick x 1000mm high	m	1.10	55.90	13.18	9.73	78.81	13.794	*-5.302*
G3.8.3.04 Extra Smooth Finish; 250 mm thick x 1000mm high	m	1.12	56.92	13.65	10.88	81.45	10.544	*-3.818*
G3.8.3.05 Rough Board Finish; 400mm thick x 1000mm high	m	1.10	55.90	13.18	9.73	78.81	13.794	*-5.302*
G3.8.3.06 Extra Smooth Finish; 400mm thick x 1000mm high	m	1.10	55.90	13.18	10.88	79.96	10.500	*-3.818*

G3.8.4 Other members

G3.8.4.01 Rough Board Finish; Box culverts; Internal dimensions 1 x 1m; wall thickness 250 mm	m	4.46	226.86	42.68	41.46	311.00	56.902	*-22.596*
G3.8.4.02 Extra Smooth Finish; Box culverts; Internal dimensions 1 x 1m; wall thickness 250 mm	m	4.46	226.86	42.68	46.37	315.91	42.864	*-16.269*
G3.8.4.03 Rough Board Finish; Box culverts; Internal dimensions 2 x 2m; wall thickness 250 mm	m	10.32	524.41	119.56	124.31	768.28	168.948	*-67.756*
G3.8.4.04 Extra Smooth Finish; Box culverts; Internal dimensions 2 x 2m; wall thickness 250 mm	m	10.32	524.41	119.56	139.04	783.01	126.854	*-48.786*
G3.8.4.05 Rough Board Finish; Box culverts; Internal dimensions 2 x 2m; wall thickness 400mm	m	10.86	551.85	122.72	131.95	806.52	178.720	*-71.920*
G3.8.4.06 Extra Smooth Finish; Box culverts; Internal dimensions 2 x 2m; wall thickness 400mm	m	10.86	551.85	122.72	147.58	822.15	134.039	*-51.784*

G3.8.5 Projections

G3.8.5.01 Extra Smooth Finish; Nibs; 50 x 50mm deep	m	0.36	18.30	2.96	3.20	24.46	3.114	*-1.123*
G3.8.5.02 Extra Smooth Finish; Nibs; 100 x 100mm deep	m	0.36	18.30	2.96	5.92	27.18	5.232	*-2.077*

G3.8.6 Intrusions

G3.8.6.01 Extra Smooth Finish; Rebates 50 x 50mm deep	m	0.36	18.30	1.40	3.20	22.90	2.786	*-1.123*
G3.8.6.02 Extra Smooth Finish; Rebates 100 x 100mm deep	m	0.36	18.30	1.40	3.20	22.90	2.786	*-1.123*

G5 Reinforcement

G5.1 Plain round steel bars

G5.1.1 Nominal size 6mm								
G5.1.1.01 Standard lengths	t	16.96	808.48	196.33	654.87	1,659.68	621.911	
G5.1.1.02 Bent and cut to length	t	16.96	808.48	196.33	732.04	1,736.85	621.911	
G5.1.2 Nominal size 8mm								
G5.1.2.01 Standard lengths	t	16.96	808.48	196.23	636.88	1,641.59	621.893	
G5.1.2.02 Bent and cut to length	t	16.96	808.48	196.23	711.88	1,716.59	621.893	
G5.1.3 Nominal size 10mm								
G5.1.3.01 Standard lengths	t	13.65	650.70	157.93	611.30	1,419.93	615.379	
G5.1.3.02 Bent and cut to length	t	13.65	650.70	157.93	683.48	1,492.11	615.379	
G5.1.4 Nominal size 12mm								
G5.1.4.01 Standard lengths	t	11.82	563.46	136.76	596.34	1,296.56	611.778	

G5   Reinforcement continued...

	Unit	Labour Hours	Labour Net £	Plant Net £	Materials Net £	Unit Net £	CO$_2$e Kg
G5.1	**Plain round steel bars**						
G5.1.4	Nominal size 12mm						
G5.1.4.02 Bent and cut to length	t	11.82	563.46	136.76	666.79	1,367.01	611.778
G5.1.5	Nominal size 16mm						
G5.1.5.01 Standard lengths	t	10.28	490.05	118.94	563.61	1,172.60	608.747
G5.1.5.02 Bent and cut to length	t	10.28	490.05	118.94	630.16	1,239.15	608.747
G5.1.6	Nominal size 20mm						
G5.1.6.01 Standard lengths	t	9.59	457.16	110.96	563.61	1,131.73	607.389
G5.1.6.02 Bent and cut to length	t	9.59	457.16	110.96	630.16	1,198.28	607.389
G5.1.7	Nominal size 25mm						
G5.1.7.01 Standard lengths	t	8.79	419.02	101.70	563.61	1,084.33	605.815
G5.1.7.02 Bent and cut to length	t	8.79	419.02	101.70	629.72	1,150.44	605.815
G5.1.8	Nominal size 32mm or greater						
G5.1.8.01 32mm; Standard lengths	t	8.35	398.04	96.61	566.64	1,061.29	604.949
G5.1.8.02 32mm; Bent and cut to length	t	8.35	398.04	96.61	633.41	1,128.06	604.949
G5.1.8.03 40mm; Standard lengths	t	7.85	374.21	90.82	570.76	1,035.79	603.965
G5.1.8.04 40mm; Bent and cut to length	t	7.85	374.21	90.82	640.78	1,105.81	603.965
G5.2	**Deformed high yield steel bars**						
G5.2.1	Nominal size 6mm						
G5.2.1.01 Standard lengths	t	16.96	808.48	196.23	674.52	1,679.23	621.893
G5.2.1.02 Bent and cut to length	t	16.96	808.48	196.23	754.00	1,758.71	621.893
G5.2.2	Nominal size 8mm						
G5.2.2.01 Standard lengths	t	16.96	808.48	196.23	655.98	1,660.69	621.893
G5.2.2.02 Bent and cut to length	t	16.96	808.48	196.23	733.24	1,737.95	621.893
G5.2.3	Nominal size 10mm						
G5.2.3.01 Standard lengths	t	13.65	650.70	157.93	629.64	1,438.27	615.379
G5.2.3.02 Bent and cut to length	t	13.65	650.70	157.93	703.99	1,512.62	615.379
G5.2.4	Nominal size 12mm						
G5.2.4.01 Standard lengths	t	11.82	563.46	136.76	614.23	1,314.45	611.778
G5.2.4.02 Bent and cut to length	t	11.82	563.46	136.76	686.80	1,387.02	611.778
G5.2.5	Nominal size 16mm						
G5.2.5.01 Standard lengths	t	10.28	490.05	118.94	580.52	1,189.51	608.747
G5.2.5.02 Bent and cut to length	t	10.28	490.05	118.94	649.06	1,258.05	608.747
G5.2.6	Nominal size 20mm						
G5.2.6.01 Standard lengths	t	9.59	457.16	110.96	580.52	1,148.64	607.389
G5.2.6.02 Bent and cut to length	t	9.59	457.16	110.96	649.06	1,217.18	607.389
G5.2.7	Nominal size 25mm						
G5.2.7.01 Standard lengths	t	8.79	419.02	101.70	580.52	1,101.24	605.815
G5.2.7.02 Bent and cut to length	t	8.79	419.02	101.70	649.06	1,169.78	605.815
G5.2.8	Nominal size 32mm or greater						
G5.2.8.01 32mm; Standard lengths	t	8.35	398.04	96.61	583.64	1,078.29	604.949
G5.2.8.02 32mm; Bent and cut to length	t	8.35	398.04	96.61	652.41	1,147.06	604.949
G5.2.8.03 40mm; Standard lengths	t	7.85	374.21	90.82	587.88	1,052.91	603.965
G5.2.8.04 40mm; Bent and cut to length	t	7.85	374.21	90.82	660.00	1,125.03	603.965

G5 Reinforcement continued...

	Unit	Labour Hours	Labour Net £	Plant Net £	Materials Net £	Unit Net £	CO$_2$e Kg
G5.3 **Stainless steel bars of stated quality**							
G5.3.3 Nominal size 10mm							
G5.3.3.01 Stainless steel bars; type 316 S66 stainless steel and warm worked; standard lengths	t	13.65	650.70	157.93	3,287.72	**4,096.35**	6,142.276
G5.3.3.02 Stainless steel bars; type 316 S66 stainless steel and warm worked; bent and cut to length	t	13.65	650.70	157.93	3,420.91	**4,229.54**	6,142.276
G5.3.4 Nominal size 12mm							
G5.3.4.01 Stainless steel bars; type 316 S66 stainless steel and warm worked; standard lengths	t	11.82	563.46	136.76	2,853.10	**3,553.32**	6,138.675
G5.3.4.02 Stainless steel bars; type 316 S66 stainless steel and warm worked; bent and cut to length	t	11.82	563.46	136.76	2,972.62	**3,672.84**	6,138.675
G5.3.5 Nominal size 16mm							
G5.3.5.01 Stainless steel bars; type 316 S66 stainless steel and warm worked; standard lengths	t	10.28	490.05	118.94	2,949.13	**3,558.12**	6,135.644
G5.3.5.02 Stainless steel bars; type 316 S66 stainless steel and warm worked; bent and cut to length	t	10.28	490.05	118.94	3,038.86	**3,647.85**	6,135.644
G5.3.6 Nominal size 20mm							
G5.3.6.01 Stainless steel bars; type 316 S66 stainless steel and warm worked; standard lengths	t	9.59	457.16	110.96	3,715.33	**4,283.45**	6,134.286
G5.3.6.02 Stainless steel bars; type 316 S66 stainless steel and warm worked; bent and cut to length	t	9.59	457.16	110.96	3,817.33	**4,385.45**	6,134.286
G5.3.7 Nominal size 25mm							
G5.3.7.01 Stainless steel bars; type 316 S66 stainless steel and warm worked; standard lengths	t	8.79	419.02	101.70	4,006.25	**4,526.97**	6,132.712
G5.3.7.02 Stainless steel bars; type 316 S66 stainless steel and warm worked; bent and cut to length	t	8.79	419.02	101.70	4,096.68	**4,617.40**	6,132.712
G5.3.8 Nominal size 32mm or greater							
G5.3.8.01 Stainless steel bars; type 316 S66 stainless steel and warm worked; standard lengths	t	8.35	398.04	96.61	3,902.50	**4,397.15**	6,131.846
G5.3.8.02 Stainless steel bars; type 316 S66 stainless steel and warm worked; bent and cut to length	t	8.35	398.04	96.61	3,993.28	**4,487.93**	6,131.846
G5.5 **Special joints**							
G5.5.0 Generally							
G5.5.0.01 Plain round steel bars, BS 4449 welded to mild steel; including wire brush cleaning and preparing surfaces. (Cost of bars not included); nominal size: 6mm	Nr	0.70	34.66	2.63	0.04	**37.33**	1.514

G5 Reinforcement continued...

	Unit	Labour Hours	Labour Net £	Plant Net £	Materials Net £	Unit Net £	CO_2e Kg
G5.5	**Special joints**						
G5.5.0	Generally						
G5.5.0.02	Plain round steel bars, BS 4449 welded to mild steel; including wire brush cleaning and preparing surfaces. (Cost of bars not included); nominal size: 8mm	Nr	0.70	34.66	2.63	0.04	37.33	1.514
G5.5.0.03	Plain round steel bars, BS 4449 welded to mild steel; including wire brush cleaning and preparing surfaces. (Cost of bars not included); nominal size: 20mm	Nr	1.50	74.28	5.64	0.11	80.03	3.257
G5.5.0.04	Deformed high yield bars, BS 4449 welded to mild steel; including wire brush cleaning and preparing surfaces. (Cost of bars not included); nominal size: 6mm	Nr	0.70	34.66	2.63	0.04	37.33	1.514
G5.5.0.05	Deformed high yield bars, BS 4449 welded to mild steel; including wire brush cleaning and preparing surfaces. (Cost of bars not included); nominal size: 8mm	Nr	0.70	34.66	2.63	0.04	37.33	1.514
G5.5.0.06	Deformed high yield bars, BS 4449 welded to mild steel; including wire brush cleaning and preparing surfaces. (Cost of bars not included); nominal size: 20mm	Nr	1.50	74.28	5.64	0.11	80.03	3.257
G5.5.0.07	Stainless steel bars type 316, S66 welded to mild steel; including wire brush cleaning and preparing surfaces. (Cost of bars not included); nominal size: 6mm	Nr	0.80	39.62	3.01	0.17	42.80	1.867
G5.5.0.08	Stainless steel bars type 316, S66 welded to mild steel; including wire brush cleaning and preparing surfaces. (Cost of bars not included); nominal size: 8mm	Nr	0.80	39.62	3.01	0.17	42.80	1.867
G5.5.0.09	Stainless steel bars type 316, S66 welded to mild steel; including wire brush cleaning and preparing surfaces. (Cost of bars not included); nominal size: 20mm	Nr	1.60	79.23	6.02	0.50	85.75	3.886
G5.6	**Steel fabric, BS 4483**						
G5.6.1	Nominal mass: not exceeding 2kg/m^2						
G5.6.1.01	0.77kg/m^2; D49	m^2	0.06	2.86	-	1.39	4.25	0.453
G5.6.1.02	1.54kg/m^2; A98	m^2	0.06	2.86	-	1.52	4.38	0.906
G5.6.2	Nominal mass: 2 - 3kg/m^2						
G5.6.2.01	2.22kg/m^2; A142	m^2	0.06	2.86	-	1.47	4.33	1.307
G5.6.2.02	2.61kg/m^2; C283	m^2	0.06	2.86	-	1.91	4.77	1.536
G5.6.3	Nominal mass: 3 - 4kg/m^2						
G5.6.3.01	3.02kg/m^2; A193	m^2	0.06	2.86	-	2.04	4.90	1.777
G5.6.3.02	3.05kg/m^2; B196	m^2	0.07	3.34	-	3.73	7.07	1.795
G5.6.3.03	3.41kg/m^2; C385	m^2	0.07	3.34	-	2.30	5.64	2.007
G5.6.3.04	3.73kg/m^2; B283	m^2	0.07	3.34	-	2.49	5.83	2.195
G5.6.3.05	3.95kg/m^2; A252	m^2	0.09	4.29	-	2.69	6.98	2.325

G5 Reinforcement continued...

	Unit	Labour Hours	Labour Net £	Plant Net £	Materials Net £	Unit Net £	CO_2e Kg
G5.6	**Steel fabric, BS 4483**						
G5.6.4	Nominal mass: 4 - 5kg/m²						
G5.6.4.01	4.34kg/m²; C503	m²	0.09	4.29	-	2.95	7.24	2.554
G5.6.4.02	4.53kg/m²; B385	m²	0.10	4.77	-	3.03	7.80	2.666
G5.6.5	Nominal mass: 5 - 6kg/m²						
G5.6.5.01	5.55kg/m²; C636	m²	0.10	4.77	-	3.77	8.54	3.266
G5.6.5.02	5.93kg/m²; B503	m²	0.10	4.77	-	3.96	8.73	3.490
G5.6.6	Nominal mass: 6 - 7kg/m²						
G5.6.6.01	6.16kg/m²; A393	m²	0.13	6.20	1.50	4.12	11.82	3.881
G5.6.6.02	6.72kg/m²; C785	m²	0.13	6.20	1.50	4.14	11.84	4.211
G5.6.8	Nominal mass: stated exceeding 8kg/m²						
G5.6.8.01	8.14kg/m2; B785	m²	0.13	6.20	1.50	5.38	13.08	5.047
G5.6.8.02	10.90kg/m²; B1131	m²	0.14	6.67	1.62	7.24	15.53	6.708

G6 Joints

	Unit	Labour Hours	Labour Net £	Plant Net £	Materials Net £	Unit Net £	CO_2e Kg
G6.1	**Open surface plain**						
G6.1.1	Average width: not exceeding 0.5m						
G6.1.1.01	Generally	m²	0.10	2.89	0.98	-	3.87	0.659
G6.1.2	Average width: 0.5 - 1m						
G6.1.2.01	Generally	m²	0.09	2.60	0.88	-	3.48	0.593
G6.1.3	Average width: stated exceeding 1m						
G6.1.3.01	Generally	m²	0.08	2.31	0.78	-	3.09	0.527
G6.2	**Open surface with filler**						
G6.2.1	Average width: not exceeding 0.5m						
G6.2.1.01	13mm Flexcell joint filler	m²	0.23	11.69	-	7.47	19.16	1.208
G6.2.1.02	19mm Flexcell joint filler	m²	0.23	11.69	-	11.46	23.15	1.740
G6.2.1.03	25mm Flexcell joint filler	m²	0.23	11.69	-	13.51	25.20	2.417
G6.2.2	Average width: 0.5 - 1m						
G6.2.2.01	13mm Flexcell joint filler	m²	0.20	10.16	-	7.47	17.63	1.208
G6.2.2.02	19mm Flexcell joint filler	m²	0.20	10.16	-	11.46	21.62	1.740
G6.2.2.03	25mm Flexcell joint filler	m²	0.20	10.16	-	13.51	23.67	2.417
G6.2.3	Average width: stated exceeding 1m						
G6.2.3.01	13mm Flexcell joint filler	m²	0.17	8.64	-	7.47	16.11	1.208
G6.2.3.02	19mm Flexcell joint filler	m²	0.17	8.64	-	11.46	20.10	1.740
G6.2.3.03	25mm Flexcell joint filler	m²	0.17	8.64	-	13.51	22.15	2.417
G6.3	**Formed surface plain (including formwork)**						
G6.3.1	Average width: not exceeding 0.5m						
G6.3.1.01	Generally; including formwork	m²	0.51	25.82	6.02	6.13	37.97	7.164	-2.807
G6.3.2	Average width: 0.5 - 1m						
G6.3.2.01	Generally; including formwork	m²	0.69	35.07	8.06	6.13	49.26	7.495	-2.807
G6.3.3	Average width: stated exceeding 1m						
G6.3.3.01	Generally; including formwork	m²	0.75	38.01	8.49	6.13	52.63	7.594	-2.807

G6 Joints continued...

Code	Description	Unit	Labour Hours	Labour Net £	Plant Net £	Materials Net £	**Unit Net £**	*CO_2e Kg*	
G6.4	**Formed surface with filler**								
G6.4.1	Average width: not exceeding 0.5m								
G6.4.1.01	13mm Flexcell joint filler	m^2	0.63	31.91	4.19	13.61	**49.71**	8.319	*-2.807*
G6.4.1.02	19mm Flexcell joint filler	m^2	0.63	31.91	6.02	17.59	**55.52**	8.904	*-2.807*
G6.4.1.03	25mm Flexcell joint filler	m^2	0.63	31.91	6.02	19.65	**57.58**	9.581	*-2.807*
G6.4.2	Average width: 0.5 - 1m								
G6.4.2.01	13mm Flexcell joint filler	m^2	0.69	35.07	4.60	13.61	**53.28**	8.405	*-2.807*
G6.4.2.02	19mm Flexcell joint filler	m^2	0.69	35.27	6.63	17.59	**59.49**	8.998	*-2.807*
G6.4.2.03	25mm Flexcell joint filler	m^2	0.69	35.27	6.63	19.65	**61.55**	9.675	*-2.807*
G6.4.3	Average width: stated exceeding 1m								
G6.4.3.01	13mm Flexcell joint filler	m^2	0.75	38.12	5.01	13.61	**56.74**	8.492	*-2.807*
G6.4.3.02	19mm Flexcell joint filler	m^2	0.75	38.01	7.29	17.59	**62.89**	9.099	*-2.807*
G6.4.3.03	25mm Flexcell joint filler	m^2	0.75	38.01	7.29	19.65	**64.95**	9.776	*-2.807*
G6.5	**Plastics or rubber waterstops**								
G6.5.1	Average width not exceeding 150mm								
G6.5.1.01	PVC Flat "X" dumbell junction pieces; width 100mm	Nr	0.55	27.95	-	27.80	**55.75**	9.500	
G6.5.1.02	Rubber flat dumbell; width 150mm	Nr	0.21	10.67	-	37.97	**48.64**	5.390	
G6.5.1.03	Rubber flat "L" dumbell junction piece; width 150mm	Nr	0.31	15.75	-	91.04	**106.79**	9.917	
G6.5.1.04	Rubber vertical "L" dumbell junction piece; width 150mm	Nr	0.50	25.41	-	91.04	**116.45**	9.917	
G6.5.1.05	Heavy duty PVC flat "T" dumbell junction piece; width 150mm	Nr	0.50	25.41	-	93.06	**118.47**	10.620	
G6.5.1.06	Heavy duty PVC vertical "T" dumbell junction piece; width 150mm	Nr	0.55	27.95	-	93.06	**121.01**	10.620	
G6.5.1.07	Heavy duty PVC flat "X" dumbell junction piece; width 150mm	Nr	0.60	30.49	-	112.38	**142.87**	10.832	
G6.5.2	Average width 150 - 200mm								
G6.5.2.01	PVC Flat centre bulb; width 150mm	Nr	0.55	27.95	-	27.80	**55.75**	9.500	
G6.5.2.02	PVC Flat "L" centre bulb junction (2 way); width 150mm	Nr	0.31	15.75	-	13.33	**29.08**	10.310	
G6.5.2.03	PVC Vertical "L" centre bulb junction (2 way); width 150mm	Nr	0.50	25.41	-	11.99	**37.40**	10.310	
G6.5.2.04	PVC Flat "T" centre bulb junction (3 way); width 150mm	Nr	0.50	25.41	-	26.03	**51.44**	10.620	
G6.5.2.05	PVC Vertical "T" centre bulb junction (3 way); width 150mm	Nr	0.55	27.95	-	27.48	**55.43**	10.620	
G6.5.2.06	PVC Flat "X" dumbell junction pieces; width 170mm	Nr	0.60	30.49	-	31.70	**62.19**	10.832	
G6.5.3	Average width 200 - 300mm								
G6.5.3.01	PVC Flat centre bulb; width 200mm	Nr	0.26	13.21	-	10.69	**23.90**	5.155	
G6.5.3.02	PVC Flat centre bulb; width 250 mm	Nr	0.31	15.75	-	13.67	**29.42**	6.444	
G6.5.3.03	PVC Flat "L" centre bulb junction (2 way); width 200mm	Nr	0.36	18.30	-	19.61	**37.91**	10.524	
G6.5.3.04	PVC Flat "L" centre bulb junction (2 way); width 250 mm	Nr	0.36	18.30	-	25.09	**43.39**	10.740	
G6.5.3.05	PVC Vertical "L" centre bulb junction (2 way); width 200mm	Nr	0.55	27.95	-	20.94	**48.89**	10.524	

G6 Joints continued...

	Unit	Labour Hours	Labour Net £	Plant Net £	Materials Net £	Unit Net £	CO_2e Kg
G6.5	**Plastics or rubber waterstops**						
G6.5.3	Average width 200 - 300mm						
G6.5.3.06	PVC Vertical "L" centre bulb junction (2 way); width 250 mm	Nr	0.60	30.49	-	22.98	53.47	10.740
G6.5.3.07	PVC Flat "T" centre bulb junction (3 way); width 200mm	Nr	0.55	27.95	-	30.87	58.82	10.836
G6.5.3.08	PVC Flat "T" centre bulb junction (3 way); width 250 mm	Nr	0.60	30.49	-	37.67	68.16	11.062
G6.5.3.09	PVC Vertical "T" centre bulb junction (3 way); width 200mm	Nr	0.60	30.49	-	31.70	62.19	10.836
G6.5.3.10	PVC Vertical "T" centre bulb junction (3 way); width 250 mm	Nr	0.60	30.49	-	38.56	69.05	11.062
G6.5.3.11	PVC Flat "X" dumbell junction pieces; width 210mm	Nr	0.65	33.03	-	38.17	71.20	11.032
G6.5.3.12	PVC Flat "X" dumbell junction pieces; width 250 mm	Nr	0.70	35.57	-	46.32	81.89	11.283
G6.5.3.13	Rubber flat dumbell; width 230mm	m	0.26	13.21	-	55.04	68.25	8.264
G6.5.3.14	Rubber flat "L" dumbell junction piece; width 230mm	Nr	0.36	18.30	-	105.73	124.03	10.330
G6.5.3.15	Rubber vertical "L" dumbell junction piece; width 230mm	Nr	0.60	30.49	-	105.73	136.22	10.330
G6.5.3.16	Heavy duty PVC flat "T" dumbell junction piece; width 230mm	Nr	0.60	30.49	-	116.68	147.17	11.062
G6.5.3.17	Heavy duty PVC vertical "T" dumbell junction piece; width 230mm	Nr	0.65	33.03	-	116.67	149.70	11.062
G6.5.3.18	Heavy duty PVC flat "X" dumbell junction piece; width 230mm	Nr	0.70	35.57	-	131.03	166.60	11.283
G6.7	**Sealed rebates or grooves**						
G6.7.0	Cold poured Expandite Colpor 200 joint sealing compound; forming groove						
G6.7.0.01	5 x 15mm	m	0.15	4.33	-	0.28	4.61	0.005
G6.7.0.02	10 x 10mm	m	0.15	4.33	-	0.47	4.80	0.009
G6.7.0.03	10 x 15mm	m	0.15	4.33	-	0.56	4.89	0.011
G6.7.0.04	10 x 20mm	m	0.15	4.33	-	0.75	5.08	0.014
G6.7.0.05	10 x 25mm	m	0.15	4.33	-	1.13	5.46	0.021
G6.7.0.06	15 x 20mm	m	0.15	4.33	-	1.22	5.55	0.023
G6.7.0.07	15 x 25mm	m	0.15	4.33	-	1.32	5.65	0.025
G6.7.0.08	15 x 40mm	m	0.25	7.22	-	2.82	10.04	0.053
G6.7.0.09	20 x 25mm	m	0.25	7.22	-	2.35	9.57	0.044
G6.7.0.10	20 x 40mm	m	0.25	7.22	-	3.76	10.98	0.070
G6.7.0.11	20 x 50mm	m	0.25	7.22	-	4.70	11.92	0.088
G6.7.0.12	25 x 30mm	m	0.25	7.22	-	3.38	10.60	0.063
G6.7.0.13	25 x 40mm	m	0.25	7.22	-	4.70	11.92	0.088
G6.7.0.14	25 x 50mm	m	0.35	10.11	-	5.26	15.37	0.099
G6.7.0.15	30 x 40mm	m	0.35	10.11	-	5.64	15.75	0.106
G6.7.0.16	30 x 50mm	m	0.35	10.11	-	7.14	17.25	0.134
G6.7.0.17	50 x 50mm	m	0.45	13.00	-	11.28	24.28	0.211
G6.8	**Dowels**						
G6.8.1	Plain or greased						
G6.8.1.01	Plain mild steel dowel bars with cages 0.5m long; cast into one side of joint; 12mm diameter	Nr	0.10	4.77	-	0.15	4.92	0.519
G6.8.1.02	Plain mild steel dowel bars with cages 0.5m long; cast into one side of joint; 16mm diameter	Nr	0.10	4.77	-	0.26	5.03	0.922
G6.8.1.03	Plain mild steel dowel bars with cages 0.5m long; cast into one side of joint; 20mm diameter	Nr	0.13	5.96	-	0.39	6.35	1.441
G6.8.1.04	Plain mild steel dowel bars with cages 0.5m long; cast into one side of joint; 25mm diameter	Nr	0.13	5.96	-	0.55	6.51	2.252

G6 Joints continued...

	Unit	Labour Hours	Labour Net £	Plant Net £	Materials Net £	Unit Net £	CO_2e Kg
G6.8	**Dowels**						
G6.8.1	Plain or greased						
G6.8.1.05	Plain mild steel dowel bars with cages 1m long; cast into one side of joint; 12mm diameter	Nr	0.13	5.96	-	0.27	6.23	0.519
G6.8.1.06	Plain mild steel dowel bars with cages 1m long; cast into one side of joint; 16mm diameter	Nr	0.13	5.96	-	0.47	6.43	0.922
G6.8.1.07	Plain mild steel dowel bars with cages 1m long; cast into one side of joint; 20mm diameter	Nr	0.15	7.15	-	0.70	7.85	1.441
G6.8.1.08	Plain mild steel dowel bars with cages 1m long; cast into one side of joint; 25mm diameter	Nr	0.15	7.15	-	0.99	8.14	2.252
G6.8.1.09	Plain mild steel dowel bars with cages 1m long; cast into one side of joint and debonding for a length of 500mm; 12mm diameter	Nr	0.15	7.15	-	0.27	7.42	0.519
G6.8.1.10	Plain mild steel dowel bars with cages 1m long; cast into one side of joint and debonding for a length of 500mm; 16mm diameter	Nr	0.15	7.15	-	0.47	7.62	0.922
G6.8.1.11	Plain mild steel dowel bars with cages 1m long; cast into one side of joint and debonding for a length of 500mm; 20mm diameter	Nr	0.17	7.87	-	0.70	8.57	1.441
G6.8.1.12	Plain mild steel dowel bars with cages 1m long; cast into one side of joint and debonding for a length of 500mm; 25mm diameter	Nr	0.17	7.87	-	0.99	8.86	2.252
G6.8.2	Sleeved or capped						
G6.8.2.01	Capped mild steel dowel bars with cages 1m long; cast into one side of joint and debonding for a length of 500mm and capped with PVC dowel caps; 12mm diameter	Nr	0.17	7.87	-	0.27	8.14	0.519
G6.8.2.02	16mm diameter	Nr	0.17	7.87	-	0.47	8.34	0.922
G6.8.2.03	20mm diameter	Nr	0.18	8.58	-	0.70	9.28	1.441
G6.8.2.04	25mm diameter	Nr	0.18	8.58	-	0.99	9.57	2.252

G7 Post-Tensioned Prestressing

Note(s): The following prices are guide prices only for the use in preliminary approximate estimating.

Prestressing tendons vary greatly in size, for example the K-Range system provides tendons with capacities in the range of 1060kN to 10230kN depending on the number of prestressing strands incorporated in the tendon. The most reasonable method of assessing prestressing costs is probably that based on the tonnage of prestressing steel required, although prices, even on the basis, vary considerably depending on the individual tendon capacity, tendon lengths, total quantities and the type of structure. Tonnage can be assessed on the basis of 4.5kg of steel per metre per 1000kN of Tendon characteristic strength (Tendon working loads specified are normally around 70% of this value). Prices are around £2750 ± 30% per tonne of prestressing steel. Small works containing less than 15 tonnes of steel would fall at the top end of this price range with the average around 40-50 tonnes. Unit prices bottom out at 100 tonnes and over. Tendons less than 15m long are rarely formed from prestressing steel as for these lengths bar tendons are usually cheaper. EXAMPLE: 12Nr. Tendons on each 3 structures, tendons 32m long with specified force of 3535kN at 70% characteristic strength.

100% characteristic strength = 3525 / 0.7 = 5036kN
Tonnage = 12 x 3 x 32 x 5036 x 4.5 / 1000 = 26.1 tonnes.

This is a relatively small job so price will be above the median, say £2750 + 15% = £3162.50/tonne x 26.1 tonnes = £82541 and the unit tendon price / 3 / 12 = £2293 per tendon. The rates include supplying and threading the prestressing cable, the supply and fixing of steel ducts and anchorages and the stressing and grouting of the tendons.

*It is usual for the main contractor to fix in position in their shutter the load distribution trumpet component of the anchorage assembly.

G7.1	**Horizontal internal tendons in in situ concrete**						
G7.1.6	Length; 20 - 25m						
G7.1.6.01	Length: 20 - 25m	Nr	6.00	368.34	361.32	1,739.15	2,468.81	344.272

G7 Post-Tensioned Prestressing continued...

	Unit	Labour Hours	Labour Net £	Plant Net £	Materials Net £	Unit Net £	CO_2e Kg
G7.1	**Horizontal internal tendons in in situ concrete**						
G7.1.8 Length; stated exceeding 30m							
G7.1.8.01 Length; 32m	Nr	6.00	368.34	361.32	1,783.46	2,513.12	361.576
G7.2	**Inclined or vertical internal tendons in in situ concrete**						
G7.2.6 Length; 20 - 25m							
G7.2.6.01 Length: 20 - 25m	Nr	8.00	491.12	481.76	1,739.15	2,712.03	432.396
G7.2.8 Length; stated exceeding 30m							
G7.2.8.01 Length; 32m	Nr	8.00	491.12	481.76	1,783.46	2,756.34	449.700

G8 Concrete Accessories

	Unit	Labour Hours	Labour Net £	Plant Net £	Materials Net £	Unit Net £	CO_2e Kg
G8.1	**Finishing of top surfaces**						
G8.1.1 Wood float							
G8.1.1.01 level	m²	0.05	1.44	-	-	1.44	-
G8.1.1.02 to crossfalls	m²	0.07	2.02	-	-	2.02	-
G8.1.2 Steel trowel							
G8.1.2.01 level	m²	0.05	1.44	-	-	1.44	-
G8.1.2.02 to crossfalls	m²	0.07	2.02	-	-	2.02	-
G8.2	**Finishing of formed surfaces**						
G8.2.2 Bush hammering							
G8.2.2.01 walls	m²	0.70	20.22	6.85	-	27.07	4.610
G8.2.2.02 beams	m²	1.00	28.89	9.79	-	38.68	6.585
G8.2.2.03 columns	m²	0.83	23.98	8.13	-	32.11	5.466
G8.2.3 Other stated surface treatment carried out after striking							
G8.2.3.01 Rubbing down concrete surfaces after striking formwork; walls	m²	0.17	4.91	-	-	4.91	-
G8.2.3.02 Rubbing down concrete surfaces after striking formwork; beams	m²	0.20	5.78	-	-	5.78	-
G8.2.3.03 Rubbing down concrete surfaces after striking formwork; columns	m²	0.20	5.78	-	-	5.78	-
G8.3	**Inserts**						
G8.3.1 Linear inserts							
G8.3.1.01 Building in pipes (supply excluded); 250 mm thick reinforced concrete; 100mm nominal bore	m	0.34	17.28	-	-	17.28	-
G8.3.1.02 Building in pipes (supply excluded); 250 mm thick reinforced concrete; 150mm nominal bore	m	0.34	17.28	-	-	17.28	-
G8.3.1.03 Building in pipes (supply excluded); 250 mm thick reinforced concrete; 300mm nominal bore	m	0.45	22.87	-	-	22.87	-
G8.3.2 Other inserts							
G8.3.2.01 Grouting in foundation bolts (supply excluded); 250 mm thick reinforced concrete; 20 x 20 x 100mm deep	Nr	0.25	12.71	-	0.10	12.81	0.254

G8   Concrete Accessories continued...

	Unit	Labour Hours	Labour Net £	Plant Net £	Materials Net £	Unit Net £	CO_2e Kg
G8.3	**Inserts**						
G8.3.2	Other inserts						
G8.3.2.02	Grouting in foundation bolts (supply excluded); 250 mm thick reinforced concrete; 25 x 25 x 175mm deep	Nr	0.25	12.71	-	0.10	12.81	0.254
G8.3.2.03	Grouting in foundation bolts (supply excluded); 250 mm thick reinforced concrete; 200 x 200 x 500mm deep	Nr	0.25	12.71	-	0.20	12.91	0.508
G8.3.2.04	Grouting in foundation bolts (supply excluded); 250 mm thick reinforced concrete; 300 x 200 x 500mm deep	Nr	0.25	12.71	-	0.31	13.02	0.762
G8.4	**Grouting under plates**						
G8.4.1	Area: not exceeding $0.1m^2$						
G8.4.1.01	Grouting in cement mortar (1:3); 25mm thick	Nr	0.34	17.28	-	0.31	17.59	0.762
G8.4.1.02	Grouting in cement mortar (1:3); 50mm thick	Nr	0.45	22.87	-	0.61	23.48	1.524
G8.4.2	Area: $0.1 - 0.5m^2$						
G8.4.2.01	Grouting in cement mortar (1:3); 25mm thick	Nr	0.34	17.28	-	0.61	17.89	1.524
G8.4.2.02	Grouting in cement mortar (1:3); 50mm thick	Nr	0.45	22.87	-	1.22	24.09	3.049
G8.4.3	Area: $0.5 - 1m^2$						
G8.4.3.01	Grouting in cement mortar (1:3); 25mm thick	Nr	0.45	22.87	-	1.02	23.89	2.541
G8.4.3.02	Grouting in cement mortar (1:3); 50mm thick	Nr	0.60	30.49	-	1.02	31.51	2.541
G8.4.4	Area stated exceeding $1m^2$						
G8.4.4.01	Grouting in cement mortar (1:3); 25mm thick	Nr	0.50	25.41	-	5.10	30.51	12.703
G8.4.4.02	Grouting in cement mortar (1:3); 50mm thick	Nr	0.60	30.49	-	5.10	35.59	12.703

CLASS H:
PRECAST CONCRETE

Calculations used throughout Class H - Precast Concrete

Labour

			Qty		Rate		Total
L G0001ICE	**Precast Concrete Install Labour Gang**						
	L A9003ICE	Labourer (Skill Rate 3)	2	x	14.86	=	£29.72
	L A9016ICE	Ganger	1	x	17.64	=	£17.64
	L A9015ICE	Banksman	1	x	14.03	=	£14.03
		Total hourly cost of gang				=	**£61.39**

Plant

			Qty		Rate		Total
P G0001ICE	**Precast Concrete Install Plant Gang**						
	P A0481ICE	Cranes Transit - 25t	1	x	60.22	=	£60.22
		Total hourly cost of gang				=	**£60.22**

Precast Concrete

Note(s):   NOTE: The following prices regarding bridge beams has been derived from historical specialist sub-contractor prices and are guide prices for approximate estimating purpose only. The prices are for supply and deliver only.

H2   Prestressed Pre-Tensioned Beams

		Unit	Labour Hours	Labour Net £	Plant Net £	Materials Net £	Unit Net £	CO_2e Kg
H2.3	**Length: 7 - 10m**							
H2.3.4	Mass: 1 - 2t							
H2.3.4.01	Bridge Beams; Inverted "T" Beams; Length 8m; top flange width 205mm; bottom flange width 495mm; Section T1; depth 380mm; cross-sectional area 98000mm^2; mass 1.88 t	Nr	-	-	-	737.76	737.76	340.298
H2.3.5	Mass: 2 - 5t							
H2.3.5.01	Bridge Beams; Inverted "T" Beams; Length 8m; top flange width 205mm; bottom flange width 495mm; Section T2; depth 420mm; cross-sectional area 106200mm^2; mass 2.04 t	Nr	-	-	-	843.15	843.15	369.260
H2.3.5.02	Bridge Beams; Inverted "T" Beams; Length 8m; top flange width 205mm; bottom flange width 495mm; Section T3; depth 535mm; cross-sectional area 114275mm^2; mass 2.19 t	Nr	-	-	-	948.54	948.54	396.411
H2.4	**Length: 10 - 15m**							
H2.4.5	Mass: 2 - 5t							
H2.4.5.01	Bridge Beams; Inverted "T" Beams; Length 12m; top flange width 205mm; bottom flange width 495mm; Section T4; depth 575mm; cross-sectional area 122475mm^2; mass 3.54 t	Nr	-	-	-	1,053.94	1,053.94	640.774
H2.4.5.02	Bridge Beams; Inverted "T" Beams; Length 12m; top flange width 205mm; bottom flange width 495mm; Section T5; depth 615mm; cross-sectional area 130675mm^2; mass 3.76 t	Nr	-	-	-	1,264.73	1,264.73	680.596
H2.4.5.03	Bridge Beams; Inverted "T" Beams; Length 12m; top flange width 205mm; bottom flange width 495mm; Section T6; depth 655mm; cross-sectional area 138875mm^2; mass 4 t	Nr	-	-	-	1,370.12	1,370.12	724.038
H2.4.5.04	Bridge Beams; Inverted "T" Beams; Length 12m; top flange width 205mm; bottom flange width 495mm; Section T7; depth 695mm; cross-sectional area 147075mm^2; mass 4.25 t	Nr	-	-	-	1,580.91	1,580.91	769.291

H2   Prestressed Pre-Tensioned Beams continued...

		Unit	Labour Hours	Labour Net £	Plant Net £	Materials Net £	Unit Net £	CO_2e Kg
H2.4	**Length: 10 - 15m**							
H2.4.5	Mass: 2 - 5t							
H2.4.5.05	Bridge Beams; Inverted "T" Beams; Length 12m; top flange width 205mm; bottom flange width 495mm; Section T8; depth 735mm; cross-sectional area 155160mm²; mass 4.48 t	Nr	-	-	-	1,791.70	1,791.70	810.923
H2.5	**Length: 15 - 20m**							
H2.5.6	Mass: 5 - 10t							
H2.5.6.01	Bridge Beams; Inverted "T" Beams; Length 18m; top flange width 400mm; bottom flange width 970mm; Section T9; depth 775mm; cross-sectional area 163 360mm²; mass 7.06 t	Nr	-	-	-	1,897.09	1,897.09	1,277.928
H2.5.6.02	Bridge Beams; Inverted "T" Beams; Length 18m; top flange width 400mm; bottom flange width 970mm; Section T10; depth 815mm; cross-sectional area 171560mm²; mass 7.43 t	Nr	-	-	-	2,107.88	2,107.88	1,344.901
H2.5.7	Mass: 10 - 20t							
H2.5.7.01	Bridge Beams; "M" Beams; Length 18m; base width 970mm; leg thickness 165mm; Section M²; depth 720mm; cross-sectional area 316650mm2; mass 13.7 t	Nr	-	-	-	4,848.12	4,848.12	2,479.831
H2.5.7.02	Bridge Beams; "M" Beams; Length 18m; base width 970mm; leg thickness 165mm; Section M³; depth 800mm; cross-sectional area 348650mm²; mass 15.1 t	Nr	-	-	-	5,480.48	5,480.48	2,733.244
H2.5.7.03	Bridge Beams; "M" Beams; Length 18m; base width 970mm; leg thickness 165mm; Section M4; depth 880mm; cross-sectional area 355050mm²; mass 15.37 t	Nr	-	-	-	6,534.42	6,534.42	2,782.117
H2.5.7.04	Bridge Beams; "Y" Beams; Length 16m; base width 750mm; Section Y1; depth 700mm; cross sectional area 309202mm²; mass 12.4t	Nr	-	-	-	4,848.12	4,848.12	2,244.519
H2.5.7.05	Bridge Beams; "Y" Beams; Length 16m; base width 750mm; Section Y2; depth 800mm; cross sectional area 339882mm²; mass 13.6t	Nr	-	-	-	5,691.27	5,691.27	2,461.730
H2.5.8	Mass: stated exceeding 20t							
H2.5.8.01	Bridge Beams; "U" Beams; Section U1; depth 800mm; cross-sectional area 466450mm²; mass 20.2 t	Nr	-	-	-	8,536.90	8,536.90	3,656.393
H2.5.8.02	Bridge Beams; "U" Beams; Section U3; depth 900mm; cross-sectional area 499450mm²; mass 21.64 t	Nr	-	-	-	10,117.81	10,117.81	3,917.047

H2 Prestressed Pre-Tensioned Beams continued...

	Unit	Labour Hours	Labour Net £	Plant Net £	Materials Net £	Unit Net £	CO₂e Kg

	Unit	Labour Hours	Labour Net £	Plant Net £	Materials Net £	Unit Net £	CO_2e Kg
H2.5 **Length: 15 - 20m**							
H2.5.8 Mass: stated exceeding 20t							
H2.5.8.03 Bridge Beams; "U" Beams; Section U5; depth 1000mm; cross-sectional area; 532450mm²; mass 23.06 t	Nr	-	-	-	11,804.11	11,804.11	4,174.081
H2.5.8.04 Bridge Beams; "U" Beams; Section U7; depth 1100mm; cross-sectional area; 565450mm2; mass 24.48 t	Nr	-	-	-	14,122.77	14,122.77	4,431.114
H2.6 **Length: 20 - 30m**							
H2.6.7 Mass: 10 - 20t							
H2.6.7.01 Bridge Beams; "Y" Beams; Length 21m; base width 750mm; Section Y3; depth 900mm; cross sectional area 373444mm²; mass 19.6t	Nr	-	-	-	6,955.99	6,955.99	3,547.787
H2.6.7.02 Bridge Beams; "Y" Beams; Length 21m; base width 750mm; Section Y4; depth 1000mm; cross sectional area 409890 mm²; mass 21.5t	Nr	-	-	-	8,009.93	8,009.93	3,891.706
H2.6.8 Mass: stated exceeding 20t							
H2.6.8.01 Bridge Beams; "M" Beams; Length 22m; top flange width 400mm; bottom flange width 970mm; Section M5; depth 960mm; cross-sectional area 380650mm²; mass 20.13 t	Nr	-	-	-	6,745.21	6,745.21	3,707.076
H2.6.8.02 Bridge Beams; "M" Beams; Length 22m; top flange width 400mm; bottom flange width 970mm; Section M6; depth 1040mm; cross-sectional area 387050mm²; mass 20.48 t	Nr	-	-	-	7,588.36	7,588.36	3,707.076
H2.6.8.03 Bridge Beams; "M" Beams; Length 22m; top flange width 400mm; bottom flange width 970mm; Section M7; depth 1120mm; cross-sectional area 393450mm²; mass 20.83 t	Nr	-	-	-	8,853.08	8,853.08	3,770.429
H2.6.8.04 Bridge Beams; "M" Beams; Length 26m; top flange width 400mm; bottom flange width 970mm; Section M8; depth 1200mm; cross-sectional area 419050mm²; mass 26.21 t	Nr	-	-	-	8,536.90	8,536.90	4,744.261
H2.6.8.05 Bridge Beams; "M" Beams; Length 26m; top flange width 400mm; bottom flange width 970mm; Section M9; depth 1280mm; cross-sectional area 425450mm²; mass 26.60 t	Nr	-	-	-	9,696.23	9,696.23	4,814.854

H2 Prestressed Pre-Tensioned Beams continued...

	Unit	Labour Hours	Labour Net £	Plant Net £	Materials Net £	Unit Net £	CO_2e Kg
H2.6	**Length: 20 - 30m**						
H2.6.8	Mass: stated exceeding 20t						
H2.6.8.06	Bridge Beams; "M" Beams; Length 26m; top flange width 400mm; bottom flange width 970mm; Section M10; depth 1360mm; cross-sectional area 457450mm^2; mass 28.63 t	Nr	-	-	-	10,644.78	10,644.78	5,182.304
H2.6.8.07	Bridge Beams; "U" Beams; Length 22m; base width 970mm; leg thickness 165mm; Section U8; depth 1200mm; cross-sectional area 598450mm^2; mass 31.66 t	Nr	-	-	-	16,125.26	16,125.26	5,730.763
H2.6.8.08	Bridge Beams; "U" Beams; Length 22m; base width 970mm; leg thickness 165mm; Section U9; depth 1300mm; cross-sectional area 631450mm^2; mass 33.42 t	Nr	-	-	-	18,233.13	18,233.13	6,049.340
H2.6.8.09	Bridge Beams; "U" Beams; Length 30m; base width 970mm; leg thickness 165mm; Section U10; depth 1400mm; cross-sectional area 664450mm^2; mass 47.94 t	Nr	-	-	-	20,551.80	20,551.80	8,670.358
H2.6.8.10	Bridge Beams; "U" Beams; Length 30m; base width 970mm; leg thickness 165mm; Section U11; depth 1500mm; cross-sectional area 697450mm^2; mass 50.34 t	Nr	-	-	-	22,238.10	22,238.10	9,112.021
H2.6.8.11	Bridge Beams; "U" Beams; Length 30m; base width 970mm; leg thickness 165mm; Section U12; depth 1600mm; cross-sectional area 730450mm^2; mass 52.71 t	Nr	-	-	-	24,662.16	24,662.16	9,541.014
H2.6.8.12	Bridge Beams; "Y" Beams; Length 25m; base width 750mm; Section Y5; depth 1100mm; cross sectional area 449220 mm^2; mass 28.1t	Nr	-	-	-	9,169.26	9,169.26	5,086.369
H2.6.8.13	Bridge Beams; "Y" Beams; Length 25m; base width 750mm; Section Y6; depth 1200mm; cross sectional area 491433mm^2; mass 30.7t	Nr	-	-	-	10,328.60	10,328.60	5,556.994
H2.6.8.14	Bridge Beams; "Y" Beams; Length 30m; base width 750mm; Section Y7; depth 1300mm; cross sectional area 536530mm^2; mass 40.2t	Nr	-	-	-	11,698.72	11,698.72	7,276.584
H2.6.8.15	Bridge Beams; "Y" Beams; Length 30m; base width 750mm; Section Y8; depth 1400mm; cross sectional area 584708mm^2; mass 43.9t	Nr	-	-	-	12,647.26	12,647.26	7,946.320

H5 Slabs

Note(s): NOTE: Prestressed precast concrete floors or roofs; hoisting and fixing on prepared bearings; grouting joints; designed to allow 1.5kN/m2 for screeds.

	Unit	Labour Hours	Labour Net £	Plant Net £	Materials Net £	Unit Net £	CO_2e Kg
H5.3 **Area: 4 - 15m2**							
H5.3.5 Mass: 2 - 5t							
H5.3.5.01 Prestressed precast concrete floors or roofs; hoisting and fixing on prepared bearings; grouting joints; designed to allow 1.5kN/m^2 for screeds; Superimposed loads maximum 5kN/m2; 6m x 1.2m x 250 mm deep; mass 4.25 t	Nr	1.00	61.39	60.22	303.53	425.14	813.353
H5.3.6 Mass: 5 - 10t							
H5.3.6.01 Prestressed precast concrete floors or roofs; hoisting and fixing on prepared bearings; grouting joints; designed to allow 1.5kN/m^2 for screeds; Superimposed loads maximum 5kN/m2; 9m x 1.2m x 250 mm deep; mass 6.4 t	Nr	1.00	61.39	60.22	455.30	576.91	1,202.523
H5.3.6.02 Prestressed precast concrete floors or roofs; hoisting and fixing on prepared bearings; grouting joints; designed to allow 1.5kN/m^2 for screeds; Superimposed loads maximum 5kN/m2; 12m x 1.2m x 250 mm deep; mass 8.5 t	Nr	1.00	61.39	60.22	682.95	804.56	1,582.643
H5.3.7 Mass: 10 - 20t							
H5.3.7.01 Prestressed precast concrete floors or roofs; hoisting and fixing on prepared bearings; grouting joints; designed to allow 1.5kN/m^2 for screeds; Superimposed loads maximum 10kN/m2; 6m x 2.4m x 300mm deep; mass 10.2 t	Nr	1.00	61.39	60.22	607.07	728.68	1,890.360
H5.4 **Area: 15 - 50m^2**							
H5.4.8 Mass: stated exceeding 20t							
H5.4.8.01 Prestressed precast concrete floors or roofs; hoisting and fixing on prepared bearings; grouting joints; designed to allow 1.5kN/m^2 for screeds; Superimposed loads maximum 10kN/m2; 9m x 2.4m x 400mm deep; mass 20.4 t	Nr	1.00	61.39	60.22	1,138.25	1,259.86	3,736.657
H5.4.8.02 Prestressed precast concrete floors or roofs; hoisting and fixing on prepared bearings; grouting joints; designed to allow 1.5kN/m^2 for screeds; Superimposed loads maximum 10kN/m2; 12m x 2.4m x 600mm deep; mass 40.8 t	Nr	1.00	61.39	60.22	1,821.21	1,942.82	7,429.252

H7 Units For Subways, Culverts And Ducts

Note(s): NOTES: Culverts
The following prices are based on historical specialist sub-contractor prices and provided as a guide for approximate estimating purpose only. The prices are for supply and delivery only and are applicable to sites within 100 miles of the point of supply for minimum quantities of 30 linear metres which are not subjected to deep fill conditions. Conditions of exposure assumed to be severe. Standard cross-sections are available in many sizes and firm prices can only be quoted against full loading information and design specification.

	Unit	Labour Hours	Labour Net £	Plant Net £	Materials Net £	Unit Net £	CO_2e Kg
H7.0	**Standard Box Culvert Units**						
H7.0.0	Precast concrete; internal cross-sectional dimensions						
H7.0.0.01 1200mm width x 600mm height	m	-	-	-	232.39	232.39	240.743
H7.0.0.02 1800mm width x 1200mm height	m	-	-	-	617.34	617.34	438.043
H7.0.0.03 2400mm width x 1800mm height	m	-	-	-	867.92	867.92	919.529
H7.0.0.04 3000mm width x 1800mm height	m	-	-	-	1,179.36	1,179.36	1,171.132

CLASS I:
PIPEWORK - PIPES

Calculations used throughout Class I - Pipework - Pipes

Labour

			Qty		Rate		Total
L A0184ICE	**Small bore pipes in shallow trenches Labour Gang**						
	L A9015ICE	Banksman	1	x	14.03	=	£14.03
	L A9016ICE	Ganger	1	x	17.64	=	£17.64
	L A9017ICE	Pipelayer (standard rate)	1	x	14.86	=	£14.86
	L A9002ICE	Labourer (General Operative)	5	x	13.02	=	£65.08
		Total hourly cost of gang				=	**£111.61**
L A0185ICE	**Small bore pipes in deep trenches Labour Gang**						
	L A9015ICE	Banksman	2	x	14.03	=	£28.06
	L A9016ICE	Ganger	1	x	17.64	=	£17.64
	L A9017ICE	Pipelayer (standard rate)	1	x	14.86	=	£14.86
	L A9002ICE	Labourer (General Operative)	8	x	13.02	=	£104.12
	L A9026ICE	Timberman	1	x	14.86	=	£14.86
		Total hourly cost of gang				=	**£179.54**
L A0186ICE	**Large bore pipes in shallow trench Labour Gang**						
	L A9016ICE	Ganger	1	x	17.64	=	£17.64
	L A9018ICE	Pipelayer (large pipes)	1	x	16.51	=	£16.51
	L A9002ICE	Labourer (General Operative)	5	x	13.02	=	£65.08
	L A9015ICE	Banksman	1	x	14.03	=	£14.03
		Total hourly cost of gang				=	**£113.26**
L A0187ICE	**Large bore pipes in deep trench Labour Gang**						
	L A9016ICE	Ganger	1	x	17.64	=	£17.64
	L A9018ICE	Pipelayer (large pipes)	1	x	16.51	=	£16.51
	L A9002ICE	Labourer (General Operative)	8	x	13.02	=	£104.12
	L A9015ICE	Banksman	2	x	14.03	=	£28.06
	L A9026ICE	Timberman	1	x	14.86	=	£14.86
		Total hourly cost of gang				=	**£181.19**
L A0188ICE	**Small bore steel pipes in shallow trench Labour Gang**						
	L A9016ICE	Ganger	1	x	17.64	=	£17.64
	L A9021ICE	Fitters and Welders	1	x	17.33	=	£17.33
	L A9003ICE	Labourer (Skill Rate 3)	3	x	14.86	=	£44.58
	L A9015ICE	Banksman	1	x	14.03	=	£14.03
	L A9002ICE	Labourer (General Operative)	5	x	13.02	=	£65.08
		Total hourly cost of gang				=	**£158.66**
L A0189ICE	**Small bore steel pipes in deep trench Labour Gang**						
	L A9016ICE	Ganger	1	x	17.64	=	£17.64
	L A9002ICE	Labourer (General Operative)	8	x	13.02	=	£104.12
	L A9003ICE	Labourer (Skill Rate 3)	3	x	14.86	=	£44.58
	L A9021ICE	Fitters and Welders	1	x	17.33	=	£17.33
	L A9015ICE	Banksman	1	x	14.03	=	£14.03
	L A9026ICE	Timberman	1	x	14.86	=	£14.86
		Total hourly cost of gang				=	**£212.56**

L A0190ICE   **Large bore steel pipes in shallow trench Labour Gang**

L A9021ICE	Fitters and Welders	2	x	17.33	=	£34.66
L A9003ICE	Labourer (Skill Rate 3)	4	x	14.86	=	£59.45
L A9002ICE	Labourer (General Operative)	5	x	13.02	=	£65.08
L A9016ICE	Ganger	1	x	17.64	=	£17.64
L A9015ICE	Banksman	1	x	14.03	=	£14.03
	Total hourly cost of gang				=	**£190.86**

L A0191ICE   **Large bore steel pipes in deep trench Labour Gang**

L A9016ICE	Ganger	1	x	17.64	=	£17.64
L A9015ICE	Banksman	1	x	14.03	=	£14.03
L A9021ICE	Fitters and Welders	2	x	17.33	=	£34.66
L A9026ICE	Timberman	1	x	14.86	=	£14.86
L A9003ICE	Labourer (Skill Rate 3)	4	x	14.86	=	£59.45
L A9002ICE	Labourer (General Operative)	8	x	13.02	=	£104.12
	Total hourly cost of gang				=	**£244.76**

L A0192ICE   **Small bore pipes in shallow trench in french drain Labour Gang**

L A9015ICE	Banksman	1	x	14.03	=	£14.03
L A9016ICE	Ganger	1	x	17.64	=	£17.64
L A9017ICE	Pipelayer (standard rate)	1	x	14.86	=	£14.86
L A9002ICE	Labourer (General Operative)	4	x	13.02	=	£52.06
	Total hourly cost of gang				=	**£98.59**

L A0193ICE   **Small bore iron pipes not in trenches Labour Gang**

L A9016ICE	Ganger	1	x	17.64	=	£17.64
L A9017ICE	Pipelayer (standard rate)	1	x	14.86	=	£14.86
L A9002ICE	Labourer (General Operative)	2	x	13.02	=	£26.03
L A9015ICE	Banksman	1	x	14.03	=	£14.03
	Total hourly cost of gang				=	**£72.56**

L A0194ICE   **Large bore iron pipes not in trenches Labour Gang**

L A9016ICE	Ganger	1	x	17.64	=	£17.64
L A9018ICE	Pipelayer (large pipes)	1	x	16.51	=	£16.51
L A9002ICE	Labourer (General Operative)	2	x	13.02	=	£26.03
L A9015ICE	Banksman	1	x	14.03	=	£14.03
	Total hourly cost of gang				=	**£74.21**

L A0195ICE   **Small bore steel pipes not in trenches Labour Gang**

L A9021ICE	Fitters and Welders	1	x	17.33	=	£17.33
L A9003ICE	Labourer (Skill Rate 3)	2	x	14.86	=	£29.72
L A9002ICE	Labourer (General Operative)	2	x	13.02	=	£26.03
L A9015ICE	Banksman	1	x	14.03	=	£14.03
L A9016ICE	Ganger	1	x	17.64	=	£17.64
	Total hourly cost of gang				=	**£104.75**

L A0196ICE   **Large bore steel pipes not in trenches Labour Gang**

L A9021ICE	Fitters and Welders	1	x	17.33	=	£17.33
L A9003ICE	Labourer (Skill Rate 3)	2	x	14.86	=	£29.72
L A9002ICE	Labourer (General Operative)	5	x	13.02	=	£65.08
L A9015ICE	Banksman	1	x	14.03	=	£14.03
L A9016ICE	Ganger	1	x	17.64	=	£17.64
	Total hourly cost of gang				=	**£143.80**

Plant

P A1184ICE	**Small bore pipes in shallow trench Plant Gang**						
	P A0592ICE	Hydraulic Excavator - Cat 320 96kW	1	x	36.36	=	£36.36
	P A0811ICE	Pump - Godwin ET50 23m3/h 4 inches	1	x	2.94	=	£2.94
	P A0853ICE	Vibrating Plate Diesel 24kN	1	x	2.53	=	£2.53
	P A0803ICE	Trench Sheets	90	x	0.09	=	£8.10
	P A0804ICE	Acrow Props	70	x	0.09	=	£6.30
	P A1023ICE	Landrover 4WD	1	x	16.16	=	£16.16
		Total hourly cost of gang				=	**£72.39**

P A1185ICE	**Small bore pipes in deep trench Plant Gang**						
	P A0605ICE	Hydraulic Excavator - Cat 166kW	1	x	44.31	=	£44.31
	P A0812ICE	Pump - Godwin ET75 74m3/h 4 inches	2	x	4.40	=	£8.80
	P A0854ICE	Vibrating Plate Diesel 33.5kN	1	x	2.95	=	£2.95
	P A0803ICE	Trench Sheets	150	x	0.09	=	£13.50
	P A0804ICE	Acrow Props	100	x	0.09	=	£9.00
	P A0805ICE	Timber Baulks	3	x	0.65	=	£1.95
	P A1023ICE	Landrover 4WD	1	x	16.16	=	£16.16
		Total hourly cost of gang				=	**£96.67**

P A1186ICE	**Large bore in shallow trench Plant Gang**						
	P A0605ICE	Hydraulic Excavator - Cat 166kW	1	x	44.31	=	£44.31
	P A0811ICE	Pump - Godwin ET50 23m3/h 4 inches	1	x	2.94	=	£2.94
	P A0853ICE	Vibrating Plate Diesel 24kN	1	x	2.53	=	£2.53
	P A0803ICE	Trench Sheets	90	x	0.09	=	£8.10
	P A0804ICE	Acrow Props	70	x	0.09	=	£6.30
	P A0511ICE	Cranes Crawler - 22RB - 15t	1	x	41.06	=	£41.06
	P A1023ICE	Landrover 4WD	1	x	16.16	=	£16.16
		Total hourly cost of gang				=	**£121.40**

P A1187ICE	**Large bore pipes in deep trench Plant Gang**						
	P A0605ICE	Hydraulic Excavator - Cat 166kW	1	x	44.31	=	£44.31
	P A0812ICE	Pump - Godwin ET75 74m3/h 4 inches	2	x	4.40	=	£8.80
	P A0854ICE	Vibrating Plate Diesel 33.5kN	1	x	2.95	=	£2.95
	P A0803ICE	Trench Sheets	150	x	0.09	=	£13.50
	P A0804ICE	Acrow Props	100	x	0.09	=	£9.00
	P A0805ICE	Timber Baulks	3	x	0.65	=	£1.95
	P A0511ICE	Cranes Crawler - 22RB - 15t	1	x	41.06	=	£41.06
	P A1023ICE	Landrover 4WD	1	x	16.16	=	£16.16
		Total hourly cost of gang				=	**£137.73**

P A1188ICE	**Small bore steel pipes in shallow trench Plant Gang**						
	P A0592ICE	Hydraulic Excavator - Cat 320 96kW	1	x	36.36	=	£36.36
	P A0811ICE	Pump - Godwin ET50 23m3/h 4 inches	1	x	2.94	=	£2.94
	P A0853ICE	Vibrating Plate Diesel 24kN	1	x	2.53	=	£2.53
	P A0803ICE	Trench Sheets	90	x	0.09	=	£8.10
	P A0804ICE	Acrow Props	70	x	0.09	=	£6.30
	P A1023ICE	Landrover 4WD	1	x	16.16	=	£16.16
	P A0991ICE	Welding Set - 300 amp Diesel Electric Start Sil	1	x	4.48	=	£4.48
	P A0946ICE	Crawler Tractor / Dozer - Cat 561 Sideboom	1	x	44.35	=	£44.35
		Total hourly cost of gang				=	**£121.22**

P A1189ICE **Small bore steel pipes in deep trench Plant Gang**

P A0605ICE	Hydraulic Excavator - Cat 166kW	1	x	44.31	=	£44.31
P A0812ICE	Pump - Godwin ET75 74m3/h 4 inches	2	x	4.40	=	£8.80
P A0854ICE	Vibrating Plate Diesel 33.5kN	1	x	2.95	=	£2.95
P A0803ICE	Trench Sheets	150	x	0.09	=	£13.50
P A0804ICE	Acrow Props	100	x	0.09	=	£9.00
P A0805ICE	Timber Baulks	3	x	0.65	=	£1.95
P A1023ICE	Landrover 4WD	1	x	16.16	=	£16.16
P A0991ICE	Welding Set - 300 amp Diesel Electric Start Sil	1	x	4.48	=	£4.48
P A0511ICE	Cranes Crawler - 22RB - 15t	1	x	41.06	=	£41.06

Total hourly cost of gang = **£142.21**

P A1190ICE **Large bore steel pipes in shallow trench Plant Gang**

P A0605ICE	Hydraulic Excavator - Cat 166kW	1	x	44.31	=	£44.31
P A0811ICE	Pump - Godwin ET50 23m3/h 4 inches	1	x	2.94	=	£2.94
P A0853ICE	Vibrating Plate Diesel 24kN	1	x	2.53	=	£2.53
P A0803ICE	Trench Sheets	90	x	0.09	=	£8.10
P A0804ICE	Acrow Props	70	x	0.09	=	£6.30
P A1023ICE	Landrover 4WD	1	x	16.16	=	£16.16
P A0991ICE	Welding Set - 300 amp Diesel Electric Start Sil	1	x	4.48	=	£4.48
P A0511ICE	Cranes Crawler - 22RB - 15t	1	x	41.06	=	£41.06

Total hourly cost of gang = **£125.88**

P A1191ICE **Large bore steel pipes in deep trench Plant Gang**

P A0605ICE	Hydraulic Excavator - Cat 166kW	1	x	44.31	=	£44.31
P A0812ICE	Pump - Godwin ET75 74m3/h 4 inches	2	x	4.40	=	£8.80
P A0854ICE	Vibrating Plate Diesel 33.5kN	1	x	2.95	=	£2.95
P A0803ICE	Trench Sheets	150	x	0.09	=	£13.50
P A0804ICE	Acrow Props	100	x	0.09	=	£9.00
P A0805ICE	Timber Baulks	3	x	0.65	=	£1.95
P A1023ICE	Landrover 4WD	1	x	16.16	=	£16.16
P A0991ICE	Welding Set - 300 amp Diesel Electric Start Sil	2	x	4.48	=	£8.96
P A0505ICE	Cranes Crawler - NCK 406C - 30t	1	x	51.96	=	£51.96

Total hourly cost of gang = **£157.59**

P A1194ICE **Placing materials in bed, haunches and surrounds Plant Gang**

P A0854ICE	Vibrating Plate Diesel 33.5kN	1	x	2.95	=	£2.95
P A0939ICE	Crawler Tractor / Dozer - Cat 941	0.5	x	37.84	=	£18.92

Total hourly cost of gang = **£21.87**

P A1195ICE **Large bore iron pipes not in trenches Plant Gang**

P A0511ICE	Cranes Crawler - 22RB - 15t	1	x	41.06	=	£41.06

Total hourly cost of gang = **£41.06**

P A1196ICE **Steel pipes not in trenches Plant Gang**

P A0991ICE	Welding Set - 300 amp Diesel Electric Start Sil	2	x	4.48	=	£8.96
P A0946ICE	Crawler Tractor / Dozer - Cat 561 Sideboom	1	x	44.35	=	£44.35

Total hourly cost of gang = **£53.31**

P A1197ICE   **Small bore MDPE in shallow trench Plant Gang**

P A0592ICE	Hydraulic Excavator - Cat 320 96kW	2	x	36.36	=	£72.72
P A0811ICE	Pump - Godwin ET50 23m3/h 4 inches	1	x	2.94	=	£2.94
P A0853ICE	Vibrating Plate Diesel 24kN	1	x	2.53	=	£2.53
P A0803ICE	Trench Sheets	90	x	0.09	=	£8.10
P A0804ICE	Acrow Props	70	x	0.09	=	£6.30
P A1023ICE	Landrover 4WD	1	x	16.16	=	£16.16
P A0994ICE	Welding Set - Plastic Pipe Welding (Small)	2	x	6.84	=	£13.68

                                    Total hourly cost of gang   =   **£122.43**

Pipework - Pipes

Note(s): NOTES: Unit prices for this section are based upon the following assumptions:
1) Excavations require a minimum of pumping.
2) No allowance has been made for testing pipework.
3) Prices include for backfilling trenches with excavated material and spreading surplus at sides of trench.
4) No allowance has been made for Double Handling of pipes along easement.
5) Back-fill hard rammed around small sizes up to 150mm above crown then in layers ompacted with plate vibrator.
6) Deep dig prices allow for some support but exclude close sheeting or driven piling. Prices also assume adequate working space to batter sides and store excavated material.
7) No allowance has been made for encountering cross services.
8) Steel pipe items are shown with and without linings.
9) Backfill excluded from prices for pipes in French Drains.
10) Gang build ups assume 'Deep Trench' = 3 - 3.5m deep or over and 'Large bore = 600mm diameter or over.

I1   Clay Pipes

		Unit	Labour Hours	Labour Net £	Plant Net £	Materials Net £	Unit Net £	CO_2e Kg
I1.1	**Nominal bore: not exceeding 200mm**							
I1.1.2	In trenches, depth: not exceeding 1.5m							
I1.1.2.01	Vitrified clay pipes, BS 65, "Extra Strength" spigot & socket flexible joints; nominal bore 100mm	m	0.06	6.70	4.30	16.84	27.84	11.499
I1.1.2.02	as above; nominal bore 150mm	m	0.08	8.48	5.44	20.28	34.20	16.878
I1.1.2.03	Vitrified clay pipes, BS 65, "Extra Strength": plain end with sleeve joints; nominal bore 100mm	m	0.06	6.70	4.30	4.89	15.89	6.122
I1.1.2.04	as above; nominal bore 150mm	m	0.08	8.48	5.44	15.27	29.19	10.003
I1.1.2.05	as above; nominal bore 200mm	m	0.09	9.93	6.37	31.36	47.66	13.806
I1.1.2.06	Vitrified clay pipes, BS 65, "Surface Water" quality: spigot & socket cement joints; nominal bore 100mm	m	0.06	6.70	4.30	7.80	18.80	11.499
I1.1.2.07	as above; nominal bore 150mm	m	0.08	8.48	5.44	19.32	33.24	16.878
I1.1.2.08	Vitrified clay pipes, BS 65, British Standard "Normal" quality: spigot & socket cement joints; nominal bore 100mm	m	0.06	6.70	4.30	17.04	28.04	11.499
I1.1.2.09	as above; nominal bore 150mm	m	0.08	8.48	5.44	22.39	36.31	16.878
I1.1.2.10	Vitrified clay pipes, BS 65, "Perforated" plain end with sleeve joints; nominal bore 100mm	m	0.06	6.70	4.30	12.99	23.99	6.122
I1.1.2.11	as above; nominal bore 150mm	m	0.08	8.48	5.44	17.75	31.67	10.003
I1.1.2.12	Clay land drains; BS 1196; plain butt joints; nominal bore 75mm	m	0.04	4.46	2.86	4.67	11.99	4.782
I1.1.2.13	as above; nominal bore 75mm; in french drains	m	0.04	3.94	2.86	4.67	11.47	4.782
I1.1.2.14	as above; nominal bore 100mm	m	0.05	5.02	3.22	7.21	15.45	6.229
I1.1.2.15	as above; nominal bore 100mm; in french drains	m	0.05	4.44	3.22	7.21	14.87	6.229
I1.1.2.16	as above; nominal bore 150mm	m	0.05	5.58	3.58	14.59	23.75	11.162
I1.1.2.17	as above; nominal bore 150mm; in french drains	m	0.05	4.93	3.58	14.59	23.10	11.162
I1.1.3	In trenches, depth: 1.5 - 2m							
I1.1.3.01	Vitrified clay pipes, BS 65, "Extra Strength" spigot & socket flexible joints; nominal bore 100mm	m	0.08	8.48	5.44	16.84	30.76	11.920
I1.1.3.02	as above; nominal bore 150mm	m	0.09	9.71	6.23	20.28	36.22	17.167
I1.1.3.03	Vitrified clay pipes, BS 65, "Extra Strength": plain end with sleeve joints; nominal bore 100mm	m	0.08	8.48	5.44	4.89	18.81	6.543
I1.1.3.04	as above; nominal bore 150mm	m	0.09	9.71	6.23	15.27	31.21	10.292
I1.1.3.05	as above; nominal bore 200mm	m	0.10	10.83	6.95	31.36	49.14	14.016
I1.1.3.06	Vitrified clay pipes, BS 65, "Surface Water" quality: spigot & socket cement joints; nominal bore 100mm	m	0.08	8.48	5.44	7.80	21.72	11.920
I1.1.3.07	as above; nominal bore 150mm	m	0.09	9.71	6.23	19.32	35.26	17.167
I1.1.3.08	Vitrified clay pipes, BS 65, British Standard "Normal" quality: spigot & socket cement joints; nominal bore 100mm	m	0.08	8.48	5.44	17.04	30.96	11.920
I1.1.3.09	as above; nominal bore 150mm	m	0.09	9.71	6.23	22.39	38.33	17.167

I1 Clay Pipes continued...

	Unit	Labour Hours	Labour Net £	Plant Net £	Materials Net £	Unit Net £	CO$_2$e Kg
I1.1	**Nominal bore: not exceeding 200mm**						
I1.1.3	In trenches, depth: 1.5 - 2m						
I1.1.3.10	Vitrified clay pipes, BS 65, "Perforated" plain end with sleeve joints; nominal bore 100mm	m	0.08	8.48	5.44	12.99	26.91	6.543
I1.1.3.11	as above; nominal bore 150mm	m	0.09	9.71	6.23	17.75	33.69	10.292
I1.1.3.12	Clay land drains; BS 1196; plain butt joints; nominal bore 75mm	m	0.06	6.70	4.30	4.67	15.67	5.308
I1.1.3.13	as above; nominal bore 75mm; in french drains	m	0.06	5.92	4.30	4.67	14.89	5.308
I1.1.3.14	as above; nominal bore 100mm	m	0.07	7.25	4.65	7.21	19.11	6.755
I1.1.3.15	as above; nominal bore 100mm; in french drains	m	0.07	6.41	4.65	7.21	18.27	6.755
I1.1.3.16	as above; nominal bore 150mm	m	0.07	7.81	5.01	14.59	27.41	11.688
I1.1.3.17	as above; nominal bore 150mm; in french drains	m	0.07	6.90	5.01	14.59	26.50	11.688
I1.1.4	In trenches, depth: 2 - 2.5m						
I1.1.4.01	Vitrified clay pipes, BS 65, "Extra Strength" spigot & socket flexible joints; nominal bore 100mm	m	0.11	11.94	7.66	16.84	36.44	12.735
I1.1.4.02	as above; nominal bore 150mm	m	0.12	13.28	8.52	20.28	42.08	18.008
I1.1.4.03	Vitrified clay pipes, BS 65, "Extra Strength": plain end with sleeve joints; nominal bore 100mm	m	0.11	11.94	7.66	4.89	24.49	7.358
I1.1.4.04	as above; nominal bore 150mm	m	0.12	13.28	8.52	15.27	37.07	11.133
I1.1.4.05	as above; nominal bore 200mm	m	0.13	14.62	9.38	31.36	55.36	14.910
I1.1.4.06	Vitrified clay pipes, BS 65, "Surface Water" quality: spigot & socket cement joints; nominal bore 100mm	m	0.11	11.94	7.66	7.80	27.40	12.735
I1.1.4.07	as above; nominal bore 150mm	m	0.12	13.28	8.52	19.32	41.12	18.008
I1.1.4.08	Vitrified clay pipes, BS 65, British Standard "Normal" quality: spigot & socket cement joints; nominal bore 100mm	m	0.11	11.94	7.66	17.04	36.64	12.735
I1.1.4.09	as above; nominal bore 150mm	m	0.12	13.28	8.52	22.39	44.19	18.008
I1.1.4.10	Vitrified clay pipes, BS 65, "Perforated" plain end with sleeve joints; nominal bore 100mm	m	0.11	11.94	7.66	12.99	32.59	7.358
I1.1.4.11	as above; nominal bore 150mm	m	0.12	13.28	8.52	17.75	39.55	11.133
I1.1.4.12	Clay land drains; BS 1196; plain butt joints; nominal bore 150mm	m	0.10	11.16	7.16	14.59	32.91	12.477
I1.1.4.13	as previous item; nominal bore 150mm; in french drains	m	0.10	9.86	7.16	14.59	31.61	12.477
I1.1.5	In trenches, depth: 2.5 - 3m						
I1.1.5.01	Vitrified clay pipes, BS 65, "Extra Strength" spigot & socket flexible joints; nominal bore 100mm	m	0.13	14.84	9.52	16.84	41.20	13.418
I1.1.5.02	as above; nominal bore 150mm	m	0.15	16.52	10.60	20.28	47.40	18.770
I1.1.5.03	Vitrified clay pipes, BS 65, "Extra Strength": plain end with sleeve joints; nominal bore 100mm	m	0.13	14.84	9.52	4.89	29.25	8.041
I1.1.5.04	as above; nominal bore 150mm	m	0.15	16.52	10.60	15.27	42.39	11.895
I1.1.5.05	as above; nominal bore 200mm	m	0.16	17.86	11.46	31.36	60.68	15.672
I1.1.5.06	Vitrified clay pipes, BS 65, "Surface Water" quality: spigot & socket cement joints; nominal bore 100mm	m	0.13	14.84	9.52	7.80	32.16	13.418
I1.1.5.07	as above; nominal bore 150mm	m	0.15	16.52	10.60	19.32	46.44	18.770

I1 Clay Pipes continued...

	Unit	Labour Hours	Labour Net £	Plant Net £	Materials Net £	Unit Net £	CO$_2$e Kg
I1.1	**Nominal bore: not exceeding 200mm**						
I1.1.5	In trenches, depth: 2.5 - 3m						
I1.1.5.08	Vitrified clay pipes, BS 65, British Standard "Normal" quality: spigot & socket cement joints; nominal bore 100mm	m	0.13	14.84	9.52	17.04	41.40	13.418
I1.1.5.09	as above; nominal bore 150mm	m	0.15	16.52	10.60	22.39	49.51	18.770
I1.1.5.10	Vitrified clay pipes, BS 65, "Perforated" plain end with sleeve joints; nominal bore 100mm	m	0.13	14.51	9.31	12.99	36.81	7.962
I1.1.5.11	as above; nominal bore 150mm	m	0.15	16.52	10.60	17.75	44.87	11.895
I1.1.6	In trenches, depth: 3 - 3.5m						
I1.1.6.01	Vitrified clay pipes, BS 65, "Extra Strength" spigot & socket flexible joints; nominal bore 100mm	m	0.18	31.96	16.99	16.84	65.79	16.274
I1.1.6.02	as above; nominal bore 150mm	m	0.20	35.01	18.61	20.28	73.90	21.838
I1.1.6.03	Vitrified clay pipes, BS 65, "Extra Strength": plain end with sleeve joints; nominal bore 100mm	m	0.18	31.96	16.99	4.89	53.84	10.897
I1.1.6.04	as previous item; nominal bore 150mm	m	0.20	35.01	18.61	15.27	68.89	14.963
I1.1.6.05	as above; nominal bore 200mm	m	0.20	35.91	19.09	31.36	86.36	18.603
I1.1.6.06	Vitrified clay pipes, BS 65, "Surface Water" quality: spigot & socket cement joints; nominal bore 100mm	m	0.18	31.96	16.99	7.80	56.75	16.274
I1.1.6.07	as above; nominal bore 150mm	m	0.20	35.01	18.61	19.32	72.94	21.838
I1.1.6.08	Vitrified clay pipes, BS 65, British Standard "Normal" quality: spigot & socket cement joints; nominal bore 100mm	m	0.18	31.96	16.99	17.04	65.99	16.274
I1.1.6.09	as above; nominal bore 150mm	m	0.20	35.01	18.61	22.39	76.01	21.838
I1.1.6.10	Vitrified clay pipes, BS 65, "Perforated" plain end with sleeve joints; nominal bore 100mm	m	0.18	31.96	16.99	12.99	61.94	10.897
I1.1.6.11	as above; nominal bore 150mm	m	0.20	35.01	18.61	17.75	71.37	14.963
I1.1.7	In trenches, depth: 3.5 - 4m						
I1.1.7.01	Vitrified clay pipes, BS 65, "Extra Strength" spigot & socket flexible joints; nominal bore 100mm	m	0.22	38.78	20.62	16.84	76.24	17.630
I1.1.7.02	as above; nominal bore 150mm	m	0.24	42.19	22.43	20.28	84.90	23.266
I1.1.7.03	Vitrified clay pipes, BS 65, "Extra Strength": plain end with sleeve joints; nominal bore 100mm	m	0.22	38.78	20.62	4.89	64.29	12.253
I1.1.7.04	as above; nominal bore 150mm	m	0.24	42.19	22.43	15.27	79.89	16.391
I1.1.7.05	as above; nominal bore 200mm	m	0.25	44.88	23.86	31.36	100.10	20.387
I1.1.7.06	Vitrified clay pipes, BS 65, "Surface Water" quality: spigot & socket cement joints; nominal bore 100mm	m	0.22	38.78	20.62	7.80	67.20	17.630
I1.1.7.07	as above; nominal bore 150mm	m	0.24	42.19	22.43	19.32	83.94	23.266
I1.1.7.08	Vitrified clay pipes, BS 65, British Standard "Normal" quality: spigot & socket cement joints; nominal bore 100mm	m	0.22	38.78	20.62	17.04	76.44	17.630
I1.1.7.09	as above; nominal bore 150mm	m	0.24	42.19	22.43	22.39	87.01	23.266
I1.1.7.10	Vitrified clay pipes, BS 65, "Perforated" plain end with sleeve joints; nominal bore 100mm	m	0.22	38.78	20.62	12.99	72.39	12.253
I1.1.7.11	as above; nominal bore 150mm	m	0.24	42.19	22.43	17.75	82.37	16.391

I1 Clay Pipes continued...

	Unit	Labour Hours	Labour Net £	Plant Net £	Materials Net £	Unit Net £	CO_2e Kg
I1.1	**Nominal bore: not exceeding 200mm**						
I1.1.8	In trenches, depth: exceeding 4m						
I1.1.8.01	Vitrified clay pipes, BS 65, "Extra Strength" spigot & socket flexible joints; nominal bore 100mm; depth: 4 - 4.5m	m	0.40	71.82	38.18	16.84	126.84	24.195
I1.1.8.02	as above; depth: 4.5 - 5m	m	0.53	95.69	50.88	16.84	163.41	28.941
I1.1.8.03	as above; nominal bore 150mm; depth: 4 - 4.5m	m	0.42	75.59	40.19	20.28	136.06	29.903
I1.1.8.04	as above; depth: 4.5 - 5m	m	0.57	102.52	54.51	20.28	177.31	35.255
I1.1.8.05	Vitrified clay pipes, BS 65, "Extra Strength": plain end with sleeve joints; nominal bore 100mm; depth: 4 - 4.5m	m	0.40	71.82	38.18	4.89	114.89	18.818
I1.1.8.06	as above; depth: 4.5 - 5m	m	0.53	95.69	50.88	4.89	151.46	23.564
I1.1.8.07	as above; nominal bore 150mm; depth: 4 - 4.5m	m	0.42	75.59	40.19	15.27	131.05	23.028
I1.1.8.08	as above; depth: 4.5 - 5m	m	0.57	102.52	54.51	15.27	172.30	28.380
I1.1.8.09	as above; nominal bore 200mm; depth: 4 - 4.5m	m	0.44	79.72	42.38	31.36	153.46	27.309
I1.1.8.10	as above; depth: 4.5 - 5m	m	0.62	110.42	58.71	31.36	200.49	33.411
I1.1.8.11	Vitrified clay pipes, BS 65, "Surface Water" quality: spigot & socket cement joints; nominal bore 100mm; depth: 4 - 4.5m	m	0.40	71.82	38.18	7.80	117.80	24.195
I1.1.8.12	as above; depth: 4.5 - 5m	m	0.53	95.69	50.88	7.80	154.37	28.941
I1.1.8.13	as above; nominal bore 150mm; depth: 4 - 4.5m	m	0.42	75.59	40.19	19.32	135.10	29.903
I1.1.8.14	as above; depth: 4.5 - 5m	m	0.57	102.52	54.51	19.32	176.35	35.255
I1.1.8.15	Vitrified clay pipes, BS 65, British Standard "Normal" quality: spigot & socket cement joints; nominal bore 100mm; depth: 4 - 4.5m	m	0.40	71.82	38.18	17.04	127.04	24.195
I1.1.8.16	as above; depth: 4.5 - 5m	m	0.53	95.69	50.88	17.04	163.61	28.941
I1.1.8.17	as above; nominal bore 150mm; depth: 4 - 4.5m	m	0.42	75.59	40.19	22.39	138.17	29.903
I1.1.8.18	as above; depth: 4.5 - 5m	m	0.57	102.52	54.51	22.39	179.42	35.255
I1.1.8.19	Vitrified clay pipes, BS 65, "Perforated" plain end with sleeve joints; nominal bore 100mm; depth: 4 - 4.5m	m	0.40	71.82	38.18	12.99	122.99	18.818
I1.1.8.20	as above; depth: 4.5 - 5m	m	0.53	95.69	50.88	12.99	159.56	23.564
I1.1.8.21	as above; nominal bore 150mm; depth: 4 - 4.5m	m	0.42	75.59	40.19	17.75	133.53	23.028
I1.1.8.22	as above; depth: 4.5 - 5m	m	0.57	102.52	54.51	17.75	174.78	28.380
I1.2	**nominal bore: 200 - 300mm**						
I1.2.2	In trenches, depth: not exceeding 1.5m						
I1.2.2.01	Vitrified clay pipes, BS 65, "Extra Strength" spigot & socket flexible joints; nominal bore 225mm	m	0.09	10.49	6.73	39.15	56.37	23.261
I1.2.2.02	as above; nominal bore 300mm	m	0.11	12.72	8.16	61.19	82.07	39.039
I1.2.2.03	Vitrified clay pipes, BS 65, "Extra Strength": plain end with sleeve joints; nominal bore 225mm	m	0.09	10.49	6.73	38.35	55.57	21.510
I1.2.2.04	Vitrified clay pipes, BS 65, "Surface Water" quality: spigot & socket cement joints; nominal bore 225mm	m	0.09	10.49	6.73	45.25	62.47	23.261
I1.2.2.05	as above; nominal bore 300mm	m	0.11	12.28	7.88	70.69	90.85	38.934

I1   Clay Pipes continued...

	Unit	Labour Hours	Labour Net £	Plant Net £	Materials Net £	Unit Net £	CO$_2$e Kg
I1.2 **nominal bore: 200 - 300mm**							
I1.2.2 In trenches, depth: not exceeding 1.5m							
I1.2.2.06 Vitrified clay pipes, BS 65, British Standard "Normal" quality: spigot & socket cement joints; nominal bore 225mm	m	0.09	10.49	6.73	43.31	60.53	23.261
I1.2.2.07 as above; nominal bore 300mm	m	0.11	12.72	8.16	68.27	89.15	39.039
I1.2.2.08 Vitrified clay pipes, BS 65, "Perforated" plain end with sleeve joints; nominal bore 225mm	m	0.09	10.49	6.73	34.67	51.89	21.510
I1.2.2.09 as above; nominal bore 300mm	m	0.11	12.28	7.88	62.24	82.40	32.964
I1.2.2.10 Clay land drains; BS 1196; plain butt joints; nominal bore 225mm	m	0.07	7.81	5.01	36.93	49.75	17.405
I1.2.3 In trenches, depth: 1.5 - 2m							
I1.2.3.01 Vitrified clay pipes, BS 65, "Extra Strength" spigot & socket flexible joints; nominal bore 225mm	m	0.10	11.27	7.23	39.15	57.65	23.445
I1.2.3.02 as above; nominal bore 300mm	m	0.12	13.73	8.81	61.19	83.73	39.275
I1.2.3.03 Vitrified clay pipes, BS 65, "Extra Strength": plain end with sleeve joints; nominal bore 225mm	m	0.10	11.27	7.23	38.35	56.85	21.694
I1.2.3.04 Vitrified clay pipes, BS 65, "Surface Water" quality: spigot & socket cement joints; nominal bore 225mm	m	0.10	11.27	7.23	45.25	63.75	23.445
I1.2.3.05 as above; nominal bore 300mm	m	0.12	13.73	8.81	70.69	93.23	39.275
I1.2.3.06 Vitrified clay pipes, BS 65, British Standard "Normal" quality: spigot & socket cement joints; nominal bore 225mm	m	0.10	11.27	7.23	43.31	61.81	23.445
I1.2.3.07 as above; nominal bore 300mm	m	0.12	13.73	8.81	68.27	90.81	39.275
I1.2.3.08 Vitrified clay pipes, BS 65, "Perforated" plain end with sleeve joints; nominal bore 225mm	m	0.10	11.27	7.23	34.67	53.17	21.694
I1.2.3.09 as above; nominal bore 300mm	m	0.12	13.73	8.81	62.24	84.78	33.305
I1.2.3.10 Clay land drains; BS 1196; plain butt joints; nominal bore 225mm	m	0.09	10.04	6.44	36.93	53.41	17.931
I1.2.4 In trenches, depth: 2 - 2.5m							
I1.2.4.01 Vitrified clay pipes, BS 65, "Extra Strength" spigot & socket flexible joints; nominal bore 225mm	m	0.14	15.18	9.74	39.15	64.07	24.365
I1.2.4.02 as above; nominal bore 300mm	m	0.15	16.18	10.38	61.19	87.75	39.854
I1.2.4.03 Vitrified clay pipes, BS 65, "Extra Strength": plain end with sleeve joints; nominal bore 225mm	m	0.14	15.18	9.74	38.35	63.27	22.614
I1.2.4.04 Vitrified clay pipes, BS 65, "Surface Water" quality: spigot & socket cement joints; nominal bore 225mm	m	0.14	15.18	9.74	45.25	70.17	24.365
I1.2.4.05 as above; nominal bore 300mm	m	0.15	16.18	10.38	70.69	97.25	39.854
I1.2.4.06 Vitrified clay pipes, BS 65, British Standard "Normal" quality: spigot & socket cement joints; nominal bore 225mm	m	0.14	15.18	9.74	43.31	68.23	24.365
I1.2.4.07 as above; nominal bore 300mm	m	0.15	16.18	10.38	68.27	94.83	39.854
I1.2.4.08 Vitrified clay pipes, BS 65, "Perforated" plain end with sleeve joints; nominal bore 225mm	m	0.14	15.18	9.74	34.67	59.59	22.614
I1.2.4.09 as above; nominal bore 300mm	m	0.15	16.18	10.38	62.24	88.80	33.884

I1   Clay Pipes continued...

	Unit	Labour Hours	Labour Net £	Plant Net £	Materials Net £	Unit Net £	CO_2e Kg
I1.2 **nominal bore: 200 - 300mm**							
I1.2.4 In trenches, depth: 2 - 2.5m							
I1.2.4.10 Clay land drains; BS 1196; plain butt joints; nominal bore 225mm	m	0.13	13.95	8.95	36.93	59.83	18.851
I1.2.5 In trenches, depth: 2.5 - 3m							
I1.2.5.01 Vitrified clay pipes, BS 65, "Extra Strength" spigot & socket flexible joints; nominal bore 225mm	m	0.17	18.64	11.96	39.15	69.75	25.180
I1.2.5.02 as above; nominal bore 300mm	m	0.19	21.32	13.68	61.19	96.19	41.063
I1.2.5.03 Vitrified clay pipes, BS 65, "Extra Strength": plain end with sleeve joints; nominal bore 225mm	m	0.17	18.64	11.96	38.35	68.95	23.429
I1.2.5.04 Vitrified clay pipes, BS 65, "Surface Water" quality: spigot & socket cement joints; nominal bore 225mm	m	0.17	18.64	11.96	45.25	75.85	25.180
I1.2.5.05 as above; nominal bore 300mm	m	0.19	21.32	13.68	70.69	105.69	41.063
I1.2.5.06 Vitrified clay pipes, BS 65, British Standard "Normal" quality: spigot & socket cement joints; nominal bore 225mm	m	0.17	18.64	11.96	43.31	73.91	25.180
I1.2.5.07 as above; nominal bore 300mm	m	0.19	21.32	13.68	68.27	103.27	41.063
I1.2.5.08 Vitrified clay pipes, BS 65, "Perforated" plain end with sleeve joints; nominal bore 225mm	m	0.17	18.64	11.96	34.67	65.27	23.429
I1.2.5.09 as above; nominal bore 300mm	m	0.19	21.32	13.68	70.69	105.69	41.063
I1.2.6 In trenches, depth: 3 - 3.5m							
I1.2.6.01 Vitrified clay pipes, BS 65, "Extra Strength" spigot & socket flexible joints; nominal bore 225mm	m	0.22	38.78	20.62	39.15	98.55	28.498
I1.2.6.02 as above; nominal bore 300mm	m	0.24	43.45	23.10	61.19	127.74	44.677
I1.2.6.03 Vitrified clay pipes, BS 65, "Extra Strength": plain end with sleeve joints; nominal bore 225mm	m	0.22	38.78	20.62	38.35	97.75	26.747
I1.2.6.04 Vitrified clay pipes, BS 65, "Surface Water" quality: spigot & socket cement joints; nominal bore 225mm	m	0.22	38.78	20.62	45.25	104.65	28.498
I1.2.6.05 as above; nominal bore 300mm	m	0.24	43.45	23.10	70.69	137.24	44.677
I1.2.6.06 Vitrified clay pipes, BS 65, British Standard "Normal" quality: spigot & socket cement joints; nominal bore 225mm	m	0.22	38.78	20.62	43.31	102.71	28.498
I1.2.6.07 as above; nominal bore 300mm	m	0.24	43.45	23.10	68.27	134.82	44.677
I1.2.6.08 Vitrified clay pipes, BS 65, "Perforated" plain end with sleeve joints; nominal bore 225mm	m	0.22	38.78	20.62	34.67	94.07	26.747
I1.2.6.09 as above; nominal bore 300mm	m	0.24	43.45	23.10	62.24	128.79	38.707
I1.2.7 In trenches, depth: 3.5 - 4m							
I1.2.7.01 Vitrified clay pipes, BS 65, "Extra Strength" spigot & socket flexible joints; nominal bore 225mm	m	0.27	47.94	25.49	39.15	112.58	30.317
I1.2.7.02 as above; nominal bore 300mm	m	0.31	55.30	29.40	61.19	145.89	47.032
I1.2.7.03 Vitrified clay pipes, BS 65, "Extra Strength": plain end with sleeve joints; nominal bore 225mm	m	0.28	49.55	26.35	38.35	114.25	28.888

I1   Clay Pipes continued...

	Unit	Labour Hours	Labour Net £	Plant Net £	Materials Net £	Unit Net £	CO₂e Kg

I1.2 **nominal bore: 200 - 300mm**							
I1.2.7 In trenches, depth: 3.5 - 4m							
I1.2.7.04 Vitrified clay pipes, BS 65, "Surface Water" quality: spigot & socket cement joints; nominal bore 225mm	m	0.27	47.94	25.49	45.25	118.68	30.317
I1.2.7.05 as above; nominal bore 300mm	m	0.31	55.30	29.40	70.69	155.39	47.032
I1.2.7.06 Vitrified clay pipes, BS 65, British Standard "Normal" quality: spigot & socket cement joints; nominal bore 225mm	m	0.27	47.94	25.49	43.31	116.74	30.317
I1.2.7.07 as above; nominal bore 300mm	m	0.31	55.30	29.40	68.27	152.97	47.032
I1.2.7.08 Vitrified clay pipes, BS 65, "Perforated" plain end with sleeve joints; nominal bore 225mm	m	0.27	47.94	25.49	34.67	108.10	28.566
I1.2.7.09 as above; nominal bore 300mm	m	0.31	55.30	29.40	62.24	146.94	41.062
I1.2.8 In trenches, depth: exceeding 4m							
I1.2.8.01 Vitrified clay pipes, BS 65, "Extra Strength" spigot & socket flexible joints; nominal bore 225mm; depth: 4 - 4.5m	m	0.47	84.56	44.96	39.15	168.67	37.597
I1.2.8.02 as above; depth: 4.5 - 5m	m	0.62	110.42	58.71	39.15	208.28	42.735
I1.2.8.03 as above; nominal bore 300mm; depth: 4 - 4.5m	m	0.50	89.77	47.73	61.19	198.69	53.884
I1.2.8.04 as above; depth: 4.5 - 5m	m	0.67	120.29	63.96	61.19	245.44	59.950
I1.2.8.05 Vitrified clay pipes, BS 65, "Extra Strength": plain end with sleeve joints; nominal bore 225mm; depth: 4 - 4.5m	m	0.47	84.56	44.96	38.35	167.87	35.846
I1.2.8.06 as above; depth: 4.5 - 5m	m	0.62	110.42	58.71	38.35	207.48	40.984
I1.2.8.07 Vitrified clay pipes, BS 65, "Surface Water" quality: spigot & socket cement joints; nominal bore 225mm; depth: 4 - 4.5m	m	0.47	84.56	44.96	45.25	174.77	37.597
I1.2.8.08 as above; depth: 4.5 - 5m	m	0.62	110.42	58.71	45.25	214.38	42.735
I1.2.8.09 as above; nominal bore 300mm; depth: 4 - 4.5m	m	0.50	89.77	47.73	70.69	208.19	53.884
I1.2.8.10 as above; depth: 4.5 - 5m	m	0.67	120.29	63.96	70.69	254.94	59.950
I1.2.8.11 Vitrified clay pipes, BS 65, British Standard "Normal" quality: spigot & socket cement joints; nominal bore 225mm; depth: 4 - 4.5m	m	0.47	84.56	44.96	43.31	172.83	37.597
I1.2.8.12 as above; depth: 4.5 - 5m	m	0.62	110.42	58.71	43.31	212.44	42.735
I1.2.8.13 as above; nominal bore 300mm; depth: 4 - 4.5m	m	0.50	89.77	47.73	68.27	205.77	53.884
I1.2.8.14 as above; depth: 4.5 - 5m	m	0.67	120.29	63.96	68.27	252.52	59.950
I1.2.8.15 Vitrified clay pipes, BS 65, "Perforated" plain end with sleeve joints; nominal bore 225mm; depth: 4 - 4.5m	m	0.47	84.56	44.96	34.67	164.19	35.846
I1.2.8.16 as above; depth: 4.5 - 5m	m	0.62	110.42	58.71	34.67	203.80	40.984
I1.2.8.17 Vitrified clay pipes, BS 65, "Perforated" plain end with sleeve joints; nominal bore 300mm; depth: 4 - 4.5m	m	0.50	89.77	47.73	62.24	199.74	47.914
I1.2.8.18 as above; depth: 4.5 - 5m	m	0.67	120.29	63.96	62.24	246.49	53.980

I1　Clay Pipes continued...

		Unit	Labour Hours	Labour Net £	Plant Net £	Materials Net £	Unit Net £	CO_2e Kg
I1.3	**nominal bore: 300 - 600mm**							
I1.3.2	In trenches, depth: not exceeding 1.5m							
I1.3.2.01	Vitrified clay pipes, BS 65, "Extra Strength" spigot & socket flexible joints; nominal bore 375mm	m	0.12	13.73	8.81	148.24	170.78	55.754
I1.3.2.02	as above; nominal bore 450mm	m	0.16	17.86	11.46	196.11	225.43	80.798
I1.3.2.03	Vitrified clay pipes, BS 65, "Surface Water" quality: spigot & socket cement joints; nominal bore 375mm	m	0.12	13.73	8.81	164.49	187.03	55.754
I1.3.2.04	as above; nominal bore 450mm	m	0.16	17.86	11.46	209.51	238.83	80.798
I1.3.2.05	Vitrified clay pipes, BS 65, British Standard "Normal" quality: spigot & socket cement joints; nominal bore 375mm	m	0.12	13.73	8.81	109.48	132.02	55.754
I1.3.2.06	as above; nominal bore 450mm	m	0.16	17.86	11.46	180.37	209.69	80.798
I1.3.3	In trenches, depth: 1.5 - 2m							
I1.3.3.01	Vitrified clay pipes, BS 65, "Extra Strength" spigot & socket flexible joints; nominal bore 375mm	m	0.13	14.84	9.52	148.24	172.60	56.017
I1.3.3.02	as above; nominal bore 450mm	m	0.20	22.32	14.32	196.11	232.75	81.849
I1.3.3.03	Vitrified clay pipes, BS 65, "Surface Water" quality: spigot & socket cement joints; nominal bore 375mm	m	0.13	14.84	9.52	164.49	188.85	56.017
I1.3.3.04	as above; nominal bore 450mm	m	0.20	22.32	14.32	209.51	246.15	81.849
I1.3.3.05	Vitrified clay pipes, BS 65, British Standard "Normal" quality: spigot & socket cement joints; nominal bore 375mm	m	0.13	14.84	9.52	109.48	133.84	56.017
I1.3.3.06	as above; nominal bore 450mm	m	0.20	22.32	14.32	180.37	217.01	81.849
I1.3.4	In trenches, depth: 2 - 2.5m							
I1.3.4.01	Vitrified clay pipes, BS 65, "Extra Strength" spigot & socket flexible joints; nominal bore 375mm	m	0.19	20.76	13.32	148.24	182.32	57.410
I1.3.4.02	as above; nominal bore 450mm	m	0.23	25.56	16.40	196.11	238.07	82.612
I1.3.4.03	Vitrified clay pipes, BS 65, "Surface Water" quality: spigot & socket cement joints; nominal bore 375mm	m	0.19	20.76	13.32	164.49	198.57	57.410
I1.3.4.04	as above; nominal bore 450mm	m	0.23	25.56	16.40	209.51	251.47	82.612
I1.3.4.05	Vitrified clay pipes, BS 65, British Standard "Normal" quality: spigot & socket cement joints; nominal bore 375mm	m	0.19	20.76	13.32	109.48	143.56	57.410
I1.3.4.06	as above; nominal bore 450mm	m	0.23	25.56	16.40	180.37	222.33	82.612
I1.3.5	In trenches, depth: 2.5 - 3m							
I1.3.5.01	Vitrified clay pipes, BS 65, "Extra Strength" spigot & socket flexible joints; nominal bore 375mm	m	0.22	24.78	15.90	148.24	188.92	58.357
I1.3.5.02	as above; nominal bore 450mm	m	0.27	29.80	19.12	196.11	245.03	83.611
I1.3.5.03	Vitrified clay pipes, BS 65, "Surface Water" quality: spigot & socket cement joints; nominal bore 375mm	m	0.22	24.78	15.90	164.49	205.17	58.357
I1.3.5.04	as above; nominal bore 450mm	m	0.27	29.80	19.12	209.51	258.43	83.611

I1 Clay Pipes continued...

	Unit	Labour Hours	Labour Net £	Plant Net £	Materials Net £	Unit Net £	CO_2e Kg
I1.3	**nominal bore: 300 - 600mm**						
I1.3.5	In trenches, depth: 2.5 - 3m						
I1.3.5.05	Vitrified clay pipes, BS 65, British Standard "Normal" quality: spigot & socket cement joints; nominal bore 375mm	m	0.22	24.78	15.90	109.48	150.16	58.357
I1.3.5.06	as above; nominal bore 450mm	m	0.27	29.80	19.12	180.37	229.29	83.611
I1.3.6	In trenches, depth: 3 - 3.5m						
I1.3.6.01	Vitrified clay pipes, BS 65, "Extra Strength" spigot & socket flexible joints; nominal bore 375mm	m	0.28	49.55	26.35	148.24	224.14	62.370
I1.3.6.02	as above; nominal bore 450mm	m	0.32	57.45	30.55	196.11	284.11	88.011
I1.3.6.03	Vitrified clay pipes, BS 65, "Surface Water" quality: spigot & socket cement joints; nominal bore 375mm	m	0.28	49.55	26.35	164.49	240.39	62.370
I1.3.6.04	as previous item; nominal bore 450mm	m	0.32	57.45	30.55	209.51	297.51	88.011
I1.3.6.05	Vitrified clay pipes, BS 65, British Standard "Normal" quality: spigot & socket cement joints; nominal bore 375mm	m	0.28	49.55	26.35	109.48	185.38	62.370
I1.3.6.06	as above; nominal bore 450mm	m	0.32	57.45	30.55	180.37	268.37	88.011
I1.3.7	In trenches, depth: 3.5 - 4m						
I1.3.7.01	Vitrified clay pipes, BS 65, "Extra Strength" spigot & socket flexible joints; nominal bore 375mm	m	0.36	65.35	34.75	148.24	248.34	65.510
I1.3.7.02	as above; nominal bore 450mm	m	0.44	79.72	42.38	196.11	318.21	92.435
I1.3.7.03	Vitrified clay pipes, BS 65, "Surface Water" quality: spigot & socket cement joints; nominal bore 375mm	m	0.36	65.35	34.75	164.49	264.59	65.510
I1.3.7.04	as above; nominal bore 450mm	m	0.44	79.72	42.38	209.51	331.61	92.435
I1.3.7.05	Vitrified clay pipes, BS 65, British Standard "Normal" quality: spigot & socket cement joints; nominal bore 375mm	m	0.36	65.35	34.75	109.48	209.58	65.510
I1.3.7.06	as above; nominal bore 450mm	m	0.44	79.72	42.38	180.37	302.47	92.435
I1.3.8	In trenches, depth: exceeding 4m						
I1.3.8.01	Vitrified clay pipes, BS 65, "Extra Strength" spigot & socket flexible joints; nominal bore 375mm; depth: 4 - 4.5m	m	0.57	102.52	54.51	148.24	305.27	72.896
I1.3.8.02	as above; depth: 4.5 - 5m	m	0.73	130.53	69.40	148.24	348.17	78.463
I1.3.8.03	as above; nominal bore 450mm; In trenches; depth: 4 - 4.5m	m	0.62	110.42	58.71	196.11	365.24	98.537
I1.3.8.04	as above; depth: 4.5 - 5m	m	0.80	143.63	76.37	196.11	416.11	105.138
I1.3.8.05	Vitrified clay pipes, BS 65, "Surface Water" quality: spigot & socket cement joints; nominal bore 375mm; depth: 4 - 4.5m	m	0.57	102.52	54.51	164.49	321.52	72.896
I1.3.8.06	as above; depth: 4.5 - 5m	m	0.73	130.53	69.40	164.49	364.42	78.463
I1.3.8.07	as above; nominal bore 450mm; depth: 4 - 4.5m	m	0.62	110.42	58.71	209.51	378.64	98.537
I1.3.8.08	as above; depth: 4.5 - 5m	m	0.80	143.63	76.37	209.51	429.51	105.138
I1.3.8.09	Vitrified clay pipes, BS 65, British Standard "Normal" quality: spigot & socket cement joints; nominal bore 375mm; depth: 4 - 4.5m	m	0.57	102.52	54.51	109.48	266.51	72.896

I1 Clay Pipes continued...

	Unit	Labour Hours	Labour Net £	Plant Net £	Materials Net £	Unit Net £	CO_2e Kg
I1.3	**nominal bore: 300 - 600mm**						
I1.3.8	In trenches, depth: exceeding 4m						
I1.3.8.10	as above; depth: 4.5 - 5m	m	0.73	130.53	69.40	109.48	309.41	78.463
I1.3.8.11	as above; nominal bore 450mm; depth: 4 - 4.5m	m	0.62	110.42	58.71	180.37	349.50	98.537
I1.3.8.12	as above; depth: 4.5 - 5m	m	0.80	143.63	76.37	180.37	400.37	105.138

I2 Concrete Pipes

	Unit	Labour Hours	Labour Net £	Plant Net £	Materials Net £	Unit Net £	CO_2e Kg
I2.1	**nominal bore: not exceeding 200mm**						
I2.1.2	In trenches, depth: not exceeding 1.5m						
I2.1.2.01	Concrete porous pipes, BS 5911, Ogee joints; nominal bore 150mm	m	0.08	8.93	5.73	5.12	19.78	7.949
I2.1.3	In trenches, depth: 1.5 - 2m						
I2.1.3.01	Concrete porous pipes, BS 5911, Ogee joints; nominal bore 150mm	m	0.09	10.04	6.44	5.12	21.60	8.212
I2.1.4	In trenches, depth: 2 - 2.5m						
I2.1.4.01	Concrete porous pipes, BS 5911, Ogee joints; nominal bore 150mm	m	0.11	12.28	7.88	5.12	25.28	8.738
I2.1.5	In trenches, depth: 2.5 - 3m						
I2.1.5.01	Concrete porous pipes, BS 5911, Ogee joints; nominal bore 150mm	m	0.15	16.74	10.74	5.12	32.60	9.789
I2.1.6	In trenches, depth: 3 - 3.5m						
I2.1.6.01	Concrete porous pipes, BS 5911, Ogee joints; nominal bore 150mm	m	0.18	32.32	17.18	5.12	54.62	12.269
I2.1.7	In trenches, depth: 3.5 - 4m						
I2.1.7.01	Concrete porous pipes, BS 5911, Ogee joints; nominal bore 150mm	m	0.20	35.91	19.09	5.12	60.12	12.983
I2.1.8	In trenches, depth: exceeding 4m						
I2.1.8.01	Concrete porous pipes, BS 5911, Ogee joints; nominal bore 150mm; depth: 4 - 4.5m	m	0.27	48.48	25.77	5.12	79.37	15.480
I2.1.8.02	as above; depth: 4.5 - 5m	m	0.32	57.45	30.55	5.12	93.12	17.265
I2.2	**nominal bore: 200 - 300mm**						
I2.2.2	In trenches, depth: not exceeding 1.5m						
I2.2.2.01	Concrete pipes, BS 5911, Class "120"; rebated flexible joints with mastic sealant to internal faces; nominal bore 300mm	m	0.14	15.51	9.95	16.10	41.56	29.026
I2.2.2.02	Concrete porous pipes, BS 5911, Ogee joints; nominal bore 225mm	m	0.09	10.04	6.44	6.03	22.51	11.135
I2.2.2.03	as above, Ogee joints; nominal bore 300mm	m	0.11	12.28	7.88	9.22	29.38	16.046
I2.2.3	In trenches, depth: 1.5 - 2m						
I2.2.3.01	Concrete pipes, BS 5911, Class "120"; rebated flexible joints with mastic sealant to internal faces; nominal bore 300mm	m	0.16	17.86	11.46	16.10	45.42	29.578
I2.2.3.02	Concrete porous pipes, BS 5911, Ogee joints; nominal bore 225mm	m	0.11	12.28	7.88	6.03	26.19	11.661

I2 Concrete Pipes continued...

	Unit	Labour Hours	Labour Net £	Plant Net £	Materials Net £	Unit Net £	CO_2e Kg
I2.2 **nominal bore: 200 - 300mm**							
I2.2.3 In trenches, depth: 1.5 - 2m							
I2.2.3.03 as above, Ogee joints; nominal bore 300mm	m	0.13	14.51	9.31	9.22	33.04	16.571
I2.2.4 In trenches, depth: 2 - 2.5m							
I2.2.4.01 Concrete pipes, BS 5911, Class "120"; rebated flexible joints with mastic sealant to internal faces; nominal bore 300mm	m	0.18	19.87	12.74	16.10	48.71	30.051
I2.2.4.02 Concrete porous pipes, BS 5911, Ogee joints; nominal bore 225mm	m	0.12	13.39	8.59	6.03	28.01	11.923
I2.2.4.03 as above, Ogee joints; nominal bore 300mm	m	0.14	15.63	10.02	9.22	34.87	16.834
I2.2.5 In trenches, depth: 2.5 - 3m							
I2.2.5.01 Concrete pipes, BS 5911, Class "120"; rebated flexible joints with mastic sealant to internal faces; nominal bore 300mm	m	0.20	22.32	14.32	16.10	52.74	30.629
I2.2.5.02 Concrete porous pipes, BS 5911, Ogee joints; nominal bore 225mm	m	0.16	17.86	11.46	6.03	35.35	12.975
I2.2.5.03 as above, Ogee joints; nominal bore 300mm	m	0.17	18.97	12.17	9.22	40.36	17.623
I2.2.6 In trenches, depth: 3 - 3.5m							
I2.2.6.01 Concrete pipes, BS 5911, Class "120"; rebated flexible joints with mastic sealant to internal faces; nominal bore 300mm	m	0.20	35.91	19.09	16.10	71.10	32.509
I2.2.6.02 Concrete porous pipes, BS 5911, Ogee joints; nominal bore 225mm	m	0.20	35.91	19.09	6.03	61.03	15.906
I2.2.6.03 as above, Ogee joints; nominal bore 300mm	m	0.22	39.50	21.00	9.22	69.72	21.004
I2.2.7 In trenches, depth: 3.5 - 4m							
I2.2.7.01 Concrete pipes, BS 5911, Class "120"; rebated flexible joints with mastic sealant to internal faces; nominal bore 300mm	m	0.23	41.11	21.86	16.10	79.07	33.543
I2.2.7.02 Concrete porous pipes, BS 5911, Ogee joints; nominal bore 225mm	m	0.23	41.29	21.96	6.03	69.28	16.976
I2.2.7.03 as above, Ogee joints; nominal bore 300mm	m	0.26	46.68	24.82	9.22	80.72	22.432
I2.2.8 In trenches, depth: exceeding 4m							
I2.2.8.01 Concrete pipes, BS 5911, Class "120"; rebated flexible joints with mastic sealant to internal faces; nominal bore 300mm; depth: 4 - 4.5m	m	0.29	52.25	27.78	16.10	96.13	35.756
I2.2.8.02 as above; depth: 4.5 - 5m	m	0.36	63.92	33.98	16.10	114.00	38.075
I2.2.8.03 as above; depth: 5 - 5.5m	m	0.40	71.82	38.18	16.10	126.10	39.645
I2.2.8.04 as above; depth: 5.5 - 6m	m	0.53	95.69	50.88	16.10	162.67	44.391
I2.2.8.05 Concrete porous pipes, BS 5911; Ogee joints; nominal bore 225mm; depth: 4 - 4.5m	m	0.30	53.86	28.64	6.03	88.53	19.474

I2 Concrete Pipes continued...

	Unit	Labour Hours	Labour Net £	Plant Net £	Materials Net £	Unit Net £	CO_2e Kg
I2.2 **nominal bore: 200 - 300mm**							
I2.2.8 In trenches, depth: exceeding 4m							
I2.2.8.06 as previous item; depth: 4.5 - 5m	m	0.44	79.00	42.00	6.03	127.03	24.470
I2.2.8.07 as above; nominal bore 300mm; depth: 4 - 4.5m	m	0.32	57.45	30.55	9.22	97.22	24.573
I2.2.8.08 as above; depth: 4.5 - 5m	m	0.47	52.46	33.65	9.22	95.33	25.509
I2.3 **nominal bore: 300 - 600mm**							
I2.3.2 In trenches, depth: not exceeding 1.5m							
I2.3.2.01 Concrete pipes, BS 5911, Class "120"; rebated flexible joints with mastic sealant to internal faces; nominal bore 375mm	m	0.15	16.18	10.38	20.09	46.65	33.627
I2.3.2.02 as above; nominal bore 450mm	m	0.16	17.86	11.46	23.95	53.27	45.421
I2.3.2.03 Concrete porous pipes, BS 5911, Ogee joints; nominal bore 375mm	m	0.13	14.51	9.31	14.95	38.77	20.955
I2.3.2.04 as above; nominal bore 450mm	m	0.16	17.86	11.46	16.74	46.06	29.783
I2.3.2.05 as above; nominal bore 525mm	m	0.16	18.12	19.30	19.35	56.77	32.278
I2.3.2.06 as above; nominal bore 600mm	m	0.18	20.39	21.71	29.15	71.25	41.793
I2.3.3 In trenches, depth: 1.5 - 2m							
I2.3.3.01 Concrete pipes, BS 5911, Class "120"; rebated flexible joints with mastic sealant to internal faces; nominal bore 375mm	m	0.17	18.75	12.03	20.09	50.87	34.231
I2.3.3.02 as above; nominal bore 450mm	m	0.19	20.98	13.46	23.95	58.39	46.157
I2.3.3.03 as above; nominal bore 600mm	m	0.18	20.39	21.71	38.56	80.66	77.455
I2.3.3.04 Concrete porous pipes, BS 5911, Ogee joints; nominal bore 375mm	m	0.16	17.86	11.46	14.95	44.27	21.744
I2.3.3.05 as above; nominal bore 450mm	m	0.18	20.09	12.89	16.74	49.72	30.309
I2.3.3.06 as above; nominal bore 525mm	m	0.18	20.39	21.71	19.35	61.45	33.024
I2.3.3.07 as above; nominal bore 600mm	m	0.20	22.65	24.12	29.15	75.92	42.539
I2.3.4 In trenches, depth: 2 - 2.5m							
I2.3.4.01 Concrete pipes, BS 5911, Class "120"; rebated flexible joints with mastic sealant to internal faces; nominal bore 375mm	m	0.19	20.98	13.46	20.09	54.53	34.757
I2.3.4.02 as above; nominal bore 450mm	m	0.21	23.77	15.25	23.95	62.97	46.814
I2.3.4.03 as above; nominal bore 600mm	m	0.20	22.65	24.12	38.56	85.33	78.201
I2.3.4.04 Concrete porous pipes, BS 5911, Ogee joints; nominal bore 375mm	m	0.18	20.09	12.89	14.95	47.93	22.270
I2.3.4.05 as above; nominal bore 450mm	m	0.20	22.32	14.32	16.74	53.38	30.834
I2.3.4.06 as above; nominal bore 525mm	m	0.20	22.65	24.12	19.35	66.12	33.770
I2.3.4.07 as above; nominal bore 600mm	m	0.22	24.92	26.53	29.15	80.60	43.286
I2.3.5 In trenches, depth: 2.5 - 3m							
I2.3.5.01 Concrete pipes, BS 5911, Class "120"; rebated flexible joints with mastic sealant to internal faces; nominal bore 375mm	m	0.21	23.77	15.25	20.09	59.11	35.414
I2.3.5.02 as above; nominal bore 450mm	m	0.25	27.46	17.61	23.95	69.02	47.682
I2.3.5.03 as above; nominal bore 600mm	m	0.23	26.05	27.74	38.56	92.35	79.321
I2.3.5.04 Concrete porous pipes, BS 5911, Ogee joints; nominal bore 375mm	m	0.20	22.32	14.32	14.95	51.59	22.795
I2.3.5.05 as above; nominal bore 450mm	m	0.23	25.67	16.47	16.74	58.88	31.623

I2   Concrete Pipes continued...

	Unit	Labour Hours	Labour Net £	Plant Net £	Materials Net £	Unit Net £	CO_2e Kg
I2.3 **nominal bore: 300 - 600mm**							
I2.3.5 In trenches, depth: 2.5 - 3m							
I2.3.5.06 as previous item; nominal bore 525mm	m	0.23	26.05	27.74	19.35	73.14	34.890
I2.3.5.07 as above; nominal bore 600mm	m	0.24	27.18	28.95	29.15	85.28	44.032
I2.3.6 In trenches, depth: 3 - 3.5m							
I2.3.6.01 Concrete pipes, BS 5911, Class "120"; rebated flexible joints with mastic sealant to internal faces; nominal bore 375mm	m	0.21	38.24	20.33	20.09	78.66	37.415
I2.3.6.02 as above; nominal bore 450mm	m	0.25	44.17	23.48	23.95	91.60	49.993
I2.3.6.03 as above; nominal bore 600mm	m	0.23	41.67	31.40	38.56	111.63	79.807
I2.3.6.04 Concrete porous pipes, BS 5911, Ogee joints; nominal bore 375mm	m	0.25	44.88	23.86	14.95	83.69	26.459
I2.3.6.05 as above; nominal bore 450mm	m	0.27	48.48	25.77	16.74	90.99	35.211
I2.3.6.06 as above; nominal bore 525mm	m	0.27	48.92	36.86	19.35	105.13	36.953
I2.3.6.07 as above; nominal bore 600mm	m	0.29	52.55	39.59	29.15	121.29	46.511
I2.3.7 In trenches, depth: 3.5 - 4m							
I2.3.7.01 Concrete pipes, BS 5911, Class "120"; rebated flexible joints with mastic sealant to internal faces; nominal bore 375mm	m	0.25	44.17	23.48	20.09	87.74	38.593
I2.3.7.02 as above; nominal bore 450mm	m	0.29	52.25	27.78	23.95	103.98	51.599
I2.3.7.03 as above; nominal bore 600mm	m	0.29	52.55	39.59	38.56	130.70	82.173
I2.3.7.04 Concrete porous pipes, BS 5911, Ogee joints; nominal bore 375mm	m	0.29	52.07	27.68	14.95	94.70	27.886
I2.3.7.05 as above; nominal bore 450mm	m	0.32	57.45	30.55	16.74	104.74	36.996
I2.3.7.06 as above; nominal bore 525mm	m	0.32	57.98	43.69	19.35	121.02	38.925
I2.3.7.07 as above; nominal bore 600mm	m	0.35	63.42	47.78	29.15	140.35	48.877
I2.3.8 In trenches, depth: exceeding 4m							
I2.3.8.01 Concrete pipes, BS 5911, Class "120"; rebated flexible joints with mastic sealant to internal faces; nominal bore 375mm; depth: 4 - 4.5m	m	0.32	57.45	30.55	20.09	108.09	41.234
I2.3.8.02 as above; depth: 4.5 - 5m	m	0.40	71.82	38.18	20.09	130.09	44.088
I2.3.8.03 as above; depth: 5 - 5.5m	m	0.43	76.66	40.76	20.09	137.51	45.052
I2.3.8.04 as above; depth: 5.5 - 6m	m	0.56	100.72	53.55	20.09	174.36	49.833
I2.3.8.05 as above; nominal bore 450mm; depth: 4 - 4.5m	m	0.34	60.50	32.17	23.95	116.62	53.240
I2.3.8.06 as above; depth: 4.5 - 5m	m	0.43	76.66	40.76	23.95	141.37	56.452
I2.3.8.07 as above; depth: 5 - 5.5m	m	0.46	82.05	43.63	23.95	149.63	57.522
I2.3.8.08 as above; depth: 5.5 - 6m	m	0.58	104.49	55.56	23.95	184.00	61.983
I2.3.8.09 as above; nominal bore 600mm; depth: 4 - 4.5m	m	0.34	61.60	46.42	38.56	146.58	84.145
I2.3.8.10 as above; depth: 4.5 - 5m	m	0.43	77.91	58.70	38.56	175.17	87.693
I2.3.8.11 as above; depth: 5 - 5.5m	m	0.46	83.35	62.80	38.56	184.71	88.876
I2.3.8.12 as above; depth: 5.5 - 6m	m	0.58	105.09	79.18	38.56	222.83	93.608
I2.3.8.13 Concrete porous pipes, BS 5911, Ogee joints; nominal bore 375mm; depth: 4 - 4.5m	m	0.35	62.84	33.41	14.95	111.20	30.027
I2.3.8.14 as above; depth: 4.5 - 5m	m	0.50	89.77	47.73	14.95	152.45	35.380
I2.3.8.15 as above; nominal bore 450mm; depth: 4 - 4.5m	m	0.40	71.82	38.18	16.74	126.74	39.850
I2.3.8.16 as above; depth: 4.5 - 5m	m	0.53	95.16	50.59	16.74	162.49	44.489
I2.3.8.17 as above; nominal bore 525mm; depth: 4 - 4.5m	m	0.40	72.48	54.61	19.35	146.44	42.079

I2 Concrete Pipes continued...

		Unit	Labour Hours	Labour Net £	Plant Net £	Materials Net £	Unit Net £	CO_2e Kg
I2.3	**nominal bore: 300 - 600mm**							
I2.3.8	In trenches, depth: exceeding 4m							
I2.3.8.18	as previous item; depth: 4.5 - 5m	m	0.53	96.03	72.36	19.35	187.74	47.205
I2.3.8.19	as above; depth: 5 - 5.5m	m	0.67	121.40	91.47	19.35	232.22	52.726
I2.3.8.20	as above; depth: 5.5 - 6m	m	0.80	144.95	109.22	19.35	273.52	57.852
I2.3.8.21	as above; nominal bore 600mm; depth: 4 - 4.5m	m	0.42	76.10	57.34	29.15	162.59	51.637
I2.3.8.22	as above; depth: 4.5 - 5m	m	0.57	103.28	77.82	29.15	210.25	57.552
I2.3.8.23	as above; depth: 5 - 5.5m	m	0.73	132.27	99.66	29.15	261.08	63.861
I2.3.8.24	as above; depth: 5.5 - 6m	m	0.89	161.26	121.50	29.15	311.91	70.170
I2.4	**nominal bore: 600 - 900mm**							
I2.4.3	In trenches, depth: 1.5 - 2m							
I2.4.3.01	Concrete pipes, BS 5911, Class "120"; rebated flexible joints with mastic sealant to internal faces; nominal bore 750mm	m	0.23	26.05	27.74	79.75	133.54	92.767
I2.4.3.02	as above; nominal bore 900mm	m	0.27	30.58	32.56	107.05	170.19	122.320
I2.4.4	In trenches, depth: 2 - 2.5m							
I2.4.4.01	Concrete pipes, BS 5911, Class "120"; rebated flexible joints with mastic sealant to internal faces; nominal bore 750mm	m	0.25	28.32	30.15	79.75	138.22	93.513
I2.4.4.02	as above; nominal bore 900mm	m	0.29	32.85	34.98	107.05	174.88	123.067
I2.4.5	In trenches, depth: 2.5 - 3m							
I2.4.5.01	Concrete pipes, BS 5911, Class "120"; rebated flexible joints with mastic sealant to internal faces; nominal bore 750mm	m	0.27	30.58	32.56	79.75	142.89	94.259
I2.4.5.02	as above; nominal bore 900mm	m	0.32	36.24	38.60	107.05	181.89	124.186
I2.4.6	In trenches, depth: 3 - 3.5m							
I2.4.6.01	Concrete pipes, BS 5911, Class "120"; rebated flexible joints with mastic sealant to internal faces; nominal bore 750mm	m	0.27	48.92	36.86	79.75	165.53	94.830
I2.4.6.02	as above; nominal bore 900mm	m	0.32	57.98	43.69	107.05	208.72	124.863
I2.4.7	In trenches, depth: 3.5 - 4m							
I2.4.7.01	Concrete pipes, BS 5911, Class "120"; rebated flexible joints with mastic sealant to internal faces; nominal bore 750mm	m	0.32	57.98	43.69	79.75	181.42	96.802
I2.4.7.02	as above; nominal bore 900mm	m	0.36	65.23	49.15	107.05	221.43	126.440
I2.4.8	In trenches, depth: exceeding 4m							
I2.4.8.01	Concrete pipes, BS 5911, Class "120"; rebated flexible joints with mastic sealant to internal faces; nominal bore 750mm; depth: 4 - 4.5m	m	0.36	65.23	49.15	79.75	194.13	98.379
I2.4.8.02	as above; depth: 4.5 - 5m	m	0.46	83.35	62.80	79.75	225.90	102.322
I2.4.8.03	as above; depth: 5 - 5.5m	m	0.53	96.03	72.36	79.75	248.14	105.082
I2.4.8.04	as above; depth: 5.5 - 6m	m	0.64	115.96	87.37	79.75	283.08	109.420
I2.4.8.05	as above; nominal bore 900mm; depth: 4 - 4.5m	m	0.40	72.48	54.61	107.05	234.14	128.017
I2.4.8.06	as above; depth: 4.5 - 5m	m	0.53	96.03	72.36	107.05	275.44	133.143
I2.4.8.07	as above; depth: 5 - 5.5m	m	0.58	105.09	79.18	107.05	291.32	135.115

I2   Concrete Pipes continued...

		Unit	Labour Hours	Labour Net £	Plant Net £	Materials Net £	Unit Net £	CO_2e Kg
I2.4	**nominal bore: 600 - 900mm**							
I2.4.8	In trenches, depth: exceeding 4m							
I2.4.8.08	as previous item; depth: 5.5 - 6m	m	0.71	128.64	96.93	107.05	332.62	140.241
I2.5	**nominal bore: 900 - 1200mm**							
I2.5.3	In trenches, depth: 1.5 - 2m							
I2.5.3.01	Concrete pipes, BS 5911, Class "120"; rebated flexible joints with mastic sealant to internal faces; nominal bore 1200mm	m	0.32	36.24	38.60	183.82	258.66	219.477
I2.5.4	In trenches, depth: 2 - 2.5m							
I2.5.4.01	Concrete pipes, BS 5911, Class "120"; rebated flexible joints with mastic sealant to internal faces; nominal bore 1200mm	m	0.36	40.77	43.42	183.82	268.01	220.970
I2.5.5	In trenches, depth: 2.5 - 3m							
I2.5.5.01	Concrete pipes, BS 5911, Class "120"; rebated flexible joints with mastic sealant to internal faces; nominal bore 1200mm	m	0.40	45.30	48.24	183.82	277.36	222.462
I2.5.6	In trenches, depth: 3 - 3.5m							
I2.5.6.01	Concrete pipes, BS 5911, Class "120"; rebated flexible joints with mastic sealant to internal faces; nominal bore 1200mm	m	0.40	72.48	54.61	183.82	310.91	223.308
I2.5.7	In trenches, depth: 3.5 - 4m							
I2.5.7.01	Concrete pipes, BS 5911, Class "120"; rebated flexible joints with mastic sealant to internal faces; nominal bore 1200mm	m	0.46	83.35	62.80	183.82	329.97	225.674
I2.5.8	In trenches, depth: exceeding 4m							
I2.5.8.01	Concrete pipes, BS 5911, Class "120"; rebated flexible joints with mastic sealant to internal faces; nominal bore 1200mm; depth: 4 - 4.5m	m	0.53	96.03	72.36	183.82	352.21	228.434
I2.5.8.02	as above; depth: 4.5 - 5m	m	0.64	115.96	87.37	183.82	387.15	232.772
I2.5.8.03	as above; depth: 5 - 5.5m	m	0.71	128.64	96.93	183.82	409.39	235.532
I2.5.8.04	as above; depth: 5.5 - 6m	m	0.80	144.95	109.22	183.82	437.99	239.081
I2.6	**nominal bore: 1200 - 1500mm**							
I2.6.3	In trenches, depth: 1.5 - 2m							
I2.6.3.01	Concrete pipes, BS 5911, Class "120"; rebated flexible joints with mastic sealant to internal faces; nominal bore 1500mm	m	0.40	45.30	48.24	323.17	416.71	320.676
I2.6.4	In trenches, depth: 2 - 2.5m							
I2.6.4.01	Concrete pipes, BS 5911, Class "120"; rebated flexible joints with mastic sealant to internal faces; nominal bore 1500mm	m	0.46	52.10	55.48	323.17	430.75	322.915

I2 Concrete Pipes continued...

		Unit	Labour Hours	Labour Net £	Plant Net £	Materials Net £	Unit Net £	CO_2e Kg
I2.6	**nominal bore: 1200 - 1500mm**							
I2.6.5 I2.6.5.01	In trenches, depth: 2.5 - 3m Concrete pipes, BS 5911, Class "120"; rebated flexible joints with mastic sealant to internal faces; nominal bore 1500mm	m	0.53	60.03	63.92	323.17	447.12	325.527
I2.6.6 I2.6.6.01	In trenches, depth: 3 - 3.5m Concrete pipes, BS 5911, Class "120"; rebated flexible joints with mastic sealant to internal faces; nominal bore 1500mm	m	0.53	96.03	72.36	323.17	491.56	326.648
I2.6.7 I2.6.7.01	In trenches, depth: 3.5 - 4m Concrete pipes, BS 5911, Class "120"; rebated flexible joints with mastic sealant to internal faces; nominal bore 1500mm	m	0.58	105.09	79.18	323.17	507.44	328.620
I2.6.8 I2.6.8.01	In trenches, depth: exceeding 4m Concrete pipes, BS 5911, Class "120"; rebated flexible joints with mastic sealant to internal faces; nominal bore 1500mm; depth: 4 - 4.5m	m	0.64	115.96	87.37	323.17	526.50	330.986
I2.6.8.02	as above; depth: 4.5 - 5m	m	0.71	128.64	96.93	323.17	548.74	333.746
I2.6.8.03	as above; depth: 5 - 5.5m	m	0.80	144.95	109.22	323.17	577.34	337.295
I2.6.8.04	as above; depth: 5.5 - 6m	m	1.07	193.87	146.08	323.17	663.12	347.941
I2.7	**nominal bore: 1500 - 1800mm**							
I2.7.3 I2.7.3.01	In trenches, depth: 1.5 - 2m Concrete pipes, BS 5911, Class "120"; rebated flexible joints with mastic sealant to internal faces; nominal bore 1800mm	m	0.53	60.03	63.92	450.06	574.01	437.772
I2.7.4 I2.7.4.01	In trenches, depth: 2 - 2.5m Concrete pipes, BS 5911, Class "120"; rebated flexible joints with mastic sealant to internal faces; nominal bore 1800mm	m	0.58	65.69	69.95	450.06	585.70	439.638
I2.7.5 I2.7.5.01	In trenches, depth: 2.5 - 3m Concrete pipes, BS 5911, Class "120"; rebated flexible joints with mastic sealant to internal faces; nominal bore 1800mm	m	0.60	67.96	72.37	450.06	590.39	440.385
I2.7.6 I2.7.6.01	In trenches, depth: 3 - 3.5m Concrete pipes, BS 5911, Class "120"; rebated flexible joints with mastic sealant to internal faces; nominal bore 1800mm	m	0.62	112.34	84.64	450.06	647.04	442.442
I2.7.7 I2.7.7.01	In trenches, depth: 3.5 - 4m Concrete pipes, BS 5911, Class "120"; rebated flexible joints with mastic sealant to internal faces; nominal bore 1800mm	m	0.65	117.77	88.74	450.06	656.57	443.625

I2 Concrete Pipes continued...

		Unit	Labour Hours	Labour Net £	Plant Net £	Materials Net £	Unit Net £	CO_2e Kg
I2.7	**nominal bore: 1500 - 1800mm**							
I2.7.8	In trenches, depth: exceeding 4m							
I2.7.8.01	Concrete pipes, BS 5911, Class "120"; rebated flexible joints with mastic sealant to internal faces; nominal bore 1800mm; depth: 4 - 4.5m	m	0.75	135.89	102.39	450.06	688.34	447.568
I2.7.8.02	as above; depth: 4.5 - 5m	m	1.07	193.87	146.08	450.06	790.01	460.186
I2.7.8.03	as above; depth: 5 - 5.5m	m	1.15	208.37	157.00	450.06	815.43	463.341
I2.7.8.04	as above; depth: 5.5 - 6m	m	1.30	235.55	177.48	450.06	863.09	469.255
I2.8	**nominal bore: exceeding 1800mm**							
I2.8.6	In trenches, depth: 3 - 3.5m							
I2.8.6.01	Concrete pipes, BS 5911, Class "120"; rebated flexible joints with mastic sealant to internal faces; nominal bore 2100mm	m	0.64	115.96	87.37	450.06	653.39	443.231
I2.8.6.02	as above; nominal bore 2400mm	m	0.69	125.02	94.20	450.06	669.28	445.202
I2.8.7	In trenches, depth: 3.5 - 4.0m							
I2.8.7.01	Concrete pipes, BS 5911, Class "120"; rebated flexible joints with mastic sealant to internal faces; nominal bore 2100mm	m	0.71	128.64	96.93	450.06	675.63	445.991
I2.8.7.02	as above; nominal bore 2400mm	m	0.71	128.64	96.93	450.06	675.63	445.991
I2.8.8	In trenches, exceeding 4.0m							
I2.8.8.01	Concrete pipes, BS 5911, Class "120"; rebated flexible joints with mastic sealant to internal faces; nominal bore 2100mm; depth: 4 - 4.5m	m	0.80	144.95	109.22	450.06	704.23	449.540
I2.8.8.02	as above; depth: 4.5 - 5m	m	1.07	193.87	146.08	450.06	790.01	460.186
I2.8.8.03	as above; depth: 5 - 5.5m	m	1.25	226.49	170.65	450.06	847.20	467.284
I2.8.8.04	as above; depth: 5.5 - 6m	m	1.60	289.90	218.43	450.06	958.39	481.085
I2.8.8.05	as above; nominal bore 2400mm; depth: 4 - 4.5m	m	0.80	144.95	109.22	450.06	704.23	449.540
I2.8.8.06	as above; depth: 4.5 - 5m	m	1.07	193.87	146.08	450.06	790.01	460.186
I2.8.8.07	as above; depth: 5 - 5.5m	m	1.25	226.49	170.65	450.06	847.20	467.284
I2.8.8.08	as above; depth: 5.5 - 6m	m	1.60	289.90	218.43	450.06	958.39	481.085

I3 Iron Pipes

		Unit	Labour Hours	Labour Net £	Plant Net £	Materials Net £	Unit Net £	CO_2e Kg
I3.1	**nominal bore: not exceeding 200mm**							
I3.1.1	Not in trenches							
I3.1.1.01	Ductile spun iron pipes; concrete lined, BS EN 598, spigot & socket Tyton joints; nominal bore 100mm; above ground on pipe supports (supports priced elsewhere)	m	0.08	5.80	-	46.49	52.29	30.179
I3.1.1.02	as above; nominal bore 150mm; above ground on pipe supports (supports priced elsewhere)	m	0.10	7.26	-	54.23	61.49	48.391

I3 Iron Pipes continued...

		Unit	Labour Hours	Labour Net £	Plant Net £	Materials Net £	Unit Net £	CO_2e Kg
I3.1	**nominal bore: not exceeding 200mm**							
I3.1.2	In trenches, depth: not exceeding 1.5m							
I3.1.2.01	Ductile spun iron pipes; concrete lined, BS EN 598, spigot & socket Tyton joints; nominal bore 100mm	m	0.08	8.93	5.73	46.49	61.15	32.282
I3.1.2.02	as above; nominal bore 150mm	m	0.08	7.89	5.73	54.23	67.85	50.494
I3.1.3	In trenches, depth: 1.5 - 2m							
I3.1.3.01	Ductile spun iron pipes; concrete lined, BS EN 598, spigot & socket Tyton joints; nominal bore 100mm	m	0.08	9.38	6.01	46.49	61.88	32.387
I3.1.3.02	as above; nominal bore 150mm	m	0.09	8.87	6.44	54.23	69.54	50.757
I3.1.4	In trenches, depth: 2 - 2.5m							
I3.1.4.01	Ductile spun iron pipes; concrete lined, BS EN 598, spigot & socket Tyton joints; nominal bore 100mm	m	0.09	10.49	6.73	46.49	63.71	32.650
I3.1.4.02	as above; nominal bore 150mm	m	0.11	10.84	7.88	54.23	72.95	51.283
I3.1.5	In trenches, depth: 2.5 - 3m							
I3.1.5.01	Ductile spun iron pipes; concrete lined, BS EN 598, spigot & socket Tyton joints; nominal bore 100mm	m	0.11	12.72	8.16	46.49	67.37	33.176
I3.1.5.02	as above; nominal bore 150mm	m	0.14	15.63	10.02	54.23	79.88	52.071
I3.1.6	In trenches, depth: 3 - 3.5m							
I3.1.6.01	Ductile spun iron pipes; concrete lined, BS EN 598, spigot & socket Tyton joints; nominal bore 100mm	m	0.11	20.47	10.88	46.49	77.84	34.247
I3.1.6.02	as above; nominal bore 150mm	m	0.14	25.14	13.36	54.23	92.73	53.387
I3.1.7	In trenches, depth: 3.5 - 4m							
I3.1.7.01	Ductile spun iron pipes; concrete lined, BS EN 598, spigot & socket Tyton joints; nominal bore 100mm	m	0.13	23.88	12.70	46.49	83.07	34.925
I3.1.7.02	as above; nominal bore 150mm	m	0.17	30.52	16.23	54.23	100.98	54.457
I3.1.8	In trenches, depth: exceeding 4m							
I3.1.8.01	Ductile spun iron pipes; concrete lined, BS EN 598, spigot & socket Tyton joints; nominal bore 100mm; depth: 4 - 4.5m	m	0.17	30.16	16.04	46.49	92.69	36.174
I3.1.8.02	as above; depth; depth: 4.5 - 5m	m	0.20	35.91	19.09	46.49	101.49	37.316
I3.1.8.03	as above; depth; depth: 5 - 5.5m	m	0.27	47.94	25.49	46.49	119.92	39.706
I3.1.8.04	as above; depth; depth: 5.5 - 6m	m	0.36	65.35	34.75	46.49	146.59	43.168
I3.1.8.05	as above; nominal bore 150mm; depth: 4 - 4.5m	m	0.20	35.91	19.09	54.23	109.23	55.528
I3.1.8.06	as above; depth: 4.5 - 5m	m	0.25	44.88	23.86	54.23	122.97	57.312
I3.1.8.07	as above; depth: 5 - 5.5m	m	0.29	52.07	27.68	54.23	133.98	58.739
I3.1.8.08	as above; depth: 5.5 - 6m	m	0.40	71.82	38.18	54.23	164.23	62.664

I3 Iron Pipes continued...

		Unit	Labour Hours	Labour Net £	Plant Net £	Materials Net £	Unit Net £	CO_2e Kg
I3.2	**nominal bore: 200 - 300mm**							
I3.2.1	Not in trenches							
I3.2.1.01	Ductile spun iron pipes; concrete lined, BS EN 598, spigot & socket Tyton joints; nominal bore 300mm; above ground on pipe supports (supports priced elsewhere)	m	0.13	9.43	5.34	101.92	116.69	129.009
I3.2.2	In trenches, depth: not exceeding 1.5m							
I3.2.2.01	Ductile spun iron pipes; concrete lined, BS EN 598, spigot & socket Tyton joints; nominal bore 300mm	m	0.13	14.51	9.31	101.92	125.74	131.939
I3.2.3	In trenches, depth: 1.5 - 2m							
I3.2.3.01	Ductile spun iron pipes; concrete lined, BS EN 598, spigot & socket Tyton joints; nominal bore 300mm	m	0.14	15.63	10.02	101.92	127.57	132.202
I3.2.4	In trenches, depth: 2 - 2.5m							
I3.2.4.01	Ductile spun iron pipes; concrete lined, BS EN 598, spigot & socket Tyton joints; nominal bore 300mm	m	0.16	17.86	11.46	101.92	131.24	132.728
I3.2.5	In trenches, depth: 2.5 - 3m							
I3.2.5.01	Ductile spun iron pipes; concrete lined, BS EN 598, spigot & socket Tyton joints; nominal bore 300mm	m	0.19	21.21	13.60	101.92	136.73	133.517
I3.2.6	In trenches, depth: 3 - 3.5m							
I3.2.6.01	Ductile spun iron pipes; concrete lined, BS EN 598, spigot & socket Tyton joints; nominal bore 300mm	m	0.19	34.11	18.14	101.92	154.17	135.302
I3.2.7	In trenches, depth: 3.5 - 4m							
I3.2.7.01	Ductile spun iron pipes; concrete lined, BS EN 598, spigot & socket Tyton joints; nominal bore 300mm	m	0.21	37.70	20.05	101.92	159.67	136.015
I3.2.8	In trenches, depth: exceeding 4m							
I3.2.8.01	Ductile spun iron pipes; concrete lined, BS EN 598, spigot & socket Tyton joints; nominal bore 300mm; depth: 4 - 4.5m	m	0.29	52.25	27.78	101.92	181.95	138.906
I3.2.8.02	as above; depth: 4.5 - 5m	m	0.36	63.92	33.98	101.92	199.82	141.225
I3.2.8.03	as above; depth: 5 - 5.5m	m	0.40	71.82	38.18	101.92	211.92	142.795
I3.2.8.04	as above; depth: 5.5 - 6m	m	0.53	95.69	50.88	101.92	248.49	147.541

I3   Iron Pipes continued...

	Unit	Labour Hours	Labour Net £	Plant Net £	Materials Net £	Unit Net £	CO_2e Kg
I3.3	**nominal bore: 300 - 600mm**						
I3.3.1	**Not in trenches**						
I3.3.1.01	Ductile spun iron pipes; concrete lined, BS EN 598, spigot & socket Tyton joints; nominal bore 600mm; above ground on pipe supports (supports priced elsewhere)	m	0.18	13.36	-	314.98	328.34	332.232
I3.3.1.02	as above; nominal bore 450mm; above ground on pipe supports (supports priced elsewhere)	m	0.16	11.61	-	186.11	197.72	216.458
I3.3.2	**In trenches, depth: not exceeding 1.5m**						
I3.3.2.01	Ductile spun iron pipes; concrete lined, BS EN 598, spigot & socket Tyton joints; nominal bore 600mm	m	0.16	18.19	11.67	314.98	344.84	336.517
I3.3.2.02	as above; nominal bore 450mm	m	0.15	16.18	10.38	186.11	212.67	220.270
I3.3.3	**In trenches, depth: 1.5 - 2m**						
I3.3.3.01	Ductile spun iron pipes; concrete lined, BS EN 598, spigot & socket Tyton joints; nominal bore 600mm	m	0.19	20.98	13.46	314.98	349.42	337.174
I3.3.3.02	as above; nominal bore 450mm	m	0.17	18.75	12.03	186.11	216.89	220.874
I3.3.4	**In trenches, depth: 2 - 2.5m**						
I3.3.4.01	Ductile spun iron pipes; concrete lined, BS EN 598, spigot & socket Tyton joints; nominal bore 600mm	m	0.21	23.77	15.25	314.98	354.00	337.831
I3.3.4.02	as previous item; nominal bore 450mm	m	0.19	20.98	13.46	186.11	220.55	221.400
I3.3.5	**In trenches, depth: 2.5 - 3m**						
I3.3.5.01	Ductile spun iron pipes; concrete lined, BS EN 598, spigot & socket Tyton joints; nominal bore 600mm	m	0.25	27.46	17.61	314.98	360.05	338.699
I3.3.5.02	as above; nominal bore 450mm	m	0.21	23.77	15.25	186.11	225.13	222.057
I3.3.6	**In trenches, depth: 3 - 3.5m**						
I3.3.6.01	Ductile spun iron pipes; concrete lined, BS EN 598, spigot & socket Tyton joints; nominal bore 600mm	m	0.25	44.17	23.48	314.98	382.63	341.010
I3.3.6.02	as above; nominal bore 450mm	m	0.21	38.24	20.33	186.11	244.68	224.058
I3.3.7	**In trenches, depth: 3.5 - 4m**						
I3.3.7.01	Ductile spun iron pipes; concrete lined, BS EN 598, spigot & socket Tyton joints; nominal bore 600mm	m	0.29	52.25	27.78	314.98	395.01	342.616
I3.3.7.02	as above; nominal bore 450mm	m	0.25	44.17	23.48	186.11	253.76	225.236
I3.3.8	**In trenches, depth: exceeding 4m**						
I3.3.8.01	Ductile spun iron pipes; concrete lined, BS EN 598, spigot & socket Tyton joints; nominal bore 450mm; depth: 4 - 4.5m	m	0.32	57.45	30.55	186.11	274.11	227.877
I3.3.8.02	as above; depth: 4.5 - 5m	m	0.40	71.82	38.18	186.11	296.11	230.731
I3.3.8.03	as above; depth: 5 - 5.5m	m	0.43	76.66	40.76	186.11	303.53	231.695

I3   Iron Pipes continued...

	Unit	Labour Hours	Labour Net £	Plant Net £	Materials Net £	Unit Net £	CO_2e Kg
I3.3	**nominal bore: 300 - 600mm**						
I3.3.8	**In trenches, depth: exceeding 4m**						
I3.3.8.04	as previous item; depth: 5.5 - 6m	m	0.56	100.72	53.55	186.11	340.38	236.476
I3.3.8.05	as above; nominal bore 600mm; depth: 4 - 4.5m	m	0.34	60.50	32.17	314.98	407.65	344.257
I3.3.8.06	as above; depth: 4.5 - 5m	m	0.43	76.66	40.76	314.98	432.40	347.469
I3.3.8.07	as above; depth: 5 - 5.5m	m	0.46	82.05	43.63	314.98	440.66	348.539
I3.3.8.08	as above; depth: 5.5 - 6m	m	0.58	104.49	55.56	314.98	475.03	353.000
I3.4	**nominal bore: 600 - 900mm**						
I3.4.1	**Not in trenches**						
I3.4.1.01	Ductile spun iron pipes; concrete lined, BS EN 598, Stantyte joints with rubber gasket; nominal bore 800mm; above ground on pipe supports (supports priced elsewhere)	m	0.23	17.07	-	377.22	394.29	423.811
I3.4.1.02	as above; nominal bore 900mm; above ground on pipe supports (supports priced elsewhere)	m	0.27	20.04	-	449.85	469.89	487.361
I3.4.3	**In trenches, depth: 1.5 - 2m**						
I3.4.3.01	Ductile spun iron pipes; concrete lined, BS EN 598, Stantyte joints with rubber gasket; nominal bore 800mm	m	0.23	25.94	27.62	377.22	430.78	432.356
I3.4.3.02	as above; nominal bore 900mm	m	0.27	30.24	32.20	449.85	512.29	497.324
I3.4.4	**In trenches, depth: 2 - 2.5m**						
I3.4.4.01	Ductile spun iron pipes; concrete lined, BS EN 598, Stantyte joints with rubber gasket; nominal bore 800mm	m	0.25	27.86	29.67	377.22	434.75	432.991
I3.4.4.02	as above; nominal bore 900mm	m	0.29	32.96	35.10	449.85	517.91	498.220
I3.4.5	**In trenches, depth: 2.5 - 3m**						
I3.4.5.01	Ductile spun iron pipes; concrete lined, BS EN 598, Stantyte joints with rubber gasket; nominal bore 800mm	m	0.27	30.24	32.20	377.22	439.66	433.774
I3.4.5.02	as above; nominal bore 900mm	m	0.32	36.24	38.60	449.85	524.69	499.302
I3.4.6	**In trenches, depth: 3 - 3.5m**						
I3.4.6.01	Ductile spun iron pipes; concrete lined, BS EN 598, Stantyte joints with rubber gasket; nominal bore 800mm	m	0.27	48.38	36.45	377.22	462.05	434.339
I3.4.6.02	as above; nominal bore 900mm	m	0.32	57.98	43.69	449.85	551.52	499.979
I3.4.7	**In trenches, depth: 3.5 - 4m**						
I3.4.7.01	Ductile spun iron pipes; concrete lined, BS EN 598, Stantyte joints with rubber gasket; nominal bore 800mm	m	0.32	57.98	43.69	377.22	478.89	436.429
I3.4.7.02	as above; nominal bore 900mm	m	0.36	64.50	48.60	449.85	562.95	501.398
I3.4.8	**In trenches, depth: exceeding 4m**						
I3.4.8.01	Ductile spun iron pipes; concrete lined, BS EN 598, Stantyte joints with rubber gasket; nominal bore 800mm; depth: 4 - 4.5m	m	0.36	64.50	48.60	377.22	490.32	437.848

I3   Iron Pipes continued...

	Unit	Labour Hours	Labour Net £	Plant Net £	Materials Net £	Unit Net £	CO_2e Kg
I3.4	**nominal bore: 600 - 900mm**						
I3.4.8	In trenches, depth: exceeding 4m						
I3.4.8.02	as previous item; depth: 4.5 - 5m	m	0.46	82.80	62.39	377.22	522.41	441.831
I3.4.8.03	as above; depth: 5 - 5.5m	m	0.53	96.57	72.77	377.22	546.56	444.828
I3.4.8.04	as above; depth: 5.5 - 6m	m	0.64	115.96	87.37	377.22	580.55	449.047
I3.4.8.05	as above; nominal bore 900mm; depth: 4 - 4.5m	m	0.40	72.48	54.61	449.85	576.94	503.133
I3.4.8.06	as above; depth: 4.5 - 5m	m	0.53	96.57	72.77	449.85	619.19	508.378
I3.4.8.07	as above; depth: 5 - 5.5m	m	0.58	105.45	79.45	449.85	634.75	510.310
I3.4.8.08	as above; depth: 5.5 - 6m	m	0.71	128.83	97.07	449.85	675.75	515.396
I3.5	**nominal bore: 900 - 1200mm**						
I3.5.1	Not in trenches						
I3.5.1.01	Ductile spun iron pipes; concrete lined, BS EN 598, Stantyte joints with rubber gasket; nominal bore 1000mm; above ground on pipe supports (supports priced elsewhere)	m	0.29	21.52	-	573.48	595.00	570.436
I3.5.1.02	as above; nominal bore 1200mm; above ground on pipe supports (supports priced elsewhere)	m	0.36	26.72	-	784.80	811.52	653.510
I3.5.3	In trenches, depth: 1.5 - 2m						
I3.5.3.01	Ductile spun iron pipes; concrete lined, BS EN 598, Stantyte joints with rubber gasket; nominal bore 1000mm	m	0.32	36.24	38.60	573.48	648.32	582.377
I3.5.3.02	as above; nominal bore 1200mm	m	0.32	36.24	38.60	784.80	859.64	665.451
I3.5.4	In trenches, depth: 2 - 2.5m						
I3.5.4.01	Ductile spun iron pipes; concrete lined, BS EN 598, Stantyte joints with rubber gasket; nominal bore 1000mm	m	0.36	40.43	43.06	573.48	656.97	583.758
I3.5.4.02	as above; nominal bore 1200mm	m	0.36	40.43	43.06	784.80	868.29	666.832
I3.5.5	In trenches, depth: 2.5 - 3m						
I3.5.5.01	Ductile spun iron pipes; concrete lined, BS EN 598, Stantyte joints with rubber gasket; nominal bore 1000mm	m	0.40	45.30	48.24	573.48	667.02	585.362
I3.5.5.02	as above; nominal bore 1200mm	m	0.40	45.30	48.24	784.80	878.34	668.436
I3.5.6	In trenches, depth: 3 - 3.5m						
I3.5.6.01	Ductile spun iron pipes; concrete lined, BS EN 598, Stantyte joints with rubber gasket; nominal bore 1000mm	m	0.40	72.48	54.61	573.48	700.57	586.208
I3.5.6.02	as above; nominal bore 1200mm	m	0.40	72.48	54.61	784.80	911.89	669.282
I3.5.7	In trenches, depth: 3.5 - 4m						
I3.5.7.01	Ductile spun iron pipes; concrete lined, BS EN 598, Stantyte joints with rubber gasket; nominal bore 1000mm	m	0.46	82.80	62.39	573.48	718.67	588.456
I3.5.7.02	as above; nominal bore 1200mm	m	0.46	82.80	62.39	784.80	929.99	671.530

I3   Iron Pipes continued...

	Unit	Labour Hours	Labour Net £	Plant Net £	Materials Net £	Unit Net £	CO_2e Kg
I3.5	**nominal bore: 900 - 1200mm**						
I3.5.8	In trenches, depth: exceeding 4m						
I3.5.8.01	Ductile spun iron pipes; concrete lined, BS EN 598, Stantyte joints with rubber gasket; nominal bore 1000mm; depth: 4 - 4.5m	m	0.53	96.57	72.77	573.48	742.82	591.453
I3.5.8.02	as above; depth: 4.5 - 5m	m	0.64	115.96	87.37	573.48	776.81	595.672
I3.5.8.03	as above; depth: 5 - 5.5m	m	0.71	128.83	97.07	573.48	799.38	598.471
I3.5.8.04	as above; depth: 5.5 - 6m	m	0.80	144.95	109.22	573.48	827.65	601.981
I3.5.8.05	as above; nominal bore 1200mm; depth: 4 - 4.5m	m	0.53	96.57	72.77	784.80	954.14	674.527
I3.5.8.06	as above; depth: 4.5 - 5m	m	0.64	115.96	87.37	784.80	988.13	678.746
I3.5.8.07	as above; depth: 5 - 5.5m	m	0.71	128.83	97.07	784.80	1,010.70	681.545
I3.5.8.08	as above; depth: 5.5 - 6m	m	0.80	144.95	109.22	784.80	1,038.97	685.055
I3.7	**nominal bore: 1500 - 1800mm**						
I3.7.1	Not in trenches						
I3.7.1.01	Ductile spun iron pipes; concrete lined, BS EN 598, Stantyte joints with rubber gasket; nominal bore 1600mm; above ground on pipe supports (supports priced elsewhere)	m	0.49	36.36	-	1,531.32	1,567.68	736.585
I3.7.2	In trenches, depth: not exceeding 1.5m						
I3.7.2.01	Ductile spun iron pipes; concrete lined, BS EN 598, Stantyte joints with rubber gasket; nominal bore 1600mm	m	0.42	47.91	51.02	1,531.32	1,630.25	752.370
I3.7.3	In trenches, depth: 1.5 - 2m						
I3.7.3.01	Ductile spun iron pipes; concrete lined, BS EN 598, Stantyte joints with rubber gasket; nominal bore 1600mm	m	0.49	55.50	59.10	1,531.32	1,645.92	754.870
I3.7.4	In trenches, depth: 2 - 2.5m						
I3.7.4.01	Ductile spun iron pipes; concrete lined, BS EN 598, Stantyte joints with rubber gasket; nominal bore 1600mm	m	0.58	65.69	69.95	1,531.32	1,666.96	758.228
I3.7.5	In trenches, depth: 2.5 - 3m						
I3.7.5.01	Ductile spun iron pipes; concrete lined, BS EN 598, Stantyte joints with rubber gasket; nominal bore 1600mm	m	0.73	82.68	88.05	1,531.32	1,702.05	763.826
I3.7.6	In trenches, depth: 3 - 3.5m						
I3.7.6.01	Ductile spun iron pipes; concrete lined, BS EN 598, Stantyte joints with rubber gasket; nominal bore 1600mm	m	0.73	132.27	99.66	1,531.32	1,763.25	765.370
I3.7.7	In trenches, depth: 3.5 - 4m						
I3.7.7.01	Ductile spun iron pipes; concrete lined, BS EN 598, Stantyte joints with rubber gasket; nominal bore 1600mm	m	0.80	144.95	109.22	1,531.32	1,785.49	768.130

I3 Iron Pipes continued...

	Unit	Labour Hours	Labour Net £	Plant Net £	Materials Net £	Unit Net £	CO$_2$e Kg

I3.7 nominal bore: 1500 - 1800mm

I3.7.8 In trenches, depth: exceeding 4m

I3.7.8.01 Ductile spun iron pipes; concrete lined, BS EN 598, Stantyte joints with rubber gasket; nominal bore 1600mm; depth: 4 - 4.5m

	Unit	Labour Hours	Labour Net £	Plant Net £	Materials Net £	Unit Net £	CO$_2$e Kg
I3.7.8.01	m	0.80	144.95	109.22	1,531.32	1,785.49	768.130
I3.7.8.02 as above; depth: 4.5 - 5m	m	1.07	193.33	145.67	1,531.32	1,870.32	778.658
I3.7.8.03 as above; depth: 5 - 5.5m	m	1.25	226.49	170.65	1,531.32	1,928.46	785.874
I3.7.8.04 as above; depth: 5.5 - 6m	m	1.60	289.90	218.43	1,531.32	2,039.65	799.675

I4 Steel Pipes

I4.1 nominal bore: not exceeding 200mm

I4.1.1 Not in trenches

I4.1.1.01 Steel pipes, BS EN 10216, ends bevelled; 9 m average length; welded joints, internal and external grit blasting; external coating on one coat bituminous primer; three coats bituminous enamel; inner wrap of fibreglass; outer wrap of bitumen impregnated fibreglass to BS 534, one coat of solar reflecting Vynamatt SP; internal lining of one coat bituminous primer, spun lining of bituminous enamel to BS 534, cold applied bitumen base tape on PVC carrier and primer to joint; 55% overlap; nominal bore 101.7mm; above ground on pipe supports (supports priced elsewhere)

	Unit	Labour Hours	Labour Net £	Plant Net £	Materials Net £	Unit Net £	CO$_2$e Kg
I4.1.1.01	m	0.09	9.43	5.86	50.06	65.35	31.165
I4.1.1.02 as above; nominal bore 154.1mm; above ground on pipe supports (supports priced elsewhere)	m	0.09	9.43	4.80	53.12	67.35	43.977

I4.1.2 In trenches, depth: not exceeding 1.5m

I4.1.2.01 Steel pipes, BS EN 10216, ends bevelled; 9 m average length; welded joints, internal and external grit blasting; external coating on one coat bituminous primer; three coats bituminous enamel; inner wrap of fibreglass; outer wrap of bitumen impregnated fibreglass to BS 534, one coat of solar reflecting Vynamatt SP; internal lining of one coat bituminous primer, spun lining of bituminous enamel to BS 534, cold applied bitumen base tape on PVC carrier and primer to joint; 55% overlap; nominal bore 101.7mm

	Unit	Labour Hours	Labour Net £	Plant Net £	Materials Net £	Unit Net £	CO$_2$e Kg
I4.1.2.01	m	0.09	14.28	10.84	46.80	71.92	32.847
I4.1.2.02 as above; nominal bore 154.1mm	m	0.09	14.28	10.84	53.13	78.25	46.122

I4   Steel Pipes continued...

	Unit	Labour Hours	Labour Net £	Plant Net £	Materials Net £	Unit Net £	CO$_2$e Kg
I4.1 **nominal bore: not exceeding 200mm**							
I4.1.3 In trenches, depth: 1.5 - 2m							
I4.1.3.01 Steel pipes, BS EN 10216, ends bevelled; 9 m average length; welded joints, internal and external grit blasting; external coating on one coat bituminous primer; three coats bituminous enamel; inner wrap of fibreglass; outer wrap of bitumen impregnated fibreglass to BS 534, one coat of solar reflecting Vynamatt SP; internal lining of one coat bituminous primer, spun lining of bituminous enamel to BS 534, cold applied bitumen base tape on PVC carrier and primer to joint; 55% overlap; nominal bore 101.7mm	m	0.09	14.28	10.84	46.80	**71.92**	32.847
I4.1.3.02 as above; nominal bore 154.1mm	m	0.10	15.87	12.04	53.13	**81.04**	46.587
I4.1.4 In trenches, depth: 2 - 2.5m							
I4.1.4.01 Steel pipes, BS EN 10216, ends bevelled; 9 m average length; welded joints, internal and external grit blasting; external coating on one coat bituminous primer; three coats bituminous enamel; inner wrap of fibreglass; outer wrap of bitumen impregnated fibreglass to BS 534, one coat of solar reflecting Vynamatt SP; internal lining of one coat bituminous primer, spun lining of bituminous enamel to BS 534, cold applied bitumen base tape on PVC carrier and primer to joint; 55% overlap; nominal bore 101.7mm	m	0.09	14.28	10.84	46.80	**71.92**	32.847
I4.1.4.02 as above; nominal bore 154.1mm	m	0.11	17.45	13.25	53.13	**83.83**	47.052
I4.1.5 In trenches, depth: 2.5 - 3m							
I4.1.5.01 Steel pipes, BS EN 10216, ends bevelled; 9 m average length; welded joints, internal and external grit blasting; external coating on one coat bituminous primer; three coats bituminous enamel; inner wrap of fibreglass; outer wrap of bitumen impregnated fibreglass to BS 534, one coat of solar reflecting Vynamatt SP; internal lining of one coat bituminous primer, spun lining of bituminous enamel to BS 534, cold applied bitumen base tape on PVC carrier and primer to joint; 55% overlap; nominal bore 101.7mm	m	0.09	14.28	10.84	46.80	**71.92**	32.847
I4.1.5.02 as above; nominal bore 154.1mm	m	0.13	19.83	15.05	53.13	**88.01**	47.750

I4 Steel Pipes continued...

		Unit	Labour Hours	Labour Net £	Plant Net £	Materials Net £	Unit Net £	CO_2e Kg
I4.1	**nominal bore: not exceeding 200mm**							
I4.1.6	In trenches, depth: 3 - 3.5m							
I4.1.6.01	Steel pipes, BS EN 10216, ends bevelled; 9 m average length; welded joints, internal and external grit blasting; external coating on one coat bituminous primer; three coats bituminous enamel; inner wrap of fibreglass; outer wrap of bitumen impregnated fibreglass to BS 534, one coat of solar reflecting Vynamatt SP; internal lining of one coat bituminous primer, spun lining of bituminous enamel to BS 534, cold applied bitumen base tape on PVC carrier and primer to joint; 55% overlap; nominal bore 101.7mm	m	0.13	28.48	18.89	46.80	94.17	34.274
I4.1.6.02	as above; nominal bore 154.1mm	m	0.13	28.48	18.89	53.13	100.50	47.548
I4.1.7	In trenches, depth: 3.5 - 4m							
I4.1.7.01	Steel pipes, BS EN 10216, ends bevelled; 9 m average length; welded joints, internal and external grit blasting; external coating on one coat bituminous primer; three coats bituminous enamel; inner wrap of fibreglass; outer wrap of bitumen impregnated fibreglass to BS 534, one coat of solar reflecting Vynamatt SP; internal lining of one coat bituminous primer, spun lining of bituminous enamel to BS 534, cold applied bitumen base tape on PVC carrier and primer to joint; 55% overlap; nominal bore 101.7mm	m	0.16	34.01	22.56	46.80	103.37	35.363
I4.1.7.02	as above; nominal bore 154.1mm	m	0.16	34.01	22.56	53.13	109.70	48.637
I4.1.8	In trenches, depth: exceeding 4m							
I4.1.8.01	Steel pipes, BS EN 10216, ends bevelled; 9 m average length; welded joints, internal and external grit blasting; external coating on one coat bituminous primer; three coats bituminous enamel; inner wrap of fibreglass; outer wrap of bitumen impregnated fibreglass to BS 534, one coat of solar reflecting Vynamatt SP; internal lining of one coat bituminous primer, spun lining of bituminous enamel to BS 534, cold applied bitumen base tape on PVC carrier and primer to joint; 55% overlap; nominal bore 101.7mm; depth: 4 - 4.5m	m	0.30	64.62	42.86	46.80	154.28	41.395
I4.1.8.02	as above; depth: 4.5 - 5m	m	0.25	53.14	35.25	46.80	135.19	39.133

I4 Steel Pipes continued...

	Unit	Labour Hours	Labour Net £	Plant Net £	Materials Net £	Unit Net £	CO_2e Kg
I4.1	**nominal bore: not exceeding 200mm**						
I4.1.8	In trenches, depth: exceeding 4m						
I4.1.8.03	as previous item; depth: 5 - 5.5m	m	0.36	76.52	50.76	46.80	174.08	43.741
I4.1.8.04	as above; nominal bore 154.1mm; depth: 4 - 4.5m	m	0.21	43.57	28.90	53.13	125.60	50.523
I4.1.8.05	as above; depth: 4.5 - 5m	m	0.25	53.14	35.25	53.13	141.52	52.408
I4.1.8.06	as above; depth: 5 - 5.5m	m	0.36	76.52	50.76	53.13	180.41	57.016
I4.3	**nominal bore: 300 - 600mm**						
I4.3.1	Not in trenches						
I4.3.1.01	Steel pipes, BS EN 10216, ends bevelled; 9 m average length; welded joints, internal and external grit blasting; external coating on one coat bituminous primer; three coats bituminous enamel; inner wrap of fibreglass; outer wrap of bitumen impregnated fibreglass to BS 534, one coat of solar reflecting Vynamatt SP; internal lining of one coat bituminous primer, spun lining of bituminous enamel to BS 534, cold applied bitumen base tape on PVC carrier and primer to joint; 55% overlap; nominal bore 304.9mm; above ground on pipe supports (supports priced elsewhere)	m	0.16	16.76	8.53	108.30	133.59	141.380
I4.3.1.02	as above; nominal bore 438.2mm; above ground on pipe supports (supports priced elsewhere)	m	0.18	18.86	9.59	225.60	254.05	203.093
I4.3.2	In trenches, depth: not exceeding 1.5m						
I4.3.2.01	Steel pipes, BS EN 10216, ends bevelled; 9 m average length; welded joints, internal and external grit blasting; external coating on one coat bituminous primer; three coats bituminous enamel; inner wrap of fibreglass; outer wrap of bitumen impregnated fibreglass to BS 534, one coat of solar reflecting Vynamatt SP; internal lining of one coat bituminous primer, spun lining of bituminous enamel to BS 534, cold applied bitumen base tape on PVC carrier and primer to joint; 55% overlap; nominal bore 304.9mm	m	0.13	20.63	15.65	108.32	144.60	143.796
I4.3.2.02	as above; nominal bore 438.2mm	m	0.13	20.63	15.65	225.62	261.90	205.055

I4 Steel Pipes continued...

	Unit	Labour Hours	Labour Net £	Plant Net £	Materials Net £	Unit Net £	CO_2e Kg
I4.3	**nominal bore: 300 - 600mm**						
I4.3.3	**In trenches, depth: 1.5 - 2m**						
I4.3.3.01	Steel pipes, BS EN 10216, ends bevelled; 9 m average length; welded joints, internal and external grit blasting; external coating on one coat bituminous primer; three coats bituminous enamel; inner wrap of fibreglass; outer wrap of bitumen impregnated fibreglass to BS 534, one coat of solar reflecting Vynamatt SP; internal lining of one coat bituminous primer, spun lining of bituminous enamel to BS 534, cold applied bitumen base tape on PVC carrier and primer to joint; 55% overlap; nominal bore 304.9mm	m	0.15	23.80	18.06	108.32	150.18	144.727
I4.3.3.02	as above; nominal bore 438.2mm	m	0.15	23.80	18.06	225.62	267.48	205.986
I4.3.4	**In trenches, depth: 2 - 2.5m**						
I4.3.4.01	Steel pipes, BS EN 10216, ends bevelled; 9 m average length; welded joints, internal and external grit blasting; external coating on one coat bituminous primer; three coats bituminous enamel; inner wrap of fibreglass; outer wrap of bitumen impregnated fibreglass to BS 534, one coat of solar reflecting Vynamatt SP; internal lining of one coat bituminous primer, spun lining of bituminous enamel to BS 534, cold applied bitumen base tape on PVC carrier and primer to joint; 55% overlap; nominal bore 304.9mm	m	0.18	28.56	21.68	108.32	158.56	146.122
I4.3.4.02	as above; nominal bore 438.2mm	m	0.18	28.56	21.68	225.62	275.86	207.381
I4.3.5	**In trenches, depth: 2.5 - 3m**						
I4.3.5.01	Steel pipes, BS EN 10216, ends bevelled; 9 m average length; welded joints, internal and external grit blasting; external coating on one coat bituminous primer; three coats bituminous enamel; inner wrap of fibreglass; outer wrap of bitumen impregnated fibreglass to BS 534, one coat of solar reflecting Vynamatt SP; internal lining of one coat bituminous primer, spun lining of bituminous enamel to BS 534, cold applied bitumen base tape on PVC carrier and primer to joint; 55% overlap; nominal bore 304.9mm	m	0.22	34.11	25.89	108.32	168.32	147.751
I4.3.5.02	as above; nominal bore 438.2mm	m	0.22	34.11	25.89	225.62	285.62	209.010

I4 Steel Pipes continued...

	Unit	Labour Hours	Labour Net £	Plant Net £	Materials Net £	Unit Net £	CO_2e Kg
I4.3	**nominal bore: 300 - 600mm**						
I4.3.6	**In trenches, depth: 3 - 3.5m**						
I4.3.6.01	Steel pipes, BS EN 10216, ends bevelled; 9 m average length; welded joints, internal and external grit blasting; external coating on one coat bituminous primer; three coats bituminous enamel; inner wrap of fibreglass; outer wrap of bitumen impregnated fibreglass to BS 534, one coat of solar reflecting Vynamatt SP; internal lining of one coat bituminous primer, spun lining of bituminous enamel to BS 534, cold applied bitumen base tape on PVC carrier and primer to joint; 55% overlap; nominal bore 304.9mm	m	0.22	47.61	31.58	108.32	187.51	147.132
I4.3.6.02	as above; nominal bore 438.2mm	m	0.22	47.61	31.58	225.62	304.81	208.391
I4.3.7	**In trenches, depth: 3.5 - 4m**						
I4.3.7.01	Steel pipes, BS EN 10216, ends bevelled; 9 m average length; welded joints, internal and external grit blasting; external coating on one coat bituminous primer; three coats bituminous enamel; inner wrap of fibreglass; outer wrap of bitumen impregnated fibreglass to BS 534, one coat of solar reflecting Vynamatt SP; internal lining of one coat bituminous primer, spun lining of bituminous enamel to BS 534, cold applied bitumen base tape on PVC carrier and primer to joint; 55% overlap; nominal bore 304.9mm	m	0.25	53.99	35.81	108.32	198.12	148.389
I4.3.7.02	as above; nominal bore 438.2mm	m	0.25	53.99	35.81	225.62	315.42	209.648
I4.3.8	**In trenches, depth: exceeding 4m**						
I4.3.8.01	Steel pipes, BS EN 10216, ends bevelled; 9 m average length; welded joints, internal and external grit blasting; external coating on one coat bituminous primer; three coats bituminous enamel; inner wrap of fibreglass; outer wrap of bitumen impregnated fibreglass to BS 534, one coat of solar reflecting Vynamatt SP; internal lining of one coat bituminous primer, spun lining of bituminous enamel to BS 534, cold applied bitumen base tape on PVC carrier and primer to joint; 55% overlap; nominal bore 304.9mm; depth: 4 - 4.5m	m	0.30	64.62	42.86	108.32	215.80	150.483
I4.3.8.02	as above; depth: 4.5 - 5m	m	0.36	76.52	50.76	108.32	235.60	152.829

I4 Steel Pipes continued...

	Unit	Labour Hours	Labour Net £	Plant Net £	Materials Net £	Unit Net £	CO_2e Kg
I4.3	**nominal bore: 300 - 600mm**						
I4.3.8	In trenches, depth: exceeding 4m						
I4.3.8.03	as previous item; depth: 5 - 5.5m	m	0.44	93.53	62.04	108.32	263.89	156.181
I4.3.8.04	as above; nominal bore 438.2mm; depth: 4 - 4.5m	m	0.30	64.62	42.86	225.62	333.10	211.742
I4.3.8.05	as above; depth: 4.5 - 5m	m	0.36	76.52	50.76	225.62	352.90	214.088
I4.3.8.06	as above; depth: 5 - 5.5m	m	0.44	93.53	62.04	225.62	381.19	217.440
I4.4	**nominal bore: 600 - 900mm**						
I4.4.1	Not in trenches						
I4.4.1.01	Steel pipes, BS EN 10216, ends bevelled; 9 m average length; welded joints, internal and external grit blasting; external coating on one coat bituminous primer; three coats bituminous enamel; inner wrap of fibreglass; outer wrap of bitumen impregnated fibreglass to BS 534, one coat of solar reflecting Vynamatt SP; internal lining of one coat bituminous primer, spun lining of bituminous enamel to BS 534, cold applied bitumen base tape on PVC carrier and primer to joint; 55% overlap; nominal bore 641mm; above ground on pipe supports (supports priced elsewhere)	m	0.23	33.07	12.26	312.97	358.30	295.941
I4.4.1.02	as above; nominal bore 692mm (supports priced elsewhere)	m	0.27	38.83	14.39	326.25	379.47	327.303
I4.4.1.03	as above; nominal bore 794mm (supports priced elsewhere)	m	0.29	41.70	15.46	373.49	430.65	373.793
I4.4.1.04	as above; nominal bore 895mm (supports priced elsewhere)	m	0.22	31.64	11.73	432.26	475.63	418.235
I4.4.3	In trenches, depth: 1.5 - 2m						
I4.4.3.01	Steel pipes, BS EN 10216, ends bevelled; 9 m average length; welded joints, internal and external grit blasting; external coating on one coat bituminous primer; three coats bituminous enamel; inner wrap of fibreglass; outer wrap of bitumen impregnated fibreglass to BS 534, one coat of solar reflecting Vynamatt SP; internal lining of one coat bituminous primer, spun lining of bituminous enamel to BS 534, cold applied bitumen base tape on PVC carrier and primer to joint; 55% overlap; nominal bore 641mm	m	0.16	30.54	20.01	312.99	363.54	297.085
I4.4.3.02	as above; nominal bore 692mm	m	0.17	32.44	21.27	326.28	379.99	327.937
I4.4.3.03	as above; nominal bore 794mm	m	0.17	32.44	21.27	373.52	427.23	373.972
I4.4.3.04	as above; nominal bore 895mm	m	0.22	41.99	27.52	432.29	501.80	421.992

I4 Steel Pipes continued...

	Unit	Labour Hours	Labour Net £	Plant Net £	Materials Net £	Unit Net £	CO_2e Kg
I4.4	**nominal bore: 600 - 900mm**						
I4.4.4	In trenches, depth: 2 - 2.5m						
I4.4.4.01	Steel pipes, BS EN 10216, ends bevelled; 9 m average length; welded joints, internal and external grit blasting; external coating on one coat bituminous primer; three coats bituminous enamel; inner wrap of fibreglass; outer wrap of bitumen impregnated fibreglass to BS 534, one coat of solar reflecting Vynamatt SP; internal lining of one coat bituminous primer, spun lining of bituminous enamel to BS 534, cold applied bitumen base tape on PVC carrier and primer to joint; 55% overlap; nominal bore 641mm	m	0.20	38.17	25.02	312.99	376.18	298.676
I4.4.4.02	as above; nominal bore 692mm	m	0.21	40.08	26.27	326.28	392.63	329.528
I4.4.4.03	as above; nominal bore 794mm	m	0.21	40.08	26.27	373.52	439.87	375.563
I4.4.4.04	as above; nominal bore 895mm	m	0.26	49.62	32.52	432.29	514.43	423.583
I4.4.5	In trenches, depth: 2.5 - 3m						
I4.4.5.01	Steel pipes, BS EN 10216, ends bevelled; 9 m average length; welded joints, internal and external grit blasting; external coating on one coat bituminous primer; three coats bituminous enamel; inner wrap of fibreglass; outer wrap of bitumen impregnated fibreglass to BS 534, one coat of solar reflecting Vynamatt SP; internal lining of one coat bituminous primer, spun lining of bituminous enamel to BS 534, cold applied bitumen base tape on PVC carrier and primer to joint; 55% overlap; nominal bore 641mm	m	0.27	50.58	33.15	312.99	396.72	301.262
I4.4.5.02	as above; nominal bore 692mm	m	0.28	52.48	34.40	326.28	413.16	332.113
I4.4.5.03	as above; nominal bore 794mm	m	0.28	52.48	34.40	373.52	460.40	378.149
I4.4.5.04	as above; nominal bore 895mm	m	0.30	56.30	36.90	432.29	525.49	424.976

I4   Steel Pipes continued...

	Unit	Labour Hours	Labour Net £	Plant Net £	Materials Net £	Unit Net £	CO_2e Kg
I4.4 **nominal bore: 600 - 900mm**							
I4.4.6 In trenches, depth: 3 - 3.5m							
I4.4.6.01 Steel pipes, BS EN 10216, ends bevelled; 9 m average length; welded joints, internal and external grit blasting; external coating on one coat bituminous primer; three coats bituminous enamel; inner wrap of fibreglass; outer wrap of bitumen impregnated fibreglass to BS 534, one coat of solar reflecting Vynamatt SP; internal lining of one coat bituminous primer, spun lining of bituminous enamel to BS 534, cold applied bitumen base tape on PVC carrier and primer to joint; 55% overlap; nominal bore 641mm	m	0.27	67.06	42.85	312.99	422.90	302.910
I4.4.6.02 as above; nominal bore 692mm	m	0.28	69.51	44.41	326.28	440.20	333.809
I4.4.6.03 as above; nominal bore 794mm	m	0.28	69.51	44.41	373.52	487.44	379.845
I4.4.6.04 as above; nominal bore 895mm	m	0.30	74.40	47.54	432.29	554.23	426.766
I4.4.7 In trenches, depth: 3.5 - 4m							
I4.4.7.01 Steel pipes, BS EN 10216, ends bevelled; 9 m average length; welded joints, internal and external grit blasting; external coating on one coat bituminous primer; three coats bituminous enamel; inner wrap of fibreglass; outer wrap of bitumen impregnated fibreglass to BS 534, one coat of solar reflecting Vynamatt SP; internal lining of one coat bituminous primer, spun lining of bituminous enamel to BS 534, cold applied bitumen base tape on PVC carrier and primer to joint; 55% overlap; nominal bore 641mm	m	0.28	69.26	44.25	312.99	426.50	303.311
I4.4.7.02 as above; nominal bore 692mm	m	0.30	73.42	46.91	326.28	446.61	334.521
I4.4.7.03 as above; nominal bore 794mm	m	0.30	73.42	46.91	373.52	493.85	380.557
I4.4.7.04 as previous item; nominal bore 895mm	m	0.35	85.66	54.73	432.29	572.68	428.812

I4   Steel Pipes continued...

	Unit	Labour Hours	Labour Net £	Plant Net £	Materials Net £	Unit Net £	CO$_2$e Kg
I4.4 **nominal bore: 600 - 900mm**							
I4.4.8 In trenches, depth: exceeding 4m							
I4.4.8.01 Steel pipes, BS EN 10216, ends bevelled; 9 m average length; welded joints, internal and external grit blasting; external coating on one coat bituminous primer; three coats bituminous enamel; inner wrap of fibreglass; outer wrap of bitumen impregnated fibreglass to BS 534, one coat of solar reflecting Vynamatt SP; internal lining of one coat bituminous primer, spun lining of bituminous enamel to BS 534, cold applied bitumen base tape on PVC carrier and primer to joint; 55% overlap; nominal bore 641mm; depth: 4 - 4.5m	m	0.31	75.87	48.47	312.99	437.33	304.512
I4.4.8.02 as above; depth: 4.5 - 5m	m	0.37	90.56	57.86	312.99	461.41	307.182
I4.4.8.03 as above; depth: 5 - 5.5m	m	0.50	122.38	78.19	312.99	513.56	312.965
I4.4.8.04 as above; depth: 5.5 - 6m	m	0.57	139.51	89.13	312.99	541.63	316.080
I4.4.8.05 Steel pipes, BS EN 10216, ends bevelled; 9 m average length; welded joints, internal and external grit blasting; external coating on one coat bituminous primer; three coats bituminous enamel; inner wrap of fibreglass; outer wrap of bitumen impregnated fibreglass to BS 534, one coat of solar reflecting Vynamatt SP; internal lining of one coat bituminous primer, spun lining of bituminous enamel to BS 534, cold applied bitumen base tape on PVC carrier and primer to joint; 55% overlap; nominal bore 692mm; depth: 4 - 4.5m	m	0.33	80.77	51.60	326.28	458.65	335.856
I4.4.8.06 as previous item; depth: 4.5 - 5m	m	0.40	97.90	62.55	326.28	486.73	338.970
I4.4.8.07 as above; depth: 5 - 5.5m	m	0.53	129.72	82.88	326.28	538.88	344.754
I4.4.8.08 as above; depth: 5.5 - 6m	m	0.62	151.75	96.95	326.28	574.98	348.758
I4.4.8.09 Steel pipes, BS EN 10216, ends bevelled; 9 m average length; welded joints, internal and external grit blasting; external coating on one coat bituminous primer; three coats bituminous enamel; inner wrap of fibreglass; outer wrap of bitumen impregnated fibreglass to BS 534, one coat of solar reflecting Vynamatt SP; internal lining of one coat bituminous primer, spun lining of bituminous enamel to BS 534, cold applied bitumen base tape on PVC carrier and primer to joint; 55% overlap; nominal bore 794mm; depth: 4 - 4.5m	m	0.33	80.77	51.60	373.52	505.89	381.892
I4.4.8.10 as above; depth: 4.5 - 5m	m	0.40	97.90	62.55	373.52	533.97	385.006

I4 Steel Pipes continued...

	Unit	Labour Hours	Labour Net £	Plant Net £	Materials Net £	Unit Net £	CO_2e Kg
I4.4	**nominal bore: 600 - 900mm**						
I4.4.8	In trenches, depth: exceeding 4m						
I4.4.8.11	as previous item; depth: 5 - 5.5m	m	0.53	129.72	82.88	373.52	586.12	390.790
I4.4.8.12	as above; depth: 5.5 - 6m	m	0.62	151.75	96.95	373.52	622.22	394.794
I4.4.8.13	Steel pipes, BS EN 10216, ends bevelled; 9 m average length; welded joints, internal and external grit blasting; external coating on one coat bituminous primer; three coats bituminous enamel; inner wrap of fibreglass; outer wrap of bitumen impregnated fibreglass to BS 534, one coat of solar reflecting Vynamatt SP; internal lining of one coat bituminous primer, spun lining of bituminous enamel to BS 534, cold applied bitumen base tape on PVC carrier and primer to joint; 55% overlap; nominal bore 895mm; depth: 4 - 4.5m	m	0.40	97.90	62.55	432.29	592.74	431.037
I4.4.8.14	as above; depth: 4.5 - 5m	m	0.47	115.03	73.49	432.29	620.81	434.151
I4.4.8.15	as above; depth: 5 - 5.5m	m	0.62	151.75	96.95	432.29	680.99	440.825
I4.4.8.16	as above; depth: 5.5 - 6m	m	0.73	178.67	114.15	432.29	725.11	445.719
I4.5	**nominal bore: 900 - 1200mm**						
I4.5.1	Not in trenches						
I4.5.1.01	Steel pipes, BS EN 10216, ends bevelled; 9 m average length; welded joints, internal and external grit blasting; external coating on one coat bituminous primer; three coats bituminous enamel; inner wrap of fibreglass; outer wrap of bitumen impregnated fibreglass to BS 534, one coat of solar reflecting Vynamatt SP; internal lining of one coat bituminous primer, spun lining of bituminous enamel to BS 534, cold applied bitumen base tape on PVC carrier and primer to joint; 55% overlap; nominal bore 997mm; above ground on pipe supports (supports priced elsewhere)	m	0.24	34.51	12.79	494.86	542.16	464.722
I4.5.1.02	as above; nominal bore 1195mm; above ground on pipe supports (supports priced elsewhere)	m	0.27	38.83	14.39	588.08	641.30	724.610

I4   Steel Pipes continued...

	Unit	Labour Hours	Labour Net £	Plant Net £	Materials Net £	Unit Net £	CO_2e Kg
I4.5	**nominal bore: 900 - 1200mm**						
I4.5.3	In trenches, depth: 1.5 - 2m						
I4.5.3.01	Steel pipes, BS EN 10216, ends bevelled; 9 m average length; welded joints, internal and external grit blasting; external coating on one coat bituminous primer; three coats bituminous enamel; inner wrap of fibreglass; outer wrap of bitumen impregnated fibreglass to BS 534, one coat of solar reflecting Vynamatt SP; internal lining of one coat bituminous primer, spun lining of bituminous enamel to BS 534, cold applied bitumen base tape on PVC carrier and primer to joint; 55% overlap; nominal bore 997mm	m	0.24	45.80	30.02	494.89	570.71	468.821
I4.5.3.02	as above; nominal bore 1195mm	m	0.27	51.53	33.77	588.12	673.42	729.222
I4.5.4	In trenches, depth: 2 - 2.5m						
I4.5.4.01	Steel pipes, BS EN 10216, ends bevelled; 9 m average length; welded joints, internal and external grit blasting; external coating on one coat bituminous primer; three coats bituminous enamel; inner wrap of fibreglass; outer wrap of bitumen impregnated fibreglass to BS 534, one coat of solar reflecting Vynamatt SP; internal lining of one coat bituminous primer, spun lining of bituminous enamel to BS 534, cold applied bitumen base tape on PVC carrier and primer to joint; 55% overlap; nominal bore 997mm	m	0.28	52.48	34.40	494.89	581.77	470.213
I4.5.4.02	as above; nominal bore 1195mm	m	0.30	57.25	37.53	588.12	682.90	730.415
I4.5.5	In trenches, depth: 2.5 - 3m						
I4.5.5.01	Steel pipes, BS EN 10216, ends bevelled; 9 m average length; welded joints, internal and external grit blasting; external coating on one coat bituminous primer; three coats bituminous enamel; inner wrap of fibreglass; outer wrap of bitumen impregnated fibreglass to BS 534, one coat of solar reflecting Vynamatt SP; internal lining of one coat bituminous primer, spun lining of bituminous enamel to BS 534, cold applied bitumen base tape on PVC carrier and primer to joint; 55% overlap; nominal bore 997mm	m	0.31	58.21	38.15	494.89	591.25	471.406
I4.5.5.02	as above; nominal bore 1195mm	m	0.33	62.03	40.65	588.12	690.80	731.410

I4   Steel Pipes continued...

	Unit	Labour Hours	Labour Net £	Plant Net £	Materials Net £	Unit Net £	CO_2e Kg
I4.5	**nominal bore: 900 - 1200mm**						
I4.5.6	**In trenches, depth: 3 - 3.5m**						
I4.5.6.01	Steel pipes, BS EN 10216, ends bevelled; 9 m average length; welded joints, internal and external grit blasting; external coating on one coat bituminous primer; three coats bituminous enamel; inner wrap of fibreglass; outer wrap of bitumen impregnated fibreglass to BS 534, one coat of solar reflecting Vynamatt SP; internal lining of one coat bituminous primer, spun lining of bituminous enamel to BS 534, cold applied bitumen base tape on PVC carrier and primer to joint; 55% overlap; nominal bore 997mm	m	0.31	76.85	49.10	494.89	620.84	473.244
I4.5.6.02	as above; nominal bore 1195mm	m	0.33	81.75	52.23	588.12	722.10	733.341
I4.5.7	**In trenches, depth: 3.5 - 4m**						
I4.5.7.01	Steel pipes, BS EN 10216, ends bevelled; 9 m average length; welded joints, internal and external grit blasting; external coating on one coat bituminous primer; three coats bituminous enamel; inner wrap of fibreglass; outer wrap of bitumen impregnated fibreglass to BS 534, one coat of solar reflecting Vynamatt SP; internal lining of one coat bituminous primer, spun lining of bituminous enamel to BS 534, cold applied bitumen base tape on PVC carrier and primer to joint; 55% overlap; nominal bore 997mm	m	0.38	93.00	59.42	494.89	647.31	476.180
I4.5.7.02	as above; nominal bore 1195mm	m	0.44	107.69	68.80	588.12	764.61	738.057
I4.5.8	**In trenches, depth: exceeding 4m**						
I4.5.8.01	Steel pipes, BS EN 10216, ends bevelled; 9 m average length; welded joints, internal and external grit blasting; external coating on one coat bituminous primer; three coats bituminous enamel; inner wrap of fibreglass; outer wrap of bitumen impregnated fibreglass to BS 534, one coat of solar reflecting Vynamatt SP; internal lining of one coat bituminous primer, spun lining of bituminous enamel to BS 534, cold applied bitumen base tape on PVC carrier and primer to joint; 55% overlap; nominal bore 997mm; depth: 4 - 4.5m	m	0.44	107.69	68.80	494.89	671.38	478.849
I4.5.8.02	as above; depth: 4.5 - 5m	m	0.53	129.72	82.88	494.89	707.49	482.853

I4 Steel Pipes continued...

	Unit	Labour Hours	Labour Net £	Plant Net £	Materials Net £	Unit Net £	CO$_2$e Kg
I4.5	**nominal bore: 900 - 1200mm**						
I4.5.8	In trenches, depth: exceeding 4m						
I4.5.8.03	as previous item; depth: 5 - 5.5m	m	0.67	163.98	104.77	494.89	763.64	489.082
I4.5.8.04	as above; depth: 5.5 - 6m	m	0.80	195.80	125.10	494.89	815.79	494.866
I4.5.8.05	as above; nominal bore 1195mm; depth: 4 - 4.5m	m	0.53	129.72	82.88	588.12	800.72	742.061
I4.5.8.06	as above; depth: 4.5 - 5m	m	0.67	163.98	104.77	588.12	856.87	748.290
I4.5.8.07	as above; depth: 5 - 5.5m	m	0.89	217.83	139.17	588.12	945.12	758.078
I4.5.8.08	as above; depth: 5.5 - 6m	m	1.00	244.75	156.37	588.12	989.24	762.971
I4.7	**nominal bore: 1500 - 1800mm**						
I4.7.1	Not in trenches						
I4.7.1.01	Steel pipes, BS EN 10216, ends bevelled; 9 m average length; welded joints, internal and external grit blasting; external coating on one coat bituminous primer; three coats bituminous enamel; inner wrap of fibreglass; outer wrap of bitumen impregnated fibreglass to BS 534, one coat of solar reflecting Vynamatt SP; internal lining of one coat bituminous primer, spun lining of bituminous enamel to BS 534, cold applied bitumen base tape on PVC carrier and primer to joint; 55% overlap; nominal bore 1595mm; above ground on pipe supports (supports priced elsewhere)	m	0.36	51.77	19.19	1,240.08	1,311.04	970.989
I4.7.2	In trenches, depth: not exceeding 1.5m						
I4.7.2.01	Steel pipes, BS EN 10216, ends bevelled; 9 m average length; welded joints, internal and external grit blasting; external coating on one coat bituminous primer; three coats bituminous enamel; inner wrap of fibreglass; outer wrap of bitumen impregnated fibreglass to BS 534, one coat of solar reflecting Vynamatt SP; internal lining of one coat bituminous primer, spun lining of bituminous enamel to BS 534, cold applied bitumen base tape on PVC carrier and primer to joint; 55% overlap; nominal bore 1595mm	m	0.21	40.08	26.27	1,240.13	1,306.48	971.170

I4 Steel Pipes continued...

		Unit	Labour Hours	Labour Net £	Plant Net £	Materials Net £	Unit Net £	CO_2e Kg
I4.7	**nominal bore: 1500 - 1800mm**							
I4.7.3	In trenches, depth: 1.5 - 2m							
I4.7.3.01	Steel pipes, BS EN 10216, ends bevelled; 9 m average length; welded joints, internal and external grit blasting; external coating on one coat bituminous primer; three coats bituminous enamel; inner wrap of fibreglass; outer wrap of bitumen impregnated fibreglass to BS 534, one coat of solar reflecting Vynamatt SP; internal lining of one coat bituminous primer, spun lining of bituminous enamel to BS 534, cold applied bitumen base tape on PVC carrier and primer to joint; 55% overlap; nominal bore 1595mm	m	0.28	53.44	35.03	1,240.13	1,328.60	973.955
I4.7.4	In trenches, depth: 2 - 2.5m							
I4.7.4.01	Steel pipes, BS EN 10216, ends bevelled; 9 m average length; welded joints, internal and external grit blasting; external coating on one coat bituminous primer; three coats bituminous enamel; inner wrap of fibreglass; outer wrap of bitumen impregnated fibreglass to BS 534, one coat of solar reflecting Vynamatt SP; internal lining of one coat bituminous primer, spun lining of bituminous enamel to BS 534, cold applied bitumen base tape on PVC carrier and primer to joint; 55% overlap; nominal bore 1595mm	m	0.36	68.71	45.03	1,240.13	1,353.87	977.137
I4.7.5	In trenches, depth: 2.5 - 3m							
I4.7.5.01	Steel pipes, BS EN 10216, ends bevelled; 9 m average length; welded joints, internal and external grit blasting; external coating on one coat bituminous primer; three coats bituminous enamel; inner wrap of fibreglass; outer wrap of bitumen impregnated fibreglass to BS 534, one coat of solar reflecting Vynamatt SP; internal lining of one coat bituminous primer, spun lining of bituminous enamel to BS 534, cold applied bitumen base tape on PVC carrier and primer to joint; 55% overlap; nominal bore 1595mm	m	0.44	83.02	54.41	1,240.13	1,377.56	980.121

I4 Steel Pipes continued...

	Unit	Labour Hours	Labour Net £	Plant Net £	Materials Net £	Unit Net £	CO_2e Kg
I4.7	**nominal bore: 1500 - 1800mm**						
I4.7.6	**In trenches, depth: 3 - 3.5m**						
I4.7.6.01	Steel pipes, BS EN 10216, ends bevelled; 9 m average length; welded joints, internal and external grit blasting; external coating on one coat bituminous primer; three coats bituminous enamel; inner wrap of fibreglass; outer wrap of bitumen impregnated fibreglass to BS 534, one coat of solar reflecting Vynamatt SP; internal lining of one coat bituminous primer, spun lining of bituminous enamel to BS 534, cold applied bitumen base tape on PVC carrier and primer to joint; 55% overlap; nominal bore 1595mm	m	0.44	108.67	69.43	1,240.13	1,418.23	982.570
I4.7.7	**In trenches, depth: 3.5 - 4m**						
I4.7.7.01	Steel pipes, BS EN 10216, ends bevelled; 9 m average length; welded joints, internal and external grit blasting; external coating on one coat bituminous primer; three coats bituminous enamel; inner wrap of fibreglass; outer wrap of bitumen impregnated fibreglass to BS 534, one coat of solar reflecting Vynamatt SP; internal lining of one coat bituminous primer, spun lining of bituminous enamel to BS 534, cold applied bitumen base tape on PVC carrier and primer to joint; 55% overlap; nominal bore 1595mm	m	0.53	129.72	82.88	1,240.13	1,452.73	986.397
I4.7.8	**In trenches, depth: exceeding 4m**						
I4.7.8.01	Steel pipes, BS EN 10216, ends bevelled; 9 m average length; welded joints, internal and external grit blasting; external coating on one coat bituminous primer; three coats bituminous enamel; inner wrap of fibreglass; outer wrap of bitumen impregnated fibreglass to BS 534, one coat of solar reflecting Vynamatt SP; internal lining of one coat bituminous primer, spun lining of bituminous enamel to BS 534, cold applied bitumen base tape on PVC carrier and primer to joint; 55% overlap; nominal bore 1595mm; depth: 4 - 4.5m	m	0.67	163.98	104.77	1,240.13	1,508.88	992.625
I4.7.8.02	as above; depth: 4.5 - 5m	m	0.89	217.83	139.17	1,240.13	1,597.13	1,002.413
I4.7.8.03	as above; depth: 5 - 5.5m	m	1.33	325.52	207.97	1,240.13	1,773.62	1,021.989
I4.7.8.04	as above; depth: 5.5 - 6m	m	1.60	391.60	250.19	1,240.13	1,881.92	1,034.001

I5 Polyvinyl Chloride Pipes

	Unit	Labour Hours	Labour Net £	Plant Net £	Materials Net £	Unit Net £	CO$_2$e Kg
I5.1	**nominal bore: not exceeding 200mm**						
I5.1.2	In trenches, depth: not exceeding 1.5m						
I5.1.2.01	Unplasticised PVC pipes; 6m lengths; compression joints with rubber ring; nominal bore 50mm	m	0.08	8.93	5.73	7.09	21.75	5.187
I5.1.2.02	as above; nominal bore 100mm	m	0.09	10.04	6.44	19.43	35.91	6.522
I5.1.2.03	as above; nominal bore 150mm	m	0.08	8.93	5.73	42.22	56.88	7.921
I5.1.3	In trenches, depth: 1.5 - 2m						
I5.1.3.01	Unplasticised PVC pipes; compression joints with rubber ring; nominal bore 50mm	m	0.09	10.04	6.44	7.09	23.57	5.450
I5.1.3.02	as above; nominal bore 100mm	m	0.10	11.16	7.16	19.43	37.75	6.785
I5.1.3.03	Unplasticised PVC pipes; 6m lengths; compression joints with rubber ring; nominal bore 150mm	m	0.11	12.28	7.88	42.22	62.38	8.710
I5.1.4	In trenches, depth: 2 - 2.5m						
I5.1.4.01	Unplasticised PVC pipes; 6m lengths; compression joints with rubber ring; nominal bore 50mm	m	0.10	15.87	7.16	7.09	30.12	5.713
I5.1.4.02	as above; nominal bore 100mm	m	0.11	12.28	7.88	19.43	39.59	7.048
I5.1.4.03	as above; nominal bore 150mm	m	0.13	14.51	9.31	42.22	66.04	9.235
I5.1.5	In trenches, depth: 2.5 - 3m						
I5.1.5.01	Unplasticised PVC pipes; compression joints with rubber ring; nominal bore 50mm	m	0.11	12.28	7.88	7.09	27.25	5.976
I5.1.5.02	as above; nominal bore 100mm	m	0.13	14.51	9.31	19.43	43.25	7.573
I5.1.5.03	as above; nominal bore 150mm	m	0.15	16.74	10.74	42.22	69.70	9.761
I5.1.6	In trenches, depth: 3 - 3.5m						
I5.1.6.01	Unplasticised PVC pipes; 6m lengths; compression joints with rubber ring; nominal bore 50mm	m	0.12	13.39	11.46	7.09	31.94	7.366
I5.1.6.02	as above; nominal bore 100mm	m	0.15	26.93	14.32	19.43	60.68	9.508
I5.1.6.03	as above; nominal bore 150mm	m	0.15	26.93	14.32	42.22	83.47	11.170
I5.1.7	In trenches, depth: 3.5 - 4m						
I5.1.7.01	Unplasticised PVC pipes; 6m lengths; compression joints with rubber ring; nominal bore 50mm	m	0.14	15.63	13.36	7.09	36.08	8.080
I5.1.7.02	as above; nominal bore 100mm	m	0.18	32.32	17.18	19.43	68.93	10.579
I5.1.7.03	as above; nominal bore 150mm	m	0.18	32.32	17.18	42.22	91.72	12.241
I5.1.8	In trenches, depth: exceeding 4m						
I5.1.8.01	Unplasticised PVC pipes; 6m lengths; compression joints with rubber ring; nominal bore 50mm; depth: 4 - 4.5m	m	0.18	20.09	17.18	7.09	44.36	9.507
I5.1.8.02	as above; depth: 4.5 - 5m	m	0.22	24.55	21.00	7.09	52.64	10.934
I5.1.8.03	as above; nominal bore 100mm; depth: 4 - 4.5m	m	0.22	39.50	21.00	19.43	79.93	12.006
I5.1.8.04	as above; depth: 4.5 - 5m	m	0.25	44.88	23.86	19.43	88.17	13.077
I5.1.8.05	as above; nominal bore 150mm; depth: 4 - 4.5m	m	0.22	39.50	21.00	42.22	102.72	13.668
I5.1.8.06	as above; depth: 4.5 - 5m	m	0.30	53.86	28.64	42.22	124.72	16.523

I5 Polyvinyl Chloride Pipes continued...

	Unit	Labour Hours	Labour Net £	Plant Net £	Materials Net £	Unit Net £	CO_2e Kg
I5.2	**nominal bore: 200 - 300mm**						
I5.2.2	In trenches, depth: not exceeding 1.5m						
I5.2.2.01	Unplasticised PVC pipes; 6m lengths; compression joints with rubber ring; nominal bore 300mm	m	0.11	12.28	7.88	145.40	165.56	20.997
I5.2.3	In trenches, depth: 1.5 - 2m						
I5.2.3.01	Unplasticised PVC pipes; 6m lengths; compression joints with rubber ring; nominal bore 300mm	m	0.13	14.51	9.31	145.40	169.22	21.522
I5.2.4	In trenches, depth: 2 - 2.5m						
I5.2.4.01	Unplasticised PVC pipes; 6m lengths; compression joints with rubber ring; nominal bore 300mm	m	0.15	16.74	10.74	145.40	172.88	22.048
I5.2.5	In trenches, depth: 2.5 - 3m						
I5.2.5.01	Unplasticised PVC pipes; 6m lengths; compression joints with rubber ring; nominal bore 300mm	m	0.18	20.09	12.89	145.40	178.38	22.837
I5.2.6	In trenches, depth: 3 - 3.5m						
I5.2.6.01	Unplasticised PVC pipes; 6m lengths; compression joints with rubber ring; nominal bore 300mm	m	0.18	32.32	17.18	145.40	194.90	24.528
I5.2.7	In trenches, depth: 3.5 - 4m						
I5.2.7.01	Unplasticised PVC pipes; 6m lengths; compression joints with rubber ring; nominal bore 300mm	m	0.22	39.50	21.00	145.40	205.90	25.955
I5.2.8	In trenches, depth: exceeding 4m						
I5.2.8.01	Unplasticised PVC pipes; 6m lengths; compression joints with rubber ring; nominal bore 300mm; depth: 4 - 4.5m	m	0.30	53.86	28.64	145.40	227.90	28.810
I5.2.8.02	as above; depth: 4.5 - 5m	m	0.44	79.00	42.00	145.40	266.40	33.806

I8 Medium Density Polyethylene Pipes

	Unit	Labour Hours	Labour Net £	Plant Net £	Materials Net £	Unit Net £	CO_2e Kg
I8.1	**nominal bore: not exceeding 200mm**						
I8.1.2	In trenches, depth: not exceeding 1.5m						
I8.1.2.01	Blue MDPE (SDR 11) water supply pipe systems, 6m lengths, butt welded joints; Outside diameter 90 mm	m	0.06	6.70	7.30	5.11	19.11	5.667
I8.1.2.02	as above; Outside diameter 125mm	m	0.07	7.81	8.52	9.80	26.13	8.813
I8.1.2.03	as above; Outside diameter 160mm	m	0.08	8.93	9.73	16.00	34.66	12.833
I8.1.2.04	as above; Outside diameter 180mm	m	0.09	10.04	10.95	20.27	41.26	15.721
I8.1.3	In trenches, depth: 1.5 - 2m						
I8.1.3.01	Blue MDPE (SDR 11) water supply pipe systems, 6m lengths, butt welded joints; Outside diameter 90 mm	m	0.07	7.81	8.52	5.11	21.44	6.130
I8.1.3.02	as above; Outside diameter 125mm	m	0.09	10.04	10.95	9.80	30.79	9.739

18 Medium Density Polyethylene Pipes continued...

	Unit	Labour Hours	Labour Net £	Plant Net £	Materials Net £	Unit Net £	CO_2e Kg
18.1	**nominal bore: not exceeding 200mm**						
18.1.3	In trenches, depth: 1.5 - 2m						
18.1.3.03	as previous item; Outside diameter 160mm	m	0.11	12.28	13.38	16.00	41.66	14.222
18.1.3.04	as above; Outside diameter 180mm	m	0.11	12.28	13.38	20.27	45.93	16.647
18.1.4	In trenches, depth: 2 - 2.5m						
18.1.4.01	Blue MDPE (SDR 11) water supply pipe systems, 6m lengths, butt welded joints; Outside diameter 90 mm	m	0.10	11.16	12.17	5.11	28.44	7.519
18.1.4.02	as above; Outside diameter 125mm	m	0.12	13.39	14.60	9.80	37.79	11.128
18.1.4.03	as above; Outside diameter 160mm	m	0.13	14.51	15.81	16.00	46.32	15.148
18.1.4.04	as above; Outside diameter 180mm	m	0.13	14.51	15.81	20.27	50.59	17.573
18.1.5	In trenches, depth: 2.5 - 3m						
18.1.5.01	Blue MDPE (SDR 11) water supply pipe systems, 6m lengths, butt welded joints; Outside diameter 90 mm	m	0.12	13.39	14.60	5.11	33.10	8.445
18.1.5.02	as above; Outside diameter 125mm	m	0.14	15.63	17.03	9.80	42.46	12.054
18.1.5.03	as above; Outside diameter 160mm	m	0.15	16.74	18.25	16.00	50.99	16.075
18.1.5.04	as above; Outside diameter 180mm	m	0.15	16.74	18.25	20.27	55.26	18.500
18.1.6	In trenches, depth: 3 - 3.5m						
18.1.6.01	Blue MDPE (SDR 11) water supply pipe systems, 6m lengths, butt welded joints; Outside diameter 90 mm	m	0.12	21.54	14.60	5.11	41.25	8.445
18.1.6.02	as above; Outside diameter 125mm	m	0.14	25.14	17.03	9.80	51.97	12.054
18.1.6.03	as above; Outside diameter 160mm	m	0.15	26.93	18.25	16.00	61.18	16.075
18.1.6.04	as above; Outside diameter 180mm	m	0.15	26.93	18.25	20.27	65.45	18.500
18.1.7	In trenches, depth: 3.5 - 4m						
18.1.7.01	Blue MDPE (SDR 11) water supply pipe systems, 6m lengths, butt welded joints; Outside diameter 90 mm	m	0.14	25.14	17.03	5.11	47.28	9.371
18.1.7.02	as above; Outside diameter 125mm	m	0.16	28.73	19.46	9.80	57.99	12.981
18.1.7.03	as above; Outside diameter 160mm	m	0.18	32.32	21.90	16.00	70.22	17.464
18.1.7.04	as above; Outside diameter 180mm	m	0.19	34.11	23.11	20.27	77.49	20.352
18.1.8	In trenches, depth: exceeding 4m						
18.1.8.01	Blue MDPE (SDR 11) water supply pipe systems, 6m lengths, butt welded joints; Outside diameter 90 mm; depth: 4 - 4.5m	m	0.17	30.52	20.68	5.11	56.31	10.761
18.1.8.02	as above; depth: 4.5 - 5m	m	0.21	37.70	25.55	5.11	68.36	12.613

I8   Medium Density Polyethylene Pipes continued...

	Unit	Labour Hours	Labour Net £	Plant Net £	Materials Net £	Unit Net £	CO_2e Kg
I8.1 **nominal bore: not exceeding 200mm**							
I8.1.8 In trenches, depth: exceeding 4m							
I8.1.8.03 as previous item; Outside diameter 125mm; depth: 4 - 4.5m	m	0.20	35.91	24.33	9.80	70.04	14.833
I8.1.8.04 as above; depth: 4.5 - 5m	m	0.26	46.68	31.63	9.80	88.11	17.612
I8.1.8.05 as above; Outside diameter 160mm; depth: 4 - 4.5m	m	0.23	41.29	27.98	16.00	85.27	19.779
I8.1.8.06 as above; depth: 4.5 - 5m	m	0.31	55.66	37.71	16.00	109.37	23.484
I8.1.8.07 as above; Outside diameter 180mm; depth: 4 - 4.5m	m	0.24	43.09	29.20	20.27	92.56	22.667
I8.1.8.08 as above; depth: 4.5 - 5m	m	0.33	59.25	40.14	20.27	119.66	26.835
I8.2 **nominal bore: 200 - 300mm**							
I8.2.2 In trenches, depth: not exceeding 1.5m							
I8.2.2.01 Blue MDPE (SDR 11) water supply pipe systems, 6m lengths, butt welded joints; Outside diameter 250 mm	m	0.10	11.16	12.17	38.94	62.27	26.917
I8.2.3 In trenches, depth: 1.5 - 2m							
I8.2.3.01 Blue MDPE (SDR 11) water supply pipe systems, 6m lengths, butt welded joints; Outside diameter 250 mm	m	0.12	13.39	14.60	38.94	66.93	27.843
I8.2.4 In trenches, depth: 2 - 2.5m							
I8.2.4.01 Blue MDPE (SDR 11) water supply pipe systems, 6m lengths, butt welded joints; Outside diameter 250 mm	m	0.14	15.63	17.03	38.94	71.60	28.769
I8.2.5 In trenches, depth: 2.5 - 3m							
I8.2.5.01 Blue MDPE (SDR 11) water supply pipe systems, 6m lengths, butt welded joints; Outside diameter 250 mm	m	0.17	18.97	20.68	38.94	78.59	30.159
I8.2.6 In trenches, depth: 3 - 3.5m							
I8.2.6.01 Blue MDPE (SDR 11) water supply pipe systems, 6m lengths, butt welded joints; Outside diameter 250 mm	m	0.17	30.52	20.68	38.94	90.14	30.159
I8.2.7 In trenches, depth: 3.5 - 4m							
I8.2.7.01 Blue MDPE (SDR 11) water supply pipe systems, 6m lengths, butt welded joints; Outside diameter 250 mm	m	0.21	37.70	25.55	38.94	102.19	32.011
I8.2.8 In trenches, depth: exceeding 4m							
I8.2.8.01 Blue MDPE (SDR 11) water supply pipe systems, 6m lengths, butt welded joints; Outside diameter 250 mm; depth: 4 - 4.5m	m	0.27	48.48	32.85	38.94	120.27	34.790
I8.2.8.02 as above; depth: 4.5 - 5m	m	0.39	70.02	47.44	38.94	156.40	40.347

I8   Medium Density Polyethylene Pipes continued...

	Unit	Labour Hours	Labour Net £	Plant Net £	Materials Net £	Unit Net £	CO_2e Kg
I8.3	**nominal bore: 300 - 600mm**						
I8.3.2	**In trenches, depth: not exceeding 1.5m**						
I8.3.2.01	Blue MDPE (SDR 11) water supply pipe systems, 6m lengths, butt welded joints; Outside diameter 315mm m	0.11	12.28	13.38	61.77	87.43	40.700
I8.3.2.02	as above; Outside diameter 335mm m	0.12	13.39	14.60	78.51	106.50	51.769
I8.3.2.03	as above; Outside diameter 400mm m	0.13	14.51	15.81	99.63	129.95	63.072
I8.3.3	**In trenches, depth: 1.5 - 2m**						
I8.3.3.01	Blue MDPE (SDR 11) water supply pipe systems, 6m lengths, butt welded joints; Outside diameter 315mm m	0.13	14.51	15.81	61.77	92.09	41.626
I8.3.3.02	as above; Outside diameter 335mm m	0.13	14.51	15.81	78.51	108.83	52.232
I8.3.3.03	as above; Outside diameter 400mm m	0.14	15.63	17.03	99.63	132.29	63.535
I8.3.4	**In trenches, depth: 2 - 2.5m**						
I8.3.4.01	Blue MDPE (SDR 11) water supply pipe systems, 6m lengths, butt welded joints; Outside diameter 315mm m	0.15	16.74	18.25	61.77	96.76	42.553
I8.3.4.02	as above; Outside diameter 335mm m	0.16	17.86	19.46	78.51	115.83	53.622
I8.3.4.03	as above; Outside diameter 400mm m	0.16	17.86	19.46	99.63	136.95	64.462
I8.3.5	**In trenches, depth: 2.5 - 3m**						
I8.3.5.01	Blue MDPE (SDR 11) water supply pipe systems, 6m lengths, butt welded joints; Outside diameter 315mm m	0.18	20.09	21.90	61.77	103.76	43.942
I8.3.5.02	as above; Outside diameter 335mm m	0.19	21.21	23.11	78.51	122.83	55.011
I8.3.5.03	as above; Outside diameter 400mm m	0.20	22.32	24.33	99.63	146.28	66.314
I8.3.6	**In trenches, depth: 3 - 3.5m**						
I8.3.6.01	Blue MDPE (SDR 11) water supply pipe systems, 6m lengths, butt welded joints; Outside diameter 315mm m	0.18	32.32	21.90	61.77	115.99	43.942
I8.3.6.02	as above; Outside diameter 335mm m	0.19	34.11	23.11	78.51	135.73	55.011
I8.3.6.03	as above; Outside diameter 400mm m	0.20	35.91	24.33	99.63	159.87	66.314
I8.3.7	**In trenches, depth: 3.5 - 4m**						
I8.3.7.01	Blue MDPE (SDR 11) water supply pipe systems, 6m lengths, butt welded joints; Outside diameter 315mm m	0.23	41.29	27.98	61.77	131.04	46.257
I8.3.7.02	as above; Outside diameter 335mm m	0.24	43.09	29.20	78.51	150.80	57.326
I8.3.7.03	as above; Outside diameter 400mm m	0.25	44.88	30.41	99.63	174.92	68.630

18   Medium Density Polyethylene Pipes continued...

	Unit	Labour Hours	Labour Net £	Plant Net £	Materials Net £	Unit Net £	CO_2e Kg
18.3	**nominal bore: 300 - 600mm**						
18.3.8	In trenches, depth: exceeding 4m						
18.3.8.01	Blue MDPE (SDR 11) water supply pipe systems, 6m lengths, butt welded joints; Outside diameter 315mm; depth: 4 - 4.5m	m	0.30	53.86	36.49	61.77	152.12	49.499
18.3.8.02	as above; depth: 4.5 - 5m	m	0.46	82.59	55.96	61.77	200.32	56.909
18.3.8.03	as above; Outside diameter 335mm; depth: 4 - 4.5m	m	0.32	57.45	38.93	78.51	174.89	61.031
18.3.8.04	as previous item; depth: 4.5 - 5m	m	0.50	89.77	60.83	78.51	229.11	69.367
18.3.8.05	as above; Outside diameter 400mm; depth: 4 - 4.5m	m	0.35	62.84	42.58	99.63	205.05	73.261
18.3.8.06	as above; depth: 4.5 - 5m	m	0.54	96.95	65.69	99.63	262.27	82.059

CLASS J:
PIPEWORK - FITTINGS AND VALVES

Calculations used throughout Class J - Pipework - Fittings and Valves

Labour

			Qty		Rate		Total
L A0183ICE	**Lay and joint pipe fitting Labour Gang**						
	L A9002ICE	Labourer (General Operative)	1	x	13.02	=	£13.02
	L A9017ICE	Pipelayer (standard rate)	1	x	14.86	=	£14.86
		Total hourly cost of gang				=	**£27.88**
L A0184ICE	**Small bore pipes in shallow trenches Labour Gang**						
	L A9015ICE	Banksman	1	x	14.03	=	£14.03
	L A9016ICE	Ganger	1	x	17.64	=	£17.64
	L A9017ICE	Pipelayer (standard rate)	1	x	14.86	=	£14.86
	L A9002ICE	Labourer (General Operative)	5	x	13.02	=	£65.08
		Total hourly cost of gang				=	**£111.61**
L A0186ICE	**Large bore pipes in shallow trench Labour Gang**						
	L A9016ICE	Ganger	1	x	17.64	=	£17.64
	L A9018ICE	Pipelayer (large pipes)	1	x	16.51	=	£16.51
	L A9002ICE	Labourer (General Operative)	5	x	13.02	=	£65.08
	L A9015ICE	Banksman	1	x	14.03	=	£14.03
		Total hourly cost of gang				=	**£113.26**
L A0188ICE	**Small bore steel pipes in shallow trench Labour Gang**						
	L A9016ICE	Ganger	1	x	17.64	=	£17.64
	L A9021ICE	Fitters and Welders	1	x	17.33	=	£17.33
	L A9003ICE	Labourer (Skill Rate 3)	3	x	14.86	=	£44.58
	L A9015ICE	Banksman	1	x	14.03	=	£14.03
	L A9002ICE	Labourer (General Operative)	5	x	13.02	=	£65.08
		Total hourly cost of gang				=	**£158.66**

Plant

			Qty		Rate		Total
P A1184ICE	**Small bore pipes in shallow trench Plant Gang**						
	P A0592ICE	Hydraulic Excavator - Cat 320 96kW	1	x	36.36	=	£36.36
	P A0811ICE	Pump - Godwin ET50 23m3/h 4 inches	1	x	2.94	=	£2.94
	P A0853ICE	Vibrating Plate Diesel 24kN	1	x	2.53	=	£2.53
	P A0803ICE	Trench Sheets	90	x	0.09	=	£8.10
	P A0804ICE	Acrow Props	70	x	0.09	=	£6.30
	P A1023ICE	Landrover 4WD	1	x	16.16	=	£16.16
		Total hourly cost of gang				=	**£72.39**
P A1186ICE	**Large bore in shallow trench Plant Gang**						
	P A0605ICE	Hydraulic Excavator - Cat 166kW	1	x	44.31	=	£44.31
	P A0811ICE	Pump - Godwin ET50 23m3/h 4 inches	1	x	2.94	=	£2.94
	P A0853ICE	Vibrating Plate Diesel 24kN	1	x	2.53	=	£2.53
	P A0803ICE	Trench Sheets	90	x	0.09	=	£8.10
	P A0804ICE	Acrow Props	70	x	0.09	=	£6.30
	P A0511ICE	Cranes Crawler - 22RB - 15t	1	x	41.06	=	£41.06
	P A1023ICE	Landrover 4WD	1	x	16.16	=	£16.16
		Total hourly cost of gang				=	**£121.40**

P A1188ICE   **Small bore steel pipes in shallow trench Plant Gang**

P A0592ICE	Hydraulic Excavator - Cat 320 96kW	1	x	36.36	=	£36.36
P A0811ICE	Pump - Godwin ET50 23m3/h 4 inches	1	x	2.94	=	£2.94
P A0853ICE	Vibrating Plate Diesel 24kN	1	x	2.53	=	£2.53
P A0803ICE	Trench Sheets	90	x	0.09	=	£8.10
P A0804ICE	Acrow Props	70	x	0.09	=	£6.30
P A1023ICE	Landrover 4WD	1	x	16.16	=	£16.16
P A0991ICE	Welding Set - 300 amp Diesel Electric Start Sil	1	x	4.48	=	£4.48
P A0946ICE	Crawler Tractor / Dozer - Cat 561 Sideboom	1	x	44.35	=	£44.35

Total hourly cost of gang   =   **£121.22**

Pipeworks - Fittings and Valves

Note(s):   The rates included under Class J are full value and the user must take this into account when pricing fittings and valves in conjunction with lengths of pipes in trenches measured under Class I Measurement Rule M3.

J1   Clay Pipe Fittings

		Unit	Labour Hours	Labour Net £	Plant Net £	Materials Net £	Unit Net £	CO_2e Kg
J1.1	**Bends**							
J1.1.1	nominal bore: not exceeding 200mm							
J1.1.1.01	Vitrified clay pipe fittings, BS 65, spigot & socket flexible joints; nominal bore 100mm; 45 degree	Nr	0.08	2.23	-	34.56	36.79	4.241
J1.1.1.02	as above; 11.25 degree bends	Nr	0.08	2.23	-	34.56	36.79	3.794
J1.1.1.03	as above; Rest bends	Nr	0.15	4.18	-	41.48	45.66	6.361
J1.1.1.04	as above; nominal bore 150mm; 45 degree bends	Nr	0.10	2.79	-	38.39	41.18	5.654
J1.1.1.05	as above; 11.25 degree bends	Nr	0.10	2.79	-	38.39	41.18	5.059
J1.1.1.06	as above; Rest bends	Nr	0.18	5.02	-	45.78	50.80	8.482
J1.1.1.07	Vitrified clay pipe fittings, BS 65, "extra strength" spigot and socket flexible joints; nominal bore 100mm; 45 degree bends	Nr	0.08	2.23	-	9.12	11.35	1.727
J1.1.1.08	as above; 11.25 degree bends	Nr	0.08	2.23	-	9.12	11.35	1.591
J1.1.1.09	as above; Rest bends	Nr	0.15	4.18	-	15.21	19.39	1.609
J1.1.1.10	as above; nominal bore 150mm; 45 degree bends	Nr	0.10	2.79	-	23.09	25.88	3.682
J1.1.1.11	as above; 11.25 degree bends	Nr	0.10	2.79	-	23.09	25.88	2.802
J1.1.1.12	as above; Rest bends	Nr	0.18	5.02	-	29.67	34.69	3.058
J1.1.1.13	as above; nominal bore 200mm; 45 degree bends	Nr	0.13	3.62	-	76.67	80.29	8.462
J1.1.1.14	as above; 11.25 degree bends	Nr	0.13	3.62	-	76.67	80.29	8.293
J1.1.1.15	Vitrified clay pipe fittings, BS 65, "surface water" spigot and socket cement joints; nominal bore 100mm; 45 degree bends	Nr	0.25	6.97	-	22.74	29.71	3.502
J1.1.1.16	as above; 11.25 degree bends	Nr	0.25	6.97	-	22.76	29.73	3.502
J1.1.1.17	as above; nominal bore 150mm; 45 degree bends	Nr	0.17	4.74	-	22.74	27.48	3.502
J1.1.1.18	as above; 11.25 degree bends	Nr	0.17	4.74	-	22.74	27.48	4.205
J1.1.1.19	Vitrified clay pipe fittings, BS 65, "British Standard Tested" quality; spigot and socket cement joints; nominal bore 100mm; 45 degree bends	Nr	0.25	6.97	-	4.89	11.86	2.334
J1.1.1.20	as above; 11.25 degree bends	Nr	0.25	6.97	-	4.89	11.86	2.334
J1.1.1.21	as above; nominal bore 150mm; 45 degree bends	Nr	0.06	1.67	-	8.74	10.41	3.502
J1.1.1.22	as above; 11.25 degree bends	Nr	0.06	1.67	-	22.74	24.41	3.502
J1.1.1.23	Vitrified clay pipe fittings, BS 65, "Perforated"; spigot and socket flexible joints; nominal bore 100mm; 45 degree bends	Nr	0.08	2.23	-	13.31	15.54	2.334
J1.1.1.24	as above; 22.5 degree bends	Nr	0.10	2.79	-	13.31	16.10	2.334
J1.1.1.25	as above; nominal bore 150mm; 45 degree bends	Nr	0.12	3.35	-	22.74	26.09	3.502
J1.1.1.26	as above; 22.5 degree bends	Nr	0.18	5.02	-	22.74	27.76	3.502

J1 Clay Pipe Fittings continued...

	Unit	Labour Hours	Labour Net £	Plant Net £	Materials Net £	Unit Net £	CO_2e Kg
J1.1	**Bends**						
J1.1.2	**nominal bore: 200 - 300mm**						
J1.1.2.01	Vitrified clay pipe fittings, BS 65, spigot & socket flexible joints; nominal bore 225mm; 45 degree	Nr	0.14	3.90	-	78.29	82.19	10.291
J1.1.2.02	as above; Rest bends	Nr	0.25	6.97	-	109.08	116.05	12.266
J1.1.2.03	as above; nominal bore 300mm; 45 degree bends	Nr	0.18	5.02	-	154.04	159.06	23.108
J1.1.2.04	as above; 22.5 degree bends	Nr	0.18	5.02	-	154.04	159.06	18.207
J1.1.2.05	Vitrified clay pipe fittings, BS 65, "extra strength" spigot and socket flexible joints; nominal bore 225mm; 45 degree bends	Nr	0.23	6.41	-	369.98	376.39	33.088
J1.1.2.06	as above; 22.5 degree bends	Nr	0.25	6.97	-	369.98	376.95	37.290
J1.1.2.07	as above; 45 degree bends	Nr	0.30	8.36	-	78.29	86.65	10.291
J1.1.2.08	as above; 11.25 degree bends	Nr	0.30	8.36	-	78.29	86.65	8.940
J1.1.2.09	as above; Rest bends	Nr	0.13	3.62	-	109.08	112.70	12.266
J1.1.2.10	Vitrified clay pipe fittings, BS 65, "surface water" spigot and socket cement joints; nominal bore 300mm; 45 degree bends	Nr	0.13	3.62	-	129.59	133.21	5.836
J1.1.2.11	as above; 11.25 degree bends	Nr	0.14	3.90	-	129.59	133.49	7.008
J1.1.2.12	as above; Rest bends	Nr	0.14	3.90	-	303.70	307.60	9.202
J1.1.2.13	Vitrified clay pipe fittings, BS 65, "British Standard Tested" quality; spigot and socket cement joints; nominal bore 225mm; 45 degree bends	Nr	0.35	9.76	-	71.31	81.07	4.669
J1.1.2.14	as above; 11.25 degree bends	Nr	0.35	9.76	-	71.31	81.07	4.669
J1.1.2.15	as above; Rest bends	Nr	0.45	12.55	-	40.18	52.73	7.256
J1.1.2.16	as above; nominal bore 300mm; 45 degree bends	Nr	0.50	13.94	-	50.14	64.08	5.836
J1.1.2.17	as above; 11.25 degree bends	Nr	0.50	13.94	-	146.82	160.76	5.836
J1.1.2.18	as above; Rest bends	Nr	0.65	18.12	-	187.87	205.99	9.202
J1.1.2.19	Vitrified clay pipe fittings, BS 65, "Perforated"; spigot and socket flexible joints; nominal bore 225mm; 45 degree bends	Nr	1.00	27.88	-	71.31	99.19	4.669
J1.1.2.20	as above; 22.5 degree bends	Nr	0.75	20.91	-	71.31	92.22	4.669
J1.1.2.21	as above; nominal bore 300mm; 45 degree bends	Nr	0.60	16.73	-	129.59	146.32	5.836
J1.1.2.22	as above; 22.5 degree bends	Nr	0.17	4.74	-	129.59	134.33	5.836
J1.1.2.23	Vitrified clay pipe fittings, BS 65, spigot & socket flexible joints; nominal bore 225mm; 22.5 degree	Nr	0.14	3.90	-	78.29	82.19	8.940
J1.1.3	**nominal bore: 300 - 600mm**						
J1.1.3.01	Vitrified clay pipe fittings, BS 65, spigot & socket flexible joints; nominal bore 375mm; 45 degree	Nr	0.23	6.41	-	369.98	376.39	33.088
J1.1.3.02	as above; 22.5 degree bends	Nr	0.23	6.41	-	369.98	376.39	37.290
J1.1.3.03	Vitrified clay pipe fittings, BS 65, spigot & socket flexible joints; nominal bore 450mm; 45 degree	Nr	0.30	8.36	-	551.09	559.45	60.508
J1.1.3.04	as above; 22.5 degree bends	Nr	0.30	8.36	-	551.09	559.45	49.402

J1 Clay Pipe Fittings continued...

	Unit	Labour Hours	Labour Net £	Plant Net £	Materials Net £	Unit Net £	CO$_2$e Kg
J1.1	**Bends**						
J1.1.3	nominal bore: 300 - 600mm						
J1.1.3.05	Vitrified clay pipe fittings, BS 65, "surface water" spigot and socket cement joints; nominal bore 375mm; 45 degree bends	Nr	0.65	18.12	-	364.19	382.31	6.420
J1.1.3.06	as above; 11.25 degree bends	Nr	0.65	18.12	-	371.89	390.01	6.420
J1.1.3.07	as above; nominal bore 450mm; 45 degree bends	Nr	0.75	20.91	-	595.14	616.05	6.741
J1.1.3.08	as above; 11.25 degree bends	Nr	0.75	20.91	-	595.14	616.05	10.785
J1.1.3.09	Vitrified clay pipe fittings, BS 65, "British Standard Tested" quality; spigot and socket cement joints; nominal bore 375mm; 45 degree bends	Nr	0.65	18.12	-	129.59	147.71	6.420
J1.1.3.10	as above; 11.25 degree bends	Nr	0.65	18.12	-	129.59	147.71	6.420
J1.1.3.11	as above; nominal bore 450mm; 45 degree bends	Nr	0.75	20.91	-	173.48	194.39	6.741
J1.1.3.12	as above; 11.25 degree bends	Nr	0.75	20.91	-	261.86	282.77	10.785
J1.2	**Junctions and Branches**						
J1.2.1	nominal bore: not exceeding 200mm						
J1.2.1.01	Vitrified clay pipe fittings, BS 65, spigot & socket flexible joints; nominal bore 100mm; Single	Nr	0.15	4.18	-	34.56	38.74	5.859
J1.2.1.02	as above; Double junctions (special)	Nr	0.20	5.58	-	57.72	63.30	6.696
J1.2.1.03	as above; Oblique saddles	Nr	0.75	20.91	-	38.23	59.14	6.696
J1.2.1.04	as above; nominal bore 150mm; Single junctions	Nr	0.18	5.02	-	50.17	55.19	7.812
J1.2.1.05	as above; Double junctions (special)	Nr	0.25	6.97	-	134.04	141.01	8.928
J1.2.1.06	as above; Oblique saddles	Nr	1.00	27.88	-	38.11	65.99	5.208
J1.2.1.07	Vitrified clay pipe fittings, BS 65, "extra strength" spigot and socket flexible joints; nominal bore 100mm; Single junctions	Nr	0.15	4.18	-	19.34	23.52	2.818
J1.2.1.08	as above; nominal bore 150mm; Single junctions	Nr	0.18	5.02	-	33.83	38.85	6.404
J1.2.1.09	Vitrified clay pipe fittings, BS 65, "surface water" spigot and socket cement joints; nominal bore 100mm; Single junctions	Nr	0.25	6.97	-	27.06	34.03	2.803
J1.2.1.10	Vitrified clay pipe fittings, BS 65, "extra strength" spigot and socket flexible joints; nominal bore 150mm; Double junctions (special)	Nr	0.25	6.97	-	48.32	55.29	3.735
J1.2.1.11	as above; Oblique saddles	Nr	0.33	9.20	-	24.11	33.31	3.630
J1.2.1.12	Vitrified clay pipe fittings, BS 65, "surface water" spigot and socket cement joints; nominal bore 150mm; 11.25 degree bends	Nr	0.75	20.91	-	44.79	65.70	5.603
J1.2.1.13	as above; Rest bends	Nr	0.35	9.76	-	110.95	120.71	5.603
J1.2.1.14	as above; 45 degree bends	Nr	0.45	12.55	-	19.99	32.54	5.445

J1 Clay Pipe Fittings continued...

	Unit	Labour Hours	Labour Net £	Plant Net £	Materials Net £	Unit Net £	CO$_2$e Kg
J1.2	**Junctions and Branches**						
J1.2.1	nominal bore: not exceeding 200mm						
J1.2.1.15	Vitrified clay pipe fittings, BS 65, "British Standard Tested" quality; spigot and socket cement joints; nominal bore 100mm; Single junctions	Nr	1.00	27.88	-	11.71	39.59	2.803
J1.2.1.16	as above; Double junctions (special)	Nr	0.25	6.97	-	22.20	29.17	3.735
J1.2.1.17	as above; Oblique saddles	Nr	0.33	9.20	-	11.71	20.91	3.630
J1.2.1.18	as above; nominal bore 150mm; Single junctions	Nr	0.75	20.91	-	19.99	40.90	4.205
J1.2.1.19	as above; Double junctions (special)	Nr	0.35	9.76	-	37.53	47.29	5.603
J1.2.1.20	as above; Oblique saddles	Nr	0.45	12.55	-	19.99	32.54	5.445
J1.2.1.21	Vitrified clay pipe fittings, BS 65, "Perforated"; spigot and socket flexible joints; nominal bore 100mm; Square junctions	Nr	0.13	3.62	-	24.43	28.05	3.735
J1.2.1.22	as above; Oblique junctions	Nr	0.12	3.35	-	24.43	27.78	3.630
J1.2.1.23	as above; nominal bore 150mm; Oblique junctions	Nr	0.65	18.12	-	41.78	59.90	5.445
J1.2.1.24	Unglazed clay pipe fittings, BS 1196, plain butt joints; nominal bore 75mm; Single junctions	Nr	0.08	2.23	-	14.66	16.89	2.801
J1.2.1.25	as above; nominal bore 100mm; Single junctions	Nr	0.13	3.62	-	19.47	23.09	3.735
J1.2.1.26	as above; nominal bore 150mm; Single junctions	Nr	0.25	6.97	-	23.98	30.95	5.603
J1.2.2	nominal bore: 200 - 300mm						
J1.2.2.01	Unglazed clay pipe fittings, BS 1196, plain butt joints; nominal bore 225mm; Single junctions	Nr	0.10	2.79	-	63.00	65.79	7.470
J1.2.2.02	Vitrified clay pipe fittings, BS 65, spigot & socket flexible joints; nominal bore 225mm; Single	Nr	0.25	6.97	-	117.78	124.75	17.359
J1.2.2.03	as above; Double junctions (special)	Nr	0.35	9.76	-	289.89	299.65	18.919
J1.2.2.04	as above; Oblique saddles	Nr	1.50	41.82	-	91.12	132.94	7.276
J1.2.2.05	as above; nominal bore 300mm; Single junctions	Nr	0.35	9.76	-	242.78	252.54	38.514
J1.2.2.06	as above; Double junctions (special)	Nr	0.50	13.94	-	584.59	598.53	45.516
J1.2.2.07	as above; Oblique saddles	Nr	2.00	55.76	-	238.76	294.52	12.254
J1.2.2.08	Vitrified clay pipe fittings, BS 65, "extra strength" spigot and socket flexible joints; nominal bore 225mm; Single junctions	Nr	0.45	12.55	-	519.67	532.22	65.651
J1.2.2.09	Vitrified clay pipe fittings, BS 65, "surface water" spigot and socket cement joints; nominal bore 225mm; Single junctions	Nr	0.50	13.94	-	124.44	138.38	5.606
J1.2.2.10	as above; Double junctions (special)	Nr	0.60	16.73	-	289.89	306.62	18.919
J1.2.2.11	as above; Oblique saddles	Nr	0.25	6.97	-	91.12	98.09	7.276
J1.2.2.12	as above; nominal bore 300mm; Single junctions	Nr	0.50	13.94	-	124.44	138.38	5.606

J1 Clay Pipe Fittings continued...

		Unit	Labour Hours	Labour Net £	Plant Net £	Materials Net £	Unit Net £	CO$_2$e Kg
J1.2	**Junctions and Branches**							
J1.2.2	nominal bore: 200 - 300mm							
J1.2.2.13	as previous item; Double junctions (special)	Nr	0.60	16.73	-	255.18	271.91	7.470
J1.2.2.14	as above; Oblique saddles	Nr	1.50	41.82	-	127.80	169.62	7.256
J1.2.2.15	Vitrified clay pipe fittings, BS 65, "British Standard Tested" quality; spigot and socket cement joints; nominal bore 225mm; Single junctions	Nr	0.65	18.12	-	55.87	73.99	5.606
J1.2.2.16	as above; Double junctions (special)	Nr	0.80	22.30	-	114.22	136.52	7.470
J1.2.2.17	as above; Oblique saddles	Nr	2.00	55.76	-	96.29	152.05	7.362
J1.2.2.18	as previous item; nominal bore 300mm; Single junctions	Nr	0.85	23.70	-	146.82	170.52	7.008
J1.2.2.19	as above; Double junctions (special)	Nr	1.00	27.88	-	185.11	212.99	9.075
J1.2.2.20	as above; Oblique saddles	Nr	2.50	69.70	-	171.62	241.32	9.070
J1.2.2.21	Vitrified clay pipe fittings, BS 65, "Perforated"; spigot and socket flexible joints; nominal bore 225mm; Square junctions	Nr	0.75	20.91	-	95.18	116.09	7.470
J1.2.2.22	as above; Oblique junctions	Nr	0.50	13.94	-	95.18	109.12	7.260
J1.2.2.23	as above; nominal bore 300mm; Square junctions	Nr	0.24	6.69	-	95.18	101.87	7.470
J1.2.2.24	as above; Oblique junctions	Nr	0.24	6.69	-	95.18	101.87	7.260
J1.2.3	nominal bore: 300 - 600mm							
J1.2.3.01	Vitrified clay pipe fittings, BS 65, "surface water" spigot and socket cement joints; nominal bore 375mm; Single junctions	Nr	0.85	23.70	-	364.19	387.89	9.812
J1.2.3.02	as above; Double junctions (special)	Nr	1.00	27.88	-	768.99	796.87	10.442
J1.2.3.03	as above; Oblique saddles	Nr	2.50	69.70	-	461.78	531.48	9.977
J1.2.3.04	Vitrified clay pipe fittings, BS 65, spigot & socket flexible joints; nominal bore 375mm; Single	Nr	0.45	12.55	-	519.67	532.22	65.651
J1.2.3.05	as above; Double junctions (special)	Nr	0.65	18.12	-	825.24	843.36	78.782
J1.2.3.06	as above; nominal bore 450mm; Single junctions	Nr	0.60	16.73	-	660.59	677.32	107.229
J1.2.3.07	Vitrified clay pipe fittings, BS 65, "surface water" spigot and socket cement joints; nominal bore 450mm; Single junctions	Nr	1.00	27.88	-	651.79	679.67	10.481
J1.2.3.08	as above; nominal bore 450mm; Oblique saddles	Nr	3.00	83.64	-	651.79	735.43	10.476
J1.2.3.09	Vitrified clay pipe fittings, BS 65, "British Standard Tested" quality; spigot and socket cement joints; nominal bore 375mm; Single junctions	Nr	0.85	23.70	-	157.59	181.29	9.812
J1.2.3.10	as above; Double junctions (special)	Nr	1.00	27.88	-	207.68	235.56	10.442
J1.2.3.11	as above; Oblique saddles	Nr	2.50	69.70	-	158.94	228.64	9.977
J1.2.3.12	as above; nominal bore 450mm; Single junctions	Nr	1.00	27.88	-	286.78	314.66	10.481
J1.2.3.13	as above; Oblique saddles	Nr	3.00	83.64	-	286.78	370.42	10.476

J1   Clay Pipe Fittings continued...

	Unit	Labour Hours	Labour Net £	Plant Net £	Materials Net £	Unit Net £	CO$_2$e Kg
J1.3 **Tapers**							
J1.3.1 nominal bore: not exceeding 200mm							
J1.3.1.01 Vitrified clay pipe fittings, BS 65, spigot & socket flexible joints; nominal bore 150mm; Tapers	Nr	0.10	2.79	-	90.58	93.37	4.464
J1.3.1.02 Vitrified clay pipe fittings, BS 65, "extra strength" spigot and socket flexible joints; nominal bore 150mm;	Nr	0.10	2.79	-	77.91	80.70	3.042
J1.3.1.03 as above; nominal bore 200mm; Tapers	Nr	0.13	3.62	-	84.77	88.39	6.431
J1.3.1.04 Vitrified clay pipe fittings, BS 65, "surface water" spigot and socket cement joints; nominal bore 150mm; Tapers	Nr	0.25	6.97	-	26.20	33.17	5.603
J1.3.1.05 Vitrified clay pipe fittings, BS 65, "British Standard Tested" quality; spigot and socket cement joints; nominal bore 150mm; Tapers	Nr	0.25	6.97	-	26.20	33.17	5.603
J1.3.2 nominal bore: 200 - 300mm							
J1.3.2.01 Vitrified clay pipe fittings, BS 65, spigot & socket flexible joints; nominal bore 225mm; Tapers	Nr	0.14	3.90	-	303.38	307.28	9.355
J1.3.2.02 as above; nominal bore 300mm; Tapers	Nr	0.18	5.02	-	288.09	293.11	17.506
J1.3.2.03 Vitrified clay pipe fittings, BS 65, "extra strength" spigot and socket flexible joints; nominal bore 225mm;	Nr	0.23	6.41	-	369.06	375.47	28.887
J1.3.2.04 Vitrified clay pipe fittings, BS 65, "surface water" spigot and socket cement joints; nominal bore 225mm; Tapers	Nr	0.13	3.62	-	303.38	307.00	9.355
J1.3.2.05 as above; nominal bore 300mm; Tapers	Nr	0.14	3.90	-	230.00	233.90	8.979
J1.3.2.06 Vitrified clay pipe fittings, BS 65, "British Standard Tested" quality; spigot and socket cement joints; nominal bore 225mm; Tapers	Nr	0.35	9.76	-	65.26	75.02	7.260
J1.3.2.07 as above; nominal bore 300mm; Tapers	Nr	0.50	13.94	-	105.64	119.58	8.979
J1.3.3 nominal bore: 300 - 600mm							
J1.3.3.01 Vitrified clay pipe fittings, BS 65, spigot & socket flexible joints; nominal bore 375mm; Tapers	Nr	0.23	6.41	-	369.06	375.47	28.887
J1.3.3.02 Vitrified clay pipe fittings, BS 65, "surface water" spigot and socket cement joints; nominal bore 375mm; Tapers	Nr	0.65	18.12	-	367.60	385.72	9.877
J1.3.3.03 as above; nominal bore 450mm; Tapers	Nr	0.75	20.91	-	676.64	697.55	10.371
J1.3.3.04 Vitrified clay pipe fittings, BS 65, "British Standard Tested" quality; spigot and socket cement joints; nominal bore 375mm; Tapers	Nr	0.65	18.12	-	141.64	159.76	9.877
J1.3.3.05 as above; nominal bore 450mm; Tapers	Nr	0.75	20.91	-	297.71	318.62	10.371

J2   Concrete Pipe Fittings

	Unit	Labour Hours	Labour Net £	Plant Net £	Materials Net £	Unit Net £	CO_2e Kg
J2.1	**Bends**						
J2.1.2	**nominal bore: 200 - 300mm**						
J2.1.2.01	BS 5911, Class "120"; rebated flexible joints with mastic sealant to internal faces; nominal bore 300mm; Bends	Nr	0.30	8.36	-	129.45	137.81	14.908
J2.1.3	**nominal bore: 300 - 600mm**						
J2.1.3.01	BS 5911, Class "120"; rebated flexible joints with mastic sealant to internal faces; nominal bore 375mm; Bends	Nr	0.36	10.04	-	150.54	160.58	17.801
J2.1.3.02	as above; nominal bore 450mm; Bends	Nr	0.50	13.94	-	173.43	187.37	24.904
J2.1.3.03	as above; nominal bore 600mm; Bends	Nr	0.10	11.33	12.06	251.25	274.64	46.350
J2.1.4	**nominal bore: 600 - 900mm**						
J2.1.4.01	BS 5911, Class "120"; rebated flexible joints with mastic sealant to internal faces; nominal bore 750mm; Bends	Nr	0.13	14.72	15.68	819.06	849.46	92.542
J2.1.4.02	as above; nominal bore 900mm; Bends	Nr	0.20	22.65	24.12	1,099.95	1,146.72	125.846
J2.1.5	**nominal bore: 900 - 1200mm**						
J2.1.5.01	BS 5911, Class "120"; rebated flexible joints with mastic sealant to internal faces; nominal bore 1200mm; Bends	Nr	0.26	29.45	31.36	1,890.44	1,951.25	265.468
J2.1.6	**nominal bore: 1200 - 1500mm**						
J2.1.6.01	BS 5911, Class "120"; rebated flexible joints with mastic sealant to internal faces; nominal bore 1500mm; Bends	Nr	0.32	36.24	38.60	3,328.51	3,403.35	399.244
J2.1.7	**nominal bore: 1500 - 1800mm**						
J2.1.7.01	BS 5911, Class "120"; rebated flexible joints with mastic sealant to internal faces; nominal bore 1800mm; Bends	Nr	0.40	45.30	48.24	4,627.32	4,720.86	533.766
J2.2	**Junctions and Branches**						
J2.2.1	**nominal bore: not exceeding 200mm**						
J2.2.1.01	Concrete porous pipe fittings, BS 5911, Ogee joints; nominal bore 100mm; Single junctions	Nr	0.20	5.58	-	23.64	29.22	2.181
J2.2.1.02	Concrete porous pipe fittings, BS 5911, Ogee joints; nominal bore 150mm; Single junctions	Nr	0.30	8.36	-	34.71	43.07	2.566
J2.2.2	**nominal bore: 200 - 300mm**						
J2.2.2.01	BS 5911, Class "120"; rebated flexible joints with mastic sealant to internal faces; nominal bore 300mm; Single junction 150mm	Nr	0.45	12.55	-	84.70	97.25	7.702
J2.2.2.02	as above; Double junctions	Nr	0.55	15.33	-	259.76	275.09	14.908

J2 Concrete Pipe Fittings continued...

	Unit	Labour Hours	Labour Net £	Plant Net £	Materials Net £	Unit Net £	CO$_2$e Kg
J2.2	**Junctions and Branches**						
J2.2.3	nominal bore: 300 - 600mm						
J2.2.3.01	BS 5911, Class "120"; rebated flexible joints with mastic sealant to internal faces; nominal bore 375mm; Single junction 150mm	Nr	0.50	13.94	-	87.77	101.71	10.384
J2.2.3.02	as above; Double junctions	Nr	0.65	18.12	-	188.27	206.39	20.768
J2.2.3.03	as above; nominal bore 450mm; Single junction 150mm	Nr	0.75	20.91	-	110.20	131.11	14.943
J2.2.3.04	as above; nominal bore 600mm; Single junction 150mm	Nr	0.15	16.99	18.09	143.60	178.68	32.588
J2.2.4	nominal bore: 600 - 900mm						
J2.2.4.01	BS 5911, Class "120"; rebated flexible joints with mastic sealant to internal faces; nominal bore 750mm; Single junction 150mm	Nr	0.20	22.65	24.12	806.43	853.20	95.154

J3 Iron Or Steel Pipe Fittings

	Unit	Labour Hours	Labour Net £	Plant Net £	Materials Net £	Unit Net £	CO$_2$e Kg
J3.1	**Bends**						
J3.1.1	nominal bore: not exceeding 200mm						
J3.1.1.01	Ductile spun iron fittings, BS EN 598, spigot and socket joints; nominal bore 100mm; 22.5 degree bends	Nr	0.20	5.58	-	23.32	28.90	16.156
J3.1.1.02	as above; 45 degree bends	Nr	0.25	6.97	-	28.41	35.38	16.156
J3.1.1.03	as above; 90 degree bends	Nr	0.35	9.76	-	30.22	39.98	16.859
J3.1.1.04	as above; 90 degree flanged bends	Nr	0.60	16.73	-	35.08	51.81	16.859
J3.1.1.05	Ductile spun iron fittings, BS EN 598, spigot and socket joints; nominal bore 150mm; 22.5 degree bends	Nr	0.30	8.36	-	80.05	88.41	25.288
J3.1.1.06	as above; 45 degree bends	Nr	0.35	9.76	-	87.44	97.20	25.991
J3.1.1.07	as above; 90 degree bends	Nr	0.45	12.55	-	136.57	149.12	31.610
J3.1.1.08	as above; 90 degree double flanged bends	Nr	0.80	22.30	-	101.68	123.98	31.610
J3.1.1.09	Carbon steel pipe fittings, butt welded joints, BS 5135; Not in trenches; fixed to pipes on pipe supports; nominal bore 100mm; 45 degree elbows	Nr	0.28	44.42	33.72	11.72	89.86	16.888
J3.1.1.10	as above; 90 degree elbows	Nr	0.29	46.01	34.92	14.07	95.00	21.214
J3.1.1.11	as above; nominal bore 150mm; 45 degree elbows	Nr	0.42	66.64	50.58	28.95	146.17	30.603
J3.1.1.12	as above; 90 degree elbows	Nr	0.44	69.81	52.98	34.29	157.08	42.595
J3.1.2	nominal bore: 200 - 300mm						
J3.1.2.01	Ductile spun iron fittings, BS EN 598, spigot and socket joints; nominal bore 300mm; 22.5 degree bends	Nr	0.10	11.33	12.06	323.27	346.66	82.406
J3.1.2.02	as above; 45 degree bends	Nr	0.11	12.46	13.27	367.72	393.45	86.994
J3.1.2.03	as above; 90 degree bends	Nr	0.12	13.59	14.47	526.59	554.65	101.416
J3.1.2.04	as above; 90 degree flanged bends	Nr	0.21	23.78	25.33	436.49	485.60	104.774

J3 Iron Or Steel Pipe Fittings continued...

	Unit	Labour Hours	Labour Net £	Plant Net £	Materials Net £	Unit Net £	CO_2e Kg
J3.1	**Bends**						
J3.1.3	nominal bore: 300 - 600mm						
J3.1.3.01	Ductile spun iron fittings, BS EN 598, spigot and socket joints; nominal bore 450mm; 22.5 degree bends	Nr	0.18	20.39	21.71	1,008.97	1,051.07	307.365
J3.1.3.02	as above; 45 degree bends	Nr	0.20	22.65	24.12	1,098.93	1,145.70	305.301
J3.1.3.03	as above; 90 degree bends	Nr	0.22	24.92	26.53	1,368.33	1,419.78	185.227
J3.1.3.04	as above; 90 degree flanged bends	Nr	0.36	40.77	43.42	1,469.40	1,553.59	190.451
J3.1.3.05	as above; nominal bore 600mm; 22.5 degree bends	Nr	0.23	26.05	27.74	1,418.98	1,472.77	383.691
J3.1.3.06	as above; 45 degree bends	Nr	0.25	28.32	30.15	1,584.11	1,642.58	380.222
J3.1.3.07	as above; 90 degree bends	Nr	0.27	30.58	32.56	2,606.00	2,669.14	465.262
J3.1.3.08	as above; 90 degree flanged bends	Nr	0.47	53.23	56.69	2,654.11	2,764.03	472.726
J3.1.4	nominal bore: 600 - 900mm						
J3.1.4.01	Ductile spun iron fittings, BS EN 598, spigot and socket joints; nominal bore 700mm; 22.5 degree bends	Nr	0.28	31.71	33.77	2,736.92	2,802.40	490.923
J3.1.4.02	as above; 45 degree bends	Nr	0.30	33.98	36.18	3,921.52	3,991.68	481.836
J3.1.4.03	as above; 90 degree bends	Nr	0.33	37.38	39.80	5,185.67	5,262.85	606.586
J3.1.4.04	as above; 90 degree double flanged bends	Nr	0.57	64.56	68.75	2,828.60	2,961.91	615.542
J3.1.4.05	as above; nominal bore 800mm; 22.5 degree bends	Nr	0.34	38.51	41.01	3,175.77	3,255.29	642.081
J3.1.4.06	as previous item; 45 degree bends	Nr	0.37	41.91	44.63	4,532.82	4,619.36	637.582
J3.1.4.07	as above; 90 degree bends	Nr	0.41	46.44	49.45	9,805.35	9,901.24	817.497
J3.1.4.08	as above; 90 degree double flanged bends	Nr	0.70	79.28	84.43	5,493.84	5,657.55	828.318
J3.1.4.09	as above; nominal bore 900mm; 22.5 degree bends	Nr	0.40	45.30	48.24	6,138.57	6,232.11	841.006
J3.1.4.10	as above; 45 degree bends	Nr	0.43	48.70	51.86	7,191.14	7,291.70	825.267
J3.1.4.11	as above; 90 degree bends	Nr	0.48	54.36	57.89	11,134.57	11,246.82	1,071.585
J3.1.4.12	as above; 90 degree flanged bends	Nr	0.82	92.87	98.90	5,997.78	6,189.55	1,084.272
J3.1.5	nominal bore: 900 - 1200mm						
J3.1.5.01	Ductile spun iron fittings, BS EN 598, spigot and socket joints; nominal bore 1000mm; 22.5 degree bends	Nr	0.47	53.23	56.69	6,720.49	6,830.41	884.923
J3.1.5.02	as above; 45 degree bends	Nr	0.50	56.63	60.31	7,931.70	8,048.64	868.340
J3.1.5.03	as above; 90 degree bends	Nr	0.56	63.43	67.54	12,701.05	12,832.02	1,127.254
J3.1.5.04	as above; 90 degree flanged bends	Nr	0.96	108.73	115.79	6,724.54	6,949.06	1,142.180
J3.1.5.05	as above; nominal bore 1200mm; 22.5 degree bends	Nr	0.68	77.02	82.01	15,399.99	15,559.02	1,692.989
J3.1.5.06	as above; 45 degree bends	Nr	0.76	86.08	91.66	18,813.79	18,991.53	2,238.264
J3.1.5.07	as above; 90 degree bends	Nr	1.30	147.24	156.79	16,513.20	16,817.23	2,258.415
J3.1.5.08	as above; 90 degree double flanged bends	Nr	0.76	86.08	91.66	20,104.51	20,282.25	2,915.425
J3.2	**Junctions and Branches**						
J3.2.1	nominal bore: not exceeding 200mm						
J3.2.1.01	Ductile spun iron fittings, BS EN 598, spigot and socket joints; nominal bore 100mm; Equal tees	Nr	0.50	13.94	-	40.58	54.52	17.561
J3.2.1.02	as previous item; Flanged equal tees	Nr	1.00	27.88	-	86.51	114.39	26.693

J3 Iron Or Steel Pipe Fittings continued...

	Unit	Labour Hours	Labour Net £	Plant Net £	Materials Net £	Unit Net £	CO_2e Kg
J3.2	**Junctions and Branches**						
J3.2.1	nominal bore: not exceeding 200mm						
J3.2.1.03	as previous item; nominal bore 150mm; Equal tees	Nr	0.65	18.12	-	165.68	183.80	41.444
J3.2.1.04	as above; Flanged equal tees	Nr	1.40	39.03	-	159.94	198.97	41.444
J3.2.1.05	Carbon steel pipe fittings, butt welded joints, BS EN 10253; Not in trenches; fixed to pipes on pipe supports; nominal bore 100mm; Equal tees	Nr	0.38	60.29	45.76	60.60	166.65	29.263
J3.2.1.06	as above; nominal bore 150mm; Equal tees	Nr	0.57	90.44	68.64	133.48	292.56	59.706
J3.2.2	nominal bore: 200 - 300mm						
J3.2.2.01	Ductile spun iron fittings, BS EN 598, spigot and socket joints; nominal bore 300mm; Equal tees	Nr	0.12	13.59	14.47	557.26	585.32	123.894
J3.2.2.02	as above; Flanged equal tees	Nr	0.30	33.98	36.18	581.47	651.63	130.611
J3.2.3	nominal bore: 300 - 600mm						
J3.2.3.01	Ductile spun iron fittings, BS EN 598, spigot and socket joints; nominal bore 450mm; Equal tees	Nr	0.22	24.92	26.53	1,864.37	1,915.82	280.760
J3.2.3.02	as above; Flanged equal tees	Nr	0.45	50.97	54.27	2,142.20	2,247.44	289.342
J3.2.3.03	as above; nominal bore 600mm; Equal tees	Nr	0.27	30.58	32.56	3,394.31	3,457.45	638.064
J3.2.3.04	as above; Flanged equal tees	Nr	0.70	79.28	84.43	3,025.35	3,189.06	654.110
J3.2.4	nominal bore: 600 - 900mm						
J3.2.4.01	Ductile spun iron fittings, BS EN 598, spigot and socket joints; nominal bore 700mm; Equal tees	Nr	0.33	37.38	39.80	4,871.91	4,949.09	782.198
J3.2.4.02	as above; Flanged equal tees	Nr	0.85	96.27	102.52	4,368.49	4,567.28	801.603
J3.2.4.03	as above; nominal bore 800mm; Equal tees	Nr	0.41	46.44	49.45	6,220.52	6,316.41	823.678
J3.2.4.04	as above; Flanged equal tees	Nr	1.05	118.92	126.64	5,933.71	6,179.27	847.560
J3.2.4.05	as above; nominal bore 900mm; Equal tees	Nr	0.48	54.36	57.89	14,386.31	14,498.56	1,386.282
J3.2.4.06	as above; Flanged equal tees	Nr	1.23	139.31	148.35	11,737.76	12,025.42	1,414.269
J3.2.5	nominal bore: 900 - 1200mm						
J3.2.5.01	Ductile spun iron fittings, BS EN 598, spigot and socket joints; nominal bore 1000mm; Equal tees	Nr	0.56	63.43	67.54	14,472.15	14,603.12	1,457.686
J3.2.5.02	as above; Flanged equal tees	Nr	1.44	163.09	173.68	13,204.73	13,541.50	1,490.524
J3.2.5.03	as above; nominal bore 1200mm; Equal tees	Nr	1.95	220.86	120.61	20,104.51	20,445.98	2,924.381
J3.2.5.04	as above; Flanged equal tees	Nr	1.95	220.86	235.19	22,450.44	22,906.49	2,959.831
J3.3	**Tapers**						
J3.3.1	nominal bore: not exceeding 200mm						
J3.3.1.01	Carbon steel pipe fittings, butt welded joints, BS EN 10253; Not in trenches; fixed to pipes on pipe supports; nominal bore 100mm; Concentric reducers, 100 - 50mm	Nr	0.27	42.84	32.51	21.96	97.31	20.501
J3.3.1.02	as above; Eccentric reducers, 100 - 50mm	Nr	0.27	42.84	32.51	36.10	111.45	20.501

J3   Iron Or Steel Pipe Fittings continued...

		Unit	Labour Hours	Labour Net £	Plant Net £	Materials Net £	Unit Net £	CO_2e Kg
J3.3	**Tapers**							
J3.3.1	nominal bore: not exceeding 200mm							
J3.3.1.03	as previous item; nominal bore 150mm; Concentric reducers, 100 - 50mm	Nr	0.41	65.05	49.37	27.80	142.22	27.780
J3.3.1.04	as above; Eccentric reducers, 100 - 50mm	Nr	0.41	65.05	49.37	51.95	166.37	28.547
J3.5	**Adaptors**							
J3.5.1	nominal bore: not exceeding 200mm							
J3.5.1.01	Ductile spun iron fittings, BS EN 598, spigot and socket joints; nominal bore 100mm; Flange and socket pieces	Nr	0.40	11.15	-	53.21	64.36	18.615
J3.5.1.02	as above; Flange and spigot pieces	Nr	0.70	19.52	-	141.58	161.10	43.903
J3.5.1.03	as above; nominal bore 150mm; Flange and socket pieces	Nr	0.55	15.33	-	84.65	99.98	26.693
J3.5.1.04	as above; Flange and spigot pieces	Nr	0.95	26.49	-	178.83	205.32	55.494
J3.5.2	nominal bore: 200 - 300mm							
J3.5.2.01	Ductile spun iron fittings, BS EN 598, spigot and socket joints; nominal bore 300mm; Flange and socket pieces	Nr	0.20	22.65	24.12	259.70	306.47	70.683
J3.5.2.02	as above; Flange and spigot pieces	Nr	0.20	22.65	24.12	249.03	295.80	66.469
J3.5.3	nominal bore: 300 - 600mm							
J3.5.3.01	Ductile spun iron fittings, BS EN 598, spigot and socket joints; nominal bore 450mm; Flange and socket pieces	Nr	0.27	30.58	32.56	581.34	644.48	199.736
J3.5.3.02	as above; Flange and spigot pieces	Nr	0.27	30.58	32.56	526.56	589.70	199.736
J3.5.3.03	as above; nominal bore 600mm; Flange and socket pieces	Nr	0.35	39.64	42.21	996.03	1,077.88	350.236
J3.5.3.04	as above; Flange and spigot pieces	Nr	0.35	39.64	42.21	1,180.67	1,262.52	350.236
J3.5.4	nominal bore: 600 - 900mm							
J3.5.4.01	Ductile spun iron fittings, BS EN 598, spigot and socket joints; nominal bore 700mm; Flange and socket pieces	Nr	0.42	47.57	50.66	2,452.16	2,550.39	518.626
J3.5.4.02	as above; Flange and spigot pieces	Nr	0.42	47.57	50.66	1,629.19	1,727.42	518.626
J3.5.4.03	as previous item; nominal bore 800mm; Flange and socket pieces	Nr	0.52	58.90	62.72	4,482.14	4,603.76	547.505
J3.5.4.04	as above; Flange and spigot pieces	Nr	0.52	58.90	62.72	1,764.33	1,885.95	547.505
J3.5.4.05	as above; nominal bore 900mm; Flange and socket pieces	Nr	0.61	69.09	73.57	4,965.86	5,108.52	941.566
J3.5.4.06	as above; Flange and spigot pieces	Nr	0.61	69.09	73.57	3,261.36	3,404.02	941.566
J3.5.5	nominal bore: 900 - 1200mm							
J3.5.5.01	Ductile spun iron fittings, BS EN 598, spigot and socket joints; nominal bore 1000mm; Flange and socket pieces	Nr	0.71	80.41	85.63	5,453.07	5,619.11	991.237
J3.5.5.02	as above; Flange and spigot pieces	Nr	0.71	80.41	85.63	4,355.32	4,521.36	991.237
J3.5.5.03	as above; nominal bore 1200mm; Flange and socket pieces	Nr	0.96	108.73	115.79	7,869.65	8,094.17	1,250.940
J3.5.5.04	as above; Flange and spigot pieces	Nr	0.64	72.49	77.19	9,906.90	10,056.58	1,728.023

J4 Polyvinyl Chloride Pipe Fittings

	Unit	Labour Hours	Labour Net £	Plant Net £	Materials Net £	Unit Net £	CO_2e Kg
J4.1 **Bends**							
J4.1.1 nominal bore: not exceeding 200mm							
J4.1.1.01 Unplasticised PVC pipe fittings; solvent cement joints; nominal bore 50mm; 45 degree elbows	Nr	0.20	5.58	-	23.32	28.90	16.156
J4.1.1.02 as above90 degree elbows	Nr	0.40	11.15	-	53.21	64.36	18.615
J4.1.1.03 as abovenominal bore 100mm; 45 degree elbows	Nr	0.30	8.36	-	80.05	88.41	25.288
J4.1.1.04 as above; 90 degree elbows	Nr	0.55	15.33	-	84.65	99.98	26.693
J4.1.1.05 Unplasticised PVC pipe fittings; solvent cement joints; Not in trenches; fixed to pipes on pipe supports; nominal bore 50mm; 45 degree elbows	Nr	0.25	6.97	-	28.41	35.38	16.156
J4.1.1.06 as above; 90 degree elbows	Nr	0.35	9.76	-	30.22	39.98	16.859
J4.1.1.07 as above; nominal bore 100mm; 45 degree elbows	Nr	0.35	9.76	-	87.44	97.20	25.991
J4.1.1.08 as above; 90 degree elbows	Nr	0.45	12.55	-	136.57	149.12	31.610
J4.3 **Tapers**							
J4.3.1 nominal bore: not exceeding 200mm							
J4.3.1.01 Unplasticised PVC pipe fittings; solvent cement joints; nominal bore 50mm; Reducing sockets	Nr	0.60	16.73	-	35.08	51.81	16.859
J4.3.1.02 as above; nominal bore 100mm; Reducing sockets	Nr	0.80	22.30	-	101.68	123.98	31.610
J4.3.1.03 Unplasticised PVC pipe fittings; solvent cement joints; Not in trenches; fixed to pipes on pipe supports; nominal bore 50mm; Reducing sockets	Nr	0.40	11.15	-	55.91	67.06	17.210
J4.3.1.04 as above; nominal bore 100mm; Reducing sockets	Nr	0.55	15.33	-	64.82	80.15	24.586

J7 Medium Density Polyethylene Pipe Fittings

	Unit	Labour Hours	Labour Net £	Plant Net £	Materials Net £	Unit Net £	CO_2e Kg
J7.1 **Bends**							
J7.1.1 nominal bore: not exceeding 200mm							
J7.1.1.01 Blue MDPE (SDR 11) water supply pupped fittings; 90 mm outside diameter; 45 degree pupped elbow	Nr	0.14	15.63	-	27.76	43.39	1.169
J7.1.1.02 as above; 90 degree pupped elbow	Nr	0.14	15.63	-	29.36	44.99	1.169
J7.1.1.03 as above; 125 mm outside diameter; 45 degree pupped elbow	Nr	0.17	18.97	-	41.50	60.47	2.244
J7.1.1.04 as above; 90 degree pupped elbow	Nr	0.17	18.97	-	48.22	67.19	2.244
J7.1.1.05 as above; 180 mm outside diameter; 45 degree pupped elbow	Nr	0.23	25.67	-	85.11	110.78	4.641
J7.1.1.06 as previous item; 90 degree pupped elbow	Nr	0.23	25.67	-	130.82	156.49	4.641

J7   Medium Density Polyethylene Pipe Fittings continued...

	Unit	Labour Hours	Labour Net £	Plant Net £	Materials Net £	Unit Net £	CO_2e Kg
J7.1 **Bends**							
J7.1.2 nominal bore: 200 - 300mm							
J7.1.2.01 Blue MDPE (SDR 11) water supply pupped fittings; 250 mm outside diameter; 45 degree pupped elbow	Nr	0.30	33.48	-	128.72	162.20	8.909
J7.1.2.02 as above; 90 degree pupped elbow	Nr	0.30	33.48	-	192.44	225.92	8.909
J7.1.3 nominal bore: 300 - 600mm							
J7.1.3.01 Blue MDPE (SDR 11) water supply pupped fittings; 315 mm outside diameter; 45 degree pupped elbow	Nr	0.37	41.30	-	225.15	266.45	14.137
J7.1.3.02 as above; 90 degree pupped elbow	Nr	0.37	41.30	-	390.77	432.07	14.137
J7.2 **Junctions and Branches**							
J7.2.1 nominal bore: not exceeding 200mm							
J7.2.1.01 Blue MDPE (SDR 11) water supply pupped fittings; 90 mm outside diameter; 90 degree pupped equal tee	Nr	0.14	15.63	-	37.57	53.20	1.636
J7.2.1.02 as above; 90 degree pupped reducing branch tee (90 x 63)	Nr	0.14	15.63	-	55.11	70.74	1.636
J7.2.1.03 as above; 125 mm outside diameter; 90 degree pupped equal tee	Nr	0.17	18.97	-	69.60	88.57	3.141
J7.2.1.04 as above; 90 degree pupped reducing branch tee (125 x 90)	Nr	0.17	18.97	-	102.26	121.23	3.141
J7.2.1.05 as previous item; 180 mm outside diameter; 90 degree pupped equal tee	Nr	0.23	25.67	-	129.56	155.23	6.498
J7.2.1.06 as above; 90 degree pupped reducing branch tee (180 x 125)	Nr	0.23	25.67	-	183.68	209.35	6.498
J7.2.2 nominal bore: 200 - 300mm							
J7.2.2.01 Blue MDPE (SDR 11) water supply pupped fittings; 250 mm outside diameter; 90 degree pupped equal tee	Nr	0.30	33.48	-	234.26	267.74	12.473
J7.2.2.02 as above; 90 degree pupped reducing branch tee (250 x 180)	Nr	0.30	33.48	-	343.13	376.61	12.473
J7.2.3 nominal bore: 300 - 600mm							
J7.2.3.01 Blue MDPE (SDR 11) water supply pupped fittings; 315 mm outside diameter; 90 degree pupped equal tee	Nr	0.37	41.30	-	586.15	627.45	19.792
J7.2.3.02 as above; 90 degree pupped reducing branch tee (315 x 250)	Nr	0.37	41.30	-	677.85	719.15	19.792
J7.3 **Tapers**							
J7.3.1 nominal bore: not exceeding 200mm							
J7.3.1.01 Blue MDPE (SDR 11) water supply pupped fittings; 90 mm outside diameter; Pupped reducer (90 x 63)	Nr	0.14	15.63	-	22.80	38.43	1.536

J7 Medium Density Polyethylene Pipe Fittings continued...

		Unit	Labour Hours	Labour Net £	Plant Net £	Materials Net £	Unit Net £	CO₂e Kg

		Unit	Labour Hours	Labour Net £	Plant Net £	Materials Net £	Unit Net £	CO_2e Kg
J7.3	**Tapers**							
J7.3.1	nominal bore: not exceeding 200mm							
J7.3.1.02	as previous item; 125 mm outside diameter; Pupped reducer (125 x 90)	Nr	0.17	18.97	-	21.22	40.19	2.064
J7.3.1.03	as above; 180 mm outside diameter; Pupped reducer (180 x 125)	Nr	0.23	25.67	-	62.39	88.06	4.270
J7.3.2	nominal bore: 200 - 300mm							
J7.3.2.01	Blue MDPE (SDR 11) water supply pupped fittings; 250 mm outside diameter; Pupped reducer (250 x 180)	Nr	0.30	33.48	-	103.72	137.20	8.196
J7.3.3	nominal bore: 300 - 600mm							
J7.3.3.01	Blue MDPE (SDR 11) water supply pupped fittings; 315 mm outside diameter; Pupped reducer (315 x 250)	Nr	0.37	41.30	-	145.49	186.79	13.006

J8 Valves And Penstocks

		Unit	Labour Hours	Labour Net £	Plant Net £	Materials Net £	Unit Net £	CO_2e Kg
J8.1	**Gate valves: hand operated**							
J8.1.1	nominal bore: not exceeding 200mm							
J8.1.1.01	Flanged Ductile iron valves; right hand closing; nominal bore; 80mm; with cap	Nr	0.15	16.74	10.74	146.14	173.62	34.851
J8.1.1.02	as above; with handwheel	Nr	0.18	20.09	12.89	171.56	204.54	37.185
J8.1.1.03	as above; 100mm; with cap	Nr	0.19	21.21	13.60	175.55	210.36	45.737
J8.1.1.04	as above; with handwheel	Nr	0.22	24.55	15.75	195.14	235.44	48.562
J8.1.1.05	as above; 150mm; with cap	Nr	0.34	37.95	24.34	267.26	329.55	80.588
J8.1.1.06	as above; with handwheel	Nr	0.37	41.30	26.49	286.86	354.65	84.958
J8.1.1.07	as above; 200mm; with cap	Nr	0.46	51.34	32.94	524.04	608.32	135.723
J8.1.1.08	as above; with handwheel	Nr	0.50	55.81	35.80	550.26	641.87	142.957
J8.1.2	nominal bore: 200 - 300mm							
J8.1.2.01	Flanged Ductile iron valves; right hand closing; nominal bore; 250 mm; with cap	Nr	0.58	64.73	41.53	759.87	866.13	225.981
J8.1.2.02	as above; with handwheel	Nr	0.62	69.20	44.39	797.73	911.32	237.569
J8.1.2.03	as above; 300mm; with cap	Nr	0.70	78.13	50.12	919.45	1,047.70	306.405
J8.1.2.04	as above; with handwheel	Nr	0.75	83.71	53.70	957.34	1,094.75	322.119
J8.1.3	nominal bore: 300 - 600mm							
J8.1.3.01	Flanged Cast iron wedge valve, rating having gunmetal faces to body and wedge, stainless steel non-rising screw stem working in a gunmetal nut housed in the wedge; end flanges faced and drilled, body tested 24 bars; seat test and maximum working pressure 16 bar; clockwise closing; complete with cast iron cap for key operation or cast iron handwheel direct on valve; prices exclude forming holes and grouting in; nominal bore: 350mm; with cap	Nr	0.84	93.75	60.14	2,474.99	2,628.88	363.116
J8.1.3.02	as above; with handwheel	Nr	0.90	100.45	64.44	2,409.00	2,573.89	358.006
J8.1.3.03	as above; nominal bore: 400mm; with cap	Nr	1.00	111.61	71.60	2,789.06	2,972.27	423.661
J8.1.3.04	as above; with handwheel	Nr	1.06	118.31	75.90	2,937.00	3,131.21	445.106
J8.1.3.05	as above; nominal bore: 450mm; with cap	Nr	1.15	128.35	82.34	3,817.02	4,027.71	490.629

J8　Valves And Penstocks continued

	Unit	Labour Hours	Labour Net £	Plant Net £	Materials Net £	Unit Net £	CO_2e Kg
J8.1	**Gate valves: hand operated**						
J8.1.3.06	Flanged Cast iron wedge valve, rating having gunmetal faces to body and wedge, stainless steel non-rising screw stem working in a gunmetal nut housed in the wedge; end flanges faced and drilled, seat test and maximum working pressure 16 bar; clockwise closing; complete with cast iron cap for key operation or cast iron handwheel direct on valve; prices exclude forming holes and grouting in; nominal bore: 450mm; with handwheel	Nr	1.22	136.16	87.35	3,987.04	4,210.55	512.337
J8.1.3.07	as above; nominal bore: 500mm; with cap	Nr	1.27	143.84	153.17	4,676.69	4,973.70	570.815
J8.1.3.08	as above; with handwheel	Nr	1.34	151.77	161.62	4,877.84	5,191.23	593.295
J8.1.3.09	as above; nominal bore: 600mm; with cap	Nr	1.60	181.22	192.98	5,806.95	6,181.15	646.155
J8.1.3.10	as above; with handwheel	Nr	1.70	192.54	205.04	5,921.89	6,319.47	669.755
J8.8	**Penstocks**						
J8.8.1	nominal bore: not exceeding 200mm						
J8.8.1.01	Circular pattern cast iron penstocks suitable for on-seat pressure each comprising of cast iron frame, door and adjustable wedges; frame with flat back for wall fixing; door and frame having copper alloy sealing faces; stainless steel non-rising screw stem working in a gunmetal nut housed in door; complete with cast iron capor handwheel mounted direct on penstock framework and necessary mild steel indented type foundation bolts; prices exclude forming holes and grouting in; nominal bore: 100mm; with cap	Nr	0.20	22.32	14.32	580.70	617.34	66.370
J8.8.1.02	as above; with handwheel	Nr	0.23	25.67	16.47	771.07	813.21	70.215
J8.8.1.03	as above; nominal bore: 150mm; with cap	Nr	0.33	36.83	23.63	687.26	747.72	116.150
J8.8.1.04	as above; with handwheel	Nr	0.36	40.18	25.78	899.18	965.14	122.311
J8.8.1.05	as above; nominal bore: 200mm; with cap	Nr	0.50	55.81	35.80	820.11	911.72	202.021

J8   Valves And Penstocks continued...

	Unit	Labour Hours	Labour Net £	Plant Net £	Materials Net £	Unit Net £	CO_2e Kg
J8.8 **Penstocks**							
J8.8.1 nominal bore: not exceeding 200mm							
J8.8.1.06 as previous item; with handwheel	Nr	0.54	60.27	38.66	1,027.29	1,126.22	208.914
J8.8.2 nominal bore: 200 - 300mm							
J8.8.2.01 Circular pattern cast iron penstocks suitable for on-seat pressure each comprising of cast iron frame, door and adjustable wedges; frame with flat back for wall fixing; door and frame having copper alloy sealing faces; stainless steel non-rising screw stem working in a gunmetal nut housed in door; complete with cast iron capor handwheel mounted direct on penstock framework and necessary mild steel indented type foundation bolts; prices exclude forming holes and grouting in; nominal bore: 225mm; with cap	Nr	0.58	64.73	41.53	821.35	927.61	299.738
J8.8.2.02 as above; with handwheel	Nr	0.62	69.20	44.39	1,093.14	1,206.73	315.021
J8.8.2.03 as above; nominal bore: 250mm; with cap	Nr	0.68	75.89	48.69	899.18	1,023.76	333.977
J8.8.2.04 as above; with handwheel	Nr	0.73	81.48	52.27	1,144.63	1,278.38	351.097
J8.8.2.05 as previous item; nominal bore: 300mm; with cap	Nr	0.80	89.29	57.28	1,005.74	1,152.31	453.036
J8.8.2.06 as above; with handwheel	Nr	0.85	94.87	60.86	1,279.92	1,435.65	475.950
J8.8.3 nominal bore: 300 - 600mm							
J8.8.3.01 Circular pattern cast iron penstocks suitable for on-seat pressure each comprising of cast iron frame, door and adjustable wedges; frame with flat back for wall fixing; door and frame having copper alloy sealing faces; stainless steel non-rising screw stem working in a gunmetal nut housed in door; complete with cast iron capor handwheel mounted direct on penstock framework and necessary mild steel indented type foundation bolts; prices exclude forming holes and grouting in; nominal bore: 350mm; with cap	Nr	1.00	111.61	71.60	1,111.10	1,294.31	510.731
J8.8.3.02 as above; with handwheel	Nr	1.06	118.31	75.90	1,415.22	1,609.43	586.838
J8.8.3.03 as above; nominal bore: 400mm; with cap	Nr	1.20	133.93	85.92	1,204.49	1,424.34	695.885
J8.8.3.04 as above; with handwheel	Nr	1.27	141.74	90.93	1,520.58	1,753.25	730.942
J8.8.3.05 as above; nominal bore: 500mm; with cap	Nr	1.60	181.22	192.98	1,391.27	1,765.47	934.782
J8.8.3.06 as above; with handwheel	Nr	1.70	192.54	205.04	1,729.32	2,126.90	971.730
J8.8.3.07 as above; nominal bore: 600mm; with cap	Nr	1.95	220.86	235.19	1,936.73	2,392.78	1,056.016
J8.8.3.08 as above; with handwheel	Nr	2.05	232.18	247.25	1,939.64	2,419.07	1,090.158

J8   Valves And Penstocks continued...

	Unit	Labour Hours	Labour Net £	Plant Net £	Materials Net £	Unit Net £	CO_2e Kg
J8.8 **Penstocks**							
J8.8.4 nominal bore: 600 - 900mm							
J8.8.4.01 Circular pattern cast iron penstocks suitable for on-seat pressure each comprising of cast iron frame, door and adjustable wedges; frame with flat back for wall fixing; door and frame having copper alloy sealing faces; stainless steel non-rising screw stem working in a gunmetal nut housed in door; complete with cast iron capor handwheel mounted direct on penstock framework and necessary mild steel indented type foundation bolts; prices exclude forming holes and grouting in; nominal bore: 700mm; with cap	Nr	2.30	260.50	277.40	1,764.53	2,302.43	1,043.736
J8.8.4.02 as above; with handwheel	Nr	2.40	271.82	289.46	2,149.17	2,710.45	1,153.901
J8.8.4.03 as above; nominal bore: 750mm; with cap	Nr	2.53	286.55	305.14	1,857.63	2,449.32	1,100.213
J8.8.4.04 as above; with handwheel	Nr	2.65	300.14	319.62	2,178.50	2,798.26	1,216.447
J8.8.4.05 as above; nominal bore: 800mm; with cap	Nr	2.65	300.14	319.62	1,951.02	2,570.78	1,154.981
J8.8.4.06 as above; with handwheel	Nr	2.77	313.73	334.09	2,359.30	3,007.12	1,276.803
J8.8.4.07 as above; nominal bore: 900mm; with cap	Nr	3.00	339.78	361.83	2,137.20	2,838.81	1,220.847
J8.8.4.08 as above; with handwheel	Nr	3.15	356.77	379.92	2,568.83	3,305.52	1,349.655
J8.8.5 nominal bore: 900 - 1200mm							
J8.8.5.01 Circular pattern cast iron penstocks suitable for on-seat pressure each comprising of cast iron frame, door and adjustable wedges; frame with flat back for wall fixing; door and frame having copper alloy sealing faces; stainless steel non-rising screw stem working in a gunmetal nut housed in door; complete with cast iron capor handwheel mounted direct on penstock framework and necessary mild steel indented type foundation bolts; prices exclude forming holes and grouting in; nominal bore: 1000mm; with cap	Nr	3.35	379.42	404.04	2,324.28	3,107.74	1,267.175
J8.8.5.02 as above; with handwheel	Nr	3.50	396.41	422.13	2,778.96	3,597.50	1,399.679

CLASS K:
PIPEWORK - MANHOLES
AND PIPEWORK ANCILLARIES

Calculations used throughout Class K - Pipework - Manholes and Pipework Ancillaries

Labour

			Qty		Rate		Total
L K0200ICE	**Manholes Labour Gang**						
	L A9016ICE	Ganger	1	x	17.64	=	£17.64
	L A9002ICE	Labourer (General Operative)	3	x	13.02	=	£39.05
	L A9000ICE	Craftsman WRA	1	x	17.33	=	£17.33
		Total hourly cost of gang				=	**£74.02**
L K0202ICE	**Gullies Labour Gang**						
	L A9016ICE	Ganger	1	x	17.64	=	£17.64
	L A9002ICE	Labourer (General Operative)	2	x	13.02	=	£26.03
		Total hourly cost of gang				=	**£43.67**
L K0203ICE	**French and rubble drains Labour Gang**						
	L A9016ICE	Ganger	1	x	17.64	=	£17.64
	L A9002ICE	Labourer (General Operative)	3	x	13.02	=	£39.05
	L A9017ICE	Pipelayer (standard rate)	1	x	14.86	=	£14.86
		Total hourly cost of gang				=	**£71.55**
L K0205ICE	**Topsoil stripping and reinstatement Labour Gang**						
	L A9002ICE	Labourer (General Operative)	1	x	13.02	=	£13.02
		Total hourly cost of gang				=	**£13.02**
L K0206ICE	**Building in pipes to manholes etc. Labour Gang**						
	L A9016ICE	Ganger	1	x	17.64	=	£17.64
	L A9002ICE	Labourer (General Operative)	1	x	13.02	=	£13.02
		Total hourly cost of gang				=	**£30.66**

Plant

			Qty		Rate		Total
P K1200ICE	**Precast concrete manholes (shallow) Plant Gang**						
	P A0592ICE	Hydraulic Excavator - Cat 320 96kW	1	x	36.36	=	£36.36
	P A0811ICE	Pump - Godwin ET50 23m3/h 4 inches	1	x	2.94	=	£2.94
	P A0911ICE	Wheeled Tractor / Grader - Ford 3190H	1	x	25.17	=	£25.17
	P A0961ICE	Trailer - Massey Tipping	1	x	1.81	=	£1.81
	P A0544ICE	Dumper - 1.50t 2WD	1	x	2.94	=	£2.94
	P A0803ICE	Trench Sheets	72	x	0.09	=	£6.48
	P A0804ICE	Acrow Props	50	x	0.09	=	£4.50
	P A0722ICE	Concrete Mixer - 4/3 Petrol	1	x	2.18	=	£2.18
		Total hourly cost of gang				=	**£82.38**

P K1201ICE   **Precast concrete manholes (deep) Plant Gang**

P A0605ICE	Hydraulic Excavator - Cat 166kW	1	x	44.31	=	£44.31
P A0812ICE	Pump - Godwin ET75 74m3/h 4 inches	1	x	4.40	=	£4.40
P A0911ICE	Wheeled Tractor / Grader - Ford 3190H	1	x	25.17	=	£25.17
P A0961ICE	Trailer - Massey Tipping	1	x	1.81	=	£1.81
P A0544ICE	Dumper - 1.50t 2WD	1	x	2.94	=	£2.94
P A0803ICE	Trench Sheets	96	x	0.09	=	£8.64
P A0804ICE	Acrow Props	60	x	0.09	=	£5.40
P A0722ICE	Concrete Mixer - 4/3 Petrol	1	x	2.18	=	£2.18

    Total hourly cost of gang   = **£94.85**

P K1202ICE   **Gullies Plant Gang**

P A0594ICE	Hydraulic Excavator - JCB 3CX Sitemaster	1	x	28.91	=	£28.91
P A0544ICE	Dumper - 1.50t 2WD	1	x	2.94	=	£2.94
P A0722ICE	Concrete Mixer - 4/3 Petrol	1	x	2.18	=	£2.18

    Total hourly cost of gang   = **£34.03**

P K1203ICE   **French and rubble drains**

P A0592ICE	Hydraulic Excavator - Cat 320 96kW	1	x	36.36	=	£36.36
P A0811ICE	Pump - Godwin ET50 23m3/h 4 inches	1	x	2.94	=	£2.94
P A0544ICE	Dumper - 1.50t 2WD	1	x	2.94	=	£2.94
P A0854ICE	Vibrating Plate Diesel 33.5kN	1	x	2.95	=	£2.95

    Total hourly cost of gang   = **£45.19**

P K1204ICE   **Breaking up and temporary reinstatement of roads and footpaths Plant Gang**

P A0452ICE	Compressor - 2-Tool (Complete)	1	x	5.20	=	£5.20
P A0401ICE	Thor 16D / Maco SK8 Medium Duty Breaker	2	x	0.82	=	£1.64
P A0544ICE	Dumper - 1.50t 2WD	1	x	2.94	=	£2.94
P A0863ICE	Roller - 28 inch Vibratory Roller 0.37t	1	x	2.19	=	£2.19
P A1030ICE	Rammer Benjo	1	x	1.59	=	£1.59
P A0971ICE	Traffic Lights - Main Generator 2-Way with 100m Cables	1	x	3.21	=	£3.21

    Total hourly cost of gang   = **£16.77**

P K1205ICE   **Topsoil, stripping and reinstatement Plant Gang**

P A0934ICE	Crawler Tractor / Dozer - Cat D6R LGP 138kW	1	x	48.97	=	£48.97

    Total hourly cost of gang   = **£48.97**

Pipework - Manholes And Pipework Ancillaries

K1 Manholes

	Unit	Labour Hours	Labour Net £	Plant Net £	Materials Net £	Unit Net £	CO$_2$e Kg
K1.1 **Brick**							
K1.1.1 **Depth: not exceeding 1.5m**							
K1.1.1.01 Brick manholes 900 x 675mm internally with 150mm in situ concrete, grade 20 base with fair faced engineering Class "B" brick walls 215mm thick in cement mortar (1:3), 150mm precast concrete cover slab on 3 course brick kerb, 150mm vitrified clay straight main channel and 1nr 3/4 section branch, concrete benching and 600 x 450mm grade "B" single seal cast iron cover and frame including excavating in firm ground by machine and filling by hand and spreading surplus excavated material by hand on site average 25m distance from excavation; Depth: 1m; total cost of one manhole	Nr	7.85	580.98	641.97	479.49	1,702.44	822.866
K1.1.2 **Depth: 1.5 - 2m**							
K1.1.2.01 Brick manholes 900 x 675mm internally with 150mm in situ concrete, grade 20 base with fair faced engineering Class "B" brick walls 215mm thick in cement mortar (1:3), 150mm precast concrete cover slab on 3 course brick kerb, 150mm vitrified clay straight main channel and 1nr 3/4 section branch, concrete benching and 600 x 450mm grade "B" single seal cast iron cover and frame including excavating in firm ground by machine and filling by hand and spreading surplus excavated material by hand on site average 25m distance from excavation; Depth: 2m; total cost of one manhole	Nr	12.90	954.73	1,054.96	694.39	2,704.08	1,350.531

K1 Manholes continued...

		Unit	Labour Hours	Labour Net £	Plant Net £	Materials Net £	Unit Net £	CO_2e Kg
K1.2	**Brick with backdrop**							
K1.2.2	Depth: 1.5 - 2m							
K1.2.2.01	Brick manholes 1200 x 1050mm internally with 150mm in situ concrete, grade 20 base with fair faced engineering Class "B" brick walls 215mm thick in cement mortar (1:3), 150mm precast concrete cover slab on 3 course brick kerb, 150mm diameter vitrified clay pipe and 1nr 150 x 150mm junction forming backdrop commencing 1m above invert encased in concrete, grade 20, with cast iron cover and frame and 1nr 3/4 section branch channel bend, concrete benching and 600 x 600mm grade "B" single seal cast iron cover and frame, including excavating and filling in firm ground by machine, removing surplus from site; Depth 2m; total cost of one manhole	Nr	18.47	1,366.96	1,510.48	1,024.93	3,902.37	1,980.626
K1.2.4	Depth: 2.5 - 3m							
K1.2.4.01	Brick manholes 1200 x 1050mm internally with 150mm in situ concrete, grade 20 base with fair faced engineering Class "B" brick walls 215mm thick in cement mortar (1:3), 150mm precast concrete cover slab on 3 course brick kerb, 150mm diameter vitrified clay pipe and 1nr 150 x 150mm junction forming backdrop commencing 1m above invert encased in concrete, grade 20, with cast iron cover and frame and 1nr 3/4 section branch channel bend, concrete benching and 600 x 600mm grade "B" single seal cast iron cover and frame, including excavating and filling in firm ground by machine, removing surplus from site; Depth; 4m, backdrop commencing 2.5m above invert, total cost of one manhole	Nr	32.10	2,375.72	2,625.14	1,631.34	6,632.20	3,501.344
K1.5	**Precast concrete**							
K1.5.1	Depth: not exceeding 1.5m							
K1.5.1.01	Precast concrete manholes; excavation and backfilling in natural material; in situ concrete grade 20 in 300mm base; 300/150mm branch junction channel; 150mm diameter half round bend channel; cement mortar benching (1:3); precast concrete manhole rings and cover slab, BS 5911; 150mm in situ concrete grade 20 surround to rings; 600mm diameter heavy duty cast iron access cover and frame; galvanised malleable step irons BS EN13101 at 305mm centres cast into sides of chamber rings; 675mm nominal internal diameter manhole rings; Depth: 1.5m; total cost of one manhole	Nr	3.95	292.34	310.76	829.45	1,432.55	768.024

K1 Manholes continued...

	Unit	Labour Hours	Labour Net £	Plant Net £	Materials Net £	Unit Net £	CO_2e Kg
K1.5 **Precast concrete**							
K1.5.1 Depth: not exceeding 1.5m							
K1.5.1.02 as previous item; 900mm nominal internal diameter manhole rings; Depth: 1.5m; total cost of one manhole	Nr	4.41	326.38	348.38	669.32	1,344.08	825.965
K1.5.1.03 as previous item; 1050mm nominal internal diameter manhole rings; Depth: 1.5m; total cost of one manhole	Nr	5.32	393.73	422.80	676.20	1,492.73	927.783
K1.5.1.04 as above; 1200mm nominal internal diameter manhole rings; Depth: 1.5m; total cost of one manhole	Nr	6.24	461.82	498.04	786.77	1,746.63	1,205.079
K1.5.1.05 as above; 1350mm nominal internal diameter manhole rings; Depth: 1.5m; total cost of one manhole	Nr	7.18	531.39	574.91	894.33	2,000.63	1,406.401
K1.5.1.06 as above; 1500mm nominal internal diameter manhole rings; Depth: 1.5m; total cost of one manhole	Nr	8.14	602.44	751.78	958.47	2,312.69	1,642.488
K1.5.1.07 as above; 1800mm nominal internal diameter manhole rings; Depth: 1.5m; total cost of one manhole	Nr	10.08	746.02	934.31	1,210.81	2,891.14	2,002.225
K1.5.2 Depth: 1.5 - 2m							
K1.5.2.01 Precast concrete manholes; excavation and backfilling in natural material; in situ concrete grade 20 in 300mm base; 300/150mm branch junction channel; 150mm diameter half round bend channel; cement mortar benching (1:3); precast concrete manhole rings and cover slab, BS 5911; 150mm in situ concrete grade 20 surround to rings; 600mm diameter heavy duty cast iron access cover and frame; galvanised malleable step irons BS EN13101 at 305mm centres cast into sides of chamber rings; 675mm nominal internal diameter manhole rings; Depth: 2m; total cost of one manhole	Nr	3.75	277.54	294.41	653.40	1,225.35	770.124
K1.5.2.02 as above; 900mm nominal internal diameter manhole rings; Depth: 2m; total cost of one manhole	Nr	5.21	385.59	413.81	732.95	1,532.35	979.156
K1.5.2.03 as above; 1050mm nominal internal diameter manhole rings; Depth: 2m; total cost of one manhole	Nr	6.33	468.48	505.40	771.42	1,745.30	1,161.165
K1.5.2.04 as above; 1200mm nominal internal diameter manhole rings; Depth: 2m; total cost of one manhole	Nr	7.49	554.33	600.27	853.97	2,008.57	1,399.664

K1   Manholes continued...

	Unit	Labour Hours	Labour Net £	Plant Net £	Materials Net £	Unit Net £	CO_2e Kg
K1.5 **Precast concrete**							
K1.5.2 Depth: 1.5 - 2m							
K1.5.2.05 as previous item; 1350mm nominal internal diameter manhole rings; Depth: 2m; total cost of one manhole	Nr	8.68	642.41	697.58	999.65	2,339.64	1,677.233
K1.5.2.06 as above; 1500mm nominal internal diameter manhole rings; Depth: 2m; total cost of one manhole	Nr	9.89	731.96	916.44	1,129.02	2,777.42	1,978.906
K1.5.2.07 as above; 1800mm nominal internal diameter manhole rings; Depth: 2m; total cost of one manhole	Nr	12.32	911.80	1,145.08	1,367.75	3,424.63	2,404.872
K1.5.2.08 as above; 1200mm nominal internal diameter manhole chamber rings tapering to 900mm nominal internal diameter shaft rings at 1.5m above channel invert; Depth: 2m; total cost of one manhole	Nr	7.02	519.55	561.83	933.86	2,015.24	1,367.493
K1.5.2.09 as above; 1350mm nominal internal diameter manhole chamber rings tapering to 900mm nominal internal diameter shaft rings at 1.5m above channel invert; Depth: 2m; total cost of one manhole	Nr	7.92	586.16	635.43	1,019.59	2,241.18	1,540.505
K1.5.2.10 as previous item; 1500mm nominal internal diameter manhole chamber rings tapering to 900mm nominal internal diameter shaft rings at 1.5m above channel invert; Depth: 2m; total cost of one manhole	Nr	9.13	675.71	844.93	1,150.70	2,671.34	1,842.031
K1.5.2.11 as above; 1800mm nominal internal diameter manhole chamber rings tapering to 900mm nominal internal diameter shaft rings at 1.5m above channel invert; Depth: 2m; total cost of one manhole	Nr	11.24	831.87	1,043.46	1,358.22	3,233.55	2,226.814

K1   Manholes continued...

	Unit	Labour Hours	Labour Net £	Plant Net £	Materials Net £	Unit Net £	CO_2e Kg
KI.5 **Precast concrete**							
KI.5.4 Depth: 2.5 - 3m							
KI.5.4.01 Precast concrete manholes; excavation and backfilling in natural material; in situ concrete grade 20 in 300mm base; 300/150mm branch junction channel; 150mm diameter half round bend channel; cement mortar benching (1:3); precast concrete manhole rings and cover slab, BS 5911; 150mm in situ concrete grade 20 surround to rings; 600mm diameter heavy duty cast iron access cover and frame; galvanised malleable step irons BS EN13101 at 305mm centres cast into sides of chamber rings; 675mm nominal internal diameter manhole rings; Depth: 3m; total cost of one manhole	Nr	4.81	355.99	377.01	769.17	1,502.17	1,017.932
KI.5.4.02 as above; 900mm nominal internal diameter manhole rings; Depth: 3m; total cost of one manhole	Nr	6.97	515.85	553.65	870.15	1,939.65	1,296.102
KI.5.4.03 as above; 1050mm nominal internal diameter manhole rings; Depth: 3m; total cost of one manhole	Nr	8.62	637.97	688.59	921.62	2,248.18	1,548.266
KI.5.4.04 as above; 1200mm nominal internal diameter manhole rings; Depth: 3m; total cost of one manhole	Nr	10.19	754.16	816.98	1,037.21	2,608.35	1,891.069
KI.5.4.05 as above; 1200mm nominal internal diameter manhole chamber rings tapering to 900mm nominal internal diameter shaft rings at 1.5m above channel invert; Depth: 3m; total cost of one manhole	Nr	7.84	580.24	624.80	871.75	2,076.79	1,090.371
KI.5.4.06 as above; 1350mm nominal internal diameter manhole rings; Depth: 3m; total cost of one manhole	Nr	11.97	885.90	962.55	1,220.23	3,068.68	2,233.041
KI.5.4.07 as above; 1350mm nominal internal diameter manhole chamber rings tapering to 900mm nominal internal diameter shaft rings at 1.5m above channel invert; Depth: 3m; total cost of one manhole	Nr	9.91	733.44	794.08	1,201.82	2,729.34	1,947.233
KI.5.4.08 as above; 1500mm nominal internal diameter manhole rings; Depth: 3m; total cost of one manhole	Nr	13.59	1,005.80	1,259.87	1,368.05	3,633.72	2,619.624

K1   Manholes continued...

	Unit	Labour Hours	Labour Net £	Plant Net £	Materials Net £	Unit Net £	CO$_2$e Kg

K1.5   Precast concrete

K1.5.4   Depth: 2.5 - 3m

	Unit	Labour Hours	Labour Net £	Plant Net £	Materials Net £	Unit Net £	CO$_2$e Kg
K1.5.4.09 as previous item; 1500mm nominal internal diameter manhole chamber rings tapering to 900mm nominal internal diameter shaft rings at 1.5m above channel invert; Depth: 3m; total cost of one manhole	Nr	11.10	821.51	1,025.58	1,348.70	3,195.79	2,245.903
K1.5.4.10 as above; 1800mm nominal internal diameter manhole rings; Depth: 3m; total cost of one manhole	Nr	16.94	1,253.73	1,582.59	1,681.92	4,518.24	3,196.045
K1.5.4.11 as above; 1800mm nominal internal diameter manhole chamber rings tapering to 900mm nominal internal diameter shaft rings at 1.5m above channel invert; Depth: 3m; total cost of one manhole	Nr	13.53	1,001.36	1,254.22	1,514.11	3,769.69	2,615.773

K1.5.6   Depth: 3.5 - 4m

	Unit	Labour Hours	Labour Net £	Plant Net £	Materials Net £	Unit Net £	CO$_2$e Kg
K1.5.6.01 Precast concrete manholes; excavation and backfilling in natural material; in situ concrete grade 20 in 300mm base; 300/150mm branch junction channel; 150mm diameter half round bend channel; cement mortar benching (1:3); precast concrete manhole rings and cover slab, BS 5911; 150mm in situ concrete grade 20 surround to rings; 600mm diameter heavy duty cast iron access cover and frame; galvanised malleable step irons BS EN13101 at 305mm centres cast into sides of chamber rings; 675mm nominal internal diameter manhole rings; Depth: 4m; total cost of one manhole	Nr	5.90	436.66	457.97	879.97	1,774.60	1,261.423
K1.5.6.02 as above; 900mm nominal internal diameter manhole rings; Depth: 4m; total cost of one manhole	Nr	8.71	644.63	687.77	1,002.39	2,334.79	1,607.353
K1.5.6.03 as above; 1050mm nominal internal diameter manhole rings; Depth: 4m; total cost of one manhole	Nr	10.72	793.39	852.15	1,071.68	2,717.22	1,938.830
K1.5.6.04 as above; 1200mm nominal internal diameter manhole rings; Depth: 4m; total cost of one manhole	Nr	12.89	953.99	1,029.61	1,201.01	3,184.61	2,335.811
K1.5.6.05 as above; 1200mm nominal internal diameter manhole chamber rings tapering to 900mm nominal internal diameter shaft rings at 1.5m above channel invert; Depth: 4m; total cost of one manhole	Nr	10.02	741.58	794.90	1,197.06	2,733.54	1,967.341

K1 Manholes continued...

	Unit	Labour Hours	Labour Net £	Plant Net £	Materials Net £	Unit Net £	CO_2e Kg
K1.5	**Precast concrete**						
K1.5.6	Depth: 3.5 - 4m						
K1.5.6.06	as previous item; 1350mm nominal internal diameter manhole rings; Depth: 4m; total cost of one manhole	Nr	15.74	1,164.92	1,443.34	1,537.63	4,145.89	3,175.730
K1.5.6.07	as above; 1350mm nominal internal diameter manhole chamber rings tapering to 900mm nominal internal diameter shaft rings at 1.5m above channel invert; Depth: 4m; total cost of one manhole	Nr	11.37	841.49	905.30	1,324.40	3,071.19	2,223.090
K1.5.6.08	as above; 1500mm nominal internal diameter manhole rings; Depth: 4m; total cost of one manhole	Nr	17.29	1,279.63	1,598.59	1,602.12	4,480.34	3,254.804
K1.5.6.09	as above; 1500mm nominal internal diameter manhole chamber rings tapering to 900mm nominal internal diameter shaft rings at 1.5m above channel invert; Depth: 4m; total cost of one manhole	Nr	12.64	935.49	1,161.07	1,471.28	3,567.84	2,535.336
K1.5.6.10	as above; 1800mm nominal internal diameter manhole rings; Depth: 4m; total cost of one manhole	Nr	21.72	1,607.50	2,015.41	1,991.13	5,614.04	3,981.680
K1.5.6.11	as above; 1800mm nominal internal diameter manhole chamber rings tapering to 900mm nominal internal diameter shaft rings at 1.5m above channel invert; Depth: 4m; total cost of one manhole	Nr	15.25	1,128.65	1,406.65	1,648.57	4,183.87	2,929.326
K1.5.7	Depth: stated exceeding 4m						
K1.5.7.01	Precast concrete manholes; excavation and backfilling in natural material; in situ concrete grade 20 in 300mm base; 300/150mm branch junction channel; 150mm diameter half round bend channel; cement mortar benching (1:3); precast concrete manhole rings and cover slab, BS 5911; 150mm in situ concrete grade 20 surround to rings; 600mm diameter heavy duty cast iron access cover and frame; galvanised malleable step irons BS EN13101 at 305mm centres cast into sides of chamber rings; 900mm nominal internal diameter manhole rings; Depth: 5m; total cost of one manhole	Nr	10.46	774.14	946.55	1,135.89	2,856.58	2,003.405
K1.5.7.02	as above; 1050mm nominal internal diameter manhole rings; Depth: 5m; total cost of one manhole	Nr	13.08	968.05	1,193.06	1,216.90	3,378.01	2,422.849

K1   Manholes continued...

	Unit	Labour Hours	Labour Net £	Plant Net £	Materials Net £	Unit Net £	CO_2e Kg
K1.5 **Precast concrete**							
K1.5.7 Depth: stated exceeding 4m							
K1.5.7.03 as previous item; 1200mm nominal internal diameter manhole rings; Depth: 5m; total cost of one manhole	Nr	15.74	1,164.92	1,443.34	1,374.46	3,982.72	2,933.505
K1.5.7.04 as above; 1200mm nominal internal diameter manhole chamber rings tapering to 900mm nominal internal diameter shaft rings at 1.5m above channel invert; Depth: 5m; total cost of one manhole	Nr	11.45	847.41	1,039.69	1,271.37	3,158.47	2,191.235
K1.5.7.05 as above; 1350mm nominal internal diameter manhole rings; Depth: 5m; total cost of one manhole	Nr	18.42	1,363.26	1,695.50	1,461.95	4,520.71	3,400.484
K1.5.7.06 as above; 1350mm nominal internal diameter manhole chamber rings tapering to 900mm nominal internal diameter shaft rings at 1.5m above channel invert; Depth: 5m; total cost of one manhole	Nr	12.86	951.77	1,172.36	1,446.98	3,571.11	2,598.171
K1.5.7.07 as above; 1500mm nominal internal diameter manhole rings; Depth: 5m; total cost of one manhole	Nr	21.10	1,561.61	1,947.66	1,841.01	5,350.28	3,907.723
K1.5.7.08 as above; 1500mm nominal internal diameter manhole chamber rings tapering to 900mm nominal internal diameter shaft rings at 1.5m above channel invert; Depth: 5m; total cost of one manhole	Nr	14.23	1,053.16	1,301.26	1,593.86	3,948.28	2,826.542
K1.5.7.09 as above; 1800mm nominal internal diameter manhole rings; Depth: 5m; total cost of one manhole	Nr	26.52	1,962.75	2,457.63	2,309.99	6,730.37	4,798.540
K1.5.7.10 as above; 1800mm nominal internal diameter manhole chamber rings tapering to 900mm nominal internal diameter shaft rings at 1.5m above channel invert; Depth: 5m; total cost of one manhole	Nr	17.01	1,258.91	1,562.83	1,771.15	4,592.89	3,226.557
K3 **Gullies**							
K3.1 **Clay**							
K3.1.0 Vitrified clay road gully; concrete grade 20 surround 150mm thick; 3 courses class 'B' engineering brick in cement mortar (1:3); 400 x 349mm cast iron grating and frame; BS EN 124, grade 'A' bedded in cement mortar (1:3)							
K3.1.0.01 450mm diameter x 900mm deep	Nr	2.03	88.65	69.08	317.01	474.74	274.369

K3 Gullies continued...

		Unit	Labour Hours	Labour Net £	Plant Net £	Materials Net £	Unit Net £	CO_2e Kg
K3.6	**Precast concrete trapped**							
K3.6.0	Precast concrete gully, BS 5911; 150mm trapped outlet; rodding eye with stopper and galvanised chain; concrete grade 20 surround 150mm thick; 3 courses class 'B' engineering brick in cement mortar (1:3); 400 x 349mm cast iron grating and frame, BS EN 124, grade 'A' bedded in cement mortar (1:3)							
K3.6.0.01	375mm diameter x 900mm deep	Nr	1.68	73.37	57.17	190.08	320.62	283.420
K3.7	**Plastics**							
K3.7.0	Polypropylene road gully; concrete grade 20 surround 150mm thick; 3 courses class 'B' engineering bricks in cement mortar (1:3); 502 x 349mm cast iron grating and frame, BS EN 124, grade 'A' bedded in cement mortar (1:3)							
K3.7.0.01	500mm diameter x 900mm deep	Nr	2.25	98.26	76.57	235.35	410.18	316.713

K4 French Drains, Rubble Drains, Ditches And Trenches

Note(s): NOTE: Items for Filling of French and rubble drains are exclusive of disposal of excavated material which is included in class 'I': Pipes.

		Unit	Labour Hours	Labour Net £	Plant Net £	Materials Net £	Unit Net £	CO_2e Kg
K4.1	**Filling french and rubble drains with graded material**							
K4.1.0	14mm limestone aggregate fill							
K4.1.0.01	Generally	m^3	0.18	12.88	8.13	22.10	43.11	13.027
K4.2	**Filling french and rubble drains with rubble**							
K4.2.0	Broken brick fill							
K4.2.0.01	Generally	m^3	0.20	14.31	9.04	12.56	35.91	13.819
K4.3	**Trenches for unpiped rubble drains**							
K4.3.1	Cross-sectional area: not exceeding $0.25m^2$							
K4.3.1.01	Generally	m	0.03	1.79	2.26	3.62	7.67	1.120
K4.3.2	Cross-sectional area: $0.25 - 0.5m^2$							
K4.3.2.01	Generally	m	0.04	2.86	3.62	5.47	11.95	1.776
K4.3.3	Cross-sectional area: $0.5 - 0.75m^2$							
K4.3.3.01	Generally	m	0.06	4.29	5.42	9.09	18.80	2.709
K4.3.4	Cross-sectional area: $0.75 - 1m^2$							
K4.3.4.01	Generally	m	0.09	6.08	7.68	12.71	26.47	3.829
K4.3.5	Cross-sectional area: $1 - 1.5m^2$							
K4.3.5.01	Generally	m	0.11	7.51	9.49	18.08	35.08	4.853
K4.3.6	Cross-sectional area: $1.5 - 2m^2$							
K4.3.6.01	Generally	m	0.13	8.94	11.30	25.31	45.55	5.973
K4.3.7	Cross-sectional area: $2 - 3m^2$							
K4.3.7.01	Generally	m	0.19	13.59	17.17	36.16	66.92	8.959
K4.3.8	Cross-sectional area: exceeding $3m^2$							
K4.3.8.01	Generally; cross-sectional area: $5m^2$	m	0.34	23.97	30.28	72.31	126.56	16.239
K4.4	**Rectangular section ditches: unlined**							
K4.4.1	Cross-sectional area: not exceeding $0.25m^2$							
K4.4.1.01	Generally	m	0.03	1.93	2.49	3.62	8.04	1.213
K4.4.2	Cross-sectional area: $0.25 - 0.5m^2$							
K4.4.2.01	Generally	m	0.04	3.15	3.98	5.47	12.60	1.925

K4 French Drains, Rubble Drains, Ditches And Trenches continued...

		Unit	Labour Hours	Labour Net £	Plant Net £	Materials Net £	Unit Net £	CO_2e Kg
K4.4	**Rectangular section ditches: unlined**							
K4.4.3	Cross-sectional area: 0.5 - 0.75m²							
K4.4.3.01	Generally	m	0.07	4.72	5.97	9.09	19.78	2.933
K4.4.4	Cross-sectional area: 0.75 - 1m²							
K4.4.4.01	Generally	m	0.09	6.73	8.45	12.71	27.89	4.146
K4.4.5	Cross-sectional area: 1 - 1.5m²							
K4.4.5.01	Generally	m	0.12	8.30	10.44	18.08	36.82	5.244
K4.4.6	Cross-sectional area: 1.5 - 2m²							
K4.4.6.01	Generally	m	0.14	9.87	12.43	25.31	47.61	6.439
K4.4.7	Cross-sectional area: 2 - 3m²							
K4.4.7.01	Generally	m	0.21	14.95	18.89	36.16	70.00	9.668
K4.4.8	Cross-sectional area: exceeding 3m²							
K4.4.8.01	Generally; cross-sectional area: 5m²	m	0.37	26.33	33.31	72.31	131.95	17.489
K4.5	**Rectangular section ditches: lined**							
K4.5.1	Cross-sectional area: not exceeding 0.25m²							
K4.5.1.01	100mm thick weak mix concrete grade 15, 20mm aggregate; Generally	m	0.06	3.94	4.50	17.66	26.10	40.923
K4.5.2	Cross-sectional area: 0.25 - 0.5m²							
K4.5.2.01	100mm thick weak mix concrete grade 15, 20mm aggregate; Generally	m	0.09	6.30	7.20	23.26	36.76	52.387
K4.5.3	Cross-sectional area: 0.5 - 0.75m²							
K4.5.3.01	100mm thick weak mix concrete grade 15, 20mm aggregate; Generally	m	0.13	9.44	10.79	31.55	51.78	66.860
K4.5.4	Cross-sectional area: 0.75 - 1m²							
K4.5.4.01	100mm thick weak mix concrete grade 15, 20mm aggregate; Generally	m	0.19	13.38	15.29	38.91	67.58	79.021
K4.5.5	Cross-sectional area: 1 - 1.5m²							
K4.5.5.01	100mm thick weak mix concrete grade 15, 20mm aggregate; Generally	m	0.23	16.53	18.89	49.90	85.32	96.199
K4.5.6	Cross-sectional area: 1.5 - 2m²							
K4.5.6.01	100mm thick weak mix concrete grade 15, 20mm aggregate; Generally	m	0.28	19.68	22.49	62.75	104.92	113.473
K4.5.7	Cross-sectional area: 2 - 3m²							
K4.5.7.01	100mm thick weak mix concrete grade 15, 20mm aggregate; Generally	m	0.42	29.91	34.18	80.15	144.24	136.276
K4.5.8	Cross-sectional area: exceeding 3m²							
K4.5.8.01	100mm thick weak mix concrete grade 15, 20mm aggregate; Generally	m	0.74	52.73	60.27	135.02	248.02	199.228

K4 French Drains, Rubble Drains, Ditches And Trenches continued...

		Unit	Labour Hours	Labour Net £	Plant Net £	Materials Net £	Unit Net £	CO_2e Kg
K4.6	**Vee section ditches: unlined**							
K4.6.1	Cross-sectional area: not exceeding 0.25m²							
K4.6.1.01	Generally	m	0.03	2.15	4.91	3.62	10.68	1.840
K4.6.2	Cross-sectional area: 0.25 - 0.5m²							
K4.6.2.01	Generally	m	0.05	3.43	7.85	5.47	16.75	2.928
K4.6.3	Cross-sectional area: 0.5 - 0.75m²							
K4.6.3.01	Generally	m	0.07	5.15	11.78	9.09	26.02	4.437
K4.6.4	Cross-sectional area: 0.75 - 1m²							
K4.6.4.01	Generally	m	0.10	7.30	16.68	12.71	36.69	6.277
K4.6.5	Cross-sectional area: 1 - 1.5m²							
K4.6.5.01	Generally	m	0.13	9.02	20.61	18.08	47.71	7.877
K4.6.6	Cross-sectional area: 1.5 - 2m²							
K4.6.6.01	Generally	m	0.15	10.73	24.53	25.31	60.57	9.573
K4.6.7	Cross-sectional area: 2 - 3m²							
K4.6.7.01	Generally	m	0.23	16.31	37.29	36.16	89.76	14.431
K4.6.8	Cross-sectional area: exceeding 3m²							
K4.6.8.01	Generally	m	0.40	28.76	65.75	72.31	166.82	25.887
K4.7	**Vee section ditches: lined**							
K4.7.1	Cross-sectional area: not exceeding 0.25m²							
K4.7.1.01	100mm thick weak mix concrete grade 15, 20mm aggregate; Generally	m	0.06	4.29	4.91	17.66	26.86	41.061
K4.7.2	Cross-sectional area: 0.25 - 0.5m²							
K4.7.2.01	100mm thick weak mix concrete grade 15, 20mm aggregate; Generally	m	0.10	6.87	7.85	23.26	37.98	52.607
K4.7.3	Cross-sectional area: 0.5 - 0.75m²							
K4.7.3.01	100mm thick weak mix concrete grade 15, 20mm aggregate; Generally	m	0.14	10.30	11.78	31.55	53.63	67.190
K4.7.4	Cross-sectional area: 0.75 - 1m²							
K4.7.4.01	100mm thick weak mix concrete grade 15, 20mm aggregate; Generally	m	0.20	14.60	16.68	38.91	70.19	79.489
K4.7.5	Cross-sectional area: 1 - 1.5m²							
K4.7.5.01	100mm thick weak mix concrete grade 15, 20mm aggregate; Generally	m	0.25	18.03	20.61	49.90	88.54	96.777
K4.7.6	Cross-sectional area: 1.5 - 2m²							
K4.7.6.01	100mm thick weak mix concrete grade 15, 20mm aggregate; Generally	m	0.30	21.47	24.53	62.75	108.75	114.162
K4.7.7	Cross-sectional area: 2 - 3m²							
K4.7.7.01	100mm thick weak mix concrete grade 15, 20mm aggregate; Generally	m	0.46	32.63	37.29	80.15	150.07	137.323

K4 French Drains, Rubble Drains, Ditches And Trenches continued...

	Unit	Labour Hours	Labour Net £	Plant Net £	Materials Net £	Unit Net £	CO_2e Kg
K4.7	**Vee section ditches: lined**						
K4.7.8	Cross-sectional area: exceeding 3m^2						
K4.7.8.01	100mm thick weak mix concrete grade 15, 20mm aggregate; Generally	m	0.80	57.53	65.75	135.02	258.30	201.074
K4.8	**Trenches for pipes or cables not to be laid by the contractor**						
K4.8.1	Cross-sectional area: not exceeding 0.25m^2						
K4.8.1.01	100mm thick weak mix concrete grade 15, 20mm aggregate; Generally	m	0.03	2.15	4.91	3.62	10.68	1.840
K4.8.2	Cross-sectional area: 0.25 - 0.5m^2						
K4.8.2.01	100mm thick weak mix concrete grade 15, 20mm aggregate; Generally	m	0.05	3.43	7.85	5.47	16.75	2.928
K4.8.3	Cross-sectional area: 0.5 - 0.75m^2						
K4.8.3.01	100mm thick weak mix concrete grade 15, 20mm aggregate; Generally	m	0.07	5.15	11.78	9.09	26.02	4.437
K4.8.4	Cross-sectional area: 0.75 - 1m^2						
K4.8.4.01	100mm thick weak mix concrete grade 15, 20mm aggregate; Generally	m	0.10	7.30	16.68	12.71	36.69	6.277
K4.8.5	Cross-sectional area: 1 - 1.5m^2						
K4.8.5.01	100mm thick weak mix concrete grade 15, 20mm aggregate; Generally	m	0.13	9.02	20.61	18.08	47.71	7.877
K4.8.6	Cross-sectional area: 1.5 - 2m^2						
K4.8.6.01	100mm thick weak mix concrete grade 15, 20mm aggregate; Generally	m	0.15	10.73	24.53	25.31	60.57	9.573
K4.8.7	Cross-sectional area: 2 - 3m^2						
K4.8.7.01	100mm thick weak mix concrete grade 15, 20mm aggregate; Generally	m	0.23	16.31	37.29	36.16	89.76	14.431
K4.8.8	Cross-sectional area: exceeding 3m^2						
K4.8.8.01	100mm thick weak mix concrete grade 15, 20mm aggregate; Generally	m	0.40	28.76	65.75	72.31	166.82	25.887

K5 Dust And Metal Culverts

	Unit	Labour Hours	Labour Net £	Plant Net £	Materials Net £	Unit Net £	CO_2e Kg
K5.1	**Cable ducts: 1 way**						
K5.1.2	In trenches, depth: not exceeding 1.5m						
K5.1.2.01	clay ducts, BS 65; 100mm diameter	m	0.08	5.72	3.62	8.32	17.66	1.992
K5.1.2.02	as above; 150mm diameter	m	0.09	6.44	4.07	18.16	28.67	2.677
K5.1.3	In trenches, depth: 1.5 - 2m						
K5.1.3.01	clay ducts, BS 65; 100mm diameter	m	0.12	8.59	5.42	8.32	22.33	2.738
K5.1.3.02	as above; 150mm diameter	m	0.14	10.02	6.33	18.16	34.51	3.610

K5   Dust And Metal Culverts continued...

	Unit	Labour Hours	Labour Net £	Plant Net £	Materials Net £	Unit Net £	CO_2e Kg
K5.1	**Cable ducts: 1 way**						
K5.1.4	In trenches, depth: 2 - 2.5m						
K5.1.4.01	clay ducts, BS 65; 100mm diameter	m	0.18	12.88	8.13	8.32	29.33	3.857
K5.1.4.02	as above; 150mm diameter	m	0.20	14.31	9.04	18.16	41.51	4.730
K5.1.5	In trenches, depth: 2.5 - 3m						
K5.1.5.01	clay ducts, BS 65; 100mm diameter	m	0.25	17.89	11.30	8.32	37.51	5.164
K5.1.5.02	as above; 150mm diameter	m	0.28	20.03	12.65	18.16	50.84	6.222
K5.1.6	In trenches, depth: 3 - 3.5m						
K5.1.6.01	clay ducts, BS 65; 100mm diameter	m	0.33	23.61	14.91	8.32	46.84	6.656
K5.1.6.02	as above; 150mm diameter	m	0.37	26.47	16.72	18.16	61.35	7.901
K5.1.7	In trenches, depth: 3.5 - 4m						
K5.1.7.01	clay ducts, BS 65; 100mm diameter	m	0.42	30.05	18.98	8.32	57.35	8.335
K5.1.7.02	as above; 150mm diameter	m	0.47	33.63	21.24	18.16	73.03	9.767

K5.5	**Sectional corrugated metal culverts, nominal internal diameter: not exceeding**							
K5.5.1	Not in trenches							
K5.5.1.01	Sections corrugated, galvanised and bitumen coated steel culverts; 1.5mm thickness; 3m length jointed with coupling bands; nominal internal diameter 0.5m	m	0.08	3.49	1.36	-	4.85	0.465
K5.5.2	In trenches, depth: not exceeding 1.5m							
K5.5.2.01	Sections corrugated, galvanised and bitumen coated steel culverts; 1.5mm thickness; 3m length jointed with coupling bands; nominal internal diameter 0.5m	m	0.21	9.17	7.15	85.49	101.81	121.580
K5.5.3	In trenches, depth: 1.5 - 2m							
K5.5.3.01	Sections corrugated, galvanised and bitumen coated steel culverts; 1.5mm thickness; 3m length jointed with coupling bands; nominal internal diameter 0.5m	m	0.25	10.92	8.51	85.49	104.92	122.045
K5.5.4	In trenches, depth: 2 - 2.5m							
K5.5.4.01	Sections corrugated, galvanised and bitumen coated steel culverts; 1.5mm thickness; 3m length jointed with coupling bands; nominal internal diameter 0.5m	m	0.30	13.10	10.21	85.49	108.80	122.626
K5.5.5	In trenches, depth: 2.5 - 3m							
K5.5.5.01	Sections corrugated, galvanised and bitumen coated steel culverts; 1.5mm thickness; 3m length jointed with coupling bands; nominal internal diameter 0.5m	m	0.36	15.72	12.25	85.49	113.46	123.323

K5   Dust And Metal Culverts continued...

	Unit	Labour Hours	Labour Net £	Plant Net £	Materials Net £	Unit Net £	CO_2e Kg
K5.5	**Sectional corrugated metal culverts, nominal internal diameter: not exceeding**						
K5.5.6 In trenches, depth: 3 - 3.5m							
K5.5.6.01 Sections corrugated, galvanised and bitumen coated steel culverts; 1.5mm thickness; 3m length jointed with coupling bands; nominal internal diameter 0.5m	m	0.44	19.21	14.97	85.49	119.67	124.253
K5.5.7 In trenches, depth: 3.5 - 4m							
K5.5.7.01 Sections corrugated, galvanised and bitumen coated steel culverts; 1.5mm thickness; 3m length jointed with coupling bands; nominal internal diameter 0.5m	m	0.54	23.58	18.38	85.49	127.45	125.415
K5.6	**Sectional corrugated metal culverts, nominal internal diameter: 0.5 - 1m**						
K5.6.1 Not in trenches							
K5.6.1.01 Sections corrugated, galvanised and bitumen coated steel culverts; 1.5mm thickness; 3m length jointed with coupling bands; nominal internal diameter 1.0m	m	0.12	5.24	2.04	170.98	178.26	238.978
K5.6.2 In trenches, depth: not exceeding 1.5m							
K5.6.2.01 Sections corrugated, galvanised and bitumen coated steel culverts; 1.5mm thickness; 3m length jointed with coupling bands; nominal internal diameter 1.0m	m	0.25	10.92	8.51	170.98	190.41	241.185
K5.6.3 In trenches, depth: 1.5 - 2m							
K5.6.3.01 Sections corrugated, galvanised and bitumen coated steel culverts; 1.5mm thickness; 3m length jointed with coupling bands; nominal internal diameter 1.0m	m	0.32	13.97	10.89	170.98	195.84	241.999
K5.6.4 In trenches, depth: 2 - 2.5m							
K5.6.4.01 Sections corrugated, galvanised and bitumen coated steel culverts; 1.5mm thickness; 3m length jointed with coupling bands; nominal internal diameter 1.0m	m	0.41	17.90	13.95	170.98	202.83	243.045
K5.6.5 In trenches, depth: 2.5 - 3m							
K5.6.5.01 Sections corrugated, galvanised and bitumen coated steel culverts; 1.5mm thickness; 3m length jointed with coupling bands; nominal internal diameter 1.0m	m	0.51	22.27	17.36	140.78	180.41	232.968
K5.6.6 In trenches, depth: 3 - 3.5m							
K5.6.6.01 Sections corrugated, galvanised and bitumen coated steel culverts; 1.5mm thickness; 3m length jointed with coupling bands; nominal internal diameter 1.0m	m	0.62	27.08	21.10	170.98	219.16	245.485

K5   Dust And Metal Culverts continued...

	Unit	Labour Hours	Labour Net £	Plant Net £	Materials Net £	Unit Net £	CO$_2$e Kg
K5.6 **Sectional corrugated metal culverts, nominal internal diameter: 0.5 - 1m**							
K5.6.7 In trenches, depth: 3.5 - 4m							
K5.6.7.01 Sections corrugated, galvanised and bitumen coated steel culverts; 1.5mm thickness; 3m length jointed with coupling bands; nominal internal diameter 1.0m	m	0.74	32.32	25.18	170.98	228.48	246.879
K5.7 **Sectional corrugated metal culverts, nominal internal diameter: 1 - 1.5m**							
K5.7.1 Not in trenches							
K5.7.1.01 Sections corrugated, galvanised and bitumen coated steel culverts; 1.5mm thickness; 3m length jointed with coupling bands; nominal internal diameter 1.5m	m	0.20	8.73	3.40	317.14	329.27	358.584
K5.7.2 In trenches, depth: not exceeding 1.5m							
K5.7.2.01 Sections corrugated, galvanised and bitumen coated steel culverts; 1.5mm thickness; 3m length jointed with coupling bands; nominal internal diameter 1.5m	m	0.49	21.40	16.67	317.14	355.21	363.116
K5.7.3 In trenches, depth: 1.5 - 2m							
K5.7.3.01 Sections corrugated, galvanised and bitumen coated steel culverts; 1.5mm thickness; 3m length jointed with coupling bands; nominal internal diameter 1.5m	m	0.41	17.90	13.95	317.14	348.99	362.186
K5.7.4 In trenches, depth: 2 - 2.5m							
K5.7.4.01 Sections corrugated, galvanised and bitumen coated steel culverts; 1.5mm thickness; 3m length jointed with coupling bands; nominal internal diameter 1.5m	m	0.63	27.51	21.44	317.14	366.09	364.743
K5.7.5 In trenches, depth: 2.5 - 3m							
K5.7.5.01 Sections corrugated, galvanised and bitumen coated steel culverts; 1.5mm thickness; 3m length jointed with coupling bands; nominal internal diameter 1.5m	m	0.77	33.63	26.20	317.14	376.97	366.369
K5.7.6 In trenches, depth: 3 - 3.5m							
K5.7.6.01 Sections corrugated, galvanised and bitumen coated steel culverts; 1.5mm thickness; 3m length jointed with coupling bands; nominal internal diameter 1.5m	m	0.91	39.74	30.97	317.14	387.85	367.996
K5.7.7 In trenches, depth: 3.5 - 4m							
K5.7.7.01 Sections corrugated, galvanised and bitumen coated steel culverts; 1.5mm thickness; 3m length jointed with coupling bands; nominal internal diameter 1.5m	m	1.05	45.85	35.73	317.14	398.72	369.623

K5   Dust And Metal Culverts continued...

	Unit	Labour Hours	Labour Net £	Plant Net £	Materials Net £	Unit Net £	CO_2e Kg
K5.8	**Sectional corrugated metal culverts, nominal internal diameter: exceeding 1.5m**						
K5.8.1	**Not in trenches**						
K5.8.1.01	Sections corrugated, galvanised and bitumen coated steel culverts; 1.5mm thickness; 3m length jointed with coupling bands; nominal internal diameter 2.0m	m	0.27	11.79	4.59	455.02	471.40	478.131
K5.8.2	**In trenches, depth: not exceeding 1.5m**						
K5.8.2.01	Sections corrugated, galvanised and bitumen coated steel culverts; 1.5mm thickness; 3m length jointed with coupling bands; nominal internal diameter 2.0m	m	0.36	15.72	12.25	455.02	482.99	480.745
K5.8.3	**In trenches, depth: 1.5 - 2m**						
K5.8.3.01	Sections corrugated, galvanised and bitumen coated steel culverts; 1.5mm thickness; 3m length jointed with coupling bands; nominal internal diameter 2.0m	m	0.47	20.52	15.99	455.02	491.53	482.024
K5.8.4	**In trenches, depth: 2 - 2.5m**						
K5.8.4.01	Sections corrugated, galvanised and bitumen coated steel culverts; 1.5mm thickness; 3m length jointed with coupling bands; nominal internal diameter 2.0m	m	0.62	27.08	21.10	455.02	503.20	483.767
K5.8.5	**In trenches, depth: 2.5 - 3m**						
K5.8.5.01	Sections corrugated, galvanised and bitumen coated steel culverts; 1.5mm thickness; 3m length jointed with coupling bands; nominal internal diameter 2.0m	m	0.79	34.50	26.88	455.02	516.40	485.742
K5.8.6	**In trenches, depth: 3 - 3.5m**						
K5.8.6.01	Sections corrugated, galvanised and bitumen coated steel culverts; 1.5mm thickness; 3m length jointed with coupling bands; nominal internal diameter 2.0m	m	0.95	41.49	32.33	455.02	528.84	487.601
K5.8.7	**In trenches, depth: 3.5 - 4m**						
K5.8.7.01	Sections corrugated, galvanised and bitumen coated steel culverts; 1.5mm thickness; 3m length jointed with coupling bands; nominal internal diameter 2.0m	m	1.12	48.91	38.11	455.02	542.04	489.577

K7 Reinstatement

Note(s): NOTE: Reinstatement of Roads and Footpaths - The following items are provided as a guide only. Many highway authorities do not permit contractors to carry out permanent reinstatement of trenches; they carry this out themselves or use their own contractors and charge the cost to Permanent reinstatement. The rates quoted should be used with care as they are for a median situation - continuous trench work in urban areas will be cheaper because of the greater lengths involved; crossings of narrow country roads will be dearer because of the mobilisation and set up costs for a short measured length. Reinstatement of land - Because of the very wide range of types of terrain likely to be encountered in the construction of a pipeline it is not possible to give examples of cost. In addition, the cost will be dependant on the time of year and construction methods used. The following, however, are the factors to be considered:

1) recovery of all surplus construction materials, consumables and rubbish
2) restoration of original ground levels and profiles (ridge and furrow fields can prove particularly expensive)
3) Cleaning, recovery of any bridging or flushing pipes, and the restoration of banks to field ditches and other minor water courses
4) Removal of any temporary fances and gates (if not measured elsewhere)
5) Liaison with farmers and landowners

	Unit	Labour Hours	Labour Net £	Plant Net £	Materials Net £	Unit Net £	CO_2e Kg
K7.1	**Breaking up and temporary reinstatement of roads**						
K7.1.1	Pipe bore: not exceeding 300mm						
K7.1.1.01	100mm macadam base course; 50mm bituminous macadam wearing course; temporary reinstatement with 150mm lean mix; 40mm bituminous macadam base course (20mm aggregate) and 10mm bituminous macadam wearing course (10mm aggregate); pipe nominal bore; 300mm	m	0.62	27.08	10.40	23.36	60.84	45.120
K7.1.1.02	200mm granular sub-base; 100mm bituminous macadam base course; 25mm wearing course; temporary reinstatement with 200mm lean mix; 85mm bituminous macadam base course (40mm aggregate) and 15mm bituminous macadam wearing course (10mm aggregate); pipe nominal bore: 300mm	m	0.91	39.74	15.26	38.59	93.59	64.869
K7.1.1.03	100mm granular sub-base; 150mm concrete pavement; temporary reinstatement with 150mm lean mix; 150mm concrete grade 20; pipe nominal bore; 300mm	m	0.81	35.37	13.58	18.84	67.79	64.757
K7.1.1.04	150mm granular sub-base; 150mm reinforced concrete pavement; temporary reinstatement with 200mm lean mix; 150mm concrete grade 20; 1 layer mesh reinforcement (ref A252); pipe nominal bore; 300mm	m	0.96	41.92	16.10	22.29	80.31	73.405

K7   Reinstatement continued...

	Unit	Labour Hours	Labour Net £	Plant Net £	Materials Net £	Unit Net £	CO_2e Kg

K7.1   Breaking up and temporary reinstatement of roads

K7.1.2   Pipe bore: 300 - 900mm

	Unit	Labour Hours	Labour Net £	Plant Net £	Materials Net £	Unit Net £	CO_2e Kg
K7.1.2.01 100mm macadam base course; 50mm bituminous macadam wearing course; temporary reinstatement with 150mm lean mix; 40mm bituminous macadam base course (20mm aggregate) and 10mm bituminous macadam wearing course (10mm aggregate); pipe nominal bore; 600mm	m	0.71	31.01	11.91	30.57	73.49	59.445
K7.1.2.02 as above; pipe nominal bore; 900mm	m	0.83	36.25	13.92	37.77	87.94	74.175
K7.1.2.03 200mm granular sub-base; 100mm bituminous macadam base course; 25mm wearing course; temporary reinstatement with 200mm lean mix; 85mm bituminous macadam base course (40mm aggregate) and 15mm bituminous macadam wearing course (10mm aggregate); pipe nominal nominal bore 600mm	m	1.03	44.98	17.27	53.55	115.80	89.469
K7.1.2.04 as above; pipe nominal nominal bore 900mm	m	1.24	54.15	20.79	66.81	141.75	113.957
K7.1.2.05 100mm granular sub-base; 150mm concrete pavement; temporary reinstatement with 150mm lean mix; 150mm concrete grade 20; pipe nominal bore; 600mm	m	0.94	41.05	15.76	26.38	83.19	88.042
K7.1.2.06 as above; pipe nominal bore; 900mm	m	1.14	49.78	19.12	33.91	102.81	112.271
K7.1.2.07 150mm granular sub-base; 150mm reinforced concrete pavement; temporary reinstatement with 200mm lean mix; 150mm concrete grade 20; 1 layer mesh reinforcement (ref A252); pipe nominal bore; 600mm	m	1.12	48.91	18.78	31.67	99.36	102.324
K7.1.2.08 as above; pipe nominal bore; 900mm	m	1.36	59.39	22.81	41.04	123.24	132.323

K7.1.3   Pipe bore: 900 - 1800mm

	Unit	Labour Hours	Labour Net £	Plant Net £	Materials Net £	Unit Net £	CO_2e Kg
K7.1.3.01 100mm macadam base course; 50mm bituminous macadam wearing course; temporary reinstatement with 150mm lean mix; 40mm bituminous macadam base course (20mm aggregate) and 10mm bituminous macadam wearing course (10mm aggregate); pipe nominal bore; 1200mm	m	0.95	41.49	15.93	45.90	103.32	91.519

K7   Reinstatement continued...

	Unit	Labour Hours	Labour Net £	Plant Net £	Materials Net £	Unit Net £	CO_2e Kg
K7.1	**Breaking up and temporary reinstatement of roads**						
K7.1.3	Pipe bore: 900 - 1800mm						
K7.1.3.02	200mm granular sub-base; 100mm bituminous macadam base course; 25mm wearing course; temporary reinstatement with 200mm lean mix; 85mm bituminous macadam base course (40mm aggregate) and 15mm bituminous macadam wearing course (10mm aggregate); pipe nominal bore; 1200mm	m	1.45	63.32	24.32	80.85	168.49	139.108
K7.1.3.03	100mm granular sub-base; 150mm concrete pavement; temporary reinstatement with 150mm lean mix; 150mm concrete grade 20; pipe nominal bore; 1200mm	m	1.33	58.08	22.30	43.33	123.71	141.748
K7.1.3.04	150mm granular sub-base; 150mm reinforced concrete pavement; temporary reinstatement with 200mm lean mix; 150mm concrete grade 20; 1 layer mesh reinforcement (ref A252); pipe nominal bore; 1200mm	m	1.60	69.87	26.83	51.38	148.08	165.089
K7.2	**Breaking up and temporary reinstatement of footpaths**						
K7.2.1	Pipe bore: not exceeding 300mm						
K7.2.1.01	100mm granular base; 50mm wearing course; temporary reinstatement with 50mm bituminous macadam; pipe nominal bore; 300mm	m	0.52	22.71	8.72	11.98	43.41	15.634
K7.2.1.02	100mm granular base; 50mm precast concrete paving slabs; pipe nominal bore; 300mm	m	0.52	22.71	-	-	22.71	-
K7.2.2	Pipe bore: 300 - 900mm						
K7.2.2.01	100mm granular base; 50mm wearing course; temporary reinstatement with 50mm bituminous macadam; pipe nominal bore; 600mm	m	0.61	26.64	10.23	16.58	53.45	20.163
K7.2.2.02	as above; pipe nominal bore; 900mm	m	0.72	31.44	12.07	21.19	64.70	24.962
K7.2.2.03	100mm granular base; 50mm precast concrete paving slabs; pipe nominal bore; 600mm	m	0.61	26.64	10.23	-	36.87	8.231
K7.2.2.04	as above; pipe nominal bore; 900mm	m	0.72	31.44	12.07	-	43.51	9.715
K7.2.3	Pipe bore: 900 - 1800mm						
K7.2.3.01	100mm granular base; 50mm wearing course; temporary reinstatement with 50mm bituminous macadam; pipe nominal bore; 1200mm	m	0.84	36.68	14.09	26.71	77.48	30.559

K7 Reinstatement continued...

	Unit	Labour Hours	Labour Net £	Plant Net £	Materials Net £	Unit Net £	CO_2e Kg
K7.2 **Breaking up and temporary reinstatement of footpaths**							
K7.2.3 Pipe bore: 900 - 1800mm							
K7.2.3.02 100mm granular base; 50mm precast concrete paving slabs; pipe nominal bore; 1200mm	m	0.84	36.68	14.09	-	50.77	11.334
K7.6 **Strip topsoil from easement and reinstate**							
K7.6.0 Storing and protecting for re-use (225mm depth)							
K7.6.0.01 Width: 5m	m	0.02	0.27	1.03	-	1.30	0.499
K7.6.0.02 Width: 10m	m	0.04	0.51	1.91	-	2.42	0.927

K8 Other Pipework Ancillaries

	Unit	Labour Hours	Labour Net £	Plant Net £	Materials Net £	Unit Net £	CO_2e Kg
K8.2 **Marker posts**							
K8.2.0 Fix only							
K8.2.0.01 Precast concrete marker post including excavation and setting base in concrete	Nr	1.25	54.59	-	-	54.59	-
K8.5 **Connection to existing manholes and other chambers**							
K8.5.1 Pipe bore: not exceeding 200mm							
K8.5.1.01 Building in ends of pipes to existing 215mm brick manhole in engineering class "B" bricks and make good with cement mortar (1:3); 2m below ground level; pipe nominal bore; 150mm	Nr	2.50	76.65	-	3.17	79.82	12.703
K8.5.1.02 as above; 4m below ground level; pipe nominal bore; 150mm	Nr	3.00	91.98	-	3.17	95.15	12.703
K8.5.1.03 Building in ends of pipes to existing 675mm precast concrete manhole chamber and make good with cement mortar (1:3); 2m below ground level; pipe nominal bore; 150mm	Nr	1.00	30.66	-	3.17	33.83	12.703
K8.5.1.04 as above; 4m below ground level; pipe nominal bore; 150mm	Nr	1.20	36.79	-	3.17	39.96	12.703
K8.5.2 Pipe bore: 200 - 300mm							
K8.5.2.01 Building in ends of pipes to existing 215mm brick manhole in engineering class "B" bricks and make good with cement mortar (1:3); 2m below ground level; pipe nominal bore; 300mm	Nr	4.00	122.64	-	3.81	126.45	15.244
K8.5.2.02 as above; 4m below ground level; pipe nominal bore; 300mm	Nr	4.80	147.17	-	3.81	150.98	15.244
K8.5.2.03 Building in ends of pipes to existing 675mm precast concrete manhole chamber and make good with cement mortar (1:3); 2m below ground level; pipe nominal bore; 300mm	Nr	2.00	61.32	-	3.81	65.13	15.244
K8.5.2.04 as above; 4m below ground level; pipe nominal bore; 300mm	Nr	2.40	73.58	-	3.81	77.39	15.244

K8 Other Pipework Ancillaries continued...

	Unit	Labour Hours	Labour Net £	Plant Net £	Materials Net £	Unit Net £	CO_2e Kg
K8.5	**Connection to existing manholes and other chambers**						
K8.5.3	Pipe bore: 300 - 600mm						
K8.5.3.01	Building in ends of pipes to existing 215mm brick manhole in engineering class "B" bricks and make good with cement mortar (1:3); 2m below ground level; pipe nominal bore; 450mm	Nr	6.50	199.29	-	4.44	203.73	17.784
K8.5.3.02	as above; pipe nominal bore; 600mm	Nr	9.50	291.27	-	6.35	297.62	25.406
K8.5.3.03	as above; 4m below ground level; pipe nominal bore; 450mm	Nr	7.80	239.15	-	4.44	243.59	17.784
K8.5.3.04	as above; pipe nominal bore; 600mm	Nr	11.40	349.52	-	6.35	355.87	25.406
K8.5.3.05	Building in ends of pipes to existing 675mm precast concrete manhole chamber and make good with cement mortar (1:3); 2m below ground level; pipe nominal bore; 450mm	Nr	2.75	84.31	-	4.44	88.75	17.784
K8.5.3.06	as above; pipe nominal bore; 600mm	Nr	3.50	107.31	-	6.35	113.66	25.406
K8.5.3.07	as above; 4m below ground level; pipe nominal bore; 450mm	Nr	3.30	101.18	-	4.44	105.62	17.784
K8.5.3.08	as above; pipe nominal bore; 600mm	Nr	4.20	128.77	-	6.35	135.12	25.406
K8.6	**Connection to existing pipes, ducts and culverts**						
K8.6.1	Pipe bore: not exceeding 200mm						
K8.6.1.01	Building new clay pipe into existing 450mm diameter clay main; including all cutting installation of saddle building in and making good upon completion; pipe nominal bore; 150mm	Nr	1.00	30.66	-	27.47	58.13	4.884
K8.6.1.02	Building in new concrete pipe into existing 750mm diameter concrete main; including all cutting, building in and making good upon completion; pipe nominal bore; 150mm	Nr	1.50	45.99	-	1.27	47.26	5.081
K8.6.2	Pipe bore: 200 - 300mm						
K8.6.2.01	Building new clay pipe into existing 450mm diameter clay main; including all cutting installation of saddle building in and making good upon completion; pipe nominal bore; 225mm	Nr	1.50	45.99	-	96.55	142.54	9.750
K8.6.2.02	as above; pipe nominal bore; 300mm	Nr	2.10	64.39	-	168.98	233.37	19.706
K8.6.2.03	Building in new concrete pipe into existing 750mm diameter concrete main; including all cutting, building in and making good upon completion; pipe nominal bore; 225mm	Nr	2.10	64.39	-	3.17	67.56	12.703

K8 Other Pipework Ancillaries continued...

	Unit	Labour Hours	Labour Net £	Plant Net £	Materials Net £	Unit Net £	CO_2e Kg
K8.6 **Connection to existing pipes, ducts and culverts**							
K8.6.3 Pipe bore: 300 - 600mm						218	
K8.6.3.01 Building in new concrete pipe into existing 750mm diameter concrete main; including all cutting, building in and making good upon completion; pipe nominal bore; 375mm	Nr	2.80	85.85	-	3.81	89.66	15.244
K8.6.3.02 as above; pipe nominal bore; 450mm	Nr	3.50	107.31	-	5.08	112.39	20.325
K8.6.3.03 as above; pipe nominal bore; 525mm	Nr	4.25	130.31	-	6.35	136.66	25.406

CLASS L:
PIPEWORK - SUPPORTS AND PROTECTION, ANCILLARIES TO LAYING AND EXCAVATION

Calculations used throughout Class L - Pipework - Supports and Protection, Ancillaries to Laying and Excavation

Labour

			Qty		Rate		Total
L A0184ICE	**Small bore pipes in shallow trenches Labour Gang**						
	L A9015ICE	Banksman	1	x	14.03	=	£14.03
	L A9016ICE	Ganger	1	x	17.64	=	£17.64
	L A9017ICE	Pipelayer (standard rate)	1	x	14.86	=	£14.86
	L A9002ICE	Labourer (General Operative)	5	x	13.02	=	£65.08
		Total hourly cost of gang				=	**£111.61**
L A0212ICE	**Breaking out rock, concrete etc. Labour Gang**						
	L A9002ICE	Labourer (General Operative)	2	x	13.02	=	£26.03
		Total hourly cost of gang				=	**£26.03**
L A0213ICE	**Breaking out rock, concrete etc in headings and pipe jacking**						
	L A9035ICE	Miners	2	x	41.27	=	£82.54
		Total hourly cost of gang				=	**£82.54**
L A0214ICE	**Placing materials in beds, haunches and surrounds Labour Gang**						
	L A9002ICE	Labourer (General Operative)	2	x	13.02	=	£26.03
	L A9016ICE	Ganger	0.5	x	17.64	=	£8.82
		Total hourly cost of gang				=	**£34.85**
L L0001ICE	**Pipework Jacking Labour Gang**						
	L A9003ICE	Labourer (Skill Rate 3)	2	x	14.86	=	£29.72
	L A9016ICE	Ganger	1	x	17.64	=	£17.64
	L A9015ICE	Banksman	1	x	14.03	=	£14.03
	L A9035ICE	Miners	2	x	41.27	=	£82.54
		Total hourly cost of gang				=	**£143.93**
L L0002ICE	**Pipework Thrust Boring Labour Gang**						
	L A9003ICE	Labourer (Skill Rate 3)	2	x	14.86	=	£29.72
	L A9016ICE	Ganger	1	x	17.64	=	£17.64
	L A9015ICE	Banksman	1	x	14.03	=	£14.03
	L A9035ICE	Miners	2	x	41.27	=	£82.54
	L A9021ICE	Fitters and Welders	1	x	17.33	=	£17.33
		Total hourly cost of gang				=	**£161.26**

Plant

			Qty		Rate		Total
P A1184ICE	**Small bore pipes in shallow trench Plant Gang**						
	P A0592ICE	Hydraulic Excavator - Cat 320 96kW	1	x	36.36	=	£36.36
	P A0811ICE	Pump - Godwin ET50 23m3/h 4 inches	1	x	2.94	=	£2.94
	P A0853ICE	Vibrating Plate Diesel 24kN	1	x	2.53	=	£2.53
	P A0803ICE	Trench Sheets	90	x	0.09	=	£8.10
	P A0804ICE	Acrow Props	70	x	0.09	=	£6.30
	P A1023ICE	Landrover 4WD	1	x	16.16	=	£16.16
		Total hourly cost of gang				=	**£72.39**

P A1192ICE	**Breaking out rock, concrete etc. Plant Gang**						
P A0456ICE	Compressor - 250 cfm	1	x	11.57	=	£11.57
P A0401ICE	Thor 16D / Maco SK8 Medium Duty Breaker	2	x	0.82	=	£1.64
	Total hourly cost of gang				=	**£13.21**

P A1194ICE	**Placing materials in bed, haunches and surrounds Plant Gang**						
P A0854ICE	Vibrating Plate Diesel 33.5kN	1	x	2.95	=	£2.95
P A0939ICE	Crawler Tractor / Dozer - Cat 941	0.5	x	37.84	=	£18.92
	Total hourly cost of gang				=	**£21.87**

Pipework - Supports and Protection, Ancillaries To Laying and Excavation

Note(s): 1) Prices for imported materials in trenches do not allow for double handling by off highway dumpers. Good access for road vehicles is assumed.
2) Price for items dealing with 'rock' are guide prices only and may vary considerably.
3) Granular material shrinkage and compaction; the conversion factor from tonne to cubic metres for granular materials allows for shrinkage, compaction and high wastage to take account of 'ragged' trenches. The conversion for sand has been calculated as follows:

tonnes per cubic metres	1.70
shrinkage and compaction 10%	0.17
wastage 15%	1.87
conversion factor for sand	2.15t per m³

4) Prices include for spreading surplus excavated material at sides of excavation. No allowance has been made for disposal of excavated rock, concrete or the like from site.

L1   Extras To Excavation And Backfilling

		Unit	Labour Hours	Labour Net £	Plant Net £	Materials Net £	Unit Net £	CO_2e Kg
L1.1	**In pipe trenches**							
L1.1.1	Excavation of rock							
L1.1.1.01	Generally	m³	2.00	52.06	26.42	-	78.48	30.466
L1.1.2	Excavation of mass concrete							
L1.1.2.01	Generally	m³	2.75	71.58	36.33	-	107.91	41.891
L1.1.3	Excavation of reinforced concrete							
L1.1.3.01	Generally	m³	3.85	100.22	50.86	-	151.08	58.647
L1.1.4	Excavation of other artificial hard material							
L1.1.4.01	Old brick foundations	m³	1.25	32.54	16.51	-	49.05	19.041
L1.1.5	Backfilling above the Final Surface with concrete							
L1.1.5.01	Grade 15, 20mm aggregate	m³	0.05	5.58	3.58	85.10	94.26	262.787
L1.1.6	Backfilling above the Final Surface with stated material other than concrete							
L1.1.6.01	Pea gravel	m³	0.05	5.58	3.58	41.70	50.86	12.519
L1.1.7	Excavation of natural material below the Final Surface and backfilling with concrete							
L1.1.7.01	Grade 15, 20mm aggregate	m³	0.12	13.39	8.59	85.10	107.08	264.627
L1.1.8	Excavation of natural material below the Final Surface and backfilling with stated material other than concrete							
L1.1.8.01	Pea gravel	m³	0.12	13.39	8.59	41.70	63.68	14.359
L1.2	**In manholes and other chambers**							
L1.2.1	Excavation of rock							
L1.2.1.01	Generally	m³	2.00	52.06	26.42	-	78.48	30.466
L1.2.2	Excavation of mass concrete							
L1.2.2.01	Generally	m³	2.75	71.58	36.33	-	107.91	41.891
L1.2.3	Excavation of reinforced concrete							
L1.2.3.01	Generally	m³	3.85	100.22	50.86	-	151.08	58.647
L1.2.4	Excavation of other artificial hard material							
L1.2.4.01	Old brick foundations	m³	1.25	32.54	16.51	-	49.05	19.041
L1.2.5	Backfilling above the Final Surface with concrete							
L1.2.5.01	Grade 15, 20mm aggregate	m³	0.06	6.70	4.30	85.10	96.10	263.050

L1 Extras To Excavation And Backfilling continued...

	Unit	Labour Hours	Labour Net £	Plant Net £	Materials Net £	Unit Net £	CO_2e Kg
L1.2	**In manholes and other chambers**						
L1.2.6	Backfilling above the Final Surface with stated material other than concrete						
L1.2.6.01	Pea gravel	m³	0.06	6.70	4.30	41.70	52.70	12.782
L1.2.7	Excavation of natural material below the Final Surface and backfilling with concrete						
L1.2.7.01	Grade 15, 20mm aggregate	m³	0.13	14.51	9.31	85.10	108.92	264.890
L1.2.8	Excavation of natural material below the Final Surface and backfilling with stated material other than concrete						
L1.2.8.01	Pea gravel	m³	0.13	14.51	9.31	41.70	65.52	14.622
L1.3	**In headings**						
L1.3.1	Excavation of rock						
L1.3.1.01	Generally	m³	6.00	495.24	79.26	-	574.50	91.398
L1.3.5	Backfilling above the Final Surface with concrete						
L1.3.5.01	Grade 15, 20mm aggregate	m³	2.00	165.08	26.42	85.10	276.60	291.939
L1.3.6	Backfilling above the Final Surface with stated material other than concrete						
L1.3.6.01	Pea gravel	m³	2.00	165.08	26.42	41.70	233.20	41.671
L1.3.7	Excavation of natural material below the Final Surface and backfilling with concrete						
L1.3.7.01	Grade 15, 20mm aggregate	m³	4.00	330.16	52.84	85.10	468.10	322.405
L1.3.8	Excavation of natural material below the Final Surface and backfilling with stated material other than concrete						
L1.3.8.01	Pea gravel	m³	4.00	330.16	52.84	41.70	424.70	72.137
L1.5	**In pipe jacking**						
L1.5.1	Excavation of rock						
L1.5.1.01	Generally	m³	6.00	495.24	79.26	-	574.50	91.398
L1.5.5	Backfilling above the Final Surface with concrete						
L1.5.5.01	Grade 15, 20mm aggregate	m³	2.00	165.08	26.42	85.10	276.60	291.939
L1.5.6	Backfilling above the Final Surface with stated material other than concrete						
L1.5.6.01	Pea gravel	m³	2.00	165.08	26.42	41.70	233.20	41.671
L1.5.7	Excavation of natural material below the Final Surface and backfilling with concrete						
L1.5.7.01	Grade 15, 20mm aggregate	m³	4.00	330.16	52.84	85.10	468.10	322.405
L1.5.8	Excavation of natural material below the Final Surface and backfilling with stated material other than concrete						
L1.5.8.01	Pea gravel	m³	4.00	330.16	52.84	41.70	424.70	72.137

L2   Special Pipelaying Methods

Note(s):   The following prices for guide purposes only and assume a typical bore length of 25 metres.
Rates do not include allowances for:
1) Removal of top soil; locating services; cutting, welding and wrapping pipe; backfilling or reinstatement of pits; specialised water pumping, grouting or supports; or fencing around pits;
2) Mobilisation and demobilisation;
3) Working and reception pits;
4) Aborted bores due to excessively bad ground conditions will be charged at S/E cost, pit cost, intersite move cost, sleeve cost, and 2/3 of length to be bored;
5) Daywork or standing time charged at £125 per hour;
6) Trench sheets and supports left in place for more than two weeks will be charged at cost plus 15%;
7) Main contractor to allow for maintaining pits, withdrawing support and holding upon completion;
8) As ground conditions are a major contributor to prices it is recommended that advice on prices is sought from a specialist sub-contractor.

	Unit	Labour Hours	Labour Net £	Plant Net £	Materials Net £	Unit Net £	CO_2e Kg
L2.2 **Thrust boring**							
L2.2.2 nominal bore: 200 - 300mm							
L2.2.2.01 Installing steel pipes by auger thrust boring; Steel pipes, BS EN 10126; ends bevelled; 7.3m average length; welded joints to, BS 5135; minimum 9.5mm wall thickness; nominal bore: 250 mm	m	0.16	25.80	163.06	75.38	264.24	117.805
L2.2.2.02 Installing steel pipes by auger thrust boring; Steel pipes, BS EN 10126; ends bevelled; 7.3m average length; welded joints to, BS 5135; minimum 9.5mm wall thickness; nominal bore: 300mm	m	0.16	25.80	163.06	109.12	297.98	135.157
L2.2.3 nominal bore: 300 - 600mm							
L2.2.3.01 Installing steel pipes by auger thrust boring; Steel pipes, BS EN 10126; ends bevelled; 7.3m average length; welded joints to, BS 5135; minimum 9.5mm wall thickness; nominal bore: 400mm	m	0.16	25.80	163.06	181.18	370.04	152.553
L2.2.3.02 Installing steel pipes by auger thrust boring; Steel pipes, BS EN 10126; ends bevelled; 7.3m average length; welded joints to, BS 5135; minimum 9.5mm wall thickness; nominal bore: 450mm	m	0.16	25.80	163.06	349.81	538.67	169.949
L2.2.3.03 Installing steel pipes by auger thrust boring; Steel pipes, BS EN 10126; ends bevelled; 7.3m average length; welded joints to, BS 5135; minimum 9.5mm wall thickness; nominal bore: 600mm	m	0.16	25.80	163.06	382.63	571.49	256.625
L2.2.4 nominal bore: 600 - 900mm							
L2.2.4.01 Installing steel pipes by auger thrust boring; Steel pipes, BS EN 10126; ends bevelled; 7.3m average length; welded joints to, BS 5135; minimum 9.5mm wall thickness; nominal bore: 750mm	m	0.16	25.80	163.06	612.43	801.29	350.545
L2.2.4.02 Installing steel pipes by auger thrust boring; Steel pipes, BS EN 10126; ends bevelled; 7.3m average length; welded joints to, BS 5135; minimum 9.5mm wall thickness; nominal bore: 900mm	m	0.16	25.80	163.06	753.77	942.63	406.941

L2 Special Pipelaying Methods continued...

	Unit	Labour Hours	Labour Net £	Plant Net £	Materials Net £	Unit Net £	CO_2e Kg
L2.3 **Pipe jacking**							
L2.3.4 nominal bore: 600 - 900mm							
L2.3.4.01 Concrete pipes, BS 5911, Part 1: Class H; 2.44m long; rebated joint with rubber sealing ring; nominal bore: 900mm	m	2.11	304.12	495.82	216.06	1,016.00	264.056
L2.3.5 nominal bore: 900 - 1200mm							
L2.3.5.01 Concrete pipes, BS 5911, Part 1: Class H; 2.44m long; rebated joint with rubber sealing ring; nominal bore: 1050mm	m	2.11	304.12	495.82	284.56	1,084.50	307.171
L2.3.5.02 Concrete pipes, BS 5911, Part 1: Class H; 2.44m long; rebated joint with rubber sealing ring; nominal bore: 1200mm	m	2.11	304.12	495.82	353.07	1,153.01	350.286
L2.3.6 nominal bore: 1200 - 1500mm							
L2.3.6.01 Concrete pipes, BS 5911, Part 1: Class H; 2.44m long; rebated joint with rubber sealing ring; nominal bore: 1350mm	m	2.11	304.12	495.82	421.58	1,221.52	402.901
L2.3.6.02 Concrete pipes, BS 5911, Part 1: Class H; 2.44m long; rebated joint with rubber sealing ring; nominal bore: 1500mm	m	2.11	304.12	495.82	490.08	1,290.02	455.515
L2.3.7 nominal bore: 1500 - 1800mm							
L2.3.7.01 Concrete pipes, BS 5911, Part 1: Class H; 2.44m long; rebated joint with rubber sealing ring; nominal bore: 1650mm	m	2.11	304.12	495.82	558.59	1,358.53	508.130
L2.3.7.02 Concrete pipes, BS 5911, Part 1: Class H; 2.44m long; rebated joint with rubber sealing ring; nominal bore: 1800mm	m	2.11	304.12	495.82	627.09	1,427.03	539.991
L2.3.8 nominal bore: stated exceeding 1800mm							
L2.3.8.01 Concrete pipes, BS 5911, Part 1: Class H; 2.44m long; rebated joint with rubber sealing ring; nominal bore: 2100mm	m	2.11	304.12	495.82	685.06	1,485.00	560.745

L3 Beds

	Unit	Labour Hours	Labour Net £	Plant Net £	Materials Net £	Unit Net £	CO_2e Kg
L3.1 **Sand**							
L3.1.1 nominal bore: not exceeding 200mm							
L3.1.1.01 Depth 100mm; nominal bore: 100mm	m	0.06	2.09	1.31	2.81	6.21	1.411
L3.1.1.02 Depth 100mm; nominal bore: 150mm	m	0.07	2.44	1.53	3.02	6.99	1.559
L3.1.1.03 Depth 150mm; nominal bore: 100mm	m	0.09	3.14	1.97	4.10	9.21	2.079
L3.1.1.04 Depth 150mm; nominal bore: 150mm	m	0.10	3.49	2.19	4.53	10.21	2.301
L3.1.2 nominal bore: 200 - 300mm							
L3.1.2.01 Depth 100mm; nominal bore: 225mm	m	0.08	2.79	1.75	3.45	7.99	1.782

L3 Beds continued...

	Unit	Labour Hours	Labour Net £	Plant Net £	Materials Net £	Unit Net £	CO_2e Kg
L3.1 **Sand**							
L3.1.2 nominal bore: 200 - 300mm							
L3.1.2.02 Depth 100mm; nominal bore: 300mm	m	0.09	3.14	1.97	3.67	8.78	1.930
L3.1.2.03 Depth 150mm; nominal bore: 225mm	m	0.12	4.18	2.62	4.96	11.76	2.598
L3.1.2.04 Depth 150mm; nominal bore: 300mm	m	0.13	4.53	2.84	5.61	12.98	2.895
L3.1.3 nominal bore: 300 - 600mm							
L3.1.3.01 Depth 100mm; nominal bore: 375mm	m	0.10	3.49	2.19	4.32	10.00	2.227
L3.1.3.02 Depth 100mm; nominal bore: 450mm	m	0.11	3.83	2.41	4.75	10.99	2.450
L3.1.3.03 Depth 100mm; nominal bore: 600mm	m	0.13	4.53	2.84	5.18	12.55	2.747
L3.1.3.04 Depth 150mm; nominal bore: 375mm	m	0.15	5.23	3.28	6.04	14.55	3.192
L3.1.3.05 Depth 150mm; nominal bore: 450mm	m	0.16	5.58	3.50	6.69	15.77	3.489
L3.1.3.06 Depth 150mm; nominal bore: 600mm	m	0.20	6.97	4.37	7.55	18.89	4.083
L3.1.4 nominal bore: 600 - 900mm							
L3.1.4.01 Depth 100mm; nominal bore: 675mm	m	0.14	4.88	3.06	5.39	13.33	2.895
L3.1.4.02 Depth 100mm; nominal bore: 750mm	m	0.15	5.23	3.28	5.83	14.34	3.118
L3.1.4.03 Depth 100mm; nominal bore: 800mm	m	0.15	5.23	3.28	6.04	14.55	3.192
L3.1.4.04 Depth 100mm; nominal bore: 900mm	m	0.17	5.92	3.72	6.47	16.11	3.489
L3.1.4.05 Depth 150mm; nominal bore: 675mm	m	0.21	7.32	4.59	8.20	20.11	4.380
L3.1.4.06 Depth 150mm; nominal bore: 750mm	m	0.23	8.02	5.03	8.63	21.68	4.677
L3.1.4.07 Depth 150mm; nominal bore: 800mm	m	0.24	8.36	5.25	9.06	22.67	4.899
L3.1.4.08 Depth 150mm; nominal bore: 900mm	m	0.25	8.71	5.47	9.71	23.89	5.196
L3.1.5 nominal bore: 900 - 1200mm							
L3.1.5.01 Depth 100mm; nominal bore: 1200mm	m	0.20	6.97	4.37	9.06	20.40	4.603
L3.1.5.02 Depth 150mm; nominal bore: 1200mm	m	0.30	10.46	6.56	13.60	30.62	6.904
L3.1.6 nominal bore: 1200 - 1500mm							
L3.1.6.01 Depth 100mm; nominal bore: 1500mm	m	0.23	8.02	5.03	10.36	23.41	5.271
L3.1.6.02 Depth 150mm; nominal bore: 1500mm	m	0.34	11.85	7.44	15.75	35.04	7.944
L3.1.7 nominal bore: 1500 - 1800mm							
L3.1.7.01 Depth 100mm; nominal bore: 1800mm	m	0.26	9.06	5.69	11.87	26.62	6.013
L3.1.7.02 Depth 150mm; nominal bore: 1800mm	m	0.39	13.59	8.53	17.70	39.82	8.983

L3 Beds continued...

	Unit	Labour Hours	Labour Net £	Plant Net £	Materials Net £	Unit Net £	CO_2e Kg
L3.2	**Selected excavated granular material**						
L3.2.1	nominal bore: not exceeding 200mm						
L3.2.1.01	Depth 100mm; nominal bore: 100mm	m	0.09	3.14	1.97	-	5.11	0.667
L3.2.1.02	Depth 100mm; nominal bore: 150mm	m	0.10	3.49	2.19	-	5.68	0.742
L3.2.1.03	Depth 150mm; nominal bore: 100mm	m	0.13	4.53	2.84	-	7.37	0.964
L3.2.1.04	Depth 150mm; nominal bore: 150mm	m	0.15	5.23	3.28	-	8.51	1.112
L3.2.2	nominal bore: 200 - 300mm						
L3.2.2.01	Depth 100mm; nominal bore: 225mm	m	0.12	4.18	2.62	-	6.80	0.890
L3.2.2.02	Depth 100mm; nominal bore: 300mm	m	0.13	4.53	2.84	-	7.37	0.964
L3.2.2.03	Depth 150mm; nominal bore: 225mm	m	0.16	5.58	3.50	-	9.08	1.186
L3.2.2.04	Depth 150mm; nominal bore: 300mm	m	0.19	6.62	4.16	-	10.78	1.409
L3.2.3	nominal bore: 300 - 600mm						
L3.2.3.01	Depth 100mm; nominal bore: 375mm	m	0.15	5.23	3.28	-	8.51	1.112
L3.2.3.02	Depth 100mm; nominal bore: 450mm	m	0.16	5.58	3.50	-	9.08	1.186
L3.2.3.03	Depth 100mm; nominal bore: 600mm	m	0.20	6.97	4.37	-	11.34	1.483
L3.2.3.04	Depth 150mm; nominal bore: 375mm	m	0.22	7.67	4.81	-	12.48	1.631
L3.2.3.05	Depth 150mm; nominal bore: 450mm	m	0.24	8.36	5.25	-	13.61	1.780
L3.2.3.06	Depth 150mm; nominal bore: 600mm	m	0.30	10.46	6.56	-	17.02	2.225
L3.2.4	nominal bore: 600 - 900mm						
L3.2.4.01	Depth 100mm; nominal bore: 675mm	m	0.21	7.32	4.59	-	11.91	1.557
L3.2.4.02	Depth 100mm; nominal bore: 750mm	m	0.23	8.02	5.03	-	13.05	1.705
L3.2.4.03	Depth 100mm; nominal bore: 800mm	m	0.24	8.36	5.25	-	13.61	1.780
L3.2.4.04	Depth 100mm; nominal bore: 900mm	m	0.25	8.71	5.47	-	14.18	1.854
L3.2.4.05	Depth 150mm; nominal bore: 675mm	m	0.32	11.15	7.00	-	18.15	2.373
L3.2.4.06	Depth 150mm; nominal bore: 750mm	m	0.35	12.20	7.65	-	19.85	2.595
L3.2.4.07	Depth 150mm; nominal bore: 800mm	m	0.36	12.55	7.87	-	20.42	2.669
L3.2.4.08	Depth 150mm; nominal bore: 900mm	m	0.38	13.24	8.31	-	21.55	2.818
L3.2.5	nominal bore: 900 - 1200mm						
L3.2.5.01	Depth 100mm; nominal bore: 1200mm	m	0.30	10.46	6.56	-	17.02	2.225
L3.2.5.02	Depth 150mm; nominal bore: 1200mm	m	0.45	15.68	9.84	-	25.52	3.337

L3 Beds continued...

	Unit	Labour Hours	Labour Net £	Plant Net £	Materials Net £	Unit Net £	CO_2e Kg
L3.2	**Selected excavated granular material**						
L3.2.6	nominal bore: 1200 - 1500mm						
L3.2.6.01	Depth 100mm; nominal bore: 1500mm	m	0.34	11.85	7.44	-	19.29	2.521
L3.2.6.02	Depth 150mm; nominal bore: 1500mm	m	0.51	17.77	11.15	-	28.92	3.782
L3.2.7	nominal bore: 1500 - 1800mm						
L3.2.7.01	Depth 100mm; nominal bore: 1800mm	m	0.39	13.59	8.53	-	22.12	2.892
L3.2.7.02	Depth 150mm; nominal bore: 1800mm	m	0.59	20.56	12.90	-	33.46	4.375
L3.3	**Imported granular material**						
L3.3.1	nominal bore: not exceeding 200mm						
L3.3.1.01	DTp clause nr. 503; Depth 100mm; nominal bore: 100mm	m	0.06	2.09	1.31	2.54	5.94	1.355
L3.3.1.02	DTp clause nr. 503; Depth 100mm; nominal bore: 150mm	m	0.07	2.44	1.53	2.74	6.71	1.499
L3.3.1.03	DTp clause nr. 503; Depth 150mm; nominal bore: 100mm	m	0.09	3.14	1.97	3.71	8.82	1.998
L3.3.1.04	DTp clause nr. 503; Depth 150mm; nominal bore: 150mm	m	0.10	3.49	2.19	4.11	9.79	2.212
L3.3.2	nominal bore: 200 - 300mm						
L3.3.2.01	DTp clause nr. 503; Depth 100mm; nominal bore: 225mm	m	0.08	2.79	1.75	3.13	7.67	1.714
L3.3.2.02	DTp clause nr. 503; Depth 100mm; nominal bore: 300mm	m	0.09	3.14	1.97	3.32	8.43	1.858
L3.3.2.03	DTp clause nr. 503; Depth 150mm; nominal bore: 225mm	m	0.12	4.18	2.62	4.50	11.30	2.500
L3.3.2.04	DTp clause nr. 503; Depth 150mm; nominal bore: 300mm	m	0.13	4.53	2.84	5.08	12.45	2.785
L3.3.3	nominal bore: 300 - 600mm						
L3.3.3.01	DTp clause nr. 503; Depth 100mm; nominal bore: 375mm	m	0.10	3.49	2.19	3.71	9.39	2.072
L3.3.3.02	DTp clause nr. 503; Depth 100mm; nominal bore: 450mm	m	0.11	3.83	2.41	3.91	10.15	2.216
L3.3.3.03	DTp clause nr. 503; Depth 100mm; nominal bore: 600mm	m	0.13	4.53	2.84	4.69	12.06	2.645
L3.3.3.04	DTp clause nr. 503; Depth 150mm; nominal bore: 375mm	m	0.15	5.23	3.28	5.47	13.98	3.073
L3.3.3.05	DTp clause nr. 503; Depth 150mm; nominal bore: 450mm	m	0.16	5.58	3.50	6.06	15.14	3.357
L3.3.3.06	DTp clause nr. 503; Depth 150mm; nominal bore: 600mm	m	0.20	6.97	4.37	6.84	18.18	3.934
L3.3.4	nominal bore: 600 - 900mm						
L3.3.4.01	DTp clause nr. 503; Depth 100mm; nominal bore: 675mm	m	0.14	4.88	3.06	4.89	12.83	2.789
L3.3.4.02	DTp clause nr. 503; Depth 100mm; nominal bore: 750mm	m	0.15	5.23	3.28	5.28	13.79	3.003
L3.3.4.03	DTp clause nr. 503; Depth 100mm; nominal bore: 800mm	m	0.15	5.23	3.28	5.47	13.98	3.073
L3.3.4.04	DTp clause nr. 503; Depth 100mm; nominal bore: 900mm	m	0.17	5.92	3.72	5.87	15.51	3.361
L3.3.4.05	DTp clause nr. 503; Depth 150mm; nominal bore: 675mm	m	0.21	7.32	4.59	7.43	19.34	4.218

L3 Beds continued...

	Unit	Labour Hours	Labour Net £	Plant Net £	Materials Net £	Unit Net £	CO$_2$e Kg
L3.3	**Imported granular material**						
L3.3.4	**nominal bore: 600 - 900mm**						
L3.3.4.06	DTp clause nr. 503; Depth 150mm; nominal bore: 750mm	m	0.23	8.02	5.03	7.82	20.87	4.507
L3.3.4.07	DTp clause nr. 503; Depth 150mm; nominal bore: 800mm	m	0.24	8.36	5.25	8.21	21.82	4.721
L3.3.4.08	DTp clause nr. 503; Depth 150mm; nominal bore: 900mm	m	0.25	8.71	5.47	8.80	22.98	5.005
L3.3.5	**nominal bore: 900 - 1200mm**						
L3.3.5.01	DTp clause nr. 503; Depth 100mm; nominal bore: 1200mm	m	0.20	6.97	4.37	8.21	19.55	4.424
L3.3.5.02	DTp clause nr. 503; Depth 150mm; nominal bore: 1200mm	m	0.30	10.46	6.56	12.32	29.34	6.636
L3.3.6	**nominal bore: 1200 - 1500mm**						
L3.3.6.01	DTp clause nr. 503; Depth 100mm; nominal bore: 1500mm	m	0.23	8.02	5.03	9.38	22.43	5.067
L3.3.6.02	DTp clause nr. 503; Depth 150mm; nominal bore: 1500mm	m	0.34	11.85	7.44	14.27	33.56	7.633
L3.3.7	**nominal bore: 1500 - 1800mm**						
L3.3.7.01	DTp clause nr. 503; Depth 100mm; nominal bore: 1800mm	m	0.26	9.06	5.69	10.75	25.50	5.780
L3.3.7.02	DTp clause nr. 503; Depth 150mm; nominal bore: 1800mm	m	0.39	13.59	8.53	16.03	38.15	8.634
L3.4	**Mass Concrete**						
L3.4.1	**nominal bore: not exceeding 200mm**						
L3.4.1.01	Grade 15, 20mm aggregate; Depth 100mm; nominal bore: 100mm	m	0.07	2.44	1.53	5.11	9.08	16.207
L3.4.1.02	Grade 15, 20mm aggregate; Depth 100mm; nominal bore: 150mm	m	0.09	3.14	1.97	5.53	10.64	17.663
L3.4.1.03	Grade 15, 20mm aggregate; Depth 150mm; nominal bore: 100mm	m	0.11	3.83	2.41	7.66	13.90	24.348
L3.4.1.04	Grade 15, 20mm aggregate; Depth 150mm; nominal bore: 150mm	m	0.12	4.18	2.62	8.51	15.31	27.037
L3.4.2	**nominal bore: 200 - 300mm**						
L3.4.2.01	Grade 15, 20mm aggregate; Depth 100mm; nominal bore: 225mm	m	0.10	3.49	2.19	6.21	11.89	19.829
L3.4.2.02	Grade 15, 20mm aggregate; Depth 100mm; nominal bore: 300mm	m	0.12	4.18	2.62	6.81	13.61	21.808
L3.4.2.03	Grade 15, 20mm aggregate; Depth 150mm; nominal bore: 225mm	m	0.15	5.23	3.28	9.28	17.79	29.613
L3.4.2.04	Grade 15, 20mm aggregate; Depth 150mm; nominal bore: 300mm	m	0.16	5.58	3.50	10.21	19.29	32.563
L3.4.3	**nominal bore: 300 - 600mm**						
L3.4.3.01	Grade 15, 20mm aggregate; Depth 100mm; nominal bore: 375mm	m	0.13	4.53	2.84	7.49	14.86	23.974

L3 Beds continued...

	Unit	Labour Hours	Labour Net £	Plant Net £	Materials Net £	Unit Net £	CO_2e Kg
L3.4 **Mass Concrete**							
L3.4.3 nominal bore: 300 - 600mm							
L3.4.3.02 Grade 15, 20mm aggregate; Depth 100mm; nominal bore: 450mm	m	0.14	4.88	3.06	8.08	16.02	25.878
L3.4.3.03 Grade 15, 20mm aggregate; Depth 100mm; nominal bore: 600mm	m	0.16	5.58	3.50	9.36	18.44	29.948
L3.4.3.04 Grade 15, 20mm aggregate; Depth 150mm; nominal bore: 375mm	m	0.18	6.27	3.94	11.23	21.44	35.849
L3.4.3.05 Grade 15, 20mm aggregate; Depth 150mm; nominal bore: 450mm	m	0.20	6.97	4.37	12.17	23.51	38.874
L3.4.3.06 Grade 15, 20mm aggregate; Depth 150mm; nominal bore: 600mm	m	0.25	8.71	5.47	14.04	28.22	44.997
L3.4.4 nominal bore: 600 - 900mm							
L3.4.4.01 Grade 15, 20mm aggregate; Depth 100mm; nominal bore: 675mm	m	0.17	5.92	3.72	10.04	19.68	32.114
L3.4.4.02 Grade 15, 20mm aggregate; Depth 100mm; nominal bore: 750mm	m	0.18	6.27	3.94	10.64	20.85	34.019
L3.4.4.03 Grade 15, 20mm aggregate; Depth 100mm; nominal bore: 800mm	m	0.19	6.62	4.16	11.06	21.84	35.400
L3.4.4.04 Grade 15, 20mm aggregate; Depth 100mm; nominal bore: 900mm	m	0.21	7.32	4.59	11.91	23.82	38.163
L3.4.4.05 Grade 15, 20mm aggregate; Depth 150mm; nominal bore: 675mm	m	0.26	9.06	5.69	15.06	29.81	48.209
L3.4.4.06 Grade 15, 20mm aggregate; Depth 150mm; nominal bore: 750mm	m	0.29	10.11	6.34	16.00	32.45	51.307
L3.4.4.07 Grade 15, 20mm aggregate; Depth 150mm; nominal bore: 800mm	m	0.30	10.46	6.56	16.59	33.61	53.212
L3.4.4.08 Grade 15, 20mm aggregate; Depth 150mm; nominal bore: 900mm	m	0.32	11.15	7.00	17.87	36.02	57.282
L3.4.5 nominal bore: 900 - 1200mm							
L3.4.5.01 Grade 15, 20mm aggregate; Depth 100mm; nominal bore: 1200mm	m	0.25	8.71	5.47	16.59	30.77	52.841
L3.4.5.02 Grade 15, 20mm aggregate; Depth 150mm; nominal bore: 1200mm	m	0.37	12.89	8.09	24.93	45.91	79.355
L3.4.6 nominal bore: 1200 - 1500mm							
L3.4.6.01 Grade 15, 20mm aggregate; Depth 100mm; nominal bore: 1500mm	m	0.29	10.11	6.34	19.15	35.60	60.982
L3.4.6.02 Grade 15, 20mm aggregate; Depth 150mm; nominal bore: 1500mm	m	0.42	14.64	9.19	28.76	52.59	91.492

L3   Beds continued...

	Unit	Labour Hours	Labour Net £	Plant Net £	Materials Net £	Unit Net £	CO_2e Kg
L3.4 **Mass Concrete**							
L3.4.7 nominal bore: 1500 - 1800mm							
L3.4.7.01 Grade 15, 20mm aggregate; Depth 100mm; nominal bore: 1800mm	m	0.33	11.50	7.22	21.70	40.42	69.123
L3.4.7.02 Grade 15, 20mm aggregate; Depth 150mm; nominal bore: 1800mm	m	0.50	17.43	10.94	32.59	60.96	103.852
L3.5 **Reinforced concrete**							
L3.5.1 nominal bore: not exceeding 200mm							
L3.5.1.01 Grade 25, 20mm aggregate; 1 layer A252 mesh reinforcement; nominal bore: 100mm	m	0.10	3.49	2.19	7.04	12.72	18.744
L3.5.1.02 Grade 25, 20mm aggregate; 1 layer A252 mesh reinforcement; nominal bore: 150mm	m	0.12	4.18	2.62	7.62	14.42	20.393
L3.5.1.03 Grade 25, 20mm aggregate; 1 layer A252 mesh reinforcement; nominal bore: 100mm	m	0.14	4.88	3.06	9.75	17.69	27.345
L3.5.1.04 Grade 25, 20mm aggregate; 1 layer A252 mesh reinforcement; nominal bore: 150mm	m	0.15	5.23	3.28	10.79	19.30	30.303
L3.5.2 nominal bore: 200 - 300mm							
L3.5.2.01 Grade 25, 20mm aggregate; 1 layer A252 mesh reinforcement; nominal bore: 225mm	m	0.14	4.88	3.06	8.56	16.50	22.942
L3.5.2.02 Grade 25, 20mm aggregate; 1 layer A252 mesh reinforcement; nominal bore: 300mm	m	0.16	5.58	3.50	9.38	18.46	25.190
L3.5.2.03 Grade 25, 20mm aggregate; 1 layer A252 mesh reinforcement; nominal bore: 225mm	m	0.19	6.62	4.16	11.81	22.59	33.277
L3.5.2.04 Grade 25, 20mm aggregate; 1 layer A252 mesh reinforcement; nominal bore: 300mm	m	0.20	6.97	4.37	13.00	24.34	36.559
L3.5.3 nominal bore: 300 - 600mm							
L3.5.3.01 Grade 25, 20mm aggregate; 1 layer A252 mesh reinforcement; Depth 100mm; nominal bore: 375mm	m	0.17	5.92	3.72	10.32	19.96	27.665
L3.5.3.02 Grade 25, 20mm aggregate; 1 layer A252 mesh reinforcement; Depth 100mm; nominal bore: 450mm	m	0.19	6.62	4.16	11.14	21.92	29.914
L3.5.3.03 Grade 25, 20mm aggregate; 1 layer A252 mesh reinforcement; Depth 100mm; nominal bore: 600mm	m	0.22	7.67	4.81	-	12.48	1.631
L3.5.3.04 Grade 25, 20mm aggregate; 1 layer A252 mesh reinforcement; Depth 150mm; nominal bore: 375mm	m	0.22	7.67	4.81	14.30	26.78	40.215
L3.5.3.05 Grade 25, 20mm aggregate; 1 layer A252 mesh reinforcement; nominal bore: 450mm	m	0.25	8.71	5.47	15.48	29.66	43.645

L3   Beds continued...

	Unit	Labour Hours	Labour Net £	Plant Net £	Materials Net £	Unit Net £	CO$_2$e Kg
L3.5 **Reinforced concrete**							
L3.5.3 nominal bore: 300 - 600mm							
L3.5.3.06 Grade 25, 20mm aggregate; 1 layer A252 mesh reinforcement; nominal bore: 600mm	m	0.31	10.80	6.78	17.87	35.45	50.528
L3.5.4 nominal bore: 600 - 900mm							
L3.5.4.01 Grade 25, 20mm aggregate; 1 layer A252 mesh reinforcement; Depth 100mm; nominal bore: 675mm	m	0.23	8.02	5.03	13.84	26.89	37.111
L3.5.4.02 Grade 25, 20mm aggregate; 1 layer A252 mesh reinforcement; Depth 100mm; nominal bore: 750mm	m	0.24	8.36	5.25	14.66	28.27	39.286
L3.5.4.03 Grade 25, 20mm aggregate; 1 layer A252 mesh reinforcement; Depth 100mm; nominal bore: 800mm	m	0.26	9.06	5.69	15.25	30.00	40.934
L3.5.4.04 Grade 25, 20mm aggregate; 1 layer A252 mesh reinforcement; Depth 100mm; nominal bore: 900mm	m	0.28	9.76	6.12	16.42	32.30	44.083
L3.5.4.05 Grade 25, 20mm aggregate; 1 layer A252 mesh reinforcement; Depth 150mm; nominal bore: 675mm	m	0.32	11.15	7.00	19.17	37.32	54.110
L3.5.4.06 Grade 25, 20mm aggregate; 1 layer A252 mesh reinforcement; Depth 150mm; nominal bore: 750mm	m	0.35	12.20	7.65	20.03	39.88	57.261
L3.5.4.07 Grade 25, 20mm aggregate; 1 layer A252 mesh reinforcement; Depth 150mm; nominal bore: 800mm	m	0.37	12.89	8.09	21.12	42.10	59.742
L3.5.4.08 Grade 25, 20mm aggregate; 1 layer A252 mesh reinforcement; Depth 150mm; nominal bore: 900mm	m	0.39	13.59	8.53	18.98	41.10	61.020
L3.5.5 nominal bore: 900 - 1200mm							
L3.5.5.01 Grade 25, 20mm aggregate; 1 layer A252 mesh reinforcement; Depth 100mm; nominal bore: 1200mm	m	0.35	12.20	7.65	22.87	42.72	61.105
L3.5.5.02 Grade 25, 20mm aggregate; 1 layer A252 mesh reinforcement; Depth 150mm; nominal bore: 1200mm	m	0.47	16.38	10.28	31.72	58.38	89.121
L3.5.6 nominal bore: 1200 - 1500mm							
L3.5.6.01 Grade 25, 20mm aggregate; 1 layer A252 mesh reinforcement; Depth 100mm; nominal bore: 1500mm	m	0.40	13.94	8.75	26.39	49.08	70.477
L3.5.6.02 Grade 25, 20mm aggregate; 1 layer A252 mesh reinforcement; Depth 150mm; nominal bore: 1500mm	m	0.53	18.47	11.59	36.60	66.66	102.719

L3 Beds continued...

	Unit	Labour Hours	Labour Net £	Plant Net £	Materials Net £	Unit Net £	CO_2e Kg
L3.5	**Reinforced concrete**						
L3.5.7	nominal bore: 1500 - 1800mm						
L3.5.7.01	Grade 25, 20mm aggregate; 1 layer A252 mesh reinforcement; Depth 100mm; nominal bore: 1800mm	m	0.46	16.03	10.06	29.90	55.99	79.923
L3.5.7.02	Grade 25, 20mm aggregate; 1 layer A252 mesh reinforcement; Depth 150mm; nominal bore: 1800mm	m	0.63	21.96	13.78	41.47	77.21	116.614

L4 Haunches

	Unit	Labour Hours	Labour Net £	Plant Net £	Materials Net £	Unit Net £	CO_2e Kg
L4.4	**Mass Concrete**						
L4.4.1	nominal bore: not exceeding 200mm						
L4.4.1.01	Including beds; grade 15, 20mm aggregate; Depth 100mm; nominal bore: 100mm	m	0.12	4.18	2.62	8.51	15.31	27.037
L4.4.1.02	Including beds; grade 15, 20mm aggregate; Depth 100mm; nominal bore: 150mm	m	0.15	5.23	3.28	9.36	17.87	29.874
L4.4.1.03	Including beds; grade 15, 20mm aggregate; Depth 150mm; nominal bore: 100mm	m	0.15	5.23	3.28	11.06	19.57	35.104
L4.4.1.04	Including beds; grade 15, 20mm aggregate; Depth 150mm; nominal bore: 150mm	m	0.19	6.62	4.16	12.17	22.95	38.799
L4.4.2	nominal bore: 200 - 300mm						
L4.4.2.01	Including beds; grade 15, 20mm aggregate; Depth 100mm; nominal bore: 225mm	m	0.18	6.27	3.94	10.64	20.85	34.019
L4.4.2.02	Including beds; grade 15, 20mm aggregate; Depth 100mm; nominal bore: 300mm	m	0.22	7.67	4.81	12.34	24.82	39.545
L4.4.2.03	Including beds; grade 15, 20mm aggregate; Depth 150mm; nominal bore: 225mm	m	0.22	7.67	4.81	14.47	26.95	46.082
L4.4.2.04	Including beds; grade 15, 20mm aggregate; Depth 150mm; nominal bore: 300mm	m	0.27	9.41	5.90	15.74	31.05	50.375
L4.4.3	nominal bore: 300 - 600mm						
L4.4.3.01	Including beds; grade 15, 20mm aggregate; Depth 100mm; nominal bore: 375mm	m	0.27	9.41	5.90	15.32	30.63	49.067
L4.4.3.02	Including beds; grade 15, 20mm aggregate; Depth 100mm; nominal bore: 450mm	m	0.32	11.15	7.00	18.72	36.87	59.897
L4.4.3.03	Including beds; grade 15, 20mm aggregate; Depth 100mm; nominal bore: 600mm	m	0.40	13.94	8.75	29.78	52.47	94.482
L4.4.3.04	Including beds; grade 15, 20mm aggregate; Depth 150mm; nominal bore: 375mm	m	0.33	11.50	7.22	19.06	37.78	61.017
L4.4.3.05	Including beds; grade 15, 20mm aggregate; Depth 150mm; nominal bore: 450mm	m	0.38	13.24	8.31	22.81	44.36	72.892

L4 Haunches continued...

	Unit	Labour Hours	Labour Net £	Plant Net £	Materials Net £	Unit Net £	CO_2e Kg
L4.4	**Mass Concrete**						
L4.4.3	nominal bore: 300 - 600mm						
L4.4.3.06	Including beds; grade 15, 20mm aggregate; Depth 150mm; nominal bore: 600mm	m	0.46	16.03	10.06	34.47	60.56	109.307
L4.4.4	nominal bore: 600 - 900mm						
L4.4.4.01	Including beds; grade 15, 20mm aggregate; Depth 100mm; nominal bore: 675mm	m	0.50	17.43	10.94	33.19	61.56	105.682
L4.4.4.02	Including beds; grade 15, 20mm aggregate; Depth 100mm; nominal bore: 750mm	m	0.69	24.05	15.09	40.85	79.99	130.623
L4.4.4.03	Including beds; grade 15, 20mm aggregate; Depth 100mm; nominal bore: 800mm	m	0.85	29.62	18.59	49.36	97.57	157.957
L4.4.4.04	Including beds; grade 15, 20mm aggregate; Depth 100mm; nominal bore: 900mm	m	0.95	33.11	20.78	53.61	107.50	171.772
L4.4.4.05	Including beds; grade 15, 20mm aggregate; Depth 150mm; nominal bore: 675mm	m	0.57	19.86	12.47	38.72	71.05	123.197
L4.4.4.06	Including beds; grade 15, 20mm aggregate; Depth 150mm; nominal bore: 750mm	m	0.77	26.83	16.84	46.21	89.88	147.689
L4.4.4.07	Including beds; grade 15, 20mm aggregate; Depth 150mm; nominal bore: 800mm	m	0.95	33.11	20.78	54.89	108.78	175.694
L4.4.4.08	Including beds; grade 15, 20mm aggregate; Depth 150mm; nominal bore: 900mm	m	1.05	36.59	22.96	59.57	119.12	190.817
L4.4.5	nominal bore: 900 - 1200mm						
L4.4.5.01	Including beds; grade 15, 20mm aggregate; Depth 100mm; nominal bore: 1200mm	m	1.15	40.08	25.15	76.59	141.82	243.853
L4.4.5.02	Including beds; grade 15, 20mm aggregate; Depth 150mm; nominal bore: 1200mm	m	1.28	44.61	27.99	85.10	157.70	270.964
L4.4.6	nominal bore: 1200 - 1500mm						
L4.4.6.01	Including beds; grade 15, 20mm aggregate; Depth 100mm; nominal bore: 1500mm	m	1.38	48.09	30.18	91.06	169.33	290.009
L4.4.6.02	Including beds; grade 15, 20mm aggregate; Depth 150mm; nominal bore: 1500mm	m	1.42	49.49	31.06	100.67	181.22	319.852
L4.4.7	nominal bore: 1500 - 1800mm						
L4.4.7.01	Including beds; grade 15, 20mm aggregate; Depth 100mm; nominal bore: 1800mm	m	1.60	55.76	34.99	110.63	201.38	351.779
L4.4.7.02	Including beds; grade 15, 20mm aggregate; Depth 150mm; nominal bore: 1800mm	m	1.76	61.34	38.49	121.52	221.35	386.434

L4   Haunches continued...

	Unit	Labour Hours	Labour Net £	Plant Net £	Materials Net £	Unit Net £	CO_2e Kg
L4.5	**Reinforced concrete**							
L4.5.1	nominal bore: not exceeding 200mm						
L4.5.1.01	Including beds; grade 25, 20mm aggregate; 1 layer A252 mesh reinforcement; Depth 100mm; nominal bore: 100mm	m	0.15	5.23	3.28	10.65	19.16	30.187
L4.5.1.02	Including beds; grade 25, 20mm aggregate; 1 layer A252 mesh reinforcement; Depth 100mm; nominal bore: 150mm	m	0.18	6.27	3.94	11.69	21.90	33.294
L4.5.1.03	Including beds; grade 25, 20mm aggregate; 1 layer A252 mesh reinforcement; Depth 150mm; nominal bore: 100mm	m	0.18	6.27	3.94	13.36	23.57	38.714
L4.5.1.04	Including beds; grade 25, 20mm aggregate; 1 layer A252 mesh reinforcement; Depth 150mm; nominal bore: 150mm	m	0.22	7.67	4.81	12.92	25.40	41.214
L4.5.2	nominal bore: 200 - 300mm						
L4.5.2.01	Including beds; grade 25, 20mm aggregate; 1 layer A252 mesh reinforcement; Depth 100mm; nominal bore: 225mm	m	0.22	7.67	4.81	13.98	26.46	40.143
L4.5.2.02	Including beds; grade 25, 20mm aggregate; 1 layer A252 mesh reinforcement; Depth 100mm; nominal bore: 300mm	m	0.26	9.06	5.69	15.26	30.01	43.924
L4.5.2.03	Including beds; grade 25, 20mm aggregate; 1 layer A252 mesh reinforcement; Depth 150mm; nominal bore: 225mm	m	0.26	9.06	5.69	17.33	32.08	50.681
L4.5.2.04	Including beds; grade 25, 20mm aggregate; 1 layer A252 mesh reinforcement; Depth 150mm; nominal bore: 300mm	m	0.31	10.80	6.78	18.87	36.45	55.366
L4.5.3	nominal bore: 300 - 600mm						
L4.5.3.01	Including beds; grade 25, 20mm aggregate; 1 layer A252 mesh reinforcement; Depth 100mm; nominal bore: 375mm	m	0.31	10.80	6.78	18.63	36.21	54.168
L4.5.3.02	Including beds; grade 25, 20mm aggregate; 1 layer A252 mesh reinforcement; Depth 100mm; nominal bore: 450mm	m	0.37	12.89	8.09	22.44	43.42	65.848
L4.5.3.03	Including beds; grade 25, 20mm aggregate; 1 layer A252 mesh reinforcement; Depth 100mm; nominal bore: 600mm	m	0.46	16.03	10.06	34.59	60.68	102.848
L4.5.3.04	Including beds; grade 25, 20mm aggregate; 1 layer A252 mesh reinforcement; Depth 150mm; nominal bore: 375mm	m	0.37	12.89	8.09	22.61	43.59	66.793
L4.5.3.05	Including beds; grade 25, 20mm aggregate; 1 layer A252 mesh reinforcement; Depth 150mm; nominal bore: 450mm	m	0.43	14.99	9.40	26.77	51.16	79.579

L4 Haunches continued...

	Unit	Labour Hours	Labour Net £	Plant Net £	Materials Net £	Unit Net £	CO_2e Kg
L4.5	**Reinforced concrete**						
L4.5.3	nominal bore: 300 - 600mm						
L4.5.3.06	Including beds; grade 25, 20mm aggregate; I layer A252 mesh reinforcement; Depth 150mm; nominal bore: 600mm	m	0.52	18.12	11.37	39.56	69.05	118.517
L4.5.4	nominal bore: 600 - 900mm						
L4.5.4.01	Including beds; grade 25, 20mm aggregate; I layer A252 mesh reinforcement; Depth 100mm; nominal bore: 675mm	m	0.56	19.52	12.25	38.42	70.19	114.848
L4.5.4.02	Including beds; grade 25, 20mm aggregate; I layer A252 mesh reinforcement; Depth 100mm; nominal bore: 750mm	m	0.75	26.14	16.40	46.74	89.28	141.331
L4.5.4.03	Including beds; grade 25, 20mm aggregate; I layer A252 mesh reinforcement; Depth 100mm; nominal bore: 800mm	m	0.92	32.06	20.12	55.91	108.09	170.388
L4.5.4.04	Including beds; grade 25, 20mm aggregate; I layer A252 mesh reinforcement; Depth 100mm; nominal bore: 900mm	m	1.02	35.55	22.31	60.70	118.56	185.202
L4.5.4.05	Including beds; grade 25, 20mm aggregate; I layer A252 mesh reinforcement; Depth 150mm; nominal bore: 675mm	m	0.63	21.96	13.78	43.75	79.49	131.698
L4.5.4.06	Including beds; grade 25, 20mm aggregate; I layer A252 mesh reinforcement; Depth 150mm; nominal bore: 750mm	m	0.83	28.93	18.15	52.43	99.51	159.363
L4.5.4.07	Including beds; grade 25, 20mm aggregate; I layer A252 mesh reinforcement; Depth 150mm; nominal bore: 800mm	m	1.02	35.55	22.31	61.79	119.65	189.121
L4.5.4.08	Including beds; grade 25, 20mm aggregate; I layer A252 mesh reinforcement; Depth 150mm; nominal bore: 900mm	m	1.12	39.03	24.49	67.03	130.55	205.319
L4.5.5	nominal bore: 900 - 1200mm						
L4.5.5.01	Including beds; grade 25, 20mm aggregate; I layer A252 mesh reinforcement; Depth 100mm; nominal bore: 1200mm	m	1.25	43.56	27.34	86.58	157.48	262.922
L4.5.5.02	Including beds; grade 25, 20mm aggregate; I layer A252 mesh reinforcement; Depth 150mm; nominal bore: 1200mm	m	1.38	48.09	30.18	95.62	173.89	291.565
L4.5.6	nominal bore: 1200 - 1500mm						
L4.5.6.01	Including beds; grade 25, 20mm aggregate; I layer A252 mesh reinforcement; Depth 100mm; nominal bore: 1500mm	m	1.49	51.93	32.59	102.75	187.27	312.455
L4.5.6.02	Including beds; grade 25, 20mm aggregate; I layer A252 mesh reinforcement; Depth 150mm; nominal bore: 1500mm	m	1.53	53.32	33.46	112.96	199.74	344.029

L4   Haunches continued...

	Unit	Labour Hours	Labour Net £	Plant Net £	Materials Net £	Unit Net £	CO_2e Kg
L4.5	**Reinforced concrete**						
L4.5.7	nominal bore: 1500 - 1800mm						
L4.5.7.01	Including beds; grade 25, 20mm aggregate; 1 layer A252 mesh reinforcement; Depth 100mm; nominal bore: 1800mm	m	1.73	60.29	37.84	110.44	208.57	336.773
L4.5.7.02	Including beds; grade 25, 20mm aggregate; 1 layer A252 mesh reinforcement; Depth 150mm; nominal bore: 1800mm	m	1.89	65.87	41.33	135.91	243.11	415.212

L5   Surrounds

	Unit	Labour Hours	Labour Net £	Plant Net £	Materials Net £	Unit Net £	CO_2e Kg
L5.1	**Sand**						
L5.1.1	nominal bore: not exceeding 200mm						
L5.1.1.01	Including beds; Thickness 100mm; Depth 100mm; nominal bore: 100mm	m	0.15	5.23	3.28	7.88	16.39	3.823
L5.1.1.02	Including beds; Thickness 100mm; Depth 100mm; nominal bore: 150mm	m	0.18	6.27	3.94	9.73	19.94	4.685
L5.1.1.03	Including beds; Thickness 100mm; Depth 150mm; nominal bore: 100mm	m	0.19	6.62	4.16	10.66	21.44	5.078
L5.1.1.04	Including beds; Thickness 100mm; Depth 150mm; nominal bore: 150mm	m	0.23	8.02	5.03	12.75	25.80	6.095
L5.1.2	nominal bore: 200 - 300mm						
L5.1.2.01	Including beds; Thickness 100mm; Depth 100mm; nominal bore: 225mm	m	0.20	6.97	4.37	12.47	23.81	5.776
L5.1.2.02	Including beds; Thickness 100mm; Depth 100mm; nominal bore: 300mm	m	0.23	8.02	5.03	15.11	28.16	6.905
L5.1.2.03	Including beds; Thickness 100mm; Depth 150mm; nominal bore: 225mm	m	0.25	8.71	5.47	15.84	30.02	7.306
L5.1.2.04	Including beds; Thickness 100mm; Depth 150mm; nominal bore: 300mm	m	0.28	9.76	6.12	18.82	34.70	8.553
L5.1.3	nominal bore: 300 - 600mm						
L5.1.3.01	Including beds; Thickness 100mm; Depth 100mm; nominal bore: 375mm	m	0.25	8.71	5.47	18.21	32.39	8.123
L5.1.3.02	Including beds; Thickness 100mm; Depth 100mm; nominal bore: 450mm	m	0.27	9.41	5.90	21.28	36.59	9.326
L5.1.3.03	Including beds; Thickness 100mm; Depth 100mm; nominal bore: 600mm	m	0.33	11.50	7.22	25.90	44.62	11.361
L5.1.3.04	Including beds; Thickness 100mm; Depth 150mm; nominal bore: 375mm	m	0.31	10.80	6.78	22.27	39.85	9.964
L5.1.3.05	Including beds; Thickness 100mm; Depth 150mm; nominal bore: 450mm	m	0.33	11.50	7.22	25.68	44.40	11.286

L5 Surrounds continued...

	Unit	Labour Hours	Labour Net £	Plant Net £	Materials Net £	Unit Net £	CO_2e Kg
L5.1 **Sand**							
L5.1.3 nominal bore: 300 - 600mm							
L5.1.3.06 Including beds; Thickness 100mm; Depth 150mm; nominal bore: 600mm	m	0.38	13.24	8.31	32.95	54.50	14.160
L5.1.4 nominal bore: 600 - 900mm							
L5.1.4.01 Including beds; Thickness 100mm; Depth 100mm; nominal bore: 675mm	m	0.35	12.20	7.65	31.14	50.99	13.314
L5.1.4.02 Including beds; Thickness 100mm; Depth 100mm; nominal bore: 750mm	m	0.38	13.24	8.31	34.59	56.14	14.725
L5.1.4.03 Including beds; Thickness 100mm; Depth 100mm; nominal bore: 800mm	m	0.40	13.94	8.75	37.12	59.81	15.742
L5.1.4.04 Including beds; Thickness 100mm; Depth 100mm; nominal bore: 900mm	m	0.43	14.99	9.40	41.65	66.04	17.524
L5.1.4.05 Including beds; Thickness 100mm; Depth 150mm; nominal bore: 675mm	m	0.40	13.94	8.75	36.60	59.29	15.564
L5.1.4.06 Including beds; Thickness 100mm; Depth 150mm; nominal bore: 750mm	m	0.43	14.99	9.40	40.40	64.79	17.094
L5.1.4.07 Including beds; Thickness 100mm; Depth 150mm; nominal bore: 800mm	m	0.46	16.03	10.06	43.16	69.25	18.267
L5.1.4.08 Including beds; Thickness 100mm; Depth 150mm; nominal bore: 900mm	m	0.49	17.08	10.72	48.12	75.92	20.198
L5.1.5 nominal bore: 900 - 1200mm							
L5.1.5.01 Including beds; Thickness 100mm; Depth 100mm; nominal bore: 1200mm	m	0.50	17.43	10.94	74.24	102.61	29.260
L5.1.5.02 Including beds; Thickness 100mm; Depth 150mm; nominal bore: 1200mm	m	0.56	19.52	12.25	83.28	115.05	32.817
L5.1.6 nominal bore: 1200 - 1500mm							
L5.1.6.01 Including beds; Thickness 100mm; Depth 100mm; nominal bore: 1500mm	m	0.58	20.21	12.68	96.03	128.92	37.355
L5.1.6.02 Including beds; Thickness 100mm; Depth 150mm; nominal bore: 1500mm	m	0.63	21.96	13.78	116.19	151.93	44.664
L5.1.7 nominal bore: 1500 - 1800mm							
L5.1.7.01 Including beds; Thickness 100mm; Depth 100mm; nominal bore: 1800mm	m	0.65	22.65	14.22	118.69	155.56	45.674
L5.1.7.02 Including beds; Thickness 100mm; Depth 150mm; nominal bore: 1800mm	m	0.72	25.09	15.75	130.52	171.36	50.263

L5   Surrounds continued...

	Unit	Labour Hours	Labour Net £	Plant Net £	Materials Net £	Unit Net £	CO_2e Kg
L5.2	**Selected excavated granular material**						
L5.2.1	nominal bore: not exceeding 200mm						
L5.2.1.01	Including beds; thickness 100mm; Depth 100mm; nominal bore: 100mm	m	0.22	7.67	4.81	-	12.48	1.631
L5.2.1.02	Including beds; thickness 100mm; Depth 100mm; nominal bore: 150mm	m	0.27	9.41	5.90	-	15.31	2.002
L5.2.1.03	Including beds; thickness 100mm; Depth 150mm; nominal bore: 100mm	m	0.29	10.11	6.34	-	16.45	2.150
L5.2.1.04	Including beds; thickness 100mm; Depth 150mm; nominal bore: 150mm	m	0.35	12.20	7.65	-	19.85	2.595
L5.2.2	nominal bore: 200 - 300mm						
L5.2.2.01	Including beds; thickness 100mm; Depth 100mm; nominal bore: 225mm	m	0.30	10.46	6.56	-	17.02	2.225
L5.2.2.02	Including beds; thickness 100mm; Depth 100mm; nominal bore: 300mm	m	0.35	12.20	7.65	-	19.85	2.595
L5.2.2.03	Including beds; thickness 100mm; Depth 150mm; nominal bore: 225mm	m	0.38	13.24	8.31	-	21.55	2.818
L5.2.2.04	Including beds; thickness 100mm; Depth 150mm; nominal bore: 300mm	m	0.42	14.64	9.19	-	23.83	3.114
L5.2.3	nominal bore: 300 - 600mm						
L5.2.3.01	Including beds; thickness 100mm; Depth 100mm; nominal bore: 375mm	m	0.38	13.24	8.31	-	21.55	2.818
L5.2.3.02	Including beds; thickness 100mm; Depth 100mm; nominal bore: 450mm	m	0.40	13.94	8.75	-	22.69	2.966
L5.2.3.03	Including beds; thickness 100mm; Depth 100mm; nominal bore: 600mm	m	0.50	17.43	10.94	-	28.37	3.708
L5.2.3.04	Including beds; thickness 100mm; Depth 150mm; nominal bore: 375mm	m	0.47	16.38	10.28	-	26.66	3.485
L5.2.3.05	Including beds; thickness 100mm; Depth 150mm; nominal bore: 450mm	m	0.50	17.43	10.94	-	28.37	3.708
L5.2.3.06	Including beds; thickness 100mm; Depth 150mm; nominal bore: 600mm	m	0.57	19.86	12.47	-	32.33	4.227
L5.2.4	nominal bore: 600 - 900mm						
L5.2.4.01	Including beds; thickness 100mm; Depth 100mm; nominal bore: 675mm	m	0.53	18.47	11.59	-	30.06	3.930
L5.2.4.02	Including beds; thickness 100mm; Depth 100mm; nominal bore: 750mm	m	0.57	19.86	12.47	-	32.33	4.227
L5.2.4.03	Including beds; thickness 100mm; Depth 100mm; nominal bore: 800mm	m	0.60	20.91	13.12	-	34.03	4.449

L5 Surrounds continued...

	Unit	Labour Hours	Labour Net £	Plant Net £	Materials Net £	Unit Net £	CO$_2$e Kg
L5.2	**Selected excavated granular material**						
L5.2.4	**nominal bore: 600 - 900mm**						
L5.2.4.04	Including beds; thickness 100mm; Depth 100mm; nominal bore: 900mm	m	0.65	22.65	14.22	-	36.87	4.820
L5.2.4.05	Including beds; thickness 100mm; Depth 150mm; nominal bore: 675mm	m	0.60	20.91	13.12	-	34.03	4.449
L5.2.4.06	Including beds; thickness 100mm; Depth 150mm; nominal bore: 750mm	m	0.65	22.65	14.22	-	36.87	4.820
L5.2.4.07	Including beds; thickness 100mm; Depth 150mm; nominal bore: 800mm	m	0.69	24.05	15.09	-	39.14	5.116
L5.2.4.08	Including beds; thickness 100mm; Depth 150mm; nominal bore: 900mm	m	0.74	25.79	16.18	-	41.97	5.487
L5.2.5	**nominal bore: 900 - 1200mm**						
L5.2.5.01	Including beds; thickness 100mm; Depth 100mm; nominal bore: 1200mm	m	0.75	26.14	16.40	-	42.54	5.561
L5.2.5.02	Including beds; thickness 100mm; Depth 150mm; nominal bore: 1200mm	m	0.84	29.27	18.37	-	47.64	6.229
L5.2.6	**nominal bore: 1200 - 1500mm**						
L5.2.6.01	Including beds; thickness 100mm; Depth 100mm; nominal bore: 1500mm	m	0.87	30.32	19.03	-	49.35	6.451
L5.2.6.02	Including beds; thickness 100mm; Depth 150mm; nominal bore: 1500mm	m	0.95	33.11	20.78	-	53.89	7.044
L5.2.7	**nominal bore: 1500 - 1800mm**						
L5.2.7.01	Including beds; thickness 100mm; Depth 100mm; nominal bore: 1800mm	m	0.98	34.15	21.43	-	55.58	7.267
L5.2.7.02	Including beds; thickness 100mm; Depth 150mm; nominal bore: 1800mm	m	1.08	37.64	23.62	-	61.26	8.008
L5.3	**Imported granular material**						
L5.3.1	**nominal bore: not exceeding 200mm**						
L5.3.1.01	Including beds; DTp clause nr. 503; thickness 100mm; Depth 100mm; nominal bore: 100mm	m	0.15	5.23	3.28	7.14	15.65	3.668
L5.3.1.02	Including beds; DTp clause nr. 503; thickness 100mm; Depth 100mm; nominal bore: 150mm	m	0.18	6.27	3.94	8.82	19.03	4.493
L5.3.1.03	Including beds; DTp clause nr. 503; thickness 100mm; Depth 150mm; nominal bore: 100mm	m	0.19	6.62	4.16	9.66	20.44	4.868
L5.3.1.04	Including beds; DTp clause nr. 503; thickness 100mm; Depth 150mm; nominal bore: 150mm	m	0.23	8.02	5.03	12.02	25.07	6.012

L5 Surrounds continued...

	Unit	Labour Hours	Labour Net £	Plant Net £	Materials Net £	Unit Net £	CO_2e Kg
L5.3 **Imported granular material**							
L5.3.2 nominal bore: 200 - 300mm							
L5.3.2.01 Including beds; DTp clause nr. 503; thickness 100mm; Depth 100mm; nominal bore: 225mm	m	0.20	6.97	4.37	11.30	22.64	5.531
L5.3.2.02 Including beds; DTp clause nr. 503; thickness 100mm; Depth 100mm; nominal bore: 300mm	m	0.23	8.02	5.03	13.69	26.74	6.608
L5.3.2.03 Including beds; DTp clause nr. 503; thickness 100mm; Depth 150mm; nominal bore: 225mm	m	0.25	8.71	5.47	14.35	28.53	6.994
L5.3.2.04 Including beds; DTp clause nr. 503; thickness 100mm; Depth 150mm; nominal bore: 300mm	m	0.28	9.76	6.12	17.05	32.93	8.183
L5.3.2.05 Including beds; DTp clause nr. 503; thickness 100mm; Depth 100mm; nominal bore: 375mm	m	0.25	8.71	5.47	16.50	30.68	7.764
L5.3.3 nominal bore: 300 - 600mm							
L5.3.3.01 Including beds; DTp clause nr. 503; thickness 100mm; Depth 100mm; nominal bore: 450mm	m	0.27	9.41	5.90	19.28	34.59	8.907
L5.3.3.02 Including beds; DTp clause nr. 503; thickness 100mm; Depth 100mm; nominal bore: 600mm	m	0.33	11.50	7.22	25.22	43.94	11.481
L5.3.3.03 Including beds; DTp clause nr. 503; thickness 100mm; Depth 150mm; nominal bore: 375mm	m	0.31	10.80	6.78	20.18	37.76	9.526
L5.3.3.04 Including beds; DTp clause nr. 503; thickness 100mm; Depth 150mm; nominal bore: 450mm	m	0.33	11.50	7.22	23.26	41.98	10.781
L5.3.3.05 Including beds; DTp clause nr. 503; thickness 100mm; Depth 150mm; nominal bore: 600mm	m	0.38	13.24	8.31	29.85	51.40	13.511
L5.3.4 nominal bore: 600 - 900mm							
L5.3.4.01 Including beds; DTp clause nr. 503; thickness 100mm; Depth 100mm; nominal bore: 675mm	m	0.35	12.20	7.65	28.21	48.06	12.701
L5.3.4.02 Including beds; DTp clause nr. 503; thickness 100mm; Depth 100mm; nominal bore: 750mm	m	0.38	13.24	8.31	31.34	52.89	14.044
L5.3.4.03 Including beds; DTp clause nr. 503; thickness 100mm; Depth 100mm; nominal bore: 800mm	m	0.40	13.94	8.75	33.63	56.32	15.011
L5.3.4.04 Including beds; DTp clause nr. 503; thickness 100mm; Depth 100mm; nominal bore: 900mm	m	0.43	14.99	9.40	44.96	69.35	19.295
L5.3.4.05 Including beds; DTp clause nr. 503; thickness 100mm; Depth 150mm; nominal bore: 675mm	m	0.40	13.94	8.75	33.16	55.85	14.843
L5.3.4.06 Including beds; DTp clause nr. 503; thickness 100mm; Depth 150mm; nominal bore: 750mm	m	0.43	14.99	9.40	36.60	60.99	16.298
L5.3.4.07 Including beds; DTp clause nr. 503; thickness 100mm; Depth 150mm; nominal bore: 800mm	m	0.46	16.03	10.06	39.10	65.19	17.417
L5.3.4.08 Including beds; DTp clause nr. 503; thickness 100mm; Depth 150mm; nominal bore: 900mm	m	0.49	17.08	10.72	43.60	71.40	19.250

L5 Surrounds continued...

	Unit	Labour Hours	Labour Net £	Plant Net £	Materials Net £	Unit Net £	CO_2e Kg
L5.3 **Imported granular material**							
L5.3.5 nominal bore: 900 - 1200mm							
L5.3.5.01 Including beds; DTp clause nr. 503; thickness 100mm; Depth 100mm; nominal bore: 1200mm	m	0.50	17.43	10.94	67.25	95.62	27.798
L5.3.5.02 Including beds; DTp clause nr. 503; thickness 100mm; Depth 150mm; nominal bore: 1200mm	m	0.56	19.52	12.25	75.44	107.21	31.177
L5.3.6 nominal bore: 1200 - 1500mm							
L5.3.6.01 Including beds; DTp clause nr. 503; thickness 100mm; Depth 100mm; nominal bore: 1500mm	m	0.58	20.21	12.68	95.80	128.69	38.615
L5.3.6.02 Including beds; DTp clause nr. 503; thickness 100mm; Depth 150mm; nominal bore: 1500mm	m	0.63	21.96	13.78	105.26	141.00	42.376
L5.3.7 nominal bore: 1500 - 1800mm							
L5.3.7.01 Including beds; DTp clause nr. 503; thickness 100mm; Depth 100mm; nominal bore: 1800mm	m	0.65	22.65	14.22	107.53	144.40	43.336
L5.3.7.02 Including beds; DTp clause nr. 503; thickness 100mm; Depth 150mm; nominal bore: 1800mm	m	0.72	25.09	15.75	118.24	159.08	47.693
L5.4 **Mass Concrete**							
L5.4.1 nominal bore: not exceeding 200mm							
L5.4.1.01 Including beds; grade 15, 20mm aggregate; thickness 100mm; Depth 100mm; nominal bore: 100mm	m	0.12	4.18	2.62	14.47	21.27	45.340
L5.4.1.02 Including beds; grade 15, 20mm aggregate; thickness 100mm; Depth 100mm; nominal bore: 150mm	m	0.15	5.23	3.28	17.87	26.38	56.022
L5.4.1.03 Including beds; grade 15, 20mm aggregate; thickness 100mm; Depth 150mm; nominal bore: 100mm	m	0.15	5.23	3.28	19.57	28.08	61.251
L5.4.1.04 Including beds; grade 15, 20mm aggregate; thickness 100mm; Depth 150mm; nominal bore: 150mm	m	0.19	6.62	4.16	23.40	34.18	73.314
L5.4.2 nominal bore: 200 - 300mm							
L5.4.2.01 Including beds; grade 15, 20mm aggregate; thickness 100mm; Depth 100mm; nominal bore: 225mm	m	0.18	6.27	3.94	22.89	33.10	71.671
L5.4.2.02 Including beds; grade 15, 20mm aggregate; thickness 100mm; Depth 100mm; nominal bore: 300mm	m	0.22	7.67	4.81	27.74	40.22	86.871
L5.4.2.03 Including beds; grade 15, 20mm aggregate; thickness 100mm; Depth 150mm; nominal bore: 225mm	m	0.22	7.67	4.81	29.02	41.50	90.794

L5 Surrounds continued...

	Unit	Labour Hours	Labour Net £	Plant Net £	Materials Net £	Unit Net £	CO_2e Kg
L5.4 **Mass Concrete**							
L5.4.2 nominal bore: 200 - 300mm							
L5.4.2.04 Including beds; grade 15, 20mm aggregate; thickness 100mm; Depth 150mm; nominal bore: 300mm	m	0.27	9.41	5.90	34.55	49.86	108.160
L5.4.3 nominal bore: 300 - 600mm							
L5.4.3.01 Including beds; grade 15, 20mm aggregate; thickness 100mm; Depth 100mm; nominal bore: 375mm	m	0.27	9.41	5.90	33.44	48.75	104.761
L5.4.3.02 Including beds; grade 15, 20mm aggregate; thickness 100mm; Depth 100mm; nominal bore: 450mm	m	0.32	11.15	7.00	39.06	57.21	122.389
L5.4.3.03 Including beds; grade 15, 20mm aggregate; thickness 100mm; Depth 100mm; nominal bore: 600mm	m	0.40	13.94	8.75	51.06	73.75	159.850
L5.4.3.04 Including beds; grade 15, 20mm aggregate; thickness 100mm; Depth 150mm; nominal bore: 375mm	m	0.33	11.50	7.22	40.85	59.57	127.954
L5.4.3.05 Including beds; grade 15, 20mm aggregate; thickness 100mm; Depth 150mm; nominal bore: 450mm	m	0.38	13.24	8.31	47.06	68.61	147.412
L5.4.3.06 Including beds; grade 15, 20mm aggregate; thickness 100mm; Depth 150mm; nominal bore: 600mm	m	0.46	16.03	10.06	60.42	86.51	189.057
L5.4.4 nominal bore: 600 - 900mm							
L5.4.4.01 Including beds; grade 15, 20mm aggregate; thickness 100mm; Depth 100mm; nominal bore: 675mm	m	0.50	17.43	10.94	57.02	85.39	178.894
L5.4.4.02 Including beds; grade 15, 20mm aggregate; thickness 100mm; Depth 100mm; nominal bore: 750mm	m	0.69	24.05	15.09	63.82	102.96	201.221
L5.4.4.03 Including beds; grade 15, 20mm aggregate; thickness 100mm; Depth 100mm; nominal bore: 800mm	m	0.85	29.62	18.59	68.08	116.29	215.481
L5.4.4.04 Including beds; grade 15, 20mm aggregate; thickness 100mm; Depth 100mm; nominal bore: 900mm	m	0.95	33.11	20.78	93.61	147.50	294.665
L5.4.4.05 Including beds; grade 15, 20mm aggregate; thickness 100mm; Depth 150mm; nominal bore: 675mm	m	0.57	19.86	12.47	67.14	99.47	210.529
L5.4.4.06 Including beds; grade 15, 20mm aggregate; thickness 100mm; Depth 150mm; nominal bore: 750mm	m	0.77	26.83	16.84	74.12	117.79	233.453

L5 Surrounds continued...

		Unit	Labour Hours	Labour Net £	Plant Net £	Materials Net £	Unit Net £	CO_2e Kg
L5.4	**Mass Concrete**							
L5.4.4	nominal bore: 600 - 900mm							
L5.4.4.07	Including beds; grade 15, 20mm aggregate; thickness 100mm; Depth 150mm; nominal bore: 800mm	m	0.95	33.11	20.78	79.14	133.03	250.214
L5.4.4.08	Including beds; grade 15, 20mm aggregate; thickness 100mm; Depth 150mm; nominal bore: 900mm	m	1.05	36.59	22.96	88.25	147.80	278.933
L5.4.5	nominal bore: 900 - 1200mm							
L5.4.5.01	Including beds; grade 15, 20mm aggregate; thickness 100mm; Depth 100mm; nominal bore: 1200mm	m	1.15	40.08	25.15	136.16	201.39	426.884
L5.4.5.02	Including beds; grade 15, 20mm aggregate; thickness 100mm; Depth 150mm; nominal bore: 1200mm	m	1.28	44.61	27.99	152.75	225.35	478.835
L5.4.6	nominal bore: 1200 - 1500mm							
L5.4.6.01	Including beds; grade 15, 20mm aggregate; thickness 100mm; Depth 100mm; nominal bore: 1500mm	m	1.38	48.09	30.18	194.03	272.30	606.391
L5.4.6.02	Including beds; grade 15, 20mm aggregate; thickness 100mm; Depth 150mm; nominal bore: 1500mm	m	1.42	49.49	31.06	213.09	293.64	665.258
L5.4.7	nominal bore: 1500 - 1800mm							
L5.4.7.01	Including beds; grade 15, 20mm aggregate; thickness 100mm; Depth 100mm; nominal bore: 1800mm	m	1.60	55.76	34.99	217.86	308.61	681.235
L5.4.7.02	Including beds; grade 15, 20mm aggregate; thickness 100mm; Depth 150mm; nominal bore: 1800mm	m	1.76	61.34	38.49	239.39	339.22	748.574
L5.5	**Reinforced concrete**							
L5.5.1	nominal bore: not exceeding 200mm							
L5.5.1.01	Including beds; grade 25, 20mm aggregate; 1 layer A252 mesh reinforcement; thickness 100mm; Depth 100mm; nominal bore: 100mm	m	0.15	5.23	3.28	16.98	25.49	49.563
L5.5.1.02	Including beds; grade 25, 20mm aggregate; 1 layer A252 mesh reinforcement; thickness 100mm; Depth 100mm; nominal bore: 150mm	m	0.18	6.27	3.94	20.73	30.94	60.974
L5.5.1.03	Including beds; grade 25, 20mm aggregate; 1 layer A252 mesh reinforcement; thickness 100mm; Depth 150mm; nominal bore: 100mm	m	0.18	6.27	3.94	22.40	32.61	66.393

L5 Surrounds continued...

	Unit	Labour Hours	Labour Net £	Plant Net £	Materials Net £	Unit Net £	CO_2e Kg
L5.5 **Reinforced concrete**							
L5.5.1 nominal bore: not exceeding 200mm							
L5.5.1.04 Including beds; grade 25, 20mm aggregate; 1 layer A252 mesh reinforcement; thickness 100mm; Depth 150mm; nominal bore: 150mm	m	0.22	7.67	4.81	26.60	39.08	79.262
L5.5.2 nominal bore: 200 - 300mm							
L5.5.2.01 Including beds; grade 25, 20mm aggregate; 1 layer A252 mesh reinforcement; thickness 100mm; Depth 100mm; nominal bore: 225mm	m	0.22	7.67	4.81	26.27	38.75	77.787
L5.5.2.02 Including beds; grade 25, 20mm aggregate; 1 layer A252 mesh reinforcement; thickness 100mm; Depth 100mm; nominal bore: 300mm	m	0.26	9.06	5.69	31.61	46.36	94.024
L5.5.2.03 Including beds; grade 25, 20mm aggregate; 1 layer A252 mesh reinforcement; thickness 100mm; Depth 150mm; nominal bore: 225mm	m	0.26	9.06	5.69	32.78	47.53	98.014
L5.5.2.04 Including beds; grade 25, 20mm aggregate; 1 layer A252 mesh reinforcement; thickness 100mm; Depth 150mm; nominal bore: 300mm	m	0.31	10.80	6.78	38.84	56.42	116.539
L5.5.3 nominal bore: 300 - 600mm							
L5.5.3.01 Including beds; grade 25, 20mm aggregate; 1 layer A252 mesh reinforcement; thickness 100mm; Depth 100mm; nominal bore: 375mm	m	0.31	10.80	6.78	37.88	55.46	113.127
L5.5.3.02 Including beds; grade 25, 20mm aggregate; 1 layer A252 mesh reinforcement; thickness 100mm; Depth 100mm; nominal bore: 450mm	m	0.37	12.89	8.09	44.04	65.02	132.003
L5.5.3.03 Including beds; grade 25, 20mm aggregate; 1 layer A252 mesh reinforcement; thickness 100mm; Depth 100mm; nominal bore: 600mm	m	0.46	16.03	10.06	57.18	83.27	172.048
L5.5.3.04 Including beds; grade 25, 20mm aggregate; 1 layer A252 mesh reinforcement; thickness 100mm; Depth 150mm; nominal bore: 375mm	m	0.37	12.89	8.09	45.74	66.72	137.653
L5.5.3.05 Including beds; grade 25, 20mm aggregate; 1 layer A252 mesh reinforcement; thickness 100mm; Depth 150mm; nominal bore: 450mm	m	0.43	14.99	9.40	52.53	76.92	158.467

L5   Surrounds continued...

	Unit	Labour Hours	Labour Net £	Plant Net £	Materials Net £	Unit Net £	CO_2e Kg
L5.5	**Reinforced concrete**						
L5.5.3	**nominal bore: 300 - 600mm**						
L5.5.3.06	Including beds; grade 25, 20mm aggregate; 1 layer A252 mesh reinforcement; thickness 100mm; Depth 150mm; nominal bore: 600mm	m	0.52	18.12	11.37	67.12	96.61	202.941
L5.5.4	**nominal bore: 600 - 900mm**						
L5.5.4.01	Including beds; grade 25, 20mm aggregate; 1 layer A252 mesh reinforcement; thickness 100mm; Depth 100mm; nominal bore: 675mm	m	0.56	19.52	12.25	64.12	95.89	193.689
L5.5.4.02	Including beds; grade 25, 20mm aggregate; 1 layer A252 mesh reinforcement; thickness 100mm; Depth 100mm; nominal bore: 750mm	m	0.77	26.83	16.84	71.14	114.81	216.215
L5.5.4.03	Including beds; grade 25, 20mm aggregate; 1 layer A252 mesh reinforcement; thickness 100mm; Depth 100mm; nominal bore: 800mm	m	0.92	32.06	20.12	75.79	127.97	231.284
L5.5.4.04	Including beds; grade 25, 20mm aggregate; 1 layer A252 mesh reinforcement; thickness 100mm; Depth 100mm; nominal bore: 900mm	m	1.02	35.55	22.31	85.10	142.96	259.937
L5.5.4.05	Including beds; grade 25, 20mm aggregate; 1 layer A252 mesh reinforcement; thickness 100mm; Depth 150mm; nominal bore: 675mm	m	0.63	21.96	13.78	74.48	110.22	225.809
L5.5.4.06	Including beds; grade 25, 20mm aggregate; 1 layer A252 mesh reinforcement; thickness 100mm; Depth 150mm; nominal bore: 750mm	m	0.83	28.93	18.15	82.07	129.15	250.153
L5.5.4.07	Including beds; grade 25, 20mm aggregate; 1 layer A252 mesh reinforcement; thickness 100mm; Depth 150mm; nominal bore: 800mm	m	1.02	35.55	22.31	87.54	145.40	268.009
L5.5.4.08	Including beds; grade 25, 20mm aggregate; 1 layer A252 mesh reinforcement; thickness 100mm; Depth 150mm; nominal bore: 900mm	m	1.12	39.03	24.49	97.48	161.00	298.600
L5.5.5	**nominal bore: 900 - 1200mm**						
L5.5.5.01	Including beds; grade 25, 20mm aggregate; 1 layer A252 mesh reinforcement; thickness 100mm; Depth 100mm; nominal bore: 1200mm	m	1.25	43.56	27.34	149.84	220.74	456.681

L5   Surrounds continued...

		Unit	Labour Hours	Labour Net £	Plant Net £	Materials Net £	Unit Net £	CO$_2$e Kg
L5.5	**Reinforced concrete**							
L5.5.5	nominal bore: 900 - 1200mm							
L5.5.5.02	Including beds; grade 25, 20mm aggregate; 1 layer A252 mesh reinforcement; thickness 100mm; Depth 150mm; nominal bore: 1200mm	m	1.38	48.09	30.18	167.46	245.73	511.621
L5.5.6	nominal bore: 1200 - 1500mm							
L5.5.6.01	Including beds; grade 25, 20mm aggregate; 1 layer A252 mesh reinforcement; thickness 100mm; Depth 100mm; nominal bore: 1500mm	m	1.49	51.93	32.59	212.10	296.62	647.381
L5.5.6.02	Including beds; grade 25, 20mm aggregate; 1 layer A252 mesh reinforcement; thickness 100mm; Depth 150mm; nominal bore: 1500mm	m	1.53	53.32	33.46	207.13	293.91	632.454
L5.5.7	nominal bore: 1500 - 1800mm							
L5.5.7.01	Including beds; grade 25, 20mm aggregate; 1 layer A252 mesh reinforcement; thickness 100mm; Depth 100mm; nominal bore: 1800mm	m	1.73	60.29	37.84	238.21	336.34	727.362
L5.5.7.02	Including beds; grade 25, 20mm aggregate; 1 layer A252 mesh reinforcement; thickness 100mm; Depth 150mm; nominal bore: 1800mm	m	1.89	65.87	41.33	261.07	368.27	798.579

L7   Concrete Stools And Thrust Blocks

		Unit	Labour Hours	Labour Net £	Plant Net £	Materials Net £	Unit Net £	CO$_2$e Kg
L7.1	**Volume: not exceeding 0.1m^3**							
L7.1.1	nominal bore: not exceeding 200mm							
L7.1.1.01	Generally	Nr	0.25	8.71	5.47	9.04	23.22	29.534
L7.2	**Volume: 0.1 – 0.2m^3**							
L7.2.1	nominal bore: not exceeding 200mm							
L7.2.1.01	Generally	Nr	0.40	13.94	8.75	18.07	40.76	58.326
L7.3	**Volume: 0.2 – 0.5m^3**							
L7.3.1	nominal bore: not exceeding 200mm							
L7.3.1.01	Generally	Nr	0.70	24.40	15.31	45.19	84.90	143.590
L7.4	**Volume: 0.5 – 1m^3**							
L7.4.1	nominal bore: not exceeding 200mm							
L7.4.1.01	Generally	Nr	1.00	34.85	21.87	90.37	147.09	284.214
L7.5	**Volume: 1 – 2m^3**							
L7.5.1	nominal bore: not exceeding 200mm							
L7.5.1.01	Generally	Nr	2.00	69.70	43.74	180.74	294.18	568.428

L7 Concrete Stools And Thrust Blocks continued...

	Unit	Labour Hours	Labour Net £	Plant Net £	Materials Net £	Unit Net £	CO_2e Kg
L7.6	**Volume: 2 – 4m³**						
L7.6.1	nominal bore: not exceeding 200mm						
L7.6.1.01 Generally	Nr	3.50	121.98	76.55	361.48	560.01	1,133.149
L7.8	**Volume: stated exceeding 6m³**						
L7.8.1	nominal bore: not exceeding 200mm						
L7.8.1.01 Volume: 8m³	Nr	6.00	209.10	131.22	722.96	1,063.28	2,258.882
L7.8.1.02 Volume: 25m³	Nr	15.00	522.75	328.05	2,259.25	3,110.05	7,031.200

CLASS M:
STRUCTURAL METALWORK

Calculations used throughout Class M - Structural Metalwork

Labour

			Qty		Rate		Total
L A0315ICE	**Paint Spray Labour Gang**						
	L A9029ICE	Spray painter	1	x	17.33	=	£17.33
		Total hourly cost of gang				=	**£17.33**
L M0001ICE	**Shot Blasting Labour Gang**						
	L A9016ICE	Ganger	1	x	17.64	=	£17.64
	L A9002ICE	Labourer (General Operative)	2	x	13.02	=	£26.03
		Total hourly cost of gang				=	**£43.67**
L M0002ICE	**Structural Steel Erection Labour Gang**						
	L A9003ICE	Labourer (Skill Rate 3)	1	x	14.86	=	£14.86
	L A9021ICE	Fitters and Welders	2	x	17.33	=	£34.66
		Total hourly cost of gang				=	**£49.52**
L M0003ICE	**Structural Steel Fabrication Labour Gang**						
	L A9003ICE	Labourer (Skill Rate 3)	1	x	14.86	=	£14.86
	L A9021ICE	Fitters and Welders	2	x	17.33	=	£34.66
		Total hourly cost of gang				=	**£49.52**

Plant

			Qty		Rate		Total
P M0004ICE	**Shot Blasting Plant Gang**						
	P A0746ICE	Shotblast Equipment (3.0 Bag Pot) with 250 cfm Compressor	1	x	3.59	=	£3.59
		Total hourly cost of gang				=	**£3.59**
P M0005ICE	**Structural Steel Erection Plant Gang**						
	P A0991ICE	Welding Set - 300 amp Diesel Electric Start Sil	1	x	4.48	=	£4.48
	P A1031ICE	Cutting and Burning Gear	1	x	3.77	=	£3.77
	P A0564ICE	Fork Lift Truck - 2.5t 4WD	1	x	28.68	=	£28.68
	P A0481ICE	Cranes Transit - 25t	1	x	60.22	=	£60.22
		Total hourly cost of gang				=	**£97.15**

Structural Metalwork

Note(s):   NOTE: The following prices are guide prices for approximate estimating.

M3   Fabrication Of Members For Frames

		Unit	Labour Hours	Labour Net £	Plant Net £	Materials Net £	Unit Net £	CO_2e Kg
M3.1	**Columns**							
M3.1.1	Straight on plan							
M3.1.1.01	Universal Column section; 254 UC to 305 UC	t	5.00	247.60	502.24	843.12	1,592.96	1,398.986
M3.2	**Beams**							
M3.2.1	Straight on plan							
M3.2.1.01	Universal Beam section; 356 UB to 533 UB; not exceeding 7m long	t	5.00	247.60	502.24	843.12	1,592.96	1,398.986
M3.3	**Portal Frames**							
M3.3.1	Straight on plan							
M3.3.1.01	Span: 15m; 203 UB to 254 UB	t	7.00	346.64	696.54	843.12	1,886.30	1,511.590
M3.3.1.02	Span: 20m; 406 UB to 457 UB	t	6.50	321.88	647.97	843.12	1,812.97	1,483.439
M3.3.1.03	Span: 30m; 533 UB to 610 UB	t	5.00	247.60	502.24	843.12	1,592.96	1,398.986
M3.4	**Trestles, towers and built-up columns**							
M3.4.1	Straight on plan							
M3.4.1.01	bolted angle construction, piece small	t	8.00	396.16	793.69	843.12	2,032.97	1,567.892
M3.5	**Trusses and built-up girders**							
M3.5.1	Straight on plan							
M3.5.1.01	15m bolted angle construction	t	9.00	445.68	890.84	843.12	2,179.64	1,624.194
M3.5.1.02	20m structural tee beam internal angles welded	t	8.00	396.16	793.69	843.12	2,032.97	1,567.892
M3.5.1.03	30m structural tee beam internal angles welded	t	8.00	396.16	793.69	843.12	2,032.97	1,567.892
M3.6	**Bracings, purlins and cladding rails**							
M3.6.1	Straight on plan							
M3.6.1.01	Generally	t	9.00	445.68	890.84	843.12	2,179.64	1,624.194
M3.8	**Anchorages and holding down bolt assemblies**							
M3.8.0	Holding down bolts and plates (per tonne of base)							
M3.8.0.01	Generally	t	12.00	594.24	1,182.29	843.12	2,619.65	1,793.100

M4   Fabrication Of Other Members

		Unit	Labour Hours	Labour Net £	Plant Net £	Materials Net £	Unit Net £	CO_2e Kg
M4.1	**Columns**							
M4.1.1	Straight on plan							
M4.1.1.01	Universal Column section; 254 UC to 305 UC	t	5.00	247.60	502.24	843.12	1,592.96	1,398.986
M4.2	**Beams**							
M4.2.1	Straight on plan							
M4.2.1.01	Universal Beam section; 356 UB to 533 UB; not exceeding 7m long	t	5.00	247.60	502.24	843.12	1,592.96	1,398.986

M4 Fabrication Of Other Members continued...

		Unit	Labour Hours	Labour Net £	Plant Net £	Materials Net £	Unit Net £	CO_2e Kg
M4.3	**Portal frames**							
M4.3.1	Straight on plan							
M4.3.1.01	Universal Beam section; minimum section 457 UB	t	6.50	321.88	647.97	843.12	1,812.97	1,483.439
M4.4	**Trestles, towers and built-up columns**							
M4.4.1	Straight on plan							
M4.4.1.01	bolted angle construction, piece small	t	8.00	396.16	793.69	843.12	2,032.97	1,567.892
M4.5	**Trusses and built-up girders**							
M4.5.1	Straight on plan							
M4.5.1.01	not exceeding 20m long	t	8.00	396.16	793.69	843.12	2,032.97	1,567.892
M4.6	**Bracings, purlins and cladding rails**							
M4.6.1	Straight on plan							
M4.6.1.01	Generally	t	9.00	445.68	890.84	843.12	2,179.64	1,624.194
M4.8	**Anchorages and holding down bolt assemblies**							
M4.8.0	Holding down bolts and plates (per tonne of base)							
M4.8.0.01	Generally	t	12.00	594.24	1,182.29	843.12	2,619.65	1,793.100

M6 Erection Of Members For Frames

		Unit	Labour Hours	Labour Net £	Plant Net £	Materials Net £	Unit Net £	CO_2e Kg
M6.2	**Permanent erection**							
M6.2.0	Steel members							
M6.2.0.01	Generally	t	3.00	148.56	291.45	–	440.01	168.906

M7 Erection Of Other Members

		Unit	Labour Hours	Labour Net £	Plant Net £	Materials Net £	Unit Net £	CO_2e Kg
M7.2	**Permanent erection**							
M7.2.0	Steel members							
M7.2.0.01	Generally	t	3.00	148.56	291.45	–	440.01	168.906
M7.3	**Site bolts: black**							
M7.3.3	Diameter: 20-24 mm							
M7.3.3.01	24mm diameter	Nr	0.01	0.25	0.49	2.71	3.45	0.552
M7.4	**Site bolts: HSFG general grade**							
M7.4.3	Diameter: 20-24 mm							
M7.4.3.01	24mm diameter	Nr	0.01	0.35	0.68	3.03	4.06	0.685
M7.5	**Site bolts: HSFG higher grade**							
M7.5.3	Diameter: 20-24 mm							
M7.5.3.01	24mm diameter	Nr	0.01	0.40	0.78	3.03	4.21	0.741
M7.6	**Site bolts: HSFG load indicating or load limit types, general grade**							
M7.6.3	Diameter: 20-24 mm							
M7.6.3.01	24mm diameter	Nr	0.01	0.64	1.26	3.81	5.71	1.012

M7 Erection Of Other Members continued...

	Unit	Labour Hours	Labour Net £	Plant Net £	Materials Net £	Unit Net £	CO_2e Kg
M7.7	**Site bolts: HSFG load indicating or load limit types, higher grade**						
M7.7.3 Diameter: 20-24 mm							
M7.7.3.01 24mm diameter	Nr	0.01	0.64	1.26	3.81	5.71	1.012

M8 Off Site Surface Treatment

	Unit	Labour Hours	Labour Net £	Plant Net £	Materials Net £	Unit Net £	CO_2e Kg
M8.1	**Blast cleaning**						
M8.1.0 Blast cleaning							
M8.1.0.01 Generally	m²	0.05	2.18	0.18	2.21	4.57	0.750
M8.6	**Galvanising**						
M8.6.0 Generally							
M8.6.0.01 Galvanising to BS EN 1461: 600g/ m²; Average: 20m2/ tonne	m²	-	-	-	-	26.62	-
M8.7	**Painting**						
M8.7.0 Generally							
M8.7.0.01 High build Epoxy zinc phosphate	m²	0.17	2.89	0.51	1.72	5.12	0.402

CLASS N:
MISCELLANEOUS METALWORK

Calculations used throughout Class N - Miscellaneous Metalwork

Labour

			Qty		Rate		Total
L M0003ICE	**Structural Steel Fabrication Labour Gang**						
	L A9003ICE	Labourer (Skill Rate 3)	1	x	14.86	=	£14.86
	L A9021ICE	Fitters and Welders	2	x	17.33	=	£34.66
		Total hourly cost of gang				=	**£49.52**
L N0001ICE	**Motorway Barrier Install Labour Gang**						
	L A9016ICE	Ganger	2	x	17.64	=	£35.28
	L A9014ICE	Labourer (Skill Rate 4)	2	x	14.03	=	£28.06
	L A9002ICE	Labourer (General Operative)	4	x	13.02	=	£52.06
	L A9025ICE	Plant Operator (Class 4)	2	x	15.30	=	£30.61
		Total hourly cost of gang				=	**£146.01**

Plant

			Qty		Rate		Total
P M0006ICE	**Structural Steel Fabrication Plant Gang**						
	P A0991ICE	Welding Set - 300 amp Diesel Electric Start Sil	1	x	4.48	=	£4.48
	P A1031ICE	Cutting and Burning Gear	1	x	3.77	=	£3.77
	P A0564ICE	Fork Lift Truck - 2.5t 4WD	1	x	28.68	=	£28.68
	P A0481ICE	Cranes Transit - 25t	1	x	60.22	=	£60.22
		Total hourly cost of gang				=	**£97.15**
P N0002ICE	**Motorway Barrier Install Plant Gang**						
	P A1001ICE	Tipping Waggon - 10t 4-Wheel (16t Gr)	1	x	44.88	=	£44.88
	P A1059ICE	Agricultural Tractor: Fencing Auger	1	x	25.08	=	£25.08
	P A0452ICE	Compressor - 2-Tool (Complete)	1	x	5.20	=	£5.20
	P A0461ICE	Hose & Breaker & 3 Steels	1	x	0.38	=	£0.38
	P A0554ICE	Dumper - 2.0t 4WD Swivel Skip	1	x	22.56	=	£22.56
	P A0573ICE	Generator - 3kvA Petrol	1	x	1.67	=	£1.67
	P A0573ICE	Generator - 3kvA Petrol	1	x	1.67	=	£1.67
		Total hourly cost of gang				=	**£101.44**

Miscellaneous Metalwork

Note(s): NOTE: The following prices are guide prices for approximate estimating. All materials are mild steel unless stated otherwise.

N1   Miscellaneous Metalwork

		Unit	Labour Hours	Labour Net £	Plant Net £	Materials Net £	Unit Net £	CO_2e Kg
N1.1	**Stairways and landings**							
N1.1.0	Mild steel fabrication, fixings included							
N1.1.0.01	Generally	t	15.00	742.80	1,622.19	1,411.92	3,776.91	4,843.497
N1.2	**Walkways and platforms**							
N1.2.0	Mild steel fabrication, fixings included							
N1.2.0.01	Generally	t	15.00	742.80	1,622.19	1,411.92	3,776.91	4,843.497
N1.3	**Ladders**							
N1.3.0	Ladders; 400mm wide; 65 x 10mm stringers; 20mm diameter rungs							
N1.3.0.01	Generally	m	1.00	49.52	103.37	55.41	208.30	228.514
N1.5	**Bridge parapets**							
N1.5.0	Mild steel fabrication, fixing included; DTp specification for railings							
N1.5.0.01	Type P1; 3 rails; posts at 3m centres	m	0.63	91.26	63.41	71.92	226.59	144.641
N1.5.0.02	Type P4; pedestrian guard railing; posts at 2m centres	m	0.63	91.26	63.41	71.92	226.59	144.641
N1.5.0.03	Type P6; sound screen; solid infill; 2m high	m	0.67	97.39	67.67	150.02	315.08	151.327
N1.6	**Miscellaneous Framing**							
N1.6.1	Angle section							
N1.6.1.01	76 x 64mm x 9.68kg/m; galvanised steel walkway supports	m	0.08	4.11	24.56	13.85	42.52	50.898
N1.6.1.02	102 x 65mm x 13.4kg/m; galvanised steel toe plate	m	0.17	8.27	32.72	18.05	59.04	67.315
N1.6.2	Channel section							
N1.6.2.01	Galvanised walkway supports; Fixing 3m above ground level; 102 x 51mm x 10.42kg/m	m	0.08	4.11	24.56	13.89	42.56	50.976
N1.6.2.02	Galvanised walkway supports; Fixing 3m above ground level; 152 x 76mm x 17.88kg/m	m	0.17	8.27	32.72	24.96	65.95	86.616
N1.6.2.03	Galvanised walkway supports; Fixing 3m above ground level; 178 x 89mm x 26.81kg/m	m	0.25	12.38	40.78	37.43	90.59	126.120
N1.6.2.04	Galvanised walkway supports; Fixing 3m above ground level; 203 x 89mm x 29.78kg/m	m	0.33	16.49	48.84	41.59	106.92	142.404
N1.6.2.05	Galvanised walkway supports; Fixing 3m above ground level; 254 x 76mm x 28.29kg/m	m	0.33	16.49	48.84	38.85	104.18	134.716
N1.6.2.06	Galvanised walkway supports; Fixing 6m above ground level; 102 x 51mm x 10.42kg/m	m	0.17	8.27	32.72	13.89	54.88	55.705
N1.6.2.07	Galvanised walkway supports; Fixing 6m above ground level; 152 x 76mm x 17.88kg/m	m	0.33	16.49	48.84	24.96	90.29	95.962
N1.6.2.08	Galvanised walkway supports; Fixing 6m above ground level; 178 x 89mm x 26.81kg/m	m	0.50	24.76	65.07	37.43	127.26	140.196

N1 Miscellaneous Metalwork continued...

	Unit	Labour Hours	Labour Net £	Plant Net £	Materials Net £	Unit Net £	CO_2e Kg
N1.6 **Miscellaneous Framing**							
N1.6.2 Channel section							
N1.6.2.09 Galvanised walkway supports; Fixing 6m above ground level; 203 x 89mm x 29.78kg/m	m	0.67	33.03	81.29	41.59	155.91	161.209
N1.6.2.10 Galvanised walkway supports; Fixing 6m above ground level; 254 x 76mm x 28.29kg/m	m	0.67	33.03	81.29	38.85	153.17	153.521
N1.6.2.11 Galvanised steel cable tray supports; Section size: 305 x 89mm x 41.69kg/m	m	1.00	49.52	113.64	58.23	221.39	226.399

	Unit	Labour Hours	Labour Net £	Plant Net £	Materials Net £	Unit Net £	CO_2e Kg

Calculations used throughout Class O - Timber

Labour

			Qty		Rate		Total
L A0240ICE	**Timber Labour Gang**						
	L A9026ICE	Timberman	1	x	14.86	=	£14.86
	L A9003ICE	Labourer (Skill Rate 3)	1	x	14.86	=	£14.86
	L A9002ICE	Labourer (General Operative)	1	x	13.02	=	£13.02
		Total hourly cost of gang				=	**£42.74**

Plant

			Qty		Rate		Total
P A1240ICE	**Timber Plant Gang**						
	P A0452ICE	Compressor - 2-Tool (Complete)	1	x	5.20	=	£5.20
	P A0702ICE	Kango Type Tool - Electric Power Drill 19mm	2	x	1.12	=	£2.24
	P A0705ICE	Kango Type Tool - Electric Screwdriver	1	x	0.52	=	£0.52
	P A0704ICE	Kango Type Tool - Electric Nut Runner	1	x	0.52	=	£0.52
	P A0507ICE	Cranes Crawler - NCK 305B - 20t	0.5	x	48.55	=	£24.28
		Total hourly cost of gang				=	**£32.76**

Timber

Note(s): 1) The hardness and toughness of species of timber varies considerably. The labour rates shown are based on working an average timber grade such as Douglas Fir. It is required more time to work on the higher grades. For example, boring Greenheart would take three times as long as boring pine.

2) The very hard woods are difficult to shape truly, this makes it difficult to marry up bolt holes so it is recommended that structural members are constructed as modules with a minimum of in-place boring.

3) Most timber is imported into this country; thus the price is largely influenced by the currency rates of exchange, so it is essential to get up to date quotations for timber particularly the hardwoods, much of which comes from the Americas and Canada

O1   Hardwood Components

	Unit	Labour Hours	Labour Net £	Plant Net £	Materials Net £	Unit Net £	CO_2e Kg	
O1.1 Cross sectional area: not exceeding 0.01m²								
O1.1.1 Length: not exceeding 1.5m								
O1.1.1.01 Sawn finish; Greenheart; 100 x 75mm	m	0.30	12.82	9.82	7.49	30.13	3.956	-4.197
O1.1.2 Length: 1.5 - 3m								
O1.1.2.01 Sawn finish; Greenheart; 100 x 75mm	m	0.28	11.97	9.17	7.49	28.63	3.737	-4.197
O1.1.3 Length: 3 - 5m								
O1.1.3.01 Sawn finish; Greenheart; 100 x 75mm	m	0.26	11.11	8.51	7.49	27.11	3.517	-4.197
O1.1.4 Length: 5 - 8m								
O1.1.4.01 Sawn finish; Greenheart; 100 x 75mm	m	0.24	10.26	7.86	7.49	25.61	3.298	-4.197
O1.2 Cross sectional area: 0.01 - 0.02m²								
O1.2.1 Length: not exceeding 1.5m								
O1.2.1.01 Sawn finish; Greenheart; 150 x 75mm; runners in marine work	m	0.32	13.68	10.48	9.36	33.52	4.342	-5.247
O1.2.1.02 Sawn finish; Greenheart; 75 x 225mm; runners in marine work	m	0.31	13.25	10.15	15.92	39.32	4.812	-8.919
O1.2.2 Length: 1.5 - 3m								
O1.2.2.01 Sawn finish; Greenheart; 150 x 75mm; runners in marine work	m	0.28	11.97	9.17	9.36	30.50	3.903	-5.247
O1.2.2.02 Sawn finish; Greenheart; 75 x 225mm; runners in marine work	m	0.29	12.39	9.49	15.92	37.80	4.593	-8.919
O1.2.3 Length: 3 - 5m								
O1.2.3.01 Sawn finish; Greenheart; 150 x 75mm; runners in marine work	m	0.26	11.11	8.51	9.36	28.98	3.683	-5.247
O1.2.3.02 Sawn finish; Greenheart; 75 x 225mm; runners in marine work	m	0.27	11.54	8.84	15.92	36.30	4.373	-8.919
O1.2.4 Length: 5 - 8m								
O1.2.4.01 Sawn finish; Greenheart; 150 x 75mm; runners in marine work	m	0.24	10.26	7.86	9.36	27.48	3.464	-5.247
O1.2.4.02 Sawn finish; Greenheart; 75 x 225mm; runners in marine work	m	0.25	10.69	8.19	15.92	34.80	4.153	-8.919
O1.3 Cross sectional area: 0.02 - 0.04m²								
O1.3.1 Length: not exceeding 1.5m								
O1.3.1.01 Sawn finish; Greenheart; 200 x 200mm; braces in marine work	m	0.58	24.79	18.99	37.45	81.23	9.682	-20.986

O1 Hardwood Components continued...

	Unit	Labour Hours	Labour Net £	Plant Net £	Materials Net £	Unit Net £	CO_2e Kg	
O1.3	**Cross sectional area: 0.02 - 0.04m²**							
O1.3.2	**Length: 1.5 - 3m**							
O1.3.2.01	Sawn finish; Greenheart; 200 x 200mm; braces in marine work	m	0.55	23.51	18.01	37.45	78.97	9.353	-20.986
O1.3.3	**Length: 3 - 5m**							
O1.3.3.01	Sawn finish; Greenheart; 200 x 200mm; braces in marine work	m	0.52	22.22	17.02	37.45	76.69	9.024	-20.986
O1.3.4	**Length: 5 - 8m**							
O1.3.4.01	Sawn finish; Greenheart; 200 x 200mm; braces in marine work	m	0.49	20.94	16.04	37.45	74.43	8.694	-20.986
O1.4	**Cross sectional area: 0.04 - 0.1m²**							
O1.4.2	**Length: 1.5 - 3m**							
O1.4.2.01	Sawn finish; Greenheart; 300 x 150mm; braces between piles in marine work	m	0.65	27.78	21.28	42.14	91.20	10.865	-23.610
O1.4.2.02	Sawn finish; Greenheart; 200 x 300mm; fenders in marine work	m	0.65	27.78	21.28	56.18	105.24	12.109	-31.480
O1.4.2.03	Sawn finish; Greenheart; 300 x 300mm; fenders in marine work	m	0.65	27.78	21.28	84.27	133.33	14.595	-47.219
O1.4.3	**Length: 3 - 5m**							
O1.4.3.01	Sawn finish; Greenheart; 300 x 150mm; braces between piles in marine work	m	0.62	26.50	20.30	42.14	88.94	10.536	-23.610
O1.4.3.02	Sawn finish; Greenheart; 200 x 300mm; fenders in marine work	m	0.62	26.50	20.30	56.18	102.98	11.779	-31.480
O1.4.3.03	Sawn finish; Greenheart; 300 x 300mm; fenders in marine work	m	0.62	26.50	20.30	84.27	131.07	14.266	-47.219
O1.4.4	**Length: 5 - 8m**							
O1.4.4.01	Sawn finish; Greenheart; 300 x 150mm; braces between piles in marine work	m	0.63	26.93	20.63	42.14	89.70	10.646	-23.610
O1.4.4.02	Sawn finish; Greenheart; 300 x 150mm; braces between piles in marine work; fixed between high water level ordinary spring tides (assumed 8m above ordnance datum) and low water level ordinary spring tides (assumed 3.5m above ordnance datum)	m	1.12	47.87	36.67	42.14	126.68	16.025	-23.610
O1.4.4.03	Sawn finish; Greenheart; 200 x 300mm; fenders in marine work	m	0.63	26.93	20.63	56.18	103.74	11.889	-31.480
O1.4.4.04	Sawn finish; Greenheart; 300 x 300mm; fenders in marine work	m	0.61	26.07	19.97	84.27	130.31	14.156	-47.219
O1.5	**Cross sectional area: 0.1 - 0.2m²**							
O1.5.3	**Length: 3 - 5m**							
O1.5.3.01	Sawn finish; Greenheart; 600 x 300mm; baulk in marine work	m	0.98	41.89	32.09	168.54	242.52	25.677	-94.439
O1.5.4	**Length: 5 - 8m**							
O1.5.4.01	Sawn finish; Greenheart; 600 x 300mm; baulk in marine work	m	0.90	38.47	29.47	168.54	236.48	24.799	-94.439

O1 Hardwood Components continued...

	Unit	Labour Hours	Labour Net £	Plant Net £	Materials Net £	Unit Net £	CO$_2$e	Kg
O1.6 **Cross sectional area: 0.2 - 0.4m^2**								
O1.6.3 Length: 3 - 5m								
O1.6.3.01 Sawn finish; Greenheart; 600 x 600mm; baulk in marine work	m	2.01	85.91	65.81	337.08	488.80	51.903	-188.878
O1.6.4 Length: 5 - 8m								
O1.6.4.01 Sawn finish; Greenheart; 600 x 600mm; baulk in marine work	m	1.21	51.72	39.62	337.08	428.42	43.120	-188.878
O1.7 **Cross sectional area: exceeding 0.4m^2**								
O1.7.3 Length: 3 - 5m								
O1.7.3.01 Sawn finish; Greenheart; 750 x 600mm; baulk in marine work	m	2.01	85.91	65.81	421.35	573.07	59.362	-236.097
O1.7.4 Length: 5 - 8m								
O1.7.4.01 Sawn finish; Greenheart; 750 x 600mm; baulk in marine work	m	1.21	51.72	39.62	421.35	512.69	50.579	-236.097

O2 Softwood Components

	Unit	Labour Hours	Labour Net £	Plant Net £	Materials Net £	Unit Net £	CO$_2$e	Kg
O2.1 **Cross sectional area: not exceeding 0.01m^2**								
O2.1.1 Length: not exceeding 1.5m								
O2.1.1.01 Wrot finish; Douglas fir; 100 x 75mm	m	0.27	11.54	8.84	4.38	24.76	3.544	-3.673
O2.1.2 Length: 1.5 - 3m								
O2.1.2.01 Wrot finish; Douglas fir; 100 x 75mm	m	0.27	11.54	8.84	4.38	24.76	3.544	-3.673
O2.1.3 Length: 3 - 5m								
O2.1.3.01 Wrot finish; Douglas fir; 100 x 75mm	m	0.23	9.83	7.53	4.38	21.74	3.105	-3.673
O2.2 **Cross sectional area: 0.01 - 0.02m^2**								
O2.2.2 Length: 1.5 - 3m								
O2.2.2.01 Wrot finish; Douglas fir; 150 x 75mm; runners in marine work	m	0.25	10.69	8.19	6.89	25.77	3.656	-5.771
O2.2.2.02 Wrot finish; Douglas fir; 75 x 225mm; runners in marine work	m	0.25	10.69	8.19	10.65	29.53	4.153	-8.919
O2.2.3 Length: 3 - 5m								
O2.2.3.01 Wrot finish; Douglas fir; 150 x 75mm; runners in marine work	m	0.20	8.55	6.55	6.89	21.99	3.107	-5.771
O2.2.3.02 Wrot finish; Douglas fir; 75 x 225mm; runners in marine work	m	0.21	8.98	6.88	10.65	26.51	3.714	-8.919
O2.3 **Cross sectional area: 0.02 - 0.04m^2**								
O2.3.2 Length: 1.5 - 3m								
O2.3.2.01 Wrot finish; Douglas fir; 200 x 200mm; braces in marine work	m	1.00	42.74	32.74	25.05	100.53	14.293	-20.986
O2.3.3 Length: 3 - 5m								
O2.3.3.01 Wrot finish; Douglas fir; 200 x 200mm; braces in marine work	m	0.93	39.66	30.38	25.05	95.09	13.503	-20.986

O2 Softwood Components continued...

	Unit	Labour Hours	Labour Net £	Plant Net £	Materials Net £	Unit Net £	CO_2e Kg	
O2.4	**Cross sectional area: 0.04 - 0.1m²**							
O2.4.2	**Length: 1.5 - 3m**							
O2.4.2.01 Wrot finish; Douglas fir; 300 × 150mm; braces between piles in marine work	m	1.04	44.28	33.92	28.19	106.39	15.103	-23.610
O2.4.2.02 Wrot finish; Douglas fir; 200 × 300mm; fenders in marine work	m	0.78	33.42	25.60	37.58	96.60	13.558	-31.480
O2.4.2.03 Wrot finish; Douglas fir; 300 × 300mm; fenders in marine work	m	0.99	42.31	32.41	56.37	131.09	18.327	-47.219
O2.4.3	**Length: 3 - 5m**							
O2.4.3.01 Wrot finish; Douglas fir; 300 × 150mm; braces between piles in marine work	m	0.96	40.86	31.30	28.19	100.35	14.225	-23.610
O2.4.3.02 Wrot finish; Douglas fir; 200 × 300mm; fenders in marine work	m	0.65	27.70	21.22	37.58	86.50	12.087	-31.480
O2.4.3.03 Wrot finish; Douglas fir; 300 × 300mm; fenders in marine work	m	0.90	38.47	29.79	56.37	124.63	17.449	-47.219
O2.4.4	**Length: 5 - 8m**							
O2.4.4.01 Wrot finish; Douglas fir; 300 × 150mm; braces between piles in marine work	m	0.88	37.44	28.68	28.19	94.31	13.346	-23.610
O2.4.4.02 Wrot finish; Douglas fir; 300 × 150mm; braces between piles in marine work; fixed between high water level ordinary spring tides (assumed 8m above ordnance datum) and low water level ordinary spring tides (assumed 3.5m above ordnance datum)	m	1.52	64.96	49.76	28.19	142.91	20.416	-23.610
O2.4.4.03 Wrot finish; Douglas fir; 200 × 300mm; fenders in marine work	m	0.60	25.73	19.71	37.58	83.02	11.582	-31.480
O2.4.4.04 Wrot finish; Douglas fir; 300 × 300mm; fenders in marine work	m	0.87	37.18	28.48	56.37	122.03	17.010	-47.219
O2.5	**Cross sectional area: 0.1 - 0.2m²**							
O2.5.3	**Length: 3 - 5m**							
O2.5.3.01 Wrot finish; Douglas fir; 600 × 300mm; baulk in marine work	m	1.08	46.16	35.36	112.74	194.26	26.775	-94.439
O2.5.4	**Length: 5 - 8m**							
O2.5.4.01 Wrot finish; Douglas fir; 600 × 300mm; baulk in marine work	m	1.02	43.59	33.39	112.74	189.72	26.116	-94.439
O2.6	**Cross sectional area: 0.2 - 0.4m2**							
O2.6.3	**Length: 3 - 5m**							
O2.6.3.01 Wrot finish; Douglas fir; 600 × 600mm; baulk in marine work	m	1.41	60.26	46.16	225.49	331.91	45.316	-188.878
O2.6.4	**Length: 5 - 8m**							
O2.6.4.01 Wrot finish; Douglas fir; 600 × 600mm; baulk in marine work	m	1.00	42.74	32.74	225.49	300.97	40.815	-188.878
O2.7	**Cross sectional area: exceeding 0.4m²**							
O2.7.3	**Length: 3 - 5m**							
O2.7.3.01 Wrot finish; Douglas fir; 750 × 600mm; baulk in marine work	m	1.44	61.55	47.15	281.86	390.56	53.104	-236.097

O2 Softwood Components continued...

	Unit	Labour Hours	Labour Net £	Plant Net £	Materials Net £	Unit Net £	CO$_2$e	Kg

O2.7 Cross sectional area: exceeding 0.4m²

O2.7.4 Length: 5 - 8m
O2.7.4.01 Wrot finish; Douglas fir; 750 x 600mm; baulk in marine work

	Unit	Labour Hours	Labour Net £	Plant Net £	Materials Net £	Unit Net £	CO$_2$e Kg
O2.7.4.01 Wrot finish; Douglas fir; 750 x 600mm; baulk in marine work	m	1.00	42.74	32.74	281.86	357.34	48.274	-236.097

O3 Hardwood Decking

O3.2 Thickness: 25 - 50mm

	Unit	Labour Hours	Labour Net £	Plant Net £	Materials Net £	Unit Net £	CO$_2$e Kg
O3.2.0 Generally							
O3.2.0.01 Generally	m²	0.60	25.64	19.64	46.82	92.10	10.731	-26.233

O3.3 Thickness: 50 - 75mm

	Unit	Labour Hours	Labour Net £	Plant Net £	Materials Net £	Unit Net £	CO$_2$e Kg
O3.3.0 Generally							
O3.3.0.01 Generally	m²	0.90	38.47	29.47	70.23	138.17	16.096	-39.350

O3.4 Thickness: 75 - 100mm

	Unit	Labour Hours	Labour Net £	Plant Net £	Materials Net £	Unit Net £	CO$_2$e Kg
O3.4.0 Generally							
O3.4.0.01 Generally	m²	1.20	51.29	39.29	93.63	184.21	21.462	-52.466

O3.5 Thickness: 100 - 125mm

	Unit	Labour Hours	Labour Net £	Plant Net £	Materials Net £	Unit Net £	CO$_2$e Kg
O3.5.0 Generally							
O3.5.0.01 Generally	m²	1.50	64.11	49.11	117.04	230.26	26.827	-65.583

O3.6 Thickness: 125 - 150mm

	Unit	Labour Hours	Labour Net £	Plant Net £	Materials Net £	Unit Net £	CO$_2$e Kg
O3.6.0 Generally							
O3.6.0.01 Generally	m²	1.80	76.93	58.93	140.45	276.31	32.192	-78.699

O3.7 Thickness: exceeding 150mm

	Unit	Labour Hours	Labour Net £	Plant Net £	Materials Net £	Unit Net £	CO$_2$e Kg
O3.7.0 Generally							
O3.7.0.01 Generally	m²	2.20	94.03	72.03	163.86	329.92	38.656	-91.816

O4 Softwood Decking

O4.2 Thickness: 25 - 50mm

	Unit	Labour Hours	Labour Net £	Plant Net £	Materials Net £	Unit Net £	CO$_2$e Kg
O4.2.0 Generally							
O4.2.0.01 Generally	m²	0.50	21.37	16.37	31.32	69.06	9.633	-26.233

O4.3 Thickness: 50 - 75mm

	Unit	Labour Hours	Labour Net £	Plant Net £	Materials Net £	Unit Net £	CO$_2$e Kg
O4.3.0 Generally							
O4.3.0.01 Generally	m²	0.60	25.64	19.64	46.98	92.26	12.803	-39.350

O4.4 Thickness: 75 - 100mm

	Unit	Labour Hours	Labour Net £	Plant Net £	Materials Net £	Unit Net £	CO$_2$e Kg
O4.4.0 Generally							
O4.4.0.01 Generally	m²	0.90	38.47	29.47	62.64	130.58	18.168	-52.466

O4.5 Thickness: 100 - 125mm

	Unit	Labour Hours	Labour Net £	Plant Net £	Materials Net £	Unit Net £	CO$_2$e Kg
O4.5.0 Generally							
O4.5.0.01 Generally	m²	1.20	51.29	39.29	78.30	168.88	23.534	-65.583

O4 Softwood Decking continued...

	Unit	Labour Hours	Labour Net £	Plant Net £	Materials Net £	Unit Net £		CO$_2$e Kg

O4.6 Thickness: 125 - 150mm

| O4.6.0 | Generally | | | | | | | | |
| O4.6.0.01 | Generally | m² | 1.50 | 64.11 | 49.11 | 93.95 | 207.17 | 28.899 | -78.699 |

O4.7 Thickness: exceeding 150mm

| O4.7.0 | Generally | | | | | | | | |
| O4.7.0.01 | Generally | m² | 1.80 | 76.93 | 58.93 | 109.61 | 245.47 | 34.264 | -91.816 |

O5 Fittings And Fastenings

Note(s): 1) Electric power drill and attachments are deemed to have been included in the general items
2) It has been assumed that the timber is no harder than Douglas Fir

O5.1 Straps

O5.1.0	Galvanised mild steel straps; fixing with screws							
O5.1.0.01	30 x 2.5 x 400mm girth	Nr	0.32	13.68	-	0.27	13.95	0.434
O5.1.0.02	30 x 2.5 x 600mm girth	Nr	0.35	14.96	-	0.44	15.40	0.651
O5.1.0.03	30 x 2.5 x 800mm girth	Nr	0.38	16.24	-	0.58	16.82	0.868
O5.1.0.04	30 x 2.5 x 1000mm girth	Nr	0.47	20.09	-	0.72	20.81	1.085
O5.1.0.05	30 x 2.5 x 1200mm girth	Nr	0.50	21.37	-	1.08	22.45	1.627

O5.2 Spikes

O5.2.0	Generally							
O5.2.0.01	Mild steel Rosehead spikes, including drilling holes; 12.5 x 12.5 x 100mm girth	Nr	0.17	7.27	-	1.34	8.61	0.029
O5.2.0.02	Mild steel Rosehead spikes, including drilling holes; 14 x 14 x 275mm girth	Nr	0.25	10.69	-	2.41	13.10	0.059
O5.2.0.03	Mild steel dogs, including drilling holes; 12.5 x 12.5 x 250 mm long with 12.5 x 12.5 x 75mm spikes each end	Nr	0.30	12.82	-	2.74	15.56	0.059

O5.3 Coach Screws

O5.3.0	Generally							
O5.3.0.01	6mm diameter; 50mm long	Nr	0.03	1.41	-	0.05	1.46	0.006
O5.3.0.02	6mm diameter; 75mm long	Nr	0.03	1.41	-	0.12	1.53	0.010
O5.3.0.03	6mm diameter; 100mm long	Nr	0.04	1.58	-	0.14	1.72	0.013
O5.3.0.04	6mm diameter; 125mm long	Nr	0.04	1.58	-	0.17	1.75	0.016
O5.3.0.05	8mm diameter; 50mm long	Nr	0.03	1.41	-	0.16	1.57	0.012
O5.3.0.06	8mm diameter; 75mm long	Nr	0.03	1.41	-	0.25	1.66	0.017
O5.3.0.07	8mm diameter; 100mm long	Nr	0.04	1.58	-	0.29	1.87	0.023
O5.3.0.08	8mm diameter; 125mm long	Nr	0.04	1.58	-	0.38	1.96	0.029
O5.3.0.09	8mm diameter; 150mm long	Nr	0.04	1.84	-	0.52	2.36	0.035
O5.3.0.10	9.5mm diameter; 50mm long	Nr	0.04	1.58	-	0.30	1.88	0.016
O5.3.0.11	9.5mm diameter; 75mm long	Nr	0.04	1.58	-	0.43	2.01	0.024
O5.3.0.12	9.5mm diameter; 100mm long	Nr	0.04	1.84	-	0.51	2.35	0.033
O5.3.0.13	9.5mm diameter; 125mm long	Nr	0.04	1.84	-	0.65	2.49	0.041
O5.3.0.14	9.5mm diameter; 150mm long	Nr	0.05	2.22	-	0.91	3.13	0.049
O5.3.0.15	12.5mm diameter; 50mm long	Nr	0.04	1.84	-	0.45	2.29	0.029
O5.3.0.16	12.5mm diameter; 75mm long	Nr	0.04	1.84	-	0.64	2.48	0.043
O5.3.0.17	12.5mm diameter; 100mm long	Nr	0.05	2.22	-	0.72	2.94	0.057
O5.3.0.18	12.5mm diameter; 125mm long	Nr	0.05	2.22	-	0.96	3.18	0.072
O5.3.0.19	12.5mm diameter; 150mm long	Nr	0.07	2.86	-	1.18	4.04	0.086

O5   Fittings And Fastenings continued...

	Unit	Labour Hours	Labour Net £	Plant Net £	Materials Net £	Unit Net £	CO_2e Kg
O5.4 **Bolts**							
O5.4.0 Generally							
O5.4.0.01 Black hexagonal head bolts and nuts grade 4.6, BS 4190 including one nut and one washer; drilling holes; M6; 25mm long	Nr	0.05	2.14	-	0.06	2.20	0.003
O5.4.0.02 Black hexagonal head bolts and nuts grade 4.6, BS 4190 including one nut and one washer; drilling holes; M6; 50mm long	Nr	0.05	2.14	-	0.06	2.20	0.006
O5.4.0.03 Black hexagonal head bolts and nuts grade 4.6, BS 4190 including one nut and one washer; drilling holes; Black hexagonal head bolts and nuts grade 4.6, BS 4190 including one nut and one washer; drilling holes; M6; 80mm long	Nr	0.05	2.14	-	0.09	2.23	0.010
O5.4.0.04 Black hexagonal head bolts and nuts grade 4.6, BS 4190 including one nut and one washer; drilling holes; M6; 100mm long	Nr	0.06	2.56	-	0.11	2.67	0.013
O5.4.0.05 Black hexagonal head bolts and nuts grade 4.6, BS 4190 including one nut and one washer; drilling holes; M8; 25mm long	Nr	0.05	2.14	-	0.07	2.21	0.006
O5.4.0.06 Black hexagonal head bolts and nuts grade 4.6, BS 4190 including one nut and one washer; drilling holes; M8; 50mm long	Nr	0.05	2.14	-	0.09	2.23	0.012
O5.4.0.07 Black hexagonal head bolts and nuts grade 4.6, BS 4190 including one nut and one washer; drilling holes; M10; 50mm	Nr	0.05	2.14	-	0.16	2.30	0.018
O5.4.0.08 Black hexagonal head bolts and nuts grade 4.6, BS 4190 including one nut and one washer; drilling holes; M10; 80mm long	Nr	0.06	2.56	-	0.20	2.76	0.029
O5.4.0.09 Black hexagonal head bolts and nuts grade 4.6, BS 4190 including one nut and one washer; drilling holes; M10; 100mm long	Nr	0.08	3.42	-	0.24	3.66	0.036
O5.4.0.10 Black hexagonal head bolts and nuts grade 4.6, BS 4190 including one nut and one washer; drilling holes; M10; 140 mm long	Nr	0.08	3.42	-	0.32	3.74	0.050
O5.4.0.11 Black hexagonal head bolts and nuts grade 4.6, BS 4190 including one nut and one washer; drilling holes; M12; 50mm long	Nr	0.06	2.56	-	0.20	2.76	0.026
O5.4.0.12 Black hexagonal head bolts and nuts grade 4.6, BS 4190 including one nut and one washer; drilling holes; M12; 80mm long	Nr	0.07	2.99	-	0.24	3.23	0.042
O5.4.0.13 Black hexagonal head bolts and nuts grade 4.6, BS 4190 including one nut and one washer; drilling holes; M12; 100mm long	Nr	0.07	2.99	-	0.28	3.27	0.052

O5 Fittings And Fastenings continued...

	Unit	Labour Hours	Labour Net £	Plant Net £	Materials Net £	Unit Net £	CO$_2$e Kg
O5.4	**Bolts**						
O5.4.0	Generally						
O5.4.0.14 Black hexagonal head bolts and nuts grade 4.6, BS 4190 including one nut and one washer; drilling holes; M12; 140 mm long	Nr	0.08	3.42	-	0.41	3.83	0.073
O5.4.0.15 Black hexagonal head bolts and nuts grade 4.6, BS 4190 including one nut and one washer; drilling holes; M12; 200mm long	Nr	0.08	3.42	-	0.78	4.20	0.104
O5.4.0.16 Black hexagonal head bolts and nuts grade 4.6, BS 4190 including one nut and one washer; drilling holes; M12; 260mm long	Nr	0.10	4.27	-	0.89	5.16	0.135
O5.4.0.17 Black hexagonal head bolts and nuts grade 4.6, BS 4190 including one nut and one washer; drilling holes; M12; 300mm long	Nr	0.10	4.27	-	1.20	5.47	0.156
O5.4.0.18 Black hexagonal head bolts and nuts grade 4.6, BS 4190 including one nut and one washer; drilling holes; M16; 50mm long	Nr	0.07	2.99	-	0.37	3.36	0.046
O5.4.0.19 Black hexagonal head bolts and nuts grade 4.6, BS 4190 including one nut and one washer; drilling holes; M16; 80mm long	Nr	0.07	2.99	-	0.45	3.44	0.074
O5.4.0.20 Black hexagonal head bolts and nuts grade 4.6, BS 4190 including one nut and one washer; drilling holes; M16; 100mm long	Nr	0.08	3.42	-	0.50	3.92	0.092
O5.4.0.21 Black hexagonal head bolts and nuts grade 4.6, BS 4190 including one nut and one washer; drilling holes; M^{2}0; 50mm long	Nr	0.08	3.42	-	0.53	3.95	0.072
O5.4.0.22 Black hexagonal head bolts and nuts grade 4.6, BS 4190 including one nut and one washer; drilling holes; M20; 100mm long	Nr	0.08	3.42	-	0.79	4.21	0.144
O5.4.0.23 Black cup square hexagon carriage bolts and nuts, grade 4.6, BS 4933 including one nut and one washer; drilling holes; M12; 75mm long	Nr	0.08	3.42	-	0.33	3.75	0.039
O5.4.0.24 Black cup square hexagon carriage bolts and nuts, grade 4.6, BS 4933 including one nut and one washer; drilling holes; M12; 100mm long	Nr	0.10	4.27	-	0.39	4.66	0.052
O5.4.0.25 Black cup square hexagon carriage bolts and nuts, grade 4.6, BS 4933 including one nut and one washer; drilling holes; M12; 150mm long	Nr	0.10	4.27	-	0.56	4.83	0.078
O5.5	**Plates**						
O5.5.0	Generally						
O5.5.0.01 Galvanised mild steel single sided round toothed plate connectors, BS EN 912; 38mm diameter for 9mm bolt	Nr	0.05	2.14	-	0.18	2.32	0.047

O5   Fittings And Fastenings continued...

	Unit	Labour Hours	Labour Net £	Plant Net £	Materials Net £	Unit Net £	CO_2e Kg
O5.5 **Plates**							
O5.5.0 Generally							
O5.5.0.02 Galvanised mild steel single sided round toothed plate connectors, BS EN 912; 57mm diameter for 12mm bolt	Nr	0.05	2.14	-	0.22	2.36	0.047
O5.5.0.03 Galvanised mild steel single sided round toothed plate connectors, BS EN 912; 63mm diameter for 12mm bolt	Nr	0.06	2.56	-	0.31	2.87	0.047
O5.5.0.04 Galvanised mild steel single sided round toothed plate connectors, BS EN 912; 75mm diameter for 12mm bolt	Nr	0.06	2.56	-	0.34	2.90	0.047
O5.5.0.05 Galvanised mild steel double sided round toothed plate connectors, BS EN 912; 38mm diameter for 9mm bolt	Nr	0.07	2.99	-	0.20	3.19	0.047
O5.5.0.06 Galvanised mild steel double sided round toothed plate connectors, BS EN 912; 50mm diameter for 12mm bolt	Nr	0.07	2.99	-	0.21	3.20	0.047
O5.5.0.07 Galvanised mild steel double sided round toothed plate connectors, BS EN 912; 63mm diameter for 12mm bolt	Nr	0.08	3.42	-	0.34	3.76	0.047
O5.5.0.08 Galvanised mild steel double sided round toothed plate connectors, BS EN 912; 75mm diameter for 12mm bolt	Nr	0.08	3.42	-	0.36	3.78	0.047
O5.5.0.09 Galvanised mild steel split ring connectors, BS EN 912; 50mm diameter; parallel sides	Nr	0.17	7.27	-	0.71	7.98	0.074
O5.5.0.10 Galvanised mild steel split ring connectors, BS EN 912; 64mm diameter; parallel sides	Nr	0.19	8.12	-	0.77	8.89	0.089
O5.5.0.11 Galvanised mild steel shear plate connectors, BS EN 912; 67mm diameter pressed steel	Nr	0.17	7.27	-	1.20	8.47	0.085
O5.5.0.12 Galvanised mild steel shear plate connectors, BS EN 912; 102mm diameter malleable cast iron	Nr	0.19	8.12	-	17.47	25.59	4.896

CLASS P:
PILES

Calculations used throughout Class P - Piles

Labour

			Qty		Rate		Total
L C0002ICE	**Bored Piling Labour Gang**						
	L A9016ICE	Ganger	2	x	17.64	=	£35.28
	L A9002ICE	Labourer (General Operative)	2	x	13.02	=	£26.03
	L A9000ICE	Craftsman WRA	2	x	17.33	=	£34.66
		Total hourly cost of gang				=	**£95.97**
L C0012ICE	**Piling Labour Gang (Mobilise / Demobilise / Move)**						
	L A9016ICE	Ganger	16	x	17.64	=	£282.24
	L A9002ICE	Labourer (General Operative)	24	x	13.02	=	£312.37
		Total hourly cost of gang				=	**£594.61**
L P0001ICE	**Driven Cast in Place Piling Labour Gang**						
	L A9016ICE	Ganger	2	x	17.64	=	£35.28
	L A9002ICE	Labourer (General Operative)	2	x	13.02	=	£26.03
	L A9000ICE	Craftsman WRA	3	x	17.33	=	£51.98
		Total hourly cost of gang				=	**£113.29**
L P0002ICE	**Driven Piling Labour Gang**						
	L A9016ICE	Ganger	1	x	17.64	=	£17.64
	L A9002ICE	Labourer (General Operative)	1	x	13.02	=	£13.02
	L A9000ICE	Craftsman WRA	1	x	17.33	=	£17.33
		Total hourly cost of gang				=	**£47.99**

Plant

			Qty		Rate		Total
P C0002ICE	**Bored Piling (Medium Rig) Plant Gang**						
	P A1063ICE	Bored Piling Rig Medium	1	x	165.10	=	£165.10
	P A0737ICE	Concrete Mixer - Schwing BP1000R Concrete Pumps	1	x	46.33	=	£46.33
		Total hourly cost of gang				=	**£211.43**
P C0003ICE	**Bored Piling (Medium) Plant Gang (Mobilise / Demobilise / Move)**						
	P A1063ICE	Bored Piling Rig Medium	8	x	165.10	=	£1,320.80
	P A0737ICE	Concrete Mixer - Schwing BP1000R Concrete Pumps	8	x	46.33	=	£370.64
	P A1044ICE	Articulated Lorry	8	x	68.51	=	£548.08
	P A1045ICE	Low Loader Trailer	8	x	13.96	=	£111.68
	P A1063ICE	Bored Piling Rig Medium	10	x	165.10	=	£1,651.00
	P A0737ICE	Concrete Mixer - Schwing BP1000R Concrete Pumps	10	x	46.33	=	£463.30
		Total hourly cost of gang				=	**£4,465.50**
P N0001ICE	**Sheet Piling Plant Gang Large (Mobilise / Demobilse / Move)**						
	P A1079ICE	Sheet Piling Rig Large	10	x	219.73	=	£2,197.30
	P A0458ICE	Compressor - 480 cfm	10	x	20.17	=	£201.70
	P A1044ICE	Articulated Lorry	10	x	68.51	=	£685.10
	P A1045ICE	Low Loader Trailer	10	x	13.96	=	£139.60
	P A0458ICE	Compressor - 480 cfm	10	x	20.17	=	£201.70
	P A0992ICE	Welding Set - 250 amp Diesel Electric Start Sil	10	x	3.76	=	£37.60
	P A1031ICE	Cutting and Burning Gear	10	x	3.77	=	£37.70
		Total hourly cost of gang				=	**£3,500.70**

P P0001ICE **Bored Piling (Large Rig) Plant Gang**

P A1062ICE	Bored Piling Rig Large	1	x	219.73	=	£219.73
P A0737ICE	Concrete Mixer - Schwing BP1000R Concrete Pumps	1	x	46.33	=	£46.33

Total hourly cost of gang = **£266.06**

P P0002ICE **Bored Piling (Large) Plant Gang (Mobilise / Demobilise / Move)**

P A1062ICE	Bored Piling Rig Large	10	x	219.73	=	£2,197.30
P A0737ICE	Concrete Mixer - Schwing BP1000R Concrete Pumps	10	x	46.33	=	£463.30
P A1044ICE	Articulated Lorry	10	x	68.51	=	£685.10
P A1045ICE	Low Loader Trailer	10	x	13.96	=	£139.60
P A1063ICE	Bored Piling Rig Medium	10	x	165.10	=	£1,651.00
P A0737ICE	Concrete Mixer - Schwing BP1000R Concrete Pumps	10	x	46.33	=	£463.30

Total hourly cost of gang = **£5,599.60**

P P0003ICE **Driven Cast in Place Piling (Large) Plant Gang (Mobilise / Demobilise / Move)**

P A1066ICE	Driven Piling Rig Large	10	x	219.73	=	£2,197.30
P A0737ICE	Concrete Mixer - Schwing BP1000R Concrete Pumps	10	x	46.33	=	£463.30
P A1044ICE	Articulated Lorry	10	x	68.51	=	£685.10
P A1045ICE	Low Loader Trailer	10	x	13.96	=	£139.60
P A1066ICE	Driven Piling Rig Large	10	x	219.73	=	£2,197.30
P A0737ICE	Concrete Mixer - Schwing BP1000R Concrete Pumps	10	x	46.33	=	£463.30
P A0992ICE	Welding Set - 250 amp Diesel Electric Start Sil	10	x	3.76	=	£37.60
P A1031ICE	Cutting and Burning Gear	10	x	3.77	=	£37.70

Total hourly cost of gang = **£6,221.20**

P P0004ICE **Driven Cast in Place Piling (Medium) Plant Gang (Mobilise / Demobilise / Move)**

P A1067ICE	Driven Piling Rig Medium	8	x	165.10	=	£1,320.80
P A0737ICE	Concrete Mixer - Schwing BP1000R Concrete Pumps	1	x	46.33	=	£46.33
P A1044ICE	Articulated Lorry	8	x	68.51	=	£548.08
P A1045ICE	Low Loader Trailer	8	x	13.96	=	£111.68
P A1063ICE	Bored Piling Rig Medium	10	x	165.10	=	£1,651.00
P A0737ICE	Concrete Mixer - Schwing BP1000R Concrete Pumps	10	x	46.33	=	£463.30
P A0992ICE	Welding Set - 250 amp Diesel Electric Start Sil	10	x	3.76	=	£37.60
P A1031ICE	Cutting and Burning Gear	10	x	3.77	=	£37.70

Total hourly cost of gang = **£4,216.49**

P P0005ICE **Driven Piling (Large) Plant Gang**

P A1066ICE	Driven Piling Rig Large	1	x	219.73	=	£219.73

Total hourly cost of gang = **£219.73**

P P0006ICE **Driven Piling (Large) Plant Gang (Mobilise / Demobilise / Move)**

P A1066ICE	Driven Piling Rig Large	10	x	219.73	=	£2,197.30
P A1044ICE	Articulated Lorry	10	x	68.51	=	£685.10
P A1045ICE	Low Loader Trailer	10	x	13.96	=	£139.60
P A1066ICE	Driven Piling Rig Large	10	x	219.73	=	£2,197.30

Total hourly cost of gang = **£5,219.30**

P P0007ICE **Driven Piling (Medium) Plant Gang**

P A1063ICE	Bored Piling Rig Medium	1	x	165.10	=	£165.10

Total hourly cost of gang = **£165.10**

P P0008ICE | **Driven Piling (Medium) Plant Gang (Mobilise / Demobilise / Move)**
P A1067ICE	Driven Piling Rig Medium	8	x	165.10	=	£1,320.80
P A1044ICE	Articulated Lorry	8	x	68.51	=	£548.08
P A1045ICE	Low Loader Trailer	8	x	13.96	=	£111.68
P A1063ICE	Bored Piling Rig Medium	10	x	165.10	=	£1,651.00

Total hourly cost of gang = **£3,631.56**

P P0009ICE | **Driven Cast in place Piling Plant Gang Medium**
P A1063ICE	Bored Piling Rig Medium	1	x	165.10	=	£165.10
P A0737ICE	Concrete Mixer - Schwing BP1000R Concrete Pumps	1	x	46.33	=	£46.33
P A0992ICE	Welding Set - 250 amp Diesel Electric Start Sil	1	x	3.76	=	£3.76
P A1031ICE	Cutting and Burning Gear	1	x	3.77	=	£3.77

Total hourly cost of gang = **£218.96**

P P0011ICE | **Sheet Piling (Large) Plant Gang**
P A1079ICE	Sheet Piling Rig Large	1	x	219.73	=	£219.73
P A0458ICE	Compressor - 480 cfm	1	x	20.17	=	£20.17
P A0992ICE	Welding Set - 250 amp Diesel Electric Start Sil	1	x	3.76	=	£3.76
P A1031ICE	Cutting and Burning Gear	1	x	3.77	=	£3.77

Total hourly cost of gang = **£247.43**

P P0012ICE | **Sheet Piling (Medium) Plant Gang**
P A1080ICE	Sheet Piling Rig Medium	1	x	165.10	=	£165.10
P A0458ICE	Compressor - 480 cfm	1	x	20.17	=	£20.17
P A0992ICE	Welding Set - 250 amp Diesel Electric Start Sil	1	x	3.76	=	£3.76
P A1031ICE	Cutting and Burning Gear	1	x	3.77	=	£3.77

Total hourly cost of gang = **£192.80**

P P0013ICE | **Sheet Piling (Medium) Plant Gang (Mobilise / Demobilise / Move)**
P A1080ICE	Sheet Piling Rig Medium	8	x	165.10	=	£1,320.80
P A0458ICE	Compressor - 480 cfm	8	x	20.17	=	£161.36
P A1044ICE	Articulated Lorry	8	x	68.51	=	£548.08
P A1045ICE	Low Loader Trailer	8	x	13.96	=	£111.68
P A0458ICE	Compressor - 480 cfm	8	x	20.17	=	£161.36
P A0992ICE	Welding Set - 250 amp Diesel Electric Start Sil	8	x	3.76	=	£30.08
P A1031ICE	Cutting and Burning Gear	8	x	3.77	=	£30.16

Total hourly cost of gang = **£2,363.52**

Piles

Note(s):   NOTE: The following prices are based on historical sub-contractor guide prices for approximate estimating only and assume not less than 50 piles will be cast on a level site. The cost of bringing to and from site the plant equipment and labour required including setting up and removal of the piling rig over each pile position is included in the items for 'number of piles'.

P1   Bored Cast In Place Concrete Piles

		Unit	Labour Hours	Labour Net £	Plant Net £	Materials Net £	Unit Net £	CO_2e Kg
P1.1	**Diameter: 300mm or 350mm**							
P1.1.1	Number of piles							
P1.1.1.01	Reinforced in situ concrete, BS EN 206 prescribed mix, C20/25, 20mm aggregate; Diameter: 300mm	Nr	0.02	11.89	89.31	-	101.20	82.044
P1.1.2	Concreted length							
P1.1.2.01	Reinforced in situ concrete, BS EN 206 prescribed mix, C20/25, 20mm aggregate; Diameter: 300mm	m	-	-	-	5.84	5.84	17.230
P1.1.3	Depth bored or driven to stated maximum depth							
P1.1.3.01	Reinforced in situ concrete, BS EN 206 prescribed mix, C20/25, 20mm aggregate; Diameter: 300mm; Depth bored: 10m	m	0.10	9.60	21.14	6.93	37.67	27.586
P1.1.3.02	Reinforced in situ concrete, BS EN 206 prescribed mix, C20/25, 20mm aggregate; Diameter: 300mm; Depth bored: 15m	m	0.15	14.40	31.71	6.93	53.04	38.142
P1.1.3.03	Reinforced in situ concrete, BS EN 206 prescribed mix, C20/25, 20mm aggregate; Diameter: 300mm; Depth bored: 20m	m	0.15	14.40	39.91	6.93	61.24	38.142
P1.4	**Diameter: 600mm or 750mm**							
P1.4.1	Number of piles							
P1.4.1.01	Reinforced in situ concrete, BS EN 206 prescribed mix, C20/25, 20mm aggregate; Diameter: 600mm	Nr	0.02	11.89	89.31	-	101.20	82.044
P1.4.2	Concreted length							
P1.4.2.01	Reinforced in situ concrete, BS EN 206 prescribed mix, C20/25, 20mm aggregate; Diameter: 600mm	m	-	-	-	23.60	23.60	69.660
P1.4.3	Depth bored or driven to stated maximum depth							
P1.4.3.01	Reinforced in situ concrete, BS EN 206 prescribed mix, C20/25, 20mm aggregate; Diameter: 600mm; Depth bored: 10m	m	0.05	4.80	10.57	26.47	41.84	35.274
P1.4.3.02	Reinforced in situ concrete, BS EN 206 prescribed mix, C20/25, 20mm aggregate; Diameter: 600mm; Depth bored: 15m	m	0.10	9.60	21.14	26.47	57.21	45.830
P1.4.3.03	Reinforced in situ concrete, BS EN 206 prescribed mix, C20/25, 20mm aggregate; Diameter: 600mm; Depth bored: 20m	m	0.15	14.40	39.91	26.47	80.78	56.386

P1 Bored Cast In Place Concrete Piles continued...

	Unit	Labour Hours	Labour Net £	Plant Net £	Materials Net £	Unit Net £	CO_2e Kg
P1.5 **Diameter: 900mm or 1050mm**							
P1.5.1 Number of piles							
P1.5.1.01 Reinforced in situ concrete, BS EN 206 prescribed mix, C20/25, 20mm aggregate; Diameter: 900mm	Nr	0.02	11.89	111.99	-	123.88	91.998
P1.5.2 Concreted length							
P1.5.2.01 Reinforced in situ concrete, BS EN 206 prescribed mix, C20/25, 20mm aggregate; Diameter: 900mm	m	-	-	-	53.36	53.36	157.534
P1.5.3 Depth bored or driven to stated maximum depth							
P1.5.3.01 Reinforced in situ concrete, BS EN 206 prescribed mix, C20/25, 20mm aggregate; Diameter: 900mm; Depth bored: 10m	m	0.05	4.80	13.30	63.02	81.12	69.408
P1.5.3.02 Reinforced in situ concrete, BS EN 206 prescribed mix, C20/25, 20mm aggregate; Diameter: 900mm; Depth bored: 15m	m	0.10	9.60	26.61	63.02	99.23	79.964
P1.5.3.03 Reinforced in situ concrete, BS EN 206 prescribed mix, C20/25, 20mm aggregate; Diameter: 900mm; Depth bored: 20m	m	0.15	14.40	39.91	63.02	117.33	90.520
P1.6 **Diameter: 1200mm or 1350mm**							
P1.6.1 Number of piles							
P1.6.1.01 Reinforced in situ concrete, BS EN 206 prescribed mix, C20/25, 20mm aggregate; Diameter: 1200mm	Nr	0.02	11.89	111.99	-	123.88	91.998
P1.6.2 Concreted length							
P1.6.2.01 Reinforced in situ concrete, BS EN 206 prescribed mix, C20/25, 20mm aggregate; Diameter: 1200mm	m	-	-	-	94.22	94.22	278.146
P1.6.3 Depth bored or driven to stated maximum depth							
P1.6.3.01 Reinforced in situ concrete, BS EN 206 prescribed mix, C20/25, 20mm aggregate; Diameter: 1200mm; Depth bored: 10m	m	0.30	28.79	79.82	107.13	215.74	163.385
P1.6.3.02 Reinforced in situ concrete, BS EN 206 prescribed mix, C20/25, 20mm aggregate; Diameter: 1200mm; Depth bored: 15m	m	0.35	33.59	93.12	107.13	233.84	173.941
P1.6.3.03 Reinforced in situ concrete, BS EN 206 prescribed mix, C20/25, 20mm aggregate; Diameter: 1200mm; Depth bored: 20m	m	0.40	38.39	106.42	107.13	251.94	184.497

P2   Driven Cast In Place Concrete Piles

	Unit	Labour Hours	Labour Net £	Plant Net £	Materials Net £	Unit Net £	CO_2e Kg
P2.1 Diameter: 300mm or 350mm							
P2.1.1 Number of piles							
P2.1.1.01 Diameter: 300mm; Design load 30t	Nr	0.02	11.89	84.33	-	96.22	82.066
P2.1.2 Concreted length							
P2.1.2.01 Diameter: 300mm; Design load 30t	m	0.08	9.06	-	41.33	50.39	108.438
P2.1.3 Depth bored or driven to stated maximum depth							
P2.1.3.01 Diameter: 300mm; Design load 30t; Depth bored: 10m	m	0.05	6.00	11.60	-	17.60	11.384
P2.1.3.02 Diameter: 300mm; Design load 30t; Depth bored: 15m	m	0.08	9.06	17.52	-	26.58	17.184
P2.1.3.03 Diameter: 300mm; Design load 30t; Depth bored: 20m	m	0.10	11.33	21.90	-	33.23	21.480
P2.4 Diameter: 600mm or 750mm							
P2.4.1 Number of piles							
P2.4.1.01 Diameter: 600mm; Design load 60t	Nr	0.02	11.89	84.33	-	96.22	82.066
P2.4.2 Concreted length							
P2.4.2.01 Diameter: 600mm; Design load 60t	m	0.08	9.06	-	108.21	117.27	266.873
P2.4.3 Depth bored or driven to stated maximum depth							
P2.4.3.01 Diameter: 600mm; Design load 60t; Depth bored: 10m	m	0.07	7.59	14.67	-	22.26	14.391
P2.4.3.02 Diameter: 600mm; Design load 60t; Depth bored: 15m	m	0.08	9.06	17.52	-	26.58	17.184
P2.4.3.03 Diameter: 600mm; Design load 60t; Depth bored: 20m	m	0.10	11.33	21.90	-	33.23	21.480
P2.5 Diameter: 900mm or 1050mm							
P2.5.1 Number of piles							
P2.5.1.01 Diameter: 900mm; Design load 90t	Nr	0.02	11.89	124.42	-	136.31	92.733
P2.5.2 Concreted length							
P2.5.2.01 Diameter: 900mm; Design load 90t	m	0.08	9.06	-	225.57	234.63	545.436
P2.5.3 Depth bored or driven to stated maximum depth							
P2.5.3.01 Diameter: 900mm; Design load 90t; Depth bored: 10m	m	0.05	6.00	14.50	-	20.50	**11.384**
P2.5.3.02 Diameter: 900mm; Design load 90t; Depth bored: 15m	m	0.08	9.06	21.89	-	30.95	17.184
P2.5.3.03 Diameter: 900mm; Design load 90t; Depth bored: 20m	m	0.10	11.33	27.36	-	38.69	21.480
P2.6 Diameter: 1200mm or 1350mm							
P2.6.1 Number of piles							
P2.6.1.01 Diameter: 1200mm; Design load 120t	Nr	0.02	11.89	124.42	-	136.31	92.733
P2.6.2 Concreted length							
P2.6.2.01 Diameter: 1200mm; Design load 120t	m	0.08	9.06	-	350.50	359.56	820.695

P2 Driven Cast In Place Concrete Piles continued...

	Unit	Labour Hours	Labour Net £	Plant Net £	Materials Net £	Unit Net £	CO_2e Kg	
P2.6	**Diameter: 1200mm or 1350mm**							
P2.6.3	Depth bored or driven to stated maximum depth							
P2.6.3.01	Diameter: 1200mm; Design load 120t; Depth bored: 10m	m	0.05	6.00	14.50	-	20.50	11.384
P2.6.3.02	Diameter: 1200mm; Design load 120t; Depth bored: 15m	m	0.08	9.06	21.89	-	30.95	17.184
P2.6.3.03	Diameter: 1200mm; Design load 120t; Depth bored: 20m	m	0.10	11.33	27.36	-	38.69	21.480

P3 Preformed Concrete Piles

	Unit	Labour Hours	Labour Net £	Plant Net £	Materials Net £	Unit Net £	CO_2e Kg	
P3.2	**Cross-sectional area: 0.025 - 0.05m²**							
P3.2.1	Number of piles of stated length							
P3.2.1.01	Cross sectional area: 0.05m²; Length 10m	Nr	0.82	50.28	204.71	105.39	360.38	444.154
P3.2.1.02	Cross sectional area: 0.05m²; Length 15m	Nr	1.02	59.87	237.73	158.09	455.69	584.917
P3.2.1.03	Cross sectional area: 0.05m²; Length 20m	Nr	1.52	83.86	320.28	210.79	614.93	787.491
P3.2.2	Depth driven							
P3.2.2.01	Cross sectional area: 0.05m²	m	0.04	1.92	6.60	-	8.52	8.242
P3.3	**Cross-sectional area: 0.05 - 0.1m²**							
P3.3.1	Number of piles of stated length							
P3.3.1.01	Cross-sectional area: 0.075m²; Length 10m	Nr	0.82	50.28	280.17	158.09	488.54	553.460
P3.3.1.02	Cross-sectional area: 0.075m²; Length 15m	Nr	1.02	59.87	324.12	237.14	621.13	744.001
P3.3.1.03	Cross-sectional area: 0.075m²; Length 20m	Nr	1.52	83.86	433.98	316.18	834.02	996.353
P3.3.2	Depth driven							
P3.3.2.01	Cross-sectional area: 0.075m²	m	0.04	1.92	8.79	-	10.71	8.242

P5 Timber Piles

	Unit	Labour Hours	Labour Net £	Plant Net £	Materials Net £	Unit Net £	CO_2e Kg	
P5.3	**Cross-sectional area: 0.05 - 0.1m²**							
P5.3.1	Number of piles of stated length							
P5.3.1.01	Douglas fir; 225 x 225 mm, length 10m	Nr	0.62	40.68	171.69	503.20	715.57	245.796 -265.610
P5.3.1.02	Douglas fir; 225 x 225 mm, length 15m	Nr	0.77	47.88	237.73	754.80	1,040.41	328.211 -398.415
P5.3.1.03	Douglas fir; 225 x 225 mm, length 20m	Nr	1.02	59.87	237.73	1,006.40	1,304.00	328.211 -531.220
P5.3.2	Depth driven							
P5.3.2.01	Cross-sectional area: 0.05 - 0.1m²	m	0.04	1.92	6.60	-	8.52	8.242

P6 Isolated Steel Piles

		Unit	Labour Hours	Labour Net £	Plant Net £	Materials Net £	Unit Net £	CO_2e Kg
P6.2	**Mass: 15 - 30Kg/m**							
P6.2.1	Number of piles of stated length							
P6.2.1.01	I Sections; Mass: 25kg/m; Length 10m	Nr	1.02	59.87	237.73	309.61	607.21	578.453
P6.2.1.02	I Sections; Mass: 25kg/m; Length 15m	Nr	1.52	83.86	320.28	463.82	867.96	827.018
P6.2.1.03	I Sections; Mass: 25kg/m; Length 20m	Nr	2.02	107.85	402.83	618.04	1,128.72	1,075.583
P6.2.2	Depth driven							
P6.2.2.01	I Sections; Mass: 25kg/m	m	0.04	1.92	6.60	-	8.52	8.242
P6.3	**Mass: 30 - 60Kg/m**							
P6.3.1	Number of piles of stated length							
P6.3.1.01	I Sections; Mass: 50Kg/m; Length 10m	Nr	1.02	59.87	237.73	618.04	915.64	869.545
P6.3.1.02	I Sections; Mass: 50Kg/m; Length 15m	Nr	1.52	83.86	320.28	927.64	1,331.78	1,264.767
P6.3.1.03	I Sections; Mass: 50Kg/m; Length 20m	Nr	2.02	107.85	402.83	1,236.07	1,746.75	1,658.878
P6.3.2	Depth driven							
P6.3.2.01	I Sections; Mass: 50Kg/m	m	0.04	1.92	6.60	-	8.52	8.242
P6.4	**Mass: 60 - 120Kg/m**							
P6.4.1	Number of piles of stated length							
P6.4.1.01	I Sections; Mass: 100Kg/m; Length 10m	Nr	1.52	83.86	320.28	1,236.07	1,640.21	1,555.859
P6.4.1.02	I Sections; Mass: 100Kg/m; Length 15m	Nr	2.02	107.85	402.83	1,854.11	2,364.79	2,242.172
P6.4.1.03	I Sections; Mass: 100Kg/m; Length 20m	Nr	2.52	131.84	517.14	2,472.14	3,121.12	2,938.237
P6.4.2	Depth driven							
P6.4.2.01	I Sections; Mass: 100Kg/m	m	0.04	1.92	6.60	-	8.52	8.242
P6.5	**Mass: 120 - 250Kg/m**							
P6.5.1	Number of piles of stated length							
P6.5.1.01	I Sections; Mass: 250Kg/m; Length 10m	Nr	2.52	131.84	653.71	3,090.18	3,875.73	3,521.531
P6.5.1.02	I Sections; Mass: 250Kg/m; Length 15m	Nr	3.52	179.82	682.24	4,635.85	5,497.91	5,186.361
P6.5.1.03	I Sections; Mass: 250Kg/m; Length 20m	Nr	4.52	227.80	847.34	6,180.35	7,255.49	6,850.079
P6.5.2	Depth driven							
P6.5.2.01	I Sections; Mass: 250Kg/m	m	0.04	1.92	6.60	-	8.52	8.242

P7 Interlocking Steel Piles

		Unit	Labour Hours	Labour Net £	Plant Net £	Materials Net £	Unit Net £	CO_2e Kg
P7.2	**Section modulus: 500 - 800cm^3/m**							
P7.2.1	Length of special piles							
P7.2.1.01	Piles weldable structural steel, BS EN 10248; Section modulus: 601 cm^3/m; Length of corners	m	0.10	7.50	19.28	52.55	79.33	35.545

P7 Interlocking Steel Piles continued...

	Unit	Labour Hours	Labour Net £	Plant Net £	Materials Net £	Unit Net £	CO_2e Kg
P7.2 **Section modulus: 500 - 800cm3/m**							
P7.2.1 Length of special piles							
P7.2.1.02 Piles weldable structural steel, BS EN 10248; Section modulus: 601 cm³/m; Length of closure piles	m	0.15	11.25	28.92	52.55	92.72	47.197
P7.2.1.03 Piles weldable structural steel, BS EN 10248; Section modulus: 601 cm³/m; Length of taper piles	m	0.25	18.76	48.20	52.55	119.51	60.484
P7.2.2 Driven area							
P7.2.2.01 Piles weldable structural steel, BS EN 10248; Section modulus: 601 cm³/m	m²	0.10	7.50	19.28	-	26.78	23.306
P7.2.3 Area of piles of length: not exceeding 14m							
P7.2.3.01 Piles weldable structural steel, BS EN 10248; Section modulus: 601 cm³/m; Area of piles of length: 10m	m²	0.02	11.89	47.27	87.35	146.51	207.116
P7.3 **Section modulus: 800 - 1200cm³/m**							
P7.3.1 Length of special piles							
P7.3.1.01 Piles weldable structural steel, BS EN 10248; Section modulus: 1199cm³/m; Length of corners	m	0.15	11.25	28.92	52.55	92.72	47.197
P7.3.1.02 Piles weldable structural steel, BS EN 10248; Section modulus: 1199cm³/m; Length of closure piles	m	0.20	15.01	38.56	52.55	106.12	58.850
P7.3.1.03 Piles weldable structural steel, BS EN 10248; Section modulus: 1199cm³/m; Length of taper piles	m	0.25	18.76	48.20	52.55	119.51	60.484
P7.3.2 Driven area							
P7.3.2.01 Piles weldable structural steel, BS EN 10248; Section modulus: 1199cm³/m	m²	0.10	7.50	19.28	-	26.78	23.306
P7.3.3 Area of piles of length: not exceeding 14m							
P7.3.3.01 Piles weldable structural steel, BS EN 10248; Section modulus: 1199cm³/m; Area of piles of length: 10m	m²	0.02	11.89	47.27	118.18	177.34	263.606
P7.3.4 Area of piles of length: 14 - 24m							
P7.3.4.01 Piles weldable structural steel, BS EN 10248; Section modulus: 1199cm³/m; Area of piles of length: 20m	m²	0.02	11.89	47.27	118.18	177.34	**263.606**
P7.5 **Section modulus: 2000 - 3000cm³/m**							
P7.5.1 Length of special piles							
P7.5.1.01 Piles weldable structural steel, BS EN 10248; Section modulus: 2009cm³/m; Length of corners	m	0.20	15.01	38.56	52.55	106.12	58.850

P7   Interlocking Steel Piles continued...

	Unit	Labour Hours	Labour Net £	Plant Net £	Materials Net £	Unit Net £	CO_2e Kg
P7.5 **Section modulus: 2000 - 3000cm3/m**							
P7.5.1 Length of special piles							
P7.5.1.02 Piles weldable structural steel, BS EN 10248; Section modulus: 2009cm^3/m; Length of closure piles	m	0.25	18.76	48.20	52.55	119.51	70.503
P7.5.1.03 Piles weldable structural steel, BS EN 10248; Section modulus: 2009cm^3/m; Length of taper piles	m	0.30	22.51	57.84	52.55	132.90	72.137
P7.5.2 Driven area							
P7.5.2.01 Piles weldable structural steel, BS EN 10248; Section modulus: 2009cm^3/m	m^2	0.15	11.25	28.92	-	40.17	34.958
P7.5.3 Area of piles of length: not exceeding 14m							
P7.5.3.01 Piles weldable structural steel, BS EN 10248; Section modulus: 2009cm^3/m; Area of piles of length: 10m	m^2	0.02	11.89	47.27	151.06	210.22	323.861
P7.5.4 Area of piles of length: 14 - 24m							
P7.5.4.01 Piles weldable structural steel, BS EN 10248; Section modulus: 2009cm^3/m; Area of piles of length: 20m	m^2	0.02	11.89	47.27	151.06	210.22	323.861
P7.5.5 Area of piles of length: exceeding 24m							
P7.5.5.01 Piles weldable structural steel, BS EN 10248; Section modulus: 2009cm^3/m; Area of piles of length: 30m	m^2	0.02	11.89	47.27	151.06	210.22	323.861
P7.7 **Section modulus: 4000 - 5000cm^3/m**							
P7.7.1 Length of special piles							
P7.7.1.01 Piles weldable structural steel, BS EN 10248; Section modulus: 4200cm^3/m; Length of corners	m	0.10	7.50	24.74	52.55	84.79	35.545
P7.7.1.02 Piles weldable structural steel, BS EN 10248; Section modulus: 4200cm^3/m; Length of closure piles	m	0.15	11.25	37.11	52.55	100.91	47.197
P7.7.1.03 Piles weldable structural steel, BS EN 10248; Section modulus: 4200cm^3/m; Length of taper piles	m	0.20	15.01	49.49	52.55	117.05	48.831
P7.7.2 Driven area							
P7.7.2.01 Piles weldable structural steel, BS EN 10248; Section modulus: 4200cm^3/m	m^2	0.10	7.50	24.74	-	32.24	23.306
P7.7.3 Area of piles of length: not exceeding 14m							
P7.7.3.01 Piles weldable structural steel, BS EN 10248; Section modulus: 4200cm^3/m; Area of piles of length: 10m	m^2	0.02	11.89	70.02	298.01	379.92	604.893

P7 Interlocking Steel Piles continued...

	Unit	Labour Hours	Labour Net £	Plant Net £	Materials Net £	Unit Net £	CO_2e Kg
P7.7	**Section modulus: 4000 - 5000cm3/m**						
P7.7.4 Area of piles of length: 14 - 24m							
P7.7.4.01 Piles weldable structural steel, BS EN 10248; Section modulus: 4200cm³/m; Area of piles of length: 20m	m²	0.02	11.89	70.02	298.01	379.92	604.893
P7.7.5 Area of piles of length: exceeding 24m							
P7.7.5.01 Piles weldable structural steel, BS EN 10248; Section modulus: 4200cm³/m; Area of piles of length: 30m	m²	0.02	11.89	70.02	298.01	379.92	604.893

CLASS Q:
PILING ANCILLARIES

Calculations used throughout Class Q - Piling Ancillaries

Labour

			Qty		Rate		Total
L A0140ICE	**Steel fixing Labour Gang**						
	L A9000ICE	Craftsman WRA	2	x	17.33	=	£34.66
	L A9002ICE	Labourer (General Operative)	1	x	13.02	=	£13.02
		Total hourly cost of gang				=	**£47.68**
L A0145ICE	**Welding Labour Gang**						
	L A9003ICE	Labourer (Skill Rate 3)	1	x	14.86	=	£14.86
	L A9021ICE	Fitters and Welders	2	x	17.33	=	£34.66
		Total hourly cost of gang				=	**£49.52**
L C0012ICE	**Piling Labour Gang (Mobilise / Demobilise / Move)**						
	L A9016ICE	Ganger	16	x	17.64	=	£282.24
	L A9002ICE	Labourer (General Operative)	24	x	13.02	=	£312.37
		Total hourly cost of gang				=	**£594.61**
L P0001ICE	**Driven Cast in Place Piling Labour Gang**						
	L A9016ICE	Ganger	2	x	17.64	=	£35.28
	L A9002ICE	Labourer (General Operative)	2	x	13.02	=	£26.03
	L A9000ICE	Craftsman WRA	3	x	17.33	=	£51.98
		Total hourly cost of gang				=	**£113.29**
L P0002ICE	**Driven Piling Labour Gang**						
	L A9016ICE	Ganger	1	x	17.64	=	£17.64
	L A9002ICE	Labourer (General Operative)	1	x	13.02	=	£13.02
	L A9000ICE	Craftsman WRA	1	x	17.33	=	£17.33
		Total hourly cost of gang				=	**£47.99**
L P0003ICE	**Sheet Piling Labour Gang**						
	L A9016ICE	Ganger	1	x	17.64	=	£17.64
	L A9002ICE	Labourer (General Operative)	2	x	13.02	=	£26.03
	L A9000ICE	Craftsman WRA	1	x	17.33	=	£17.33
	L A9015ICE	Banksman	1	x	14.03	=	£14.03
		Total hourly cost of gang				=	**£75.03**
L Q0001ICE	**Pile Testing Labour Gang**						
	L A9016ICE	Ganger	1	x	17.64	=	£17.64
	L A9002ICE	Labourer (General Operative)	2	x	13.02	=	£26.03
		Total hourly cost of gang				=	**£43.67**
L Q0002ICE	**Piling - Head Works Labour Gang**						
	L A9003ICE	Labourer (Skill Rate 3)	2	x	14.86	=	£29.72
		Total hourly cost of gang				=	**£29.72**
L Q0003ICE	**Piling Backfilling Labour Gang**						
	L A9016ICE	Ganger	1	x	17.64	=	£17.64
	L A9002ICE	Labourer (General Operative)	1	x	13.02	=	£13.02
	L A9015ICE	Banksman	1	x	14.03	=	£14.03
		Total hourly cost of gang				=	**£44.69**

Plant

P A1000ICE   **Pile Testing Rig Plant Gang (Mobilise / Demobilise / Move)**

P A1047ICE	Pile Testing Rig and Equipment	10	x	234.71	=	£2,347.10
P A1044ICE	Articulated Lorry	10	x	68.51	=	£685.10
P A1045ICE	Low Loader Trailer	10	x	13.96	=	£139.60
P A1047ICE	Pile Testing Rig and Equipment	5	x	234.71	=	£1,173.55
	Total hourly cost of gang				=	**£4,345.35**

P P0007ICE   **Driven Piling (Medium) Plant Gang**

P A1063ICE	Bored Piling Rig Medium	1	x	165.10	=	£165.10
	Total hourly cost of gang				=	**£165.10**

P P0011ICE   **Sheet Piling (Large) Plant Gang**

P A1079ICE	Sheet Piling Rig Large	1	x	219.73	=	£219.73
P A0458ICE	Compressor - 480 cfm	1	x	20.17	=	£20.17
P A0992ICE	Welding Set - 250 amp Diesel Electric Start Sil	1	x	3.76	=	£3.76
P A1031ICE	Cutting and Burning Gear	1	x	3.77	=	£3.77
	Total hourly cost of gang				=	**£247.43**

P Q0005ICE   **Piling - Head Works Plant Gang**

P A0452ICE	Compressor - 2-Tool (Complete)	1	x	5.20	=	£5.20
P A0402ICE	CP9 Air Breaker Drill	2	x	0.45	=	£0.90
P A0608ICE	Hydraulic Excavator - Komatsu PC130	1	x	45.87	=	£45.87
P A0550ICE	Dumper - 3.0t 4WD	1	x	25.44	=	£25.44
	Total hourly cost of gang				=	**£77.41**

P Q0006ICE   **Piling Backfilling Plant Gang**

P A0608ICE	Hydraulic Excavator - Komatsu PC130	1	x	45.87	=	£45.87
P A0550ICE	Dumper - 3.0t 4WD	1	x	25.44	=	£25.44
	Total hourly cost of gang				=	**£71.31**

Piling Ancillaries

Note(s):   NOTES: The following prices are Specialist sub-contractor guide prices for approximate estimating only.

Q1   Cast In Place Concrete Piles

	Unit	Labour Hours	Labour Net £	Plant Net £	Materials Net £	Unit Net £	CO_2e Kg
Q1.2	**Backfilling empty bore with stated material**						
Q1.2.1 Diameter: 300mm or 350mm							
Q1.2.1.01 Material arising from excavations; Diameter: 300mm	m	0.01	0.63	1.00	-	1.63	0.367
Q1.2.1.02 Sand; Diameter: 300mm	m	0.01	0.63	1.00	2.04	3.67	1.243
Q1.2.4 Diameter: 600mm or 750mm							
Q1.2.4.01 Material arising from excavations; Diameter: 600mm	m	0.05	2.10	3.35	-	5.45	1.230
Q1.2.4.02 Sand; Diameter: 600mm	m	0.05	2.10	3.35	8.12	13.57	4.722
Q1.2.5 Diameter: 900 - 1050mm							
Q1.2.5.01 Material arising from excavations; Diameter: 900mm	m	0.07	3.13	4.99	-	8.12	1.833
Q1.2.5.02 Sand; Diameter: 900mm	m	0.07	3.13	4.99	18.28	26.40	9.691
Q1.2.6 Diameter: 1200 - 1350mm							
Q1.2.6.01 Material arising from excavations; Diameter: 1200mm	m	0.09	4.20	6.70	-	10.90	2.461
Q1.2.6.02 Sand; Diameter: 1200mm	m	0.09	4.20	6.70	32.80	43.70	16.559
Q1.3	**Permanent casings each length: not exceeding 13m**						
Q1.3.1 Diameter: 300mm or 350mm							
Q1.3.1.01 Permanent casings each length: 10m; Diameter: 300mm	m	0.10	4.80	16.51	28.56	49.87	105.338
Q1.3.4 Diameter: 600mm or 750mm							
Q1.3.4.01 Permanent casings each length: 10m; Diameter: 600mm	m	0.10	4.80	16.51	58.39	79.70	193.838
Q1.3.5 Diameter: 900 - 1050mm							
Q1.3.5.01 Permanent casings each length: 10m; Diameter: 900mm	m	0.25	11.99	41.27	111.71	164.97	382.914
Q1.3.6 Diameter: 1200 - 1350mm							
Q1.3.6.01 Permanent casings each length: 10m; Diameter: 1200mm	m	0.40	19.19	66.04	149.16	234.39	524.916
Q1.4	**Permanent casings each length: exceeding 13m**						
Q1.4.1 Diameter: 300mm or 350mm							
Q1.4.1.01 Permanent casings each length: 20m; Diameter: 300mm	m	0.10	11.33	16.51	28.56	56.40	105.338
Q1.4.4 Diameter: 600mm or 750mm							
Q1.4.4.01 Permanent casings each length: 20m; Diameter: 600mm	m	0.10	11.33	16.51	58.39	86.23	193.838
Q1.4.5 Diameter: 900 - 1050mm							
Q1.4.5.01 Permanent casings each length: 20m; Diameter: 900mm	m	0.25	28.32	41.27	111.71	181.30	382.914
Q1.4.6 Diameter: 1200 - 1350mm							
Q1.4.6.01 Permanent casings each length: 20m; Diameter: 1200mm	m	0.40	45.32	66.04	149.16	260.52	524.916

Q1　Cast In Place Concrete Piles continued...

	Unit	Labour Hours	Labour Net £	Plant Net £	Materials Net £	Unit Net £	CO_2e Kg
Q1.5	**Enlarged bases**						
Q1.5.1　Diameter: 300mm or 350mm							
Q1.5.1.01　Diameter: 300mm	Nr	1.00	47.98	165.10	18.43	231.51	244.395
Q1.5.4　Diameter: 600mm or 750mm							
Q1.5.4.01　Diameter: 600mm	Nr	1.50	71.97	247.65	87.06	406.68	466.070
Q1.5.5　Diameter: 900 - 1050mm							
Q1.5.5.01　Diameter: 900mm	Nr	2.25	107.96	494.39	168.97	771.32	814.406
Q1.5.6　Diameter: 1200 - 1350mm							
Q1.5.6.01　Diameter: 1200mm	Nr	4.00	191.92	878.92	298.39	1,369.23	1,443.583
Q1.7	**Cutting off surplus lengths**						
Q1.7.1　Diameter: 300mm or 350mm							
Q1.7.1.01　Diameter: 300mm	m	0.17	4.96	12.93	-	17.89	5.933
Q1.7.4　Diameter: 600mm or 750mm							
Q1.7.4.01　Diameter: 600mm	m	0.25	7.43	19.36	-	26.79	8.881
Q1.7.5　Diameter: 900 - 1050mm							
Q1.7.5.01　Diameter: 900mm	m	0.33	9.90	25.78	-	35.68	11.829
Q1.7.6　Diameter: 1200 - 1350mm							
Q1.7.6.01　Diameter: 1200mm	m	0.50	14.86	38.71	-	53.57	17.762
Q1.8	**Preparing heads**						
Q1.8.1　Diameter: 300mm or 350mm							
Q1.8.1.01　Diameter: 300mm	Nr	0.33	9.90	25.78	-	35.68	11.829
Q1.8.4　Diameter: 600mm or 750mm							
Q1.8.4.01　Diameter: 600mm	Nr	0.42	12.39	32.28	-	44.67	14.814
Q1.8.5　Diameter: 900 - 1050mm							
Q1.8.5.01　Diameter: 900mm	Nr	0.50	14.86	38.71	-	53.57	17.762
Q1.8.6　Diameter: 1200 - 1350mm							
Q1.8.6.01　Diameter: 1200mm	Nr	0.75	22.29	58.06	-	80.35	26.643

Q2　Cast In Place Concrete Piles

	Unit	Labour Hours	Labour Net £	Plant Net £	Materials Net £	Unit Net £	CO_2e Kg
Q2.1	**Reinforcement**						
Q2.1.1　Straight bars, nominal size: not exceeding 25mm							
Q2.1.1.01　nominal size: 8mm	t	10.00	476.70	-	711.88	1,188.58	588.516
Q2.1.1.02　nominal size: 12mm	t	10.00	476.70	-	666.79	1,143.49	588.516
Q2.1.1.03　nominal size: 20mm	t	8.00	381.36	-	630.16	1,011.52	588.516
Q2.1.1.04　nominal size: 25mm	t	8.00	381.36	-	629.72	1,011.08	588.516
Q2.1.2　Straight bars, nominal size: exceeding 25mm							
Q2.1.2.01　nominal size: 32mm	t	8.00	381.36	-	633.41	1,014.77	588.516
Q2.1.2.02　nominal size: 40mm	t	8.00	381.36	-	640.78	1,022.14	588.516

Q3　Preformed Concrete Piles

	Unit	Labour Hours	Labour Net £	Plant Net £	Materials Net £	Unit Net £	CO_2e Kg
Q3.1	**Pre-boring**						
Q3.1.2　Cross-sectional area: 0.025 - 0.05m^2							
Q3.1.2.01　Cross sectional area: 0.05m^2	m	0.05	2.40	8.26	-	10.66	10.302

Q3 Preformed Concrete Piles continued...

	Unit	Labour Hours	Labour Net £	Plant Net £	Materials Net £	Unit Net £	CO_2e Kg
Q3.1 **Pre-boring**							
Q3.1.3 Cross-sectional area: 0.05 - 0.1m^2							
Q3.1.3.01 Cross sectional area: 0.10m^2	m	0.05	2.40	8.26	-	10.66	10.302
Q3.1.4 Cross-sectional area: 0.1 - 0.15m^2							
Q3.1.4.01 Cross sectional area: 0.15m^2	m	0.08	3.60	12.38	-	15.98	15.453
Q3.1.5 Cross-sectional area: 0.15 - 0.25m^2							
Q3.1.5.01 Cross sectional area: 0.20m^2	m	0.10	4.80	16.51	-	21.31	20.604
Q3.2 **Jetting**							
Q3.2.2 Cross-sectional area: 0.025 - 0.05m^2							
Q3.2.2.01 Cross sectional area: 0.05m^2	m	0.05	2.40	10.08	-	12.48	16.502
Q3.2.3 Cross-sectional area: 0.05 - 0.1m^2							
Q3.2.3.01 Cross sectional area: 0.10m^2	m	0.05	2.40	10.08	-	12.48	16.502
Q3.2.4 Cross-sectional area: 0.1 - 0.15m^2							
Q3.2.4.01 Cross sectional area: 0.15m^2	m	0.10	4.80	20.15	-	24.95	33.005
Q3.2.5 Cross-sectional area: 0.15 - 0.25m^2							
Q3.2.5.01 Cross sectional area: 0.20m^2	m	0.15	7.20	30.23	-	37.43	49.507
Q3.3 **Filling hollow piles with concrete**							
Q3.3.2 Cross-sectional area: 0.025 - 0.05m^2							
Q3.3.2.01 Cross sectional area: 0.05m^2	m	0.10	4.80	4.63	4.17	13.60	12.816
Q3.3.3 Cross-sectional area: 0.05 - 0.1m^2							
Q3.3.3.01 Cross sectional area: 0.10m^2	m	0.15	7.20	6.95	8.34	22.49	25.378
Q3.3.4 Cross-sectional area: 0.1 - 0.15m^2							
Q3.3.4.01 Cross sectional area: 0.15m^2	m	0.20	9.60	9.27	12.51	31.38	37.939
Q3.3.5 Cross-sectional area: 0.15 - 0.25m^2							
Q3.3.5.01 Cross sectional area: 0.20m^2	m	0.25	11.99	11.58	16.68	40.25	50.501
Q3.4 **Number of pile extensions**							
Q3.4.2 Cross-sectional area: 0.025 - 0.05m^2							
Q3.4.2.01 Cross sectional area: 0.05m^2	Nr	0.25	11.99	41.27	-	53.26	51.510
Q3.4.3 Cross-sectional area: 0.05 - 0.1m^2							
Q3.4.3.01 Cross sectional area: 0.10m^2	Nr	0.30	14.39	49.53	-	63.92	61.811
Q3.4.4 Cross-sectional area: 0.1 - 0.15m^2							
Q3.4.4.01 Cross sectional area: 0.15m^2	Nr	0.35	16.79	57.78	-	74.57	72.113
Q3.4.5 Cross-sectional area: 0.15 - 0.25m^2							
Q3.4.5.01 Cross sectional area: 0.20m^2	Nr	0.30	14.39	65.92	-	80.31	61.811
Q3.5 **Length of pile extensions, each length: 3m**							
Q3.5.2 Cross-sectional area: 0.025 - 0.05m^2							
Q3.5.2.01 Each length 3m; Cross sectional area: 0.05m^2	m	-	-	-	84.32	84.32	139.377
Q3.5.3 Cross-sectional area: 0.05 - 0.1m^2							
Q3.5.3.01 Each length 3m; Cross sectional area: 0.10m^2	m	-	-	-	94.86	94.86	159.288

Q3 Preformed Concrete Piles continued...

	Unit	Labour Hours	Labour Net £	Plant Net £	Materials Net £	Unit Net £	CO_2e Kg
Q3.5	**Length of pile extensions, each length: 3m**						
Q3.5.4	Cross-sectional area: 0.1 - 0.15m²						
Q3.5.4.01	Each length 3m; Cross sectional area: 0.15m²	m	-	-	-	115.94	115.94	298.666
Q3.5.5	Cross-sectional area: 0.15 - 0.25m²						
Q3.5.5.01	Each length 3m; Cross sectional area: 0.20m²	m	-	-	-	137.01	137.01	318.577
Q3.6	**Length of pile extensions, each length: exceeding 3m**						
Q3.6.2	Cross-sectional area: 0.025 - 0.05m²						
Q3.6.2.01	Each length 5m; Cross sectional area: 0.05m²	m	-	-	-	105.40	105.40	219.022
Q3.6.3	Cross-sectional area: 0.05 - 0.1m²						
Q3.6.3.01	Each length 5m; Cross sectional area: 0.10m²	m	-	-	-	115.94	115.94	238.933
Q3.6.4	Cross-sectional area: 0.1 - 0.15m²						
Q3.6.4.01	Each length 5m; Cross sectional area: 0.15m²	m	-	-	-	126.48	126.48	438.043
Q3.6.5	Cross-sectional area: 0.15 - 0.25m²						
Q3.6.5.01	Each length 5m; Cross sectional area: 0.20m²	m	-	-	-	126.48	126.48	438.043
Q3.7	**Cutting off surplus lengths**						
Q3.7.2	Cross-sectional area: 0.025 - 0.05m²						
Q3.7.2.01	Cross sectional area: 0.05m²	m	0.25	7.43	19.36	-	26.79	8.881
Q3.7.3	Cross-sectional area: 0.05 - 0.1m²						
Q3.7.3.01	Cross sectional area: 0.10m²	m	0.33	9.90	25.78	-	35.68	11.829
Q3.7.4	Cross-sectional area: 0.1 - 0.15m²						
Q3.7.4.01	Cross sectional area: 0.15m²	m	0.42	12.39	32.28	-	44.67	14.814
Q3.7.5	Cross-sectional area: 0.15 - 0.25m²						
Q3.7.5.01	Cross sectional area: 0.20m²	m	0.50	14.86	38.71	-	53.57	17.762
Q3.8	**Preparing heads**						
Q3.8.2	Cross-sectional area: 0.025 - 0.05m²						
Q3.8.2.01	Cross sectional area: 0.05m²	Nr	0.25	7.43	19.36	-	26.79	8.881
Q3.8.3	Cross-sectional area: 0.05 - 0.1m²						
Q3.8.3.01	Cross sectional area: 0.10m²	Nr	0.33	9.90	25.78	-	35.68	11.829
Q3.8.4	Cross-sectional area: 0.1 - 0.15m²						
Q3.8.4.01	Cross sectional area: 0.15m²	Nr	0.42	12.39	32.28	-	44.67	14.814
Q3.8.5	Cross-sectional area: 0.15 - 0.25m²						
Q3.8.5.01	Cross sectional area: 0.20m²	Nr	0.50	14.86	38.71	-	53.57	17.762

Q4 Timber Piles

		Unit	Labour Hours	Labour Net £	Plant Net £	Materials Net £	Unit Net £	CO_2e Kg
Q4.7	**Cutting off surplus lengths**							
Q4.7.3	Cross-sectional area: 0.05 - 0.1m^2							
Q4.7.3.01	Cross-sectional area: 0.1m^2	m	0.10	2.97	7.74	-	10.71	3.552
Q4.7.4	Cross-sectional area: 0.1 - 0.15m^2							
Q4.7.4.01	Cross-sectional area: 0.15m^2	m	0.12	3.57	9.29	-	12.86	4.263
Q4.7.5	Cross-sectional area: 0.15 - 0.25m^2							
Q4.7.5.01	Cross-sectional area: 0.25m^2	m	0.15	4.46	11.61	-	16.07	5.329
Q4.8	**Preparing heads**							
Q4.8.3	Cross-sectional area: 0.05 - 0.1m^2							
Q4.8.3.01	Cross-sectional area: 0.1m^2	Nr	0.10	2.97	7.74	-	10.71	3.552
Q4.8.4	Cross-sectional area: 0.1 - 0.15m^2							
Q4.8.4.01	Cross-sectional area: 0.15m^2	Nr	0.12	3.57	9.29	-	12.86	4.263
Q4.8.5	Cross-sectional area: 0.15 - 0.25m^2							
Q4.8.5.01	Cross-sectional area: 0.25m^2	Nr	0.15	4.46	11.61	-	16.07	5.329

Q5 Isolated Steel Piles

		Unit	Labour Hours	Labour Net £	Plant Net £	Materials Net £	Unit Net £	CO_2e Kg
Q5.1	**Pre-boring**							
Q5.1.1	Mass: not exceeding 15kg/m							
Q5.1.1.01	Mass: 25kg/m	m	0.05	2.40	8.26	-	10.66	10.302
Q5.1.3	Mass: 30 - 60kg/m							
Q5.1.3.01	Mass: 50kg/m	m	0.05	2.40	8.26	-	10.66	10.302
Q5.1.5	Mass: 120 - 250kg/m							
Q5.1.5.01	Mass: 100kg/m	m	0.08	3.60	12.38	-	15.98	15.453
Q5.1.6	Mass: 250 - 500kg/m							
Q5.1.6.01	Mass: 250kg/m	m	0.10	4.80	16.51	-	21.31	20.604
Q5.2	**Jetting**							
Q5.2.1	Mass: not exceeding 15kg/m							
Q5.2.1.01	Mass: 25kg/m	m	0.05	2.40	10.08	-	12.48	16.502
Q5.2.3	Mass: 30 - 60kg/m							
Q5.2.3.01	Mass: 50kg/m	m	0.05	2.40	10.08	-	12.48	16.502
Q5.2.5	Mass: 120 - 250kg/m							
Q5.2.5.01	Mass: 100kg/m	m	0.10	4.80	20.15	-	24.95	33.005
Q5.2.6	Mass: 250 - 500kg/m							
Q5.2.6.01	Mass: 250kg/m	m	0.15	7.20	30.23	-	37.43	49.507
Q5.4	**Number of pile extensions**							
Q5.4.1	Mass: not exceeding 15kg/m							
Q5.4.1.01	Mass: 25kg/m	Nr	0.50	24.38	42.39	-	66.77	52.125
Q5.4.3	Mass: 30 - 60kg/m							
Q5.4.3.01	Mass: 50kg/m	Nr	0.67	32.47	56.47	-	88.94	69.431
Q5.4.5	Mass: 120 - 250kg/m							
Q5.4.5.01	Mass: 100kg/m	Nr	1.00	48.75	84.79	-	133.54	104.251

Q5   Isolated Steel Piles continued...

	Unit	Labour Hours	Labour Net £	Plant Net £	Materials Net £	Unit Net £	CO_2e Kg
Q5.4	**Number of pile extensions**						
Q5.4.6 Mass: 250 - 500kg/m							
Q5.4.6.01 Mass: 250kg/m	Nr	1.50	73.13	127.18	-	200.31	156.376
Q5.5	**Length of pile extensions, each length: not exceeding 3m**						
Q5.5.1 Mass: not exceeding 15kg/m							
Q5.5.1.01 Length of pile extensions, each length 3m; Mass: 25kg/m	m	-	-	-	21.08	21.08	27.748
Q5.5.3 Mass: 30 - 60kg/m							
Q5.5.3.01 Length of pile extensions, each length 3m; Mass: 50kg/m	m	-	-	-	42.16	42.16	55.496
Q5.5.5 Mass: 120 - 250kg/m							
Q5.5.5.01 Length of pile extensions, each length 3m; Mass: 100kg/m	m	-	-	-	84.31	84.31	110.993
Q5.5.6 Mass: 250 - 500kg/m							
Q5.5.6.01 Length of pile extensions, each length 3m; Mass: 250kg/m	m	-	-	-	210.78	210.78	277.482
Q5.7	**Cutting off surplus lengths**						
Q5.7.1 Mass: not exceeding 15kg/m							
Q5.7.1.01 Length of pile extensions, each length 3m; Mass: 25kg/m	m	0.20	9.85	6.08	-	15.93	7.055
Q5.7.3 Mass: 30 - 60kg/m							
Q5.7.3.01 Length of pile extensions, each length 3m; Mass: 50kg/m	m	0.28	13.96	6.39	-	20.35	7.183
Q5.7.5 Mass: 120 - 250kg/m							
Q5.7.5.01 Length of pile extensions, each length 3m; Mass: 100kg/m	m	0.38	18.89	9.51	-	28.40	10.812
Q5.7.6 Mass: 250 - 500kg/m							
Q5.7.6.01 Length of pile extensions, each length 3m; Mass: 250kg/m	m	0.55	27.16	10.14	-	37.30	11.068

Q6   Interlocking Steel Piles

	Unit	Labour Hours	Labour Net £	Plant Net £	Materials Net £	Unit Net £	CO_2e Kg
Q6.1	**Pre-boring**						
Q6.1.1 Section modulus: not exceeding 500cm³/m							
Q6.1.1.01 Section modulus: 500cm³/m	m	0.05	3.75	9.64	-	13.39	11.653
Q6.1.3 Section modulus: 800 - 1200cm³/m							
Q6.1.3.01 Section modulus: 1000cm³/m	m	0.05	3.75	9.64	-	13.39	11.653
Q6.1.4 Section modulus: 1200 - 2000cm³/m							
Q6.1.4.01 Section modulus: 2000cm³/m	m	0.08	6.23	16.00	-	22.23	19.344
Q6.1.6 Section modulus: 3000 - 4000cm³/m							
Q6.1.6.01 Section modulus: 4000cm³/m	m	0.10	7.50	19.28	-	26.78	23.306
Q6.2	**Jetting**						
Q6.2.1 Section modulus: not exceeding 500cm³/m							
Q6.2.1.01 Section modulus: 500cm³/m	m	0.07	5.03	15.36	-	20.39	23.923

Q6 Interlocking Steel Piles continued...

	Unit	Labour Hours	Labour Net £	Plant Net £	Materials Net £	Unit Net £	CO_2e Kg
Q6.2 **Jetting**							
Q6.2.3 Section modulus: 800 - 1200cm^3/m							
Q6.2.3.01 Section modulus: 1000cm^3/m	m	0.07	5.03	15.36	-	20.39	23.923
Q6.2.4 Section modulus: 1200 - 2000cm^3/m							
Q6.2.4.01 Section modulus: 2000cm^3/m	m	0.10	7.50	22.92	-	30.42	35.707
Q6.2.6 Section modulus: 3000 - 4000cm^3/m							
Q6.2.6.01 Section modulus: 4000cm^3/m	m	0.17	12.53	38.28	-	50.81	59.630
Q6.4 **Number of pile extensions**							
Q6.4.1 Section modulus: not exceeding 500cm^3/m							
Q6.4.1.01 Full butt weld; Section modulus: 500cm^3/m	Nr	1.67	112.32	227.24	-	339.56	273.207
Q6.4.3 Section modulus: 800 - 1200cm^3/m							
Q6.4.3.01 Full butt weld; Section modulus: 1000cm^3/m	Nr	2.17	143.46	276.56	-	420.02	332.086
Q6.4.4 Section modulus: 1200 - 2000cm^3/m							
Q6.4.4.01 Full butt weld; Section modulus: 2000cm^3/m	Nr	2.67	174.60	325.88	-	500.48	390.966
Q6.4.6 Section modulus: 3000 - 4000cm^3/m							
Q6.4.6.01 Full butt weld; Section modulus: 4000cm^3/m	Nr	4.67	299.15	668.86	-	968.01	626.484
Q6.5 **Length of pile extensions, each length: not exceeding 3m**							
Q6.5.1 Section modulus: not exceeding 500cm^3/m							
Q6.5.1.01 Each length 3m; Section modulus: 500cm^3/m	m	0.17	12.53	32.20	46.24	90.97	123.654
Q6.5.3 Section modulus: 800 - 1200cm^3/m							
Q6.5.3.01 Each length 3m; Section modulus: 1000cm^3/m	m	0.17	12.53	32.20	65.77	110.50	159.431
Q6.5.4 Section modulus: 1200 - 2000cm^3/m							
Q6.5.4.01 Each length 3m; Section modulus: 2000cm^3/m	m	0.33	24.98	64.20	85.29	174.47	233.895
Q6.5.6 Section modulus: 3000 - 4000cm^3/m							
Q6.5.6.01 Each length 3m; Section modulus: 4000cm^3/m	m	0.50	37.52	123.72	149.01	310.25	389.560
Q6.6 **Length of pile extensions, each length: exceeding 3m**							
Q6.6.1 Section modulus: not exceeding 500cm^3/m							
Q6.6.1.01 Length of pile extensions, each length 5m; Section modulus: 500cm^3/m	m	0.10	7.50	19.28	46.24	73.02	108.040
Q6.6.3 Section modulus: 800 - 1200cm^3/m							
Q6.6.3.01 Length of pile extensions, each length 5m; Section modulus: 1000cm^3/m	m	0.10	7.50	19.28	65.77	92.55	143.816

Q6 Interlocking Steel Piles continued...

	Unit	Labour Hours	Labour Net £	Plant Net £	Materials Net £	Unit Net £	CO_2e Kg
Q6.6	**Length of pile extensions, each length: exceeding 3m**						
Q6.6.4 Section modulus: 1200 - 2000cm³/m							
Q6.6.4.01 Length of pile extensions, each length 5m; Section modulus: 2000cm³/m	m	0.30	22.51	57.84	65.77	146.12	190.427
Q6.6.6 Section modulus: 3000 - 4000cm³/m							
Q6.6.6.01 Length of pile extensions, each length 5m; Section modulus: 4000cm³/m	m	0.40	30.01	98.97	65.77	194.75	213.733
Q6.7	**Cutting off surplus lengths**						
Q6.7.1 Section modulus: not exceeding 500cm³/m							
Q6.7.1.01 Section modulus: 500cm³/m	m	0.28	14.86	7.30	-	22.16	8.074
Q6.7.3 Section modulus: 800 - 1200cm³/m							
Q6.7.3.01 Section modulus: 1000cm³/m	m	0.53	27.24	8.25	-	35.49	8.457
Q6.7.4 Section modulus: 1200 - 2000cm³/m							
Q6.7.4.01 Section modulus: 2000cm³/m	m	0.70	35.51	8.88	-	44.39	8.713
Q6.7.6 Section modulus: 3000 - 4000cm³/m							
Q6.7.6.01 Section modulus: 4000cm³/m	m	1.03	52.00	11.94	-	63.94	9.224

Q7 Obstructions

	Unit	Labour Hours	Labour Net £	Plant Net £	Materials Net £	Unit Net £	CO_2e Kg
Q7.1	**Breaking out rock**						
Q7.1.1 Encountered above the founding stratum							
Q7.1.1.01 per rig	Hr	1.00	75.03	192.80	-	267.83	233.055

Q8 Pile Tests

	Unit	Labour Hours	Labour Net £	Plant Net £	Materials Net £	Unit Net £	CO_2e Kg
Q8.1	**Maintained loading with various reactions**						
Q8.1.1 Test load: not exceeding 100t							
Q8.1.1.01 Tests by anchor piles, loads applied in 10 tonne increments at hourly intervals maintained for 24 hours; removed in equal increments over 6 hours; Test load: 50t, establish equipment and carry out test	Nr	19.20	1,092.05	9,083.91	-	10,175.96	7,904.932
Q8.1.1.02 Tests by anchor piles, loads applied in 10 tonne increments at hourly intervals maintained for 24 hours; removed in equal increments over 6 hours; Test load: 100t, establish equipment and carry out test	Nr	24.20	1,310.40	10,257.46	-	11,567.86	8,935.122
Q8.1.1.03 Tests by anchor piles, loads applied in 10 tonne increments at hourly intervals maintained for 24 hours; removed in equal increments over 6 hours; Test load: 50t; excludes establish of equipment	Nr	15.20	845.67	3,450.87	-	4,296.54	2,960.020

Q8 Pile Tests continued...

	Unit	Labour Hours	Labour Net £	Plant Net £	Materials Net £	Unit Net £	CO_2e Kg
Q8.1	**Maintained loading with various reactions**						
Q8.1.1	Test load: not exceeding 100t						
Q8.1.1.04 Tests by anchor piles, loads applied in 10 tonne increments at hourly intervals maintained for 24 hours; removed in equal increments over 6 hours; Test load: 100t; excludes establish of equipment	Nr	20.20	1,064.02	4,624.42	-	5,688.44	3,990.210
Q8.1.2	Test load: 100 - 200t						
Q8.1.2.01 Tests by anchor piles, loads applied in 10 tonne increments at hourly intervals maintained for 24 hours; removed in equal increments over 6 hours; Test load: 150t, establish equipment and carry out test	Nr	34.10	1,687.64	10,996.48	-	12,684.12	9,618.511
Q8.1.2.02 Tests by anchor piles, loads applied in 10 tonne increments at hourly intervals maintained for 24 hours; removed in equal increments over 6 hours; Test load: 200t, establish equipment and carry out test	Nr	34.10	1,687.64	12,170.03	-	13,857.67	10,648.701
Q8.1.2.03 Tests by anchor piles, loads applied in 10 tonne increments at hourly intervals maintained for 24 hours; removed in equal increments over 6 hours; Test load: 150t; excludes establish of equipment	Nr	25.20	1,282.37	5,797.97	-	7,080.34	5,020.400
Q8.1.2.04 Tests by anchor piles, loads applied in 10 tonne increments at hourly intervals maintained for 24 hours; removed in equal increments over 6 hours; Test load: 200t; excludes establish of equipment	Nr	30.20	1,500.72	6,971.52	-	8,472.24	6,050.590
Q8.2	**Constant rate of penetration**						
Q8.2.1	Test load: not exceeding 100t						
Q8.2.1.01	Test load: 50t	Nr	14.20	802.00	3,216.16	-	4,018.16	2,753.982
Q8.2.1.02	Test load: 100t	Nr	19.20	1,020.35	4,389.71	-	5,410.06	3,784.172
Q8.2.2	Test load: 100 - 200t						
Q8.2.2.01	Test load: 150t	Nr	24.20	1,238.70	5,563.26	-	6,801.96	4,814.362
Q8.2.2.02	Test load: 200t	Nr	29.20	1,457.05	6,736.81	-	8,193.86	5,844.552

Calculations used throughout Class R - Roads and Pavings

Labour

			Qty		Rate		Total
L A0140ICE	**Steel fixing Labour Gang**						
	L A9000ICE	Craftsman WRA	2	x	17.33	=	£34.66
	L A9002ICE	Labourer (General Operative)	1	x	13.02	=	£13.02
		Total hourly cost of gang				**=**	**£47.68**
L A0270ICE	**Road sub-base Labour Gang**						
	L A9016ICE	Ganger	1	x	17.64	=	£17.64
	L A9002ICE	Labourer (General Operative)	2	x	13.02	=	£26.03
		Total hourly cost of gang				**=**	**£43.67**
L A0271ICE	**Soil cement Labour Gang**						
	L A9015ICE	Banksman	1	x	14.03	=	£14.03
	L A9016ICE	Ganger	1	x	17.64	=	£17.64
	L A9002ICE	Labourer (General Operative)	2	x	13.02	=	£26.03
	L A9003ICE	Labourer (Skill Rate 3)	1	x	14.86	=	£14.86
		Total hourly cost of gang				**=**	**£72.56**
L A0274ICE	**Concrete pavement Labour Gang**						
	L A9000ICE	Craftsman WRA	1	x	17.33	=	£17.33
	L A9016ICE	Ganger	1	x	17.64	=	£17.64
	L A9002ICE	Labourer (General Operative)	2	x	13.02	=	£26.03
	L A9003ICE	Labourer (Skill Rate 3)	1	x	14.86	=	£14.86
	L A9015ICE	Banksman	1	x	14.03	=	£14.03
		Total hourly cost of gang				**=**	**£89.89**
L A0275ICE	**Concrete joints Labour Gang**						
	L A9000ICE	Craftsman WRA	2	x	17.33	=	£34.66
	L A9002ICE	Labourer (General Operative)	1	x	13.02	=	£13.02
		Total hourly cost of gang				**=**	**£47.68**
L A0276ICE	**Kerbs and paving Labour Gang**						
	L A9016ICE	Ganger	1	x	17.64	=	£17.64
	L A9002ICE	Labourer (General Operative)	2	x	13.02	=	£26.03
		Total hourly cost of gang				**=**	**£43.67**
L A0320ICE	**Waterproofing Labour Gang**						
	L A9016ICE	Ganger	1	x	17.64	=	£17.64
	L A9002ICE	Labourer (General Operative)	3	x	13.02	=	£39.05
		Total hourly cost of gang				**=**	**£56.69**

Plant

			Qty		Rate		Total
P A1140ICE	**Steel fixing Plant Gang**						
	P A0509ICE	Cranes Crawler - NCK 305C - 19t	0.25	x	46.29	=	£11.57
		Total hourly cost of gang				**=**	**£11.57**

P A1270ICE **Road sub-base Plant Gang**

| P A0932ICE | Crawler Tractor / Dozer - Cat D4C LGP 60kW | 1 | x | 38.18 | = | £38.18 |
| P A0869ICE | Roller - 800mm Ride on Roller | 1 | x | 6.49 | = | £6.49 |

| **Total hourly cost of gang** | | | | = | **£44.67** |

P A1271ICE **Soil cement Plant Gang**

P A0592ICE	Hydraulic Excavator - Cat 320 96kW	1	x	36.36	=	£36.36
P A1002ICE	Tipping Waggon - 16t 6-Wheel (24t Gr)	2	x	51.34	=	£102.68
P A0735ICE	Concrete Mixer - Liner Rolpanit	1	x	39.98	=	£39.98
P A0473ICE	Cement Silo 50t	1	x	2.64	=	£2.64
P A0902ICE	Road Plant - Barber Greene BGP 200 Paver	1	x	55.06	=	£55.06

| **Total hourly cost of gang** | | | | = | **£236.72** |

P A1272ICE **Cement bound granular material Plant Gang**

P A1002ICE	Tipping Waggon - 16t 6-Wheel (24t Gr)	2	x	51.34	=	£102.68
P A0735ICE	Concrete Mixer - Liner Rolpanit	1	x	39.98	=	£39.98
P A0473ICE	Cement Silo 50t	1	x	2.64	=	£2.64
P A0902ICE	Road Plant - Barber Greene BGP 200 Paver	1	x	55.06	=	£55.06

| **Total hourly cost of gang** | | | | = | **£200.36** |

P A1274ICE **Concrete pavement Plant Gang**

P A0902ICE	Road Plant - Barber Greene BGP 200 Paver	1	x	55.06	=	£55.06
P A0457ICE	Compressor - 375 cfm	0.5	x	16.00	=	£8.00
P A0405ICE	Air Vibrating Poker up to 75mm	2	x	1.33	=	£2.66

| **Total hourly cost of gang** | | | | = | **£65.72** |

Roads and Pavings

R1   Unbound Sub-Base

	Unit	Labour Hours	Labour Net £	Plant Net £	Materials Net £	Unit Net £	CO_2e Kg
R1.1	**Type 1 unbound mixtures**						
R1.1.3	Depth: 60 - 100mm						
R1.1.3.01	DTp clause nr. 803; Levelling and compacting with a vibratory roller; depth: 75mm	m²	0.01	0.39	0.40	2.55	3.34	1.229
R1.1.4	Depth: 100 - 150mm						
R1.1.4.01	DTp clause nr. 803; Levelling and compacting with a vibratory roller; depth: 100mm	m²	0.01	0.39	0.40	3.40	4.19	1.593
R1.1.4.02	DTp clause nr. 803; Levelling and compacting with a vibratory roller; depth: 150mm	m²	0.01	0.57	0.58	5.10	6.25	2.382
R1.1.5	Depth: 150 - 200mm						
R1.1.5.01	DTp clause nr. 803; Levelling and compacting with a vibratory roller; depth: 200mm	m²	0.02	0.79	0.80	6.80	8.39	3.186
R1.1.6	Depth: 200 - 250mm						
R1.1.6.01	DTp clause nr. 803; Levelling and compacting with a vibratory roller; depth: 250 mm	m²	0.02	0.96	0.98	8.50	10.44	3.975
R1.1.7	Depth: 250 - 300mm						
R1.1.7.01	DTp clause nr. 803; Levelling and compacting with a vibratory roller; depth: 300mm	m²	0.03	1.14	1.16	10.20	12.50	4.764
R1.1.8	Depth: exceeding 300mm						
R1.1.8.01	DTp clause nr. 803; Levelling and compacting with a vibratory roller; depth: 400mm	m²	0.04	1.53	1.56	13.60	16.69	6.357
R1.1.8.02	DTp clause nr. 803; Levelling and compacting with a vibratory roller; depth: 500mm	m²	0.04	1.92	1.97	17.00	20.89	7.950
R1.2	**Type 2 unbound mixtures**						
R1.2.3	Depth: 60 - 100mm						
R1.2.3.01	DTp clause nr. 804; Levelling and compacting with a vibratory roller; depth: 75mm	m²	0.01	0.39	0.40	2.73	3.52	1.291
R1.2.4	Depth: 100 - 150mm						
R1.2.4.01	DTp clause nr. 804; Levelling and compacting with a vibratory roller; depth: 100mm	m²	0.01	0.39	0.40	3.64	4.43	1.677
R1.2.4.02	DTp clause nr. 804; Levelling and compacting with a vibratory roller; depth: 150mm	m²	0.01	0.57	0.58	5.46	6.61	2.507
R1.2.5	Depth: 150 - 200mm						
R1.2.5.01	DTp clause nr. 804; Levelling and compacting with a vibratory roller; depth: 200mm	m²	0.02	0.79	0.80	7.28	8.87	3.353

R1 Unbound Sub-Base continued...

	Unit	Labour Hours	Labour Net £	Plant Net £	Materials Net £	Unit Net £	CO_2e Kg
R1.2 **Type 2 unbound mixtures**							
R1.2.6 Depth: 200 - 250mm							
R1.2.6.01 DTp clause nr. 804; Levelling and compacting with a vibratory roller; depth: 250 mm	m²	0.02	0.96	0.98	9.10	11.04	4.184
R1.2.7 Depth: 250 - 300mm							
R1.2.7.01 DTp clause nr. 804; Levelling and compacting with a vibratory roller; depth: 300mm	m²	0.03	1.14	1.16	10.92	13.22	5.015
R1.2.8 Depth: exceeding 300mm							
R1.2.8.01 DTp clause nr. 804; Levelling and compacting with a vibratory roller; depth: 400mm	m²	0.04	1.53	1.56	14.06	17.15	6.481
R1.2.8.02 DTp clause nr. 804; Levelling and compacting with a vibratory roller; depth: 500mm	m²	0.04	1.92	1.97	18.19	22.08	8.368
R1.3 **Type 3 (open graded) unbound mixtures**							
R1.3.3 Depth: 60 - 100mm							
R1.3.3.01 DTp clause nr. 805; Levelling and compacting with a vibratory roller; depth: 75mm	m²	0.01	0.39	0.40	2.94	3.73	1.291
R1.3.4 Depth: 100 - 150mm							
R1.3.4.01 DTp clause nr. 805; Levelling and compacting with a vibratory roller; depth: 100mm	m²	0.01	0.39	0.40	3.92	4.71	1.677
R1.3.4.02 DTp clause nr. 805; Levelling and compacting with a vibratory roller; depth: 150mm	m²	0.01	0.57	0.58	5.87	7.02	2.507
R1.3.5 Depth: 150 - 200mm							
R1.3.5.01 DTp clause nr. 805; Levelling and compacting with a vibratory roller; depth: 200mm	m²	0.02	0.79	0.80	7.83	9.42	3.353
R1.3.6 Depth: 200 - 250mm							
R1.3.6.01 DTp clause nr. 805; Levelling and compacting with a vibratory roller; depth: 250 mm	m²	0.02	0.96	0.98	9.79	11.73	4.184
R1.3.7 Depth: 250 - 300mm							
R1.3.7.01 DTp clause nr. 805; Levelling and compacting with a vibratory roller; depth: 300mm	m²	0.03	1.14	1.16	11.75	14.05	5.015
R1.3.8 Depth: exceeding 300mm							
R1.3.8.01 DTp clause nr. 805; Levelling and compacting with a vibratory roller; depth: 400mm	m²	0.04	1.53	1.56	15.13	18.22	6.481
R1.3.8.02 DTp clause nr. 805; Levelling and compacting with a vibratory roller; depth: 500mm	m²	0.04	1.92	1.97	19.58	23.47	8.368
R1.6 **Geotextiles**							
R1.6.0 Generally							
R1.6.0.01 Lotrak 10/7 ground stabilising matting; 150mm laps; horizontal	m²	0.01	0.40	-	0.79	1.19	0.229

R1 Unbound Sub-Base continued...

	Unit	Labour Hours	Labour Net £	Plant Net £	Materials Net £	Unit Net £	CO_2e Kg
R1.6 **Geotextiles**							
R1.6.0 Generally							
R1.6.0.02 Typar ground stabilising geotextile; 300 mm side and end laps; Horizontally	m²	0.01	0.40	-	1.70	2.10	0.580

R2 Cement And Other Hydralically Bound Pavements

	Unit	Labour Hours	Labour Net £	Plant Net £	Materials Net £	Unit Net £	CO_2e Kg
R2.1 **Cement bound granular mixture**							
R2.1.3 Depth: 60 - 100mm							
R2.1.3.01 100kg dry Ordinary Portland Cement per cubic metre of Type 1 sub-base granular material; depth: 75mm	m²	0.01	0.58	1.60	3.58	5.76	9.583
R2.1.4 Depth: 100 - 150mm							
R2.1.4.01 100kg dry Ordinary Portland Cement per cubic metre of Type 1 sub-base granular material; depth: 100mm	m²	0.01	0.80	2.20	4.68	7.68	12.159
R2.1.4.02 100kg dry Ordinary Portland Cement per cubic metre of Type 1 sub-base granular material; depth: 150mm	m²	0.02	1.23	3.41	7.02	11.66	18.282
R2.1.5 Depth: 150 - 200mm							
R2.1.5.01 100kg dry Ordinary Portland Cement per cubic metre of Type 1 sub-base granular material; depth: 200mm	m²	0.02	1.60	4.41	9.37	15.38	24.317
R2.1.6 Depth: 200 - 250mm							
R2.1.6.01 100kg dry Ordinary Portland Cement per cubic metre of Type 1 sub-base granular material; depth: 250mm	m²	0.03	1.96	5.41	11.71	19.08	30.352
R2.1.7 Depth: 250 - 300mm							
R2.1.7.01 100kg dry Ordinary Portland Cement per cubic metre of Type 1 sub-base granular material; depth: 300mm	m²	0.03	1.96	5.41	14.05	21.42	35.946
R2.1.8 Depth: exceeding 300mm							
R2.1.8.01 100kg dry Ordinary Portland Cement per cubic metre of Type 1 sub-base granular material; depth: 400mm	m²	0.03	2.32	6.41	18.73	27.46	47.575
R2.1.8.02 100kg dry Ordinary Portland Cement per cubic metre of Type 1 sub-base granular material; depth: 500mm	m²	0.04	2.68	7.41	23.41	33.50	59.204
R2.2 **Fly ash bound mixture 1 and hydraulic road binder bound mixture 1**							
R2.2.3 Depth: 60 - 100mm							
R2.2.3.01 Fly ash bound mixture 1; primary aggregate; 75 mm thick	m²	0.01	0.58	1.60	3.41	5.59	2.131
R2.2.3.02 Fly ash bound mixture 1; recycled aggregate; 75 mm thick	m²	0.01	0.58	1.60	3.06	5.24	2.178

R2   Cement And Other Hydralically Bound Pavements continued...

	Unit	Labour Hours	Labour Net £	Plant Net £	Materials Net £	Unit Net £	CO_2e Kg
R2.2	**Fly ash bound mixture 1 and hydraulic road binder bound mixture 1**						
R2.2.4	Depth: 100 - 150mm						
R2.2.4.01 Fly ash bound mixture 1; primary aggregate; 100 mm thick	m²	0.01	0.80	2.20	4.52	7.52	2.856
R2.2.4.02 Fly ash bound mixture 1; primary aggregate; 150 mm thick	m²	0.02	1.23	3.41	6.78	11.42	4.328
R2.2.4.03 Fly ash bound mixture 1; recycled aggregate; 100 mm thick	m²	0.01	0.80	2.20	4.05	7.05	2.920
R2.2.4.04 Fly ash bound mixture 1; recycled aggregate; 150 mm thick	m²	0.02	1.23	3.41	6.08	10.72	4.423
R2.2.5	Depth: 150 - 200mm						
R2.2.5.01 Fly ash bound mixture 1; primary aggregate; 200 mm thick	m²	0.02	1.60	4.41	9.03	15.04	5.712
R2.2.5.02 Fly ash bound mixture 1; recycled aggregate; 200 mm thick	m²	0.02	1.60	4.41	8.11	14.12	5.839
R2.2.6	Depth: 200 - 250mm						
R2.2.6.01 Fly ash bound mixture 1; primary aggregate; 250 mm thick	m²	0.03	1.96	5.41	11.29	18.66	7.096
R2.2.6.02 Fly ash bound mixture 1; recycled aggregate; 250 mm thick	m²	0.03	1.96	5.41	10.13	17.50	7.255
R2.2.7	Depth: 250 - 300mm						
R2.2.7.01 Fly ash bound mixture 1; primary aggregate; 300 mm thick	m²	0.03	1.96	5.41	13.55	20.92	8.039
R2.2.7.02 Fly ash bound mixture 1; recycled aggregate; 300 mm thick	m²	0.03	1.96	5.41	12.16	19.53	8.229
R2.2.8	Depth: exceeding 300mm						
R2.2.8.01 Fly ash bound mixture 1; primary aggregate; 400 mm thick	m²	0.03	2.32	6.41	18.07	26.80	10.365
R2.2.8.02 Fly ash bound mixture 1; primary aggregate; 500 mm thick	m²	0.04	2.68	7.41	22.59	32.68	12.692
R2.2.8.04 Fly ash bound mixture 1; recycled aggregate; 400 mm thick	m²	0.03	2.32	6.41	16.21	24.94	10.619
R2.2.8.05 Fly ash bound mixture 1; recycled aggregate; 500 mm thick	m²	0.04	2.68	7.41	20.26	30.35	13.009
R2.7	**Soil treated by cement**						
R2.7.3	Depth: 60 - 100mm						
R2.7.3.01 Soil treated by cement; 10% cement content; 75 mm thick	m²	0.01	0.58	1.60	3.27	5.45	6.978
R2.7.4	Depth: 100 - 150mm						
R2.7.4.01 Soil treated by cement, 10% cement content; 100 mm thick	m²	0.01	0.80	2.20	4.09	7.09	8.811
R2.7.4.02 Soil treated by cement, 10% cement content; 150 mm thick	m²	0.02	1.23	3.41	6.13	10.77	13.261
R2.7.5	Depth: 150 - 200mm						
R2.7.5.01 Soil treated by cement, 10% cement content;200 mm thick	m²	0.02	1.60	4.41	8.18	14.19	17.623
R2.7.6	Depth: 200 - 250mm						
R2.7.6.01 Soil treated by cement, 10% cement content; 250 mm thick	m²	0.03	1.96	5.41	10.22	17.59	21.984
R2.7.7	Depth: 250 - 300mm						
R2.7.7.01 Soil treated by cement, 10% cement content; 300 mm thick	m²	0.03	1.96	5.41	12.26	19.63	25.904

R2 Cement And Other Hydralically Bound Pavements continued...

	Unit	Labour Hours	Labour Net £	Plant Net £	Materials Net £	Unit Net £	CO_2e Kg
R2.7 **Soil treated by cement**							
R2.7.8 Depth: exceeding 300mm							
R2.7.8.01 Soil treated by cement, 10% cement content; 400 mm thick	m²	0.03	2.32	6.41	16.35	25.08	34.186
R2.7.8.02 Soil treated by cement, 10% cement content; 500 mm thick	m²	0.04	2.68	7.41	20.44	30.53	42.468

R3 Bituminous Bound Pavement

	Unit	Labour Hours	Labour Net £	Plant Net £	Materials Net £	Unit Net £	CO_2e Kg
R3.1 **Hot rolled asphalt base**							
R3.1.2 Depth: 30 - 60mm							
R3.1.2.01 Hot rolled asphalt base, 50 mm thick	m²	0.08	3.49	3.57	5.22	12.28	8.832
R3.1.3 Depth: 60 - 100mm							
R3.1.3.01 Hot rolled asphalt base, 60 mm thick	m²	0.08	3.49	3.57	6.26	13.32	10.356
R3.1.3.02 Hot rolled asphalt base, 80 mm thick	m²	0.09	3.93	4.02	8.34	16.29	13.557
R3.1.4 Depth: 100 - 150mm							
R3.1.4.01 Hot rolled asphalt base, 100 mm thick	m²	0.10	4.37	4.47	10.43	19.27	16.757
R3.1.4.02 Hot rolled asphalt base, 150 mm thick	m²	0.10	4.37	4.47	15.65	24.49	24.381
R3.2 **Hot rolled asphalt binder**							
R3.2.2 Depth: 30 - 60mm							
R3.2.2.01 Hot rolled asphalt base, 50 mm thick	m²	0.08	3.49	3.57	5.22	12.28	8.832
R3.2.3 Depth: 60 - 100mm							
R3.2.3.01 Hot rolled asphalt base, 60 mm thick	m²	0.08	3.49	3.57	6.26	13.32	10.356
R3.2.3.02 Hot rolled asphalt base, 80 mm thick	m²	0.09	3.93	4.02	8.34	16.29	13.557
R3.2.4 Depth: 100 - 150mm							
R3.2.4.01 Hot rolled asphalt base, 100 mm thick	m²	0.10	4.37	4.47	10.43	19.27	16.757
R3.2.4.02 Hot rolled asphalt base, 150 mm thick	m²	0.10	4.37	4.47	15.65	24.49	24.381
R3.4 **Dense asphalt concrete surface course**							
R3.4.2 Depth: 30 - 60mm							
R3.4.2.01 10mm nominal size aggregate; depth: 30mm	m²	0.01	1.59	1.67	4.00	7.26	5.298
R3.4.2.02 10mm nominal size aggregate; depth: 40mm	m²	0.01	1.59	1.67	5.34	8.60	6.822
R3.4.2.03 14mm nominal size aggregate; depth: 30mm	m²	0.01	1.59	1.67	4.00	7.26	5.298
R3.4.2.04 14mm nominal size aggregate; depth: 40mm	m²	0.01	1.59	1.67	5.34	8.60	6.822
R3.4.3 Depth: 60 - 100mm							
R3.4.3.01 14mm nominal size aggregate; depth: 60mm	m²	0.02	3.18	3.34	8.70	15.22	11.391

R3 Bituminous Bound Pavement continued...

	Unit	Labour Hours	Labour Net £	Plant Net £	Materials Net £	Unit Net £	CO_2e Kg
R3.4 **Dense asphalt concrete surface course**							
R3.4.3 Depth: 60 - 100mm							
R3.4.3.02 14mm nominal size aggregate; depth: 80mm	m²	0.02	3.18	3.34	10.68	17.20	13.645
R3.6 **Close graded asphalt concrete surface course**							
R3.6.2 Depth: 30 - 60mm							
R3.6.2.01 10mm nominal size aggregate; depth: 30mm	m²	0.01	1.59	1.67	4.20	7.46	0.737
R3.6.2.02 10mm nominal size aggregate; depth: 40mm	m²	0.01	1.59	1.67	5.60	8.86	0.742
R3.6.2.03 14mm nominal size aggregate; depth: 30mm	m²	0.01	1.59	1.67	4.03	7.29	0.737
R3.6.2.04 14mm nominal size aggregate; depth: 40mm	m²	0.01	1.59	1.67	5.37	8.63	0.742
R3.6.3 Depth: 60 - 100mm							
R3.6.3.01 40mm nominal size aggregate; depth: 60mm	m²	0.02	3.18	3.34	8.75	15.27	1.477
R3.6.3.02 40mm nominal size aggregate; depth: 80mm	m²	0.02	3.18	3.34	10.74	17.26	1.484
R3.7 **Fine graded asphalt concrete surface course**							
R3.7.2 Depth: 30 - 60mm							
R3.7.2.01 Depth: 30mm	m²	0.01	1.59	1.67	4.59	7.85	0.737
R3.7.2.02 Depth: 40mm	m²	0.01	1.59	1.67	6.12	9.38	0.742

R4 Bituminous Bound Pavements

	Unit	Labour Hours	Labour Net £	Plant Net £	Materials Net £	Unit Net £	CO_2e Kg
R4.0 **Surface Treatments**							
R4.0.0 Generally							
R4.0.0.01 Slurry sealing; BS 434; Bitumen emulsion, sand, cement composition (200 litres per tonne dry aggregate; 2% cement by mass of aggregate); spread by mechanical means; depth: 3mm	m²	0.00	0.32	-	1.34	1.66	1.431
R4.0.0.02 Surface dressing; Coated chippings evenly applied to macadam surfaces; rolled or pressed into surface; nominal size 10mm; at the rate of 4kg per m²	m²	0.01	0.59	0.55	0.62	1.76	0.547
R4.0.0.03 Surface dressing; Coated chippings evenly applied to macadam surfaces; rolled or pressed into surface; nominal size 10mm; at the rate of 6kg per m²	m²	0.01	0.59	0.55	0.65	1.79	0.560
R4.0.0.04 Surface dressing; Coated chippings evenly applied to macadam surfaces; rolled or pressed into surface; nominal size 10mm; at the rate of 8kg per m²	m²	0.01	0.59	0.55	0.68	1.82	0.573
R4.0.0.05 Surface dressing; Coated chippings evenly applied to macadam surfaces; rolled or pressed into surface; nominal size 14mm; at the rate of 6kg per m²	m²	0.01	0.59	0.55	0.65	1.79	0.560

R4 Bituminous Bound Pavements continued...

	Unit	Labour Hours	Labour Net £	Plant Net £	Materials Net £	Unit Net £	CO_2e Kg
R4.0	**Surface Treatments**						
R4.0.0	Generally						
R4.0.0.06	Surface dressing; Coated chippings evenly applied to macadam surfaces; rolled or pressed into surface; nominal size 14mm; at the rate of 8kg per m²	m²	0.01	0.59	0.55	0.68	1.82	0.573
R4.0.0.07	Surface dressing; Coated chippings evenly applied to macadam surfaces; rolled or pressed into surface; nominal size 14mm; at the rate of 10kg per m²	m²	0.01	0.59	0.55	0.71	1.85	0.587
R4.0.0.08	Surface dressing; Uncoated chippings evenly applied to macadam surfaces; rolled or pressed into surface; nominal size 10mm; at the rate of 4kg per m²	m²	0.01	0.59	0.55	0.06	1.20	0.336
R4.0.0.09	Surface dressing; Uncoated chippings evenly applied to macadam surfaces; rolled or pressed into surface; nominal size 10mm; at the rate of 6kg per m²	m²	0.01	0.59	0.55	0.09	1.23	0.349
R4.0.0.10	Surface dressing; Uncoated chippings evenly applied to macadam surfaces; rolled or pressed into surface; nominal size 10mm; at the rate of 8kg per m²	m²	0.01	0.59	0.55	0.13	1.27	0.362
R4.0.0.11	Surface dressing; Uncoated chippings evenly applied to macadam surfaces; rolled or pressed into surface; nominal size 14mm; at the rate of 6kg per m²	m²	0.01	0.59	0.55	0.09	1.23	0.349
R4.0.0.12	Surface dressing; Uncoated chippings evenly applied to macadam surfaces; rolled or pressed into surface; nominal size 14mm; at the rate of 8kg per m²	m²	0.01	0.59	0.55	0.12	1.26	0.362
R4.0.0.13	Surface dressing; Uncoated chippings evenly applied to macadam surfaces; rolled or pressed into surface; nominal size 14mm; at the rate of 10kg per m²	m²	0.01	0.59	0.55	0.15	1.29	0.376
R4.0.0.14	Bituminous Spray; Tack coat, Cationic bitumen emulsion tack coat applied to; new surfaces	m²	0.02	0.31	0.03	0.37	0.71	0.301
R4.0.0.15	Bituminous Spray; Tack coat, Cationic bitumen emulsion tack coat applied to; existing surfaces	m²	0.02	0.31	0.03	0.37	0.71	0.301
R4.7	**Cold milling / planning**						
R4.7.0	Generally						
R4.7.0.01	Milling pavement; 50mm deep; scarifying surface; disposal on site and re-use	m²	0.04	3.91	2.42	-	6.33	6.960
R4.7.0.02	Milling pavement; 75mm deep; scarifying surface; disposal on site and re-use	m²	0.06	5.87	3.62	-	9.49	10.440

R4 Bituminous Bound Pavements continued...

	Unit	Labour Hours	Labour Net £	Plant Net £	Materials Net £	Unit Net £	CO_2e Kg
R4.7	**Cold milling / planning**						
R4.7.0	Generally						
R4.7.0.03	Milling pavement; 50mm deep; disposal on site and re-use	m²	0.04	3.42	2.11	-	5.53	6.090
R4.7.0.04	Milling pavement; 75mm deep;disposal on site and re-use	m²	0.06	5.38	3.32	-	8.70	9.570
R4.7.0.05	Milling pavement; 100mm deep; disposal on site and re-use	m²	0.07	6.84	4.23	-	11.07	12.180

R5 Concrete Pavement

	Unit	Labour Hours	Labour Net £	Plant Net £	Materials Net £	Unit Net £	CO_2e Kg
R5.2	**Jointed reinforced concrete surface slabs**						
R5.2.3	Depth: 60 - 100mm						
R5.2.3.01	Concrete C25/30, 20mm aggregate; depth: 100mm	m²	0.02	1.80	1.31	9.28	12.39	28.299
R5.2.4	Depth: 100 - 150mm						
R5.2.4.01	Concrete C25/30, 20mm aggregate; depth: 150mm	m²	0.03	2.70	1.97	13.91	18.58	42.449
R5.2.5	Depth: 150 - 200mm						
R5.2.5.01	Concrete C25/30, 20mm aggregate;; depth: 200mm	m²	0.04	3.60	2.63	18.55	24.78	56.599
R5.2.6	Depth: 200 - 250mm						
R5.2.6.01	Concrete C25/30, 20mm aggregate; depth: 250mm	m²	0.05	4.49	3.29	23.19	30.97	70.749
R5.2.7	Depth: 250 - 300mm						
R5.2.7.01	Concrete C25/30, 20mm aggregate; depth: 300mm	m²	0.06	5.39	3.94	27.82	37.15	84.898
R5.2.8	Depth: exceeding 300mm						
R5.2.8.01	Concrete C25/30, 20mm aggregate; depth: 400mm	m²	0.07	6.29	4.60	37.10	47.99	112.888
R5.6	**Steel fabric reinforcement**						
R5.6.2	Nominal mass: 2 - 3kg/m²						
R5.6.2.01	Nominal mass: 2.22kg/m²; A142	m²	0.06	2.86	-	1.47	4.33	1.307
R5.6.2.02	Nominal mass: 2.61kg/m²; C283	m²	0.06	2.86	-	1.91	4.77	1.536
R5.6.3	Nominal mass: 3 - 4kg/m²						
R5.6.3.01	Nominal mass: 3.02kg/m²; A193	m²	0.06	2.86	-	2.04	4.90	1.777
R5.6.3.02	Nominal mass: 3.05kg/m²; B196	m²	0.07	3.34	-	3.73	7.07	1.795
R5.6.3.03	Nominal mass: 3.41kg/m²; C385	m²	0.07	3.34	-	2.30	5.64	2.007
R5.6.3.04	Nominal mass: 3.73kg/m²; B283	m²	0.07	3.34	-	2.49	5.83	2.195
R5.6.3.05	Nominal mass: 3.95kg/m²; A252	m²	0.09	4.29	-	2.69	6.98	2.325
R5.6.4	Nominal mass: 4 - 5kg/m²						
R5.6.4.01	Nominal mass: 4.34kg/m²; C503	m²	0.09	4.29	-	2.95	7.24	2.554
R5.6.4.02	Nominal mass: 4.53kg/m²; B385	m²	0.10	4.77	-	3.03	7.80	2.666
R5.6.5	Nominal mass: 5 - 6kg/m²						
R5.6.5.01	Nominal mass: 5.55kg/m²; C636	m²	0.10	4.77	-	3.77	8.54	3.266
R5.6.5.02	Nominal mass: 5.93kg/m²; B503	m²	0.10	4.77	-	3.96	8.73	3.490

R5 Concrete Pavement continued...

	Unit	Labour Hours	Labour Net £	Plant Net £	Materials Net £	Unit Net £	CO_2e Kg
R5.7	**Steel bar reinforcement**						
R5.7.1	Nominal size: 6mm						
R5.7.1.01 Plain round steel bar reinforcement to BS 4449	t	11.70	557.74	135.37	681.10	1,374.21	611.542
R5.7.1.02 Deformed high yield steel bar reinforcement to BS 4449	t	11.70	557.74	135.37	701.53	1,394.64	611.542
R5.7.2	Nominal size: 8mm						
R5.7.2.01 Plain round steel bar reinforcement to BS 4449	t	9.70	462.40	112.23	662.24	1,236.87	607.606
R5.7.2.02 Deformed high yield steel bar reinforcement to BS 4449	t	9.70	462.40	112.23	682.11	1,256.74	607.606
R5.7.3	Nominal size: 10mm						
R5.7.3.01 Plain round steel bar reinforcement to BS 4449	t	7.00	333.69	80.99	635.79	1,050.47	602.292
R5.7.3.02 Deformed high yield steel bar reinforcement to BS 4449	t	7.00	333.69	80.99	654.87	1,069.55	602.292
R5.7.4	Nominal size: 12mm						
R5.7.4.01 Plain round steel bar reinforcement to BS 4449	t	6.30	300.32	72.89	620.19	993.40	600.914
R5.7.4.02 Deformed high yield steel bar reinforcement to BS 4449	t	6.30	300.32	72.89	638.79	1,012.00	600.914
R5.7.5	Nominal size: 16mm						
R5.7.5.01 Plain round steel bar reinforcement to BS 4449	t	5.30	252.65	61.32	586.15	900.12	598.946
R5.7.5.02 Deformed high yield steel bar reinforcement to BS 4449	t	5.30	252.65	61.32	603.74	917.71	598.946
R5.7.6	Nominal size: 20mm						
R5.7.6.01 Plain round steel bar reinforcement to BS 4449	t	5.00	238.35	57.85	586.15	882.35	598.356
R5.7.6.02 Deformed high yield steel bar reinforcement to BS 4449	t	5.00	238.35	57.85	603.74	899.94	598.356
R5.8	**Separation and waterproof membranes**						
R5.8.0	Generally						
R5.8.0.01 1200 gauge	m²	0.01	0.28	-	0.83	1.11	0.520
R5.8.0.02 4000 gauge	m²	0.01	0.28	-	2.05	2.33	0.520

R6 Joint In Concrete Pavements

	Unit	Labour Hours	Labour Net £	Plant Net £	Materials Net £	Unit Net £	CO_2e Kg
R6.2	**Expansion joints**						
R6.2.4	Depth of joint: 100 - 150mm						
R6.2.4.01 Korkpak joint filler; 12mm diameter mild steel bar 1m long at 400mm centres; cold pored Expandite Colpor 200 joint sealant compound; forming groove; casting bars into side of joint and debonding for a length of 500mm and capped with PVC dowel caps; 10 x 150mm deep; 13 x 25mm sealant	m	0.54	25.88	-	16.59	42.47	5.581

R6   Joint In Concrete Pavements continued...

	Unit	Labour Hours	Labour Net £	Plant Net £	Materials Net £	Unit Net £	CO_2e Kg
R6.2	**Expansion joints**						
R6.2.4	Depth of joint: 100 - 150mm						
R6.2.4.02 Korkpak joint filler; 12mm diameter mild steel bar 1m long at 400mm centres; cold pored Expandite Colpor 200 joint sealant compound; forming groove; casting bars into side of joint and debonding for a length of 500mm and capped with PVC dowel caps; 13 x 150mm deep; 19 x 25mm sealant	m	0.54	25.88	-	18.06	43.94	5.608
R6.2.4.03 Korkpak joint filler; 12mm diameter mild steel bar 1m long at 400mm centres; cold pored Expandite Colpor 200 joint sealant compound; forming groove; casting bars into side of joint and debonding for a length of 500mm and capped with PVC dowel caps; 19 x 150mm deep; 25 x 25mm sealant	m	0.54	25.88	-	25.10	50.98	6.180
R6.2.4.04 Korkpak joint filler; 12mm diameter mild steel bar 1m long at 400mm centres; cold pored Expandite Colpor 200 joint sealant compound; forming groove; casting bars into side of joint and debonding for a length of 500mm and capped with PVC dowel caps; 25 x 100mm deep; 30 x 25mm sealant	m	0.54	25.88	-	31.13	57.01	6.908
R6.2.5	Depth of joint: 150 - 200mm						
R6.2.5.01 Korkpak joint filler; 12mm diameter mild steel bar 1m long at 400mm centres; cold pored Expandite Colpor 200 joint sealant compound; forming groove; casting bars into side of joint and debonding for a length of 500mm and capped with PVC dowel caps	m	0.54	25.88	-	20.54	46.42	7.809
R6.2.6	Depth of joint: 200 - 250mm						
R6.2.6.01 Korkpak joint filler; 12mm diameter mild steel bar 1m long at 400mm centres; cold pored Expandite Colpor 200 joint sealant compound; forming groove; casting bars into side of joint and debonding for a length of 500mm and capped with PVC dowel caps	m	0.54	25.88	-	27.36	53.24	10.805

R6 Joint In Concrete Pavements continued...

	Unit	Labour Hours	Labour Net £	Plant Net £	Materials Net £	Unit Net £	CO_2e Kg
R6.2 **Expansion joints**							
R6.2.7 Depth of joint: 250 - 300mm							
R6.2.7.01 Korkpak joint filler; 12mm diameter mild steel bar 1m long at 400mm centres; cold pored Expandite Colpor 200 joint sealant compound; forming groove; casting bars into side of joint and debonding for a length of 500mm and capped with PVC dowel caps	m	0.54	25.88	-	33.93	59.81	13.759
R6.3 **Contraction joints**							
R6.3.4 Depth of joint: 100 - 150mm							
R6.3.4.01 12mm diameter mild steel bars 1m long at 600mm centres; cold poured Expandite Colpor 200 joint sealing compound; forming groove; crack inducer; including formwork; 150mm deep; 10 x 10mm sealant	m	0.51	24.31	-	5.45	29.76	4.344
R6.3.4.02 12mm diameter mild steel bars 1m long at 600mm centres; cold poured Expandite Colpor 200 joint sealing compound; forming groove; crack inducer; including formwork; 150mm deep; 15 x 15mm sealant	m	0.51	24.31	-	7.80	32.11	4.388
R6.3.4.03 12mm diameter mild steel bars 1m long at 600mm centres; cold poured Expandite Colpor 200 joint sealing compound; forming groove; crack inducer; including formwork; 150mm deep; 20 x 20mm sealant	m	0.51	24.31	-	10.15	34.46	4.432
R6.3.5 Depth of joint: 150 - 200mm							
R6.3.5.01 12mm diameter mild steel bars 1m long at 600mm centres; cold poured Expandite Colpor 200 joint sealing compound; forming groove; crack inducer; including formwork; 200mm deep; 10 x 10mm sealant	m	0.51	24.31	-	7.06	31.37	6.194
R6.3.6 Depth of joint: 200 - 250mm							
R6.3.6.01 12mm diameter mild steel bars 1m long at 600mm centres; cold poured Expandite Colpor 200 joint sealing compound; forming groove; crack inducer; including formwork; 250mm deep; 10 x 10mm sealant	m	0.51	24.31	-	8.98	33.29	8.397

R6 Joint In Concrete Pavements continued...

	Unit	Labour Hours	Labour Net £	Plant Net £	Materials Net £	Unit Net £	CO$_2$e Kg
R6.3 Contraction joints							
R6.3.7 Depth of joint: 250 - 300mm							
R6.3.7.01 12mm diameter mild steel bars 1m long at 600mm centres; cold poured Expandite Colpor 200 joint sealing compound; forming groove; crack inducer; including formwork; 300mm deep; 10 x 10mm sealant	m	0.51	24.31	-	10.90	35.21	10.600
R6.4 Warping joints							
R6.4.4 Depth of joint: 100 - 150mm							
R6.4.4.01 12mm diameter mild steel bars 1.2m long at 400mm centres; cold poured Expandite Colpor 200 joint sealing compound; forming groove; crack inducer; including formwork; 150mm deep; 10 x 10mm sealant	m	0.51	24.31	-	5.45	29.76	4.344
R6.4.5 Depth of joint: 150 - 200mm							
R6.4.5.01 12mm diameter mild steel bars 1.2m long at 300mm centres; cold poured Expandite Colpor 200 joint sealing compound; forming groove; crack inducer; including formwork; 200mm deep; 10 x 10mm sealant	m	0.51	24.31	-	7.15	31.46	6.365
R6.4.6 Depth of joint: 200 - 250mm							
R6.4.6.01 12mm diameter mild steel bars 1.2m long at 230mm centres; cold poured Expandite Colpor 200 joint sealing compound; forming groove; crack inducer; including formwork; 250mm deep; 10 x 10mm sealant	m	0.51	24.31	-	9.18	33.49	8.781
R6.4.7 Depth of joint: 250 - 300mm							
R6.4.7.01 12mm diameter mild steel bars 1.2m long at 180mm centres; cold poured Expandite Colpor 200 joint sealing DTp clause nr. 1009/11; compound; forming groove; crack inducer; including formwork; 300mm deep; 10 x 10mm sealant	m	0.51	24.31	-	11.23	35.54	11.234
R6.5 Longitudinal joints							
R6.5.4 Depth of joint: 100 - 150mm							
R6.5.4.01 12mm diameter mild steel bars 1m long at 600mm centres; 5 x 15mm cold poured Expandite Colpor 200 joint sealing compound; forming groove; crack inducer; including formwork; 150mm deep	m	0.47	22.45	-	5.45	27.90	4.344

R6 Joint In Concrete Pavements continued...

	Unit	Labour Hours	Labour Net £	Plant Net £	Materials Net £	Unit Net £	CO$_2$e Kg
R6.5	**Longitudinal joints**						
R6.5.5	Depth of joint: 150 - 200mm						
R6.5.5.01	12mm diameter mild steel bars 1m long at 600mm centres; 5 x 15mm cold poured Expandite Colpor 200 joint sealing compound; forming groove; crack inducer; including formwork; 200mm deep	m	0.47	22.45	-	7.06	29.51	6.194
R6.5.6	Depth of joint: 200 - 250mm						
R6.5.6.01	12mm diameter mild steel bars 1m long at 600mm centres; 5 x 15mm cold poured Expandite Colpor 200 joint sealing compound; forming groove; crack inducer; including formwork; 250 mm deep	m	0.47	22.45	-	8.98	31.43	8.397
R6.5.7	Depth of joint: 250 - 300mm						
R6.5.7.01	12mm diameter mild steel bars 1m long at 600mm centres; 5 x 15mm cold poured Expandite Colpor 200 joint sealing compound; forming groove; crack inducer; including formwork; 300mm deep	m	0.47	22.45	-	10.90	33.35	10.600
R6.6	**Construction joints**						
R6.6.4	Depth of joint: 100 - 150mm						
R6.6.4.01	12mm diameter mild steel bars 1m long at 600mm centres; including formwork; 150mm deep	m	0.47	22.45	-	3.57	26.02	4.308
R6.6.5	Depth of joint: 150 - 200mm						
R6.6.5.01	12mm diameter mild steel bars 1m long at 600mm centres; including formwork; 200mm deep	m	0.47	22.45	-	5.18	27.63	6.159
R6.6.6	Depth of joint: 200 - 250mm						
R6.6.6.01	12mm diameter mild steel bars 1m long at 600mm centres; including formwork; 250mm deep	m	0.47	22.45	-	7.10	29.55	8.362
R6.6.7	Depth of joint: 250 - 300mm						
R6.6.7.01	12mm diameter mild steel bars 1m long at 600mm centres; including formwork; 300mm deep	m	0.47	22.45	-	9.02	31.47	10.565

R7 Kerbs, Channels, Edgings, Footways And Paved Areas

		Unit	Labour Hours	Labour Net £	Plant Net £	Materials Net £	Unit Net £	CO$_2$e Kg
R7.1	**Precast concrete kerbs**							
R7.1.1	Straight or curved to radius exceeding 12m							
R7.1.1.01	Kerbs; 125 x 150mm; conventional concrete bed and backing	m	0.13	5.68	-	11.65	17.33	26.649
R7.1.1.02	Kerbs; 125 x 255mm; conventional concrete bed and backing	m	0.16	6.99	-	12.65	19.64	31.230

R7 Kerbs, Channels, Edgings, Footways And Paved Areas continued...

	Unit	Labour Hours	Labour Net £	Plant Net £	Materials Net £	Unit Net £	CO$_2$e Kg
R7.1	**Precast concrete kerbs**						
R7.1.1	Straight or curved to radius exceeding 12m						
R7.1.1.03	Kerbs; 150 x 305mm; conventional concrete bed and backing	m	0.16	6.99	-	15.86	22.85	36.345
R7.1.1.04	Kerbs; 125 x 150mm; low carbon concrete bed and backing	m	0.13	5.68	-	10.27	15.95	14.787
R7.1.1.05	Kerbs; 125 x 255mm; low carbon concrete bed and backing	m	0.16	6.99	-	11.27	18.26	19.367
R7.1.1.06	Kerbs; 150 x 305mm; low carbon concrete bed and backing	m	0.16	6.99	-	14.48	21.47	24.483
R7.1.2	Curved to radius not exceeding 12m						
R7.1.2.01	Kerbs; 125 x 150mm; 4m radius; conventional concrete bed and backing	m	0.26	11.35	-	11.98	23.33	37.167
R7.1.2.02	Kerbs; 125 x 150mm; 8m radius; conventional concrete bed and backing	m	0.21	9.17	-	11.98	21.15	37.167
R7.1.2.03	Kerbs; 125 x 150mm; 12m radius; conventional concrete bed and backing	m	0.16	6.99	-	11.98	18.97	37.167
R7.1.2.04	Kerbs; 125 x 255mm; 4m radius; conventional concrete bed and backing	m	0.26	11.35	-	13.29	24.64	31.934
R7.1.2.05	Kerbs; 125 x 255mm; 8m radius; conventional concrete bed and backing	m	0.21	9.17	-	13.29	22.46	31.934
R7.1.2.06	Kerbs; 125 x 255mm; 12m radius; conventional concrete bed and backing	m	0.16	6.99	-	13.29	20.28	31.934
R7.1.2.07	Kerbs; 150 x 305mm; 4m radius; conventional concrete bed and backing	m	0.26	11.35	-	14.61	25.96	31.013
R7.1.2.08	Kerbs; 150 x 305mm; 8m radius; conventional concrete bed and backing	m	0.21	9.17	-	14.61	23.78	31.013
R7.1.2.09	Kerbs; 150 x 305mm; 12m radius; conventional concrete bed and backing	m	0.16	6.99	-	14.61	21.60	31.013
R7.1.2.10	Kerbs; 125 x 150mm; 4m radius; low carbon concrete bed and backing	m	0.26	11.35	-	10.60	21.95	25.305
R7.1.2.11	Kerbs; 125 x 150mm; 8m radius; low carbon concrete bed and backing	m	0.21	9.17	-	10.60	19.77	25.305
R7.1.2.12	Kerbs; 125 x 150mm; 12m radius; low carbon concrete bed and backing	m	0.16	6.99	-	10.60	17.59	25.305
R7.1.2.13	Kerbs; 125 x 255mm; 4m radius; low carbon concrete bed and backing	m	0.26	11.35	-	11.91	23.26	20.072
R7.1.2.14	Kerbs; 125 x 255mm; 8m radius; low carbon concrete bed and backing	m	0.21	9.17	-	11.91	21.08	20.072
R7.1.2.15	Kerbs; 125 x 255mm; 12m radius; low carbon concrete bed and backing	m	0.16	6.99	-	11.91	18.90	20.072

R7 Kerbs, Channels, Edgings, Footways And Paved Areas continued...

	Unit	Labour Hours	Labour Net £	Plant Net £	Materials Net £	Unit Net £	CO_2e Kg
R7.1	**Precast concrete kerbs**						
R7.1.2	Curved to radius not exceeding 12m						
R7.1.2.16	Kerbs; 150 x 305mm; 4m radius; low carbon concrete bed and backing	m	0.26	11.35	-	14.61	25.96	31.013
R7.1.2.17	Kerbs; 150 x 305mm; 8m radius; low carbon concrete bed and backing	m	0.21	9.17	-	13.69	22.86	23.105
R7.1.2.18	Kerbs; 150 x 305mm; 12m radius; low carbon concrete bed and backing	m	0.16	6.99	-	13.69	20.68	23.105
R7.1.3	Quadrants						
R7.1.3.01	455 x 455 x 255mm; conventional concrete bed and backing	Nr	0.16	6.99	-	18.07	25.06	37.420
R7.1.3.02	455 x 455 x 150mm; conventional concrete bed and backing	Nr	0.16	6.99	-	14.85	21.84	29.613
R7.1.3.03	305 x 305 x 255mm; conventional concrete bed and backing	Nr	0.14	6.11	-	16.00	22.11	20.826
R7.1.3.04	305 x 305 x 150mm; conventional concrete bed and backing	Nr	0.14	6.11	-	14.60	20.71	17.318
R7.1.3.05	455 x 455 x 255mm; low carbon concrete bed and backing	Nr	0.16	6.99	-	16.70	23.69	25.558
R7.1.3.06	455 x 455 x 150mm; low carbon concrete bed and backing	Nr	0.16	6.99	-	13.48	20.47	17.751
R7.1.3.07	305 x 305 x 255mm; low carbon concrete bed and backing	Nr	0.14	6.11	-	15.08	21.19	12.918
R7.1.3.08	305 x 305 x 150mm; low carbon concrete bed and backing	Nr	0.14	6.11	-	13.68	19.79	9.410
R7.1.4	Drops						
R7.1.4.01	125 x 255mm; conventional concrete bed and backing	Nr	0.14	6.11	-	10.71	16.82	21.293
R7.1.4.02	125 x 175mm; conventional concrete bed and backing	Nr	0.14	6.11	-	10.63	16.74	17.702
R7.1.4.03	125 x 150mm; conventional concrete bed and backing	Nr	0.14	6.11	-	10.50	16.61	16.580
R7.1.4.04	125 x 255mm; low carbon concrete bed and backing	Nr	0.14	6.11	-	9.97	16.08	14.966
R7.1.4.05	125 x 175mm; low carbon concrete bed and backing	Nr	0.14	6.11	-	9.89	16.00	11.375
R7.1.4.06	125 x 150mm; low carbon concrete bed and backing	Nr	0.14	6.11	-	9.76	15.87	10.253
R7.2	**Precast concrete channels**						
R7.2.1	Straight or curved to radius exceeding 12m						
R7.2.1.01	255 x 125mm	m	0.10	4.37	-	12.13	16.50	19.431
R7.2.1.02	150 x 125mm	m	0.10	4.37	-	9.84	14.21	22.561
R7.2.2	Curved to radius not exceeding 12m						
R7.2.2.01	255 x 125mm; 4m radius	m	0.17	7.42	-	12.13	19.55	19.431
R7.2.2.02	255 x 125mm; 8m radius	m	0.15	6.55	-	12.13	18.68	19.431
R7.2.2.03	255 x 125mm; 12m radius	m	0.13	5.68	-	12.13	17.81	19.431
R7.2.2.04	150 x 125mm; 4m radius	m	0.17	7.42	-	10.38	17.80	12.865
R7.2.2.05	150 x 125mm; 8m radius	m	0.15	6.55	-	10.38	16.93	12.865
R7.2.2.06	150 x 125mm; 12m radius	m	0.13	5.68	-	10.38	16.06	12.865

R7 Kerbs, Channels, Edgings, Footways And Paved Areas continued...

	Unit	Labour Hours	Labour Net £	Plant Net £	Materials Net £	Unit Net £	CO_2e Kg
R7.3	**Precast concrete edgings**						
R7.3.1 Straight or curved to radius exceeding 12m							
R7.3.1.01 50 x 150mm	m	0.09	3.84	-	5.18	9.02	10.823
R7.3.1.02 50 x 200mm	m	0.09	4.02	-	5.75	9.77	11.805
R7.3.1.03 50 x 250mm	m	0.10	4.24	-	6.26	10.50	12.886
R7.3.2 Curved to radius not exceeding 12m							
R7.3.2.01 50 x 250 mm; 8m radius	m	0.10	4.37	-	5.80	10.17	12.948
R7.5	**Precast concrete, natural stone, block and clay slabs and pavers**						
R7.5.0 Generally							
R7.5.0.01 Precast concrete flags, natural finish; on mortar dabs in cement lime mortar (1:1:6); Close butt joints; 600 x 450 x 50mm thick	m²	0.15	6.55	-	16.28	22.83	23.957
R7.5.0.02 Precast concrete flags, natural finish; on mortar dabs in cement lime mortar (1:1:6); Close butt joints; 600 x 600 x 50mm thick	m²	0.17	7.42	-	14.06	21.48	24.474
R7.5.0.03 Precast concrete flags, natural finish; on mortar dabs in cement lime mortar (1:1:6); Close butt joints; 600 x 750 x 50mm thick	m²	0.19	8.30	-	13.52	21.82	24.442
R7.5.0.04 Precast concrete flags, natural finish; on mortar dabs in cement lime mortar (1:1:6); Close butt joints; 600 x 900 x 50mm thick	m²	0.21	9.17	-	12.75	21.92	24.442
R7.5.0.05 Precast concrete flags, natural finish; on mortar dabs in cement lime mortar (1:1:6); Pointing 10mm joints in cement lime mortar (1:1:6); 600 x 450 x 50mm thick	m²	0.17	7.42	-	16.28	23.70	23.957
R7.5.0.06 Precast concrete flags, natural finish; on mortar dabs in cement lime mortar (1:1:6); Pointing 10mm joints in cement lime mortar (1:1:6); 600 x 600 x 50mm thick	m²	0.19	8.30	-	14.06	22.36	24.474
R7.5.0.07 Precast concrete flags, natural finish; on mortar dabs in cement lime mortar (1:1:6); Pointing 10mm joints in cement lime mortar (1:1:6); 600 x 750 x 50mm thick	m²	0.21	9.17	-	13.52	22.69	24.442
R7.5.0.08 Precast concrete flags, natural finish; on mortar dabs in cement lime mortar (1:1:6); Pointing 10mm joints in cement lime mortar (1:1:6); 600 x 900 x 50mm thick	m²	0.23	10.04	-	12.75	22.79	24.442
R7.5.0.09 Precast concrete flags, buff finish; on mortar dabs in cement lime mortar (1:1:6); Pointing 10mm joints in cement lime mortar (1:1:6); 600 x 450 x 50mm thick	m²	0.08	3.49	-	22.54	26.03	23.957

R7 Kerbs, Channels, Edgings, Footways And Paved Areas continued...

	Unit	Labour Hours	Labour Net £	Plant Net £	Materials Net £	Unit Net £	CO_2e Kg
R7.5	**Precast concrete, natural stone, block and clay slabs and pavers**						
R7.5.0	Generally						
R7.5.0.10	Precast concrete flags, buff finish; on mortar dabs in cement lime mortar (1:1:6); Pointing 10mm joints in cement lime mortar (1:1:6); 600 × 600 × 50mm thick	m²	0.11	4.80	-	19.93	24.73	24.474
R7.5.0.11	Precast concrete flags, buff finish; on mortar dabs in cement lime mortar (1:1:6); Pointing 10mm joints in cement lime mortar (1:1:6); 600 × 750 × 50mm thick	m²	0.12	5.24	-	18.74	23.98	24.442
R7.5.0.12	Precast concrete flags, buff finish; on mortar dabs in cement lime mortar (1:1:6); Pointing 10mm joints in cement lime mortar (1:1:6); 600 × 900 × 50mm thick	m²	0.17	7.42	-	17.08	24.50	24.442
R7.5.0.13	Precast concrete flags, natural finish; on mortar dabs in cement lime mortar (1:1:6); Close butt joints; 600 × 450 × 63mm thick	m²	0.10	4.37	-	18.29	22.66	28.496
R7.5.0.14	Precast concrete flags, natural finish; on mortar dabs in cement lime mortar (1:1:6); Close butt joints; 600 × 600 × 63mm thick	m²	0.12	5.24	-	16.04	21.28	29.147
R7.5.0.15	Precast concrete flags, natural finish; on mortar dabs in cement lime mortar (1:1:6); Close butt joints; 600 × 750 × 63mm thick	m²	0.14	6.11	-	14.86	20.97	29.106
R7.5.0.16	Precast concrete flags, natural finish; on mortar dabs in cement lime mortar (1:1:6); Close butt joints; 600 × 900 × 63mm thick	m²	0.18	7.86	-	13.84	21.70	29.105
R7.5.0.17	Precast concrete flags, natural finish; on mortar dabs in cement lime mortar (1:1:6); Pointing 10mm joints in cement lime mortar (1:1:6); 600 × 450 × 63mm thick	m²	0.12	5.24	-	18.29	23.53	28.496
R7.5.0.18	Precast concrete flags, natural finish; on mortar dabs in cement lime mortar (1:1:6); Pointing 10mm joints in cement lime mortar (1:1:6); 600 × 600 × 63mm thick	m²	0.14	6.11	-	16.04	22.15	29.147
R7.5.0.19	Precast concrete flags, natural finish; on mortar dabs in cement lime mortar (1:1:6); Pointing 10mm joints in cement lime mortar (1:1:6); 600 × 750 × 63mm thick	m²	0.16	6.99	-	14.86	21.85	29.106
R7.5.0.20	Precast concrete flags, natural finish; on mortar dabs in cement lime mortar (1:1:6); Pointing 10mm joints in cement lime mortar (1:1:6); 600 × 900 × 63mm thick	m²	0.20	8.73	-	13.84	22.57	29.105

R7 Kerbs, Channels, Edgings, Footways And Paved Areas continued...

	Unit	Labour Hours	Labour Net £	Plant Net £	Materials Net £	Unit Net £	CO_2e Kg
R7.5 Precast concrete, natural stone, block and clay slabs and pavers							
R7.5.0 Generally							
R7.5.0.21 Precast concrete flags, buff finish; on mortar dabs in cement lime mortar (1:1:6); Pointing 10mm joints in cement lime mortar (1:1:6); 600 x 450 x 63mm thick	m^2	0.10	4.37	-	29.13	33.50	28.496
R7.5.0.22 Precast concrete flags, buff finish; on mortar dabs in cement lime mortar (1:1:6); Pointing 10mm joints in cement lime mortar (1:1:6); 600 x 600 x 63mm thick	m^2	0.12	5.24	-	25.68	30.92	29.147
R7.5.0.23 Precast concrete flags, buff finish; on mortar dabs in cement lime mortar (1:1:6); Pointing 10mm joints in cement lime mortar (1:1:6); 600 x 750 x 63mm thick	m^2	0.14	6.11	-	23.34	29.45	29.106
R7.5.0.24 Precast concrete flags, buff finish; on mortar dabs in cement lime mortar (1:1:6); Pointing 10mm joints in cement lime mortar (1:1:6); 600 x 900 x 63mm thick	m^2	0.18	7.86	-	21.67	29.53	29.105

R8 Ancillaries

	Unit	Labour Hours	Labour Net £	Plant Net £	Materials Net £	Unit Net £	CO_2e Kg
R8.1 Traffic signs							
R8.1.1 Non-illuminated							
R8.1.1.01 Post mounted non-lit traffic signs; excavation, disposal and backfilling; general signage: $1.50m^2$	Nr	1.75	76.42	48.75	208.83	334.00	113.370
R8.1.1.02 Post mounted non-lit traffic signs; excavation, disposal and backfilling; general signage: $2.50m^2$	Nr	2.00	87.34	55.40	288.07	430.81	156.282
R8.1.1.03 Post mounted non-lit traffic signs; excavation, disposal and backfilling; general signage: $3.00m^2$	Nr	3.00	131.01	83.10	426.63	640.74	244.387
R8.2 Surface markings							
R8.2.2 Reflecting road studs							
R8.2.2.01 Cats-eyes; in preformed pockets	Nr	0.05	2.18	2.37	10.52	15.07	3.507
R8.2.3 Letters and shapes							
R8.2.3.01 Directional arrow 6.0m long; to tarmac; thermoplastic material	Nr	0.33	14.54	15.75	7.20	37.49	27.399
R8.2.3.02 Directional arrow 4.0m long; to tarmac; thermoplastic material	Nr	0.25	10.92	11.83	5.04	27.79	20.256
R8.2.3.03 Give way triangle; 3.75m high; thermoplastic material	Nr	0.17	7.29	7.90	4.32	19.51	14.349
R8.2.4 Continuous lines							
R8.2.4.01 To tarmac; 100mm wide	m	0.01	0.31	0.33	0.36	1.00	0.755
R8.2.4.02 To tarmac; 200mm wide	m	0.01	0.35	0.38	0.78	1.51	0.819
R8.2.4.03 To tarmac; 150mm wide	m	0.01	0.35	0.38	0.57	1.30	0.819
R8.2.4.04 To concrete; one coat bituminous tack coat; 100mm wide	m	0.01	0.31	0.33	0.38	1.02	0.764

R8 Ancillaries continued...

	Unit	Labour Hours	Labour Net £	Plant Net £	Materials Net £	Unit Net £	CO_2e Kg
R8.2 **Surface markings**							
R8.2.4 Continuous lines							
R8.2.4.05 To concrete; one coat bituminous tack coat; 200mm wide	m	-	0.35	0.38	0.80	1.53	0.828
R8.2.4.06 To concrete; one coat bituminous tack coat; 150mm wide	m	-	0.35	0.38	0.59	1.32	0.828
R8.2.5 Intermittent lines							
R8.2.5.01 To tarmac; 100mm wide	m	0.01	0.35	0.38	0.36	1.09	0.819
R8.2.5.02 To tarmac; 200mm wide	m	0.01	0.44	0.47	0.78	1.69	0.946
R8.2.5.03 To tarmac; 150mm wide	m	0.01	0.39	0.43	0.57	1.39	0.882

Calculations used throughout Class S - Rail Track

Labour

			Qty		Rate		Total
L A0320ICE	**Waterproofing Labour Gang**						
	L A9016ICE	Ganger	1	x	17.64	=	£17.64
	L A9002ICE	Labourer (General Operative)	3	x	13.02	=	£39.05
		Total hourly cost of gang				=	**£56.69**
L S2000ICE	**Bottom ballast Labour Gang**						
	L A9036ICE	Principal Track Officer	2	x	75.71	=	£151.43
	L A9037ICE	Chainman	2	x	34.20	=	£68.40
	L A9038ICE	Trackman	4	x	37.08	=	£148.31
		Total hourly cost of gang				=	**£368.14**
L S2005ICE	**Top ballast Labour Gang**						
	L A9039ICE	Track Chargeman	8	x	40.98	=	£327.87
	L A9040ICE	Leading Trackman	8	x	40.98	=	£327.87
	L A9038ICE	Trackman	32	x	37.08	=	£1,186.46
	L A9036ICE	Principal Track Officer	2	x	75.71	=	£151.43
	L A9037ICE	Chainman	2	x	34.20	=	£68.40
		Total hourly cost of gang				=	**£2,062.03**
L S0004ICE	**Rail Installation Labour Gang**						
	L A9040ICE	Leading Trackman	1	x	40.98	=	£40.98
	L A9038ICE	Trackman	2	x	37.08	=	£74.15
		Total hourly cost of gang				=	**£115.13**
L S0005ICE	**Rail removal Labour Gang**						
	L A9040ICE	Leading Trackman	1	x	40.98	=	£40.98
	L A9038ICE	Trackman	2	x	37.08	=	£74.15
		Total hourly cost of gang				=	**£115.13**
L S2071ICE	**Tamping Labour Gang**						
	L A9046ICE	Supervisor	8	x	42.26	=	£338.08
	L A9047ICE	Controller of Site Safety (COSS)	8	x	25.57	=	£204.56
	L A9048ICE	Trackman (labourer)	32	x	20.88	=	£668.16
		Total hourly cost of gang				=	**£1,210.80**
L S2072ICE	**Stone Blowing Labour Gang**						
	L A9046ICE	Supervisor	8	x	42.26	=	£338.08
		Total hourly cost of gang				=	**£338.08**
L S2073ICE	**Track Labour Gang**						
	L A9046ICE	Supervisor	1	x	42.26	=	£42.26
	L A9047ICE	Controller of Site Safety (COSS)	1	x	25.57	=	£25.57
	L A9048ICE	Trackman (labourer)	6	x	20.88	=	£125.28
		Total hourly cost of gang				=	**£193.11**

Plant			Qty		Rate		Total
P S2025ICE	**Top ballast Plant Gang**						
	P A0549ICE	Dumper - 2.5t 4WD	2	x	22.99	=	£45.98
	P A0713ICE	Lighting Tower - Lighting the Works	20	x	5.30	=	£106.00
	P A0578ICE	Generator - 10kvA Diesel	5	x	3.80	=	£19.00
		Total hourly cost of gang				=	**£170.98**
P S2200ICE	**Bottom ballast Plant Gang**						
	P A0713ICE	Lighting Tower - Lighting the Works	20	x	5.30	=	£106.00
	P A0578ICE	Generator - 10kvA Diesel	5	x	3.80	=	£19.00
	P A0943ICE	Crawler Tractor / Dozer - Cat D6 LGP 160 Hp	4	x	52.02	=	£208.08
		Total hourly cost of gang				=	**£333.08**
P S0001ICE	**Rail Installation Plant Gang**						
	P A0481ICE	Cranes Transit - 25t	1	x	60.22	=	£60.22
	P A0402ICE	CP9 Air Breaker Drill	1	x	0.45	=	£0.45
	P A0452ICE	Compressor - 2-Tool (Complete)	1	x	5.20	=	£5.20
	P A1031ICE	Cutting and Burning Gear	1	x	3.77	=	£3.77
		Total hourly cost of gang				=	**£69.64**
P S0002ICE	**Rail removal Plant Gang**						
	P A0481ICE	Cranes Transit - 25t	1	x	60.22	=	£60.22
	P A1044ICE	Articulated Lorry	1	x	68.51	=	£68.51
	P A1045ICE	Low Loader Trailer	1	x	13.96	=	£13.96
	P A0402ICE	CP9 Air Breaker Drill	2	x	0.45	=	£0.90
	P A0452ICE	Compressor - 2-Tool (Complete)	1	x	5.20	=	£5.20
	P A1031ICE	Cutting and Burning Gear	1	x	3.77	=	£3.77
		Total hourly cost of gang				=	**£152.56**
P S2201ICE	**Tamping Plant Gang**						
	P A1086ICE	Tamper	8	x	567.10	=	£4,536.80
		Total hourly cost of gang				=	**£4,536.80**
P S2202ICE	**Stone Blowing Plant Gang**						
	P A1087ICE	Stone Blower	8	x	743.10	=	£5,944.80
		Total hourly cost of gang				=	**£5,944.80**
P S2203ICE	**Track Plant Gang**						
	P A1088ICE	Road Rail Excavator	1	x	70.58	=	£70.58
	P A1089ICE	Crane Supervisor	1	x	32.32	=	£32.32
	P A1090ICE	Attachment	1	x	10.00	=	£10.00
		Total hourly cost of gang				=	**£112.90**

Rail Track

Note(s): 1) The prices are applicable to sites on the average UK where total track work involved exceeds 2000 linear metres.
2) The prices assume that the track to be railed is adjacent to an existing track that is under possession from which locomotives and other plant can operate.
3) The prices assume that the works will be carried out under a series of possession arrangements of a typical 29 hour duration Sunday 00:01 to Monday 05:00.
4) Excluded from the prices are any allowances for:-
- design of the works
- follow up tamping and removal of temporary speed restrictions
- forming access to the works and removal on completion
- site welfare and accommodation facilities
- non-working supervision staff e.g. site managers, Agents, contract manager etc.
- other items of costs normally allowed for in the General Items section of a Tender
- overheads and profit
5) Possession, protection and isolation costs have been separately identified and are not included in other items of work.
6) Tipping charges have not been included in the disposal rates due to the wide variation in prices from tip to tip.
7) Landfill Tax has not been included in the disposal rates as this is dependent on the nature of the material and statutory adjustment.
8) Tamping - The prices assumed that initial and speed off tamp in typical 29 hour possession (400m per shift per tamp) and follow up tamp in Midweek Night Possession (1,600m per shift)
9) Stone Blowing - The prices assumed that work undertaken in Midweek Night Possession, 10mm lift required per sleeper (2.60m long) and output of 400m per shift.

S1   Track Foundations

		Unit	Labour Hours	Labour Net £	Plant Net £	Materials Net £	Unit Net £	CO_2e Kg
S1.1	**Bottom ballast**							
S1.1.0	Crushed and graded;							
S1.1.0.01	Limestone	m³	0.01	4.05	8.66	26.39	39.10	18.171
S1.1.0.02	Granite	m³	0.01	4.05	8.66	31.22	43.93	18.807
S1.2	**Top ballast**							
S1.2.0	Crushed and graded;							
S1.2.0.01	Limestone	m³	0.01	22.68	4.62	26.39	53.69	15.650
S1.2.0.02	Granite	m³	0.01	22.68	4.62	31.22	58.52	16.285
S1.3	**Blinding**							
S1.3.0	Imported material							
S1.3.0.01	Sand; 100 mm thick	m²	0.01	3.68	3.33	4.75	11.76	4.377
S1.4	**Blankets and vibration mats**							
S1.4.0	Imported material							
S1.4.0.01	Granular stone thickness: 100mm	m²	0.00	0.37	2.67	3.13	6.17	3.364
S1.4.0.02	Granular stone thickness: 150mm	m²	0.00	0.74	4.00	4.70	9.44	5.050
S1.4.0.03	Granular sand thickness: 100mm	m²	0.00	0.37	2.67	2.85	5.89	3.435
S1.4.0.04	Granular sand thickness: 150mm	m²	0.00	0.74	4.00	4.28	9.02	5.156
S1.5	**Waterproof and membranes**							
S1.5.0	Polythene sheet							
S1.5.0.01	1200 gauge	m²	0.02	0.85	-	0.83	1.68	0.520
S1.7	**Tamping**							
S1.7.0	Tamping							
S1.7.0.01	Plain Line, initial tamp, speed off tamp and follow up tamp	m	0.01	7.26	27.22	-	34.48	5.898
S1.7.0.02	Extra over for typical Switches & Crossing upto C turnout (initial tamp, speed off tamp and follow up tamp)	Nr	1.25	1,513.50	5,671.02	-	7,184.52	1,228.698
S1.8	**Pneumatic ballast injection: stone blowing**							
S1.8.0	Generally							
S1.8.0.01	Stone blowing	m	0.00	1.01	17.83	2.31	21.15	3.634

S2   Taking Up

		Unit	Labour Hours	Labour Net £	Plant Net £	Materials Net £	Unit Net £	CO_2e Kg
S2.1	**Bullhead rails**							
S2.1.1	Plain track							
S2.1.1.01	Take up, strip into component parts and stack in designated area for subsequent disposal; plain fishplated; bullhead or flat bottom rails; timber sleepers	m	0.10	11.51	15.26	-	26.77	9.268

S2 Taking Up continued...

	Unit	Labour Hours	Labour Net £	Plant Net £	Materials Net £	Unit Net £	CO$_2$e Kg
S2.1	**Bullhead rails**						
S2.1.1 Plain track							
S2.1.1.02 Take up, strip into component parts and stack in designated area for subsequent disposal; plain fishplated; bullhead or flat bottom rails; concrete sleepers	m	0.11	12.67	16.78	-	29.45	10.195
S2.1.1.03 Take up, strip into component parts and stack in designated area for subsequent disposal; turnouts; bullhead or flat bottom rails; timber sleepers	Nr	10.00	1,151.40	1,525.70	-	2,677.10	926.830
S2.1.2 Turnouts							
S2.1.2.01 Take up; turnout design C20 of Bullhead on timbers, strip into component parts and stack in designated areas for disposal	Nr	20.00	2,302.80	3,051.40	-	5,354.20	1,853.660
S2.1.3 Diamond crossings							
S2.1.3.01 Take up, strip into component parts and stack in designated area for subsequent disposal; Diamond crossing of Bullhead rails on timbers	Nr	10.00	1,151.40	1,525.70	-	2,677.10	926.830
S2.6	**Sundries**						
S2.6.1 Buffer stops							
S2.6.1.01 Take up buffer stops of steel rail and timber construction, weight approx 2.5 tonnes and stack in designated areas for disposal	Nr	3.00	345.42	457.71	-	803.13	278.049
S4	**Supplying**						
S4.1	**Bullhead rails**						
S4.1.4 Mass: 40 - 50 kg/m							
S4.1.4.01 Rails for jointed or welded track; new perfect quality BS95R section Bullhead rail in main lengths of 18.288m manufactured in accordance with BS9, both ends to be drilled for fish bolts	t	-	-	-	1,014.94	1,014.94	1,109.927
S4.1.4.02 Rails for jointed or welded track; new perfect quality BS95R section Bullhead Rail in main lengths of 18.288m manufactured in accordance with BS9, ends undrilled	t	-	-	-	1,014.94	1,014.94	1,109.927
S4.2	**Flat bottom rails**						
S4.2.5 Mass: exceeding 50kg/m							
S4.2.5.01 Rails for jointed or welded track; new perfect quality BS113 "A" section Flat bottom rail in main lengths of 18.288m manufactured in accordance with BS11, both ends drilled for fish bolts	t	-	-	-	787.87	787.87	1,109.927

S4 Supplying continued...

	Unit	Labour Hours	Labour Net £	Plant Net £	Materials Net £	Unit Net £	CO₂e Kg	

S4.2 Flat bottom rails

S4.2.5	Mass: exceeding 50kg/m							
S4.2.5.02	Rails for jointed or welded track; new perfect quality BS113 "A" section Flat bottom rail in main lengths of 18.288m manufactured in accordance with BS11, ends undrilled	t	-	-	-	787.87	787.87	1,109.927
S4.2.5.03	Guard rails: new perfect quality BS113 "A" section Flat bottom rail in main lengths of 18.288m manufactured in accordance with BS11, both ends drilled for fish bolts and flange planed to allow 50mm FWC	t	-	-	-	787.87	787.87	1,109.927

S4.6 Twist rails

S4.6.5	Mass: exceeding 50kg/m							
S4.6.5.01	Twist rails for jointed track adjacent turnouts: new perfect quality BS113 "A" section Flat bottom rail in main lengths of 9.144m manufactured in accordance with BS11, both ends drilled for fish bolts (Mass exceeding 50kg/m)	t	-	-	-	787.87	787.87	1,109.927

S4.7 Sleepers

S4.7.1	Timber								
S4.7.1.01	Softwood timber sleepers, new French Maritime Pine, 250 x 125 x 2600mm long	Nr	-	-	-	25.86	25.86	6.734	*-42.629*
S4.7.1.02	Hardwood timber sleepers, new Western Australian Jarrah, 250 x 130 x 2600mm long	Nr	-	-	-	39.51	39.51	23.109	*-56.693*
S4.7.2	Concrete								
S4.7.2.01	Concrete sleepers, type "F27" including malleable iron cast in shoulders	Nr	-	-	-	63.56	63.56	0.617	

S4.8 Fittings

S4.8.1	Chairs							
S4.8.1.01	Cast iron chairs, new "S1" pattern with 3nr. 25mm dia x 160mm long galvanised chairscrews, 3 nr. plastic ferrules and new spring steel key	Nr	-	-	-	30.93	30.93	4.134
S4.8.1.02	Cast iron chairs, new "CC" pattern 50mm FWC with 4nr. 25mm dia x 160mm long galvanised chairscrews, 4 nr. plastic ferrules, new spring steel and new oak keys	Nr	-	-	-	56.70	56.70	6.303

S4 Supplying continued...

	Unit	Labour Hours	Labour Net £	Plant Net £	Materials Net £	Unit Net £	CO_2e Kg
S4.8 **Fittings**							
S4.8.2 Baseplates							
S4.8.2.01 Cast iron baseplates, new "Pan 6" pattern with 5mm thick resilient rail pad, 3nr. 25mm dia x 160mm long galvanised chairscrews, 3 nr. plastic ferrules, 2 nr. "Pandrol" rail clips and nylon insulators	Nr	-	-	-	41.24	41.24	5.357
S4.8.2.02 Cast iron baseplates, new "VN" pattern with 5mm thick resilient rail pad, 3nr. 25mm dia x 160mm long galvanised chairscrews, 3 nr. plastic ferrules, 2nr. "Pandrol" rail clips and nylon insulators	Nr	-	-	-	43.30	43.30	5.987
S4.8.2.03 Cast iron baseplates, new "C" pattern with 50mm thick resilient rail pad, 4nr. 25mm dia x 160mm long galvanised chairscrews, 4 nr. plastic ferrules, 2nr. "Pandrol" rail clips and nylon insulators	Nr	-	-	-	58.28	58.28	7.413
S4.8.3 Pandrol rail fastenings							
S4.8.3.01 New "Pandrol" rail clips and nylon insulator for F27 concrete sleepers	Nr	-	-	-	45.41	45.41	8.862
S4.8.4 Plain fishplates							
S4.8.4.01 Set comprising 2nr. steel four hole skirted pattern fishplates for BS95R section rail and 4nr. standard pattern fishbolts with nuts and washers	Nr	-	-	-	47.53	47.53	4.717
S4.8.4.02 Set comprising 2nr. steel four hole shallow section fishplates for BS113 "A" section rail and 4nr. standard pattern fishbolts with nuts and washers	Nr	-	-	-	58.09	58.09	5.550
S4.8.4.03 Set comprising 2nr. steel four hole joggled pattern fishplates for jointing BS95R and BS113 "A" section rails and 4nr. standard pattern fishbolts with nuts and washers	Nr	-	-	-	121.45	121.45	9.989
S4.8.5 Insulated fishplates							
S4.8.5.01 Set comprising 2nr. four hole steel billet type insulated fishplates for jointing BS95R section rail with 4nr. 25mm dia. high tensile steel bolts with nuts and washers and 1nr. end post	Nr	-	-	-	163.70	163.70	13.874
S4.8.5.02 Set comprising 2nr. four hole steel billet type insulated fishplates for jointing BS113 "A" section rail with 4nr. 25mm dia. high tensile steel bolts with nuts and washers and 1nr. end post	Nr	-	-	-	147.86	147.86	11.099

S4   Supplying continued...

	Unit	Labour Hours	Labour Net £	Plant Net £	Materials Net £	Unit Net £	CO_2e Kg
S4.8	**Fittings**						
S4.8.5	Insulated fishplates						
S4.8.5.03	Set comprising 2nr. four hole steel billet type insulated fishplates for jointing BS95R and BS113 "A" section rails and 4nr. 25mm dia. high tensile steel bolts with nuts and washers and 1nr. end post	Nr	-	-	-	147.86	147.86	11.099

S5   Supplying

	Unit	Labour Hours	Labour Net £	Plant Net £	Materials Net £	Unit Net £	CO_2e Kg
S5.1	**Turnouts and crossings**						
S5.1.3	Turnouts						
S5.1.3.01	Plain Line Materials; Track Assemblies; Switches and Crossings; standard assemblies complete with all associated closure and check rails and all fittings; BR design, B8, manufactured from new perfect quality BS95R Bullhead rail, set on new 300 x 150mm section Douglas Fir softwood timbers	Nr	-	-	-	25,017.74	25,017.74	1,831.380
S5.1.3.02	Plain Line Materials; Track Assemblies; Switches and Crossings; standard assemblies complete with all associated closure and check rails and all fittings; BR design, C10, manufactured from new perfect quality BS95R Bullhead rail, set on new 300 x 150mm section Douglas Fir softwood timbers	Nr	-	-	-	25,117.02	25,117.02	1,775.884
S5.1.3.03	Plain Line Materials; Track Assemblies; Switches and Crossings; standard assemblies complete with all associated closure and check rails and all fittings; BR vertical design, Bv8, manufactured from new perfect quality BS113 'A' flat bottom rail, set on 300 x 130mm section Jarrah hardwood timbers	Nr	-	-	-	34,746.86	34,746.86	2,219.855
S5.1.3.04	Plain Line Materials; Track Assemblies; Switches and Crossings; standard assemblies complete with all associated closure and check rails and all fittings; BR vertical design, Cv9.25, manufactured from new perfect quality BS113 'A' flat bottom rail, set on 300 x 130mm section Jarrah hardwood timbers	Nr	-	-	-	35,243.24	35,243.24	2,774.819

S5   Supplying continued...

	Unit	Labour Hours	Labour Net £	Plant Net £	Materials Net £	Unit Net £	CO_2e Kg

S5.1   Turnouts and crossings

S5.1.4   Diamond crossings

S5.1.4.01   Plain Line Materials; Track Assemblies; Switches and Crossings; standard assemblies complete with all associated closures and check rails and all fittings; BR design, angle 1 in 4, manufactured from new perfect quality BS95R Bullhead rail, set on new 300 x 150mm section Douglas Fir softwood timbers; Double slip, BR design, angle 1 in 8, manufactured from new perfect quality BS95R Bullhead rail, set on new 300 x 150mm section Douglas Fir softwood timbers

	Nr	-	-	-	110,197.19	110,197.19	9,156.901

S5.1.4.02   Plain Line Materials; Track Assemblies; Switches and Crossings; standard assemblies complete with all associated closure and check rails and all fittings; BR Vertical design, angle 1 in 4 manufactured from new perfect quality BS113 "A" flat bottom rail, set on 300 x 130mm section Jarrah hardwood timbers

	Nr	-	-	-	56,587.74	56,587.74	6,437.579

S5.2   Sundries

S5.2.1   Buffer stops

S5.2.1.01   Buffer Stop - Rawie Friction Type (1.6t)

	Nr	-	-	-	15,591.49	15,591.49	1,859.972

S5.2.2   Retraders

S5.2.2.01   Retraders

	Nr	-	-	-	799.97	799.97	147.249

S5.2.3   Wheel stops

S5.2.3.01   Wheel Stop (pair)

	Nr	-	-	-	1,528.95	1,528.95	176.367

S5.2.4   Lubricators

S5.2.4.01   Rail Lubricator

	Nr	-	-	-	2,804.91	2,804.91	145.021

S7   Laying

S7.1   Bullhead rails: slab track

S7.1.1   Plain track

S7.1.1.01   Plain fishplated track, type BS95R Bullhead rail on softwood timber sleepers

	m	0.50	57.57	34.82	-	92.39	26.780

S7.1.1.02   Plain welded track, type BS95R Bullhead rail on softwood timber sleepers

	m	0.75	86.36	52.23	-	138.59	40.169

S7.1.1.03   Guard rail, type BS95R Bullhead rail fixed to BS95R Bullhead track

	m	0.33	38.00	22.98	-	60.98	17.674

S7 Laying continued...

	Unit	Labour Hours	Labour Net £	Plant Net £	Materials Net £	Unit Net £	CO_2e Kg
S7.1 **Bullhead rails: slab track**							
S7.1.2 Form curve in plain track radius not exceeding 300m							
S7.1.2.01 Extra over straight track for curved plain fishplated track, type BS95R Bullhead rail on softwood timber sleepers	m	0.25	28.79	17.41	-	46.20	13.390
S7.1.2.02 Extra over straight track for curved plain welded track, type BS95R Bullhead rail on softwood timber sleepers	m	0.25	28.79	17.41	-	46.20	13.390
S7.1.4 Turnouts							
S7.1.4.01 Standard turnout, B8, type BS95R rail on softwood sleepers	Nr	20.00	2,302.80	1,392.80	-	3,695.60	1,071.180
S7.1.4.02 Standard turnout, C10, type BS95R rail on softwood sleepers	Nr	24.00	2,763.36	1,671.36	-	4,434.72	1,285.416
S7.1.5 Diamond crossings							
S7.1.5.01 Standard diamond crossing, type BS95R rail on softwood timbers	Nr	24.00	2,763.36	1,671.36	-	4,434.72	1,285.416
S7.1.5.02 Standard double slip, type BS95R rail on softwood timbers	Nr	24.00	2,763.36	1,671.36	-	4,434.72	1,285.416
S7.1.6 Welded joints							
S7.1.6.01 Thermit welds; by the S kV process BS95R Bullhead rail	Nr	-	-	-	-	450.00	-
S7.2 **Flat bottom rails: slab track**							
S7.2.1 Plain track							
S7.2.1.01 Plain fishplated track, Flat bottom rail on hardwood timber sleepers	m	0.50	57.57	34.82	-	92.39	26.780
S7.2.1.02 Plain fishplated track, Flat bottom rail on concrete sleepers	m	0.50	57.57	34.82	-	92.39	26.780
S7.2.1.03 Plain welded track, Flat bottom rail on hardwood timber sleepers	m	0.75	86.36	52.23	-	138.59	40.169
S7.2.1.04 Plain welded track, Flat bottom rail on concrete sleepers	m	0.75	86.36	52.23	-	138.59	40.169
S7.2.1.05 Guard rail, Flat bottom rail fixed to flat bottom track	m	0.33	38.00	22.98	-	60.98	17.674
S7.2.2 Form curve in plain track radius not exceeding 300m							
S7.2.2.01 Extra over straight track for curved plain fishplated track, flat bottom rail on hardwood timber or concrete sleepers	m	0.25	28.79	17.41	-	46.20	13.390
S7.2.2.02 Extra over straight track for curved plain welded track, flat bottom rail on hardwood timber or concrete sleepers	m	0.25	28.79	17.41	-	46.20	13.390
S7.2.4 Turnouts							
S7.2.4.01 Standard turnout, Bv8, rail on hardwood timbers	Nr	24.00	2,763.36	1,671.36	-	4,434.72	1,285.416
S7.2.4.02 Standard turnout, Cv9.25, rail on hardwood timbers	Nr	32.00	3,684.48	2,228.48	-	5,912.96	1,713.888
S7.2.5 Diamond crossings							
S7.2.5.01 Standard diamond crossing, angle 1 in 4, rail on hardwood timbers	Nr	32.00	3,684.48	2,228.48	-	5,912.96	1,713.888

S7 Laying continued...

	Unit	Labour Hours	Labour Net £	Plant Net £	Materials Net £	Unit Net £	CO$_2$e Kg
S7.2	**Flat bottom rails: slab track**						
S7.2.6 Welded joints							
S7.2.6.01 Thermit welds; by the SkV process flat bottom rail	Nr	-	-	-	-	250.00	-
S7.5	**Sundries**						
S7.5.1 Buffer stops							
S7.5.1.01 Buffer Stop - Rawie Friction Type (1.6t)	Nr	8.00	1,544.88	451.60	-	1,996.48	171.320
S7.5.2 Retraders							
S7.5.2.01 Retraders	Nr	1.00	193.11	28.23	-	221.34	10.708
S7.5.3 Wheel stops							
S7.5.3.01 Wheel Stop (pair)	Nr	2.00	386.22	56.45	-	442.67	21.415
S7.5.4 Lubricators							
S7.5.4.01 Rail Lubricator	Nr	2.50	482.78	112.90	-	595.68	42.830

CLASS T:
TUNNELS

Calculations used throughout Class T - Tunnels

Labour

			Qty		Rate		Total
L A0120ICE	**General Earthworks Labour Gang**						
	L A9016ICE	Ganger	1	x	17.64	=	£17.64
	L A9002ICE	Labourer (General Operative)	1	x	13.02	=	£13.02
	L A9015ICE	Banksman	1	x	14.03	=	£14.03
		Total hourly cost of gang				=	**£44.69**
L A0122ICE	**Drill and Blast Labour Gang**						
	L A9019ICE	Driller	1	x	14.86	=	£14.86
	L A9020ICE	Shot firer	0.5	x	14.86	=	£7.43
		Total hourly cost of gang				=	**£22.29**
L A0130ICE	**Formwork (Make, Fix and Strike) Labour Gang**						
	L A9000ICE	Craftsman WRA	1	x	17.33	=	£17.33
	L A9003ICE	Labourer (Skill Rate 3)	1	x	14.86	=	£14.86
	L A9001ICE	Carpenter (charge hand)	1	x	18.63	=	£18.63
		Total hourly cost of gang				=	**£50.82**
L A0155ICE	**Placing of concrete Labour Gang**						
	L A9003ICE	Labourer (Skill Rate 3)	1	x	14.86	=	£14.86
	L A9002ICE	Labourer (General Operative)	4	x	13.02	=	£52.06
	L A9015ICE	Banksman	0.25	x	14.03	=	£3.51
	L A9016ICE	Ganger	1	x	17.64	=	£17.64
	L A9000ICE	Craftsman WRA	0.5	x	17.33	=	£8.66
		Total hourly cost of gang				=	**£96.73**
L C0010ICE	**Grout Hole Labour Gang**						
	L A9016ICE	Ganger	1	x	17.64	=	£17.64
	L A9002ICE	Labourer (General Operative)	1	x	13.02	=	£13.02
	L A9000ICE	Craftsman WRA	1	x	17.33	=	£17.33
		Total hourly cost of gang				=	**£47.99**
L C0011ICE	**Grout Hole Labour Gang (Mobilise / Demobilise / Move)**						
	L A9016ICE	Ganger	20	x	17.64	=	£352.80
	L A9002ICE	Labourer (General Operative)	30	x	13.02	=	£390.46
		Total hourly cost of gang				=	**£743.26**
L M0003ICE	**Structural Steel Fabrication Labour Gang**						
	L A9003ICE	Labourer (Skill Rate 3)	1	x	14.86	=	£14.86
	L A9021ICE	Fitters and Welders	2	x	17.33	=	£34.66
		Total hourly cost of gang				=	**£49.52**
L T0001ICE	**Tunnelling Labour Gang**						
	L A9003ICE	Labourer (Skill Rate 3)	2	x	14.86	=	£29.72
	L A9016ICE	Ganger	1	x	17.64	=	£17.64
	L A9015ICE	Banksman	1	x	14.03	=	£14.03
	L A9035ICE	Miners	4	x	41.27	=	£165.08
	L A9021ICE	Fitters and Welders	1	x	17.33	=	£17.33
		Total hourly cost of gang				=	**£243.80**

Plant

P A1035ICE   **Formwork (Fix and Strike) Plant Gang**

P A0703ICE	Kango Type Tool - Electric Power Woodauger	1	x	0.61	=	£0.61
P A0704ICE	Kango Type Tool - Electric Nut Runner	1	x	0.52	=	£0.52
P A0507ICE	Cranes Crawler - NCK 305B - 20t	0.25	x	48.55	=	£12.14

Total hourly cost of gang   =   **£13.27**

P A1040ICE   **Formwork (Make) Plant Gang**

P A0891ICE	Saw Bench - 24 inch Diesel / Electric	1	x	2.03	=	£2.03
P A0703ICE	Kango Type Tool - Electric Power Woodauger	1	x	0.61	=	£0.61
P A0704ICE	Kango Type Tool - Electric Nut Runner	1	x	0.52	=	£0.52

Total hourly cost of gang   =   **£3.16**

P A1128ICE   **Drill and Blast Plant Gang**

P A0420ICE	Waggon Drill with Steel & bits	1	x	4.90	=	£4.90
P A0457ICE	Compressor - 375 cfm	1	x	16.00	=	£16.00
P A0421ICE	Exploder with Circuit Tester	1	x	2.64	=	£2.64

Total hourly cost of gang   =   **£23.54**

P A1155ICE   **Placing of concrete Plant Gang**

P A0405ICE	Air Vibrating Poker up to 75mm	3	x	1.33	=	£3.99
P A0476ICE	Concrete Skip	2	x	0.75	=	£1.50
P A0408ICE	Scabbler - Floor - 3 Headed	1	x	2.64	=	£2.64
P A0409ICE	Scabbler - Floor - 5 Headed	1	x	3.77	=	£3.77
P A0526ICE	Excavators Cable - NCK 305A 0.67m3	0.25	x	44.72	=	£11.18
P A0457ICE	Compressor - 375 cfm	1	x	16.00	=	£16.00

Total hourly cost of gang   =   **£39.08**

P C0012ICE   **Grout Hole Plant Gang (Grouting rock or other artificial hard material)**

P A1060ICE	Anchor Drilling Rig (Rock)	1	x	85.26	=	£85.26
P A0458ICE	Compressor - 480 cfm	1	x	20.17	=	£20.17
P A0442ICE	Bowsers (250 gallon water / fuel)	1	x	0.94	=	£0.94
P A1069ICE	Grouting Rig (Anchorage Items)	1	x	10.53	=	£10.53

Total hourly cost of gang   =   **£116.90**

P C0013ICE   **Grout Hole Plant Gang (Mobilise / Demobilise / Move)**

P A1061ICE	Anchor Drilling Rig (Soil)	12	x	78.91	=	£946.92
P A0458ICE	Compressor - 480 cfm	12	x	20.17	=	£242.04
P A0442ICE	Bowsers (250 gallon water / fuel)	12	x	0.94	=	£11.28
P A0737ICE	Concrete Mixer - Schwing BP1000R Concrete Pumps	12	x	46.33	=	£555.96
P A1071ICE	Pile Jetting Water Pump	12	x	522.14	=	£6,265.68
P A1044ICE	Articulated Lorry	10	x	68.51	=	£685.10
P A1045ICE	Low Loader Trailer	10	x	13.96	=	£139.60

Total hourly cost of gang   =   **£8,846.58**

P C0014ICE   **Grout Hole Plant Gang (Grouting Soils)**

P A1061ICE	Anchor Drilling Rig (Soil)	1	x	78.91	=	£78.91
P A0458ICE	Compressor - 480 cfm	1	x	20.17	=	£20.17
P A0442ICE	Bowsers (250 gallon water / fuel)	1	x	0.94	=	£0.94
P A1069ICE	Grouting Rig (Anchorage Items)	1	x	10.53	=	£10.53

Total hourly cost of gang   =   **£110.55**

P M0005ICE **Structural Steel Erection Plant Gang**

P A0991ICE	Welding Set - 300 amp Diesel Electric Start Sil	1	x	4.48	=	£4.48
P A1031ICE	Cutting and Burning Gear	1	x	3.77	=	£3.77
P A0564ICE	Fork Lift Truck - 2.5t 4WD	1	x	28.68	=	£28.68
P A0481ICE	Cranes Transit - 25t	1	x	60.22	=	£60.22

Total hourly cost of gang = **£97.15**

P M0006ICE **Structural Steel Fabrication Plant Gang**

P A0991ICE	Welding Set - 300 amp Diesel Electric Start Sil	1	x	4.48	=	£4.48
P A1031ICE	Cutting and Burning Gear	1	x	3.77	=	£3.77
P A0564ICE	Fork Lift Truck - 2.5t 4WD	1	x	28.68	=	£28.68
P A0481ICE	Cranes Transit - 25t	1	x	60.22	=	£60.22

Total hourly cost of gang = **£97.15**

Tunnels

Note(s): Tunneling works are highly specialized activities. The prices can vary considerably according to conditions prevailing on particular contracts, e.g. ground conditions, water level, hardness of rocks, lengths of tunnels, size of contract etc. The following prices are therefore provided as guide prices only for use in preliminary approximate estimating. In accordance with common practice in tunneling all diameters referred to are internal diameter of segments. Rates quoted are average rates and are for machine working, with the use of tunnel shields where necessary. The following works are not priced separately, their costs are included in the rates:

1) Temporary supports to excavations.
2) Keeping excavations free from water
3) Working around and supporting services.
4) Any shafts required
5) Use of tunnel shield where necessary
6) Interruption of works required to carry out survey and inspection checks
7) Construction and removal of temporary shaft rings
8) Breaking through linings at shaft/tunnel or shaft/heading intersections. The grouting up of tunnels and shafts is included in the items excavated surfaces. The rates quoted made allowance to cover the disposal of surplus excavated material off site. This figure excludes landfill tax which should be added at the relevant level. Compressed air working will increase the rates quoted according to the project.
9) Cabling, lighting and electric
10) Rail track and ventilation system
11) Spoil and equipment consumable transport

NOTES:

Excavation of tunnels in rock based on Tunnel Boring Machine (TBM) methods
Excavation of tunnels in other stated materials based on Sprayed Concrete Lining (SCL) methods
Excavation of shafts in rock based on Drill and Blast (D & B) methods
Excavation of shafts in other stated materials based on Sprayed Concrete Lining (SCL) methods

T1 Excavation

		Unit	Labour Hours	Labour Net £	Plant Net £	Materials Net £	Unit Net £	CO_2e Kg
T1.1	**Tunnels in rock**							
T1.1.1	Stated diameter: not exceeding 2m							
T1.1.1.01	Diameter: 1.2m	m³	0.90	293.58	338.83	8.50	640.91	295.597
T1.1.1.02	Diameter: 1.5m	m³	0.80	270.04	299.93	8.50	578.47	253.644
T1.1.1.03	Diameter: 1.8m	m³	0.70	246.50	261.03	8.50	516.03	211.691
T1.1.2	Stated diameter: 2 - 3m							
T1.1.2.01	Diameter: 3m	m³	0.62	223.58	230.48	8.50	462.56	182.228
T1.1.3	Stated diameter: 3 - 4m							
T1.1.3.01	Diameter: 4m	m³	0.55	200.97	204.09	8.50	413.56	159.009
T1.1.4	Stated diameter: 4 - 5m							
T1.1.4.01	Diameter: 5m	m³	0.50	178.99	186.05	8.50	373.54	148.280

T1 Excavation continued...

	Unit	Labour Hours	Labour Net £	Plant Net £	Materials Net £	Unit Net £	CO_2e Kg
T1.2 **Tunnels in other stated material**							
T1.2.1 Stated diameter: not exceeding 2m							
T1.2.1.01 In generally soft material; Diameter: 1.2m	m³	0.80	175.86	144.34	10.00	330.20	85.832
T1.2.1.02 In generally soft material; Diameter: 1.5m	m³	0.76	167.07	137.12	10.00	314.19	81.540
T1.2.1.03 In generally soft material; Diameter: 1.8m	m³	0.70	153.88	126.30	10.00	290.18	75.103
T1.2.2 Stated diameter: 2 - 3m							
T1.2.2.01 In generally soft material; Diameter: 3m	m³	0.60	131.89	108.26	10.00	250.15	64.374
T1.2.3 Stated diameter: 3 - 4m							
T1.2.3.01 In generally soft material; Diameter: 4m	m³	0.56	123.10	101.04	10.00	234.14	60.082
T1.2.4 Stated diameter: 4 - 5m							
T1.2.4.01 In generally soft material; Diameter: 5m	m³	0.50	109.91	90.21	10.00	210.12	53.645
T1.3 **Shafts in rock**							
T1.3.2 Stated diameter: 2 - 3m							
T1.3.2.01 Diameter: 3m	m³	2.00	66.97	178.55	-	245.52	125.334
T1.3.3 Stated diameter: 3 - 4m							
T1.3.3.01 Diameter: 3.3m	m³	1.80	60.27	160.70	-	220.97	112.801
T1.3.3.02 Diameter: 3.6m	m³	1.60	53.58	142.84	-	196.42	100.267
T1.3.4 Stated diameter: 4 - 5m							
T1.3.4.01 Diameter: 4.5m	m³	1.40	46.88	124.98	-	171.86	87.734
T1.4 **Shafts in other stated material**							
T1.4.2 Stated diameter: 2 - 3m							
T1.4.2.01 In generally soft material; Diameter: 3m	m³	0.70	31.28	108.51	-	139.79	72.479
T1.4.3 Stated diameter: 3 - 4m							
T1.4.3.01 In generally soft material; Diameter: 3.3m	m³	0.60	26.81	93.01	-	119.82	62.125

T1 Excavation continued...

	Unit	Labour Hours	Labour Net £	Plant Net £	Materials Net £	Unit Net £	CO_2e Kg
T1.4	**Shafts in other stated material**						
T1.4.3 Stated diameter: 3 - 4m							
T1.4.3.02 In generally soft material; Diameter: 3.6m	m³	0.60	26.81	93.01	-	119.82	62.125
T1.4.4 Stated diameter: 4 - 5m							
T1.4.4.01 In generally soft material; Diameter: 4.5m	m³	0.50	22.34	77.50	-	99.84	51.771
T1.5	**Other cavities in rock**						
T1.5.1 Stated diameter: not exceeding 2m							
T1.5.1.01 Diameter: 1.2m	m³	4.50	150.68	283.48	-	434.16	190.090
T1.5.2 Stated diameter: 2 - 3m							
T1.5.2.01 Diameter: 3m	m³	4.20	140.64	260.84	-	401.47	175.180
T1.5.3 Stated diameter: 3 - 4m							
T1.5.3.01 Diameter: 4m	m³	3.80	127.24	234.39	-	361.63	157.550
T1.5.4 Stated diameter: 4 - 5m							
T1.5.4.01 Diameter: 5m	m³	3.60	120.55	215.55	-	336.10	145.380
T1.6	**Other cavities in other stated material**						
T1.6.1 Stated diameter: not exceeding 2m							
T1.6.1.01 In generally soft material; Diameter: 1.2m	m³	1.50	67.02	191.11	-	258.13	116.364
T1.6.2 Stated diameter: 2 - 3m							
T1.6.2.01 Diameter: 3m	m³	1.00	44.68	164.85	-	209.53	99.901
T1.6.3 Stated diameter: 3 - 4m							
T1.6.3.01 Diameter: 4m	m³	0.85	37.98	156.97	-	194.95	94.962
T1.6.4 Stated diameter: 4 - 5m							
T1.6.4.01 In generally soft material; Diameter: 5m	m³	1.00	44.68	138.84	-	183.52	85.285
T1.7	**Excavated surfaces in rock**						
T1.7.0 Generally							
T1.7.0.01 Excavation of surfaces in rock	m²	0.25	11.34	12.56	-	23.90	17.450
T1.8	**Excavated surfaces in other stated material**						
T1.8.0 Generally							
T1.8.0.01 Excavation of surfaces in soft material	m²	0.25	11.34	12.24	-	23.58	17.450

T2 In Situ Linings To Tunnels

	Unit	Labour Hours	Labour Net £	Plant Net £	Materials Net £	Unit Net £	CO_2e Kg
T2.1	**Sprayed concrete primary**						
T2.1.1 Stated diameter: not exceeding 2m							
T2.1.1.01 Concrete C32/40; Diameter 2m	m³	1.00	63.99	85.55	172.18	321.72	351.285
T2.1.2 Stated diameter: 2 - 3m							
T2.1.2.01 Concrete C32/40; Diameter 3m	m³	0.90	57.59	79.32	172.18	309.09	347.411

T2 In Situ Linings To Tunnels continued...

		Unit	Labour Hours	Labour Net £	Plant Net £	Materials Net £	Unit Net £	CO_2e Kg	
T2.1	**Sprayed concrete primary**								
T2.1.3	Stated diameter: 3 - 4m								
T2.1.3.01	Concrete C32/40; Diameter 4 m	m³	0.80	51.19	73.08	172.18	296.45	343.537	
T2.1.4	Stated diameter: 4 - 5m								
T2.1.4.01	Concrete C32/40; Diameter 5 m	m³	0.70	44.79	66.84	172.18	283.81	339.662	
T2.2	**Sprayed concrete secondary**								
T2.2.1	Stated diameter: not exceeding 2m								
T2.2.1.01	Concrete C32/40; Diameter 2m	m³	1.30	83.19	108.91	172.18	364.28	363.416	
T2.2.2	Stated diameter: 2 - 3m								
T2.2.2.01	Concrete C32/40; Diameter 3m	m³	1.10	70.39	96.43	172.18	339.00	355.668	
T2.2.3	Stated diameter: 3 - 4m								
T2.2.3.01	Concrete C32/40; Diameter 4 m	m³	1.00	63.99	87.87	172.18	324.04	351.539	
T2.2.4	Stated diameter: 4 - 5m								
T2.2.4.01	Concrete C32/40; Diameter 5 m	m³	0.96	61.43	83.99	172.18	317.60	349.837	
T2.4	**Cast concrete secondary**								
T2.4.1	Stated diameter: not exceeding 2m								
T2.4.1.01	Concrete C25/30, 20mm aggregate; Diameter 2m	m³	1.10	106.40	93.97	92.75	293.12	313.375	
T2.5	**Formwork to stated finish**								
T2.5.1	Stated diameter: not exceeding 2m								
T2.5.1.01	Rough finish; Diameter: 2m	m²	0.52	26.27	17.36	6.94	50.57	32.793	-13.478
T2.5.1.02	Wrot finish; Diameter: 2m	m²	0.58	29.63	19.61	10.41	59.65	35.828	-14.677

T3 In Situ Linings To Shafts

		Unit	Labour Hours	Labour Net £	Plant Net £	Materials Net £	Unit Net £	CO_2e Kg	
T3.4	**Cast concrete secondary**								
T3.4.1	Stated diameter: not exceeding 2m								
T3.4.1.01	Concrete C25/30, 20mm aggregate; Diameter 2m	m³	0.90	87.06	76.89	92.75	256.70	306.725	
T3.5	**Formwork to stated finish**								
T3.5.1	Stated diameter: not exceeding 2m								
T3.5.1.01	Rough finish; Diameter: 2m	m²	0.75	38.12	25.21	6.94	70.27	34.097	-13.478
T3.5.1.02	Wrot finish; Diameter: 2m	m²	0.83	42.33	28.02	10.41	80.76	37.225	-14.677

T4 In Situ Linings To Other Cavities

		Unit	Labour Hours	Labour Net £	Plant Net £	Materials Net £	Unit Net £	CO_2e Kg	
T4.4	**Cast concrete secondary**								
T4.4.1	Stated diameter: not exceeding 2m								
T4.4.1.01	Concrete C25/30, 20mm aggregate; Diameter 2m	m³	1.10	106.40	93.97	92.75	293.12	313.375	
T4.5	**Formwork to stated finish**								
T4.5.1	Stated diameter: not exceeding 2m								
T4.5.1.01	Rough finish; Diameter: 2m	m²	0.75	38.12	25.21	6.94	70.27	34.097	-13.478
T4.5.1.02	Wrot finish; Diameter: 2m	m²	0.83	42.33	28.02	10.41	80.76	37.225	-14.677

T5   Preformed Segmental Lining To Tunnels

Note(s):   NOTE: Segment prices may vary considerably according to size of contract, location and other contracts in operation. The following prices are provided as guide prices only for use in preliminary approximate estimating. Specialist advice should always be obtained for more accurate cost data.

		Unit	Labour Hours	Labour Net £	Plant Net £	Materials Net £	Unit Net £	CO_2e Kg
T5.1	**Precast concrete bolted rings**							
T5.1.1	Stated diameter: not exceeding 2m							
T5.1.1.01	Standard Bolted Segments; Diameter: 1.52m	Nr	1.00	439.65	34.37	208.52	682.54	130.963
T5.1.1.02	Standard Bolted Segments; Diameter: 1.83m	Nr	1.00	439.65	34.37	251.07	725.09	158.114
T5.7	**Lining ancillaries**							
T5.7.1	Parallel circumferential packing							
T5.7.1.01	Generally	Nr	0.01	3.52	-	4.75	8.27	9.579
T5.7.2	Tapered circumferential packing							
T5.7.2.01	Generally	Nr	0.02	7.47	-	7.91	15.38	15.964
T5.7.4	Caulking of stated material							
T5.7.4.01	Caulking; PC4AF	m	0.02	7.47	-	5.00	12.47	0.256

T6   Preformed Segmental Lining To Shafts

		Unit	Labour Hours	Labour Net £	Plant Net £	Materials Net £	Unit Net £	CO_2e Kg
T6.1	**Precast concrete bolted rings**							
T6.1.3	Stated diameter: 3 - 4m							
T6.1.3.01	Standard Bolted Segments; Diameter: 3.05m	Nr	0.50	219.82	17.18	636.58	873.58	271.832
T6.1.3.02	Standard Bolted Segments; Diameter: 3.35m	Nr	0.75	329.74	25.78	669.78	1,025.30	307.288
T6.1.3.03	Standard Bolted Segments; Diameter: 3.66m	Nr	0.75	329.74	25.78	702.98	1,058.50	329.009
T6.1.4	Stated diameter: 4 - 5m							
T6.1.4.01	Standard Bolted Segments; Diameter: 4.57m	Nr	1.00	439.65	34.37	1,202.54	1,676.56	527.374

T7   Preformed Segmental Lining To Other Cavities

		Unit	Labour Hours	Labour Net £	Plant Net £	Materials Net £	Unit Net £	CO_2e Kg
T7.1	**Precast concrete bolted rings**							
T7.1.3	Stated diameter: 3 - 4m							
T7.1.3.01	Standard Bolted Segments; Diameter: 3.05m	Nr	0.75	329.74	25.78	636.58	992.10	272.896
T7.1.3.02	Standard Bolted Segments; Diameter: 3.35m	Nr	1.00	439.65	34.37	669.78	1,143.80	308.352
T7.1.3.03	Standard Bolted Segments; Diameter: 3.66m	Nr	1.25	549.56	42.96	702.98	1,295.50	331.137
T7.1.4	Stated diameter: 4 - 5m							
T7.1.4.01	Standard Bolted Segments; Diameter: 4.57m	Nr	1.50	659.47	51.55	1,202.54	1,913.56	529.502

T8 Support And Stabilization

		Unit	Labour Hours	Labour Net £	Plant Net £	Materials Net £	Unit Net £	CO_2e Kg
T8.1	**Rock bolts**							
T8.1.1	Mechanical							
T8.1.1.01	Generally	m	0.05	21.98	0.31	18.10	40.39	1.915
T8.1.2	Mechanical grouted							
T8.1.2.01	Generally	m	0.05	21.98	0.31	27.50	49.79	2.069
T8.1.3	Pre-grouted impacted							
T8.1.3.01	Generally	m	0.05	21.98	0.31	31.67	53.96	2.210
T8.1.4	Chemical and anchor							
T8.1.4.01	Generally	m	0.05	21.98	0.31	29.61	51.90	15.284
T8.1.5	Chemical grouted							
T8.1.5.01	Generally	m	0.05	21.98	0.31	31.39	53.68	3.010
T8.1.6	Chemically filled							
T8.1.6.01	Generally	m	0.05	21.98	0.31	41.09	63.38	2.975
T8.2	**Internal support**							
T8.2.1	Steel arches: supply							
T8.2.1.01	Generally	t	6.00	297.12	599.39	843.12	1,739.63	1,455.288
T8.2.2	Steel arches: erection							
T8.2.2.01	Generally	t	1.50	659.47	145.73	-	805.20	84.453
T8.2.3	Timber supports: supply							
T8.2.3.01	Generally	m³	-	-	-	341.89	341.89	823.315 -591.075
T8.2.4	Timber supports: erection							
T8.2.4.01	Generally	m³	2.00	490.47	16.43	-	506.90	3.460
T8.2.5	Lagging							
T8.2.5.01	Generally	m²	0.05	21.98	-	10.67	32.65	21.520
T8.2.6	Sprayed concrete; 50mm thick							
T8.2.6.01	Generally	m²	0.10	43.97	3.07	2.02	49.06	16.342
T8.2.7	Mesh or link							
T8.2.7.01	Generally	m²	0.03	14.51	-	2.69	17.20	2.325
T8.3	**Pressure grouting**							
T8.3.1	Sets of drilling and grouting plant							
T8.3.1.01	Generally	Nr	0.22	24.46	200.25	861.22	1,085.93	192.996
T8.3.2	Face packers							
T8.3.2.01	Generally	Nr	0.83	73.63	-	1.34	74.97	3.766
T8.3.3	Deep packers of stated size							
T8.3.3.01	Generally	Nr	1.38	120.37	-	2.67	123.04	7.532
T8.3.4	Drilling and flushing to stated diameter							
T8.3.4.01	40mm diameter; not exceeding 20m deep	m	0.17	8.01	17.76	-	25.77	45.318
T8.3.5	Re-drilling and flushing							
T8.3.5.01	Generally	m	0.17	8.01	17.76	-	25.77	45.318
T8.3.6	Injection of grout materials of stated composition							
T8.3.6.01	Injection of cement grout	t	1.00	47.98	106.37	591.78	746.13	872.803

T8 Support And Stabilization continued...

		Unit	Labour Hours	Labour Net £	Plant Net £	Materials Net £	Unit Net £	CO_2e Kg
T8.4	**Forward probing**							
T8.4.0	Generally							
T8.4.0.01	Generally	m	0.20	9.60	21.27	–	30.87	54.273

CLASS U:
BRICKWORK, BLOCKWORK AND MASONRY

Calculations used throughout Class U - Brickwork, Blockwork and Masonry

Labour

			Qty		Rate		Total
L A0300ICE	**Brickwork Labour Gang**						
	L A9033ICE	Bricklayer (chargehand)	1	x	18.63	=	£18.63
	L A9034ICE	Bricklayer	4	x	17.33	=	£69.31
	L A9002ICE	Labourer (General Operative)	2	x	13.02	=	£26.03
		Total hourly cost of gang				=	**£113.97**
L A0301ICE	**Blockwork Labour Gang**						
	L A9034ICE	Bricklayer	4	x	17.33	=	£69.31
	L A9002ICE	Labourer (General Operative)	3	x	13.02	=	£39.05
		Total hourly cost of gang				=	**£108.36**

Brickwork, Blockwork and Masonry

Note(s): 1) The hire, erection and dismantling of access and working scaffolding has not been allowed for in the rates because it is usually included in the preliminaries as scaffolding is frequently used by more than one trade. However if the user wishes to include for scaffolding a sub-contract price for hire, erect and dismantle should be obtained as the cost varies with the shape and height of the structure as well as the length of time the scaffolding is required.
2) Rates for brickwork assume a height not in excess of 10 metres; heights in excess of this bear the following multiples of the labour element.

Height above ground	Labour hours
metres	multiple
0-10	1.00
10-15	1.10
15-20	1.25
20-25	1.50

3) Surface features are measured as part of the wall to which they belong, the volume having been enhanced by the appropriate amount. The additional work entailed measured as a labour only item and should be added to the labour content of the wall.
4) Labour rates have been based on the following outputs. The reader will be able to increase or decrease the labour element of the rate in accordance with his knowledge of the circumstances applying.

Type of brickwork	Bricks laid per hour
Common brickwork 102.5mm thick	45
Common brickwork 215mm thick	50
Common brickwork 317.5mm thick	60
Common brickwork 440mm thick	65
Common brickwork Mass	70
Brick facings pointed as	
Work proceeds - one face	40
both faces	35
Rake out joints and point	0.75 hours per m².

5) It has been assumed that materials have been delivered to the point of placing and that water is freely available.

U1   Common Brickwork

	Unit	Labour Hours	Labour Net £	Plant Net £	Materials Net £	Unit Net £	CO$_2$e Kg
U1.1 **Thickness: not exceeding: 150mm**							
U1.1.1 Vertical straight walls							
U1.1.1.01 PC£230 per 1000 in cement mortar (1:3); 102.5mm thick	m²	0.26	29.40	-	15.54	44.94	38.457
U1.1.2 Vertical curved walls							
U1.1.2.01 PC£230 per 1000 in cement mortar (1:3); 102.5mm thick; 5m mean radius	m²	0.45	51.29	-	15.54	66.83	38.457
U1.1.2.02 as above; 10m mean radius	m²	0.39	44.22	-	15.54	59.76	38.457
U1.2 **Thickness: 150 - 250mm**							
U1.2.1 Vertical straight walls							
U1.2.1.01 PC£230 per 1000 in cement mortar (1:3); 215mm thick	m²	0.46	52.88	-	30.72	83.60	75.887
U1.2.2 Vertical curved walls							
U1.2.2.01 PC£230 per 1000 in cement mortar (1:3); 215mm thick; 5m mean radius	m²	0.81	92.32	-	30.72	123.04	75.887
U1.2.2.02 as above; 10m mean radius	m²	0.70	79.78	-	30.72	110.50	75.887
U1.3 **Thickness: 250 - 500mm**							
U1.3.1 Vertical straight walls							
U1.3.1.01 PC£230 per 1000 in cement mortar (1:3); 327.5mm thick	m²	0.59	67.24	-	44.29	111.53	106.468
U1.3.1.02 as above; 440mm thick	m²	0.79	90.04	-	61.90	151.94	152.811
U1.3.2 Vertical curved walls							
U1.3.2.01 PC£230 per 1000 in cement mortar (1:3); 327.5mm thick; 5m mean radius	m²	1.03	117.39	-	44.29	161.68	106.468
U1.3.2.02 as above; 10m mean radius	m²	0.88	100.29	-	44.29	144.58	106.468
U1.3.2.03 as above; 440mm thick; 5m mean radius	m²	1.38	157.28	-	61.90	219.18	152.811
U1.3.2.04 as above; 10m mean radius	m²	1.19	135.62	-	61.90	197.52	152.811
U1.3.5 Vertical facing to concrete							
U1.3.5.01 PC£230 per 1000 in cement mortar (1:3); 440mm thick; building in 4nr wall ties/m²; casting into concrete	m²	0.79	90.04	-	44.74	134.78	106.800

U1 Common Brickwork continued...

		Unit	Labour Hours	Labour Net £	Plant Net £	Materials Net £	Unit Net £	CO_2e Kg
U1.4	**Thickness: 500mm - 1m**							
U1.4.1	Vertical straight walls							
U1.4.1.01	PC£230 per 1000 in cement mortar (1:3); 890 mm thick	m²	1.35	153.86	-	123.80	277.66	305.622
U1.4.2	Vertical curved walls							
U1.4.2.01	PC£230 per 1000 in cement mortar (1:3); 890 mm thick; 5m mean radius	m²	2.36	268.97	-	123.80	392.77	305.622
U1.4.2.02	as above; 10m mean radius	m²	2.02	230.22	-	123.80	354.02	305.622
U1.4.3	Battered straight walls							
U1.4.3.01	PC£230 per 1000 in cement mortar (1:3); 890 mm thick; one face battered at an angle of 1:20	m²	2.41	274.67	-	123.80	398.47	305.622
U1.5	**Thickness: exceeding 1m**							
U1.5.1	Vertical straight walls							
U1.5.1.01	PC£230 per 1000 in cement mortar (1:3)	m²	1.31	149.07	-	143.18	292.25	352.636
U1.6	**Columns and piers of stated cross-sectional dimensions**							
U1.6.0	Generally							
U1.6.0.01	PC£230 per 1000 in cement mortar (1:3); 215 x 215mm	m	0.15	17.55	-	7.10	24.65	17.552
U1.6.0.02	as above; 327.5 x 215mm	m	0.23	25.99	-	10.53	36.52	26.068
U1.6.0.03	as above; 327.5 x 327.5mm	m	0.34	38.98	-	15.83	54.81	39.229
U1.6.0.04	as above; 440 x 440mm	m	0.62	70.21	-	28.27	98.48	69.698
U1.6.0.05	as above; 890 x 890mm	m	1.13	129.24	-	111.54	240.78	274.676
U1.7	**Surface features**							
U1.7.1	Copings and cills, material stated							
U1.7.1.01	Brick on-edge-coping; flush pointing top and both sides; 225mm wide	m	0.06	6.84	-	2.45	9.29	6.189
U1.7.2	Rebates and chases							
U1.7.2.01	Forming fair chase 25 x 25mm wide	m	0.20	22.79	-	-	22.79	-
U1.7.4	Band courses							
U1.7.4.01	Projecting plain band set forward 25mm from wall face; 225mm high flush pointing one side and top and bottom of projection	m	0.03	3.42	-	-	3.42	-
U1.7.6	Pilasters							
U1.7.6.01	attached to wall face; 215 x 112.5mm; flush pointing all faces	m	0.03	3.42	-	3.88	7.30	10.295
U1.7.6.02	as above; 890 x 327.5mm; flush pointing all faces	m	0.12	13.68	-	44.13	57.81	113.886
U1.7.7	Plinths							
U1.7.7.01	102.5mm thick; set forward 25mm from wall face; 450mm high; flush pointing face and top of projection	m	0.06	6.84	-	-	6.84	-
U1.7.7.02	as above; 900mm high; flush pointing face and top of projection	m	0.12	13.68	-	-	13.68	-
U1.7.8	Fair facings							
U1.7.8.01	in stretcher bond including flush pointing as the work proceeds	m²	0.06	6.84	-	-	6.84	-
U1.7.8.02	in stretcher bond including weather struck pointing as the work proceeds	m²	0.07	7.98	-	-	7.98	-
U1.7.8.03	in English bond including flush pointing as the work proceeds	m²	0.09	10.26	-	-	10.26	-

U1 Common Brickwork continued...

	Unit	Labour Hours	Labour Net £	Plant Net £	Materials Net £	Unit Net £	CO$_2$e Kg
UI.7	**Surface features**						
UI.7.8	Fair facings						
UI.7.8.04	in English bond including weatherstruck pointing as the work proceeds	m^2	0.10	11.40	-	-	11.40	-
UI.7.8.05	in Flemish bond including flush pointing as the work proceeds	m^2	0.08	9.35	-	-	9.35	-
UI.7.8.06	in Flemish bond including weather struck pointing as the work proceeds	m^2	0.08	9.57	-	-	9.57	-
UI.7.8.07	in stretcher bond including raking out joints and flush repointing with coloured gauged mortar	m^2	0.15	17.10	-	2.07	19.17	3.864
UI.7.8.08	in stretcher bond including raking out joints and weather struck repointing with coloured gauged mortar	m^2	0.16	18.24	-	2.07	20.31	3.864
UI.8	**Ancillaries**						
UI.8.1	Joint reinforcement						
UI.8.1.01	Exmet galvanised brick reinforcement; 24 gauge; horizontal;	m	0.04	4.56	-	1.06	5.62	1.883
UI.8.1.02	as above; 225mm wide	m	0.05	5.70	-	2.12	7.82	3.766
UI.8.1.03	as above; 305mm wide	m	0.06	6.84	-	3.18	10.02	5.649
UI.8.1.04	as above; 450mm wide	m	0.07	7.98	-	4.24	12.22	7.532
UI.8.1.05	as above; 900mm wide	m	0.14	15.96	-	8.48	24.44	15.064
UI.8.2	Damp proof courses						
UI.8.2.01	Bitumen damp proof course; 100mm laps in cement mortar (1:3); pointing where exposed; Hessian based; horizontal; 115mm wide	m	0.01	1.37	-	1.48	2.85	3.009
UI.8.2.02	as above; 225mm wide	m	0.02	2.28	-	2.95	5.23	6.018
UI.8.2.03	as above; 330mm wide	m	0.04	4.56	-	4.43	8.99	9.027
UI.8.2.04	as above; 450mm wide	m	0.06	6.27	-	5.90	12.17	12.036
UI.8.2.05	as above; 900mm wide	m	0.07	7.41	-	11.80	19.21	24.071
UI.8.2.06	as above; vertical; 115mm wide	m	0.02	1.71	-	1.48	3.19	3.009
UI.8.2.07	as above; 225mm wide	m	0.02	2.51	-	2.95	5.46	6.018
UI.8.2.08	as above; 330mm wide	m	0.05	5.13	-	4.43	9.56	9.027
UI.8.2.09	as above; 450mm wide	m	0.06	6.84	-	5.90	12.74	12.036
UI.8.2.10	as above; 900mm wide	m	0.07	7.98	-	11.80	19.78	24.071
UI.8.2.11	as above; Fibre based; horizontal; 115mm wide	m	0.01	0.68	-	1.17	1.85	3.009
UI.8.2.12	as above; 225mm wide	m	0.01	1.37	-	2.33	3.70	6.018
UI.8.2.13	as above; 330mm wide	m	0.02	2.05	-	3.50	5.55	9.027
UI.8.2.14	as above; 450mm wide	m	0.02	2.74	-	4.66	7.40	12.036
UI.8.2.15	as above; 900mm wide	m	0.05	5.47	-	9.32	14.79	24.071
UI.8.2.16	as above; vertical; 115mm wide	m	0.01	1.37	-	1.17	2.54	3.009
UI.8.2.17	as above; 225mm wide	m	0.02	2.74	-	2.33	5.07	6.018
UI.8.2.18	as above; 330mm wide	m	0.04	4.10	-	3.50	7.60	9.027
UI.8.2.19	as above; 450mm wide	m	0.05	5.47	-	4.66	10.13	12.036
UI.8.2.20	as above; 900mm wide	m	0.10	10.94	-	9.32	20.26	24.071
UI.8.2.21	Hyload pitch polymer damp proof course; 100mm laps, sealed with Hyload contact adhesive; in cement mortar (1:3); pointing where exposed; Horizontal; 115mm wide	m	0.01	0.68	-	1.49	2.17	3.009

U1 Common Brickwork continued...

	Unit	Labour Hours	Labour Net £	Plant Net £	Materials Net £	Unit Net £	CO_2e Kg
U1.8	**Ancillaries**						
U1.8.2	Damp proof courses						
U1.8.2.22 as previous item; 225mm wide	m	0.01	1.37	-	2.97	4.34	6.018
U1.8.2.23 as above; 330mm wide	m	0.02	2.05	-	4.46	6.51	9.027
U1.8.2.24 as above; 450mm wide	m	0.02	2.74	-	5.94	8.68	12.036
U1.8.2.25 as above; 900mm wide	m	0.05	5.47	-	11.88	17.35	24.071
U1.8.2.26 as above; Vertical; 115mm wide	m	0.01	1.37	-	1.49	2.86	3.009
U1.8.2.27 as above; 225mm wide	m	0.02	2.74	-	2.97	5.71	6.018
U1.8.2.28 as above; 330mm wide	m	0.04	4.10	-	4.46	8.56	9.027
U1.8.2.29 as above; 450mm wide	m	0.05	5.47	-	5.94	11.41	12.036
U1.8.2.30 as above; 900mm wide	m	0.10	10.94	-	11.88	22.82	24.071
U1.8.3	Movement joints						
U1.8.3.01 Expansion joints; filling with Fillcrete joint filler; 20mm thick expansion joint; vertical; 102.5mm wide	m	0.06	7.18	-	2.42	9.60	2.531
U1.8.3.02 as above; 215mm wide	m	0.10	11.85	-	4.83	16.68	5.062
U1.8.3.03 as above; 327.5mm wide	m	0.17	19.03	-	7.25	26.28	7.592
U1.8.3.04 as above; 440mm wide	m	0.25	28.61	-	9.66	38.27	10.123
U1.8.3.05 as above; 890 mm wide	m	0.42	47.87	-	19.32	67.19	20.246
U1.8.4	Bonds to existing work						
U1.8.4.01 Bonding 102.5mm brickwork to existing 215mm common brickwork; cutting and toothing alternate courses	m²	0.22	25.07	-	1.24	26.31	3.344
U1.8.4.02 Bonding 102.5mm brickwork to existing concrete wall; cutting mortice and grouting in 1nr mild steel galvanised cramp every third course	m²	0.42	47.87	-	1.67	49.54	3.659
U1.8.4.03 Bonding 215mm brickwork to existing 215mm common brickwork; cutting and toothing alternate courses	m²	0.37	42.62	-	2.05	44.67	5.407
U1.8.4.04 Bonding 215mm brickwork to existing concrete wall; cutting mortice and grouting in 1nr mild steel galvanised cramp every third course	m²	0.57	64.62	-	2.49	67.11	5.722
U1.8.5	Infills of stated thickness						
U1.8.5.01 Concrete infill; Grade 25, 20mm aggregate; 50mm thick	m²	0.20	22.79	-	4.86	27.65	13.840
U1.8.5.02 as above; 100mm thick	m²	0.30	34.19	-	9.72	43.91	27.680
U1.8.5.03 as above; 150mm thick	m²	0.40	45.59	-	14.58	60.17	41.520
U1.8.5.04 as above; 200mm thick	m²	0.50	56.98	-	19.44	76.42	55.360
U1.8.7	Built-in pipes and ducts, cross-sectional area: not exceeding 0.05m²						
U1.8.7.01 supply excluded; 102.5mm brickwork; cross-sectional area; not exceeding 0.025m²	Nr	0.12	13.68	-	0.51	14.19	2.033
U1.8.7.02 as above; 0.025 - 0.25m²	Nr	0.12	13.68	-	0.51	14.19	2.033
U1.8.7.03 150mm diameter clay pipe to, BS 65, 102.5mm brickwork; cross-sectional area; not exceeding 0 025m²	Nr	0.09	10.26	-	10.01	20.27	6.061

U1 Common Brickwork continued...

	Unit	Labour Hours	Labour Net £	Plant Net £	Materials Net £	Unit Net £	CO$_2$e Kg
U1.8 Ancillaries							
U1.8.7 Built-in pipes and ducts, cross-sectional area: not exceeding 0.05m^2							
U1.8.7.04 supply excluded; 215mm brickwork; cross-sectional area; not exceeding 0.025m^2	Nr	0.10	11.40	-	0.25	11.65	1.016
U1.8.7.05 as above; 0.025 - 0.25m^2	Nr	0.12	13.68	-	0.51	14.19	2.033
U1.8.7.06 150mm diameter clay pipe to, BS 65, 215mm brickwork; cross-sectional area; not exceeding 0.025m^2	Nr	0.10	11.40	-	15.27	26.67	10.107
U1.8.8 Built-in pipes and ducts, cross-sectional area: exceeding 0.05m^2							
U1.8.8.01 supply excluded; 102.5mm brickwork; cross-sectional area; 0.40m^2	Nr	0.16	18.24	-	0.63	18.87	2.541
U1.8.8.02 supply excluded; 215mm brickwork; cross-sectional area; 0.40m^2	Nr	0.14	15.96	-	0.70	16.66	2.795

U2 Facing Brickwork

	Unit	Labour Hours	Labour Net £	Plant Net £	Materials Net £	Unit Net £	CO$_2$e Kg
U2.1 Thickness: not exceeding: 150mm							
U2.1.1 Vertical straight walls							
U2.1.1.01 PC£325 per 1000 in cement mortar (1:3); 102.5mm thick in stretcher bond	m^2	0.33	37.84	-	21.08	58.92	38.203
U2.1.2 Vertical curved walls							
U2.1.2.01 PC£325 per 1000 in cement mortar (1:3); 102.5mm thick in stretcher bond; curved walls; 5m mean radius	m^2	0.58	66.10	-	22.58	88.68	34.907
U2.1.2.02 as above; 10m mean radius	m^2	1.05	119.67	-	44.33	164.00	71.948
U2.2 Thickness: 150 - 250mm							
U2.2.1 Vertical straight walls							
U2.2.1.01 PC£325 per 1000 in cement mortar (1:3); 215mm thick in stretcher bond	m^2	0.69	78.64	-	42.16	120.80	76.405
U2.2.1.02 as above in Flemish bond	m^2	0.63	71.80	-	42.16	113.96	76.405
U2.2.2 Vertical curved walls							
U2.2.2.01 PC£325 per 1000 in cement mortar (1:3); 215mm thick in English bond; curved walls; 5m mean radius	m^2	1.20	136.76	-	46.38	183.14	71.201
U2.2.2.02 as above; 10m mean radius	m^2	1.03	117.39	-	46.38	163.77	71.201
U2.2.2.03 as above in Flemish bond; curved walls; 5m mean radius	m^2	1.11	126.51	-	46.38	172.89	71.201
U2.2.2.04 as above; 10m mean radius	m^2	0.95	108.27	-	46.38	154.65	71.201
U2.2.5 Vertical facing to concrete							
U2.2.5.01 building in 4nr wall ties/m^2; casting into concrete	m^2	0.65	74.08	-	31.83	105.91	76.036

U2   Facing Brickwork continued...

	Unit	Labour Hours	Labour Net £	Plant Net £	Materials Net £	Unit Net £	CO_2e Kg
U2.3	**Thickness: 250 - 500mm**						
U2.3.1	Vertical straight walls						
U2.3.1.01	PC£325 per 1000 in cement mortar (1:3); 327.5mm thick in stretcher bond	m²	0.73	83.20	-	63.24	146.44	114.608
U2.3.2	Vertical curved walls						
U2.3.2.01	PC£325 per 1000 in cement mortar (1:3); 327.5mm thick in stretcher bond; curved walls; 5m mean radius	m²	1.26	143.60	-	67.85	211.45	103.655
U2.3.2.02	as above; 10m mean radius	m²	1.08	123.09	-	67.85	190.94	103.655
U2.3.5	Vertical facing to concrete						
U2.3.5.01	PC£325 per 1000 in cement mortar (1:3); 327.5mm thick; building in 4nr wall ties/m²; casting into concrete	m²	0.93	105.99	-	63.00	168.99	111.394
U2.6	**Columns and piers of stated cross-sectional dimensions**						
U2.6.0	Generally						
U2.6.0.01	PC£325 per 1000 in cement mortar (1:3); 215 x 215mm	m	0.15	17.10	-	9.03	26.13	15.011
U2.6.0.02	as above; 327.5 x 215mm	m	0.23	26.21	-	13.38	39.59	22.257
U2.6.0.03	as above; 327.5 x 327.5mm	m	0.34	38.75	-	17.86	56.61	30.012
U2.6.0.04	as above; 440 x 440mm	m	0.54	61.54	-	39.95	101.49	66.009
U2.7	**Surface features**						
U2.7.1	Copings and cills, material stated						
U2.7.1.01	Brick on edge coping; flush pointing top and both sides; 225mm wide	m	0.08	8.89	-	3.31	12.20	6.189
U2.7.2	Rebates and chases						
U2.7.2.01	Forming fair chase 25 x 25mm wide	m	0.26	29.40	-	-	29.40	-
U2.7.4	Band courses						
U2.7.4.01	Projecting plain band set forward 25mm from wall face; 225mm wide; flush pointing one side and top and bottom of projection	m	0.04	4.33	-	-	4.33	-
U2.7.6	Pilasters						
U2.7.6.01	attached to wall face; 215 x 112.5mm; flush pointing all faces	m	0.04	4.33	-	58.95	63.28	113.886
U2.7.6.02	as above; 890 x 327.5mm; flush pointing all faces	m	0.15	17.21	-	14.71	31.92	27.592
U2.7.7	Plinths						
U2.7.7.01	102.5mm thick; set forward 25mm from wall face; 450mm high; flush pointing face and top of projection	m	0.07	8.43	-	-	8.43	-
U2.7.7.02	as above; 900mm high; flush pointing face and top of projection	m	0.16	17.67	-	-	17.67	-
U2.7.8	Fair facings						
U2.7.8.01	in stretcher bond including flush pointing as the work proceeds	m²	0.08	8.78	-	-	8.78	-

U2 Facing Brickwork continued...

	Unit	Labour Hours	Labour Net £	Plant Net £	Materials Net £	Unit Net £	CO$_2$e Kg
U2.7 Surface features							
U2.7.8 Fair facings							
U2.7.8.02 in stretcher bond including weather struck pointing as the work proceeds	m²	0.08	8.78	-	-	8.78	-
U2.7.8.03 in English bond including flush pointing as the work proceeds	m²	0.12	13.22	-	-	13.22	-
U2.7.8.04 in English bond including weather struck pointing as the work proceeds	m²	0.12	13.22	-	-	13.22	-
U2.7.8.05 in Flemish bond including flush pointing as the work proceeds	m²	0.11	12.08	-	-	12.08	-
U2.7.8.06 in Flemish bond including weatherstruck pointing as the work proceeds	m²	0.11	12.08	-	-	12.08	-
U2.7.8.07 in stretcher bond including raking out joints and flush repointing with coloured gauged mortar	m²	0.09	10.03	-	2.07	12.10	3.864
U2.7.8.08 in stretcher bond including raking out joints and weather struck repointing with coloured gauged mortar	m²	0.09	10.03	-	2.07	12.10	3.864
U2.8 Ancillaries							
U2.8.1 Joint reinforcement							
U2.8.1.01 Exmet galvanised brick reinforcement; 24 gauge; horizontal; 115mm wide	m	0.04	4.56	-	1.06	5.62	1.883
U2.8.1.02 as above; 225mm wide	m	0.05	5.70	-	2.12	7.82	3.766
U2.8.1.03 as above; 305mm wide	m	0.06	6.84	-	3.18	10.02	5.649
U2.8.1.04 as above; 450mm wide	m	0.07	7.98	-	4.24	12.22	7.532
U2.8.1.05 as above; 900mm wide	m	0.14	15.96	-	8.48	24.44	15.064
U2.8.2 Damp proof courses							
U2.8.2.01 Bitumen damp proof course; 100mm laps in cement mortar (1:3); pointing where exposed; Hessian based; horizontal; 115mm wide	m	0.01	1.37	-	1.48	2.85	3.009
U2.8.2.02 as above; 225mm wide	m	0.02	2.28	-	2.95	5.23	6.018
U2.8.2.03 as above; 330mm wide	m	0.04	4.56	-	4.43	8.99	9.027
U2.8.2.04 as above; 450mm wide	m	0.06	6.27	-	5.90	12.17	12.036
U2.8.2.05 as above; 900mm wide	m	0.07	7.41	-	11.80	19.21	24.071
U2.8.2.06 as above; vertical; 115mm wide	m	0.02	1.71	-	1.48	3.19	3.009
U2.8.2.07 as above; 225mm wide	m	0.02	2.51	-	2.95	5.46	6.018
U2.8.2.08 as above; 330mm wide	m	0.05	5.13	-	4.43	9.56	9.027
U2.8.2.09 as above; 450mm wide	m	0.06	6.84	-	5.90	12.74	12.036
U2.8.2.10 as above; 900mm wide	m	0.07	7.98	-	11.80	19.78	24.071
U2.8.2.11 as above; Fibre based; horizontal; 115mm wide	m	0.01	0.68	-	1.17	1.85	3.009
U2.8.2.12 as above; 225mm wide	m	0.01	1.37	-	2.33	3.70	6.018
U2.8.2.13 as above; 330mm wide	m	0.02	2.05	-	3.50	5.55	9.027
U2.8.2.14 as above; 450mm wide	m	0.02	2.74	-	4.66	7.40	12.036
U2.8.2.15 as above; 900mm wide	m	0.05	5.47	-	9.32	14.79	24.071
U2.8.2.16 as above; vertical; 115mm wide	m	0.01	1.37	-	1.17	2.54	3.009
U2.8.2.17 as above; 225mm wide	m	0.02	2.74	-	2.33	5.07	6.018
U2.8.2.18 as above; 330mm wide	m	0.04	4.10	-	3.50	7.60	9.027
U2.8.2.19 as above; 450mm wide	m	0.05	5.47	-	4.66	10.13	12.036
U2.8.2.20 as above; 900mm wide	m	0.10	10.94	-	9.32	20.26	24.071

U2   Facing Brickwork continued...

	Unit	Labour Hours	Labour Net £	Plant Net £	Materials Net £	Unit Net £	CO_2e Kg
U2.8	**Ancillaries**						
U2.8.2	Damp proof courses						
U2.8.2.21	Hyload pitch polymer damp proof course; 100mm laps, sealed with Hyload contact adhesive; in cement mortar (1:3); pointing where exposed; Horizontal; 115mm wide	m	0.01	0.68	-	1.49	2.17	3.009
U2.8.2.22	as above; 225mm wide	m	0.01	1.37	-	2.97	4.34	6.018
U2.8.2.23	as above; 330mm wide	m	0.02	2.05	-	4.46	6.51	9.027
U2.8.2.24	as above; 450mm wide	m	0.02	2.74	-	5.94	8.68	12.036
U2.8.2.25	as above; 900mm wide	m	0.05	5.47	-	11.88	17.35	24.071
U2.8.2.26	as above; Vertical; 115mm wide	m	0.01	1.37	-	1.49	2.86	3.009
U2.8.2.27	as above; 225mm wide	m	0.02	2.74	-	2.97	5.71	6.018
U2.8.2.28	as above; 330mm wide	m	0.04	4.10	-	4.46	8.56	9.027
U2.8.2.29	as above; 450mm wide	m	0.05	5.47	-	5.94	11.41	12.036
U2.8.2.30	as above; 900mm wide	m	0.10	10.94	-	11.88	22.82	24.071
U2.8.3	Movement joints						
U2.8.3.01	Expansion joints; filling with Fillcrete joint filler; 20mm thick expansion joint; vertical; 102.5mm wide	m	0.06	7.18	-	2.42	9.60	2.531
U2.8.3.02	as above; 215mm wide	m	0.10	11.85	-	4.83	16.68	5.062
U2.8.3.03	as above; 327.5mm wide	m	0.17	19.03	-	7.25	26.28	7.592
U2.8.3.04	as above; 440mm wide	m	0.25	28.61	-	9.66	38.27	10.123
U2.8.3.05	as above; 890 mm wide	m	0.42	47.87	-	19.32	67.19	20.246
U2.8.4	Bonds to existing work						
U2.8.4.01	Bonding 102.5mm brickwork to existing 215mm common brickwork; cutting and toothing alternate courses	m²	0.24	27.58	-	1.62	29.20	3.344
U2.8.4.02	Bonding 102.5mm brickwork to existing concrete wall; cutting mortice and grouting in 1nr mild steel galvanised cramp every third course	m²	0.46	52.65	-	2.05	54.70	3.659
U2.8.4.03	Bonding 215mm brickwork to existing 215mm facing brickwork; cutting and toothing alternate courses	m²	0.41	46.73	-	1.62	48.35	3.344
U2.8.4.04	Bonding 215mm brickwork to existing concrete wall; cutting mortice and grouting in 1nr mild steel galvanised cramp every third course	m²	0.64	73.17	-	2.05	75.22	3.659
U2.8.5	Infills of stated thickness						
U2.8.5.01	Concrete infill; Grade 25, 20mm aggregate; 50mm thick	m²	0.20	22.79	-	4.86	27.65	13.840
U2.8.5.02	as above; 100mm thick	m²	0.30	34.19	-	9.72	43.91	27.680
U2.8.5.03	as above; 150mm thick	m²	0.40	45.59	-	14.58	60.17	41.520
U2.8.5.04	as above; 200mm thick	m²	0.50	56.98	-	19.44	76.42	55.360
U2.8.7	Built-in pipes and ducts, cross-sectional area: not exceeding 0.05m²						
U2.8.7.01	Built-in pipes and ducts (supply excluded); 102.5mm brickwork; cross-sectional area; not exceeding 0.025m²	Nr	0.09	10.03	-	0.38	10.41	1.524

U2 Facing Brickwork continued...

	Unit	Labour Hours	Labour Net £	Plant Net £	Materials Net £	Unit Net £	CO_2e Kg
U2.8 **Ancillaries**							
U2.8.7 Built-in pipes and ducts, cross-sectional area: not exceeding 0.05m2							
U2.8.7.02 as previous item; cross-sectional area;-0.025 0.25m^2	Nr	0.13	15.04	-	0.51	15.55	2.033
U2.8.7.03 Built-in pipes and ducts; 150mm diameter clay pipe to, BS 65, 102.5mm brickwork; cross-sectional area; not exceeding 0.025m^2	Nr	0.07	7.98	-	10.01	17.99	6.061
U2.8.7.04 Built-in pipes and ducts (supply excluded); 215mm brickwork; cross-sectional area; not exceeding 0.025m^2	Nr	0.11	12.54	-	0.25	12.79	1.016
U2.8.7.05 as above; cross-sectional area; 0.025 - 0.25m^2	Nr	0.13	14.82	-	0.51	15.33	2.033
U2.8.7.06 Built-in pipes and ducts; 150mm diameter clay pipe to, BS 65, 215mm brickwork; cross-sectional area; not exceeding 0.025m^2	Nr	0.11	12.54	-	10.52	23.06	8.093
U2.8.8 Built-in pipes and ducts, cross-sectional area: exceeding 0.05m^2							
U2.8.8.01 Built-in pipes and ducts (supply excluded); 102.5mm brickwork; cross-sectional area; 0.40m^2	Nr	0.13	15.04	-	0.51	15.55	2.033
U2.8.8.02 as above; 215mm brickwork; cross-sectional area; 0.40m^2	Nr	0.15	17.55	-	0.70	18.25	2.795

U3 Engineering Brickwork

	Unit	Labour Hours	Labour Net £	Plant Net £	Materials Net £	Unit Net £	CO_2e Kg
U3.1 **Thickness: not exceeding: 150mm**							
U3.1.1 Vertical straight walls							
U3.1.1.01 Class A Engineering bricks, PC£375 per 1000 in cement mortar (1:3); 102.5mm thick	m^2	0.28	32.37	-	24.03	56.40	38.203
U3.1.1.02 Class B Engineering bricks, PC£275 per 1000 in cement mortar (1:3); 102.5mm thick	m^2	0.28	32.37	-	20.28	52.65	35.200
U3.1.2 Vertical curved walls							
U3.1.2.01 Class A Engineering bricks, PC£375 per 1000 in cement mortar (1:3); 102.5mm thick; curved walls; 5m mean radius	m^2	0.50	56.98	-	26.18	83.16	35.200
U3.1.2.02 as above; 10m mean radius	m^2	0.43	49.01	-	26.18	75.19	35.200
U3.1.2.03 Class B Engineering bricks, PC£275 per 1000 in cement mortar (1:3); 102.5mm thick; curved walls; 5m mean radius	m^2	0.50	56.98	-	20.28	77.26	35.200
U3.1.2.04 as above; 10m mean radius	m^2	0.43	49.01	-	20.28	69.29	35.200
U3.2 **Thickness: 150 - 250mm**							
U3.2.1 Vertical straight walls							
U3.2.1.01 Class A Engineering bricks, PC£375 per 1000 in cement mortar (1:3); 215mm thick	m^2	0.51	58.12	-	48.06	106.18	76.405

U3   Engineering Brickwork continued...

	Unit	Labour Hours	Labour Net £	Plant Net £	Materials Net £	Unit Net £	CO_2e Kg
U3.2 **Thickness: 150 - 250mm**							
U3.2.1 Vertical straight walls							
U3.2.1.02 Class B Engineering bricks, PC£275 per 1000 in cement mortar (1:3); 215mm thick	m²	0.51	58.12	-	40.56	98.68	70.400
U3.2.2 Vertical curved walls							
U3.2.2.01 Class A Engineering bricks, PC£375 per 1000 in cement mortar (1:3); 215mm thick; curved walls; 5m mean radius	m²	0.89	101.43	-	52.36	153.79	70.400
U3.2.2.02 as above; 10m mean radius	m²	0.76	86.62	-	52.36	138.98	70.400
U3.2.2.03 Class B Engineering bricks, PC£275 per 1000 in cement mortar (1:3); 215mm thick; curved walls; 5m mean radius	m²	0.89	101.43	-	40.56	141.99	70.400
U3.2.2.04 as above; 10m mean radius	m²	0.76	86.62	-	40.56	127.18	70.400
U3.3 **Thickness: 250 - 500mm**							
U3.3.1 Vertical straight walls							
U3.3.1.01 Class A Engineering bricks, PC£375 per 1000 in cement mortar (1:3); 327.5mm thick	m²	0.66	75.22	-	78.54	153.76	105.600
U3.3.1.02 Class B Engineering bricks, PC£275 per 1000 in cement mortar (1:3); 327.5mm thick	m²	0.65	73.85	-	60.84	134.69	105.600
U3.3.1.03 Class A Engineering bricks, PC£375 per 1000 in cement mortar (1:3); 440mm thick	m²	0.87	98.70	-	96.12	194.82	152.811
U3.3.1.04 Class B Engineering bricks, PC£275 per 1000 in cement mortar (1:3); 440mm thick	m²	0.87	98.70	-	81.12	179.82	140.801
U3.3.2 Vertical curved walls							
U3.3.2.01 Class A Engineering bricks, PC£375 per 1000 in cement mortar (1:3); 327.5mm thick; curved walls; 5m mean radius	m²	1.13	128.79	-	78.54	207.33	105.600
U3.3.2.02 as above; 10m mean radius	m²	0.97	110.55	-	78.54	189.09	105.600
U3.3.2.03 Class B Engineering bricks, PC£275 per 1000 in cement mortar (1:3); 327.5mm thick; curved walls; 5m mean radius	m²	1.13	128.79	-	60.84	189.63	105.600
U3.3.2.04 as above; 10m mean radius	m²	0.97	110.55	-	60.84	171.39	105.600
U3.3.2.05 Class A Engineering bricks, PC£375 per 1000 in cement mortar (1:3); 440mm thick; curved walls; 5m mean radius	m²	1.52	173.23	-	104.72	277.95	140.801
U3.3.2.06 as above; 10m mean radius	m²	1.31	149.30	-	104.72	254.02	140.801
U3.3.2.07 Class B Engineering bricks, PC£275 per 1000 in cement mortar (1:3); 440mm thick; curved walls; 5m mean radius	m²	1.13	128.79	-	81.12	209.91	140.801
U3.3.2.08 as above; 10m mean radius	m²	1.31	149.30	-	81.12	230.42	140.801

U3   Engineering Brickwork continued...

	Unit	Labour Hours	Labour Net £	Plant Net £	Materials Net £	Unit Net £	CO_2e Kg
U3.3 **Thickness: 250 - 500mm**							
U3.3.5 Vertical facing to concrete							
U3.3.5.01 Class A Engineering bricks, PC£375 per 1000 in cement mortar (1:3); 327.5mm thick; building in 4nr wall ties/m²; casting into concrete	m²	0.83	94.60	-	72.54	167.14	114.940
U3.3.5.02 Class B Engineering bricks, PC£275 per 1000 in cement mortar (1:3); 327.5mm thick; building in 4nr wall ties/m²; casting into concrete	m²	0.83	94.60	-	54.84	149.44	114.940
U3.4 **Thickness: 500mm - 1m**							
U3.4.1 Vertical straight walls							
U3.4.1.01 Class A Engineering bricks, PC£375 per 1000 in cement mortar (1:3); 890 mm thick	m²	1.48	168.90	-	192.24	361.14	305.622
U3.4.1.02 Class B Engineering bricks, PC£275 per 1000 in cement mortar (1:3); 890 mm thick	m²	1.48	168.90	-	162.24	331.14	281.601
U3.4.2 Vertical curved walls							
U3.4.2.01 Class A Engineering bricks, PC£375 per 1000 in cement mortar (1:3); 890 mm thick; curved walls; 5m mean radius	m²	2.65	302.25	-	192.24	494.49	305.622
U3.4.2.02 as above; 10m mean radius	m²	2.65	302.25	-	145.04	447.29	305.622
U3.4.2.03 Class B Engineering bricks, PC£275 per 1000 in cement mortar (1:3); 890 mm thick; curved walls; 5m mean radius	m²	2.59	295.18	-	162.24	457.42	281.601
U3.4.2.04 as above; 10m mean radius	m²	2.22	253.01	-	162.24	415.25	281.601
U3.4.3 Battered straight walls							
U3.4.3.01 Class A Engineering bricks, PC£375 per 1000 in cement mortar (1:3); 890 mm thick; battered walls; one face battered at an angle of 1:20	m²	2.59	295.41	-	192.24	487.65	305.622
U3.4.3.02 Class B Engineering bricks, PC£275 per 1000 in cement mortar (1:3); 890 mm thick; battered walls; one face battered at an angle of 1:20	m²	2.65	302.02	-	162.24	464.26	281.601
U3.5 **Thickness: exceeding 1m**							
U3.5.1 Vertical straight walls							
U3.5.1.01 Class A Engineering bricks, PC£375 per 1000 in cement mortar (1:3)	m³	1.50	170.95	-	222.64	393.59	352.636
U3.5.1.02 Class B Engineering bricks, PC£275 per 1000 in cement mortar (1:3)	m³	1.52	173.01	-	167.84	340.85	352.636

U3 Engineering Brickwork continued...

	Unit	Labour Hours	Labour Net £	Plant Net £	Materials Net £	Unit Net £	CO_2e Kg
U3.6	**Columns and piers of stated cross-sectional dimensions**						
U3.6.0	Generally						
U3.6.0.01 Class A Engineering bricks, PC£375 per 1000 in cement mortar (1:3); 215 x 215mm	m	0.17	19.26	-	11.01	30.27	17.552
U3.6.0.02 Class B Engineering bricks, PC£275 per 1000 in cement mortar (1:3); 215 x 215mm	m	0.17	19.26	-	8.31	27.57	17.552
U3.6.0.03 Class A Engineering bricks, PC£375 per 1000 in cement mortar (1:3); 327.5 x 215mm	m	0.25	28.49	-	16.33	44.82	26.068
U3.6.0.04 Class B Engineering bricks, PC£275 per 1000 in cement mortar (1:3); 327.5 x 215mm	m	0.25	28.49	-	12.33	40.82	26.068
U3.6.0.05 Class A Engineering bricks, PC£375 per 1000 in cement mortar (1:3); 327.5 x 327.5mm	m	0.38	42.85	-	24.53	67.38	39.229
U3.6.0.06 Class B Engineering bricks, PC£275 per 1000 in cement mortar (1:3); 327.5 x 327.5mm	m	0.37	42.17	-	18.53	60.70	39.229
U3.6.0.07 Class A Engineering bricks, PC£375 per 1000 in cement mortar (1:3);440 x 440mm	m	0.68	77.27	-	43.93	121.20	69.698
U3.6.0.08 Class B Engineering bricks, PC£275 per 1000 in cement mortar (1:3); 440 x 440mm	m	0.68	77.27	-	52.93	130.20	107.017
U3.6.0.09 Class A Engineering bricks, PC£375 per 1000 in cement mortar (1:3); 890 x 890mm	m	2.35	267.60	-	173.46	441.06	274.676
U3.6.0.10 Class B Engineering bricks, PC£275 per 1000 in cement mortar (1:3); 890 x 890mm	m	2.35	267.60	-	130.76	398.36	274.676
U3.7	**Surface features**						
U3.7.1	Copings and cills, material stated						
U3.7.1.01 Class A Engineering bricks, PC£375 per 1000 in cement mortar (1:3); Brick on edge coping; flush pointing top and both sides; 225mm wide	m	0.07	7.52	-	3.76	11.28	6.189
U3.7.1.02 as above; 75mm wide	m	0.22	25.07	-	-	25.07	-
U3.7.1.03 Class B Engineering bricks, PC£275 per 1000 in cement mortar (1:3); Brick on edge coping; flush pointing top and both sides; 225mm wide	m	0.07	7.52	-	2.86	10.38	6.189
U3.7.1.04 as above; Cill laid flat; flush with wall; flush pointing top and one side; 75mm wide	m	0.22	25.07	-	-	25.07	-
U3.7.2	Rebates and chases						
U3.7.2.01 Class A Engineering bricks, PC£375 per 1000 in cement mortar (1:3); Forming fair chase 25 x 25mm wide	m	0.03	3.76	-	-	3.76	-
U3.7.2.02 Class B Engineering bricks, PC£275 per 1000 in cement mortar (1:3); Forming fair chase 25 x 25mm wide	m	0.03	3.76	-	-	3.76	-

U3   Engineering Brickwork continued...

	Unit	Labour Hours	Labour Net £	Plant Net £	Materials Net £	Unit Net £	CO_2e Kg
U3.7 **Surface features**							
U3.7.6 Pilasters							
U3.7.6.01 Class A Engineering bricks, PC£375 per 1000 in cement mortar (1:3); Pilaster attached to wall face; 215 x 112.5mm; flush pointing all faces	m	0.03	3.76	-	4.46	8.22	10.295
U3.7.6.02 as above; 890 x 327.5mm; flush pointing all faces	m	0.13	15.04	-	51.15	66.19	113.886
U3.7.6.03 Class B Engineering bricks, PC£275 per 1000 in cement mortar (1:3); Pilaster attached to wall face; 215 x 112.5mm; flush pointing all faces	m	0.03	3.76	-	4.46	8.22	10.295
U3.7.6.04 as above; 890 x 327.5mm; flush pointing all faces	m	0.13	15.04	-	51.15	66.19	113.886
U3.7.7 Plinths							
U3.7.7.01 Class A Engineering bricks, PC£375 per 1000 in cement mortar (1:3); Plinth 102.5mm thick; set forward 25mm from wall face; 450mm high; flush pointing face and top of projection	m	0.07	7.52	-	-	7.52	-
U3.7.7.02 as above; 900mm high; flush pointing face and top of projection	m	0.13	15.04	-	-	15.04	-
U3.7.7.03 Class B Engineering bricks, PC£275 per 1000 in cement mortar (1:3); Plinth 102.5mm thick; set forward 25mm from wall face; 450mm high; flush pointing face and top of projection	m	0.07	7.52	-	-	7.52	-
U3.7.7.04 as above; 900mm high; flush pointing face and top of projection	m	0.13	15.04	-	-	15.04	-
U3.7.8 Fair facings							
U3.7.8.01 Class A Engineering bricks, PC£375 per 1000 in cement mortar (1:3); Fair facing in stretcher bond including flush pointing as the work proceeds	m²	0.07	7.52	-	-	7.52	-
U3.7.8.02 as above; in stretcher bond including weather struck pointing as the work proceeds	m²	0.08	8.66	-	-	8.66	-
U3.7.8.03 as above; in English bond including flush pointing as the work proceeds	m²	0.10	11.17	-	-	11.17	-
U3.7.8.04 as above; in English bond including weather struck pointing as the work proceeds	m²	0.11	12.54	-	-	12.54	-
U3.7.8.05 as above; in Flemish bond including flush pointing as the work proceeds	m²	0.09	10.26	-	-	10.26	-
U3.7.8.06 as above; in Flemish bond including weather struck pointing as the work proceeds	m²	0.10	11.40	-	-	11.40	-

U3   Engineering Brickwork continued...

	Unit	Labour Hours	Labour Net £	Plant Net £	Materials Net £	Unit Net £	CO_2e Kg
U3.7	**Surface features**						
U3.7.8	Fair facings						
U3.7.8.07	as previous item; in stretcher bond including raking out joints and flush repointing with coloured gauged mortar	m²	0.16	18.69	-	2.07	20.76	3.864
U3.7.8.08	as above; in stretcher bond including raking out joints and weatherstruck repointing with coloured gauged mortar	m²	0.18	20.06	-	2.07	22.13	3.864
U3.7.8.09	Class B Engineering bricks, PC£275 per 1000 in cement mortar (1:3); Fair facing in stretcher bond including flush pointing as the work proceeds	m²	0.07	7.52	-	-	7.52	-
U3.7.8.10	as above; in stretcher bond including weatherstruck pointing as the work proceeds	m²	0.08	8.66	-	-	8.66	-
U3.7.8.11	as above; in English bond including flush pointing as the work proceeds	m²	0.10	11.17	-	-	11.17	-
U3.7.8.12	as above; in English bond including weatherstruck pointing as the work proceeds	m²	0.11	12.54	-	-	12.54	-
U3.7.8.13	as above; in Flemish bond including flush pointing as the work proceeds	m²	0.09	10.26	-	-	10.26	-
U3.7.8.14	as above; in Flemish bond including weatherstruck pointing as the work proceeds	m²	0.10	11.40	-	-	11.40	-
U3.7.8.15	as above; in stretcher bond including raking out joints and flush repointing with coloured gauged mortar	m²	0.16	18.69	-	2.07	20.76	3.864
U3.7.8.16	as above; in stretcher bond including raking out joints and weatherstruck repointing with coloured gauged mortar	m²	0.18	20.06	-	2.07	22.13	3.864
U3.8	**Ancillaries**						
U3.8.1	Joint reinforcement						
U3.8.1.01	Exmet galvanised brick reinforcement; 24 gauge; horizontal;	m	0.04	4.56	-	1.06	5.62	1.883
U3.8.1.02	as above; 225mm wide	m	0.05	5.70	-	2.12	7.82	3.766
U3.8.1.03	as above; 305mm wide	m	0.06	6.84	-	3.18	10.02	5.649
U3.8.1.04	as above; 450mm wide	m	0.07	7.98	-	4.24	12.22	7.532
U3.8.1.05	as above; 900mm wide	m	0.14	15.96	-	8.48	24.44	15.064
U3.8.2	Damp proof courses						
U3.8.2.01	Bitumen damp proof course; 100mm laps in cement mortar (1:3); pointing where exposed; Hessian based; horizontal; 115mm wide	m	0.01	1.37	-	1.48	2.85	3.009
U3.8.2.02	as above; 225mm wide	m	0.02	2.28	-	2.95	5.23	6.018
U3.8.2.03	as above; 330mm wide	m	0.04	4.56	-	4.43	8.99	9.027
U3.8.2.04	as above; 450mm wide	m	0.06	6.27	-	5.90	12.17	12.036
U3.8.2.05	as above; 900mm wide	m	0.07	7.41	-	11.80	19.21	24.071
U3.8.2.06	as above; vertical; 115mm wide	m	0.02	1.71	-	1.48	3.19	3.009

U3   Engineering Brickwork continued...

	Unit	Labour Hours	Labour Net £	Plant Net £	Materials Net £	Unit Net £	CO_2e Kg
U3.8	**Ancillaries**						
U3.8.2	Damp proof courses						
U3.8.2.07 as previous item; 225mm wide	m	0.02	2.51	-	2.95	5.46	6.018
U3.8.2.08 as above; 330mm wide	m	0.05	5.13	-	4.43	9.56	9.027
U3.8.2.09 as above; 450mm wide	m	0.06	6.84	-	5.90	12.74	12.036
U3.8.2.10 as above; 900mm wide	m	0.07	7.98	-	11.80	19.78	24.071
U3.8.2.11 as above; Fibre based; horizontal; 115mm wide	m	0.01	0.68	-	1.17	1.85	3.009
U3.8.2.12 as above; 225mm wide	m	0.01	1.37	-	2.33	3.70	6.018
U3.8.2.13 as above; 330mm wide	m	0.02	2.05	-	3.50	5.55	9.027
U3.8.2.14 as above; 450mm wide	m	0.02	2.74	-	4.66	7.40	12.036
U3.8.2.15 as above; 900mm wide	m	0.05	5.47	-	9.32	14.79	24.071
U3.8.2.16 as above; vertical; 115mm wide	m	0.01	1.37	-	1.17	2.54	3.009
U3.8.2.17 as above; 225mm wide	m	0.02	2.74	-	2.33	5.07	6.018
U3.8.2.18 as above; 330mm wide	m	0.04	4.10	-	3.50	7.60	9.027
U3.8.2.19 as above; 450mm wide	m	0.05	5.47	-	4.66	10.13	12.036
U3.8.2.20 as above; 900mm wide	m	0.10	10.94	-	9.32	20.26	24.071
U3.8.2.21 Hyload pitch polymer damp proof course; 100mm laps, sealed with Hyload contact adhesive; in cement mortar (1:3); pointing where exposed; Horizontal; 115mm wide	m	0.01	0.68	-	1.49	2.17	3.009
U3.8.2.22 as above; 225mm wide	m	0.01	1.37	-	2.97	4.34	6.018
U3.8.2.23 as above; 330mm wide	m	0.02	2.05	-	4.46	6.51	9.027
U3.8.2.24 as above; 450mm wide	m	0.02	2.74	-	5.94	8.68	12.036
U3.8.2.25 as above; 900mm wide	m	0.05	5.47	-	11.88	17.35	24.071
U3.8.2.26 as above; Vertical; 115mm wide	m	0.01	1.37	-	1.49	2.86	3.009
U3.8.2.27 as above; 225mm wide	m	0.02	2.74	-	2.97	5.71	6.018
U3.8.2.28 as above; 330mm wide	m	0.04	4.10	-	4.46	8.56	9.027
U3.8.2.29 as above; 450mm wide	m	0.05	5.47	-	5.94	11.41	12.036
U3.8.2.30 as above; 900mm wide	m	0.10	10.94	-	11.88	22.82	24.071
U3.8.3	Movement joints						
U3.8.3.01 Expansion joints; filling with Fillcrete joint filler; 20mm thick expansion joint; vertical; 102.5mm wide	m	0.06	7.18	-	2.42	9.60	2.531
U3.8.3.02 as above; 215mm wide	m	0.10	11.85	-	4.83	16.68	5.062
U3.8.3.03 as above; 327.5mm wide	m	0.17	19.03	-	7.25	26.28	7.592
U3.8.3.04 as above; 440mm wide	m	0.25	28.61	-	9.66	38.27	10.123
U3.8.3.05 as above; 890 mm wide	m	0.42	47.87	-	19.32	67.19	20.246
U3.8.4	Bonds to existing work						
U3.8.4.01 Class A Engineering bricks, PC£375 per 1000 in cement mortar (1:3); bonding 102.5mm brickwork to existing 215mm common brickwork; cutting and toothing alternate courses	m²	0.24	27.58	-	1.82	29.40	3.344
U3.8.4.02 as above to existing concrete wall; cutting mortice and grouting in 1nr mild steel galvanised cramp every third course	m²	0.46	52.43	-	2.25	54.68	3.659
U3.8.4.03 Class A Engineering bricks, PC£375 per 1000 in cement mortar (1:3); bonding 215mm brickwork to existing 215mm facing brickwork; cutting and toothing alternate courses	m²	0.41	46.73	-	3.50	50.23	5.722

U3   Engineering Brickwork continued...

	Unit	Labour Hours	Labour Net £	Plant Net £	Materials Net £	Unit Net £	CO_2e Kg
U3.8	**Ancillaries**						
U3.8.4 Bonds to existing work							
U3.8.4.04 as previous item to existing concrete wall; cutting mortice and grouting in 1nr mild steel galvanised cramp every third course	m²	0.63	72.26	-	3.50	75.76	5.722
U3.8.4.05 Class B Engineering bricks, PC£275 per 1000 in cement mortar (1:3); bonding 102.5mm brickwork to existing 215mm common brickwork; cutting and toothing alternate courses	m²	0.24	27.58	-	1.42	29.00	3.344
U3.8.4.06 as above to existing concrete wall; cutting mortice and grouting in 1nr mild steel galvanised cramp every third course	m²	0.46	52.43	-	1.85	54.28	3.659
U3.8.4.07 Class B Engineering bricks, PC£275 per 1000 in cement mortar (1:3); bonding 215mm brickwork to existing 215mm facing brickwork; cutting and toothing alternate courses	m²	0.41	46.73	-	2.80	49.53	5.722
U3.8.4.08 as above to existing concrete wall; cutting mortice and grouting in 1nr mild steel galvanised cramp every third course	m²	0.63	72.26	-	2.80	75.06	5.722
U3.8.5 Infills of stated thickness							
U3.8.5.01 Concrete infill; Grade 25, 20mm aggregate; 50mm thick	m²	0.20	22.79	-	4.86	27.65	13.840
U3.8.5.02 as above; 100mm thick	m²	0.30	34.19	-	9.72	43.91	27.680
U3.8.5.03 as above; 150mm thick	m²	0.40	45.59	-	14.58	60.17	41.520
U3.8.5.04 as above; 200mm thick	m²	0.50	56.98	-	19.44	76.42	55.360
U3.8.7 Built-in pipes and ducts, cross-sectional area: not exceeding 0.05m²							
U3.8.7.01 Class A Engineering bricks, PC£375 per 1000 in cement mortar (1:3); supply excluded; 102.5mm brickwork; cross-sectional area; not exceeding 0.025m²	Nr	0.13	15.04	-	0.38	15.42	1.524
U3.8.7.02 as above; cross-sectional area; 0.025 - 0.25m²	Nr	0.13	15.04	-	0.51	15.55	2.033
U3.8.7.03 Class A Engineering bricks, PC£375 per 1000 in cement mortar (1:3); 150mm diameter clay pipe to, BS 65, 102.5mm brickwork; cross-sectional area; not exceeding 0.025m²	Nr	0.13	15.04	-	0.38	15.42	1.524
U3.8.7.04 Class A Engineering bricks, PC£375 per 1000 in cement mortar (1:3); supply excluded; 215mm brickwork; cross-sectional area; not exceeding 0.025m²	Nr	0.13	15.04	-	0.51	15.55	2.033
U3.8.7.05 as above; cross-sectional area; 0.025 - 0.25m²	Nr	0.13	14.82	-	0.63	15.45	2.541

U3 Engineering Brickwork continued...

	Unit	Labour Hours	Labour Net £	Plant Net £	Materials Net £	Unit Net £	CO_2e Kg
U3.8	**Ancillaries**						
U3.8.7	Built-in pipes and ducts, cross-sectional area: not exceeding 0.05m²						
U3.8.7.06	Class A Engineering bricks, PC£375 per 1000 in cement mortar (1:3); 150mm diameter clay pipe to, BS 65, 215mm brickwork; cross-sectional area; not exceeding 0.025m²	Nr	0.11	12.54	-	10.46	23.00	7.839
U3.8.7.07	Class B Engineering bricks, PC£275 per 1000 in cement mortar (1:3); supply excluded; 102.5mm brickwork; cross-sectional area; not exceeding 0.025m²	Nr	0.09	10.26	-	0.32	10.58	1.270
U3.8.7.08	as above; cross-sectional area; 0.025 - 0.25m²	Nr	0.13	14.82	-	0.63	15.45	2.541
U3.8.7.09	Class B Engineering bricks, PC£275 per 1000 in cement mortar (1:3); 150mm diameter clay pipe to, BS 65, 102.5mm brickwork; cross-sectional area; not exceeding 0.025m²	Nr	0.10	11.40	-	0.95	12.35	3.811
U3.8.7.10	Class B Engineering bricks, PC£275 per 1000 in cement mortar (1:3); supply excluded; 215mm brickwork; cross-sectional area; not exceeding 0.025m²	Nr	0.11	12.54	-	0.38	12.92	1.524
U3.8.7.11	as above; cross-sectional area; 0.025 - 0.25m²	Nr	0.13	14.82	-	0.63	15.45	2.541
U3.8.7.12	Class B Engineering bricks, PC£275 per 1000 in cement mortar (1:3); 150mm diameter clay pipe to, BS 65, 215mm brickwork; cross-sectional area; not exceeding 0.025m²	Nr	0.11	12.54	-	10.77	23.31	9.109
U3.8.8	Built-in pipes and ducts, cross-sectional area: exceeding 0.05m²						
U3.8.8.01	Class A Engineering bricks, PC£375 per 1000 in cement mortar (1:3); supply excluded; 102.5mm brickwork; cross-sectional area; 0.40m²	Nr	0.18	20.06	-	0.51	20.57	2.033
U3.8.8.02	as above; 215mm brickwork; cross-sectional area; 0.40m²	Nr	0.15	17.55	-	0.70	18.25	2.795
U3.8.8.03	Class B Engineering bricks, PC£275 per 1000 in cement mortar (1:3); supply excluded; 102.5mm brickwork; cross-sectional area; 0.40m²	Nr	0.18	20.06	-	0.51	20.57	2.033
U3.8.8.04	as above; 215mm brickwork; cross-sectional area; 0.40m²	Nr	0.15	17.55	-	0.70	18.25	2.795

U4 Lightweight Blockwork

	Unit	Labour Hours	Labour Net £	Plant Net £	Materials Net £	Unit Net £	CO_2e Kg
U4.1	**Thickness: not exceeding: 150mm**						
U4.1.1	Vertical straight walls						
U4.1.1.01 Precast concrete blocks, solid; compressive strength 4.2N/mm², face size 440 x 215mm; in gauged mortar (1:2:9); 100mm thick	m²	0.16	17.34	-	9.21	26.55	16.106
U4.1.1.02 as above; hollow; compressive strength 4.2N/mm², face size 440 x 215mm; in gauged mortar (1:2:9); 100mm thick	m²	0.14	15.17	-	9.59	24.76	16.106
U4.1.1.03 as above, solid; fair faced; compressive strength 4.2N/mm², face size 440 x 215mm; in gauged mortar (1:2:9); 100mm thick	m²	0.17	18.10	-	10.16	28.26	16.106
U4.1.1.04 as above, hollow fair faced; compressive strength 4.2N/mm², face size 440 x 215mm; in gauged mortar (1:2:9); 100mm thick	m²	0.16	17.01	-	12.92	29.93	13.091
U4.1.1.05 as above, solid; compressive strength 4.2N/mm², face size 440 x 215mm; in gauged mortar (1:2:9); 140 mm thick	m²	0.18	19.50	-	13.53	33.03	25.495
U4.1.1.06 as above, hollow; compressive strength 4.2N/mm², face size 440 x 215mm; in gauged mortar (1:2:9); 140 mm thick	m²	0.17	18.42	-	16.40	34.82	25.495
U4.1.1.07 as above, solid; fair faced; compressive strength 4.2N/mm², face size 440 x 215mm; in gauged mortar (1:2:9); 140 mm thick	m²	0.20	21.67	-	14.87	36.54	25.495
U4.1.1.08 as above, hollow fair faced; compressive strength 4.2N/mm², face size 440 x 215mm; in gauged mortar (1:2:9); 140 mm thick	m²	0.26	28.17	-	17.86	46.03	20.653
U4.1.2	Vertical curved walls						
U4.1.2.01 Precast concrete blocks, solid; compressive strength 4.2N/mm², face size 440 x 215mm; in gauged mortar (1:2:9); curved walls; 5m mean radius; 140 mm thick	m²	0.27	29.26	-	9.21	38.47	16.106
U4.1.2.02 as above, ; 10m mean radius; 140 mm thick	m²	0.24	26.01	-	9.21	35.22	16.106
U4.1.2.03 Precast concrete blocks, hollow; compressive strength 4.2N/mm², face size 440 x 215mm; in gauged mortar (1:2:9); curved walls; 5m mean radius; 140 mm thick	m²	0.24	26.01	-	9.59	35.60	16.106
U4.1.2.04 as above, ; 10m mean radius; 140 mm thick	m²	0.20	21.67	-	9.59	31.26	16.106
U4.1.2.05 Precast concrete blocks, solid; fair faced; compressive strength 4.2N/mm², face size 440 x 215mm; in gauged mortar (1:2:9); curved walls; 5m mean radius; 140 mm thick	m²	0.29	31.42	-	10.16	41.58	16.106
U4.1.2.06 as above, ; 10m mean radius; 140 mm thick	m²	0.25	27.09	-	10.16	37.25	16.106

U4   Lightweight Blockwork continued...

	Unit	Labour Hours	Labour Net £	Plant Net £	Materials Net £	Unit Net £	CO_2e Kg
U4.1	**Thickness: not exceeding: 150mm**						
U4.1.2	Vertical curved walls						
U4.1.2.07	Precast concrete blocks, hollow fair faced; compressive strength 4.2N/mm², face size 440 x 215mm; in gauged mortar (1:2:9); curved walls; 5m mean radius; 140 mm thick	m²	0.23	24.92	-	12.92	37.84	13.091
U4.1.2.08	as above, ; 10m mean radius; 140 mm thick	m²	0.20	21.67	-	12.92	34.59	13.091
U4.1.2.09	Precast concrete blocks, solid; compressive strength 4.2N/mm², face size 440 x 215mm; in gauged mortar (1:2:9); curved walls; 5m mean radius; 215mm thick	m²	0.31	33.59	-	13.53	47.12	25.495
U4.1.2.10	as above, ; 10m mean radius; 215mm thick	m²	0.27	29.26	-	13.53	42.79	25.495
U4.1.2.11	Precast concrete blocks, hollow; compressive strength 4.2N/mm², face size 440 x 215mm; in gauged mortar (1:2:9); curved walls; 5m mean radius; 215mm thick	m²	0.29	31.42	-	16.40	47.82	25.495
U4.1.2.12	as above, ; 10m mean radius; 215mm thick	m²	0.25	27.09	-	16.40	43.49	25.495
U4.1.2.13	Precast concrete blocks, solid; fair faced; compressive strength 4.2N/mm², face size 440 x 215mm; in gauged mortar (1:2:9); curved walls; 5m mean radius; 215mm thick	m²	0.35	37.93	-	14.87	52.80	25.495
U4.1.2.14	as above, ; 10m mean radius; 215mm thick	m²	0.30	32.51	-	14.87	47.38	25.495
U4.1.2.15	Precast concrete blocks, hollow fair faced; compressive strength 4.2N/mm², face size 440 x 215mm; in gauged mortar (1:2:9); curved walls; 5m mean radius; 215mm thick	m²	0.33	35.76	-	17.86	53.62	20.653
U4.1.2.16	as above, ; 10m mean radius; 215mm thick	m²	0.28	30.34	-	17.86	48.20	20.653
U4.2	**Thickness: 150 - 250mm**						
U4.2.1	Vertical straight walls						
U4.2.1.01	Precast concrete blocks, solid; compressive strength 4.2N/mm², face size 440 x 215mm; in gauged mortar (1:2:9); 215mm thick	m²	0.23	24.92	-	21.32	46.24	35.269
U4.2.1.02	as above, hollow; compressive strength 4.2N/mm², face size 440 x 215mm; in gauged mortar (1:2:9); 215mm thick	m²	0.22	23.84	-	24.90	48.74	35.269
U4.2.1.03	as above, solid; fair faced; compressive strength 4.2N/mm², face size 440 x 215mm; in gauged mortar (1:2:9); 215mm thick	m²	0.23	24.92	-	23.83	48.75	35.269

U4 Lightweight Blockwork continued...

	Unit	Labour Hours	Labour Net £	Plant Net £	Materials Net £	Unit Net £	CO_2e Kg
U4.2	**Thickness: 150 - 250mm**						
U4.2.1	**Vertical straight walls**						
U4.2.1.04	as previous item, hollow fair faced; compressive strength 4.2N/mm², face size 440 x 215mm; in gauged mortar (1:2:9); 215mm thick	m²	0.21	22.65	-	27.24	49.89	28.602
U4.2.2	**Vertical curved walls**						
U4.2.2.01	Precast concrete blocks, solid; compressive strength 4.2N/mm², face size 440 x 215mm; in gauged mortar (1:2:9); 215mm thick; curved walls; 5m mean radius	m²	0.40	43.34	-	21.32	64.66	35.269
U4.2.2.02	as above; 10m mean radius	m²	0.34	36.84	-	21.32	58.16	35.269
U4.2.2.03	Precast concrete blocks, hollow; compressive strength 4.2N/mm², face size 440 x 215mm; in gauged mortar (1:2:9); 215mm thick; curved walls; 5m mean radius	m²	0.38	41.18	-	24.90	66.08	35.269
U4.2.2.04	as above; 10m mean radius	m²	0.33	35.76	-	24.90	60.66	35.269
U4.2.2.05	Precast concrete blocks, solid; fair faced; compressive strength 4.2N/mm², face size 440 x 215mm; in gauged mortar (1:2:9); 215mm thick; curved walls; 5m mean radius	m²	0.40	43.34	-	23.83	67.17	35.269
U4.2.2.06	Precast concrete blocks, solid; fair faced; compressive strength 4.2N/mm², face size 440 x 215mm; in gauged mortar (1:2:9); 215mm thick; curved walls; 10m mean radius	m²	0.34	36.84	-	23.83	60.67	35.269
U4.2.2.07	Precast concrete blocks, hollow fair faced; compressive strength 4.2N/mm², face size 440 x 215mm; in gauged mortar (1:2:9); 215mm thick; curved walls; 5m mean radius	m²	0.37	40.09	-	27.24	67.33	28.602
U4.2.2.08	Precast concrete blocks, hollow fair faced; compressive strength 4.2N/mm², face size 440 x 215mm; in gauged mortar (1:2:9); 215mm thick; curved walls; 10m mean radius	m²	0.31	33.59	-	27.24	60.83	28.602
U4.7	**Surface features**						
U4.7.6	**Pilasters**						
U4.7.6.01	Precast concrete blocks, solid; compressive strength 4.2N/mm², face size 440 x 215mm; in gauged mortar (1:2:9); Pilaster attached to wall face; 440 x 100mm; flush pointing all faces	m	0.07	7.80	-	3.38	11.18	5.919
U4.7.6.02	as above; 890 x 140 mm; flush pointing all faces	m	0.20	21.46	-	12.05	33.51	22.703

U4 Lightweight Blockwork continued...

	Unit	Labour Hours	Labour Net £	Plant Net £	Materials Net £	Unit Net £	CO_2e Kg
U4.7	**Surface features**							
U4.7.6	Pilasters						
U4.7.6.03	Precast concrete blocks, hollow; compressive strength $4.2N/mm^2$, face size 440 x 215mm; in gauged mortar (1:2:9); Pilaster attached to wall face; 440 x 100mm; flush pointing all faces	m	0.07	7.80	-	3.38	11.18	5.919
U4.7.6.04	as above; 890 x 140 mm; flush pointing all faces	m	0.20	21.46	-	12.05	33.51	22.703
U4.7.6.05	Precast concrete blocks, solid; fair faced; compressive strength $4.2N/mm^2$, face size 440 x 215mm; in gauged mortar (1:2:9); Pilaster attached to wall face; 440 x 100mm; flush pointing all faces	m	0.07	7.80	-	3.63	11.43	5.919
U4.7.6.06	as above; 890 x 140 mm; flush pointing all faces	m	0.20	21.46	-	12.88	34.34	22.703
U4.7.6.07	Precast concrete blocks, hollow fair faced; compressive strength $4.2N/mm^2$, face size 440 x 215mm; in gauged mortar (1:2:9); Pilaster attached to wall face; 440 x 100mm; flush pointing all faces	m	0.07	7.80	-	4.61	12.41	5.919
U4.7.6.08	as above; 890 x 140 mm; flush pointing all faces	m	0.20	21.46	-	15.47	36.93	22.703
U4.7.7	Plinths						
U4.7.7.01	Precast concrete blocks, solid; compressive strength $4.2N/mm^2$, face size 440 x 215mm; in gauged mortar (1:2:9); Plinth 100mm thick; set forward 25mm from wall face; 440mm high; flush pointing face and top of projection	m	0.07	7.69	-	-	7.69	-
U4.7.7.02	as above; 890 mm high; flush pointing face and top of projection	m	0.11	11.38	-	-	11.38	-
U4.7.7.03	Precast concrete blocks, hollow; compressive strength $4.2N/mm^2$, face size 440 x 215mm; in gauged mortar (1:2:9); Plinth 100mm thick; set forward 25mm from wall face; 440mm high; flush pointing face and top of projection	m	0.07	7.69	-	-	7.69	-
U4.7.7.04	as above; 890 mm high; flush pointing face and top of projection	m	0.11	11.38	-	-	11.38	-

U4 Lightweight Blockwork continued...

	Unit	Labour Hours	Labour Net £	Plant Net £	Materials Net £	Unit Net £	CO_2e Kg
U4.7	**Surface features**						
U4.7.7	Plinths						
U4.7.7.05 Precast concrete blocks, solid; fair faced; compressive strength $4.2N/mm^2$, face size 440 x 215mm; in gauged mortar (1:2:9); Plinth 100mm thick; set forward 25mm from wall face; 440mm high; flush pointing face and top of projection	m	0.07	7.69	-	-	7.69	-
U4.7.7.06 as above; 890 mm high; flush pointing face and top of projection	m	0.11	11.38	-	-	11.38	-
U4.7.7.07 Precast concrete blocks, hollow fair faced; compressive strength $4.2N/mm^2$, face size 440 x 215mm; in gauged mortar (1:2:9); Plinth 100mm thick; set forward 25mm from wall face; 440mm high; flush pointing face and top of projection	m	0.07	7.69	-	-	7.69	-
U4.7.7.08 as above; 890 mm high; flush pointing face and top of projection	m	0.11	11.38	-	-	11.38	-
U4.7.8	Fair facings						
U4.7.8.01 Precast concrete blocks, solid; compressive strength $4.2N/mm^2$, face size 440 x 215mm; in gauged mortar (1:2:9); Fair facing in stretcher bond including flush pointing as the work proceeds	m^2	0.06	6.83	-	-	6.83	-
U4.7.8.02 as above, hollow; compressive strength $4.2N/mm^2$, face size 440 x 215mm; in gauged mortar (1:2:9); Fair facing in stretcher bond including flush pointing as the work proceeds	m^2	0.06	6.83	-	-	6.83	-
U4.7.8.03 as above, solid; fair faced; compressive strength $4.2N/mm^2$, face size 440 x 215mm; in gauged mortar (1:2:9); Fair facing in stretcher bond including flush pointing as the work proceeds	m^2	0.06	6.83	-	-	6.83	-
U4.7.8.04 as above, hollow fair faced; compressive strength $4.2N/mm^2$, face size 440 x 215mm; in gauged mortar (1:2:9); Fair facing in stretcher bond including flush pointing as the work proceeds	m^2	0.06	6.83	-	-	6.83	-
U4.8	**Ancillaries**						
U4.8.1	Joint reinforcement						
U4.8.1.01 Exmet galvanised brick reinforcement; 24 gauge; horizontal; 100mm wide	m	0.04	4.56	-	1.06	5.62	1.883
U4.8.1.02 as above; 225mm wide	m	0.05	5.70	-	2.12	7.82	3.766
U4.8.1.03 as above; 305mm wide	m	0.06	6.84	-	3.18	10.02	5.649
U4.8.1.04 as above; 450mm wide	m	0.07	7.98	-	4.24	12.22	7.532
U4.8.1.05 as above; 900mm wide	m	0.14	15.96	-	8.48	24.44	15.064

U4 Lightweight Blockwork continued...

	Unit	Labour Hours	Labour Net £	Plant Net £	Materials Net £	Unit Net £	CO_2e Kg
U4.8	**Ancillaries**						
U4.8.2	Damp proof courses						
U4.8.2.01	Bitumen damp proof course; 100mm laps in cement mortar (1:3); pointing where exposed; Hessian based; horizontal; 115mm wide	m	0.01	1.37	-	1.48	2.85	3.009
U4.8.2.02	as above; 225mm wide	m	0.02	2.28	-	2.95	5.23	6.018
U4.8.2.03	as above; 330mm wide	m	0.04	4.56	-	4.43	8.99	9.027
U4.8.2.04	as above; 450mm wide	m	0.06	6.27	-	5.90	12.17	12.036
U4.8.2.05	as above; 900mm wide	m	0.07	7.41	-	11.80	19.21	24.071
U4.8.2.06	as above; vertical; 115mm wide	m	0.02	1.71	-	1.48	3.19	3.009
U4.8.2.07	as above; 225mm wide	m	0.02	2.51	-	2.95	5.46	6.018
U4.8.2.08	as above; 330mm wide	m	0.05	5.13	-	4.43	9.56	9.027
U4.8.2.09	as above; 450mm wide	m	0.06	6.84	-	5.90	12.74	12.036
U4.8.2.10	as above; 900mm wide	m	0.07	7.98	-	11.80	19.78	24.071
U4.8.2.11	as above; Fibre based; horizontal; 115mm wide	m	0.01	0.68	-	1.17	1.85	3.009
U4.8.2.12	as above; 225mm wide	m	0.01	1.37	-	2.33	3.70	6.018
U4.8.2.13	as above; 330mm wide	m	0.02	2.05	-	3.50	5.55	9.027
U4.8.2.14	as above; 450mm wide	m	0.02	2.74	-	4.66	7.40	12.036
U4.8.2.15	as above; 900mm wide	m	0.05	5.47	-	9.32	14.79	24.071
U4.8.2.16	as above; vertical; 115mm wide	m	0.01	1.37	-	1.17	2.54	3.009
U4.8.2.17	as above; 225mm wide	m	0.02	2.74	-	2.33	5.07	6.018
U4.8.2.18	as above; 330mm wide	m	0.04	4.10	-	3.50	7.60	9.027
U4.8.2.19	as above; 450mm wide	m	0.05	5.47	-	4.66	10.13	12.036
U4.8.2.20	as above; 900mm wide	m	0.10	10.94	-	9.32	20.26	24.071
U4.8.2.21	Hyload pitch polymer damp proof course; 100mm laps, sealed with Hyload contact adhesive; in cement mortar (1:3); pointing where exposed; Horizontal; 115mm wide	m	0.01	0.68	-	1.49	2.17	3.009
U4.8.2.22	as above; 225mm wide	m	0.01	1.37	-	2.97	4.34	6.018
U4.8.2.23	as above; 330mm wide	m	0.02	2.05	-	4.46	6.51	9.027
U4.8.2.24	as above; 450mm wide	m	0.02	2.74	-	5.94	8.68	12.036
U4.8.2.25	as above; 900mm wide	m	0.05	5.47	-	11.88	17.35	24.071
U4.8.2.26	as above; Vertical; 115mm wide	m	0.01	1.37	-	1.49	2.86	3.009
U4.8.2.27	as above; 225mm wide	m	0.02	2.74	-	2.97	5.71	6.018
U4.8.2.28	as above; 330mm wide	m	0.04	4.10	-	4.46	8.56	9.027
U4.8.2.29	as above; 450mm wide	m	0.05	5.47	-	5.94	11.41	12.036
U4.8.2.30	as above; 900mm wide	m	0.10	10.94	-	11.88	22.82	24.071
U4.8.3	Movement joints						
U4.8.3.01	Expansion joints; filling with Fillcrete joint filler; 20mm thick expansion joint; vertical; 102.5mm wide	m	0.06	7.18	-	2.42	9.60	2.531
U4.8.3.02	as above; 215mm wide	m	0.10	11.85	-	4.83	16.68	5.062
U4.8.3.03	as above; 327.5mm wide	m	0.17	19.03	-	7.25	26.28	7.592
U4.8.3.04	as above; 440mm wide	m	0.25	28.61	-	9.66	38.27	10.123
U4.8.3.05	as above; 890 mm wide	m	0.42	47.87	-	19.32	67.19	20.246
U4.8.4	Bonds to existing work						
U4.8.4.01	Precast concrete blocks, solid; compressive strength 4.2N/mm², face size 440 x 215mm; in gauged mortar (1:2:9); bonding 100mm blockwork to existing 215mm facing brickwork; cutting and toothing alternate courses	m²	0.07	7.69	-	0.20	7.89	0.408

U4 Lightweight Blockwork continued...

	Unit	Labour Hours	Labour Net £	Plant Net £	Materials Net £	Unit Net £	CO_2e Kg
U4.8	**Ancillaries**						
U4.8.4	**Bonds to existing work**						
U4.8.4.02	as previous item; bonding 140 mm blockwork to existing 215mm facing brickwork; cutting and toothing alternate courses	m²	0.08	9.10	-	0.24	9.34	0.500
U4.8.4.03	as above, hollow; compressive strength 4.2N/mm², face size 440 x 215mm; in gauged mortar (1:2:9); bonding 100mm blockwork to existing 215mm facing brickwork; cutting and toothing alternate courses	m²	0.07	7.69	-	0.27	7.96	0.500
U4.8.4.04	as above; bonding 140 mm blockwork to existing 215mm facing brickwork; cutting and toothing alternate courses	m²	0.08	9.10	-	0.27	9.37	0.500
U4.8.4.05	as above, solid; fair faced; compressive strength 4.2N/mm², face size 440 x 215mm; in gauged mortar (1:2:9); bonding 100mm blockwork to existing 215mm facing brickwork; cutting and toothing alternate courses	m²	0.07	7.69	-	0.21	7.90	0.408
U4.8.4.06	as above; bonding 140 mm blockwork to existing 215mm facing brickwork; cutting and toothing alternate courses	m²	0.08	9.10	-	0.25	9.35	0.500
U4.8.4.07	as above, hollow fair faced; compressive strength 4.2N/mm², face size 440 x 215mm; in gauged mortar (1:2:9); bonding 100mm blockwork to existing 215mm facing brickwork; cutting and toothing alternate courses	m²	0.07	7.69	-	0.24	7.93	0.408
U4.8.4.08	as above; bonding 140 mm blockwork to existing 215mm facing brickwork; cutting and toothing alternate courses	m²	0.08	9.10	-	0.28	9.38	0.500
U4.8.5	**Infills of stated thickness**						
U4.8.5.01	Concrete infill; Grade 25, 20mm aggregate; 50mm thick	m²	0.20	22.79	-	4.86	27.65	13.840
U4.8.5.02	as above; 100mm thick	m²	0.30	34.19	-	9.72	43.91	27.680
U4.8.5.03	as above; 150mm thick	m²	0.40	45.59	-	14.58	60.17	41.520
U4.8.5.04	as above; 200mm thick	m²	0.50	56.98	-	19.44	76.42	55.360
U4.8.7	**Built-in pipes and ducts, cross-sectional area: not exceeding 0.05m²**						
U4.8.7.01	Precast concrete blocks, solid; compressive strength 4.2N/mm², face size 440 x 215mm; in gauged mortar (1:2:9); 100mm blockwork; cross-sectional area; not exceeding 0.025m2	Nr	0.13	13.55	-	0.46	14.01	1.030
U4.8.7.02	as above; 100mm blockwork; cross-sectional area; 0.025 - 0.25m²	Nr	0.13	13.55	-	0.46	14.01	1.030

U4   Lightweight Blockwork continued...

	Unit	Labour Hours	Labour Net £	Plant Net £	Materials Net £	Unit Net £	CO_2e Kg
U4.8 **Ancillaries**							
U4.8.7 Built-in pipes and ducts, cross-sectional area: not exceeding 0.05m²							
U4.8.7.03 as previous item; 150mm diameter clay pipe to, BS 65, 100mm blockwork; cross-sectional area; not exceeding 0.025m²	Nr	0.17	18.42	-	0.57	18.99	1.288
U4.8.7.04 as above, hollow; compressive strength 4.2N/mm², face size 440 x 215mm; in gauged mortar (1:2:9); supply excluded; 100mm blockwork; cross-sectional area; not exceeding 0.025m²	Nr	0.08	8.67	-	0.57	9.24	1.288
U4.8.7.05 as above; cross-sectional area; 0.025 - 0.25m²	Nr	0.13	14.09	-	0.57	14.66	1.288
U4.8.7.06 as above; 150mm diameter clay pipe to, BS 65, 100mm blockwork; cross-sectional area; not exceeding 0.025m²	Nr	0.10	10.84	-	0.57	11.41	1.288
U4.8.7.07 as above, solid; fair faced; compressive strength 4.2N/mm², face size 440 x 215mm; in gauged mortar (1:2:9); supply excluded; 100mm blockwork; cross-sectional area; not exceeding 0.025m²	Nr	0.08	8.67	-	0.57	9.24	1.288
U4.8.7.08 as above; cross-sectional area; 0.025 - 0.25m²	Nr	0.13	14.09	-	0.57	14.66	1.288
U4.8.7.09 Precast concrete blocks, solid; fair faced; compressive strength 4.2N/mm², face size 440 x 215mm; in gauged mortar (1:2:9); 150mm diameter clay pipe to, BS 65, 100mm blockwork; cross-sectional area; not exceeding 0.025m²	Nr	0.10	10.84	-	0.57	11.41	1.288
U4.8.7.10 as above, hollow fair faced; compressive strength 4.2N/mm², face size 440 x 215mm; in gauged mortar (1:2:9); supply excluded; 100mm blockwork; cross-sectional area; not exceeding 0.025m²	Nr	0.08	8.67	-	0.57	9.24	1.288
U4.8.7.11 as above; cross-sectional area; 0.025 - 0.25m²	Nr	0.13	14.09	-	0.57	14.66	1.288
U4.8.7.12 as above; 150mm diameter clay pipe to, BS 65, 100mm blockwork; cross-sectional area; not exceeding 0.025m²	Nr	0.10	10.84	-	0.57	11.41	1.288
U4.8.8 Built-in pipes and ducts, cross-sectional area: exceeding 0.05m²							
U4.8.8.01 Precast concrete blocks, solid; compressive strength 4.2N/mm², face size 440 x 215mm; in gauged mortar (1:2:9); 100mm blockwork; cross-sectional area; 0.40m²	Nr	0.17	18.10	-	0.57	18.67	1.288

U4   Lightweight Blockwork continued...

	Unit	Labour Hours	Labour Net £	Plant Net £	Materials Net £	Unit Net £	CO_2e Kg

U4.8   Ancillaries

U4.8.8   Built-in pipes and ducts, cross-sectional area: exceeding 0.05m²

U4.8.8.02   as previous item, hollow; compressive strength 4.2N/mm², face size 440 x 215mm; in gauged mortar (1:2:9); 100mm blockwork; cross-sectional area; 0.40m2

	Nr	0.17	18.42	-	0.57	18.99	1.288

U5   Dense Concrete Blockwork

U5.1   Thickness: not exceeding: 150mm

U5.1.1   Vertical straight walls

U5.1.1.01 Precast concrete blocks, solid; compressive strength 7 N/mm², face size 440 x 215mm; in gauged mortar (1:2:9); 100mm thick	m²	0.19	20.37	-	15.27	35.64	13.969
U5.1.1.02 as above; 140 mm thick	m²	0.23	24.92	-	22.12	47.04	19.870
U5.1.1.03 as above, hollow; compressive strength 7 N/mm², face size 440 x 215mm; in gauged mortar (1:2:9); 100mm thick	m²	0.18	19.29	-	13.77	33.06	12.042
U5.1.1.04 as above; 140 mm thick	m²	0.22	23.84	-	19.19	43.03	14.364

U5.1.2   Vertical curved walls

U5.1.2.01 Precast concrete blocks, solid; compressive strength 7 N/mm², face size 440 x 215mm; in gauged mortar (1:2:9); 100mm thick; curved walls; 5m mean radius	m²	0.33	35.76	-	15.27	51.03	13.969
U5.1.2.02 as above; 10m mean radius	m²	0.28	30.34	-	15.27	45.61	13.969
U5.1.2.03 as above; 140 mm thick; curved walls; 5m mean radius	m²	0.40	43.34	-	22.12	65.46	19.870
U5.1.2.04 as above; 10m mean radius	m²	0.34	36.84	-	22.12	58.96	19.870
U5.1.2.05 as above, hollow; compressive strength 7 N/mm², face size 440 x 215mm; in gauged mortar (1:2:9); 100mm thick; curved walls; 5m mean radius	m²	0.31	33.59	-	13.77	47.36	12.042
U5.1.2.06 as above; 10m mean radius	m²	0.27	29.26	-	13.77	43.03	12.042
U5.1.2.07 as above; 140 mm thick; curved walls; 5m mean radius	m²	0.38	41.18	-	19.19	60.37	14.364
U5.1.2.08 as above; 10m mean radius	m²	0.33	35.76	-	19.19	54.95	14.364

U5.2   Thickness: 150 - 250mm

U5.2.1   Vertical straight walls

U5.2.1.01 Precast concrete blocks, solid; compressive strength 7 N/mm², face size 440 x 215mm; in gauged mortar (1:2:9); 215mm thick	m²	0.30	32.51	-	33.86	66.37	23.267
U5.2.1.02 as above, hollow; compressive strength 7 N/mm², face size 440 x 215mm; in gauged mortar (1:2:9); 215mm thick	m²	0.29	31.42	-	26.91	58.33	16.385

U5 Dense Concrete Blockwork continued...

	Unit	Labour Hours	Labour Net £	Plant Net £	Materials Net £	Unit Net £	CO_2e Kg
U5.2 **Thickness: 150 - 250mm**							
U5.2.2 Vertical curved walls							
U5.2.2.01 Precast concrete blocks, solid; compressive strength 7 N/mm^2, face size 440 x 215mm; in gauged mortar (1:2:9); 215mm thick; curved walls; 5m mean radius	m^2	0.53	57.43	-	33.86	91.29	23.267
U5.2.2.02 as above; 10m mean radius	m^2	0.45	48.76	-	33.86	82.62	23.267
U5.2.2.03 as above, hollow; compressive strength 7 N/mm^2, face size 440 x 215mm; in gauged mortar (1:2:9); 215mm thick; curved walls; 5m mean radius	m^2	0.42	45.51	-	26.91	72.42	16.385
U5.2.2.04 as above; 10m mean radius	m^2	0.39	42.26	-	26.91	69.17	16.385
U5.7 **Surface features**							
U5.7.6 Pilasters							
U5.7.6.01 Precast concrete blocks, solid; compressive strength 7 N/mm^2, face size 440 x 215mm; in gauged mortar (1:2:9); Pilaster attached to wall face; 440 x 100mm; flush pointing all faces	m	0.08	9.10	-	5.61	14.71	5.135
U5.7.6.02 as above; 890 x 140 mm; flush pointing all faces	m	0.25	27.20	-	19.70	46.90	17.697
U5.7.6.03 as above, hollow; compressive strength 7 N/mm^2, face size 440 x 215mm; in gauged mortar (1:2:9); Pilaster attached to wall face; 440 x 100mm; flush pointing all faces	m	0.08	9.10	-	5.06	14.16	4.428
U5.7.6.04 as above; 890 x 140 mm; flush pointing all faces	m	0.25	27.20	-	17.09	44.29	12.797
U5.7.7 Plinths							
U5.7.7.01 Precast concrete blocks, solid; compressive strength 7 N/mm^2, face size 440 x 215mm; in gauged mortar (1:2:9); Plinth 100mm thick; set forward 25mm from wall face; 440mm high; flush pointing face and top of projection	m	0.08	9.10	-	-	9.10	-
U5.7.7.02 as above; 890 mm high; flush pointing face and top of projection	m	0.13	13.55	-	-	13.55	-
U5.7.7.03 as above, hollow; compressive strength 7 N/mm^2, face size 440 x 215mm; in gauged mortar (1:2:9); Plinth 100mm thick; set forward 25mm from wall face; 440mm high; flush pointing face and top of projection	m	0.08	9.10	-	-	9.10	-
U5.7.7.04 as above; 890 mm high; flush pointing face and top of projection	m	0.13	13.55	-	-	13.55	-

U5 Dense Concrete Blockwork continued...

	Unit	Labour Hours	Labour Net £	Plant Net £	Materials Net £	Unit Net £	CO_2e Kg
U5.7	**Surface features**						
U5.7.8	Fair facings						
U5.7.8.01	Precast concrete blocks, solid; compressive strength 7 N/mm², face size 440 x 215mm; in gauged mortar (1:2:9); Fair facing in stretcher bond including flush pointing as the work proceeds	m²	0.08	9.10	-	-	9.10	-
U5.7.8.02	as above, hollow; compressive strength 7 N/mm², face size 440 x 215mm; in gauged mortar (1:2:9); Fair facing in stretcher bond including flush pointing as the work proceeds	m²	0.08	9.10	-	-	9.10	-
U5.8	**Ancillaries**						
U5.8.1	Joint reinforcement						
U5.8.1.01	Exmet galvanised brick reinforcement; 24 gauge; horizontal; 100mm wide	m	0.04	4.56	-	1.06	5.62	1.883
U5.8.1.02	as above; 225mm wide	m	0.05	5.70	-	2.12	7.82	3.766
U5.8.1.03	as above; 305mm wide	m	0.06	6.84	-	3.18	10.02	5.649
U5.8.1.04	as above; 450mm wide	m	0.07	7.98	-	4.24	12.22	7.532
U5.8.1.05	as above; 900mm wide	m	0.14	15.96	-	8.48	24.44	15.064
U5.8.2	Damp proof courses						
U5.8.2.01	Bitumen damp proof course; 100mm laps in cement mortar (1:3); pointing where exposed; Hessian based; horizontal; 115mm wide	m	0.01	1.37	-	1.48	2.85	3.009
U5.8.2.02	as above; 225mm wide	m	0.02	2.28	-	2.95	5.23	6.018
U5.8.2.03	as above; 330mm wide	m	0.04	4.56	-	4.43	8.99	9.027
U5.8.2.04	as above; 450mm wide	m	0.06	6.27	-	5.90	12.17	12.036
U5.8.2.05	as above; 900mm wide	m	0.07	7.41	-	11.80	19.21	24.071
U5.8.2.06	as above; vertical; 115mm wide	m	0.02	1.71	-	1.48	3.19	3.009
U5.8.2.07	as above; 225mm wide	m	0.02	2.51	-	2.95	5.46	6.018
U5.8.2.08	as above; 330mm wide	m	0.05	5.13	-	4.43	9.56	9.027
U5.8.2.09	as above; 450mm wide	m	0.06	6.84	-	5.90	12.74	12.036
U5.8.2.10	as above; 900mm wide	m	0.07	7.98	-	11.80	19.78	24.071
U5.8.2.11	as above; Fibre based; horizontal; 115mm wide	m	0.01	0.68	-	1.17	1.85	3.009
U5.8.2.12	as above; 225mm wide	m	0.01	1.37	-	2.33	3.70	6.018
U5.8.2.13	as above; 330mm wide	m	0.02	2.05	-	3.50	5.55	9.027
U5.8.2.14	as above; 450mm wide	m	0.02	2.74	-	4.66	7.40	12.036
U5.8.2.15	as above; 900mm wide	m	0.05	5.47	-	9.32	14.79	24.071
U5.8.2.16	as above; vertical; 115mm wide	m	0.01	1.37	-	1.17	2.54	3.009
U5.8.2.17	as above; 225mm wide	m	0.02	2.74	-	2.33	5.07	6.018
U5.8.2.18	as above; 330mm wide	m	0.04	4.10	-	3.50	7.60	9.027
U5.8.2.19	as above; 450mm wide	m	0.05	5.47	-	4.66	10.13	12.036
U5.8.2.20	as above; 900mm wide	m	0.10	10.94	-	9.32	20.26	24.071
U5.8.2.21	Hyload pitch polymer damp proof course; 100mm laps, sealed with Hyload contact adhesive; in cement mortar (1:3); pointing where exposed; Horizontal; 115mm wide	m	0.01	0.68	-	1.49	2.17	3.009
U5.8.2.22	as above; 225mm wide	m	0.01	1.37	-	2.97	4.34	6.018
U5.8.2.23	as above; 330mm wide	m	0.02	2.05	-	4.46	6.51	9.027
U5.8.2.24	as above; 450mm wide	m	0.02	2.74	-	5.94	8.68	12.036

U5   Dense Concrete Blockwork continued...

	Unit	Labour Hours	Labour Net £	Plant Net £	Materials Net £	Unit Net £	CO$_2$e Kg
U5.8 **Ancillaries**							
U5.8.2 Damp proof courses							
U5.8.2.25 as previous item; 900mm wide	m	0.05	5.47	-	11.88	17.35	24.071
U5.8.2.26 as above; Vertical; 115mm wide	m	0.01	1.37	-	1.49	2.86	3.009
U5.8.2.27 as above; 225mm wide	m	0.02	2.74	-	2.97	5.71	6.018
U5.8.2.28 as above; 330mm wide	m	0.04	4.10	-	4.46	8.56	9.027
U5.8.2.29 as above; 450mm wide	m	0.05	5.47	-	5.94	11.41	12.036
U5.8.2.30 as above; 900mm wide	m	0.10	10.94	-	11.88	22.82	24.071
U5.8.3 Movement joints							
U5.8.3.01 Expansion joints; filling with Fillcrete joint filler; 20mm thick expansion joint; vertical; 102.5mm wide	m	0.06	7.18	-	2.42	9.60	2.531
U5.8.3.02 as above; 215mm wide	m	0.10	11.85	-	4.83	16.68	5.062
U5.8.3.03 as above; 327.5mm wide	m	0.17	19.03	-	7.25	26.28	7.592
U5.8.3.04 as above; 440mm wide	m	0.25	28.61	-	9.66	38.27	10.123
U5.8.3.05 as above; 890 mm wide	m	0.42	47.87	-	19.32	67.19	20.246
U5.8.4 Bonds to existing work							
U5.8.4.01 Precast concrete blocks, solid; compressive strength 7 N/mm^2, face size 440 x 215mm; in gauged mortar (1:2:9); bonding 100mm blockwork to existing 215mm facing brickwork; cutting and toothing alternate courses	m^2	0.08	8.13	-	0.26	8.39	0.387
U5.8.4.02 as above; Bonding 140 mm blockwork to existing 215mm facing brickwork; cutting and toothing alternate courses	m^2	0.09	9.54	-	0.33	9.87	0.443
U5.8.4.03 as above, hollow; compressive strength 7 N/mm^2, face size 440 x 215mm; in gauged mortar (1:2:9); bonding 100mm blockwork to existing 215mm facing brickwork; cutting and toothing alternate courses	m^2	0.08	8.13	-	0.25	8.38	0.368
U5.8.4.04 as above; Bonding 140 mm blockwork to existing 215mm facing brickwork; cutting and toothing alternate courses	m^2	0.09	9.54	-	0.30	9.84	0.388
U5.8.5 Infills of stated thickness							
U5.8.5.01 Concrete infill; Grade 25, 20mm aggregate; 50mm thick	m^2	0.20	22.79	-	4.86	27.65	13.840
U5.8.5.02 as above; 100mm thick	m^2	0.30	34.19	-	9.72	43.91	27.680
U5.8.5.03 as above; 150mm thick	m^2	0.40	45.59	-	14.58	60.17	41.520
U5.8.5.04 as above; 200mm thick	m^2	0.50	56.98	-	19.44	76.42	55.360
U5.8.7 Built-in pipes and ducts, cross-sectional area: not exceeding 0.05m^2							
U5.8.7.01 Precast concrete blocks, solid; compressive strength 7 N/mm^2, face size 440 x 215mm; in gauged mortar (1:2:9); supply excluded; 100mm clockwork; cross-sectional area; not exceeding 0.025m^2	Nr	0.15	15.82	-	0.46	16.28	1.030
U5.8.7.02 as above; cross-sectional area; 0.025 - 0.25m^2	Nr	0.15	15.82	-	0.46	16.28	1.030

U5 Dense Concrete Blockwork continued...

	Unit	Labour Hours	Labour Net £	Plant Net £	Materials Net £	Unit Net £	CO_2e Kg
U5.8	**Ancillaries**						
U5.8.7	Built-in pipes and ducts, cross-sectional area: not exceeding 0.05m^2						
U5.8.7.03	as above; 150mm diameter clay pipe to, BS 65, 100mm blockwork; cross-sectional area; not exceeding 0.025m^2	Nr	0.13	14.09	-	0.57	14.66	1.288
U5.8.7.04	as above, hollow; compressive strength 7 N/mm^2, face size 440 x 215mm; in gauged mortar (1:2:9); supply excluded; 100mm clockwork; cross-sectional area; not exceeding 0.025m^2	Nr	0.09	9.75	-	0.57	10.32	1.288
U5.8.7.05	as above; cross-sectional area; 0.025 - 0.25m^2	Nr	0.15	16.25	-	0.57	16.82	1.288
U5.8.7.06	as above; 150mm diameter clay pipe to, BS 65, 100mm blockwork; cross-sectional area; not exceeding 0.025m^2	Nr	0.13	14.09	-	0.57	14.66	1.288
U5.8.8	Built-in pipes and ducts, cross-sectional area: exceeding 0.05m^2						
U5.8.8.01	Precast concrete blocks, solid; compressive strength 7 N/mm^2, face size 440 x 215mm; in gauged mortar (1:2:9); supply excluded; 100mm clockwork; cross-sectional area; 0.40m2	Nr	0.18	19.94	-	0.57	20.51	1.288
U5.8.8.02	as above, hollow; compressive strength 7 N/mm^2, face size 440 x 215mm; in gauged mortar (1:2:9); supply excluded; 100mm clockwork; cross-sectional area; 0.40m^2	Nr	0.18	19.50	-	0.57	20.07	1.288

U6 Artificial Stone Blockwork

	Unit	Labour Hours	Labour Net £	Plant Net £	Materials Net £	Unit Net £	CO_2e Kg
U6.1	**Thickness: not exceeding: 150mm**						
U6.1.1	Vertical straight walls						
U6.1.1.01	Cast stonework; Bradstone walling blocks; Cotswold shades, random course in cement lime mortar (1:2:9); traditional walling; flush pointing, as the work proceeds; 100mm thick	m^2	0.36	38.58	-	44.28	82.86	20.114
U6.1.2	Vertical curved walls						
U6.1.2.01	Cast stonework; Bradstone walling blocks; Cotswold shades, random course in cement lime mortar (1:2:9); traditional walling; flush pointing, as the work proceeds; 100mm thick; curved walls; 5m mean radius	m^2	0.62	67.18	-	44.28	111.46	20.114
U6.1.2.02	as above; 10m mean radius	m^2	0.91	98.61	-	89.01	187.62	40.561

U6 Artificial Stone Blockwork continued...

	Unit	Labour Hours	Labour Net £	Plant Net £	Materials Net £	Unit Net £	CO_2e Kg
U6.1 **Thickness: not exceeding: 150mm**							
U6.1.5 Vertical facing to concrete							
U6.1.5.01 Cast stonework; Bradstone walling blocks; Cotswold shades, random course in cement lime mortar (1:2:9); traditional walling; flush pointing, as the work proceeds; 100mm thick; facing to concrete; building in 4nr fish tailed ties/m^2; casting into concrete	m^2	0.56	60.68	-	44.73	105.41	20.446
U6.7 **Surface features**							
U6.7.2 Rebates and chases							
U6.7.2.01 Cast stonework; Bradstone walling blocks; Cotswold shades, random course in cement lime mortar (1:2:9); traditional walling; flush pointing, as the work proceeds; Forming fair rebate 25 x 10mm deep	m	0.11	11.38	-	-	11.38	-
U6.7.4 Band courses							
U6.7.4.01 Cast stonework; Bradstone walling blocks; Cotswold shades, random course in cement lime mortar (1:2:9); traditional walling; flush pointing, as the work proceeds; Projecting band set forward 25mm from wall face; one course wide; flush pointing one side and top and bottom projection	m	0.05	5.63	-	-	5.63	-
U6.7.6 Pilasters							
U6.7.6.01 Cast stonework; Bradstone walling blocks; Cotswold shades, random course in cement lime mortar (1:2:9); traditional walling; flush pointing, as the work proceeds; Pilaster attached to wall face; 300 x 110mm; flush pointing all faces	m	0.05	5.63	-	13.26	18.89	5.983
U6.7.6.02 as above; 600 x 220mm; flush pointing all faces	m	0.26	28.39	-	52.81	81.20	23.416
U6.7.7 Plinths							
U6.7.7.01 Cast stonework; Bradstone walling blocks; Cotswold shades, random course in cement lime mortar (1:2:9); traditional walling; flush pointing, as the work proceeds; Plinth 100mm thick; set forward 25mm from wall face; 450mm high; flush pointing face and top	m	0.13	13.65	-	-	13.65	-
U6.7.7.02 as above; 900mm high; flush pointing face and top of projection	m	0.27	28.82	-	-	28.82	-

U6 Artificial Stone Blockwork continued...

	Unit	Labour Hours	Labour Net £	Plant Net £	Materials Net £	Unit Net £	CO_2e Kg
U6.8	**Ancillaries**						
U6.8.1	Joint reinforcement						
U6.8.1.01	Exmet galvanised brick reinforcement; 24 gauge; horizontal;	m	0.04	4.56	-	1.06	5.62	1.883
U6.8.1.02	as above; 225mm wide	m	0.05	5.70	-	2.12	7.82	3.766
U6.8.1.03	as above; 305mm wide	m	0.06	6.84	-	3.18	10.02	5.649
U6.8.1.04	as above; 450mm wide	m	0.07	7.98	-	4.24	12.22	7.532
U6.8.1.05	as above; 900mm wide	m	0.14	15.96	-	8.48	24.44	15.064
U6.8.2	Damp proof courses						
U6.8.2.01	Bitumen damp proof course; 100mm laps in cement mortar (1:3); pointing where exposed; Hessian based; horizontal; 115mm wide	m	0.01	1.37	-	1.48	2.85	3.009
U6.8.2.02	as above; 225mm wide	m	0.02	2.28	-	2.95	5.23	6.018
U6.8.2.03	as above; 330mm wide	m	0.04	4.56	-	4.43	8.99	9.027
U6.8.2.04	as above; 450mm wide	m	0.06	6.27	-	5.90	12.17	12.036
U6.8.2.05	as above; 900mm wide	m	0.07	7.41	-	11.80	19.21	24.071
U6.8.2.06	as above; vertical; 115mm wide	m	0.02	1.71	-	1.48	3.19	3.009
U6.8.2.07	as above; 225mm wide	m	0.02	2.51	-	2.95	5.46	6.018
U6.8.2.08	as above; 330mm wide	m	0.05	5.13	-	4.43	9.56	9.027
U6.8.2.09	as above; 450mm wide	m	0.06	6.84	-	5.90	12.74	12.036
U6.8.2.10	as above; 900mm wide	m	0.07	7.98	-	11.80	19.78	24.071
U6.8.2.11	as above; Fibre based; horizontal; 115mm wide	m	0.01	0.68	-	1.17	1.85	3.009
U6.8.2.12	as above; 225mm wide	m	0.01	1.37	-	2.33	3.70	6.018
U6.8.2.13	as above; 330mm wide	m	0.02	2.05	-	3.50	5.55	9.027
U6.8.2.14	as above; 450mm wide	m	0.02	2.74	-	4.66	7.40	12.036
U6.8.2.15	as above; 900mm wide	m	0.05	5.47	-	9.32	14.79	24.071
U6.8.2.16	as above; vertical; 115mm wide	m	0.01	1.37	-	1.17	2.54	3.009
U6.8.2.17	as above; 225mm wide	m	0.02	2.74	-	2.33	5.07	6.018
U6.8.2.18	as above; 330mm wide	m	0.04	4.10	-	3.50	7.60	9.027
U6.8.2.19	as above; 450mm wide	m	0.05	5.47	-	4.66	10.13	12.036
U6.8.2.20	as above; 900mm wide	m	0.10	10.94	-	9.32	20.26	24.071
U6.8.2.21	Hyload pitch polymer damp proof course; 100mm laps, sealed with Hyload contact adhesive; in cement mortar (1:3); pointing where exposed; Horizontal; 115mm wide	m	0.01	0.68	-	1.49	2.17	3.009
U6.8.2.22	as above; 225mm wide	m	0.01	1.37	-	2.97	4.34	6.018
U6.8.2.23	as above; 330mm wide	m	0.02	2.05	-	4.46	6.51	9.027
U6.8.2.24	as above; 450mm wide	m	0.02	2.74	-	5.94	8.68	12.036
U6.8.2.25	as above; 900mm wide	m	0.05	5.47	-	11.88	17.35	24.071
U6.8.2.26	as above; Vertical; 115mm wide	m	0.01	1.37	-	1.49	2.86	3.009
U6.8.2.27	as above; 225mm wide	m	0.02	2.74	-	2.97	5.71	6.018
U6.8.2.28	as above; 330mm wide	m	0.04	4.10	-	4.46	8.56	9.027
U6.8.2.29	as above; 450mm wide	m	0.05	5.47	-	5.94	11.41	12.036
U6.8.2.30	as above; 900mm wide	m	0.10	10.94	-	11.88	22.82	24.071
U6.8.3	Movement joints						
U6.8.3.01	Expansion joints; filling with Fillcrete joint filler; 20mm thick expansion joint; vertical; 102.5mm wide	m	0.06	7.18	-	2.42	9.60	2.531
U6.8.3.02	as above; 215mm wide	m	0.10	11.85	-	4.83	16.68	5.062
U6.8.3.03	as above; 327.5mm wide	m	0.17	19.03	-	7.25	26.28	7.592
U6.8.3.04	as above; 440mm wide	m	0.25	28.61	-	9.66	38.27	10.123
U6.8.3.05	as above; 890 mm wide	m	0.42	47.87	-	19.32	67.19	20.246

U6   Artificial Stone Blockwork continued...

	Unit	Labour Hours	Labour Net £	Plant Net £	Materials Net £	Unit Net £	CO_2e Kg
U6.8	**Ancillaries**						
U6.8.4	**Bonds to existing work**						
U6.8.4.01	Cast stonework; Bradstone walling blocks; Cotswold shades, random course in cement lime mortar (1:2:9); traditional walling; flush pointing, as the work proceeds; Bonding 100mm cast stone to existing 100mm cast stone; cutting and toothing alternate course	m^2	0.10	10.84	-	0.55	11.39	0.448
U6.8.5	**Infills of stated thickness**						
U6.8.5.01	Concrete infill; Grade 25, 20mm aggregate; 50mm thick	m^2	0.20	22.79	-	4.86	27.65	13.840
U6.8.5.02	as above; 100mm thick	m^2	0.30	34.19	-	9.72	43.91	27.680
U6.8.5.03	as above; 150mm thick	m^2	0.40	45.59	-	14.58	60.17	41.520
U6.8.5.04	as above; 200mm thick	m^2	0.50	56.98	-	19.44	76.42	55.360
U6.8.7	**Built-in pipes and ducts, cross-sectional area: not exceeding 0.05m^2**						
U6.8.7.01	Cast stonework; Bradstone walling blocks; Cotswold shades, random course in cement lime mortar (1:2:9); traditional walling; flush pointing, as the work proceeds; supply excluded; 100mm cast stone; cross-sectional area; not exceeding 0.025m^2	Nr	0.09	9.75	-	0.57	10.32	1.288
U6.8.7.02	as above; cross-sectional area; 0.025 - 0.25m^2	Nr	0.15	16.25	-	0.57	16.82	1.288
U6.8.7.03	as above; 150mm diameter clay pipe to, BS 65, 100mm cross-sectional area; not exceeding 0.025m^2	Nr	0.13	14.09	-	0.57	14.66	1.288
U6.8.8	**Built-in pipes and ducts, cross-sectional area: exceeding 0.05m^2**						
U6.8.8.01	Cast stonework; Bradstone walling blocks; Cotswold shades, random course in cement lime mortar (1:2:9); traditional walling; flush pointing, as the work proceeds; Built-in pipes and ducts (supply excluded); 100mm cast stone; cross-sectional area; 0.40m^2	Nr	0.18	19.50	-	0.57	20.07	1.28$_8$

U7 Ashlar Masonry

Note(s): The following Specialist prices are for natural stonework labour and materials fixed in position and should be regarded as indicative. The prices allow for the stone to be delivered in full loads within 60 miles of the stone masons depot.

	Unit	Labour Hours	Labour Net £	Plant Net £	Materials Net £	Unit Net £	CO_2e Kg
U7.1	**Thickness: not exceeding: 150mm**						
U7.1.1	Vertical straight walls						
U7.1.1.01 Natural stonework; Portland Whitbed with one exposed face in cement lime putty (2:5:7) with crushed stone dust; smooth finish; flush pointing one side as the work proceeds; 75mm thick	m²	0.95	108.27	8.76	39.62	156.65	19.889
U7.1.1.02 as above; 100mm thick	m²	1.25	142.46	11.53	50.71	204.70	23.153
U7.1.1.03 as above; 150mm thick	m²	1.50	170.95	13.83	72.88	257.66	27.919
U7.1.2	Vertical curved walls						
U7.1.2.01 Natural stonework; Portland Whitbed with one exposed face in cement lime putty (2:5:7) with crushed stone dust; smooth finish; flush pointing one side as the work proceeds; 75mm thick; curved walls; 5m mean radius	m²	1.75	199.45	16.14	39.62	255.21	23.915
U7.1.2.02 as above; 10m mean radius	m²	1.45	165.26	13.37	39.62	218.25	22.405
U7.1.2.03 as above; 100mm thick; curved walls; 5m mean radius	m²	2.00	227.94	18.44	50.71	297.09	26.927
U7.1.2.04 as above; 10m mean radius	m²	1.75	199.45	16.14	50.71	266.30	25.669
U7.1.2.05 as above; 150mm thick; curved walls; 5m mean radius	m²	2.50	284.93	23.05	72.88	380.86	32.952
U7.1.2.06 as above; 10m mean radius	m²	2.00	227.94	18.44	72.88	319.26	30.435
U7.1.5	Vertical facing to concrete						
U7.1.5.01 Natural stonework; Portland Whitbed with one exposed face in cement lime putty (2:5:7) with crushed stone dust; smooth finish; flush pointing one side as the work proceeds; 75mm thick; facing to concrete; building in 4nr fish tailed ties/m²; casting into concrete (by others)	m²	1.50	170.95	-	39.62	210.57	15.107
U7.1.5.02 as above; 100mm thick; facing to concrete; building in 4nr fish tailed ties/m²; casting into concrete (by others)	m²	1.75	199.45	-	50.71	250.16	16.861
U7.1.5.03 as above; 150mm thick; facing to concrete; building in 4nr fish tailed ties/m²; casting into concrete (by others)	m²	2.00	227.94	-	72.88	300.82	20.369
U7.2	**Thickness: 150 - 250mm**						
U7.2.1	Vertical straight walls						
U7.2.1.01 Natural stonework; Portland Whitbed with one exposed face in cement lime putty (2:5:7) with crushed stone dust; smooth finish; flush pointing one side as the work proceeds; 200mm thick	m²	2.00	227.94	13.83	95.05	336.82	31.426

U7 Ashlar Masonry continued...

	Unit	Labour Hours	Labour Net £	Plant Net £	Materials Net £	Unit Net £	CO_2e Kg
U7.2	**Thickness: 150 - 250mm**						
U7.2.2	Vertical curved walls						
U7.2.2.01	Natural stonework; Portland Whitbed with one exposed face in cement lime putty (2:5:7) with crushed stone dust; smooth finish; flush pointing one side as the work proceeds; 200mm thick; curved walls; 5m mean radius	m²	3.00	341.91	27.66	95.05	464.62	38.976
U7.2.2.02	as above; 10m mean radius	m²	2.50	284.93	23.05	95.05	403.03	36.459
U7.2.5	Vertical facing to concrete						
U7.2.5.01	Natural stonework; Portland Whitbed with one exposed face in cement lime putty (2:5:7) with crushed stone dust; smooth finish; flush pointing one side as the work proceeds; 200mm thick; facing to concrete; building in 4nr fish tailed ties/m²; casting into concrete (by others)	m²	2.00	227.94	18.44	95.05	341.43	33.943
U7.6	**Columns and piers of stated cross-sectional dimensions**						
U7.6.0	Generally						
U7.6.0.01	Natural stonework; Portland Whitbed with one exposed face in cement lime putty (2:5:7) with crushed stone dust; smooth finish; flush pointing one side as the work proceeds; 300 x 300mm	m	2.00	227.94	18.44	46.27	292.65	26.226
U7.6.0.02	as above; 450 x 300mm	m	2.50	284.93	23.05	66.23	374.21	31.899
U7.6.0.03	as above; 450 x 450mm	m	4.00	455.88	36.88	96.38	589.14	44.219
U7.6.0.04	as above; 900 x 900mm	m	8.00	911.76	73.76	365.51	1,351.03	106.935
U7.7	**Surface features**						
U7.7.2	Rebates and chases						
U7.7.2.01	Natural stonework; Portland Whitbed with one exposed face in cement lime putty (2:5:7) with crushed stone dust; smooth finish; flush pointing one side as the work proceeds; Forming fair rebate 25 x 10m deep	m	0.15	17.10	1.38	-	18.48	0.755
U7.7.4	Band courses						
U7.7.4.01	Natural stonework; Portland Whitbed with one exposed face in cement lime putty (2:5:7) with crushed stone dust; smooth finish; flush pointing one side as the work proceeds; Projecting band set forward 25mm from wall face; one course wide; flush pointing one side on the work proceeds.	m	0.20	22.79	1.84	8.14	32.77	11.133

U7 Ashlar Masonry continued...

	Unit	Labour Hours	Labour Net £	Plant Net £	Materials Net £	Unit Net £	CO_2e Kg
U7.7	**Surface features**						
U7.7.6	Pilasters						
U7.7.6.01	Natural stonework; Portland Whitbed with one exposed face in cement lime putty (2:5:7) with crushed stone dust; smooth finish; flush pointing one side as the work proceeds; Pilaster attached to wall face; 300 x 110mm; flush pointing all faces	m	0.75	85.48	6.92	21.00	113.40	15.936
U7.7.6.02	as above; 600 x 220mm; flush pointing all faces	m	1.00	113.97	9.22	64.90	188.09	24.139
U7.7.7	Plinths						
U7.7.7.01	Natural stonework; Portland Whitbed with one exposed face in cement lime putty (2:5:7) with crushed stone dust; smooth finish; flush pointing one side as the work proceeds; Plinth 100mm thick; set forward 25mm from wall face; 450mm high; flush pointing face and top of projection	m	1.00	113.97	9.22	26.32	149.51	18.036
U7.7.7.02	as above; 900mm high; flush pointing face and top of projection	m	2.00	227.94	18.44	46.27	292.65	26.226
U7.8	**Ancillaries**						
U7.8.1	Joint reinforcement						
U7.8.1.01	Exmet galvanised brick reinforcement; 24 gauge; horizontal;	m	0.04	4.56	-	1.06	5.62	1.883
U7.8.1.02	as above; 225mm wide	m	0.05	5.70	-	2.12	7.82	3.766
U7.8.1.03	as above; 305mm wide	m	0.06	6.84	-	3.18	10.02	5.649
U7.8.1.04	as above; 450mm wide	m	0.07	7.98	-	4.24	12.22	7.532
U7.8.1.05	as above; 900mm wide	m	0.14	15.96	-	8.48	24.44	15.064
U7.8.2	Damp proof courses						
U7.8.2.01	Bitumen damp proof course; 100mm laps in cement mortar (1:3); pointing where exposed; Hessian based; horizontal; 115mm wide	m	0.01	1.37	-	1.48	2.85	3.009
U7.8.2.02	as above; 225mm wide	m	0.02	2.28	-	2.95	5.23	6.018
U7.8.2.03	as above; 330mm wide	m	0.04	4.56	-	4.43	8.99	9.027
U7.8.2.04	as above; 450mm wide	m	0.06	6.27	-	5.90	12.17	12.036
U7.8.2.05	as above; 900mm wide	m	0.07	7.41	-	11.80	19.21	24.071
U7.8.2.06	as above; vertical; 115mm wide	m	0.02	1.71	-	1.48	3.19	3.009
U7.8.2.07	as above; 225mm wide	m	0.02	2.51	-	2.95	5.46	6.018
U7.8.2.08	as above; 330mm wide	m	0.05	5.13	-	4.43	9.56	9.027
U7.8.2.09	as above; 450mm wide	m	0.06	6.84	-	5.90	12.74	12.036
U7.8.2.10	as above; 900mm wide	m	0.07	7.98	-	11.80	19.78	24.071
U7.8.2.11	as above; Fibre based; horizontal; 115mm wide	m	0.01	0.68	-	1.17	1.85	3.009
U7.8.2.12	as above; 225mm wide	m	0.01	1.37	-	2.33	3.70	6.018
U7.8.2.13	as above; 330mm wide	m	0.02	2.05	-	3.50	5.55	9.027
U7.8.2.14	as above; 450mm wide	m	0.02	2.74	-	4.66	7.40	12.036
U7.8.2.15	as above; 900mm wide	m	0.05	5.47	-	9.32	14.79	24.071
U7.8.2.16	as above; vertical; 115mm wide	m	0.01	1.37	-	1.17	2.54	3.009
U7.8.2.17	as above; 225mm wide	m	0.02	2.74	-	2.33	5.07	6.018

U7 Ashlar Masonry continued...

	Unit	Labour Hours	Labour Net £	Plant Net £	Materials Net £	Unit Net £	CO_2e Kg
U7.8 **Ancillaries**							
U7.8.2 Damp proof courses							
U7.8.2.18 as previous item; 330mm wide	m	0.04	4.10	-	3.50	7.60	9.027
U7.8.2.19 as above; 450mm wide	m	0.05	5.47	-	4.66	10.13	12.036
U7.8.2.20 as above; 900mm wide	m	0.10	10.94	-	9.32	20.26	24.071
U7.8.2.21 Hyload pitch polymer damp proof course; 100mm laps, sealed with Hyload contact adhesive; in cement mortar (1:3); pointing where exposed; Horizontal; 115mm wide	m	0.01	0.68	-	1.49	2.17	3.009
U7.8.2.22 as above; 225mm wide	m	0.01	1.37	-	2.97	4.34	6.018
U7.8.2.23 as above; 330mm wide	m	0.02	2.05	-	4.46	6.51	9.027
U7.8.2.24 as above; 450mm wide	m	0.02	2.74	-	5.94	8.68	12.036
U7.8.2.25 as above; 900mm wide	m	0.05	5.47	-	11.88	17.35	24.071
U7.8.2.26 as above; Vertical; 115mm wide	m	0.01	1.37	-	1.49	2.86	3.009
U7.8.2.27 as above; 225mm wide	m	0.02	2.74	-	2.97	5.71	6.018
U7.8.2.28 as above; 330mm wide	m	0.04	4.10	-	4.46	8.56	9.027
U7.8.2.29 as above; 450mm wide	m	0.05	5.47	-	5.94	11.41	12.036
U7.8.2.30 as above; 900mm wide	m	0.10	10.94	-	11.88	22.82	24.071
U7.8.3 Movement joints							
U7.8.3.01 Expansion joints; filling with Fillcrete joint filler; 20mm thick expansion joint; vertical; 102.5mm wide	m	0.06	7.18	-	2.42	9.60	2.531
U7.8.3.02 as above; 215mm wide	m	0.10	11.85	-	4.83	16.68	5.062
U7.8.3.03 as above; 327.5mm wide	m	0.17	19.03	-	7.25	26.28	7.592
U7.8.3.04 as above; 440mm wide	m	0.25	28.61	-	9.66	38.27	10.123
U7.8.3.05 as above; 890 mm wide	m	0.42	47.87	-	19.32	67.19	20.246
U7.8.4 Bonds to existing work							
U7.8.4.01 Bonding 75mm natural stonework to existing 200mm natural stonework; cutting and toothing alternate courses	m²	0.33	37.95	3.07	-	41.02	1.676
U7.8.4.02 Bonding 100mm natural stonework to existing 200mm natural stonework; cutting and toothing alternate courses	m²	0.33	37.95	3.07	-	41.02	1.676
U7.8.5 Infills of stated thickness							
U7.8.5.01 Concrete infill; Grade 25, 20mm aggregate; 50mm thick	m²	0.20	22.79	-	4.86	27.65	13.840
U7.8.5.02 as above; 100mm thick	m²	0.30	34.19	-	9.72	43.91	27.680
U7.8.5.03 as above; 150mm thick	m²	0.40	45.59	-	14.58	60.17	41.520
U7.8.5.04 as above; 200mm thick	m²	0.50	56.98	-	19.44	76.42	55.360
U7.8.7 Built-in pipes and ducts, cross-sectional area: not exceeding 0.05m²							
U7.8.7.01 Natural stonework; Portland Whitbed with one exposed face in cement lime putty (2:5:7) with crushed stone dust; smooth finish; flush pointing one side as the work proceeds; supply excluded; 75mm natural stonework; cross-sectional area; not exceeding 0.025m²	Nr	0.17	19.03	1.54	-	20.57	0.841
U7.8.7.02 as above; cross-sectional area; 0.025 - 0.25m²	Nr	0.17	19.03	1.54	-	20.57	0.841

U7 Ashlar Masonry continued...

	Unit	Labour Hours	Labour Net £	Plant Net £	Materials Net £	Unit Net £	CO_2e Kg
U7.8 **Ancillaries**							
U7.8.7 Built-in pipes and ducts, cross-sectional area: not exceeding 0.05m²							
U7.8.7.03 as previous item; 150mm diameter clay pipe to, BS 65, 75mm natural stonework; cross-sectional area; not exceeding 0.025m²	Nr	0.17	19.03	1.54	-	20.57	0.841
U7.8.8 Built-in pipes and ducts, cross-sectional area: exceeding 0.05m²							
U7.8.8.01 Natural stonework; Portland Whitbed with one exposed face in cement lime putty (2:5:7) with crushed stone dust; smooth finish; flush pointing one side as the work proceeds; supply excluded; 75mm natural stonework; cross-sectional area; 0.40m²	Nr	0.17	19.03	1.54	-	20.57	0.841

U8 Rubble Masonry

	Unit	Labour Hours	Labour Net £	Plant Net £	Materials Net £	Unit Net £	CO_2e Kg
U8.2 **Thickness: 150 - 250mm**							
U8.2.5 Vertical facing to concrete							
U8.2.5.01 Cotswold limestone, Farmington stone quarry uncoursed random rubble walling 100 - 200mm high with natural exposed faces; in cement lime mortar (1:2:9); facing and recessed pointing as the work proceeds; 200mm thick	m²	0.44	47.57	-	81.96	129.53	30.508
U8.3 **Thickness: 250 - 500mm**							
U8.3.1 Vertical straight walls							
U8.3.1.01 Cotswold limestone, Farmington stone quarry uncoursed random rubble walling 50 - 100mm high with natural exposed faces; laid dry; facing both sides; 375mm thick	m²	0.57	61.22	-	174.97	236.19	54.648
U8.3.1.02 as above; 412mm thick	m²	0.62	67.40	-	192.26	259.66	60.048
U8.3.1.03 as above; 450mm thick	m²	0.68	73.47	-	209.97	283.44	65.577
U8.3.1.04 as above 100 - 200mm high with natural exposed faces; in cement lime mortar (1:2:9); facing and recessed pointing as the work proceeds; 450mm thick	m²	0.67	73.03	-	84.27	157.30	35.715
U8.3.1.05 as above; 500mm thick	m²	0.73	79.32	-	93.89	173.21	40.256
U8.3.2 Vertical curved walls							
U8.3.2.01 Cotswold limestone, Farmington stone quarry uncoursed random rubble walling 50 - 100mm high with natural exposed faces; laid dry; facing both sides; 375mm thick; curved walls; 5m mean radius	m²	1.08	116.92	-	174.97	291.89	54.648
U8.3.2.02 as above; 10m mean radius	m²	0.85	91.78	-	174.97	266.75	54.648
U8.3.2.03 as above; 412mm thick; curved walls; 5m mean radius	m²	1.21	130.57	-	192.26	322.83	60.048

U8 Rubble Masonry continued...

	Unit	Labour Hours	Labour Net £	Plant Net £	Materials Net £	Unit Net £	CO_2e Kg
U8.3	**Thickness: 250 - 500mm**						
U8.3.2	Vertical curved walls						
U8.3.2.04 as previous item; 10m mean radius	m²	0.93	100.88	-	192.26	293.14	60.048
U8.3.2.05 as above; 450mm thick; curved walls; 5m mean radius	m²	0.72	77.59	-	93.74	171.33	29.276
U8.3.2.06 as above; curved walls; 10m mean radius	m²	0.74	80.29	-	93.74	174.03	29.276
U8.3.2.07 as above 100 - 200mm high with natural exposed faces; in cement lime mortar (1:2:9); facing and recessed pointing as the work proceeds; 450mm thick; curved walls; 5m mean radius	m²	0.83	89.61	-	84.27	173.88	35.715
U8.3.2.08 as above; curved walls; 10m mean radius	m²	1.01	108.90	-	84.27	193.17	35.715
U8.3.2.09 as above; 500mm thick; curved walls; 5m mean radius	m²	0.91	98.61	-	93.89	192.50	40.256
U8.3.2.10 as above; 10m mean radius	m²	1.11	119.74	-	93.89	213.63	40.256
U8.3.3	Battered straight walls						
U8.3.3.01 Cotswold limestone, Farmington stone quarry uncoursed random rubble walling 50 - 100mm high with natural exposed faces; laid dry; facing both sides; 375mm thick; battered walls; one face battered at an angle of 1:20	m²	0.60	64.58	-	174.97	239.55	54.648
U8.3.3.02 as above; both faces battered at an angle of 1:20	m²	0.62	66.86	-	174.97	241.83	54.648
U8.3.3.03 as above; 412mm thick; battered walls; one face battered at an angle of 1:20	m²	0.66	70.98	-	192.26	263.24	60.048
U8.3.3.04 as above; both faces battered at an angle of 1:20	m²	0.68	73.47	-	192.26	265.73	60.048
U8.3.3.05 as above; 450mm thick; battered walls; one face battered at an angle of 1:20	m²	1.30	140.43	-	93.74	234.17	29.276
U8.3.3.06 as above; both faces battered at an angle of 1:20	m²	1.02	110.20	-	93.74	203.94	29.276
U8.3.3.07 as above 100 - 200mm high with natural exposed faces; in cement lime mortar (1:2:9); facing and recessed pointing as the work proceeds; 450mm thick; battered walls; one face battered at angle of 1:20	m²	1.17	127.00	-	84.27	211.27	35.715
U8.3.3.08 as above; battered walls; both faces battered at an angle of 1:20	m²	1.00	108.79	-	84.27	193.06	35.715
U8.3.3.09 as above; 500mm thick; battered walls; one face battered at angle of 1:20	m²	1.28	138.70	-	93.89	232.59	40.256
U8.3.3.10 as above; battered walls; both faces battered at an angle of 1:20	m²	1.10	118.98	-	93.89	212.87	40.256

U8　Rubble Masonry continued...

	Unit	Labour Hours	Labour Net £	Plant Net £	Materials Net £	Unit Net £	CO_2e Kg
U8.4	**Thickness: 500mm - 1m**						
U8.4.1	Vertical straight walls						
U8.4.1.01 Cotswold limestone, Farmington stone quarry uncoursed random rubble walling 100 - 200mm high with natural exposed faces; in cement lime mortar (1:2:9); facing and recessed pointing as the work proceeds; facing and recessed pointing as the work proceeds; 600mm thick	m²	0.86	92.97	-	112.55	205.52	48.050
U8.4.2	Vertical curved walls						
U8.4.2.01 Cotswold limestone, Farmington stone quarry uncoursed random rubble walling 100 - 200mm high with natural exposed faces; in cement lime mortar (1:2:9); facing and recessed pointing as the work proceeds; facing and recessed pointing as the work proceeds; 600mm thick; 5m mean radius	m²	1.04	112.26	-	112.55	224.81	48.050
U8.4.2.02 as above; 10m mean radius	m²	1.23	133.39	-	112.55	245.94	48.050
U8.4.3	Battered straight walls						
U8.4.3.01 Cotswold limestone, Farmington stone quarry uncoursed random rubble walling 100 - 200mm high with natural exposed faces; in cement lime mortar (1:2:9); facing and recessed pointing as the work proceeds; facing and recessed pointing as the work proceeds; 600mm thick	m²	1.41	152.35	-	112.55	264.90	48.050
U8.4.3.02 as above; battered walls; both faces battered at an angle of 1:20	m²	1.24	133.82	-	112.55	246.37	48.050
U8.7	**Surface features**						
U8.7.2	Rebates and chases						
U8.7.2.01 Cotswold limestone, Farmington stone quarry uncoursed random rubble walling 50 - 100mm high with natural exposed faces; laid dry; facing both sides; Forming fair rebate 25 x 10mm deep	m	0.13	13.55	-	-	13.55	-
U8.7.2.02 as above 100 - 200mm high with natural exposed faces; in cement lime mortar (1:2:9); facing and recessed pointing as the work proceeds; Forming fair rebate 25 x 10mm deep	m	0.13	13.55	-	-	13.55	-

U8 Rubble Masonry continued...

	Unit	Labour Hours	Labour Net £	Plant Net £	Materials Net £	Unit Net £	CO_2e Kg
U8.7	**Surface features**						
U8.7.4	Band courses						
U8.7.4.01	Cotswold limestone, Farmington stone quarry uncoursed random rubble walling 50 - 100mm high with natural exposed faces; laid dry; facing both sides; Projecting band set forward 25mm from wall face; one course wide; flush pointing one side and top and bottom of projection	m	0.06	6.83	-	-	6.83	-
U8.7.4.02	as above 100 - 200mm high with natural exposed faces; in cement lime mortar (1:2:9); facing and recessed pointing as the work proceeds; Projecting band set forward 25mm from wall face; one course wide; flush pointing one side and top and bottom of projection	m	0.06	6.83	-	-	6.83	-
U8.7.6	Pilasters						
U8.7.6.01	Cotswold limestone, Farmington stone quarry uncoursed random rubble walling 50 - 100mm high with natural exposed faces; laid dry; facing both sides; Pilaster attached to wall face; 600 x 200mm; recessed pointing all faces	m	0.17	18.10	-	25.57	43.67	9.095
U8.7.6.02	as above; 900 x 300mm	m	0.29	31.75	-	56.81	88.56	18.853
U8.7.6.03	Cotswold limestone, Farmington stone quarry uncoursed random rubble walling 100 - 200mm high with natural exposed faces; in cement lime mortar (1:2:9); facing and recessed pointing as the work proceeds; Pilaster attached to wall face; 600 x 200mm; recessed pointing all faces	m	0.17	18.10	-	22.28	40.38	9.095
U8.7.6.04	as above; 900 x 300mm; recessed pointing all faces	m	0.29	31.75	-	49.99	81.74	20.141
U8.7.7	Plinths						
U8.7.7.01	Cotswold limestone, Farmington stone quarry uncoursed random rubble walling 50 - 100mm high with natural exposed faces; laid dry; facing both sides; Plinth one course thick; set forward 25mm from wall face; 450mm high	m	0.17	18.10	-	-	18.10	-
U8.7.7.02	as above; 900mm high	m	0.33	35.76	-	-	35.76	-

U8 Rubble Masonry continued...

	Unit	Labour Hours	Labour Net £	Plant Net £	Materials Net £	Unit Net £	CO_2e Kg
U8.7	**Surface features**						
U8.7.7	Plinths						
U8.7.7.03	Cotswold limestone, Farmington stone quarry uncoursed random rubble walling 100 - 200mm high with natural exposed faces; in cement lime mortar (1:2:9); facing and recessed pointing as the work proceeds; Plinth one course thick; set forward 25mm from wall face; 450mm high; recessed pointing face and top of projection	m	0.17	18.10	-	-	18.10	-
U8.7.7.04	Cotswold limestone, Farmington stone quarry uncoursed random rubble walling 100 - 200mm high with natural exposed faces; in cement lime mortar (1:2:9); facing and recessed pointing as the work proceeds; Plinth one course thick; set forward 25mm from wall face; 900mm high; recessed pointing face and top of projection	m	0.33	35.76	-	-	35.76	-
U8.8	**Ancillaries**						
U8.8.1	Joint reinforcement						
U8.8.1.01	Exmet galvanised brick reinforcement; 24 gauge; horizontal; 100mm wide	m	0.04	4.56	-	1.06	5.62	1.883
U8.8.1.02	as above; 225mm wide	m	0.05	5.70	-	2.12	7.82	3.766
U8.8.1.03	as above; 305mm wide	m	0.06	6.84	-	3.18	10.02	5.649
U8.8.1.04	as above; 450mm wide	m	0.07	7.98	-	4.24	12.22	7.532
U8.8.1.05	as above; 900mm wide	m	0.14	15.96	-	8.48	24.44	15.064
U8.8.2	Damp proof courses						
U8.8.2.01	Bitumen damp proof course; 100mm laps in cement mortar (1:3); pointing where exposed; Hessian based; horizontal; 115mm wide	m	0.01	1.37	-	1.48	2.85	3.009
U8.8.2.02	as above; 225mm wide	m	0.02	2.28	-	2.95	5.23	6.018
U8.8.2.03	as above; 330mm wide	m	0.04	4.56	-	4.43	8.99	9.027
U8.8.2.04	as above; 450mm wide	m	0.06	6.27	-	5.90	12.17	12.036
U8.8.2.05	as above; 900mm wide	m	0.07	7.41	-	11.80	19.21	24.071
U8.8.2.06	as above; vertical; 115mm wide	m	0.02	1.71	-	1.48	3.19	3.009
U8.8.2.07	as above; 225mm wide	m	0.02	2.51	-	2.95	5.46	6.018
U8.8.2.08	as above; 330mm wide	m	0.05	5.13	-	4.43	9.56	9.027
U8.8.2.09	as above; 450mm wide	m	0.06	6.84	-	5.90	12.74	12.036
U8.8.2.10	as above; 900mm wide	m	0.07	7.98	-	11.80	19.78	24.071
U8.8.2.11	as above; Fibre based; horizontal; 115mm wide	m	0.01	0.68	-	1.17	1.85	3.009
U8.8.2.12	as above; 225mm wide	m	0.01	1.37	-	2.33	3.70	6.018
U8.8.2.13	as above; 330mm wide	m	0.02	2.05	-	3.50	5.55	9.027
U8.8.2.14	as above; 450mm wide	m	0.02	2.74	-	4.66	7.40	12.036
U8.8.2.15	as above; 900mm wide	m	0.05	5.47	-	9.32	14.79	24.071
U8.8.2.16	as above; vertical; 115mm wide	m	0.01	1.37	-	1.17	2.54	3.009
U8.8.2.17	as above; 225mm wide	m	0.02	2.74	-	2.33	5.07	6.018
U8.8.2.18	as above; 330mm wide	m	0.04	4.10	-	3.50	7.60	9.027
U8.8.2.19	as above; 450mm wide	m	0.05	5.47	-	4.66	10.13	12.036
U8.8.2.20	as above; 900mm wide	m	0.10	10.94	-	9.32	20.26	24.071

U8 Rubble Masonry continued...

	Unit	Labour Hours	Labour Net £	Plant Net £	Materials Net £	Unit Net £	CO_2e Kg
U8.8	**Ancillaries**						
U8.8.2	Damp proof courses						
U8.8.2.21	Hyload pitch polymer damp proof course; 100mm laps, sealed with Hyload contact adhesive; in cement mortar (1:3); pointing where exposed; Horizontal; 115mm wide	m	0.01	0.68	-	1.49	2.17	3.009
U8.8.2.22	as above; 225mm wide	m	0.01	1.37	-	2.97	4.34	6.018
U8.8.2.23	as above; 330mm wide	m	0.02	2.05	-	4.46	6.51	9.027
U8.8.2.24	as above; 450mm wide	m	0.02	2.74	-	5.94	8.68	12.036
U8.8.2.25	as above; 900mm wide	m	0.05	5.47	-	11.88	17.35	24.071
U8.8.2.26	as above; Vertical; 115mm wide	m	0.01	1.37	-	1.49	2.86	3.009
U8.8.2.27	as above; 225mm wide	m	0.02	2.74	-	2.97	5.71	6.018
U8.8.2.28	as above; 330mm wide	m	0.04	4.10	-	4.46	8.56	9.027
U8.8.2.29	as above; 450mm wide	m	0.05	5.47	-	5.94	11.41	12.036
U8.8.2.30	as above; 900mm wide	m	0.10	10.94	-	11.88	22.82	24.071
U8.8.3	Movement joints						
U8.8.3.01	Expansion joints; filling with Fillcrete joint filler; 20mm thick expansion joint; vertical; 102.5mm wide	m	0.06	7.18	-	2.42	9.60	2.531
U8.8.3.02	as above; 215mm wide	m	0.10	11.85	-	4.83	16.68	5.062
U8.8.3.03	as above; 327.5mm wide	m	0.17	19.03	-	7.25	26.28	7.592
U8.8.3.04	as above; 440mm wide	m	0.25	28.61	-	9.66	38.27	10.123
U8.8.3.05	as above; 890 mm wide	m	0.42	47.87	-	19.32	67.19	20.246
U8.8.4	Bonds to existing work						
U8.8.4.01	Cotswold limestone, Farmington stone quarry uncoursed random rubble walling 50 - 100mm high with natural exposed faces; laid dry; facing both sides; bonding 450mm rubble masonry to existing 450mm rubble masonry; cutting and toothing alternate courses	m²	0.13	13.55	-	2.20	15.75	0.908
U8.8.4.02	as above; bonding 500mm rubble masonry to existing 500mm rubble masonry; cutting and toothing alternate courses	m²	0.17	18.10	-	2.25	20.35	1.037
U8.8.4.03	as above 100 - 200mm high with natural exposed faces; in cement lime mortar (1:2:9); facing and recessed pointing as the work proceeds; bonding 450mm rubble masonry to existing 450mm rubble masonry; cutting and toothing alternate courses	m²	0.13	13.55	-	1.92	15.47	0.908
U8.8.4.04	as above; bonding 500mm rubble masonry to existing 500mm rubble masonry; cutting and toothing alternate courses	m²	0.17	18.10	-	1.98	20.08	1.037
U8.8.5	Infills of stated thickness						
U8.8.5.01	Concrete infill; Grade 25, 20mm aggregate; 50mm thick	m²	0.20	22.79	-	4.86	27.65	13.840
U8.8.5.02	as above; 100mm thick	m²	0.30	34.19	-	9.72	43.91	27.680
U8.8.5.03	as above; 150mm thick	m²	0.40	45.59	-	14.58	60.17	41.520

U8 Rubble Masonry continued...

	Unit	Labour Hours	Labour Net £	Plant Net £	Materials Net £	Unit Net £	CO_2e Kg
U8.8	**Ancillaries**						
U8.8.5	Infills of stated thickness						
U8.8.5.04	as previous item; 200mm thick	m^2	0.50	56.98	-	19.44	76.42	55.360
U8.8.7	Built-in pipes and ducts, cross-sectional area: not exceeding 0.05m²						
U8.8.7.01	Cotswold limestone, Farmington stone quarry uncoursed random rubble walling 50 - 100mm high with natural exposed faces; laid dry; facing both sides; supply excluded; 450mm rubble masonry; cross-sectional area; not exceeding 0.025m²	Nr	0.21	23.08	-	0.29	23.37	0.644
U8.8.7.02	as above; 150mm diameter clay pipe to, BS 65, 450mm rubble masonry; cross-sectional area; not exceeding 0.025m²	Nr	0.19	20.37	-	0.57	20.94	1.288
U8.8.7.03	as above 100 - 200mm high with natural exposed faces; in cement lime mortar (1:2:9); facing and recessed pointing as the work proceeds; supply excluded; 450mm rubble masonry; cross-sectional area; not exceeding 0.025m²	Nr	0.17	18.10	-	0.57	18.67	1.288
U8.8.7.04	as above; 150mm diameter clay pipe to, BS 65, 450mm rubble masonry; cross-sectional area; not exceeding 0.025m²	Nr	0.19	20.37	-	0.57	20.94	1.288
U8.8.8	Built-in pipes and ducts, cross-sectional area: exceeding 0.05m²						
U8.8.8.01	Cotswold limestone, Farmington stone quarry uncoursed random rubble walling 50 - 100mm high with natural exposed faces; laid dry; facing both sides; supply excluded; 450mm rubble masonry; cross-sectional area; 0.025 - 0.25m²	Nr	0.21	23.08	-	0.29	23.37	0.644
U8.8.8.02	as above; cross-sectional area; 0.40m²	Nr	0.29	31.75	-	0.34	32.09	0.773
U8.8.8.03	as above 100 - 200mm high with natural exposed faces; in cement lime mortar (1:2:9); facing and recessed pointing as the work proceeds; supply excluded; 450mm rubble masonry; cross-sectional area; 0.025 - 0.25m²	Nr	0.21	23.08	-	0.57	23.65	1.288
U8.8.8.04	as above; cross-sectional area: 0.40m²	Nr	0.29	31.75	-	0.57	32.32	1.288

CLASS V:
PAINTING

Calculations used throughout Class V - Painting

Labour

			Qty		Rate		Total
L A0310ICE	**Painting Labour Gang**						
	L A9041ICE	Painter (chargehand)	1	x	18.63	=	£18.63
	L A9027ICE	Painter	2	x	17.33	=	£34.66
	L A9028ICE	Brush hand (labourer)	1	x	13.02	=	£13.02
		Total hourly cost of gang				=	**£66.31**

Painting

Note(s): 1) Labour rates are based on free access to the work. No allowance has been made for scaffolding or time required to gain access to the work. The rates include an element of normal clean up time.

2) All the rates shown are based on hand work; where spray painting is justified, that is where large areas have to be covered, the cost per metre can be substantially reduced. The following table shows gang hours to paint various surfaces one coat using one gun and non-lead paint.

Surface	Area Covered m^2/litre	Area m^2/Hour	Gang Hours /m^2
Concrete	10	30	0.033
Brickwork	10	30	0.033
Plasterer	14	35	0.028
Metal (smooth)	19	40	0.025
Wood (planed)	16	38	0.026

3) Priming coats or first coats include the cost of labour in preparing surfaces.

V1 Iron Or Zinc Based Primer Paint

		Unit	Labour Hours	Labour Net £	Plant Net £	Materials Net £	Unit Net £	CO_2e Kg
V1.1	**Metal, other than metal sections and pipework**							
V1.1.1	Upper surfaces inclined at an angle not exceeding 30 degrees to the horizontal							
V1.1.1.04	Iron based primer paint	m^2	0.08	4.97	-	0.44	5.41	0.124
V1.1.1.05	Zinc rich primer paint	m^2	0.08	4.97	-	0.56	5.53	0.124
V1.1.2	Upper surfaces inclined at 30 - 60 degrees to the horizontal							
V1.1.2.04	Iron based primer paint	m^2	0.08	5.17	-	0.44	5.61	0.124
V1.1.2.05	Zinc rich primer paint	m^2	0.08	5.17	-	0.56	5.73	0.124
V1.1.3	Surfaces inclined at an angle exceeding 60 degrees to the horizontal							
V1.1.3.04	Iron based primer paint	m^2	0.08	5.30	-	0.44	5.74	0.124
V1.1.3.05	Zinc rich primer paint	m^2	0.08	5.30	-	0.56	5.86	0.124
V1.1.4	Soffit surfaces and lower surfaces inclined at an angle not exceeding 60 degrees to the horizontal							
V1.1.4.04	Iron based primer paint	m^2	0.10	6.63	-	0.44	7.07	0.124
V1.1.4.05	Zinc rich primer paint	m^2	0.10	6.63	-	0.56	7.19	0.124
V1.1.6	Surfaces of width not exceeding 300mm							
V1.1.6.04	Iron based primer paint	m	0.03	1.66	-	0.13	1.79	0.037
V1.1.6.05	Zinc rich primer paint	m	0.03	1.66	-	0.17	1.83	0.037
V1.1.7	Surfaces of width 300mm - 1m							
V1.1.7.04	Iron based primer paint	m	0.04	2.52	-	0.22	2.74	0.062
V1.1.7.05	Zinc rich primer paint	m	0.04	2.52	-	0.42	2.94	0.093
V1.1.8	Isolated groups of surfaces							
V1.1.8.04	Iron based primer paint; 500 x 300mm inspection covers and frames in wall; painting one side	Nr	0.15	9.94	-	0.88	10.82	0.248
V1.1.8.05	Zinc rich primer paint; 500 x 300mm inspection covers and frames in wall; painting one side	Nr	0.15	9.94	-	0.56	10.50	0.124
V1.2	**Timber**							
V1.2.1	Upper surfaces inclined at an angle not exceeding 30 degrees to the horizontal							
V1.2.1.01	Wood primer; planed timber	m^2	0.08	4.97	-	0.66	5.63	0.124
V1.2.2	Upper surfaces inclined at 30 - 60 degrees to the horizontal							
V1.2.2.01	Wood primer; planed timber	m^2	0.08	5.17	-	0.66	5.83	0.124
V1.2.3	Surfaces inclined at an angle exceeding 60 degrees to the horizontal							
V1.2.3.01	Wood primer; planed timber	m^2	0.08	5.30	-	0.66	5.96	0.124

V1   Iron Or Zinc Based Primer Paint continued...

		Unit	Labour Hours	Labour Net £	Plant Net £	Materials Net £	Unit Net £	CO_2e Kg
V1.2	**Timber**							
V1.2.4	Soffit surfaces and lower surfaces inclined at an angle not exceeding 60 degrees to the horizontal							
V1.2.4.01	Wood primer; planed timber	m²	0.10	6.63	-	0.66	7.29	0.124
V1.2.6	Surfaces of width not exceeding 300mm							
V1.2.6.01	Wood primer; planed timber	m	0.03	1.66	-	0.20	1.86	0.037
V1.2.7	Surfaces of width 300mm - 1m							
V1.2.7.01	Wood primer; planed timber	m	0.04	2.52	-	0.33	2.85	0.062
V1.2.8	Isolated groups of surfaces							
V1.2.8.01	Wood primer; planed timber; 300 x 200mm hatch and frame in wall; painting one side	Nr	0.15	9.94	-	1.32	11.26	0.248
V1.3	**Smooth concrete**							
V1.3.1	Upper surfaces inclined at an angle not exceeding 30 degrees to the horizontal							
V1.3.1.01	Masonry sealer; Unprimed smooth concrete	m²	0.08	4.97	-	1.03	6.00	0.210
V1.3.2	Upper surfaces inclined at 30 - 60 degrees to the horizontal							
V1.3.2.01	Masonry sealer; Unprimed smooth concrete	m²	0.08	5.11	-	1.03	6.14	0.210
V1.3.3	Surfaces inclined at an angle exceeding 60 degrees to the horizontal							
V1.3.3.01	Masonry sealer; Unprimed smooth concrete	m²	0.08	5.44	-	1.03	6.47	0.210
V1.3.4	Soffit surfaces and lower surfaces inclined at an angle not exceeding 60 degrees to the horizontal							
V1.3.4.01	Masonry sealer; Unprimed smooth concrete	m²	0.10	6.63	-	1.03	7.66	0.210
V1.3.6	Surfaces of width not exceeding 300mm							
V1.3.6.01	Masonry sealer; Unprimed smooth concrete	m	0.03	1.66	-	0.36	2.02	0.074
V1.3.7	Surfaces of width 300mm - 1m							
V1.3.7.01	Masonry sealer; Unprimed smooth concrete	m	0.04	2.45	-	0.73	3.18	0.149
V1.3.8	Isolated groups of surfaces							
V1.3.8.01	Masonry sealer; Unprimed smooth concrete; stepped and splayed plinth overall size 450 x 450 x 450mm; painting all faces	Nr	0.15	9.94	-	2.07	12.01	0.421
V1.4	**Rough concrete**							
V1.4.1	Upper surfaces inclined at an angle not exceeding 30 degrees to the horizontal							
V1.4.1.01	Masonry sealer; unprimed rough concrete; sawn board finish	m²	0.08	5.44	-	1.16	6.60	0.235
V1.4.2	Upper surfaces inclined at 30 - 60 degrees to the horizontal							
V1.4.2.01	Masonry sealer; unprimed rough concrete; sawn board finish	m²	0.09	5.64	-	1.16	6.80	0.235
V1.4.3	Surfaces inclined at an angle exceeding 60 degrees to the horizontal							
V1.4.3.01	Masonry sealer; unprimed rough concrete; sawn board finish	m²	0.09	5.97	-	1.16	7.13	0.235

V1 Iron Or Zinc Based Primer Paint continued...

		Unit	Labour Hours	Labour Net £	Plant Net £	Materials Net £	Unit Net £	CO_2e Kg
V1.4	**Rough concrete**							
V1.4.4	Soffit surfaces and lower surfaces inclined at an angle not exceeding 60 degrees to the horizontal							
V1.4.4.01	Masonry sealer; unprimed rough concrete; sawn board finish	m²	0.11	7.29	-	1.16	8.45	0.235
V1.4.6	Surfaces of width not exceeding 300mm							
V1.4.6.01	Masonry sealer; unprimed rough concrete; sawn board finish	m	0.03	1.79	-	0.36	2.15	0.074
V1.4.7	Surfaces of width 300mm - 1m							
V1.4.7.01	Masonry sealer; unprimed rough concrete; sawn board finish	m	0.04	2.65	-	0.73	3.38	0.149
V1.4.8	Isolated groups of surfaces							
V1.4.8.01	Masonry sealer; unprimed rough concrete; sawn board finish; stepped and splayed plinth overall size 450 x 450 x 450mm; painting all faces	Nr	0.17	10.94	-	2.31	13.25	0.470
V1.5	**Masonry**							
V1.5.1	Upper surfaces inclined at an angle not exceeding 30 degrees to the horizontal							
V1.5.1.01	Masonry sealer; unprimed masonry	m²	0.08	5.44	-	1.16	6.60	0.235
V1.5.2	Upper surfaces inclined at 30 - 60 degrees to the horizontal							
V1.5.2.01	Masonry sealer; unprimed masonry	m²	0.09	5.64	-	1.16	6.80	0.235
V1.5.3	Surfaces inclined at an angle exceeding 60 degrees to the horizontal							
V1.5.3.01	Masonry sealer; unprimed masonry	m²	0.09	5.97	-	1.16	7.13	0.235
V1.5.4	Soffit surfaces and lower surfaces inclined at an angle not exceeding 60 degrees to the horizontal							
V1.5.4.01	Masonry sealer; unprimed masonry	m²	0.11	7.29	-	1.16	8.45	0.235
V1.5.6	Surfaces of width not exceeding 300mm							
V1.5.6.01	Masonry sealer; unprimed masonry	m	0.03	1.79	-	0.36	2.15	0.074
V1.5.7	Surfaces of width 300mm - 1m							
V1.5.7.01	Masonry sealer; unprimed masonry	m	0.04	2.65	-	0.73	3.38	0.149
V1.5.8	Isolated groups of surfaces							
V1.5.8.01	Masonry sealer; unprimed masonry; stepped and splayed plinth overall size 450 x 450 x 450mm; painting all faces	Nr	0.17	10.94	-	2.31	13.25	0.470
V1.6	**Brickwork and blockwork**							
V1.6.1	Upper surfaces inclined at an angle not exceeding 30 degrees to the horizontal							
V1.6.1.01	Masonry sealer; unprimed brickwork and blockwork	m²	0.09	5.97	-	1.34	7.31	0.272

V1 Iron Or Zinc Based Primer Paint continued...

	Unit	Labour Hours	Labour Net £	Plant Net £	Materials Net £	Unit Net £	CO_2e Kg
VI.6 Brickwork and blockwork							
VI.6.2 Upper surfaces inclined at 30 - 60 degrees to the horizontal							
VI.6.2.01 Masonry sealer; unprimed brickwork and blockwork	m²	0.10	6.30	-	1.34	7.64	0.272
VI.6.3 Surfaces inclined at an angle exceeding 60 degrees to the horizontal							
VI.6.3.01 Masonry sealer; unprimed brickwork and blockwork	m²	0.10	6.63	-	1.34	7.97	0.272
VI.6.4 Soffit surfaces and lower surfaces inclined at an angle not exceeding 60 degrees to the horizontal							
VI.6.4.01 Masonry sealer; unprimed brickwork and blockwork	m²	0.12	7.96	-	1.34	9.30	0.272
VI.6.6 Surfaces of width not exceeding 300mm							
VI.6.6.01 Masonry sealer; unprimed brickwork and blockwork	m	0.03	1.99	-	0.43	2.42	0.087
VI.6.7 Surfaces of width 300mm - Im							
VI.6.7.01 Masonry sealer; unprimed brickwork and blockwork	m	0.05	2.98	-	0.67	3.65	0.136
VI.6.8 Isolated groups of surfaces							
VI.6.8.01 Masonry sealer; unprimed brickwork and blockwork; stepped and splayed plinth overall size 450 x 450 x 450mm; painting all faces	Nr	0.18	11.93	-	2.68	14.61	0.545
VI.7 Metal sections							
VI.7.0 Generally							
VI.7.0.04 Iron based primer paint	m²	0.10	6.63	-	0.44	7.07	0.124
VI.7.0.05 Zinc rich primer paint	m²	0.08	4.97	-	0.56	5.53	0.124
VI.8 Pipework							
VI.8.0 Generally							
VI.8.0.04 Iron based primer paint	m²	0.10	6.63	-	0.44	7.07	0.124

V2 Etch Primer Paint

	Unit	Labour Hours	Labour Net £	Plant Net £	Materials Net £	Unit Net £	CO_2e Kg
V2.I Metal, other than metal sections and pipework							
V2.I.I Upper surfaces inclined at an angle not exceeding 30 degrees to the horizontal							
V2.I.I.01 Zinc coated or aluminium surfaces	m²	0.08	4.97	-	1.15	6.12	0.186
V2.I.2 Upper surfaces inclined at 30 - 60 degrees to the horizontal							
V2.I.2.01 Zinc coated or aluminium surfaces	m²	0.08	5.17	-	1.15	6.32	0.186
V2.I.3 Surfaces inclined at an angle exceeding 60 degrees to the horizontal							
V2.I.3.01 Zinc coated or aluminium surfaces	m²	0.08	5.30	-	1.15	6.45	0.186
V2.I.4 Soffit surfaces and lower surfaces inclined at an angle not exceeding 60 degrees to the horizontal							
V2.I.4.01 Zinc coated or aluminium surfaces	m²	0.10	6.63	-	1.15	7.78	0.186

V2 Etch Primer Paint continued...

		Unit	Labour Hours	Labour Net £	Plant Net £	Materials Net £	Unit Net £	CO_2e Kg
V2.1	**Metal, other than metal sections and pipework**							
V2.1.6	Surfaces of width not exceeding 300mm							
V2.1.6.01	Zinc coated or aluminium surfaces	m	0.03	1.66	-	0.38	2.04	0.062
V2.1.7	Surfaces of width 300mm - 1m							
V2.1.7.01	Zinc coated or aluminium surfaces	m	0.04	2.52	-	0.54	3.06	0.087
V2.1.8	Isolated groups of surfaces							
V2.1.8.01	Zinc coated or aluminium surfaces; 300 x 200mm access hatch and frame in wall; painting one side	Nr	0.15	9.94	-	0.23	10.17	0.037
V2.7	**Metal sections**							
V2.7.0	Generally							
V2.7.0.01	Zinc coated or aluminium surfaces	m²	0.08	5.44	-	1.15	6.59	0.186
V2.8	**Pipework**							
V2.8.0	Generally							
V2.8.0.01	Zinc coated or aluminium surfaces	m²	0.08	5.44	-	1.15	6.59	0.186

V3 Oil Paint

		Unit	Labour Hours	Labour Net £	Plant Net £	Materials Net £	Unit Net £	CO_2e Kg
V3.1	**Metal, other than metal sections and pipework**							
V3.1.1	Upper surfaces inclined at an angle not exceeding 30 degrees to the horizontal							
V3.1.1.01	in three coats; primed metal	m²	0.09	5.77	-	2.17	7.94	0.347
V3.1.2	Upper surfaces inclined at 30 - 60 degrees to the horizontal							
V3.1.2.01	in three coats; primed metal	m²	0.09	5.97	-	2.17	8.14	0.347
V3.1.3	Surfaces inclined at an angle exceeding 60 degrees to the horizontal							
V3.1.3.01	in three coats; primed metal	m²	0.10	6.30	-	2.17	8.47	0.347
V3.1.4	Soffit surfaces and lower surfaces inclined at an angle not exceeding 60 degrees to the horizontal							
V3.1.4.01	in three coats; primed metal	m²	0.12	7.76	-	2.17	9.93	0.347
V3.1.6	Surfaces of width not exceeding 300mm							
V3.1.6.01	in three coats; primed metal	m	0.03	1.99	-	0.70	2.69	0.111
V3.1.7	Surfaces of width 300mm - 1m							
V3.1.7.01	in three coats; primed metal	m	0.04	2.78	-	1.40	4.18	0.223
V3.1.8	Isolated groups of surfaces							
V3.1.8.01	in three coats; primed metal; 300 x 200mm access hatch and frame in wall; painting one side	Nr	0.18	11.93	-	4.35	16.28	0.693
V3.2	**Timber**							
V3.2.1	Upper surfaces inclined at an angle not exceeding 30 degrees to the horizontal							
V3.2.1.01	in three coats; planed, primed timber	m²	0.09	5.77	-	2.17	7.94	0.347
V3.2.2	Upper surfaces inclined at 30 - 60 degrees to the horizontal							
V3.2.2.01	in three coats; planed, primed timber	m²	0.09	5.97	-	2.17	8.14	0.347

V3 Oil Paint continued...

	Unit	Labour Hours	Labour Net £	Plant Net £	Materials Net £	Unit Net £	CO_2e Kg
V3.2 **Timber**							
V3.2.3 Surfaces inclined at an angle exceeding 60 degrees to the horizontal							
V3.2.3.01 in three coats; planed, primed timber	m²	0.10	6.30	-	2.17	8.47	0.347
V3.2.4 Soffit surfaces and lower surfaces inclined at an angle not exceeding 60 degrees to the horizontal							
V3.2.4.01 in three coats; planed, primed timber	m²	0.12	7.76	-	2.17	9.93	0.347
V3.2.6 Surfaces of width not exceeding 300mm							
V3.2.6.01 in three coats; planed, primed timber	m	0.03	1.99	-	0.70	2.69	0.111
V3.2.7 Surfaces of width 300mm - 1m							
V3.2.7.01 in three coats; planed, primed timber	m	0.04	2.78	-	1.40	4.18	0.223
V3.2.8 Isolated groups of surfaces							
V3.2.8.01 in three coats; planed, primed timber; 300 x 200mm access hatch and frame in wall; painting one side	Nr	0.18	11.93	-	4.35	16.28	0.693
V3.7 **Metal sections**							
V3.7.0 Generally							
V3.7.0.01 Oil paint in three coats; planed, primed timber	m²	0.12	7.76	-	2.17	9.93	0.347
V3.8 **Pipework**							
V3.8.0 Generally							
V3.8.0.01 Oil paint in three coats; planed, primed timber	m²	0.12	7.76	-	2.17	9.93	0.347

V4 Alkyd Gloss Paint

	Unit	Labour Hours	Labour Net £	Plant Net £	Materials Net £	Unit Net £	CO_2e Kg
V4.1 **Metal, other than metal sections and pipework**							
V4.1.1 Upper surfaces inclined at an angle not exceeding 30 degrees to the horizontal							
V4.1.1.01 in three coats; primed metal	m²	0.09	5.77	-	2.17	7.94	0.347
V4.1.2 Upper surfaces inclined at 30 - 60 degrees to the horizontal							
V4.1.2.01 in three coats; primed metal	m²	0.09	5.97	-	2.17	8.14	0.347
V4.1.3 Surfaces inclined at an angle exceeding 60 degrees to the horizontal							
V4.1.3.01 in three coats; primed metal	m²	0.10	6.30	-	2.17	8.47	0.347
V4.1.4 Soffit surfaces and lower surfaces inclined at an angle not exceeding 60 degrees to the horizontal							
V4.1.4.01 in three coats; primed metal	m²	0.12	7.76	-	2.17	9.93	0.347
V4.1.6 Surfaces of width not exceeding 300mm							
V4.1.6.01 in three coats; primed metal	m	0.03	1.99	-	0.70	2.69	0.111
V4.1.7 Surfaces of width 300mm - 1m							
V4.1.7.01 in three coats; primed metal	m	0.04	2.78	-	1.40	4.18	0.223

V4 Alkyd Gloss Paint continued...

	Unit	Labour Hours	Labour Net £	Plant Net £	Materials Net £	Unit Net £	CO_2e Kg
V4.1	**Metal, other than metal sections and pipework**						
V4.1.8 Isolated groups of surfaces							
V4.1.8.01 in three coats; primed metal; 300 x 200mm access hatch and frame in wall; painting one side	Nr	0.18	11.93	-	4.35	16.28	0.693
V4.2	**Timber**						
V4.2.1 Upper surfaces inclined at an angle not exceeding 30 degrees to the horizontal							
V4.2.1.01 in three coats; primed timber	m²	0.09	5.77	-	2.17	7.94	0.347
V4.2.2 Upper surfaces inclined at 30 - 60 degrees to the horizontal							
V4.2.2.01 in three coats; primed timber	m²	0.09	5.97	-	2.17	8.14	0.347
V4.2.3 Surfaces inclined at an angle exceeding 60 degrees to the horizontal							
V4.2.3.01 in three coats; primed timber	m²	0.10	6.30	-	2.17	8.47	0.347
V4.2.4 Soffit surfaces and lower surfaces inclined at an angle not exceeding 60 degrees to the horizontal							
V4.2.4.01 in three coats; primed timber	m²	0.12	7.76	-	2.17	9.93	0.347
V4.2.6 Surfaces of width not exceeding 300mm							
V4.2.6.01 in three coats; primed timber	m	0.03	1.99	-	0.70	2.69	0.111
V4.2.7 Surfaces of width 300mm - 1m							
V4.2.7.01 in three coats; primed timber	m	0.04	2.78	-	1.40	4.18	0.223
V4.2.8 Isolated groups of surfaces							
V4.2.8.01 in three coats; primed timber; 300 x 200mm access hatch and frame in wall; painting one side	Nr	0.18	11.93	-	4.35	16.28	0.693
V4.3	**Smooth concrete**						
V4.3.1 Upper surfaces inclined at an angle not exceeding 30 degrees to the horizontal							
V4.3.1.01 one coat sealer; two coats finish	m²	0.11	7.49	-	1.62	9.11	0.297
V4.3.2 Upper surfaces inclined at 30 - 60 degrees to the horizontal							
V4.3.2.01 one coat sealer; two coats finish	m²	0.12	7.76	-	1.62	9.38	0.297
V4.3.3 Surfaces inclined at an angle exceeding 60 degrees to the horizontal							
V4.3.3.01 one coat sealer; two coats finish	m²	0.12	8.09	-	1.62	9.71	0.297
V4.3.4 Soffit surfaces and lower surfaces inclined at an angle not exceeding 60 degrees to the horizontal							
V4.3.4.01 one coat sealer; two coats finish	m²	0.15	9.75	-	1.62	11.37	0.297
V4.3.6 Surfaces of width not exceeding 300mm							
V4.3.6.01 one coat sealer; two coats finish	m	0.04	2.45	-	0.54	2.99	0.099
V4.3.7 Surfaces of width 300mm - 1m							
V4.3.7.01 one coat sealer; two coats finish	m	0.06	3.65	-	0.81	4.46	0.149
V4.3.8 Isolated groups of surfaces							
V4.3.8.01 one coat sealer; two coats finish; 2000 x 215 x 200mm deep lintel in wall; painting one face and soffit	Nr	0.31	20.69	-	0.98	21.67	0.210

V4 Alkyd Gloss Paint continued...

		Unit	Labour Hours	Labour Net £	Plant Net £	Materials Net £	Unit Net £	CO_2e Kg
V4.4	**Rough concrete**							
V4.4.1	Upper surfaces inclined at an angle not exceeding 30 degrees to the horizontal							
V4.4.1.01	one coat sealer; two coats finish; sawn board finish	m²	0.12	7.76	-	1.62	9.38	0.297
V4.4.2	Upper surfaces inclined at 30 - 60 degrees to the horizontal							
V4.4.2.01	one coat sealer; two coats finish; sawn board finish	m²	0.12	8.15	-	1.62	9.77	0.297
V4.4.3	Surfaces inclined at an angle exceeding 60 degrees to the horizontal							
V4.4.3.01	one coat sealer; two coats finish; sawn board finish	m²	0.13	8.42	-	1.62	10.04	0.297
V4.4.4	Soffit surfaces and lower surfaces inclined at an angle not exceeding 60 degrees to the horizontal							
V4.4.4.01	one coat sealer; two coats finish; sawn board finish	m²	0.15	9.75	-	1.62	11.37	0.297
V4.4.6	Surfaces of width not exceeding 300mm							
V4.4.6.01	one coat sealer; two coats finish; sawn board finish	m	0.04	2.45	-	0.54	2.99	0.099
V4.4.7	Surfaces of width 300mm - 1m							
V4.4.7.01	one coat sealer; two coats finish; sawn board finish	m	0.06	3.78	-	0.81	4.59	0.149
V4.4.8	Isolated groups of surfaces							
V4.4.8.01	one coat sealer; two coats finish; sawn board finish; 2000 x 215 x 200mm deep lintel in wall; painting one face and soffit	Nr	0.31	20.75	-	2.30	23.05	0.421
V4.5	**Masonry**							
V4.5.1	Upper surfaces inclined at an angle not exceeding 30 degrees to the horizontal							
V4.5.1.01	one coat sealer; two coats finish	m²	0.12	7.76	-	1.62	9.38	0.297
V4.5.2	Upper surfaces inclined at 30 - 60 degrees to the horizontal							
V4.5.2.01	one coat sealer; two coats finish	m²	0.12	8.15	-	1.62	9.77	0.297
V4.5.3	Surfaces inclined at an angle exceeding 60 degrees to the horizontal							
V4.5.3.01	one coat sealer; two coats finish	m²	0.13	8.42	-	1.62	10.04	0.297
V4.5.4	Soffit surfaces and lower surfaces inclined at an angle not exceeding 60 degrees to the horizontal							
V4.5.4.01	one coat sealer; two coats finish	m²	0.15	9.75	-	1.62	11.37	0.297
V4.5.6	Surfaces of width not exceeding 300mm							
V4.5.6.01	one coat sealer; two coats finish	m	0.04	2.45	-	0.54	2.99	0.099
V4.5.7	Surfaces of width 300mm - 1m							
V4.5.7.01	one coat sealer; two coats finish	m	0.06	3.78	-	0.81	4.59	0.149
V4.5.8	Isolated groups of surfaces							
V4.5.8.01	one coat sealer; two coats finish; stepped and spayed plinth overall size 450 x 450 x 450mm; painting all faces	Nr	0.31	20.75	-	0.98	21.73	0.210

V4   Alkyd Gloss Paint continued...

	Unit	Labour Hours	Labour Net £	Plant Net £	Materials Net £	Unit Net £	CO_2e Kg
V4.6	**Brickwork and blockwork**						
V4.6.1	Upper surfaces inclined at an angle not exceeding 30 degrees to the horizontal						
V4.6.1.01	one coat sealer; two coats finish	m²	0.12	7.76	-	1.62	9.38	0.297
V4.6.2	Upper surfaces inclined at 30 - 60 degrees to the horizontal						
V4.6.2.01	one coat sealer; two coats finish	m²	0.12	8.15	-	1.62	9.77	0.297
V4.6.3	Surfaces inclined at an angle exceeding 60 degrees to the horizontal						
V4.6.3.01	one coat sealer; two coats finish	m²	0.13	8.42	-	1.62	10.04	0.297
V4.6.4	Soffit surfaces and lower surfaces inclined at an angle not exceeding 60 degrees to the horizontal						
V4.6.4.01	one coat sealer; two coats finish	m²	0.15	9.88	-	1.62	11.50	0.297
V4.6.6	Surfaces of width not exceeding 300mm						
V4.6.6.01	one coat sealer; two coats finish	m	0.04	2.45	-	0.54	2.99	0.099
V4.6.7	Surfaces of width 300mm - 1m						
V4.6.7.01	one coat sealer; two coats finish	m	0.06	3.78	-	0.81	4.59	0.149
V4.6.8	Isolated groups of surfaces						
V4.6.8.01	one coat sealer; two coats finish; stepped and spayed plinth overall size 450 x 450 x 450mm; painting all faces	Nr	0.31	20.75	-	2.30	23.05	0.421
V4.7	**Metal sections**						
V4.7.0	Generally						
V4.7.0.01	one coat sealer; two coats finish	m²	0.09	6.17	-	1.40	7.57	0.223
V4.8	**Pipework**						
V4.8.0	Generally						
V4.8.0.01	one coat sealer; two coats finish	m²	0.09	6.17	-	1.40	7.57	0.223

V5   Emulsion Paint

	Unit	Labour Hours	Labour Net £	Plant Net £	Materials Net £	Unit Net £	CO_2e Kg
V5.2	**Timber**						
V5.2.1	Upper surfaces inclined at an angle not exceeding 30 degrees to the horizontal						
V5.2.1.01	in three coats; planed, primed timber	m²	0.08	5.30	-	0.98	6.28	0.248
V5.2.2	Upper surfaces inclined at 30 - 60 degrees to the horizontal						
V5.2.2.01	in three coats; planed, primed timber	m²	0.08	5.30	-	0.98	6.28	0.248
V5.2.3	Surfaces inclined at an angle exceeding 60 degrees to the horizontal						
V5.2.3.01	in three coats; planed, primed timber	m²	0.08	5.30	-	0.98	6.28	0.248
V5.2.4	Soffit surfaces and lower surfaces inclined at an angle not exceeding 60 degrees to the horizontal						
V5.2.4.01	in three coats; planed, primed timber	m²	0.11	7.29	-	0.98	8.27	0.248
V5.2.6	Surfaces of width not exceeding 300mm						
V5.2.6.01	in three coats; planed, primed timber	m	0.02	1.33	-	0.34	1.67	0.087

V5   Emulsion Paint continued...

		Unit	Labour Hours	Labour Net £	Plant Net £	Materials Net £	Unit Net £	CO_2e Kg
V5.2	**Timber**							
V5.2.7	Surfaces of width 300mm - 1m							
V5.2.7.01	in three coats; planed, primed timber	m	0.03	1.99	-	0.49	2.48	0.124
V5.2.8	Isolated groups of surfaces							
V5.2.8.01	in three coats; planed, primed timber; 300 x 200mm access hatch and frame in wall; painting one side	Nr	0.16	10.61	-	1.97	12.58	0.495
V5.3	**Smooth concrete**							
V5.3.1	Upper surfaces inclined at an angle not exceeding 30 degrees to the horizontal							
V5.3.1.01	in three coats; primed smooth concrete	m²	0.08	5.30	-	0.49	5.79	0.124
V5.3.2	Upper surfaces inclined at 30 - 60 degrees to the horizontal							
V5.3.2.01	in three coats; primed smooth concrete	m²	0.08	5.30	-	0.49	5.79	0.124
V5.3.3	Surfaces inclined at an angle exceeding 60 degrees to the horizontal							
V5.3.3.01	in three coats; primed smooth concrete	m²	0.08	5.30	-	0.49	5.79	0.124
V5.3.4	Soffit surfaces and lower surfaces inclined at an angle not exceeding 60 degrees to the horizontal							
V5.3.4.01	in three coats; primed smooth concrete	m²	0.11	7.29	-	0.49	7.78	0.124
V5.3.6	Surfaces of width not exceeding 300mm							
V5.3.6.01	in three coats; primed smooth concrete	m	0.02	1.33	-	0.15	1.48	0.037
V5.3.7	Surfaces of width 300mm - 1m							
V5.3.7.01	in three coats; primed smooth concrete	m	0.03	1.99	-	0.25	2.24	0.062
V5.3.8	Isolated groups of surfaces							
V5.3.8.01	in three coats; primed smooth concrete; 2000 x 215 x 200mm deep lintel in wall; painting one face and soffit	Nr	0.16	10.61	-	0.74	11.35	0.186
V5.6	**Brickwork and blockwork**							
V5.6.1	Upper surfaces inclined at an angle not exceeding 30 degrees to the horizontal							
V5.6.1.01	in three coats	m²	0.11	7.09	-	0.49	7.58	0.124
V5.6.2	Upper surfaces inclined at 30 - 60 degrees to the horizontal							
V5.6.2.01	in three coats	m²	0.12	7.62	-	0.49	8.11	0.124
V5.6.3	Surfaces inclined at an angle exceeding 60 degrees to the horizontal							
V5.6.3.01	in three coats	m²	0.12	7.76	-	0.49	8.25	0.124
V5.6.4	Soffit surfaces and lower surfaces inclined at an angle not exceeding 60 degrees to the horizontal							
V5.6.4.01	in three coats	m²	0.15	9.94	-	0.59	10.53	0.149
V5.6.6	Surfaces of width not exceeding 300mm							
V5.6.6.01	in three coats	m	0.04	2.65	-	0.15	2.80	0.037

V5 Emulsion Paint continued...

		Unit	Labour Hours	Labour Net £	Plant Net £	Materials Net £	Unit Net £	CO_2e Kg
V5.6	**Brickwork and blockwork**							
V5.6.7	Surfaces of width 300mm - 1m							
V5.6.7.01	in three coats	m	0.05	3.31	-	0.30	3.61	0.074
V5.6.8	Isolated groups of surfaces							
V5.6.8.01	in three coats; stepped and spayed plinth overall size 450 x 450 x 450mm; painting all faces	Nr	0.22	14.25	-	1.18	15.43	0.297

V6 Cement Paint

		Unit	Labour Hours	Labour Net £	Plant Net £	Materials Net £	Unit Net £	CO_2e Kg
V6.3	**Smooth concrete**							
V6.3.1	Upper surfaces inclined at an angle not exceeding 30 degrees to the horizontal							
V6.3.1.01	in three coats	m²	0.06	3.98	-	1.00	4.98	0.248
V6.3.2	Upper surfaces inclined at 30 - 60 degrees to the horizontal							
V6.3.2.01	in three coats	m²	0.06	3.98	-	1.00	4.98	0.248
V6.3.3	Surfaces inclined at an angle exceeding 60 degrees to the horizontal							
V6.3.3.01	in three coats	m²	0.06	3.98	-	1.00	4.98	0.248
V6.3.4	Soffit surfaces and lower surfaces inclined at an angle not exceeding 60 degrees to the horizontal							
V6.3.4.01	in three coats	m²	0.08	5.30	-	1.00	6.30	0.248
V6.3.6	Surfaces of width not exceeding 300mm							
V6.3.6.01	in three coats	m	0.02	1.33	-	3.50	4.83	0.867
V6.3.7	Surfaces of width 300mm - 1m							
V6.3.7.01	in three coats	m	0.04	2.65	-	0.50	3.15	0.124
V6.3.8	Isolated groups of surfaces							
V6.3.8.01	in three coats; 2000 x 215 x 200mm deep lintel in wall; painting one face and soffit	Nr	0.12	7.96	-	2.00	9.96	0.495
V6.4	**Rough concrete**							
V6.4.1	Upper surfaces inclined at an angle not exceeding 30 degrees to the horizontal							
V6.4.1.01	in three coats; primed rough concrete; sawn board finish	m²	0.08	5.30	-	1.25	6.55	0.310
V6.4.2	Upper surfaces inclined at 30 - 60 degrees to the horizontal							
V6.4.2.01	in three coats; primed rough concrete; sawn board finish	m²	0.08	5.30	-	1.25	6.55	0.310
V6.4.3	Surfaces inclined at an angle exceeding 60 degrees to the horizontal							
V6.4.3.01	in three coats; primed rough concrete; sawn board finish	m²	0.08	5.30	-	1.25	6.55	0.310
V6.4.4	Soffit surfaces and lower surfaces inclined at an angle not exceeding 60 degrees to the horizontal							
V6.4.4.01	in three coats; primed rough concrete; sawn board finish	m²	0.10	6.63	-	1.25	7.88	0.310
V6.4.6	Surfaces of width not exceeding 300mm							
V6.4.6.01	in three coats; primed rough concrete; sawn board finish	m	0.03	1.99	-	0.35	2.34	0.087

V6 Cement Paint continued...

		Unit	Labour Hours	Labour Net £	Plant Net £	Materials Net £	Unit Net £	CO_2e Kg
V6.4	**Rough concrete**							
V6.4.7	Surfaces of width 300mm - 1m							
V6.4.7.01	in three coats; primed rough concrete; sawn board finish	m	0.04	2.65	-	0.60	3.25	0.149
V6.4.8	Isolated groups of surfaces							
V6.4.8.01	in three coats; primed rough concrete; sawn board finish; 2000 x 215 x 200mm deep lintel in wall; painting one face and soffit	Nr	0.16	10.61	-	2.50	13.11	0.619
V6.5	**Masonry**							
V6.5.1	Upper surfaces inclined at an angle not exceeding 30 degrees to the horizontal							
V6.5.1.01	in three coats; primed masonry	m^2	0.08	5.30	-	1.25	6.55	0.310
V6.5.2	Upper surfaces inclined at 30 - 60 degrees to the horizontal							
V6.5.2.01	in three coats; primed masonry	m^2	0.08	5.30	-	1.25	6.55	0.310
V6.5.3	Surfaces inclined at an angle exceeding 60 degrees to the horizontal							
V6.5.3.01	in three coats; primed masonry	m^2	0.08	5.30	-	1.25	6.55	0.310
V6.5.4	Soffit surfaces and lower surfaces inclined at an angle not exceeding 60 degrees to the horizontal							
V6.5.4.01	in three coats; primed masonry	m^2	0.10	6.63	-	1.25	7.88	0.310
V6.5.6	Surfaces of width not exceeding 300mm							
V6.5.6.01	in three coats; primed masonry	m	0.03	1.99	-	0.35	2.34	0.087
V6.5.7	Surfaces of width 300mm - 1m							
V6.5.7.01	in three coats; primed masonry	m	0.04	2.65	-	0.60	3.25	0.149
V6.5.8	Isolated groups of surfaces							
V6.5.8.01	in three coats; primed masonry; stepped and spayed plinth overall size 450 x 450 x 450mm; painting all faces	Nr	0.16	10.61	-	2.50	13.11	0.619
V6.6	**Brickwork and blockwork**							
V6.6.1	Upper surfaces inclined at an angle not exceeding 30 degrees to the horizontal							
V6.6.1.01	in three coats; primed brickwork and blockwork	m^2	0.08	5.50	-	1.50	7.00	0.371
V6.6.2	Upper surfaces inclined at 30 - 60 degrees to the horizontal							
V6.6.2.01	in three coats; primed brickwork and blockwork	m^2	0.09	5.64	-	1.50	7.14	0.371
V6.6.3	Surfaces inclined at an angle exceeding 60 degrees to the horizontal							
V6.6.3.01	in three coats; primed brickwork and blockwork	m^2	0.09	5.97	-	1.50	7.47	0.371
V6.6.4	Soffit surfaces and lower surfaces inclined at an angle not exceeding 60 degrees to the horizontal							
V6.6.4.01	in three coats; primed brickwork and blockwork	m^2	0.11	7.09	-	1.50	8.59	0.371
V6.6.6	Surfaces of width not exceeding 300mm							
V6.6.6.01	in three coats; primed brickwork and blockwork	m	0.03	1.79	-	0.50	2.29	0.124

V6   Cement Paint continued...

	Unit	Labour Hours	Labour Net £	Plant Net £	Materials Net £	Unit Net £	CO_2e Kg
V6.6 **Brickwork and blockwork**							
V6.6.7 Surfaces of width 300mm - 1m							
V6.6.7.01 in three coats; primed brickwork and blockwork	m	0.04	2.65	-	0.75	3.40	0.186
V6.6.8 Isolated groups of surfaces							
V6.6.8.01 in three coats; primed brickwork and blockwork; stepped and spayed plinth overall size 450 x 450 x 450mm; painting all faces	Nr	0.17	10.94	-	1.50	12.44	0.371

V7   Epoxy Or Polyurethane Paint

	Unit	Labour Hours	Labour Net £	Plant Net £	Materials Net £	Unit Net £	CO_2e Kg
V7.2 **Timber**							
V7.2.1 Upper surfaces inclined at an angle not exceeding 30 degrees to the horizontal							
V7.2.1.01 Polyurethane varnish in two coats; planed, primed timber	m²	0.09	5.97	-	1.04	7.01	0.173
V7.2.2 Upper surfaces inclined at 30 - 60 degrees to the horizontal							
V7.2.2.01 Polyurethane varnish in two coats; planed, primed timber	m²	0.09	5.97	-	1.04	7.01	0.173
V7.2.3 Surfaces inclined at an angle exceeding 60 degrees to the horizontal							
V7.2.3.01 Polyurethane varnish in two coats; planed, primed timber	m²	0.09	5.97	-	1.04	7.01	0.173
V7.2.4 Soffit surfaces and lower surfaces inclined at an angle not exceeding 60 degrees to the horizontal							
V7.2.4.01 Polyurethane varnish in two coats; planed, primed timber	m²	0.12	7.96	-	1.04	9.00	0.173
V7.2.6 Surfaces of width not exceeding 300mm							
V7.2.6.01 Polyurethane varnish in two coats; planed, primed timber	m	0.03	1.99	-	0.37	2.36	0.062
V7.2.7 Surfaces of width 300mm - 1m							
V7.2.7.01 Polyurethane varnish in two coats; planed, primed timber	m	0.04	2.65	-	0.52	3.17	0.087
V7.2.8 Isolated groups of surfaces							
V7.2.8.01 Polyurethane varnish in two coats; planed, primed timber; 300 x 200mm access hatch and frame in wall; painting one side	Nr	0.18	11.93	-	2.09	14.02	0.347

V8   Bituminous Or Coal Tar Paint

	Unit	Labour Hours	Labour Net £	Plant Net £	Materials Net £	Unit Net £	CO_2e Kg
V8.1 **Metal, other than metal sections and pipework**							
V8.1.1 Upper surfaces inclined at an angle not exceeding 30 degrees to the horizontal							
V8.1.1.01 Bituminous paint, in two coats	m²	0.09	5.97	-	0.67	6.64	0.248
V8.1.1.02 Bituminous paint, heavy duty; in two coats	m²	0.14	9.08	-	5.16	14.24	1.448
V8.1.2 Upper surfaces inclined at 30 - 60 degrees to the horizontal							
V8.1.2.01 Bituminous paint, in two coats	m²	0.09	5.97	-	0.67	6.64	0.248
V8.1.2.02 Bituminous paint, heavy duty; in two coats	m²	0.14	9.48	-	5.16	14.64	1.448

V8 Bituminous Or Coal Tar Paint continued...

	Unit	Labour Hours	Labour Net £	Plant Net £	Materials Net £	Unit Net £	CO_2e Kg
V8.1	**Metal, other than metal sections and pipework**						
V8.1.3	Surfaces inclined at an angle exceeding 60 degrees to the horizontal						
V8.1.3.01	Bituminous paint, in two coats	m²	0.09	5.97	-	0.67	6.64	0.248
V8.1.3.02	Bituminous paint, heavy duty; in two coats	m²	0.15	9.94	-	5.16	15.10	1.448
V8.1.4	Soffit surfaces and lower surfaces inclined at an angle not exceeding 60 degrees to the horizontal						
V8.1.4.01	Bituminous paint, in two coats	m²	0.12	7.96	-	0.67	8.63	0.248
V8.1.4.02	Bituminous paint, heavy duty; in two coats	m²	0.18	12.13	-	5.16	17.29	1.448
V8.1.6	Surfaces of width not exceeding 300mm						
V8.1.6.01	Bituminous paint, in two coats	m	0.02	1.52	-	0.24	1.76	0.087
V8.1.6.02	Bituminous paint, heavy duty; in two coats	m	0.05	2.98	-	1.76	4.74	0.495
V8.1.7	Surfaces of width 300mm - 1m						
V8.1.7.01	Bituminous paint, in two coats	m	0.03	2.19	-	0.51	2.70	0.186
V8.1.7.02	Bituminous paint, heavy duty; in two coats	m	0.07	4.44	-	2.65	7.09	0.743
V8.1.8	Isolated groups of surfaces						
V8.1.8.01	Bituminous paint, in two coats; rolled steel joist 2000 x 215 x 200mm deep; painting one face and soffit	Nr	0.18	11.93	-	1.35	13.28	0.495
V8.1.8.02	Bituminous paint, heavy duty; in two coats; rolled steel joist 2000 x 215 x 200mm deep; painting one face and soffit	Nr	0.28	18.23	-	10.32	28.55	2.897
V8.2	**Metal, other than metal sections and pipework**						
V8.2.1	Upper surfaces inclined at an angle not exceeding 30 degrees to the horizontal						
V8.2.1.01	Bituminous paint, in two coats; primed timber	m²	0.09	5.97	-	0.67	6.64	0.248
V8.2.1.02	Bituminous paint, heavy duty; in two coats; primed planed timber	m²	0.14	9.08	-	5.16	14.24	1.448
V8.2.2	Upper surfaces inclined at 30 - 60 degrees to the horizontal						
V8.2.2.01	Bituminous paint, in two coats; primed timber	m²	0.09	5.97	-	0.67	6.64	0.248
V8.2.2.02	Bituminous paint, heavy duty; in two coats; primed planed timber	m²	0.14	9.48	-	5.16	14.64	1.448
V8.2.3	Surfaces inclined at an angle exceeding 60 degrees to the horizontal						
V8.2.3.01	Bituminous paint, in two coats; primed timber	m²	0.09	5.97	-	0.67	6.64	0.248
V8.2.3.02	Bituminous paint, heavy duty; in two coats; primed planed timber	m²	0.15	9.94	-	5.16	15.10	1.448
V8.2.4	Soffit surfaces and lower surfaces inclined at an angle not exceeding 60 degrees to the horizontal						
V8.2.4.01	Bituminous paint, in two coats; primed timber	m²	0.12	7.96	-	0.67	8.63	0.248
V8.2.4.02	Bituminous paint, heavy duty; in two coats; primed planed timber	m²	0.18	12.13	-	5.16	17.29	1.448
V8.2.6	Surfaces of width not exceeding 300mm						
V8.2.6.01	Bituminous paint, in two coats; primed timber	m	0.03	1.99	-	0.24	2.23	0.087

V8 Bituminous Or Coal Tar Paint continued...

	Unit	Labour Hours	Labour Net £	Plant Net £	Materials Net £	Unit Net £	CO$_2$e Kg
V8.2	**Metal, other than metal sections and pipework**						
V8.2.6	Surfaces of width not exceeding 300mm						
V8.2.6.02	Bituminous paint, heavy duty; in two coats; primed planed timber	m	0.05	2.98	-	1.76	4.74	0.495
V8.2.7	Surfaces of width 300mm - 1m						
V8.2.7.01	Bituminous paint, in two coats; primed timber	m	0.04	2.65	-	0.51	3.16	0.186
V8.2.7.02	Bituminous paint, heavy duty; in two coats; primed planed timber	m	0.07	4.44	-	2.65	7.09	0.743
V8.2.8	Isolated groups of surfaces						
V8.2.8.01	Bituminous paint, in two coats; primed timber; beam 2000 x 215 x 200mm deep; painting one face and soffit	Nr	0.18	11.93	-	1.35	13.28	0.495
V8.2.8.02	Bituminous paint, heavy duty; in two coats; primed planed timber; beam 2000 x 215 x 200mm deep; painting one face and soffit	Nr	0.28	18.23	-	10.32	28.55	2.897
V8.3	**Smooth concrete**						
V8.3.1	Upper surfaces inclined at an angle not exceeding 30 degrees to the horizontal						
V8.3.1.01	Bituminous paint, in two coats	m^2	0.07	4.84	-	0.74	5.58	0.272
V8.3.1.02	Bituminous paint, heavy duty; in two coats; sealed smooth concrete	m^2	0.15	10.14	-	5.64	15.78	1.585
V8.3.2	Upper surfaces inclined at 30 - 60 degrees to the horizontal						
V8.3.2.01	Bituminous paint, in two coats	m^2	0.08	4.97	-	0.74	5.71	0.272
V8.3.2.02	Bituminous paint, heavy duty; in two coats; sealed smooth concrete	m^2	0.16	10.61	-	5.64	16.25	1.585
V8.3.3	Surfaces inclined at an angle exceeding 60 degrees to the horizontal						
V8.3.3.01	Bituminous paint, in two coats	m^2	0.08	5.30	-	0.74	6.04	0.272
V8.3.3.02	Bituminous paint, heavy duty; in two coats; sealed smooth concrete	m^2	0.17	11.07	-	5.64	16.71	1.585
V8.3.4	Soffit surfaces and lower surfaces inclined at an angle not exceeding 60 degrees to the horizontal						
V8.3.4.01	Bituminous paint, in two coats	m^2	0.10	6.63	-	0.74	7.37	0.272
V8.3.4.02	Bituminous paint, heavy duty; in two coats; sealed smooth concrete	m^2	0.23	15.25	-	5.64	20.89	1.585
V8.3.6	Surfaces of width not exceeding 300mm						
V8.3.6.01	Bituminous paint, in two coats	m	0.04	2.65	-	0.24	2.89	0.087
V8.3.6.02	Bituminous paint, heavy duty; in two coats; sealed smooth concrete	m	0.05	3.31	-	1.90	5.21	0.532
V8.3.7	Surfaces of width 300mm - 1m						
V8.3.7.01	Bituminous paint, in two coats	m	0.04	2.45	-	0.37	2.82	0.136
V8.3.7.02	Bituminous paint, heavy duty; in two coats; sealed smooth concrete	m	0.08	4.97	-	2.82	7.79	0.792

V8 Bituminous Or Coal Tar Paint continued...

		Unit	Labour Hours	Labour Net £	Plant Net £	Materials Net £	Unit Net £	CO_2e Kg
V8.3	**Smooth concrete**							
V8.3.8	Isolated groups of surfaces							
V8.3.8.01	Bituminous paint, in two coats; beam 2000 x 215 x 200mm deep, concrete lintel in wall; painting one face and soffit	Nr	0.18	11.93	-	1.48	13.41	0.545
V8.3.8.02	Bituminous paint, heavy duty; in two coats; sealed smooth concrete; beam 2000 x 215 x 200mm deep concrete lintel in wall; painting one face and soffit	Nr	0.28	18.23	-	11.29	29.52	3.169
V8.4	**Rough concrete**							
V8.4.1	Upper surfaces inclined at an angle not exceeding 30 degrees to the horizontal							
V8.4.1.01	Bituminous paint, in two coats; sawn board finish	m²	0.08	5.30	-	1.75	7.05	0.644
V8.4.1.02	Bituminous paint, heavy duty; in two coats; sealed rough concrete; sawn board finish	m²	0.17	10.94	-	5.73	16.67	1.609
V8.4.2	Upper surfaces inclined at 30 - 60 degrees to the horizontal							
V8.4.2.01	Bituminous paint, in two coats; sawn board finish	m²	0.08	5.50	-	1.75	7.25	0.644
V8.4.2.02	Bituminous paint, heavy duty; in two coats; sealed rough concrete; sawn board finish	m²	0.17	11.47	-	5.73	17.20	1.609
V8.4.3	Surfaces inclined at an angle exceeding 60 degrees to the horizontal							
V8.4.3.01	Bituminous paint, in two coats; sawn board finish	m²	0.09	5.77	-	1.75	7.52	0.644
V8.4.3.02	Bituminous paint, heavy duty; in two coats; sealed rough concrete; sawn board finish	m²	0.18	11.93	-	5.73	17.66	1.609
V8.4.4	Soffit surfaces and lower surfaces inclined at an angle not exceeding 60 degrees to the horizontal							
V8.4.4.01	Bituminous paint, in two coats; sawn board finish	m²	0.11	7.29	-	1.75	9.04	0.644
V8.4.4.02	Bituminous paint, heavy duty; in two coats; sealed rough concrete; sawn board finish	m²	0.22	14.59	-	5.73	20.32	1.609
V8.4.6	Surfaces of width not exceeding 300mm							
V8.4.6.01	Bituminous paint, in two coats; sawn board finish	m	0.03	1.66	-	0.57	2.23	0.210
V8.4.6.02	Bituminous paint, heavy duty; in two coats; sealed rough concrete; sawn board finish	m	0.06	3.65	-	1.81	5.46	0.508
V8.4.7	Surfaces of width 300mm - 1m							
V8.4.7.01	Bituminous paint, in two coats; sawn board finish	m	0.04	2.65	-	0.88	3.53	0.322
V8.4.7.02	Bituminous paint, heavy duty; in two coats; sealed rough concrete; sawn board finish	m	0.08	5.50	-	2.87	8.37	0.805

V8 Bituminous Or Coal Tar Paint continued...

	Unit	Labour Hours	Labour Net £	Plant Net £	Materials Net £	Unit Net £	CO_2e Kg
V8.4 **Rough concrete**							
V8.4.8 Isolated groups of surfaces							
V8.4.8.01 Bituminous paint, in two coats; sawn board finish; 2000 x 215 x 200mm deep, concrete beam in wall; painting one face and soffit	Nr	0.16	10.61	-	0.54	11.15	0.198
V8.4.8.02 Bituminous paint, heavy duty; in two coats; sealed rough concrete; sawn board finish; 2000 x 215 x 200mm deep concrete beam in wall; painting one face and soffit	Nr	0.33	21.88	-	11.47	33.35	3.219
V8.5 **Masonry**							
V8.5.1 Upper surfaces inclined at an angle not exceeding 30 degrees to the horizontal							
V8.5.1.01 Bituminous paint, in two coats	m²	0.08	5.30	-	1.75	7.05	0.644
V8.5.1.02 Bituminous paint, heavy duty; in two coats; sealed masonry	m²	0.17	10.94	-	5.73	16.67	1.609
V8.5.2 Upper surfaces inclined at 30 - 60 degrees to the horizontal							
V8.5.2.01 Bituminous paint, in two coats	m²	0.08	5.50	-	1.75	7.25	0.644
V8.5.2.02 Bituminous paint, heavy duty; in two coats; sealed masonry	m²	0.17	11.47	-	5.73	17.20	1.609
V8.5.3 Surfaces inclined at an angle exceeding 60 degrees to the horizontal							
V8.5.3.01 Bituminous paint, in two coats	m²	0.09	5.77	-	1.75	7.52	0.644
V8.5.3.02 Bituminous paint, heavy duty; in two coats; sealed masonry	m²	0.18	11.93	-	5.73	17.66	1.609
V8.5.4 Soffit surfaces and lower surfaces inclined at an angle not exceeding 60 degrees to the horizontal							
V8.5.4.01 Bituminous paint, in two coats	m²	0.11	7.29	-	1.75	9.04	0.644
V8.5.4.02 Bituminous paint, heavy duty; in two coats; sealed masonry	m²	0.22	14.59	-	5.73	20.32	1.609
V8.5.6 Surfaces of width not exceeding 300mm							
V8.5.6.01 Bituminous paint, in two coats	m	0.03	1.66	-	0.57	2.23	0.210
V8.5.6.02 Bituminous paint, heavy duty; in two coats; sealed masonry	m	0.06	3.65	-	1.81	5.46	0.508
V8.5.7 Surfaces of width 300mm - 1m							
V8.5.7.01 Bituminous paint, in two coats	m	0.04	2.65	-	0.88	3.53	0.322
V8.5.7.02 Bituminous paint, heavy duty; in two coats; sealed masonry	m	0.08	5.50	-	2.87	8.37	0.805
V8.5.8 Isolated groups of surfaces							
V8.5.8.01 Bituminous paint, in two coats; stepped and spayed plinth overall size 450 x 450 x 450mm; painting all faces	Nr	0.16	10.61	-	0.54	11.15	0.198
V8.5.8.02 Bituminous paint, heavy duty; in two coats; sealed masonry; stepped and spayed plinth overall size 450 x 450 x 450mm; painting all faces	Nr	0.33	21.88	-	11.47	33.35	3.219
V8.6 **Brickwork and blockwork**							
V8.6.1 Upper surfaces inclined at an angle not exceeding 30 degrees to the horizontal							
V8.6.1.01 Bituminous paint, in two coats	m²	0.10	6.63	-	0.88	7.51	0.322

V8 Bituminous Or Coal Tar Paint continued...

	Unit	Labour Hours	Labour Net £	Plant Net £	Materials Net £	Unit Net £	CO_2e Kg
V8.6 **Brickwork and blockwork**							
V8.6.1 Upper surfaces inclined at an angle not exceeding 30 degrees to the horizontal							
V8.6.1.02 Bituminous paint, heavy duty; in two coats; sealed brickwork and blockwork	m²	0.18	11.93	-	6.88	18.81	1.931
V8.6.2 Upper surfaces inclined at 30 - 60 degrees to the horizontal							
V8.6.2.01 Bituminous paint, in two coats	m²	0.01	0.66	-	0.88	1.54	0.322
V8.6.2.02 Bituminous paint, heavy duty; in two coats; sealed brickwork and blockwork	m²	0.19	12.60	-	6.88	19.48	1.931
V8.6.3 Surfaces inclined at an angle exceeding 60 degrees to the horizontal							
V8.6.3.01 Bituminous paint, in two coats	m²	0.10	6.63	-	0.88	7.51	0.322
V8.6.3.02 Bituminous paint, heavy duty; in two coats; sealed brickwork and blockwork	m²	0.20	13.06	-	6.88	19.94	1.931
V8.6.4 Soffit surfaces and lower surfaces inclined at an angle not exceeding 60 degrees to the horizontal							
V8.6.4.01 Bituminous paint, in two coats	m²	0.13	8.62	-	0.88	9.50	0.322
V8.6.4.02 Bituminous paint, heavy duty; in two coats; sealed brickwork and blockwork	m²	0.24	16.11	-	6.88	22.99	1.931
V8.6.6 Surfaces of width not exceeding 300mm							
V8.6.6.01 Bituminous paint, in two coats	m	0.03	1.99	-	0.30	2.29	0.111
V8.6.6.02 Bituminous paint, heavy duty; in two coats; sealed brickwork and blockwork	m	0.06	3.98	-	2.25	6.23	0.631
V8.6.7 Surfaces of width 300mm - 1m							
V8.6.7.01 Bituminous paint, in two coats	m	0.04	2.65	-	0.44	3.09	0.161
V8.6.7.02 Bituminous paint, heavy duty; in two coats; sealed brickwork and blockwork	m	0.09	5.97	-	3.44	9.41	0.966
V8.6.8 Isolated groups of surfaces							
V8.6.8.01 Bituminous paint, in two coats; stepped and spayed plinth overall size 450 x 450 x 450mm; painting all faces	Nr	0.20	13.26	-	1.75	15.01	0.644
V8.6.8.02 Bituminous paint, heavy duty; in two coats; sealed brickwork and blockwork; stepped and spayed plinth overall size 450 x 450 x 450mm; painting all faces	Nr	0.36	23.54	-	13.76	37.30	3.863
V8.7 **Metal sections**							
V8.7.0 Generally							
V8.7.0.01 Bituminous paint, in two coats	m²	0.10	6.63	-	0.67	7.30	0.248
V8.7.0.02 Bituminous paint, heavy duty; in two coats; sealed brickwork and blockwork	m²	0.07	4.84	-	0.67	5.51	0.248
V8.8 **Pipework**							
V8.8.0 Generally							
V8.8.0.01 Bituminous paint, in two coats	m²	0.10	6.63	-	0.67	7.30	0.248

V8   Bituminous Or Coal Tar Paint continued...

	Unit	Labour Hours	Labour Net £	Plant Net £	Materials Net £	Unit Net £	CO_2e Kg
V8.8 **Pipework**							
V8.8.0 Generally							
V8.8.0.02 Bituminous paint, heavy duty; in two coats; sealed brickwork and blockwork	m²	0.07	4.84	-	0.67	5.51	0.248

CLASS W:
WATERPROOFING

Calculations used throughout Class W - Waterproofing

Labour

			Qty		Rate		Total
L A0315ICE	**Paint spray Labour Gang**						
	L A9029ICE	Spray painter	1	x	17.33	=	£17.33
		Total hourly cost of gang				=	**£17.33**
L A0320ICE	**Waterproofing Labour Gang**						
	L A9016ICE	Ganger	1	x	17.64	=	£17.64
	L A9002ICE	Labourer (General Operative)	3	x	13.02	=	£39.05
		Total hourly cost of gang				=	**£56.69**
L A0324ICE	**Tiling Labour Gang (protective layer)**						
	L A9032ICE	Tiler	2	x	17.33	=	£34.66
	L A9002ICE	Labourer (General Operative)	1	x	13.02	=	£13.02
		Total hourly cost of gang				=	**£47.68**
L A0325ICE	**Plastering Labour Gang**						
	L A9031ICE	Plasterer	1	x	17.33	=	£17.33
	L A9002ICE	Labourer (General Operative)	1	x	13.02	=	£13.02
		Total hourly cost of gang				=	**£30.35**

Waterproofing

W1   Damp Proofing

	Unit	Labour Hours	Labour Net £	Plant Net £	Materials Net £	Unit Net £	CO_2e Kg
W1.3	**Waterproof sheeting**						
W1.3.1	Upper surfaces inclined at an angle not exceeding 30 degrees to the horizontal						
W1.3.1.01 Polythene in one layer, 1200 gauge	m²	0.00	0.17	-	0.83	1.00	0.520
W1.3.2	Upper surfaces inclined at 30 - 60 degrees to the horizontal						
W1.3.2.01 Polythene in one layer, 1200 gauge	m²	0.01	0.40	-	0.83	1.23	0.520
W1.3.3	Surfaces inclined at an angle exceeding 60 degrees to the horizontal						
W1.3.3.01 Polythene in one layer, 1200 gauge	m²	0.01	0.57	-	0.83	1.40	0.520
W1.3.4	Curved surfaces						
W1.3.4.01 Polythene in one layer, 1200 gauge	m²	0.02	1.13	-	0.83	1.96	0.520
W1.3.5	Domed surfaces						
W1.3.5.01 Polythene in one layer, 1200 gauge	m²	0.02	1.13	-	0.83	1.96	0.520
W1.3.6	Surfaces of width not exceeding 300mm						
W1.3.6.01 Polythene in one layer, 1200 gauge	m	0.00	0.17	-	0.17	0.34	0.104
W1.3.7	Surfaces of width 300mm - 1m						
W1.3.7.01 Polythene in one layer, 1200 gauge	m	0.01	0.57	-	0.58	1.15	0.364
W1.3.8	Isolated groups of surfaces						
W1.3.8.01 Polythene in one layer, 1200 gauge; sump below basement level; 600 x 400 x 400mm to sides and base	Nr	0.02	1.13	-	0.93	2.06	0.582
W1.5	**Rendering in ordinary cement mortar**						
W1.5.1	Upper surfaces inclined at an angle not exceeding 30 degrees to the horizontal						
W1.5.1.01 cement mortar (1:3); 25mm thick screed in one coat to concrete; trowelled	m²	0.23	6.89	-	5.93	12.82	19.024
W1.5.2	Upper surfaces inclined at 30 - 60 degrees to the horizontal						
W1.5.2.01 cement mortar (1:3); 25mm thick screed in one coat to concrete; trowelled	m²	0.34	10.32	-	5.93	16.25	19.024
W1.5.3	Surfaces inclined at an angle exceeding 60 degrees to the horizontal						
W1.5.3.01 cement mortar (1:3); 25mm thick in two coats to brickwork; trowelled	m²	0.58	17.45	-	5.93	23.38	19.024
W1.5.4	Curved surfaces						
W1.5.4.01 cement mortar (1:3);25mm thick in two coats to brickwork; trowelled	m²	0.86	26.15	-	5.93	32.08	19.024

W1 Damp Proofing continued...

	Unit	Labour Hours	Labour Net £	Plant Net £	Materials Net £	Unit Net £	CO_2e Kg
W1.5	**Rendering in ordinary cement mortar**						
W1.5.5 Domed surfaces							
W1.5.5.01 cement mortar (1:3); 25mm thick in two coats to brickwork; trowelled	m²	1.15	34.89	-	5.93	40.82	19.024
W1.5.6 Surfaces of width not exceeding 300mm							
W1.5.6.01 cement mortar (1:3); 25mm thick in two coats to brickwork; trowelled	m	0.18	5.31	-	1.96	7.27	6.017
W1.5.7 Surfaces of width 300mm - 1m							
W1.5.7.01 cement mortar (1:3); 25mm thick in two coats to brickwork; trowelled	m	0.35	10.62	-	3.22	13.84	12.026
W1.5.8 Isolated groups of surfaces							
W1.5.8.01 cement mortar (1:3); 25mm thick in two coats to brickwork; trowelled; stepped brick plinth; overall size 450 x 450 x 450mm; to all faces	Nr	1.15	34.89	-	10.06	44.95	38.026
W1.6	**Rendering in waterproof cement mortar**						
W1.6.3 Surfaces inclined at an angle exceeding 60 degrees to the horizontal							
W1.6.3.01 cement lime mortar (1:1:6); treated with water resistant additive; 25mm thick in two coats to brickwork; trowelled	m²	0.58	17.45	-	8.13	25.58	69.975
W1.6.4 Curved surfaces							
W1.6.4.01 cement lime mortar (1:1:6); treated with water resistant additive; 25mm thick in two coats to brickwork; trowelled	m²	0.86	26.15	-	8.13	34.28	69.975
W1.6.5 Domed surfaces							
W1.6.5.01 cement lime mortar (1:1:6); treated with water resistant additive; 25mm thick in two coats to brickwork; trowelled	m²	1.15	34.89	-	8.13	43.02	69.975
W1.6.6 Surfaces of width not exceeding 300mm							
W1.6.6.01 cement lime mortar (1:1:6); treated with water resistant additive; 25mm thick in two coats to brickwork; trowelled	m	0.18	5.31	-	2.72	8.03	26.597
W1.6.7 Surfaces of width 300mm - 1m							
W1.6.7.01 cement lime mortar (1:1:6); treated with water resistant additive; 25mm thick in two coats to brickwork; trowelled	m	0.35	10.62	-	2.72	13.34	26.597
W1.6.8 Isolated groups of surfaces							
W1.6.8.01 cement lime mortar (1:1:6); treated with water resistant additive; 25mm thick in two coats to brickwork; trowelled; stepped brick plinth; overall size 450 x 450 x 450mm; to all faces	Nr	1.15	34.89	-	6.51	41.40	46.936

W2 Tanking

	Unit	Labour Hours	Labour Net £	Plant Net £	Materials Net £	Unit Net £	CO_2e Kg
W2.1	**Asphalt**						

W2.1.1 Upper surfaces inclined at an angle not exceeding 30 degrees to the horizontal

W2.1.1.01 to BS 6925; 20mm thick in two coats to brickwork; raking out joints to form key

| | m² | 0.55 | 16.69 | - | 11.23 | 27.92 | 4.365 |

W2.1.1.02 to BS 6925; 20mm thick in two coats laid on prepared concrete surface

| | m² | 0.50 | 15.17 | - | 10.24 | 25.41 | 3.982 |

W2.1.2 Upper surfaces inclined at 30 - 60 degrees to the horizontal

W2.1.2.01 to BS 6925; 20mm thick in two coats to brickwork; raking out joints to form key

| | m² | 0.82 | 24.88 | - | 11.23 | 36.11 | 4.365 |

W2.1.2.02 to BS 6925; 20mm thick in two coats laid on prepared concrete surface

| | m² | 0.75 | 22.75 | - | 10.24 | 32.99 | 3.982 |

W2.1.3 Surfaces inclined at an angle exceeding 60 degrees to the horizontal

W2.1.3.01 to BS 6925; 20mm thick in two coats to brickwork; raking out joints to form key

| | m² | 1.39 | 42.17 | - | 11.23 | 53.40 | 4.365 |

W2.1.3.02 to BS 6925; 20mm thick in two coats laid on prepared concrete surface

| | m² | 1.30 | 39.44 | - | 10.24 | 49.68 | 3.982 |

W2.1.4 Curved surfaces

W2.1.4.01 to BS 6925; 20mm thick in two coats to brickwork; raking out joints to form key

| | m² | 1.59 | 48.24 | - | 11.23 | 59.47 | 4.365 |

W2.1.4.02 to BS 6925; 20mm thick in two coats laid on prepared concrete surface

| | m² | 1.60 | 48.54 | - | 10.24 | 58.78 | 3.982 |

W2.1.5 Domed surfaces

W2.1.5.01 to BS 6925; 20mm thick in two coats to brickwork; raking out joints to form key

| | m² | 2.18 | 66.14 | - | 11.23 | 77.37 | 4.365 |

W2.1.5.02 to BS 6925; 20mm thick in two coats laid on prepared concrete surface

| | m² | 2.20 | 66.75 | - | 10.24 | 76.99 | 3.982 |

W2.1.6 Surfaces of width not exceeding 300mm

W2.1.6.01 to BS 6925; 20mm thick in two coats to brickwork; raking out joints to form key

| | m | 0.29 | 8.65 | - | 3.74 | 12.39 | 1.455 |

W2.1.6.02 to BS 6925; 20mm thick in two coats laid on prepared concrete surface

| | m | 0.20 | 6.07 | - | 3.74 | 9.81 | 1.455 |

W2.1.7 Surfaces of width 300mm - 1m

W2.1.7.01 to BS 6925; 20mm thick in two coats to brickwork; raking out joints to form key

| | m | 0.42 | 12.74 | - | 5.71 | 18.45 | 2.221 |

W2.1.7.02 to BS 6925; 20mm thick in two coats laid on prepared concrete surface

| | m | 0.30 | 9.10 | - | 5.12 | 14.22 | 1.991 |

W2 Tanking continued...

	Unit	Labour Hours	Labour Net £	Plant Net £	Materials Net £	Unit Net £	CO_2e Kg
W2.1 **Asphalt**							
W2.1.8 Isolated groups of surfaces							
W2.1.8.01 to BS 6925; 20mm thick in two coats to brickwork; raking out joints to form key; brick sump below basement level; 600 x 400 x 400mm to sides and base	Nr	2.18	66.14	-	33.68	99.82	13.094
W2.1.8.02 to BS 6925; 20mm thick in two coats laid on prepared concrete surface; concrete sump below basement level; 600 x 400 x 400mm to sides and base	Nr	2.20	66.75	-	30.72	97.47	11.946

W3 Roofing

	Unit	Labour Hours	Labour Net £	Plant Net £	Materials Net £	Unit Net £	CO_2e Kg
W3.1 **Asphalt**							
W3.1.1 Upper surfaces inclined at an angle not exceeding 30 degrees to the horizontal							
W3.1.1.01 Mastic asphalt, BS 6925; 20mm thick in two coats laid on prepared concrete surface	m²	0.50	15.17	-	9.60	24.77	3.982
W3.1.1.02 Mastic asphalt, BS 6925; 20mm thick in two coats; heavy gauge polythene membrane; sheathing felt, BS 747, type IB, 18kg roll, expanded polystyrene insulation 25mm thick; on concrete surface	m²	0.39	22.11	-	17.43	39.54	6.593
W3.1.2 Upper surfaces inclined at 30 - 60 degrees to the horizontal							
W3.1.2.01 Mastic asphalt, BS 6925; 20mm thick in two coats laid on prepared concrete surface	m²	0.75	22.75	-	9.60	32.35	3.982
W3.1.2.02 Mastic asphalt, BS 6925; 20mm thick in two coats; heavy gauge polythene membrane; sheathing felt, BS 747, type IB, 18kg roll, expanded polystyrene insulation 25mm thick; on concrete surface	m²	0.59	33.45	-	17.43	50.88	6.593
W3.1.3 Surfaces inclined at an angle exceeding 60 degrees to the horizontal							
W3.1.3.01 Mastic asphalt, BS 6925; 20mm thick in two coats laid on prepared concrete surface	m²	1.30	39.44	-	9.60	49.04	3.982
W3.1.3.02 Mastic asphalt, BS 6925; 20mm thick in two coats; heavy gauge polythene membrane; sheathing felt, BS 747, type IB, 18kg roll, expanded polystyrene insulation 25mm thick; on concrete surface	m²	0.98	55.56	-	17.43	72.99	6.593
W3.1.4 Curved surfaces							
W3.1.4.01 Mastic asphalt, BS 6925; 20mm thick in two coats laid on prepared concrete surface	m²	1.60	48.54	-	9.60	58.14	3.982
W3.1.5 Domed surfaces							
W3.1.5.01 Mastic asphalt, BS 6925; 20mm thick in two coats laid on prepared concrete surface	m²	2.20	66.75	-	9.60	76.35	3.982

W3 Roofing continued...

	Unit	Labour Hours	Labour Net £	Plant Net £	Materials Net £	Unit Net £	CO_2e Kg

W3.1 Asphalt

W3.1.6 Surfaces of width not exceeding 300mm

	Unit	Labour Hours	Labour Net £	Plant Net £	Materials Net £	Unit Net £	CO_2e Kg
W3.1.6.01 Mastic asphalt, BS 6925; 20mm thick in two coats laid on prepared concrete surface	m	0.20	6.07	-	3.14	9.21	1.302
W3.1.6.02 Mastic asphalt, BS 6925; 20mm thick in two coats; heavy gauge polythene membrane; sheathing felt, BS 747, type IB, 18kg roll, expanded polystyrene insulation 25mm thick; on concrete surface	m	0.13	7.37	-	5.80	13.17	2.190

W3.1.7 Surfaces of width 300mm - 1m

	Unit	Labour Hours	Labour Net £	Plant Net £	Materials Net £	Unit Net £	CO_2e Kg
W3.1.7.01 Mastic asphalt, BS 6925; 20mm thick in two coats laid on prepared concrete surface	m	0.30	9.10	-	6.28	15.38	2.604
W3.1.7.02 Mastic asphalt, BS 6925; 20mm thick in two coats; heavy gauge polythene membrane; sheathing felt, BS 747, type IB, 18kg roll, expanded polystyrene insulation 25mm thick; on concrete surface	m	0.26	14.74	-	12.31	27.05	4.614

W3.1.8 Isolated groups of surfaces

	Unit	Labour Hours	Labour Net £	Plant Net £	Materials Net £	Unit Net £	CO_2e Kg
W3.1.8.01 Mastic asphalt, BS 6925; 20mm thick in two coats laid on prepared concrete surface; stepped penthouse roof; overall size 800 x 500 x 150mm high step	Nr	2.20	66.75	-	19.21	85.96	7.964
W3.1.8.02 Mastic asphalt, BS 6925; 20mm thick in two coats; heavy gauge polythene membrane; sheathing felt, BS 747, type IB, 18kg roll, expanded polystyrene insulation 25mm thick; on concrete surface; stepped penthouse roof; overall size 800 x 500 x 150mm high step	Nr	1.60	90.70	-	34.13	124.83	12.880

W3.2 Sheet metal

W3.2.6 Surfaces of width not exceeding 300mm

	Unit	Labour Hours	Labour Net £	Plant Net £	Materials Net £	Unit Net £	CO_2e Kg
W3.2.6.01 Milled lead sheet, BS EN 12588, lead wedge fixings; Flashings; 200mm girth; one edge dressed along the verge of single lap tiling; one edge wedged into raked out joints of brickwork	m	0.36	16.97	-	11.11	28.08	1.545
W3.2.6.02 Milled lead sheet, BS EN 12588, lead wedge fixings; Flashings; 300mm girth; one edge dressed along the verge of single lap tiling; one edge wedged into raked out joints of brickwork	m	0.41	19.40	-	16.62	36.02	2.318

W4 Protective Layers

	Unit	Labour Hours	Labour Net £	Plant Net £	Materials Net £	Unit Net £	CO_2e Kg
W4.2 **Flexible sheeting**							
W4.2.1 Upper surfaces inclined at an angle not exceeding 30 degrees to the horizontal							
W4.2.1.01 Polythene in one layer 4000 gauge	m²	0.01	0.40	-	2.05	2.45	0.520
W4.2.1.02 Lotrak 10/7 ground stabilising matting	m²	0.01	0.40	-	0.79	1.19	0.229
W4.2.2 Upper surfaces inclined at 30 - 60 degrees to the horizontal							
W4.2.2.01 Polythene in one layer 4000 gauge	m²	0.01	0.62	-	2.05	2.67	0.520
W4.2.2.02 Lotrak 10/7 ground stabilising matting	m²	0.01	0.62	-	0.79	1.41	0.229
W4.2.3 Surfaces inclined at an angle exceeding 60 degrees to the horizontal							
W4.2.3.01 Polythene in one layer 4000 gauge	m²	0.02	0.85	-	2.05	2.90	0.520
W4.2.3.02 Lotrak 10/7 ground stabilising matting	m²	0.02	0.85	-	0.79	1.64	0.229
W4.2.4 Curved surfaces							
W4.2.4.01 Polythene in one layer 4000 gauge	m²	0.02	0.85	-	2.05	2.90	0.520
W4.2.4.02 Lotrak 10/7 ground stabilising matting	m²	0.02	0.85	-	0.79	1.64	0.229
W4.2.5 Domed surfaces							
W4.2.5.01 Polythene in one layer 4000 gauge	m²	0.02	0.96	-	2.05	3.01	0.520
W4.2.5.02 Lotrak 10/7 ground stabilising matting	m²	0.02	1.08	-	0.79	1.87	0.229
W4.2.6 Surfaces of width not exceeding 300mm							
W4.2.6.01 Polythene in one layer 4000 gauge	m	0.01	0.40	-	0.70	1.10	0.177
W4.2.6.02 Lotrak 10/7 ground stabilising matting	m	0.01	0.57	-	0.27	0.84	0.078
W4.2.7 Surfaces of width 300mm - 1m							
W4.2.7.01 Polythene in one layer 4000 gauge	m	0.01	0.40	-	1.37	1.77	0.348
W4.2.7.02 Lotrak 10/7 ground stabilising matting	m	0.01	0.57	-	0.53	1.10	0.153
W4.2.8 Isolated groups of surfaces							
W4.2.8.01 Polythene in one layer 4000 gauge; sump below basement level; 600 × 400 × 400mm to sides and base	Nr	0.02	1.13	-	0.71	1.84	0.206
W4.2.8.02 Lotrak 10/7 ground stabilising matting; 150mm laps; sump below basement level; 600 × 400 × 400mm to sides and base	Nr	0.02	1.30	-	1.58	2.88	0.458
W4.3 **Sand**							
W4.3.1 Upper surfaces inclined at an angle not exceeding 30 degrees to the horizontal							
W4.3.1.01 25mm thick to granular base	m²	0.00	0.17	-	0.68	0.85	0.312
W4.3.8 Isolated groups of surfaces							
W4.3.8.01 to base of stepped concrete plinth 450 × 450mm	Nr	0.01	0.57	-	1.36	1.93	0.624

W4 Protective Layers continued...

	Unit	Labour Hours	Labour Net £	Plant Net £	Materials Net £	Unit Net £	CO$_2$e Kg	
W4.4 **Sand and cement screed**								
W4.4.1 Upper surfaces inclined at an angle not exceeding 30 degrees to the horizontal								
W4.4.1.01 Sand and cement (1:3) screed; treated with water resistant additive; 50mm thick to prepared concrete surface	m^2	0.56	16.99	-	8.87	25.86	38.982	
W4.4.2 Upper surfaces inclined at 30 - 60 degrees to the horizontal								
W4.4.2.01 Sand and cement (1:3) screed; treated with water resistant additive; 50mm thick to prepared concrete surface	m^2	0.84	25.49	-	8.87	34.36	38.982	
W4.4.7 Surfaces of width 300mm - 1m								
W4.4.7.01 Sand and cement (1:3) screed; treated with water resistant additive; 50mm thick to prepared concrete surface	m	0.36	10.92	-	14.29	25.21	23.945	
W4.4.8 Isolated groups of surfaces								
W4.4.8.01 Sand and cement (1:3) screed; treated with water resistant additive; 50mm thick to prepared concrete surface; to base of stepped concrete plinth 450 x 450mm	Nr	1.12	33.98	-	15.82	49.80	76.182	
W4.5 **Tiles**								
W4.5.1 Upper surfaces inclined at an angle not exceeding 30 degrees to the horizontal								
W4.5.1.01 Plain tiles; 265 x 165mm; overlapping broken bonded; twice nailed each title to timber battens; 38 x 25mm softwood battens, plugged and screwed through 1000 gauge polythene membrane to concrete background	m^2	0.69	32.89	-	36.42	69.31	28.906	-5.247
W4.5.2 Upper surfaces inclined at 30 - 60 degrees to the horizontal								
W4.5.2.01 Plain tiles; 265 x 165mm; overlapping broken bonded; twice nailed each title to timber battens; 38 x 25mm softwood battens, plugged and screwed through 1000 gauge polythene membrane to concrete background	m^2	0.74	35.04	-	36.42	71.46	28.906	-5.247
W4.5.3 Surfaces inclined at an angle exceeding 60 degrees to the horizontal								
W4.5.3.01 Plain tiles; 265 x 165mm; overlapping broken bonded; twice nailed each title to timber battens; 38 x 25mm softwood battens, plugged and screwed through 1000 gauge polythene membrane to concrete background	m^2	1.04	49.34	-	36.42	85.76	28.906	-5.247

W4 Protective Layers continued...

	Unit	Labour Hours	Labour Net £	Plant Net £	Materials Net £	Unit Net £	CO_2e Kg	

W4.5	**Tiles**								
W4.5.4	Curved surfaces								
W4.5.4.01	Plain tiles; 265 x 165mm; overlapping broken bonded; twice nailed each title to timber battens; 38 x 25mm softwood battens, plugged and screwed through 1000 gauge polythene membrane to concrete background	m²	1.04	49.34	-	36.42	85.76	28.906	*-5.247*
W4.5.6	Surfaces of width not exceeding 300mm								
W4.5.6.01	Plain tiles; 265 x 165mm; overlapping broken bonded; twice nailed each title to timber battens; 38 x 25mm softwood battens, plugged and screwed through 1000 gauge polythene membrane to concrete background	m	0.23	10.96	-	12.37	23.33	9.694	*-2.099*
W4.5.7	Surfaces of width 300mm - 1m								
W4.5.7.01	Plain tiles; 265 x 165mm; overlapping broken bonded; twice nailed each title to timber battens; 38 x 25mm softwood battens, plugged and screwed through 1000 gauge polythene membrane to concrete background	m	0.46	21.93	-	24.73	46.66	19.388	*-4.197*

W5 Sprayed Or Brushed Waterproofing

W5.0	**Two coats R.I.W. solution**							
W5.0.0	Generally							
W5.0.0.01	To concrete surfaces	m²	0.05	2.83	-	4.07	6.90	0.047

CLASS X:
MISCELLANEOUS WORK

Calculations used throughout Class X - Miscellaneous Work

Labour

			Qty		Rate		Total
L X0003ICE	**Rock Gabion Labour Gang**						
	L A9016ICE	Ganger	1	x	17.64	=	£17.64
	L A9002ICE	Labourer (General Operative)	4	x	13.02	=	£52.06
	L A9024ICE	Plant Operator (Class 3)	0.5	x	16.21	=	£8.11
		Total hourly cost of gang				=	**£77.81**
L A0332ICE	**Fencing / Gate Installation Labour Gang**						
	L A9016ICE	Ganger	0.5	x	17.64	=	£8.82
	L A9014ICE	Labourer (Skill Rate 4)	0.5	x	14.03	=	£7.01
	L A9002ICE	Labourer (General Operative)	1	x	13.02	=	£13.02
	L A9025ICE	Plant Operator (Class 4)	1	x	15.30	=	£15.30
		Total hourly cost of gang				=	**£44.15**

Plant

			Qty		Rate		Total
P A1329ICE	**Gabion Plant Gang**						
	P A0507ICE	Cranes Crawler - NCK 305B - 20t	1	x	48.55	=	£48.55
		Total hourly cost of gang				=	**£48.55**
P X0001ICE	**Fencing / Gate Installation Plant Gang**						
	P A1059ICE	Agricultural Tractor: Fencing Auger	1	x	25.08	=	£25.08
	P A0961ICE	Trailer - Massey Tipping	1	x	1.81	=	£1.81
		Total hourly cost of gang				=	**£26.89**

Miscellaneous Work

X1 Fences

Note(s): The prices for work in this section are guide prices for approximate estimating purposes only. The prices are inclusive of labour and materials and are applicable to sites over 30 miles but not exceeding 60 miles radius from the point of supply.

1) All treated posts have been kiln dried where necessary to ensure a moisture content of below 28% and treated with a copper chrome arsenic preservative to a net dry salt retention of 6.4kg per m2. Specify Jakcured treated timber.

2) The prices for fencing include all necessary post hole excavation, disposal of surplus soil and concrete bases, whererequired by specification.

3) Extras would be deemed to be included in the linear items measured for fencing under the CESMM4.

		Unit	Labour Hours	Labour Net £	Plant Net £	Materials Net £	Unit Net £	CO_2e Kg	
X1.1	**Timber post and rail**								
X1.1.1	Height: not exceeding 1m								
X1.1.1.01	Cleft chestnut pale fencing BS 1722, part 4, with intermediate posts at maximum centres and spacings as Table 1; height: 0.90m; untreated chestnut, pointed for driving	m	0.30	13.29	0.06	5.57	18.92	1.733	-10.938
X1.1.1.02	as above; extra for straining posts and struts: end post and one strut	Nr	0.49	21.63	0.12	8.55	30.30	1.729	-10.625
X1.1.1.03	as above; extra for straining posts and struts: corner post and two struts	Nr	0.58	25.39	0.18	13.52	39.09	2.662	-16.528
X1.1.1.04	as above; extra for one line of barbed wire: wooden posts	m	0.03	1.10	-	0.29	1.39	0.516	
X1.1.1.05	as above; extra for two lines of barbed wire: wooden posts	m	0.05	2.21	-	0.57	2.78	1.032	
X1.1.2	Height: 1 - 1.25m								
X1.1.2.01	Wooden palisade fencing BS 1722, part 5, with intermediate posts at 3m centres; 75 x 20mm rectangular pales with square tops spaced 75mm apart; height: 1.00m; treated wooden posts	m	0.47	20.62	0.94	22.62	44.18	9.910	-10.366
X1.1.2.02	as above; extra for end post	Nr	0.38	16.78	5.47	26.79	49.04	52.070	-11.805
X1.1.2.03	as above; extra for angle post	Nr	0.47	20.84	7.28	25.54	53.66	34.292	-17.117
X1.1.2.04	Timber post and rail fencing; BS 1722 part 7, morticed type; intermediate main posts at 2.85m centres, one prick post between each main post; height: 1.10m (3 rails); treated softwood posts and rails; prick posts pointed for driving	m	0.36	15.72	0.06	7.06	22.84	1.509	-9.553
X1.1.2.05	as above; extra for end post	Nr	0.31	13.77	0.12	18.24	32.13	3.078	-19.479
X1.1.2.06	as above; extra for angle post	Nr	0.37	16.11	0.17	26.45	42.73	4.463	-28.245
X1.1.2.07	as above; extra for intersection post	Nr	0.31	13.77	0.12	26.45	40.34	4.463	-28.245
X1.1.2.08	as above; untreated Oak posts and rails; prick posts pointed for driving	m	0.53	23.44	0.06	17.72	41.22	4.245	-10.418
X1.1.2.09	as above; extra for end post	Nr	0.31	13.77	0.12	25.62	39.51	6.153	-15.096
X1.1.2.10	as above; extra for angle post	Nr	0.37	16.11	0.17	37.15	53.43	8.922	-21.889
X1.1.2.11	as above; extra for intersection post	Nr	0.31	13.77	0.12	37.15	51.04	8.922	-21.889

X1 Fences continued...

	Unit	Labour Hours	Labour Net £	Plant Net £	Materials Net £	Unit Net £	CO$_2$e Kg	

X1.1 Timber post and rail

X1.1.2 Height: 1 - 1.25m

X1.1.2.12	as previous item: 1.10m (4 rails); treated softwood posts and rails; prick posts pointed for driving	m	0.38	16.64	0.06	8.32	25.02	1.827	-11.567
X1.1.2.13	as above; extra for end post	Nr	0.35	15.32	0.12	18.60	34.04	3.140	-19.869
X1.1.2.14	as above; extra for angle post	Nr	0.40	17.79	0.17	27.14	45.10	4.579	-28.978
X1.1.2.15	as above; extra for intersection post	Nr	0.35	15.32	0.12	27.14	42.58	4.579	-28.978
X1.1.2.16	as above; untreated Oak posts and rails; prick posts pointed for driving	m	0.41	18.01	0.06	22.06	40.13	5.284	-12.968
X1.1.2.17	as above; extra for end post	Nr	0.33	14.70	0.12	25.62	40.44	6.153	-15.096
X1.1.2.18	as above; extra for angle post	Nr	0.39	17.22	0.17	37.15	54.54	8.922	-21.889
X1.1.2.19	as above; extra for intersection post	Nr	0.33	14.70	0.12	37.15	51.97	8.922	-21.889
X1.1.2.20	Wooden post and rail fencing, BS 1722, part 7, nailed type; intermediate main posts at 1.80m centres; height: 1.10m (3 rails); treated softwood posts and rails	m	0.30	13.20	0.06	5.93	19.19	1.334	-8.443
X1.1.2.21	as above; extra for end post	Nr	0.33	14.70	0.12	9.70	24.52	1.772	-11.215
X1.1.2.22	as above; extra for angle post	Nr	0.39	17.22	0.17	14.43	31.82	2.611	-16.527
X1.1.2.23	as above; extra for intersection post	Nr	0.33	14.70	0.12	14.43	29.25	2.611	-16.527
X1.1.2.24	as above, posts pointed for driving	m	0.31	13.69	0.06	5.93	19.68	1.334	-8.443
X1.1.2.25	as above; extra for end post	Nr	0.32	14.30	0.12	9.70	24.12	1.772	-11.215
X1.1.2.26	as above; extra for angle post	Nr	0.38	16.78	0.17	14.43	31.38	2.611	-16.527
X1.1.2.27	as above; extra for intersection post	Nr	0.32	14.30	0.12	14.43	28.85	2.611	-16.527
X1.1.2.28	as above; untreated Oak posts and rails	m	0.58	25.39	0.06	17.72	43.17	4.245	-10.418
X1.1.2.29	as above; extra for end post	Nr	0.31	13.77	0.12	25.62	39.51	6.153	-15.096
X1.1.2.30	as above; extra for angle post	Nr	0.34	15.06	0.17	37.15	52.38	8.922	-21.889
X1.1.2.31	as above; extra for intersection post	Nr	0.32	14.30	0.12	37.15	51.57	8.922	-21.889
X1.1.2.32	as above; posts pointed for driving	m	0.68	30.07	0.06	17.72	47.85	4.245	-10.418
X1.1.2.33	as above; extra for end post	Nr	0.32	14.13	0.12	25.62	39.87	6.153	-15.096
X1.1.2.34	as above; extra for angle post	Nr	0.38	16.56	0.17	37.15	53.88	8.922	-21.889
X1.1.2.35	as above; extra for intersection post	Nr	0.32	14.13	0.12	37.15	51.40	8.922	-21.889
X1.1.2.36	Wooden post and rail fencing, BS 1722, part 7, nailed type; intermediate main posts at 1.80m centres; height: 1.10m (4 rails); treated softwood posts and rails	m	0.35	15.63	0.06	7.23	22.92	1.665	-10.537
X1.1.2.37	as above; extra for end post	Nr	0.35	15.32	0.12	9.70	25.14	1.772	-11.215
X1.1.2.38	as above; extra for angle post	Nr	0.40	17.79	0.17	14.67	32.63	2.705	-17.118
X1.1.2.39	as above; extra for intersection post	Nr	0.35	15.32	0.12	14.67	30.11	2.705	-17.118
X1.1.2.40	as above, posts pointed for driving	m	0.35	15.23	0.06	7.23	22.52	1.665	-10.537
X1.1.2.41	as above; extra for end post	Nr	0.34	14.97	0.12	9.70	24.79	1.772	-11.215
X1.1.2.42	as above; extra for angle post	Nr	0.39	17.35	0.17	14.67	32.19	2.705	-17.118
X1.1.2.43	as above; extra for intersection post	Nr	0.34	14.97	0.12	14.67	29.76	2.705	-17.118
X1.1.2.44	as above; untreated Oak posts and rails	m	0.50	22.16	0.06	22.06	44.28	5.284	-12.968
X1.1.2.45	as above; extra for end post	Nr	0.33	14.70	0.12	25.62	40.44	6.153	-15.096
X1.1.2.46	as above; extra for angle post	Nr	0.39	17.22	0.17	37.15	54.54	8.922	-21.889

X1 Fences continued...

	Unit	Labour Hours	Labour Net £	Plant Net £	Materials Net £	Unit Net £		CO_2e Kg

X1.1 Timber post and rail

X1.1.2 Height: 1 - 1.25m

		Unit	Labour Hours	Labour Net £	Plant Net £	Materials Net £	Unit Net £		CO_2e Kg
X1.1.2.47	as previous item; extra for intersection post	Nr	0.33	14.70	0.12	37.15	51.97	8.922	-21.889
X1.1.2.48	as above; posts pointed for driving	m	0.49	21.63	0.06	22.06	43.75	5.284	-12.968
X1.1.2.49	as above; extra for end post	Nr	0.32	14.30	0.12	25.62	40.04	6.153	-15.096
X1.1.2.50	as above; extra for angle post	Nr	0.38	16.78	0.17	37.15	54.10	8.922	-21.889
X1.1.2.51	as above; extra for intersection post	Nr	0.32	14.30	0.12	37.15	51.57	8.922	-21.889
X1.1.2.52	Wooden palisade fencing BS 1722, part 5, with intermediate posts at 3m centres; 75 x 20mm rectangular pales with square tops spaced 75mm apart; height: 1.20m; treated wooden posts	m	0.49	21.46	0.94	22.62	45.02	9.910	-10.366
X1.1.2.53	as above; extra for end post	Nr	0.40	17.44	5.47	26.79	49.70	52.070	-11.805
X1.1.2.54	as above; extra for angle post	Nr	0.49	21.63	7.28	25.54	54.45	34.292	-17.117
X1.1.2.55	Cleft chestnut pale fencing BS 1722, part 4, with intermediate posts at maximum centres and spacings as Table 1; height: 1.05m; untreated chestnut, pointed for driving	m	0.30	13.29	0.06	6.20	19.55	1.970	-12.434
X1.1.2.56	as above; extra for straining posts and struts: end post and one strut	Nr	0.49	21.63	0.12	11.13	32.88	1.915	-11.805
X1.1.2.57	as above; extra for straining posts and struts: corner post and two struts	Nr	0.57	24.99	0.18	17.21	42.38	2.941	-18.298
X1.1.2.58	as above; height: 1.20m; untreated chestnut, pointed for driving	m	0.30	13.29	0.06	6.57	19.92	2.237	-14.124
X1.1.2.59	as above; extra for straining posts and struts: end post and one strut	Nr	0.52	22.87	0.12	11.13	34.12	1.915	-11.805
X1.1.2.60	as above; extra for straining posts and struts: corner post and two struts	Nr	0.64	28.12	0.18	15.86	44.16	2.754	-17.117

X1.1.3 Height: 1.25 - 1.5m

		Unit	Labour Hours	Labour Net £	Plant Net £	Materials Net £	Unit Net £		CO_2e Kg
X1.1.3.01	BS 1722 part 7, morticed type; intermediate main posts at 2.85m centres, one prick post between each main post; height: 1.30m (4 rails); treated softwood posts and rails; prick posts pointed for driving	m	0.41	18.01	0.06	8.91	26.98	1.927	-12.196
X1.1.3.02	as above; extra for end post	Nr	0.35	15.32	0.12	21.89	37.33	3.694	-23.375
X1.1.3.03	as above; extra for angle post	Nr	0.40	17.79	0.17	31.92	49.88	5.387	-34.088
X1.1.3.04	as above; extra for intersection post	Nr	0.35	15.32	0.12	31.92	47.36	5.387	-34.088
X1.1.3.05	as above; untreated Oak posts and rails; prick posts pointed for driving	m	0.50	22.16	0.06	22.91	45.13	5.489	-13.471
X1.1.3.06	as above; extra for end post	Nr	0.35	15.32	0.12	30.74	46.18	7.384	-18.115
X1.1.3.07	as above; extra for angle post	Nr	0.40	17.79	0.17	44.41	62.37	10.665	-26.166
X1.1.3.08	as above; extra for intersection post	Nr	0.35	15.32	0.12	44.83	60.27	10.768	-26.418

X1 Fences continued...

	Unit	Labour Hours	Labour Net £	Plant Net £	Materials Net £	Unit Net £	CO$_2$e Kg	

X1.1 Timber post and rail

X1.1.3 Height: 1.25 - 1.5m

		Unit	Labour Hours	Labour Net £	Plant Net £	Materials Net £	Unit Net £	CO$_2$e Kg	
X1.1.3.09	Wooden post and rail fencing, BS 1722, part 7, nailed type; intermediate main posts at 1.80m centres; height: 1.30m (4 rails); treated softwood posts and rails	m	0.35	15.63	0.06	7.23	22.92	1.665	-10.537
X1.1.3.10	as above; extra for end post	Nr	0.35	15.32	0.12	9.70	25.14	1.772	-11.215
X1.1.3.11	as above; extra for angle post	Nr	0.40	17.79	0.17	14.67	32.63	2.705	-17.118
X1.1.3.12	as above; extra for intersection post	Nr	0.35	15.32	0.12	14.67	30.11	2.705	-17.118
X1.1.3.13	as above, posts pointed for driving	m	0.35	15.23	0.06	7.23	22.52	1.665	-10.537
X1.1.3.14	as above; extra for end post	Nr	0.34	14.97	0.12	9.70	24.79	1.772	-11.215
X1.1.3.15	as above; extra for angle post	Nr	0.39	17.35	0.17	14.67	32.19	2.705	-17.118
X1.1.3.16	as above; extra for intersection post	Nr	0.34	14.97	0.12	14.67	29.76	2.705	-17.118
X1.1.3.17	as above; untreated Oak posts and rails	m	0.50	22.16	0.06	22.91	45.13	5.489	-13.471
X1.1.3.18	as above; extra for end post	Nr	0.33	14.70	0.12	30.74	45.56	7.384	-18.115
X1.1.3.19	as above; extra for angle post	Nr	0.39	17.22	0.17	44.41	61.80	10.665	-26.166
X1.1.3.20	as above; extra for intersection post	Nr	0.33	14.70	0.12	44.83	59.65	10.768	-26.418
X1.1.3.21	as above, posts pointed for driving	m	0.49	21.63	0.06	22.91	44.60	5.489	-13.471
X1.1.3.22	as above; extra for end post	Nr	0.32	14.30	0.12	30.74	45.16	7.384	-18.115
X1.1.3.23	as above; extra for angle post	Nr	0.38	16.78	0.17	44.41	61.36	10.665	-26.166
X1.1.3.24	as above; extra for intersection post	Nr	0.32	14.30	0.12	44.83	59.25	10.768	-26.418
X1.1.3.25	Wooden palisade fencing BS 1722, part 5, with intermediate posts at 3m centres; 75 x 20mm rectangular pales with square tops spaced 75mm apart; height: 1.40m; treated wooden posts	m	0.49	21.55	0.94	28.64	51.13	10.501	-13.032
X1.1.3.26	as above; extra for end post	Nr	0.42	18.54	5.47	30.33	54.34	52.629	-15.347
X1.1.3.27	as above; extra for angle post	Nr	0.50	22.07	7.28	30.97	60.32	35.131	-22.430
X1.1.3.28	Cleft chestnut pale fencing BS 1722, part 4, with intermediate posts at maximum centres and spacings as Table 1; height: 1.35m; untreated chestnut, pointed for driving	m	0.30	13.29	0.06	10.19	23.54	2.535	-16.013
X1.1.3.29	as above; extra for straining posts and struts: end post and one strut	Nr	0.52	22.87	0.12	13.57	36.56	2.288	-14.166
X1.1.3.30	as above; extra for straining posts and struts: corner post and two struts	Nr	0.64	28.12	0.18	19.65	47.95	3.314	-20.659

X1.1.4 Height: 1.5 - 2.0m

		Unit	Labour Hours	Labour Net £	Plant Net £	Materials Net £	Unit Net £	CO$_2$e Kg	
X1.1.4.01	Wooden palisade fencing BS 1722, part 5, with intermediate posts at 3m centres; 75 x 20mm rectangular pales with square tops spaced 75mm apart; height: 1.60m; treated wooden posts	m	0.49	21.55	0.94	31.86	54.35	10.703	-14.317
X1.1.4.02	as above; extra for end post	Nr	0.47	20.75	5.47	30.33	56.55	52.629	-15.347
X1.1.4.03	as above; extra for angle post	Nr	0.60	26.49	7.28	30.97	64.74	35.131	-22.430
X1.1.4.04	as above: 1.80m; treated wooden posts	m	0.54	23.89	0.94	32.20	57.03	10.796	-14.906
X1.1.4.05	as above; extra for end post	Nr	0.45	19.87	5.47	32.67	58.01	53.189	-18.888
X1.1.4.06	as above; extra for angle post	Nr	0.53	23.22	8.18	34.64	66.04	36.302	-27.742

X1   Fences continued...

	Unit	Labour Hours	Labour Net £	Plant Net £	Materials Net £	Unit Net £	CO$_2$e Kg	
X1.1	**Timber post and rail**							
X1.1.4	Height: 1.5 - 2.0m							
X1.1.4.07	Cleft chestnut pale fencing BS 1722, part 4, with intermediate posts at maximum centres and spacings as Table 1; height: 1.80m; untreated chestnut, pointed for driving	m	0.30	13.29	0.06	12.46	25.81	3.336	-21.088
X1.1.4.08	as above; extra for straining posts and struts: end post and one strut	Nr	0.52	22.87	0.12	16.29	39.28	2.847	-17.708
X1.1.4.09	as above; extra for straining posts and struts: corner post and two struts	Nr	0.64	28.12	0.18	24.02	52.32	4.152	-25.972
X1.2	**Timber post and wire**							
X1.2.1	Height: not exceeding 1m							
X1.2.1.01	Strained wire fencing, BS 1722 part 2, with 4mm galvanised wire; intermediate posts at 3m centres; height: 0.85m (3 wires); treated round softwood posts; pointed for driving	m	0.21	9.40	0.05	1.38	10.83	0.605	-1.158
X1.2.1.02	as above; extra for straining posts and struts: end post and one strut	Nr	0.46	20.31	0.12	7.64	28.07	1.302	-6.257
X1.2.1.03	as above; extra for straining posts and struts: corner post and two struts	Nr	0.57	24.99	0.17	12.37	37.53	2.005	-9.038
X1.2.1.04	as above; sawn softwood posts; pointed for driving	m	0.21	9.40	0.05	2.18	11.63	0.733	-1.966
X1.2.1.05	as above; extra for straining posts and struts: end post and one strut	Nr	0.46	20.31	0.12	10.86	31.29	1.993	-10.625
X1.2.1.06	as above; extra for straining posts and struts: corner post and two struts	Nr	0.57	24.99	0.17	16.43	41.59	3.003	-15.347
X1.2.1.07	as above; sawn softwood posts; pointed for driving	m	0.22	9.67	0.06	1.90	11.63	1.026	-1.158
X1.2.1.08	Strained wire fencing, BS 1722 part 2, with 4mm galvanised wire; intermediate posts at 3m centres; height: 1.00m (6 wires); treated round softwood posts; pointed for driving; extra for straining posts and struts: end post and one strut	Nr	0.47	20.62	0.12	10.41	31.15	1.676	-6.951
X1.2.1.09	as above: corner post and two struts	Nr	0.58	25.65	0.17	18.11	43.93	2.913	-11.122
X1.2.1.10	as above; sawn softwood posts; pointed for driving	m	0.22	9.67	0.06	2.75	12.48	1.196	-1.966
X1.2.1.11	as above; extra for straining posts and struts: end post and one strut	Nr	0.47	20.62	0.12	14.82	35.56	2.443	-11.805
X1.2.1.12	as above; extra for straining posts and struts: corner post and two struts	Nr	0.58	25.65	0.17	26.44	52.26	4.141	-18.889
X1.2.1.13	Chain link fencing, BS 1722, part 1, with galvanised mesh, line and tying wire, intermediate posts at 3m centres; height: 0.90m (medium 2 line wires); treated softwood posts	m	0.30	13.16	0.17	3.99	17.32	3.385	-1.572

X1 Fences continued...

	Unit	Labour Hours	Labour Net £	Plant Net £	Materials Net £	Unit Net £	CO$_2$e Kg	

X1.2 Timber post and wire

X1.2.1 Height: not exceeding 1m

| X1.2.1.14 | as previous item; extra for straining posts and struts: end post and one strut | Nr | 0.85 | 37.53 | 0.12 | 10.86 | 48.51 | 1.993 | -10.625 |

X1.2.2 Height: 1 - 1.25m

| X1.2.2.01 | Chain link fencing, BS 1722, part 1, with galvanised mesh, line and tying wire, intermediate posts at 3m centres; height: 1.20m (heavy 3 line wires); treated softwood posts | m | 0.34 | 15.01 | 0.17 | 4.74 | 19.92 | 4.520 | -1.966 |
| X1.2.2.02 | as above; extra for straining posts and struts: end post and one strut | Nr | 0.85 | 37.53 | 0.12 | 14.22 | 51.87 | 2.366 | -12.986 |

X1.2.3 Height: 1.25 - 1.5m

X1.2.3.01	Strained wire fencing, BS 1722 part 2, with 4mm galvanised wire; intermediate posts at 3m centres; height: 1.40m (8 wires); treated round softwood posts; pointed for driving	m	0.23	9.98	0.06	2.57	12.61	1.377	-1.620
X1.2.3.02	as above; extra for straining posts and struts: end post and one strut	Nr	0.47	20.93	0.12	13.09	34.14	2.182	-9.035
X1.2.3.03	as above; extra for straining posts and struts: corner post and two struts	Nr	0.59	25.92	0.17	22.62	48.71	3.595	-13.205
X1.2.3.04	as above; sawn softwood posts; pointed for driving	m	0.23	9.98	0.06	3.95	13.99	1.556	-2.752
X1.2.3.05	as above; extra for straining posts and struts: end post and one strut	Nr	0.47	20.93	0.12	20.83	41.88	3.178	-15.347
X1.2.3.06	as above; extra for straining posts and struts: corner post and two struts	Nr	0.59	25.92	0.17	33.93	60.02	5.051	-22.430
X1.2.3.07	as above, dropper pattern with 4mm wire; intermediate posts at 5m centres; height: 1.40m (8 wires); treated round softwood posts, pointed for driving, cleft chestnut pale dropper	m	0.23	9.98	0.06	5.43	15.47	2.502	-8.743
X1.2.3.08	as above; treated wooden batten dropper	m	0.23	9.98	0.06	5.43	15.47	2.502	-8.743
X1.2.3.09	Chain link fencing, BS 1722, part 1, with galvanised mesh, line and tying wire, intermediate posts at 3m centres; height: 1.40m (medium 3 line wires); treated softwood posts	m	0.43	18.76	0.17	5.60	24.53	5.244	-2.359
X1.2.3.10	as above; extra for straining posts and struts: end post and one strut	Nr	0.85	37.53	0.12	16.98	54.63	2.738	-15.347

X1.2.4 Height: 1.5 - 2.0m

| X1.2.4.01 | Chain link fencing, BS 1722, part 1, with galvanised mesh, line and tying wire, intermediate posts at 3m centres; height: 1.80m (heavy 3 line wires); treated softwood posts | m | 0.45 | 19.87 | 0.17 | 8.63 | 28.67 | 6.631 | -2.752 |
| X1.2.4.02 | as above; extra for straining posts and struts: end post and one strut | Nr | 0.90 | 39.73 | 0.16 | 19.08 | 58.97 | 3.106 | -17.118 |

X1 Fences continued...

	Unit	Labour Hours	Labour Net £	Plant Net £	Materials Net £	Unit Net £	CO$_2$e Kg	
X1.2	**Timber post and wire**							
X1.2.5	Height: 2 - 2.5m							
X1.2.5.01	Chain link fencing, BS 1722, part 1, with galvanised mesh, line and tying wire, intermediate posts at 3m centres; 2.13m (heavy 3 line wires); treated softwood posts	m	0.45	19.87	0.17	9.01	29.05	7.718	-3.341
X1.2.5.02	as above; extra for straining posts and struts: end post and one strut	Nr	0.90	39.73	0.16	20.38	60.27	3.479	-19.478
X1.3	**Concrete post and wire**							
X1.3.1	Height: not exceeding 1m							
X1.3.1.01	Strained wire fencing, BS 1722, part 2, with 4mm galvanised wire; intermediate posts at 3m centres; height: 0.85m (3 wires); concrete posts	m	0.32	14.08	0.94	5.97	20.99	10.698	
X1.3.1.02	as above; extra for straining posts and struts: end post and one strut	Nr	0.62	27.37	4.45	32.88	64.70	58.168	
X1.3.1.03	as above; extra for straining posts and struts: corner post and two struts	Nr	0.83	36.64	8.45	35.94	81.03	45.050	
X1.3.1.04	as above; height: 1.00m (6 wires); concrete posts	m	0.33	14.48	0.94	7.32	22.74	11.722	
X1.3.1.05	as above; extra for straining posts and struts: end post and one strut	Nr	0.65	28.83	5.47	37.42	71.72	62.416	
X1.3.1.06	as above; extra for straining posts and struts: corner post and two struts	Nr	0.83	36.56	8.44	41.88	86.88	49.136	
X1.3.1.07	Chain link fencing, BS 1722, part 1, with galvanised mesh, line and tying wire; intermediate posts at 3m centres; height: 0.90m (medium 2 line wires); concrete posts	m	0.30	13.16	0.94	8.36	22.46	13.418	
X1.3.1.08	as above; extra for straining posts and struts: end post and one strut	Nr	0.62	27.37	4.59	33.22	65.18	59.214	
X1.3.1.09	as above; extra for straining posts and struts: corner post and two struts	Nr	0.83	36.64	8.45	35.94	81.03	45.050	
X1.3.1.10	Wooden palisade fencing BS 1722, part 5, with intermediate posts at 3m centres; 75 x 20mm rectangular pales with square tops spaced 75mm apart; height: 1.00m; concrete posts	m	0.47	20.62	0.94	23.17	44.73	11.174	-8.204
X1.3.1.11	as above; extra for end post	Nr	0.38	16.78	5.47	27.28	49.53	41.584	
X1.3.1.12	as above; extra for angle post	Nr	0.47	20.84	7.28	35.73	63.85	46.086	
X1.3.1.13	Cleft chestnut pale fencing BS 1722, part 4, with intermediate posts at maximum centres and spacings as Table 1; height: 0.90m; concrete posts	m	0.23	9.98	2.04	12.41	24.43	22.449	
X1.3.1.14	as above; extra for straining posts and struts: end post and one strut	Nr	0.58	25.39	2.26	22.02	49.67	28.364	
X1.3.1.15	as above; extra for straining posts and struts: corner post and two struts	Nr	0.63	27.95	2.26	28.49	58.70	32.013	

X1 Fences continued...

	Unit	Labour Hours	Labour Net £	Plant Net £	Materials Net £	Unit Net £	CO_2e Kg	
X1.3	**Concrete post and wire**							
X1.3.1	**Height: not exceeding 1m**							
X1.3.1.16	as previous item; untreated chestnut, pointed for driving; additional costs of barbed wire fixed above the fencing; posts increased in length by 150mm for each line of barbed wire: one line of barbed wire: concrete posts	m	0.05	2.21	-	0.29	2.50	0.516	
X1.3.1.17	as above; two lines of barbed wire: concrete posts	m	0.10	4.42	-	0.57	4.99	1.032	
X1.3.2	**Height: 1 - 1.25m**							
X1.3.2.01	Chain link fencing, BS 1722, part 1, with galvanised mesh, line and tying wire; intermediate posts at 3m centres; height: 1.20m (heavy 3 line wires); concrete posts	m	0.33	14.48	0.94	9.38	24.80	15.097	
X1.3.2.02	as above; extra for straining posts and struts: end post and one strut	Nr	0.76	33.64	5.47	37.42	76.53	62.416	
X1.3.2.03	as above; extra for straining posts and struts: corner post and two struts	Nr	0.89	39.38	4.02	41.88	85.28	47.408	
X1.3.2.04	Wooden palisade fencing BS 1722, part 6, with intermediate posts at 3m centres; 75 x 20mm rectangular pales with square tops spaced 75mm apart; height: 1.20m; concrete posts	m	0.48	21.06	0.94	26.53	48.53	12.227	-9.254
X1.3.2.05	as above; extra for end post	Nr	0.42	18.54	5.47	28.88	52.89	42.970	
X1.3.2.06	as above; extra for angle post	Nr	0.49	21.63	7.28	38.64	67.55	49.734	
X1.3.2.07	Cleft chestnut pale fencing BS 1722, part 4, with intermediate posts at maximum centres and spacings as Table 1; height: 1.05m; concrete posts	m	0.24	10.64	2.04	13.00	25.68	22.220	-10.468
X1.3.2.08	as above; extra for straining posts and struts: end post and one strut	Nr	0.60	26.67	2.26	22.91	51 84	29.276	
X1.3.2.09	as above; extra for straining posts and struts: corner post and two struts	Nr	0.75	33.20	2.26	30.27	65.73	33.837	
X1.3.2.10	as above; height: 1.20m; concrete posts	m	0.30	13.29	2.04	13.07	28.40	22.861	-11.962
X1.3.2.11	as above; extra for straining posts and struts: end post and one strut	Nr	0.60	26.67	2.26	24.16	53.09	30.189	
X1.3.2.12	as above; extra for straining posts and struts: corner post and two struts	Nr	0.89	39.07	2.32	33.36	74.75	37.519	
X1.3.3	**Height: 1.25 - 1.5m**							
X1.3.3.01	Strained wire fencing, BS 1722, part 2, with 4mm galvanised wire; intermediate posts at 3m centres; height: 1.40m (8 wires); concrete posts	m	0.34	14.79	0.94	7.77	23.50	11.882	
X1.3.3.02	as above; extra for straining posts and struts: end post and one strut	Nr	0.80	35.32	4.45	37.78	77.55	53.490	
X1.3.3.03	as above; extra for straining posts and struts: corner post and two struts	Nr	0.88	38.85	4.26	49.42	92.53	52.875	

X1 Fences continued...

		Unit	Labour Hours	Labour Net £	Plant Net £	Materials Net £	Unit Net £	CO_2e Kg
X1.3	**Concrete post and wire**							
X1.3.3	Height: 1.25 - 1.5m							
X1.3.3.04	Straining wire fencing, BS 1722, part 2, dropper pattern with 4mm wire; intermediate posts at 5m centres; concrete posts; cleft chestnut pale droppers	m	0.34	14.79	0.87	8.70	24.36	11.845 *-7.090*
X1.3.3.05	Chain link fencing, BS 1722, part 1, with galvanised mesh, line and tying wire; intermediate posts at 3m centres; height: 1.40m (medium 3 line wires); concrete posts	m	0.33	14.48	0.94	9.98	25.40	16.164
X1.3.3.06	as above; extra for straining posts and struts: end post and one strut	Nr	0.76	33.64	5.47	39.51	78.62	64.546
X1.3.3.07	as above; extra for straining posts and struts: corner post and two struts	Nr	0.89	39.38	4.02	45.22	88.62	50.451
X1.3.3.08	Wooden palisade fencing BS 1722, part 6, with intermediate posts at 3m centres; 75 x 20mm rectangular pales with square tops spaced 75mm apart; height: 1.40m; concrete posts: extra for end post	m	0.49	21.55	0.94	34.28	56.77	17.364 *-10.280*
X1.3.3.09	as above; extra for end post	Nr	0.42	18.54	5.47	37.80	61.81	64.751
X1.3.3.10	as above; extra for angle post	Nr	0.45	19.87	7.28	42.57	69.72	53.382
X1.3.3.11	Cleft chestnut pale fencing BS 1722, part 4, with intermediate posts at maximum centres and spacings as Table 1; height: 1.35m; concrete posts	m	0.30	13.29	2.04	16.33	31.66	23.097 *-13.458*
X1.3.3.12	as above; extra for straining posts and struts: end post and one strut	Nr	0.60	26.67	2.26	25.47	54.40	31.405
X1.3.3.13	as above; extra for straining posts and struts: corner post and two struts	Nr	0.89	39.07	2.32	34.67	76.06	38.735
X1.3.4	Height: 1.5 - 2.0m							
X1.3.4.01	Chain link fencing, BS 1722, part 1, with galvanised mesh, line and tying wire; intermediate posts at 3m centres; height: 1.80m (heavy 3 line wires); concrete posts	m	0.34	14.79	0.94	13.29	29.02	17.554
X1.3.4.02	as above; extra for straining posts and struts: end post and one strut	Nr	0.86	38.01	4.59	43.54	86.14	66.472
X1.3.4.03	as above; extra for straining posts and struts: corner post and two struts	Nr	1.02	44.86	8.45	51.35	104.66	55.571
X1.3.4.04	as above, 3 lines of barbed wire protection at top; intermediate posts at 3m centres; height: 1.80m (heavy 3 line wires); concrete posts; steel extension arms	m	0.37	16.42	0.94	16.90	34.26	19.647
X1.3.4.05	as above; extra for straining posts and struts: end post and one strut	Nr	0.90	39.65	5.47	45.19	90.31	67.866
X1.3.4.06	as above; extra for straining posts and struts: corner post and two struts	Nr	1.02	44.86	8.45	54.36	107.67	75.358

X1 Fences continued...

	Unit	Labour Hours	Labour Net £	Plant Net £	Materials Net £	Unit Net £	CO$_2$e Kg	
X1.3	**Concrete post and wire**							
X1.3.4	Height: 1.5 - 2.0m							
X1.3.4.07	as previous item; concrete posts with cranked tops	m	0.38	16.95	0.94	14.05	31.94	18.549	
X1.3.4.08	as above; extra for straining posts and struts: end post and one strut	Nr	0.92	40.49	5.47	42.63	88.59	66.674	
X1.3.4.09	as above; concrete posts with cranked tops; extra for straining posts and struts: corner post and two struts	Nr	1.00	44.33	8.18	48.90	101.41	55.126	
X1.3.4.10	Wooden palisade fencing BS 1722, part 6, with intermediate posts at 3m centres; 75 x 20mm rectangular pales with square tops spaced 75mm apart; height: 1.60m; concrete posts: extra for end post	m	0.50	22.03	0.94	37.90	60.87	18.870	-11.565
X1.3.4.11	as above; extra for end post	Nr	0.42	18.54	5.47	40.12	64.13	67.185	
X1.3.4.12	as above; extra for angle post	Nr	0.49	21.59	5.35	46.05	72.99	56.141	
X1.3.4.13	as above: 1.80m; concrete posts: extra for end post	m	0.54	23.89	0.94	38.25	63.08	20.171	-11.565
X1.3.4.14	as above; extra for end post	Nr	0.45	19.87	5.47	41.08	66.42	69.617	
X1.3.4.15	as above; extra for angle post	Nr	0.53	23.22	5.35	47.49	76.06	59.789	
X1.3.4.16	Cleft chestnut pale fencing BS 1722, part 4, with intermediate posts at maximum centres and spacings as Table 1; height: 1.80m; concrete posts	m	0.35	15.45	2.04	18.40	35.89	24.615	-17.943
X1.3.4.17	as above; extra for straining posts and struts: end post and one strut	Nr	0.65	28.70	2.26	27.19	58.15	33.534	
X1.3.4.18	as above; extra for straining posts and struts: corner post and two struts	Nr	0.89	39.07	2.32	36.95	78.34	41.776	
X1.3.5	Height: 2 - 2.5m							
X1.3.5.01	Chain link fencing, BS 1722, part 1, with galvanised mesh, line and tying wire; intermediate posts at 3m centres; height: 2.13m (heavy 3 line wires); concrete posts	m	0.39	17.26	0.94	15.39	33.59	19.942	
X1.3.5.02	as above; extra for straining posts and struts: end post and one strut	Nr	0.92	40.49	5.47	48.55	94.51	70.932	
X1.3.5.03	as above; extra for straining posts and struts: corner post and two struts	Nr	1.02	44.86	8.45	60.42	113.73	61.654	
X1.3.5.04	as above, 3 lines of barbed wire protection at top; intermediate posts at 3m centres; height: 2.13m (heavy 3 line wires); concrete posts; steel extension arms	m	0.40	17.79	0.94	17.66	36.39	21.046	
X1.3.5.05	as above; extra for straining posts and struts: end post and one strut	Nr	0.98	43.40	5.47	51.74	100.61	72.124	
X1.3.5.06	as above; extra for straining posts and struts: corner post and two struts	Nr	1.07	47.20	8.18	60.53	115.91	62.401	
X1.3.5.07	as above; concrete posts with cranked tops	m	0.46	20.31	0.94	14.54	35.79	18.902	
X1.3.5.08	as above; extra for straining posts and struts: end post and one strut	Nr	1.00	44.24	5.47	46.69	96.40	69.715	

X1 Fences continued...

	Unit	Labour Hours	Labour Net £	Plant Net £	Materials Net £	Unit Net £	CO_2e Kg
X1.3 **Concrete post and wire**							
X1.3.5 Height: 2 - 2.5m							
X1.3.5.09 as previous item; extra for straining posts and struts: corner post and two struts	Nr	1.17	51.57	8.18	55.48	115.23	*59.992*
X1.4 **Metal post and wire**							
X1.4.1 Height: not exceeding 1m							
X1.4.1.01 Strained wire fencing, BS 1722, part 2, with 4mm galvanised wire; intermediate posts at 3m centres; height: 0.85m (3 wires); galvanised steel angle posts pointed for driving	m	0.29	12.89	0.94	5.55	19.38	11.904
X1.4.1.02 as above; extra for straining posts and struts with welded base plates: end post and one strut	Nr	0.61	26.98	5.14	30.06	62.18	61.672
X1.4.1.03 as above; extra for straining posts and struts with welded base plates: corner post and two struts	Nr	0.68	30.07	7.52	45.08	82.67	89.146
X1.4.1.04 as above; height: 1.00m (6 wires); galvanised steel angle posts pointed for driving	m	0.29	12.89	0.94	6.06	19.89	12.321
X1.4.1.05 as above; extra for straining posts and struts with welded base plates: end post and one strut	Nr	0.65	28.74	5.14	30.06	63.94	61.672
X1.4.1.06 as above; extra for straining posts and struts with welded base plates: corner post and two struts	Nr	0.73	32.23	7.52	45.08	84.83	89.146
X1.4.1.07 Chain link fencing, BS 1722, part 1, with galvanised mesh, line and tying wire, intermediate posts at 3m centres; height: 0.90m (medium 2 line wires); galvanised steel angle posts pointed for driving	m	0.23	10.29	0.94	10.69	21.92	12.657
X1.4.1.08 as above; extra for straining posts and struts with welded base plates: end post and one strut	Nr	0.61	26.98	5.14	30.06	62.18	61.672
X1.4.1.09 as above; extra for straining posts and struts with welded base plates: corner post and two struts	Nr	0.68	30.07	7.52	45.08	82.67	89.146
X1.4.2 Height: 1 - 1.25m							
X1.4.2.01 Chain link fencing, BS 1722, part 1, with galvanised mesh, line and tying wire, intermediate posts at 3m centres; height: 1.20m (heavy 3 line wires); galvanised steel angle posts pointed for driving	m	0.31	13.73	0.94	8.30	22.97	15.879
X1.4.2.02 as above; extra for straining posts and struts with welded base plates: end post and one strut	Nr	0.67	29.45	5.14	30.06	64.65	61.672
X1.4.2.03 as above; extra for straining posts and struts with welded base plates: corner post and two struts	Nr	0.78	34.35	7.52	45.08	86.95	89.146

X1 Fences continued...

	Unit	Labour Hours	Labour Net £	Plant Net £	Materials Net £	Unit Net £	CO_2e Kg
X1.4	**Metal post and wire**						
X1.4.3	Height: 1.25 - 1.5m						
X1.4.3.01	Strained wire fencing, BS 1722, part 2, with 4mm galvanised wire; intermediate posts at 3m centres; height: 1.40m (8 wires); galvanised steel angle posts pointed for driving	m	0.27	12.01	0.94	13.82	26.77	21.467
X1.4.3.02	as above; extra for straining posts and struts with welded base plates: end post and one strut	Nr	0.65	28.74	5.14	34.23	68.11	66.659
X1.4.3.03	as above; extra for straining posts and struts with welded base plates: corner post and two struts	Nr	0.73	32.23	7.52	50.64	90.39	95.795
X1.4.3.04	Strained wire fencing, BS 1722, part 2, dropper pattern with 4mm wire; intermediate posts at 5m centres; Galvanised steel angle posts; pointed for driving: 20 x 3mm galvanised flat steel dropper	m	0.27	12.01	0.94	16.06	29.01	22.525
X1.4.3.05	Chain link fencing, BS 1722, part 1, with galvanised mesh, line and tying wire, intermediate posts at 3m centres; height: 1.40m (medium 3 line wires); galvanised steel angle posts pointed for driving	m	0.25	11.17	0.94	16.04	28.15	25.409
X1.4.3.06	as above; extra for straining posts and struts with welded base plates: end post and one strut	Nr	0.65	28.74	5.14	34.23	68.11	66.659
X1.4.3.07	as above; extra for straining posts and struts with welded base plates: corner post and two struts	Nr	0.73	32.23	7.52	50.64	90.39	95.795
X1.4.4	Height: 1.5 - 2.0m						
X1.4.4.01	Chain link fencing, BS 1722, part 1, with galvanised mesh, line and tying wire, intermediate posts at 3m centres; height: 1.80m (heavy 3 line wires); galvanised steel angle posts pointed for driving	m	0.25	11.17	0.94	20.10	32.21	28.396
X1.4.4.02	as above; extra for straining posts and struts with welded base plates: end post and one strut	Nr	0.81	35.94	5.47	38.42	79.83	71.789
X1.4.4.03	as above; extra for straining posts and struts with welded base plates: corner post and two struts	Nr	0.68	30.07	8.45	57.62	96.14	104.502
X1.4.4.04	as above, 3 lines of barbed wire protection at top; intermediate posts at 3m centres; height: 1.80m (heavy 3 line wires); galvanised steel angle posts; steel extension arms; pointed for driving	m	0.27	11.74	0.88	23.43	36.05	29.628

X1 Fences continued...

	Unit	Labour Hours	Labour Net £	Plant Net £	Materials Net £	Unit Net £	CO_2e Kg
X1.4 **Metal post and wire**							
X1.4.4 Height: 1.5 - 2.0m							
X1.4.4.05 as previous item, 3 lines of barbed wire protection at top; intermediate posts at 3m centres; height: 1.80m (heavy 3 line wires); galvanised steel angle posts pointed for driving; extra for straining posts and struts with welded base plates: end post and one strut	Nr	1.03	45.30	5.47	43.83	94.60	77.514
X1.4.4.06 as above; extra for straining posts and struts with welded base plates: corner post and two struts	Nr	1.03	45.30	5.47	54.27	105.04	89.981
X1.4.4.07 as above, 3 lines of barbed wire protection at top; intermediate posts at 3, 3 lines of barbed wire protection at top; intermediate posts at 3m centres; height: 1.80m (heavy 3 line wires); galvanised steel angle posts with cranked tops; pointed for driving	m	0.29	12.89	0.88	26.07	39.84	30.958
X1.4.4.08 as above, 3 lines of barbed wire protection at top; intermediate posts at 3m centres; height: 1.80m (heavy 3 line wires); galvanised steel angle posts pointed for driving; extra for straining posts and struts with welded base plates: end post and one strut	Nr	1.03	45.30	5.47	46.47	97.24	78.844
X1.4.4.09 as above, 3 lines of barbed wire protection at top; intermediate posts at 3: corner post and two struts	Nr	1.03	45.30	5.47	56.91	107.68	91.311
X1.4.5 Height: 2 - 2.5m							
X1.4.5.01 Chain link fencing, BS 1722, part 1, with galvanised mesh, line and tying wire, intermediate posts at 3m centres; height: 2.13m (heavy 3 line wires); galvanised steel angle posts pointed for driving	m	0.26	11.52	0.94	21.49	33.95	30.058
X1.4.5.02 as above; extra for straining posts and struts with welded base plates: end post and one strut	Nr	0.81	35.94	5.47	40.51	81.92	74.282
X1.4.5.03 as above; extra for straining posts and struts with welded base plates: corner post and two struts	Nr	0.97	42.96	8.45	60.41	111.82	107.826
X1.4.5.04 as above, 3 lines of barbed wire protection at top; intermediate posts at 3m centres; height: 2.13m (heavy 3 line wires); galvanised steel angle posts; steel extension arms; pointed for driving	m	0.27	11.74	0.88	24.86	37.48	32.284
X1.4.5.05 as above; extra for straining posts and struts with welded base plates: end post and one strut	Nr	1.03	45.30	5.47	45.92	96.69	80.007

X1 Fences continued...

	Unit	Labour Hours	Labour Net £	Plant Net £	Materials Net £	Unit Net £	CO_2e Kg
X1.4 **Metal post and wire**							
X1.4.5 **Height: 2 - 2.5m**							
X1.4.5.06 as previous item; extra for straining posts and struts with welded base plates: corner post and two struts	Nr	1.03	45.30	5.47	57.06	107.83	93.305
X1.4.5.07 as above, 3 lines of barbed wire protection at top; intermediate posts at 3m centres; height: 2.13m (heavy 3 line wires); galvanised steel angle posts with cranked tops; pointed for driving	m	0.29	12.89	0.88	27.49	41.26	33.780
X1.4.5.08 as above; extra for straining posts and struts with welded base plates: end post and one strut	Nr	1.03	45.47	5.47	51.19	102.13	82.833
X1.4.5.09 as above; extra for straining posts and struts with welded base plates: corner post and two struts	Nr	1.09	48.12	5.47	64.97	118.56	97.461
X1.4.6 **Height: 2.5 - 3m**							
X1.4.6.01 Anti-intruder chain link fencing, BS 1722, part 10, with galvanised mesh, lines and tying wire; intermediate posts at 3m centres with cranked tops or extension arms; 3 lines barbed wire protection at top; bottom of mesh buried vertically 0.30m deep including trenching; 2.9m vertical height; concrete posts; cranked top; single protective top	m	0.58	25.39	0.94	24.40	50.73	29.821
X1.4.6.02 as above; concrete posts; cranked top; single protective top; extra for straining posts and struts: end post and one strut	Nr	1.34	58.98	5.47	50.79	115.24	72.959
X1.4.6.03 as above; concrete posts; cranked top; single protective top; extra for straining posts and struts: corner post and two struts	Nr	1.46	64.46	8.18	65.88	138.52	82.585
X1.4.6.04 as above; concrete posts; steel extension arms; single protective top	m	0.58	25.39	0.94	27.59	53.92	31.013
X1.4.6.05 as above; concrete posts; steel extension arms; single protective top; extra for straining posts and struts: end post and one strut	Nr	1.40	61.81	5.47	53.98	121.26	74.151
X1.4.6.06 as above; concrete posts; steel extension arms; single protective top; extra for straining posts and struts: corner post and two struts	Nr	1.56	68.87	5.47	72.26	146.60	83.956
X1.4.6.07 as above; galvanised steel angle posts with welded extension arms; spragged; single protective top	m	0.58	25.39	0.94	22.37	48.70	31.185
X1.4.6.08 as above; galvanised steel angle posts with welded extension arms; spragged; single protective top; extra for straining posts and struts: end post and one strut, spragged; bolted tie and bars	Nr	1.48	65.34	5.47	48.71	119.52	84.994

X1 Fences continued...

	Unit	Labour Hours	Labour Net £	Plant Net £	Materials Net £	Unit Net £	CO_2e Kg
X1.4	**Metal post and wire**						
X1.4.6	Height: 2.5 - 3m						
X1.4.6.09	as previous item; galvanised steel angle posts with welded extension arms; spragged; single protective top; extra for straining posts and struts: corner post and two struts; spragged; bolted tie bars	Nr	1.56	68.87	5.47	67.83	142.17	101.548
X1.4.6.10	as above; concrete posts; cranked top; double protective top	m	0.63	27.90	0.94	24.97	53.81	30.853
X1.4.6.11	as above; concrete posts; cranked top; double protective top; extra for straining posts and struts: end post and one strut	Nr	1.40	61.81	5.47	50.79	118.07	72.959
X1.4.6.12	as above; concrete posts; cranked top; double protective top; extra for straining posts and struts: corner post and two struts	Nr	1.53	67.68	8.18	65.88	141.74	82.585
X1.4.6.13	as above; concrete posts; steel extension arms; double protective top	m	0.63	27.90	0.94	28.16	57.00	32.045
X1.4.6.14	as above; concrete posts; steel extension arms; double protective top; extra for straining posts and struts: end post and one strut	Nr	1.55	68.43	5.47	53.98	127.88	74.151
X1.4.6.15	as above; concrete posts; steel extension arms; double protective top; extra for straining posts and struts: corner post and two struts	Nr	1.72	75.94	5.47	72.26	153.67	83.956
X1.4.6.16	as above; galvanised steel angle posts with welded extension arms; spragged; double protective top	m	0.63	27.90	0.94	22.94	51.78	32.217
X1.4.6.17	as above; galvanised steel angle posts with welded extension arms; spragged; double protective top; extra for straining posts and struts: end post and one strut, spragged; bolted tie and bars	Nr	1.72	75.94	5.47	48.71	130.12	84.994
X1.4.6.18	as above; galvanised steel angle posts with welded extension arms; spragged; double protective top; extra for straining posts and struts: corner post and two struts; spragged; bolted tie bars	Nr	1.72	75.94	5.47	67.83	149.24	101.548
X1.4.6.19	Chain link fencing for tennis court surrounds, BS 1722, part 13, with galvanised mesh, line and tying wire; galvanised mild steel intermediate posts at 3m centres; Fencing 2.75m high; angle section posts pointed for driving	m	0.36	15.94	0.88	22.97	39.79	17.038
X1.4.6.20	Chain link fencing for tennis court surrounds, BS 1722, part 13, with galvanised mesh, line and tying wire; galvanised mild steel intermediate posts at 3m centres; Fencing 2.75m high; angle section posts with spragged ends	m	0.36	15.94	0.88	22.97	39.79	17.038

X1 Fences continued...

	Unit	Labour Hours	Labour Net £	Plant Net £	Materials Net £	Unit Net £	CO₂e Kg	

X1.4	**Metal post and wire**								
X1.4.6	Height: 2.5 - 3m								
X1.4.6.21	as previous item; extra for angle section straining posts and struts: one way straining post or gate post with one strut; welded base plate	Nr	1.19	52.72	5.47	46.22	104.41	85.464	
X1.4.6.22	as above; extra for angle section straining posts and struts: two way straining post or gate post with two struts; welded base plate	Nr	1.45	64.06	8.18	60.75	132.99	98.514	
X1.4.6.23	as above; extra for gates: single 1.00 x 2.00m	Nr	0.60	26.58	0.88	302.58	330.04	19.503	
X1.4.6.24	as above; extra for gates: double 1.80 x 2.00m	Nr	0.60	26.58	0.87	656.83	684.28	30.599	

X1.6	**Timber close boarded**								
X1.6.1	Height: not exceeding 1m								
X1.6.1.01	Close boarded fencing, BS 1722, part 5, softwood pressure treated pales, horizontal rails, capping and gravel boards; intermediate posts at 3m centres; centre stump in each bay; height: 1.00m; concrete posts; morticed types	m	0.39	17.00	1.13	24.92	43.05	21.402	-13.242
X1.6.1.02	as above; concrete posts; morticed types; extra for: end post	Nr	0.45	19.87	3.01	10.99	33.87	14.326	
X1.6.1.03	as above; concrete posts; morticed types; extra for: angle post	Nr	0.45	19.87	3.01	10.99	33.87	14.326	
X1.6.1.04	as above; treated sawn softwood posts	m	0.38	16.78	1.13	22.11	40.02	19.791	-15.863
X1.6.1.05	as above; treated sawn softwood posts; extra for: end post	Nr	0.24	10.55	3.01	15.29	28.85	13.544	-7.870
X1.6.1.06	as above; treated sawn softwood posts; extra for: angle post	Nr	0.27	11.74	3.01	15.29	30.04	13.544	-7.870
X1.6.1.07	Woven wood fencing, BS 1722, part 11, treated softwood panels between posts at 1.80m centres; height: 1.00m; concrete posts	m	0.40	17.84	0.75	14.82	33.41	13.638	-2.363
X1.6.1.08	as above; concrete posts; extra for: end post	Nr	0.20	9.01	2.26	14.25	25.52	24.107	
X1.6.1.09	as above; concrete posts; extra for: angle post	Nr	0.24	10.55	2.26	14.25	27.06	24.107	
X1.6.1.10	as above; treated sawn softwood posts	m	0.40	17.84	0.75	12.81	31.40	11.075	-5.316
X1.6.1.11	as above; treated sawn softwood posts; extra for: end post	Nr	0.20	9.01	2.26	10.52	21.79	19.472	-5.312
X1.6.1.12	as above; treated sawn softwood posts; extra for: angle post	Nr	0.24	10.55	2.26	10.52	23.33	19.472	-5.312
X1.6.1.13	as above; overlap panels; treated softwood panels between posts at 1.80m centres: Concrete posts	m	0.40	17.84	0.75	15.88	34.47	13.634	-2.246
X1.6.1.14	as above; overlap panels; treated softwood panels between posts at 1.80m centres: Concrete posts; extra for: end post	Nr	0.20	9.01	2.26	14.25	25.52	24.107	

X1 Fences continued...

	Unit	Labour Hours	Labour Net £	Plant Net £	Materials Net £	Unit Net £	CO₂e Kg	

X1.6	**Timber close boarded**								

X1.6.1 Height: not exceeding 1m

		Unit	Labour Hours	Labour Net £	Plant Net £	Materials Net £	Unit Net £	CO₂e Kg	
X1.6.1.15	as previous item; overlap panels; treated softwood panels between posts at 1.80m centres: Concrete posts; extra for: angle post	Nr	0.24	10.55	2.26	14.25	27.06	24.107	
X1.6.1.16	as above; overlap panels; treated softwood panels between posts at 1.80m centres: Treated softwood posts	m	0.40	17.84	0.75	13.81	32.40	11.057	-5.200
X1.6.1.17	as above; overlap panels; treated softwood panels between posts at 1.80m centres: Treated softwood posts; extra for: end post	Nr	0.20	9.01	2.26	10.52	21.79	19.472	-5.312
X1.6.1.18	as above; overlap panels; treated softwood panels between posts at 1.80m centres: Treated softwood posts; extra for: angle post	Nr	0.24	10.55	2.26	10.52	23.33	19.472	-5.312
X1.6.1.19	as above; trellis panel fencing; treated panels between posts at 1.80m centres: Concrete posts	m	0.40	17.84	0.75	17.94	36.53	13.634	-2.246
X1.6.1.20	as above; trellis panel fencing; treated panels between posts at 1.80m centres: Concrete posts; extra for: end post	Nr	0.20	9.01	2.26	14.25	25.52	24.107	
X1.6.1.21	as above; trellis panel fencing; treated panels between posts at 1.80m centres: Concrete posts; extra for: angle post	Nr	0.24	10.55	2.26	14.25	27.06	24.107	
X1.6.1.22	as above; trellis panel fencing; treated panels between posts at 1.80m centres: Treated softwood posts	m	0.40	17.84	0.75	15.87	34.46	11.057	-5.200
X1.6.1.23	as above; trellis panel fencing; treated panels between posts at 1.80m centres: Treated softwood posts; extra for: end post	Nr	0.20	9.01	2.26	10.52	21.79	19.472	-5.312
X1.6.1.24	as above; trellis panel fencing; treated panels between posts at 1.80m centres: Treated softwood posts; extra for: angle post	Nr	0.24	10.55	2.26	10.52	23.33	19.472	-5.312

X1.6.2 Height: 1 - 1.25m

		Unit	Labour Hours	Labour Net £	Plant Net £	Materials Net £	Unit Net £	CO₂e Kg	
X1.6.2.01	Close boarded fencing, BS 1722, part 5, softwood pressure treated pales, horizontal rails, capping and gravel boards; intermediate posts at 3m centres; centre stump in each bay; height: 1.20m; concrete posts; morticed types	m	0.39	17.00	1.13	28.96	47.09	22.412	-17.048
X1.6.2.02	as above; concrete posts; morticed types; extra for: end post	Nr	0.45	19.87	3.01	20.78	43.66	19.599	
X1.6.2.03	as above; concrete posts; morticed types; extra for: angle post	Nr	0.45	19.87	3.01	20.78	43.66	19.599	
X1.6.2.04	as above; treated sawn softwood posts	m	0.39	17.00	1.13	27.61	45.74	20.479	-20.193
X1.6.2.05	as above; treated sawn softwood posts; extra for: end post	Nr	0.45	19.87	3.01	16.74	39.62	13.793	-9.444

X1 Fences continued...

	Unit	Labour Hours	Labour Net £	Plant Net £	Materials Net £	Unit Net £	CO_2e Kg	

X1.6 Timber close boarded

X1.6.2 Height: 1 - 1.25m

X1.6.2.06 as previous item; treated sawn softwood posts; extra for: angle post	Nr	0.45	19.87	3.01	16.74	39.62	13.793	-9.444
X1.6.2.07 Woven wood fencing, BS 1722, part 11, treated softwood panels between posts at 1.80m centres; height: 1.20m; concrete posts	m	0.40	17.84	0.75	17.41	36.00	14.776	-3.150
X1.6.2.08 as above; concrete posts; extra for: end post	Nr	0.20	9.01	2.26	16.79	28.06	25.931	
X1.6.2.09 as above; concrete posts; extra for: angle post	Nr	0.24	10.55	2.26	16.79	29.60	25.931	
X1.6.2.10 as above; treated sawn softwood posts	m	0.40	17.84	0.75	15.35	33.94	11.407	-7.416
X1.6.2.11 as above; treated sawn softwood posts; extra for: end post	Nr	0.26	11.35	2.26	12.96	26.57	19.845	-7.673
X1.6.2.12 as above; treated sawn softwood posts; extra for: angle post	Nr	0.28	12.49	2.26	12.96	27.71	19.845	-7.673
X1.6.2.13 as above; overlap panels; treated softwood panels between posts at 1.80m centres: Concrete posts	m	0.40	17.84	0.75	17.86	36.45	14.768	-3.005
X1.6.2.14 as above; overlap panels; treated softwood panels between posts at 1.80m centres: Concrete posts; extra for: end post	Nr	0.26	11.35	2.26	16.79	30.40	25.931	
X1.6.2.15 as above; overlap panels; treated softwood panels between posts at 1.80m centres: Concrete posts; extra for: angle post	Nr	0.28	12.49	2.26	16.79	31.54	25.931	
X1.6.2.16 as above; overlap panels; treated softwood panels between posts at 1.80m centres: Treated softwood posts	m	0.40	17.84	0.75	15.73	34.32	11.384	-7.271
X1.6.2.17 as above; overlap panels; treated softwood panels between posts at 1.80m centres: Treated softwood posts; extra for: end post	Nr	0.26	11.35	2.26	12.96	26.57	19.845	-7.673
X1.6.2.18 as above; overlap panels; treated softwood panels between posts at 1.80m centres: Treated softwood posts; extra for: angle post	Nr	0.28	12.49	2.26	12.96	27.71	19.845	-7.673
X1.6.2.19 as above; trellis panel fencing; treated panels between posts at 1.80m centres: Concrete posts	m	0.40	17.84	0.75	21.13	39.72	14.768	-3.005
X1.6.2.20 as above; trellis panel fencing; treated panels between posts at 1.80m centres: Concrete posts; extra for: end post	Nr	0.26	11.35	2.26	16.79	30.40	25.931	
X1.6.2.21 as above; trellis panel fencing; treated panels between posts at 1.80m centres: Concrete posts; extra for: angle post	Nr	0.28	12.49	2.26	16.79	31.54	25.931	
X1.6.2.22 as above; trellis panel fencing; treated panels between posts at 1.80m centres: Treated softwood posts	m	0.40	17.84	0.75	19.00	37.59	11.384	-7.271
X1.6.2.23 as above; trellis panel fencing; treated panels between posts at 1.80m centres: Treated softwood posts; extra for: end post	Nr	0.26	11.35	2.26	12.96	26.57	19.845	-7.673

X1 Fences continued...

	Unit	Labour Hours	Labour Net £	Plant Net £	Materials Net £	Unit Net £	CO₂e Kg	

| **X1.6** | **Timber close boarded** | | | | | | | | |

X1.6.2 Height: 1 - 1.25m

| X1.6.2.24 | as previous item; trellis panel fencing; treated panels between posts at 1.80m centres: Treated softwood posts; extra for: angle post | Nr | 0.28 | 12.49 | 2.26 | 12.96 | 27.71 | 19.845 | -7.673 |

X1.6.3 Height: 1.25 - 1.5m

X1.6.3.01	Close boarded fencing, BS 1722, part 5, softwood pressure treated pales, horizontal rails, capping and gravel boards; intermediate posts at 3m centres; centre stump in each bay; height: 1.50m; concrete posts; morticed types	m	0.39	17.00	1.13	29.18	47.31	23.422	-20.854
X1.6.3.02	as above; concrete posts; morticed types; extra for: end post	Nr	0.47	20.88	3.01	21.46	45.35	20.816	
X1.6.3.03	as above; concrete posts; morticed types; extra for: angle post	Nr	0.47	20.88	3.01	21.46	45.35	20.816	
X1.6.3.04	as above; treated sawn softwood posts	m	0.39	17.00	1.13	28.16	46.29	21.166	-24.523
X1.6.3.05	as above; treated sawn softwood posts; extra for: end post	Nr	0.47	20.88	3.01	18.38	42.27	14.041	-11.018
X1.6.3.06	as above; treated sawn softwood posts; extra for: angle post	Nr	0.47	20.88	3.01	18.38	42.27	14.041	-11.018
X1.6.3.07	Woven wood fencing, BS 1722, part 11, treated softwood panels between posts at 1.80m centres; height: 1.50m; concrete posts	m	0.58	25.78	0.75	19.04	45.57	15.578	-3.938
X1.6.3.08	as above; concrete posts; extra for: end post	Nr	0.27	11.74	2.26	17.63	31.63	27.148	
X1.6.3.09	as above; concrete posts; extra for: angle post	Nr	0.29	12.89	2.26	17.63	32.78	27.148	
X1.6.3.10	as above; treated sawn softwood posts	m	0.48	21.10	0.75	16.86	38.71	11.591	-8.533
X1.6.3.11	as above; treated sawn softwood posts; extra for: end post	Nr	0.27	11.74	2.26	13.52	27.52	19.938	-8.264
X1.6.3.12	as above; treated sawn softwood posts; extra for: angle post	Nr	0.29	12.89	2.26	13.52	28.67	19.938	-8.264
X1.6.3.13	as above; overlap panels; treated softwood panels between posts at 1.80m centres: Concrete posts	m	0.58	25.78	0.75	18.82	45.35	15.545	-3.734
X1.6.3.14	as above; overlap panels; treated softwood panels between posts at 1.80m centres: Concrete posts; extra for: end post	Nr	0.27	11.74	2.26	17.63	31.63	27.148	
X1.6.3.15	as above; overlap panels; treated softwood panels between posts at 1.80m centres: Concrete posts; extra for: angle post	Nr	0.29	12.89	2.26	17.63	32.78	27.148	
X1.6.3.16	as above; overlap panels; treated softwood panels between posts at 1.80m centres: Treated softwood posts	m	0.48	21.10	0.75	16.64	38.49	11.559	-8.329
X1.6.3.17	as above; overlap panels; treated softwood panels between posts at 1.80m centres: Treated softwood posts; extra for: end post	Nr	0.27	11.74	2.26	13.52	27.52	19.938	-8.264

X1 Fences continued...

	Unit	Labour Hours	Labour Net £	Plant Net £	Materials Net £	Unit Net £	CO_2e Kg	
X1.6	**Timber close boarded**							
X1.6.3	Height: 1.25 - 1.5m							
X1.6.3.18	as previous item; overlap panels; treated softwood panels between posts at 1.80m centres: Treated softwood posts; extra for: angle post	Nr	0.29	12.89	2.26	13.52	28.67	19.938	-8.264
X1.6.3.19	as above; trellis panel fencing; treated panels between posts at 1.80m centres: Concrete posts	m	0.58	25.78	0.75	27.53	54.06	15.545	-3.734
X1.6.3.20	as above; trellis panel fencing; treated panels between posts at 1.80m centres: Concrete posts; extra for: end post	Nr	0.27	11.74	2.26	17.63	31.63	27.148	
X1.6.3.21	as above; trellis panel fencing; treated panels between posts at 1.80m centres: Concrete posts; extra for: angle post	Nr	0.29	12.89	2.26	17.63	32.78	27.148	
X1.6.3.22	as above; trellis panel fencing; treated panels between posts at 1.80m centres: Treated softwood posts	m	0.48	21.10	0.75	25.35	47.20	11.559	-8.329
X1.6.3.23	as above; trellis panel fencing; treated panels between posts at 1.80m centres: Treated softwood posts; extra for: end post	Nr	0.27	11.74	2.26	13.52	27.52	19.938	-8.264
X1.6.3.24	as above; trellis panel fencing; treated panels between posts at 1.80m centres: Treated softwood posts; extra for: angle post	Nr	0.29	12.89	2.26	13.52	28.67	19.938	-8.264
X1.6.4	Height: 1.5 - 2.0m							
X1.6.4.01	Close boarded fencing, BS 1722, part 5, softwood pressure treated pales, horizontal rails, capping and gravel boards; intermediate posts at 3m centres; centre stump in each bay; height: 1.60m; concrete posts; morticed types	m	0.39	17.00	1.13	30.06	48.19	23.719	-22.757
X1.6.4.02	as above; height: 1.60m; concrete posts; morticed types; extra for: end post	Nr	0.50	21.85	3.01	21.46	46.32	20.816	
X1.6.4.03	as above; height: 1.60m; concrete posts; morticed types; extra for: angle post	Nr	0.50	21.85	3.01	21.46	46.32	20.816	
X1.6.4.04	as above; height: 1.60m; treated sawn softwood posts	m	0.39	17.00	1.13	29.04	47.17	21.463	-26.426
X1.6.4.05	as above; height: 1.60m; treated sawn softwood posts; extra for: end post	Nr	0.49	21.68	3.01	18.38	43.07	14.041	-11.018
X1.6.4.06	as above; height: 1.60m; treated sawn softwood posts; extra for: angle post	Nr	0.49	21.68	3.01	19.77	44.46	14.725	-11.018
X1.6.4.07	as above; height: 1.80m; concrete posts; morticed types	m	0.39	17.00	1.13	31.04	49.17	24.421	-24.660
X1.6.4.08	as above; height: 1.80m; concrete posts; morticed types; extra for: end post	Nr	0.54	23.84	3.01	21.76	48.61	22.032	
X1.6.4.09	as above; height: 1.80m; concrete posts; morticed types; extra for: angle post	Nr	0.54	23.84	3.01	21.76	48.61	22.032	

X1 Fences continued...

	Unit	Labour Hours	Labour Net £	Plant Net £	Materials Net £	Unit Net £	CO₂e Kg	

		Unit	Labour Hours	Labour Net £	Plant Net £	Materials Net £	Unit Net £	CO₂e Kg	

X1.6	**Timber close boarded**								
X1.6.4	Height: 1.5 - 2.0m								
X1.6.4.10	as previous item; height: 1.80m; treated sawn softwood posts	m	0.39	17.00	1.13	30.46	48.59	21.843	-28.853
X1.6.4.11	as above; height: 1.80m; treated sawn softwood posts; extra for: end post	Nr	0.49	21.68	3.01	20.01	44.70	14.290	-12.592
X1.6.4.12	as above; height: 1.80m; treated sawn softwood posts; extra for: angle post	Nr	0.65	28.65	3.01	24.57	56.23	17.031	-12.592
X1.6.4.13	Woven wood fencing, BS 1722, part 11, treated softwood panels between posts at 1.80m centres; height: 1.60m; concrete posts	m	0.58	25.78	0.75	19.59	46.12	15.619	-4.201
X1.6.4.14	as above; height: 1.60m; concrete posts; extra for: end post	Nr	0.28	12.32	2.26	17.63	32.21	27.148	
X1.6.4.15	as above; height: 1.60m; concrete posts; extra for: angle post	Nr	0.31	13.55	2.26	17.63	33.44	27.148	
X1.6.4.16	as above; height: 1.60m; treated sawn softwood posts	m	0.48	21.10	0.75	17.41	39.26	11.632	-8.795
X1.6.4.17	as above; height: 1.60m; treated sawn softwood posts; extra for: end post	Nr	0.28	12.32	2.26	13.52	28.10	19.938	-8.264
X1.6.4.18	as above; height: 1.60m; treated sawn softwood posts; extra for: angle post	Nr	0.31	13.55	2.26	13.52	29.33	19.938	-8.264
X1.6.4.19	as above; height: 1.60m; overlap panels; treated softwood panels between posts at 1.80m centres: Concrete posts	m	0.58	25.78	0.75	19.10	45.63	15.586	-3.997
X1.6.4.20	as above; height: 1.60m; overlap panels; treated softwood panels between posts at 1.80m centres: Concrete posts; extra for: end post	Nr	0.28	12.32	2.26	17.63	32.21	27.148	
X1.6.4.21	as above; height: 1.60m; overlap panels; treated softwood panels between posts at 1.80m centres: Concrete posts; extra for: angle post	Nr	0.31	13.55	2.26	17.63	33.44	27.148	
X1.6.4.22	as above; height: 1.60m; overlap panels; treated softwood panels between posts at 1.80m centres: Treated softwood posts	m	0.48	21.10	0.75	16.92	38.77	11.600	-8.591
X1.6.4.23	as above; height: 1.60m; overlap panels; treated softwood panels between posts at 1.80m centres: Treated softwood posts; extra for: end post	Nr	0.28	12.32	2.26	13.52	28.10	19.938	-8.264
X1.6.4.24	as above; height: 1.60m; overlap panels; treated softwood panels between posts at 1.80m centres: Treated softwood posts; extra for: angle post	Nr	0.31	13.55	2.26	13.52	29.33	19.938	-8.264
X1.6.4.25	as above; height: 1.60m; trellis panel fencing; treated panels between posts at 1.80m centres: Concrete post	m	0.58	25.78	0.75	29.76	56.29	15.586	-3.997

X1 Fences continued...

	Unit	Labour Hours	Labour Net £	Plant Net £	Materials Net £	Unit Net £	CO$_2$e Kg	
X1.6	**Timber close boarded**							
X1.6.4	Height: 1.5 - 2.0m							
X1.6.4.26	as previous item; height: 1.60m; trellis panel fencing; treated panels between posts at 1.80m centres: Concrete posts; extra for: end post	Nr	0.28	12.32	2.26	17.63	32.21	27.148
X1.6.4.27	as above; height: 1.60m; trellis panel fencing; treated panels between posts at 1.80m centres: Concrete posts; extra for: angle post	Nr	0.31	13.55	2.26	17.63	33.44	27.148
X1.6.4.28	as above; height: 1.60m; trellis panel fencing; treated panels between posts at 1.80m centres: Treated softwood posts	m	0.48	21.10	0.75	27.58	49.43	11.600 -8.591
X1.6.4.29	as above; height: 1.60m; trellis panel fencing; treated panels between posts at 1.80m centres: Treated softwood posts; extra for: end post	Nr	0.28	12.32	2.26	13.52	28.10	19.938 -8.264
X1.6.4.30	as above; height: 1.60m; trellis panel fencing; treated panels between posts at 1.80m centres: Treated softwood posts; extra for: angle post	Nr	0.31	13.55	2.26	13.52	29.33	19.938 -8.264
X1.6.4.31	as above; height: 1.80m; concrete posts	m	0.61	26.98	0.75	21.21	48.94	16.378 -4.725
X1.6.4.32	as above; height: 1.80m; concrete posts; extra for: end post	Nr	0.28	12.32	2.26	19.56	34.14	28.364
X1.6.4.33	as above; height: 1.80m; concrete posts; extra for: angle post	Nr	0.30	13.29	2.26	19.56	35.11	28.364
X1.6.4.34	as above; height: 1.80m; treated sawn softwood posts	m	0.62	27.55	0.75	10.03	38.33	11.429 -5.574
X1.6.4.35	as above; height: 1.80m; treated sawn softwood posts; extra for: end post	Nr	0.28	12.32	2.26	14.03	28.61	20.125 -9.444
X1.6.4.36	as above; height: 1.80m; treated sawn softwood posts; extra for: angle post	Nr	0.30	13.29	2.26	14.03	29.58	20.125 -9.444
X1.6.4.37	as above; height: 1.80m; overlap panels; treated softwood panels between posts at 1.80m centres: Concrete posts	m	0.61	26.98	0.75	20.46	48.19	16.341 -4.492
X1.6.4.38	as above; height: 1.80m; overlap panels; treated softwood panels between posts at 1.80m centres: Concrete posts; extra for: end post	Nr	0.28	12.32	2.26	19.56	34.14	28.364
X1.6.4.39	as above; height: 1.80m; overlap panels; treated softwood panels between posts at 1.80m centres: Concrete posts; extra for: angle post	Nr	0.30	13.29	2.26	19.56	35.11	28.364
X1.6.4.40	as above; height: 1.80m; overlap panels; treated softwood panels between posts at 1.80m centres: Treated softwood posts	m	0.62	27.55	0.75	18.89	47.19	12.088 -9.743

X1 Fences continued...

		Unit	Labour Hours	Labour Net £	Plant Net £	Materials Net £	**Unit Net £**	*CO_2e Kg*	
X1.6	**Timber close boarded**								
X1.6.4	Height: 1.5 - 2.0m								
X1.6.4.41	as previous item; height: 1.80m; overlap panels; treated softwood panels between posts at 1.80m centres: Treated softwood posts; extra for: end post	Nr	0.28	12.32	2.26	14.03	**28.61**	20.125	-9.444
X1.6.4.42	as above; height: 1.80m; overlap panels; treated softwood panels between posts at 1.80m centres: Treated softwood posts; extra for: angle post	Nr	0.30	13.29	2.26	14.03	**29.58**	20.125	-9.444
X1.6.4.43	as above; height: 1.80m; trellis panel fencing; treated panels between posts at 1.80m centres: Concrete	m	0.56	24.72	0.75	30.83	**56.30**	16.341	-4.492
X1.6.4.44	as above; height: 1.80m; trellis panel fencing; treated panels between posts at 1.80m centres: Concrete posts; extra for: end post	Nr	0.28	12.32	2.26	19.56	**34.14**	28.364	
X1.6.4.45	as above; height: 1.80m; trellis panel fencing; treated panels between posts at 1.80m centres: Concrete posts; extra for: angle post	Nr	0.30	13.29	2.26	19.56	**35.11**	28.364	
X1.6.4.46	as above; height: 1.80m; trellis panel fencing; treated panels between posts at 1.80m centres: Treated softwood posts	m	0.56	24.72	0.75	29.27	**54.74**	12.088	-9.743
X1.6.4.47	as above; height: 1.80m; trellis panel fencing; treated panels between posts at 1.80m centres: Treated softwood posts; extra for: end post	Nr	0.28	12.32	2.26	14.03	**28.61**	20.125	-9.444
X1.6.4.48	as above; height: 1.80m; trellis panel fencing; treated panels between posts at 1.80m centres: Treated softwood posts; extra for: angle post	Nr	0.30	13.29	2.26	14.03	**29.58**	20.125	-9.444
X1.7	**Metal guard rails**								
X1.7.2	Height: 1 - 1.25m								
X1.7.2.01	Mild steel (low carbon steel) fencing with round or square vertical and flat posts and horizontals, BS 1722, part 9; welded type railing panel with flat horizontals and round bar verticals pointed at top; 102 x 44 x 7.4 kg/m R.S.J. posts for concreting in the ground; 1.20m; 16mm Bar Bluntops; Calcium plumbate primed components	m	1.24	54.70	2.26	29.64	**86.60**	44.013	
X1.7.2.02	as above; Calcium plumbate primed components; extra for: end standard	Nr	0.92	40.62	2.26	18.23	**61.11**	32.559	
X1.7.2.03	as above; Galvanised components	m	1.29	56.95	2.26	30.73	**89.94**	43.920	

X1 Fences continued...

	Unit	Labour Hours	Labour Net £	Plant Net £	Materials Net £	Unit Net £	CO_2e Kg
X1.7 **Metal guard rails**							
X1.7.2 Height: 1 - 1.25m							
X1.7.2.04 as previous item; Galvanised components; extra for: end standard	Nr	0.98	43.40	2.26	19.76	65.42	32.535
X1.7.3 Height: 1.25 - 1.5m							
X1.7.3.01 Mild steel (low carbon steel) fencing with round or square vertical and flat posts and horizontals, BS 1722, part 9; welded type railing panel with flat horizontals and round bar verticals pointed at top; 102 x 44 x 7.4 kg/m R.S.J. posts for concreting in the ground; 1.35m; 20mm Bar Bluntops; Calcium plumbate primed components	m	1.31	57.84	2.26	40.28	100.38	54.405
X1.7.3.02 as above; 1.35m; 20mm Bar Bluntops; Calcium plumbate primed components; extra for: end standard	Nr	0.92	40.62	2.26	22.57	65.45	34.202
X1.7.3.03 as above; 1.35m; 20mm Bar Bluntops; Galvanised components	m	1.48	65.34	2.26	41.74	109.34	54.294
X1.7.3.04 as above; 1.35m; 20mm Bar Bluntops; Galvanised components; extra for: end standard	Nr	0.98	43.40	2.26	24.75	70.41	34.178
X1.7.3.05 as above; 1.50m; 20mm Bar Bluntops; Calcium plumbate primed components	m	1.35	59.38	2.26	42.12	103.76	56.460
X1.7.3.06 as above; 1.50m; 20mm Bar Bluntops; Calcium plumbate primed components; extra for: end standard	Nr	0.95	41.94	2.26	23.35	67.55	35.023
X1.7.3.07 as above; 1.50m; 20mm Bar Bluntops; Galvanised components	m	1.48	65.34	2.26	43.66	111.26	56.348
X1.7.3.08 as above; 1.50m; 20mm Bar Bluntops; Galvanised components; extra for: end standard	Nr	0.99	43.53	2.26	25.65	71.44	34.999
X1.7.4 Height: 1.5 - 2.0m							
X1.7.4.01 Mild steel (low carbon steel) fencing with round or square vertical and flat posts and horizontals, BS 1722, part 9; welded type railing panel with flat horizontals and round bar verticals pointed at top; 102 x 44 x 7.4 kg/m R.S.J. posts for concreting in the ground; 1.80m; 20mm Bar Bluntops; Calcium plumbate primed components	m	1.35	59.38	2.26	47.64	109.28	62.624
X1.7.4.02 as above; Calcium plumbate primed components; extra for: end standard	Nr	0.95	41.94	2.26	25.68	69.88	37.487
X1.7.4.03 as above; Galvanised components	m	1.48	65.34	2.26	49.44	117.04	62.512
X1.7.4.04 as above; Galvanised components; extra for: end standard	Nr	0.99	43.53	2.26	28.32	74.11	37.463

X1 Fences continued...

	Unit	Labour Hours	Labour Net £	Plant Net £	Materials Net £	Unit Net £	CO_2e Kg
X1.7 **Metal guard rails**							
X1.7.4 Height: 1.5 - 2.0m							
X1.7.4.05 Steel palisade fences, BS 1722, part 12, galvanised after manufacture; fences of corrugated pales with plain tops; height: 1.80m	m	0.83	36.73	0.94	90.79	128.46	124.543
X1.7.5 Height: 2 - 2.5m							
X1.7.5.01 Mild steel (low carbon steel) fencing with round or square vertical and flat posts and horizontals, BS 1722, part 9; welded type railing panel with flat horizontals and round bar verticals pointed at top; 102 x 44 x 7.4 kg/m R.S.J. posts for concreting in the ground; 2.10m; 22mm Bar Bluntops; Calcium plumbate primed components	m	1.35	59.38	2.26	66.39	128.03	85.945
X1.7.5.02 as above; Calcium plumbate primed components; extra for: end standard	Nr	0.95	41.94	2.26	28.00	72.20	39.951
X1.7.5.03 as above; Galvanised components	m	1.48	65.34	2.26	68.45	136.05	85.833
X1.7.5.04 as above; Galvanised components; extra for: end standard	Nr	0.99	43.53	2.26	31.00	76.79	39.927
X1.7.5.05 Steel palisade fences, BS 1722, part 12, galvanised after manufacture; fences of corrugated pales with plain tops; height: 2.10m	m	0.83	36.73	0.94	99.59	137.26	134.443
X1.7.5.06 as above; height: 2.40m	m	0.83	36.73	0.94	108.31	145.98	144.273
X1.7.6 Height: 2.5 - 3m							
X1.7.6.01 Steel palisade fences, BS 1722, part 12, galvanised after manufacture; fences of corrugated pales with plain tops; height: 3.00m	m	0.83	36.73	0.94	125.82	163.49	163.969

X2 Gates And Stiles

	Unit	Labour Hours	Labour Net £	Plant Net £	Materials Net £	Unit Net £	CO_2e Kg
X2.2 **Metal gates**							
X2.2.1 Width: not exceeding 1.5m							
X2.2.1.01 Single gate width 1.00m; infilling with galvanised chain link mesh; Circular hollow section framing; fittings bolted to concrete or timber posts: bitumen coated; fence height: 0.90m	Nr	1.00	44.15	26.90	86.81	157.86	34.914
X2.2.1.02 as above: bitumen coated; fence height: 1.2m	Nr	1.00	44.15	26.90	115.75	186.80	43.839
X2.2.1.03 as above: bitumen coated; fence height: 1.8m	Nr	1.00	44.15	26.90	173.63	244.68	61.690
X2.2.1.04 as above: bitumen coated; fence height: 2.18m	Nr	1.00	44.15	26.90	205.46	276.51	71.508
X2.2.1.05 as above: galvanised; fence height: 0.90m	Nr	1.00	44.15	26.90	115.35	186.40	46.035
X2.2.1.06 as above: galvanised; fence height: 1.2m	Nr	1.00	44.15	26.90	153.80	224.85	58.666

X2 Gates And Stiles continued...

	Unit	Labour Hours	Labour Net £	Plant Net £	Materials Net £	Unit Net £	CO$_2$e Kg	
X2.2 **Metal gates**								
X2.2.1 Width: not exceeding 1.5m								
X2.2.1.07 as previous item: galvanised; fence height: 1.8m	Nr	1.00	44.15	26.90	230.70	301.75	83.932	
X2.2.1.08 as above: galvanised; fence height: 2.18m	Nr	1.00	44.15	26.90	269.15	340.20	96.563	
X2.2.1.09 Single gate width 1.00m; infilling with galvanised chain link mesh; Rectangular hollow section framing; fittings bolted to concrete or timber posts: bitumen coated; fence height: 0.90m	Nr	1.00	44.15	26.90	86.81	157.86	34.914	
X2.2.1.10 as above: bitumen coated; fence height: 1.2m	Nr	1.00	44.15	26.90	115.75	186.80	43.839	
X2.2.1.11 as above: bitumen coated; fence height: 1.8m	Nr	1.00	44.15	26.90	173.63	244.68	61.690	
X2.2.1.12 as above: bitumen coated; fence height: 2.18m	Nr	1.00	44.15	26.90	202.57	273.62	70.615	
X2.2.1.13 as above: galvanised; fence height: 0.90m	Nr	1.00	44.15	26.90	128.17	199.22	50.246	
X2.2.1.14 as above: galvanised; fence height: 1.2m	Nr	1.00	44.15	26.90	153.80	224.85	58.666	
X2.2.1.15 as above: galvanised; fence height: 1.8m	Nr	1.00	44.15	26.90	230.70	301.75	83.932	
X2.2.1.16 as above: galvanised; fence height: 2.18m	Nr	1.00	44.15	26.90	269.15	340.20	96.563	
X2.2.1.17 Single gate width 1.00m; infilling with galvanised chain link mesh; additional costs of three lines of barbed wire above gates; extended stiles and extension arms 0.33m high: bitumen coated	Nr	0.20	8.83	-	6.53	15.36	7.925	
X2.2.1.18 as above; galvanised	Nr	0.20	8.83	-	8.76	17.59	29.277	
X2.2.1.19 Single gate width 1.00m; Gates in conjunction with anti-intruder chain link fencing, Infilling with: galvanised chain link mesh or plastic coated chain link mesh	Nr	1.00	44.15	26.90	300.40	371.45	82.889	
X2.2.1.20 as above; Infilling with: galvanised mild steel wire fabric	Nr	1.00	44.15	26.90	300.40	371.45	82.889	
X2.2.2 Width: 1.5 - 2m								
X2.2.2.01 Single gate width 1.50m; Corrugated pales with plain tops; Steel gates and gate posts in conjunction with steel palisade fencing; galvanised after manufacture; Height: 1.80m	Nr	1.50	66.22	40.35	419.69	526.26	102.316	-35.415
X2.2.2.02 as above; fence height: 2.1m	Nr	1.50	66.22	40.35	437.92	544.49	106.243	-41.318
X2.2.2.03 as above; fence height: 2.4m	Nr	1.50	66.22	40.35	454.07	560.64	109.830	-47.219
X2.2.2.04 as above; fence height: 3.0m	Nr	1.50	66.22	40.35	510.64	617.21	120.988	-59.024
X2.2.3 Width: 2 - 2.5m								
X2.2.3.01 Single gate width 2.00m; infilling with galvanised chain link mesh; Circular hollow section framing; fittings bolted to concrete or timber posts: bitumen coated; fence height: 0.90m	Nr	1.00	44.15	26.90	173.63	244.68	61.690	

X2 Gates And Stiles continued...

	Unit	Labour Hours	Labour Net £	Plant Net £	Materials Net £	Unit Net £	CO$_2$e Kg
X2.2 **Metal gates**							
X2.2.3 Width: 2 - 2.5m							
X2.2.3.02 as previous item: bitumen coated; fence height: 1.2m	Nr	1.00	44.15	26.90	231.50	302.55	79.540
X2.2.3.03 as above: bitumen coated; fence height: 1.8m	Nr	1.00	44.15	26.90	347.26	418.31	115.242
X2.2.3.04 as above: bitumen coated; fence height: 2.18m	Nr	1.00	44.15	26.90	405.13	476.18	133.092
X2.2.3.05 as above: galvanised; fence height: 0.90m	Nr	1.00	44.15	26.90	230.70	301.75	83.932
X2.2.3.06 as above: galvanised; fence height: 1.2m	Nr	1.00	44.15	26.90	307.60	378.65	109.196
X2.2.3.07 as above: galvanised; fence height: 1.8m	Nr	1.00	44.15	26.90	461.40	532.45	159.725
X2.2.3.08 as above: galvanised; fence height: 2.18m	Nr	1.00	44.15	26.90	538.30	609.35	184.989
X2.2.3.09 Single gate width 2.00m; infilling with galvanised chain link mesh; Rectangular hollow section framing; fittings bolted to concrete or timber posts: bitumen coated; fence height: 0.90m	Nr	1.25	55.19	33.63	173.63	262.45	63.724
X2.2.3.10 as above: bitumen coated; fence height: 1.2m	Nr	1.25	55.19	33.63	231.50	320.32	81.575
X2.2.3.11 as above: bitumen coated; fence height: 1.8m	Nr	1.25	55.19	33.63	347.26	436.08	117.276
X2.2.3.12 as above: bitumen coated; fence height: 2.18m	Nr	1.25	55.19	33.63	405.13	493.95	135.127
X2.2.3.13 as above: galvanised; fence height: 0.90m	Nr	1.25	55.19	33.63	230.70	319.52	85.967
X2.2.3.14 as above: galvanised; fence height: 1.2m	Nr	1.25	55.19	33.63	307.60	396.42	111.231
X2.2.3.15 as above: galvanised; fence height: 1.8m	Nr	1.25	55.19	33.63	461.40	550.22	161.759
X2.2.3.16 as above: galvanised; fence height: 2.18m	Nr	1.25	55.19	33.63	538.30	627.12	187.023
X2.2.3.17 Single gate width 2.00m; infilling with galvanised chain link mesh; additional costs of three lines of barbed wire above gates; extended stiles and extension arms 0.33m high: bitumen coated	Nr	0.20	8.83	-	6.53	15.36	7.925
X2.2.3.18 as above: galvanised	Nr	0.20	8.83	-	8.76	17.59	29.277
X2.2.3.19 Single gate width 2.00m; Gates in conjunction with anti-intruder chain link fencing, Infilling with: galvanised chain link mesh or plastic coated chain link mesh	Nr	1.00	44.15	26.90	600.80	671.85	157.641
X2.2.3.20 as above: galvanised mild steel wire fabric	Nr	1.00	44.15	26.90	600.80	671.85	157.641
X2.2.5 Width: 3 - 4m							
X2.2.5.01 Single gate width 3.00m; infilling with galvanised chain link mesh; Circular hollow section framing; fittings bolted to concrete or timber posts: bitumen coated; fence height: 0.90m	Nr	2.00	88.30	53.80	260.44	402.54	96.604
X2.2.5.02 as above: bitumen coated; fence height: 1.2m	Nr	2.00	88.30	53.80	347.26	489.36	123.380

X2 Gates And Stiles continued...

	Unit	Labour Hours	Labour Net £	Plant Net £	Materials Net £	Unit Net £	CO_2e Kg	
X2.2	**Metal gates**							
X2.2.5	Width: 3 - 4m							
X2.2.5.03	as previous iitem: bitumen coated; fence height: 1.8m	Nr	2.00	88.30	53.80	520.88	662.98	176.931
X2.2.5.04	as above: bitumen coated; fence height: 2.18m	Nr	2.00	88.30	53.80	607.70	749.80	203.707
X2.2.5.05	as above: galvanised; fence height: 0.90m	Nr	2.50	110.38	67.25	345.96	523.59	133.852
X2.2.5.06	as above: galvanised; fence height: 1.2m	Nr	2.50	110.38	67.25	461.28	638.91	171.689
X2.2.5.07	as above: galvanised; fence height: 1.8m	Nr	2.50	110.38	67.25	691.92	869.55	247.359
X2.2.5.08	as above: galvanised; fence height: 2.18m	Nr	2.50	110.38	67.25	807.24	984.87	285.196
X2.2.5.09	Single gate width 3.00m; infilling with galvanised chain link mesh; Rectangular hollow section framing; fittings bolted to concrete or timber posts: bitumen coated; fence height: 0.90m	Nr	2.00	88.30	53.80	260.44	402.54	96.604
X2.2.5.10	as above: bitumen coated; fence height: 1.2m	Nr	2.00	88.30	53.80	347.26	489.36	123.380
X2.2.5.11	as above: bitumen coated; fence height: 1.8m	Nr	2.00	88.30	53.80	520.88	662.98	176.931
X2.2.5.12	as above: bitumen coated; fence height: 2.18m	Nr	2.00	88.30	53.80	607.70	749.80	203.707
X2.2.5.13	as above: galvanised; fence height: 0.90m	Nr	2.50	110.38	67.25	345.96	523.59	133.852
X2.2.5.14	as above: galvanised; fence height: 1.2m	Nr	2.50	110.38	67.25	461.28	638.91	171.689
X2.2.5.15	as above: galvanised; fence height: 1.8m	Nr	2.50	110.38	67.25	691.92	869.55	247.359
X2.2.5.16	as above: galvanised; fence height: 2.18m	Nr	2.50	110.38	67.25	807.24	984.87	285.196
X2.2.5.17	Single gate width 3.00m; infilling with galvanised chain link mesh; additional costs of three lines of barbed wire above gates; extended stiles and extension arms 0.33m high: bitumen coated	Nr	0.20	8.83	–	6.53	15.36	7.925
X2.2.5.18	as above; galvanised	Nr	0.20	8.83	–	8.76	17.59	29.277
X2.2.5.19	Single gate width 3.00m; Corrugated pales with plain tops; Steel gates and gate posts in conjunction with steel palisade fencing; galvanised after manufacture; Height: 1.80m	Nr	2.00	88.30	53.80	695.80	837.90	157.341	-70.829
X2.2.5.20	as above; fence height: 2.1m	Nr	2.00	88.30	53.80	726.17	868.27	164.196	-82.634
X2.2.5.21	as above; fence height: 2.4m	Nr	2.00	88.30	53.80	753.11	895.21	170.485	-94.439
X2.2.5.22	as above; fence height: 3.0m	Nr	2.00	88.30	53.80	847.37	989.47	189.703	-118.048
X2.2.6	Width: 4 - 5m							
X2.2.6.01	Double gates width 4.00m; infilling with galvanised chain link mesh; Circular hollow section framing; fittings bolted to concrete or timber posts: bitumen coated; fence height: 0.90m	Nr	2.50	110.38	67.25	347.26	524.89	127.449
X2.2.6.02	as above: bitumen coated; fence height: 1.2m	Nr	2.50	110.38	67.25	463.01	640.64	163.150

X2 Gates And Stiles continued...

	Unit	Labour Hours	Labour Net £	Plant Net £	Materials Net £	Unit Net £	CO$_2$e Kg	
X2.2 **Metal gates**								
X2.2.6 Width: 4 - 5m								
X2.2.6.03 as previous item; bitumen coated; fence height: 1.8m	Nr	2.50	110.38	67.25	694.51	872.14	234.552	
X2.2.6.04 as above; bitumen coated; fence height: 2.18m	Nr	2.50	110.38	67.25	810.26	987.89	270.253	
X2.2.6.05 as above; galvanised; fence height: 0.90m	Nr	3.00	132.45	80.70	461.28	674.43	175.758	
X2.2.6.06 as above; galvanised; fence height: 1.2m	Nr	3.00	132.45	80.70	615.04	828.19	226.205	
X2.2.6.07 as above; galvanised; fence height: 1.8m	Nr	3.00	132.45	80.70	922.56	1,135.71	327.100	
X2.2.6.08 as above; galvanised; fence height: 2.18m	Nr	3.00	132.45	80.70	1,076.32	1,289.47	377.547	
X2.2.6.09 Double gates width 4.00m; infilling with galvanised chain link mesh; Rectangular hollow section framing; fittings bolted to concrete or timber posts; bitumen coated; fence height: 0.90m	Nr	2.50	110.38	67.25	347.26	524.89	127.449	
X2.2.6.10 as above; bitumen coated; fence height: 1.2m	Nr	2.50	110.38	67.25	463.01	640.64	163.150	
X2.2.6.11 as above; bitumen coated; fence height: 1.8m	Nr	2.50	110.38	67.25	694.51	872.14	234.552	
X2.2.6.12 as above; bitumen coated; fence height: 2.18m	Nr	2.50	110.38	67.25	810.26	987.89	270.253	
X2.2.6.13 as above; galvanised; fence height: 0.90m	Nr	3.00	132.45	80.70	461.28	674.43	175.758	
X2.2.6.14 as above; galvanised; fence height: 1.2m	Nr	3.00	132.45	80.70	615.04	828.19	226.205	
X2.2.6.15 as above; galvanised; fence height: 1.8m	Nr	3.00	132.45	80.70	922.56	1,135.71	327.100	
X2.2.6.16 as above; galvanised; fence height: 2.18m	Nr	3.00	132.45	80.70	1,076.32	1,289.47	377.547	
X2.2.6.17 Double gates width 4.00m; infilling with galvanised chain link mesh; additional costs of three lines of barbed wire above gates; extended stiles and extension arms 0.33m high; bitumen coated	Nr	0.20	8.83	–	6.53	15.36	7.925	
X2.2.6.18 as above; galvanised	Nr	0.20	8.83	–	8.76	17.59	29.277	
X2.2.6.19 Double gates width 4.00m; Gates in conjunction with anti-intruder chain link fencing, Infilling with: galvanised chain link mesh or plastic coated chain link mesh	Nr	1.00	44.15	26.90	1,201.60	1,272.65	307.144	
X2.2.6.20 as above; galvanised mild steel wire fabric	Nr	1.00	44.15	26.90	1,201.60	1,272.65	307.144	
X2.2.6.21 Double gates width 4.50m; Corrugated pales with plain tops; Steel gates and gate posts in conjunction with steel palisade fencing; galvanised after manufacture; Height: 1.80m	Nr	2.00	88.30	53.80	923.59	1,065.69	200.360	*-106.244*
X2.2.6.22 as above fence height: 2.1m	Nr	2.00	88.30	53.80	986.62	1,128.72	213.515	*-123.952*
X2.2.6.23 as above; fence height: 2.4m	Nr	2.00	88.30	53.80	1,037.05	1,179.15	224.597	*-141.658*
X2.2.6.24 as above; fence height: 3.0m	Nr	2.00	88.30	53.80	1,214.70	1,356.80	259.377	*-177.073*

X2 Gates And Stiles continued...

	Unit	Labour Hours	Labour Net £	Plant Net £	Materials Net £	Unit Net £	CO_2e Kg
X2.2	**Metal gates**						
X2.2.7	Width: exceeding 5m						
X2.2.7.01	Double gates width 6.00m; infilling with galvanised chain link mesh; Circular hollow section framing; fittings bolted to concrete or timber posts: bitumen coated; fence height: 0.90m	Nr	3.00	132.45	80.70	520.88	734.03	185.069
X2.2.7.02	as above: bitumen coated; fence height: 1.2m	Nr	3.00	132.45	80.70	694.51	907.66	238.621
X2.2.7.03	as above: bitumen coated; fence height: 1.8m	Nr	3.00	132.45	80.70	1,041.77	1,254.92	345.725
X2.2.7.04	as above: bitumen coated; fence height: 2.18m	Nr	3.00	132.45	80.70	1,215.40	1,428.55	399.277
X2.2.7.05	as above: galvanised; fence height: 0.90m	Nr	3.50	154.53	94.15	691.92	940.60	255.497
X2.2.7.06	as above: galvanised; fence height: 1.2m	Nr	3.50	154.53	94.15	922.56	1,171.24	331.169
X2.2.7.07	as above: galvanised; fence height: 1.8m	Nr	3.50	154.53	94.15	1,383.84	1,632.52	482.513
X2.2.7.08	as above: galvanised; fence height: 2.18m	Nr	3.50	154.53	94.15	1,614.48	1,863.16	558.185
X2.2.7.09	Double gates width 6.00m; infilling with galvanised chain link mesh; Rectangular hollow section framing; fittings bolted to concrete or timber posts: bitumen coated; fence height: 0.90m	Nr	3.00	132.45	80.70	520.88	734.03	185.069
X2.2.7.10	as above: bitumen coated; fence height: 1.2m	Nr	3.00	132.45	80.70	694.51	907.66	238.621
X2.2.7.11	as above: bitumen coated; fence height: 1.8m	Nr	3.00	132.45	80.70	1,041.77	1,254.92	345.725
X2.2.7.12	as above: bitumen coated; fence height: 2.18m	Nr	3.00	132.45	80.70	1,215.40	1,428.55	399.277
X2.2.7.13	as above: galvanised; fence height: 0.90m	Nr	3.50	154.53	94.15	691.92	940.60	255.497
X2.2.7.14	as above: galvanised; fence height: 1.2m	Nr	3.50	154.53	94.15	922.56	1,171.24	331.169
X2.2.7.15	as above: galvanised; fence height: 1.8m	Nr	3.50	154.53	94.15	1,383.84	1,632.52	482.513
X2.2.7.16	as above: galvanised; fence height: 2.18m	Nr	3.50	154.53	94.15	1,614.48	1,863.16	558.185
X2.2.7.17	Double gates width 6.00m; infilling with galvanised chain link mesh; additional costs of three lines of barbed wire above gates; extended stiles and extension arms 0.33m high: bitumen coated	Nr	0.20	8.83	-	6.53	15.36	7.925
X2.2.7.18	as above; galvanised	Nr	0.20	8.83	-	8.76	17.59	29.277
X2.2.7.19	Double gates width 6.00m; Gates in conjunction with anti-intruder chain link fencing, Infilling with: galvanised chain link mesh or plastic coated chain link mesh	Nr	1.50	66.22	40.35	1,796.05	1,902.62	458.652
X2.2.7.20	as above: galvanised mild steel wire fabric	Nr	1.50	66.22	40.35	1,796.05	1,902.62	458.652

X2 Gates And Stiles continued...

	Unit	Labour Hours	Labour Net £	Plant Net £	Materials Net £	Unit Net £	CO$_2$e Kg	

X2.2 Metal gates

X2.2.7 Width: exceeding 5m

X2.2.7.21	Double gates width 6.00m; Corrugated pales with plain tops; Steel gates and gate posts in conjunction with steel palisade fencing; galvanised after manufacture; Height: 1.80m	Nr	2.00	88.30	53.80	1,037.05	1,179.15	224.597	-141.658
X2.2.7.22	as above; fence height: 2.1m	Nr	2.00	88.30	53.80	1,071.44	1,213.54	233.975	-165.268
X2.2.7.23	as above; fence height: 2.4m	Nr	2.00	88.30	53.80	1,123.01	1,265.11	246.178	-188.878
X2.2.7.24	as above; fence height: 3.0m	Nr	2.00	88.30	53.80	1,495.50	1,637.60	314.835	-236.097
X2.2.7.25	Double gates width 8.00m; infilling with galvanised chain link mesh; Circular hollow section framing; fittings bolted to concrete or timber posts: bitumen coated; fence height: 0.90m	Nr	3.50	154.53	94.15	694.51	943.19	242.690	
X2.2.7.26	as above: bitumen coated; fence height: 1.2m	Nr	3.50	154.53	94.15	926.02	1,174.70	314.093	
X2.2.7.27	as above: bitumen coated; fence height: 1.8m	Nr	3.50	154.53	94.15	1,389.02	1,637.70	456.897	
X2.2.7.28	as above: bitumen coated; fence height: 2.18m	Nr	3.50	154.53	94.15	1,620.53	1,869.21	528.300	
X2.2.7.29	as above: galvanised; fence height: 0.90m	Nr	4.00	176.60	107.60	922.56	1,206.76	335.238	
X2.2.7.30	as above: galvanised; fence height: 1.2m	Nr	4.00	176.60	107.60	1,230.08	1,514.28	436.134	
X2.2.7.31	as above: galvanised; fence height: 1.8m	Nr	4.00	176.60	107.60	1,845.12	2,129.32	637.924	
X2.2.7.32	as above: galvanised; fence height: 2.18m	Nr	4.00	176.60	107.60	2,152.64	2,436.84	738.820	
X2.2.7.33	Double gates width 8.00m; infilling with galvanised chain link mesh; Rectangular hollow section framing; fittings bolted to concrete or timber posts: bitumen coated; fence height: 0.90m	Nr	3.50	154.53	94.15	694.51	943.19	242.690	
X2.2.7.34	as above: bitumen coated; fence height: 1.2m	Nr	3.50	154.53	94.15	926.02	1,174.70	314.093	
X2.2.7.35	as above: bitumen coated; fence height: 1.8m	Nr	3.50	154.53	94.15	1,389.02	1,637.70	456.897	
X2.2.7.36	as above: bitumen coated; fence height: 2.18m	Nr	3.50	154.53	94.15	1,620.53	1,869.21	528.300	
X2.2.7.37	as above: galvanised; fence height: 0.90m	Nr	4.00	176.60	107.60	922.56	1,206.76	335.238	
X2.2.7.38	as above: galvanised; fence height: 1.2m	Nr	4.00	176.60	107.60	1,230.08	1,514.28	436.134	
X2.2.7.39	as above: galvanised; fence height: 1.8m	Nr	4.00	176.60	107.60	1,845.12	2,129.32	637.924	
X2.2.7.40	as above: galvanised; fence height: 2.18m	Nr	4.00	176.60	107.60	2,152.64	2,436.84	738.820	
X2.2.7.41	Double gates width 8.00m; infilling with galvanised chain link mesh; additional costs of three lines of barbed wire above gates; extended stiles and extension arms 0.33m high: bitumen coated	Nr	0.20	8.83	-	6.53	15.36	7.925	
X2.2.7.42	as above; galvanised	Nr	0.20	8.83	-	8.76	17.59	29.277	

X2 Gates And Stiles continued...

	Unit	Labour Hours	Labour Net £	Plant Net £	Materials Net £	Unit Net £	CO_2e Kg

X2.2 Metal gates

X2.2.7 Width: exceeding 5m

X2.2.7.43 Double gates width 8.00m; Gates in conjunction with anti-intruder chain link fencing, Infilling with: galvanised chain link mesh or plastic coated chain link mesh	Nr	1.50	66.22	40.35	2,390.49	2,497.06	606.091
X2.2.7.44 as above; Infilling with: galvanised mild steel wire fabric	Nr	1.50	66.22	40.35	2,390.49	2,497.06	606.091

X3 Drainage To Structures Above Ground

X3.2 Cast Iron

X3.2.1 Gutters

X3.2.1.01 Fixing with standard brackets at heights not exceeding 3m; half round eaves gutters; Diameter: 100mm	m	0.25	7.01	-	23.85	30.86	15.220
X3.2.1.02 as above; Diameter: 150mm	m	0.30	8.42	-	46.15	54.57	22.831
X3.2.1.03 as above; Ogee eaves gutters; Diameter: 100mm	m	0.25	7.01	-	26.76	33.77	15.541
X3.2.1.04 as above; Ogee eaves gutters; Diameter: 150mm	m	0.30	8.42	-	30.71	39.13	22.831

X3.2.2 Fittings to gutters

X3.2.2.01 Fixing with standard brackets at heights not exceeding 3m; half round eaves gutters; diameter 100mm; Stopped ends	Nr	0.03	0.84	-	4.58	5.42	44.118
X3.2.2.02 as above; square angles	Nr	0.04	1.12	-	12.91	14.03	3.529
X3.2.2.03 as above; outlets (drop end)	Nr	0.04	1.12	-	12.59	13.71	2.899
X3.2.2.04 as above; diameter 150mm; Stopped ends	Nr	0.03	0.84	-	7.80	8.64	1.134
X3.2.2.05 as above; square angles	Nr	0.04	1.12	-	28.62	29.74	5.294
X3.2.2.06 as above; outlets (drop end)	Nr	0.04	1.12	-	27.13	28.25	4.349
X3.2.2.07 as above; Ogee eaves gutters; diameter 100mm; Stopped ends	Nr	0.03	0.84	-	4.44	5.28	0.794
X3.2.2.08 as above; square angles	Nr	0.04	1.12	-	14.00	15.12	3.706
X3.2.2.09 as above; outlets (drop end)	Nr	0.04	1.12	-	13.72	14.84	3.044
X3.2.2.10 as above; diameter 150mm; Stopped ends	Nr	0.03	0.84	-	15.93	16.77	1.191
X3.2.2.11 as above; square angles	Nr	0.04	1.12	-	20.00	21.12	5.559
X3.2.2.12 as above; outlets (drop end)	Nr	0.04	1.12	-	16.05	17.17	4.566

X3.2.3 Downpipes

X3.2.3.01 Fixing with standard brackets at heights not exceeding 3m; ears cast on; fixing with pipe nails and hardwood distance pieces; nominal size; 75mm	m	0.28	7.86	-	55.13	62.99	10.185
X3.2.3.02 as above; nominal size; 100mm	m	0.28	7.86	-	70.67	78.53	14.811
X3.2.3.03 as above; nominal size; Rectangular section 100 x 75mm	m	0.28	7.86	-	181.64	189.50	17.288

X3.2.4 Fittings to downpipes

X3.2.4.01 Fixing with standard brackets at heights not exceeding 3m; nominal size 75mm; Shoes	Nr	0.18	5.05	-	29.41	34.46	2.657

X3 Drainage To Structures Above Ground continued...

	Unit	Labour Hours	Labour Net £	Plant Net £	Materials Net £	Unit Net £	CO_2e Kg
X3.2	**Cast Iron**						
X3.2.4	Fittings to downpipes						
X3.2.4.02 Fixing with standard brackets at heights not exceeding 3m; ears cast on; fixing with pipe nails and hardwood distance pieces; nominal size 75mm; Obtuse bend	Nr	0.28	7.86	-	21.85	29.71	1.686
X3.2.4.03 as above; nominal size 75mm; Swan neck offset: 150mm projection	Nr	0.28	7.86	-	27.55	35.41	3.107
X3.2.4.04 as above; nominal size 75mm; Swan neck offset: 305mm projection	Nr	0.28	7.86	-	27.55	35.41	3.107
X3.2.4.05 as above; nominal size 75mm; Single equal branch	Nr	0.28	7.86	-	27.55	35.41	3.107
X3.2.4.06 as above; nominal size 100mm; Shoes	Nr	0.18	5.05	-	39.04	44.09	3.126
X3.2.4.07 as above; nominal size 100mm; Obtuse bend	Nr	0.18	5.05	-	30.86	35.91	6.041
X3.2.4.08 as above; nominal size 100mm; Swan neck offset: 150mm projection	Nr	0.18	5.05	-	52.98	58.03	3.655
X3.2.4.09 as above; nominal size 100mm; Swan neck offset: 305mm projection	Nr	0.18	5.05	-	52.98	58.03	3.655
X3.2.4.10 as above; nominal size 100mm; Single equal branch	Nr	0.18	5.05	-	52.98	58.03	3.655
X3.2.4.11 as above; nominal size 100 x 75 rectangular section; Shoes	Nr	0.18	5.05	-	110.14	115.19	3.908
X3.2.4.12 as above; nominal size 100 x 75 rectangular section; Side bend	Nr	0.18	5.05	-	104.87	109.92	7.551
X3.2.4.13 as above; nominal size 100 x 75 rectangular section; Front plinth offset	Nr	0.18	5.05	-	140.30	145.35	4.576
X3.2.4.14 as above; nominal size 100 x 75 rectangular section; Side plinth offset	Nr	0.18	5.05	-	140.30	145.35	4.576
X3.2.4.15 as above; rainwater heads; jointing to pipes; Hopper type flat; 75mm outlet	Nr	0.18	5.05	-	52.58	57.63	2.464
X3.2.4.16 as above; rainwater heads; jointing to pipes; Rectangular, 250 x 178 x 178mm; 75mm outlet	Nr	0.18	5.05	-	79.56	84.61	6.226
X3.2.4.17 as above; rainwater heads; jointing to pipes; Rectangular, 250 x 178 x 178mm; 100mm outlet	Nr	0.18	5.05	-	100.78	105.83	8.313
X3.3	**Plastics**						
X3.3.1	Gutters						
X3.3.1.01 UPVC; fixing with standard brackets at heights not exceeding 3m; half round eaves gutters; nominal size; 112mm	m	0.18	5.05	-	10.36	15.41	4.156
X3.3.1.02 as above; nominal size; 150mm	m	0.18	5.05	-	13.24	18.29	7.611

X3 Drainage To Structures Above Ground continued...

	Unit	Labour Hours	Labour Net £	Plant Net £	Materials Net £	Unit Net £	CO_2e Kg
X3.3 Plastics							
X3.3.2 Fittings to gutters							
X3.3.2.01 UPVC; fixing with standard brackets at heights not exceeding 3m; half round eaves gutters; nominal size 112m; Stopped ends	Nr	0.05	1.40	-	4.30	5.70	0.149
X3.3.2.02 as above; half round eaves gutters; nominal size 112m; Square angles	Nr	0.14	3.93	-	18.75	22.68	1.121
X3.3.2.03 as above; half round eaves gutters; nominal size 112m; Outlets	Nr	0.05	1.40	-	15.36	16.76	0.374
X3.3.2.04 as above; half round eaves gutters; diameter 150mm; Stopped ends	Nr	0.05	1.40	-	5.85	7.25	0.284
X3.3.2.05 as above; half round eaves gutters; diameter 150mm; Square angles	Nr	0.14	3.93	-	14.80	18.73	1.370
X3.3.2.06 as above; half round eaves gutters; diameter 150mm; Outlets	Nr	0.05	1.40	-	16.20	17.60	1.245
X3.3.3 Downpipes							
X3.3.3.01 UPVC; fixing with standard brackets at heights not exceeding 3m; fixing with standard brackets; nominal size; 68mm	m	0.15	4.21	-	7.89	12.10	3.084
X3.3.3.02 as above; nominal size; 100mm	m	0.15	4.21	-	17.98	22.19	4.156
X3.3.4 Fittings to downpipes							
X3.3.4.01 UPVC; fixing with standard brackets at heights not exceeding 3m; nominal size 68mm; Shoes	Nr	0.10	2.81	-	7.51	10.32	0.182
X3.3.4.02 as above; nominal size 68mm; Bends	Nr	0.16	4.49	-	8.69	13.18	0.633
X3.3.4.03 as above; nominal size 68mm; Offsets; 150mm projection	Nr	0.18	5.05	-	4.81	9.86	0.458
X3.3.4.04 as above; nominal size 68mm; Offsets; 300mm projection	Nr	0.18	5.05	-	4.81	9.86	0.458
X3.3.4.05 as above; nominal size 68mm; Branches	Nr	0.05	1.40	-	7.17	8.57	0.384
X3.3.4.06 as above; nominal size 110mm; Shoes	Nr	0.10	2.81	-	17.23	20.04	1.121
X3.3.4.07 as above; nominal size 110mm; Bends	Nr	0.16	4.49	-	21.31	25.80	0.902
X3.3.4.08 as above; nominal size 110mm; Offsets; 150mm projection	Nr	0.18	5.05	-	20.61	25.66	1.121
X3.3.4.09 as above; nominal size 110mm; Offsets; 300mm projection	Nr	0.18	5.05	-	20.61	25.66	1.121
X3.3.4.10 as above; nominal size 110mm; Branches	Nr	0.18	5.05	-	20.94	25.99	0.115

X4   Rock Filled Gabions

	Unit	Labour Hours	Labour Net £	Plant Net £	Materials Net £	Unit Net £	CO_2e Kg
X4.1	**Box of stated size**						
X4.1.0	Generally						
X4.1.0.01 Placed on river bank above water level; Zinc wire mesh 80mm; random filled by hand with broken rock of cubic character; average mass 2 - 10kg; Size: 2 x 1 x 1m	Nr	0.80	62.25	18.35	85.74	166.34	146.847
X4.1.0.02 as above; Size: 2 x 1 x 0.5m	Nr	0.65	50.58	14.91	50.82	116.31	75.525
X4.1.0.03 PVC coated wire mesh 80mm; random filled by hand with broken rock of cubic character; average mass 2 - 10kg; Size: 2 x 1 x 1m	Nr	0.80	62.25	18.35	94.60	175.20	154.371
X4.1.0.04 as above; Size: 2 x 1 x 0.5m	Nr	0.65	50.58	14.91	58.18	123.67	79.287
X4.1.0.05 Placed below water level and filled by grab; Zinc wire mesh 80mm; random filled by machine with broken rock of cubic character; average mass 2 - 10kg; Size: 2 x 1 x 1m	Nr	0.80	62.25	59.41	85.74	207.40	150.595
X4.2	**Mattress of stated thickness**						
X4.2.0	Generally						
X4.2.0.01 Gabions filled on bank then lifted into position below water level; distance 10m from bank top; galvanised wire mesh 60mm; random filled by hand with broken rock of cubic character; average mass 2 - 10kg; Size: 6 x 2m x 230mm	m^2	0.15	11.67	3.44	20.97	36.08	20.113
X4.2.0.02 as above Size: 6 x 2m x 300mm	m^2	0.15	11.67	3.44	24.72	39.83	25.850

CLASS Y:
SEWER AND WATER MAIN RENOVATION AND ANCILLARY WORKS

Calculations used throughout Class Y - Sewer and Water Main Renovation and Ancillary Works

Labour

			Qty		Rate		Total
L Y0004ICE	**Sewer Renovation Labour Gang**						
	L A9002ICE	Labourer (General Operative)	2	x	13.02	=	£26.03
	L A9003ICE	Labourer (Skill Rate 3)	2	x	14.86	=	£29.72
	L A9034ICE	Bricklayer	3	x	17.33	=	£51.98
	L A9015ICE	Banksman	1	x	14.03	=	£14.03
		Total hourly cost of gang				=	**£121.76**

Plant

			Qty		Rate		Total
P Y0005ICE	**Sewer Renovation Plant Gang**						
	P A1083ICE	Vactor Unit for Sewer Cleaning	1	x	8.50	=	£8.50
		Total hourly cost of gang				=	**£8.50**
P Y0006ICE	**Sewer Repair Plant Gang**						
	P A1073ICE	Robotic Sewer Repair Unit	1	x	73.93	=	£73.93
		Total hourly cost of gang				=	**£73.93**

Sewer And Water Main Renovation And Ancillary Works

Y1 Preparation Of Existing Sewers

Note(s): Cleaning

Cleaning is done in two stages

To remove all silt and adhesions prior to a survey of the existing sewer to establish the size of the lining units. Immediately prior to placing lining units to clean away any recent silt or sewage deposits to ensure efficient grouting to the annulus. Stage (i) usually requires 'heavy' jetting and vacuum cleaning or dredging. There is a variety of equipment/sub-contractors available depending on the size of sewer, length between the manholes etc., but usually the more expensive equipment has a compensatingly greater output and the cost will be determined more by the degree of situation rather than the size of sewer or length. Stage (ii) costs have been included in the laying operations. Guide Prices (Sub-contractors) for man-entry sewers 1200mm maximum diameter.

Normally re-lining is only contemplated where the existing structure is considered safe to work in. There should be minimal `making good' following a cleaning operation and it would be done by the lining gang.

Plugging Laterals

The following are Specialist prices and are based on robotic methods for non man entry pipelines. Assume as a permanant plug in lieu of complete filling of a lateral with PFA/cement.

Local Internal Repairs

The following are Specialist prices and are based on robotic methods for non man entry pipelines. Including cutting and repointing.

	Unit	Labour Hours	Labour Net £	Plant Net £	Materials Net £	Unit Net £		CO_2e Kg
Y1.1	**Cleaning**							
Y1.1.0	Generally							
Y1.1.0.01	0 - 10% siltation	m	0.17	20.34	1.42	-	21.76		1.359
Y1.1.0.02	10 - 40% siltation	m	0.25	30.44	2.13	-	32.57		2.035
Y1.2	**Removing intrusions**							
Y1.2.1	Laterals, bore not exceeding 150mm							
Y1.2.1.01	GVC, concrete, cast iron or steel pipes	Nr	2.00	243.54	7.54	-	251.08		3.066
Y1.2.2	Laterals, stated profile and size exceeding 150mm in one or more dimension							
Y1.2.2.01	GVC, or concrete pipes not exceeding 300mm into the bore of the main sewer	Nr	3.00	365.31	11.31	-	376.62		4.599
Y1.2.3	Other stated artifical intrusions							
Y1.2.3.01	Mass brickwork	Nr	0.50	60.88	2.79	-	63.67		3.293
Y1.2.3.02	Mass concrete	Nr	0.75	91.33	4.19	-	95.52		4.939
Y1.2.3.03	Reinforced concrete	Nr	1.00	121.77	5.58	-	127.35		6.585
Y1.2.3.04	Isolated bricks	Nr	0.01	1.70	0.08	-	1.78		0.092
Y1.4	**Plugging laterals, material stated**							
Y1.4.1	Bore not exceeding 300mm							
Y1.4.1.01	Epoxy resin plug	Nr	1.00	121.77	-	394.35	516.12	26.886	-0.066
Y1.4.1.02	Permanent concrete plug	Nr	0.50	60.88	-	4.07	64.95	11.294	-0.066
Y1.6	**Local internal repairs**							
Y1.6.1	Area: not exceeding 0.1m²							
Y1.6.1.01	Based on robotic methods for non man entry pipelines	Nr	1.00	121.77	73.93	197.16	392.86		77.091
Y1.6.2	Area: 0.1 - 0.25m²							
Y1.6.2.01	Based on robotic methods for non man entry pipelines	Nr	1.50	182.66	110.90	394.32	687.88		122.322

Y2 Stabilisation Of Existing Sewers

Note(s): Note: Pointing to be carried out by hand for small areas and by pressure pointing for large areas. Material cost is minimal compared with raking out and pointing labour costs. Prices will vary depending upon accessibility.

	Unit	Labour Hours	Labour Net £	Plant Net £	Materials Net £	Unit Net £	CO_2e Kg
Y2.1	**Pointing, materials stated**						
Y2.1.0	Generally						
Y2.1.0.01	Pointing with cement mortar (1:3); including preparation of joints	m^2	0.25	30.44	0.70	1.40	32.54	6.038

Y3 Renovation Of Existing Sewers

Note(s): Segmental lining
Linings are usually tailor-made to suit a given sewer. This is usually determined by passing a 3-dimensional template (the length of the proposed lining units) through the existing sewer to determine what lining of constant X-section will be suitable between any two manholes. The price charged for materials will depend upon the number of units which are to be provided which can be made from one run or former. The labour/plant cost of installation will depend on a number of factors: Minimum safety cover (ie. personnel) at shaft positions. If work is taking place at shaft positions while lining is carried out, then extra men would not be required for emergency use; if an existing shaft is being used for access then extra labour should be allowed over and above the lining crew to maintain (a) air flow through system (at least two manhole covers open and attended), (b) traffic direction, (c) gas monitoring and (d) escape facilities. The output in metres/shift does not vary greatly with diameter since large units are often easier to install because of the greater working room. The factors affecting output are: (a) Length between manholes. (b) Whether linings are in one piece or have an invert and soffit section. (c) Whether fixings are required to pin the units to existing brickwork. (d) Whether units are stiff/thick enough to withstand annulus grouting without internal strutting.

		Unit	Labour Hours	Labour Net £	Plant Net £	Materials Net £	Unit Net £	CO_2e Kg
Y3.3	**Segmental lining**							
Y3.3.3	Glass reinforced plastic							
Y3.3.3.01	1050 x 750 original size; One piece unit	m	0.05	6.09	1.72	413.03	420.84	79.460
Y3.3.3.02	900 x 600 original size; One piece unit	m	0.03	4.02	1.13	335.59	340.74	63.538
Y3.3.4	Glass reinforced concrete							
Y3.3.4.01	1050 x 750 original size; Two piece unit	m	0.10	12.18	3.44	380.20	395.82	122.524
Y3.3.4.02	900 x 600 original size; Two piece unit	m	0.07	8.16	2.30	306.28	316.74	97.963
Y3.6	**Annulus grouting, materials stated**							
Y3.6.0	Generally							
Y3.6.0.01	Pozament	m^3	0.10	12.18	0.73	281.80	294.71	287.212

Y4 Laterals To Renovated Sewers

		Unit	Labour Hours	Labour Net £	Plant Net £	Materials Net £	Unit Net £	CO_2e Kg
Y4.1	**Jointing**							
Y4.1.1	Bore: not exceeding 150mm							
Y4.1.1.01	Generally	Nr	0.30	36.53	1.13	26.94	64.60	7.173
Y4.1.2	Bore: 150 - 300mm							
Y4.1.2.01	Generally	Nr	0.55	66.97	2.07	37.71	106.75	9.793
Y4.1.3	Stated profile and size exceeding 300mm in one or more dimension							
Y4.1.3.01	bore: 450mm	Nr	0.80	97.42	3.02	59.26	159.70	14.651

Y5 Water Mains Renovation And Ancillary Works

	Unit	Labour Hours	Labour Net £	Plant Net £	Materials Net £	Unit Net £	CO_2e Kg
Y5.1 **Cleaning**							
Y5.1.2 Nominal bore: 200 - 300mm							
Y5.1.2.01 Siltation: not exceeding 35%	m	0.02	2.07	0.14	-	2.21	0.138
Y5.1.2.02 Siltation: not exceeding 35 - 50%	m	0.03	3.04	0.21	-	3.25	0.203
Y5.1.2.03 Siltation: not exceeding 50 - 65%	m	0.03	4.02	0.28	-	4.30	0.269
Y5.1.2.04 Siltation: over 65%	m	0.05	6.09	0.43	-	6.52	0.407
Y5.1.3 Nominal bore: 300 - 600mm							
Y5.1.3.01 Siltation: not exceeding 35%	m	0.03	4.02	0.28	-	4.30	0.269
Y5.1.3.02 Siltation: not exceeding 35 - 50%	m	0.05	6.09	0.43	-	6.52	0.407
Y5.1.3.03 Siltation: not exceeding 50 - 65%	m	0.07	8.16	0.57	-	8.73	0.545
Y5.1.3.04 Siltation: over 65%	m	0.10	12.18	0.85	-	13.03	0.814
Y5.1.4 Nominal bore: 600 - 900mm							
Y5.1.4.01 Siltation: not exceeding 35%	m	0.08	10.11	0.71	-	10.82	0.675
Y5.1.4.02 Siltation: not exceeding 35 - 50%	m	0.13	15.22	1.06	-	16.28	1.017
Y5.1.4.03 Siltation: not exceeding 50 - 65%	m	0.17	20.34	1.42	-	21.76	1.359
Y5.1.4.04 Siltation: over 65%	m	0.25	30.44	2.13	-	32.57	2.035
Y5.1.5 Nominal bore: 900 - 1200mm							
Y5.1.5.01 Siltation: not exceeding 35%	m	0.17	20.34	1.42	-	21.76	1.359
Y5.1.5.02 Siltation: not exceeding 35 - 50%	m	0.25	30.44	2.13	-	32.57	2.035
Y5.1.5.03 Siltation: not exceeding 50-65%	m	0.33	40.55	2.83	-	43.38	2.710
Y5.1.5.04 Siltation: over 65%	m	0.40	48.71	3.40	-	52.11	3.255
Y5.2 **Removing intrusions**							
Y5.2.1 Nominal bore: not exceeding 200mm							
Y5.2.1.01 Laterals of bore not exceeding 150mm; GVC, concrete, cast iron or steel pipes	Nr	2.00	243.54	7.54	-	251.08	3.066
Y5.2.2 Nominal bore: 200 - 300mm							
Y5.2.2.01 Laterals exceeding 200mm in one or more dimension; GVC, or concrete pipes not exceeding 300mm into the bore of the main pipe	Nr	3.00	365.31	11.31	-	376.62	4.599
Y5.2.2.02 Laterals exceeding 200mm in one or more dimension; Isolated bricks	Nr	3.00	365.31	11.31	-	376.62	4.599
Y5.4 **Closed-circuit television surveys**							
Y5.4.3 Nominal bore: 300 - 600mm							
Y5.4.3.01 for cast iron pipe; Nominal bore 600mm	m	-	-	1.74	-	1.74	0.138
Y5.4.4 Nominal bore: 600 - 900mm							
Y5.4.4.01 for cast iron pipe; Nominal bore 900mm	m	-	-	1.74	-	1.74	0.138
Y5.4.5 Nominal bore: 900 - 1200mm							
Y5.4.5.01 for cast iron pipe; nominal bore 1000mm	m	-	-	1.74	-	1.74	0.138
Y5.4.5.02 for cast iron pipe; nominal bore 1200mm	m	-	-	1.74	-	1.74	0.138
Y5.4.6 Nominal bore: stated exceeding 1200mm							
Y5.4.6.01 for cast iron pipe; nominal bore 1500mm	m	-	-	1.74	-	1.74	0.138
Y5.4.6.02 for cast iron pipe; nominal bore 1800mm	m	-	-	1.74	-	1.74	0.138

Y8 Interruptions

Note(s): NOTE: Prices will depend greatly upon the amount of labour and plant that is tied up by the stopping of an operation. In certain circumstances labour and plant may be diverted on to other work.

		Unit	Labour Hours	Labour Net £	Plant Net £	Materials Net £	Unit Net £	CO_2e Kg
Y8.1	**Preparation of existing sewers**							
Y8.1.0	Cleaning							
Y8.1.0.01	Vacuum cleaning	Hr	1.00	121.77	8.50	-	130.27	8.138
Y8.2	**Stabilisation of existing sewers**							
Y8.2.0	Generally							
Y8.2.0.01	Pointing	Hr	1.00	121.77	2.80	-	124.57	1.794
Y8.3	**Renovation of existing sewers**							
Y8.3.4	Stated proprietary lining							
Y8.3.4.01	GRC and GRP linings	Hr	1.00	121.77	34.37	-	156.14	4.256

CLASS Z:
SIMPLE BUILDING WORKS INCIDENTAL TO CIVIL ENGINEERING WORK

Calculations used throughout Class Z - Simple Building Works Incidental to Civil Engineering Work

Labour

			Qty		Rate		Total
L A0120ICE	**General Earthworks Labour Gang**						
	L A9016ICE	Ganger	1	x	17.64	=	£17.64
	L A9002ICE	Labourer (General Operative)	1	x	13.02	=	£13.02
	L A9015ICE	Banksman	1	x	14.03	=	£14.03
	Total hourly cost of gang					=	**£44.69**
L A0300ICE	**Brickwork Labour Gang**						
	L A9033ICE	Bricklayer (chargehand)	1	x	18.63	=	£18.63
	L A9034ICE	Bricklayer	4	x	17.33	=	£69.31
	L A9002ICE	Labourer (General Operative)	2	x	13.02	=	£26.03
	Total hourly cost of gang					=	**£113.97**
L A0310ICE	**Painting Labour Gang**						
	L A9041ICE	Painter (chargehand)	1	x	18.63	=	£18.63
	L A9027ICE	Painter	2	x	17.33	=	£34.66
	L A9028ICE	Brush hand (labourer)	1	x	13.02	=	£13.02
	Total hourly cost of gang					=	**£66.31**
L A0325ICE	**Plastering Labour Gang**						
	L A9031ICE	Plasterer	1	x	17.33	=	£17.33
	L A9002ICE	Labourer (General Operative)	1	x	13.02	=	£13.02
	Total hourly cost of gang					=	**£30.35**
L Z0002ICE	**Patent Glazing Labour Gang**						
	L A9014ICE	Labourer (Skill Rate 4)	3	x	14.03	=	£42.08
	L A9000ICE	Craftsman WRA	1	x	17.33	=	£17.33
	Total hourly cost of gang					=	**£59.41**
L Z0005ICE	**Carpentry Labour Gang**						
	L A9000ICE	Craftsman WRA	5	x	17.33	=	£86.64
	L A9002ICE	Labourer (General Operative)	1	x	13.02	=	£13.02
	L A9016ICE	Ganger	1	x	17.64	=	£17.64
	Total hourly cost of gang					=	**£117.30**
L Z0006ICE	**Electrical Labour Gang**						
	L A9043ICE	Electrician	2	x	23.16	=	£46.33
	L A9042ICE	Apprentice	1	x	11.58	=	£11.58
	Total hourly cost of gang					=	**£57.91**
L Z0007ICE	**Floor Finishes Labour Gang**						
	L A9000ICE	Craftsman WRA	1	x	17.33	=	£17.33
	L A9002ICE	Labourer (General Operative)	1	x	13.02	=	£13.02
	Total hourly cost of gang					=	**£30.35**

L Z0008ICE	**HVAC Labour Gang**							
	L A9044ICE	HVAC Craftsman	2	x	18.63	=	£37.26
	L A9042ICE	Apprentice	1	x	11.58	=	£11.58
		Total hourly cost of gang				=	**£48.84**

L Z0010ICE	**Screeding Labour Gang**							
	L A9000ICE	Craftsman WRA	1	x	17.33	=	£17.33
	L A9002ICE	Labourer (General Operative)	1	x	13.02	=	£13.02
		Total hourly cost of gang				=	**£30.35**

L Z0011ICE	**Wall Tiling Labour Gang**							
	L A9016ICE	Ganger	1	x	17.64	=	£17.64
	L A9002ICE	Labourer (General Operative)	1	x	13.02	=	£13.02
		Total hourly cost of gang				=	**£30.66**

Plant

P A1124ICE	**General Excavation Plant Gang**							
	P A0605ICE	Hydraulic Excavator - Cat 166kW	1	x	44.31	=	£44.31
	P A0943ICE	Crawler Tractor / Dozer - Cat D6 LGP 160 Hp	0.5	x	52.02	=	£26.01
		Total hourly cost of gang				=	**£70.32**

P Z0002ICE	**Carpentry Plant Gang**							
	P A0578ICE	Generator - 10kvA Diesel	1	x	3.80	=	£3.80
	P A1058ICE	9" Circular Saw	1	x	1.04	=	£1.04
	P A1082ICE	Small Tools	1	x	1.61	=	£1.61
		Total hourly cost of gang				=	**£6.45**

P Z0004ICE	**Plastering Plant Gang**							
	P A0723ICE	Concrete Mixer - 5/3 Diesel	1	x	2.80	=	£2.80
		Total hourly cost of gang				=	**£2.80**

P Z0006ICE	**Screeding Plant Gang**							
	P A0723ICE	Concrete Mixer - 5/3 Diesel	1	x	2.80	=	£2.80
	P A1077ICE	Screed Pump	1	x	15.48	=	£15.48
		Total hourly cost of gang				=	**£18.28**

Simple Building Works Incidental To Civil Engineering Work

Note(s): NOTE: The prices contained in this class are intended to assist in the compilation of estimates for activities commonly found in simple building works incidental to civil engineering works. The items listed hereafter are not all measured strictly in accordance with Class Z of CESMM4 but are included as a representative sample based on a composite basis wherever possible, to provide an indication of the level of pricing. The prices do not include for any Preliminary type items. Attention should be given to guidance notes which follow some items.

Z1   Carpentry And Joinery

	Unit	Labour Hours	Labour Net £	Plant Net £	Materials Net £	Unit Net £	CO_2e Kg	
Z1.1	**Structural and carcassing timber**							
Z1.1.1	Floors							
Z1.1.1.01	Suspended timber floors; including structural and carcassing timbers with boarding; sawn softwood joists at 600mm centres, ends built in; sawn softwood herringbone strutting 50 x 50mm to centre line of joists span; all timbers treated; Joint sizes: 175 x 50mm; Plywood, tongued and grooved joints; 18mm thick	m²	0.32	37.54	2.06	13.19	52.79	31.668	-20.321
Z1.1.1.02	as above; Joint sizes: 175 x 50mm; Chipboard, tongued and grooved joints; 18mm thick; pre-felted	m²	0.24	28.15	1.55	10.46	40.16	13.825	-19.293
Z1.1.1.03	as above; Joint sizes: 175 x 50mm; Softwood, wrought, tongued and grooved jointed; 150mm wide, 25mm thick	m²	0.40	46.92	2.58	13.87	63.37	7.960	-23.710
Z1.1.1.04	as above; Joint sizes: 200 x 50mm; Plywood, tongued and grooved joints; 18mm thick	m²	0.32	37.54	2.06	14.05	53.65	32.038	-21.764
Z1.1.1.05	as above; Joint sizes: 200 x 50mm; Chipboard, tongued and grooved joints; 18mm thick; pre-felted	m²	0.24	28.15	1.55	11.32	41.02	14.197	-20.736
Z1.1.1.06	as above; Joint sizes: 200 x 50mm; Softwood, wrought, tongued and grooved jointed; 150mm wide, 25mm thick	m²	0.40	46.92	2.58	14.73	64.23	8.332	-25.153
Z1.1.1.07	as above; Joint sizes: 225 x 50mm; Plywood, tongued and grooved joints; 18mm thick	m²	0.32	37.54	2.06	14.91	54.51	32.408	-23.207
Z1.1.1.08	as above; Joint sizes: 225 x 50mm; Chipboard, tongued and grooved joints; 18mm thick; pre-felted	m²	0.24	28.15	1.55	12.18	41.88	14.567	-22.179
Z1.1.1.09	as above; Joint sizes: 225 x 50mm; Softwood, wrought, tongued and grooved jointed; 150mm wide, 25mm thick	m²	0.40	46.92	2.58	15.59	65.09	8.702	-26.596

Z1 Carpentry And Joinery continued...

	Unit	Labour Hours	Labour Net £	Plant Net £	Materials Net £	Unit Net £		CO$_2$e Kg

Z1.1 Structural and carcassing timber

Z1.1.2 Walls and partitions

Z1.1.2.01	Timber internal stud partitions; including structural carcassing timbers within partitions; studs at 450mm centres, noggins at 1220 mm centres; head and sole plate; all timbers treated; Stud sizes: 38 x 50mm; Hardwood, sawn	m^2	0.20	23.46	1.29	4.61	29.36	3.950	-5.470
Z1.1.2.02	as above; Stud sizes: 38 x 50mm; Softwood, sawn	m^2	0.10	11.73	0.65	2.56	14.94	1.580	-4.277
Z1.1.2.03	as above; Stud sizes: 50 x 75mm; Hardwood, sawn	m^2	0.30	35.19	1.94	8.56	45.69	6.981	-10.146
Z1.1.2.04	as above; Stud sizes: 50 x 75mm; Softwood, sawn	m^2	0.15	17.59	0.97	4.79	23.35	2.762	-7.932
Z1.1.2.05	as above; Stud sizes: 75 x 100mm; Hardwood, sawn	m^2	0.60	70.38	3.87	18.19	92.44	14.666	-21.587
Z1.1.2.06	as above; Stud sizes: 75 x 100mm; Softwood, sawn	m^2	0.30	35.19	1.94	10.15	47.28	5.784	-16.881

Z1.1.3 Flat roofs

Z1.1.3.01	as above; Joint sizes: 150 x 50mm; Plywood, butt joints; 18mm thick	m^2	0.32	37.54	2.06	13.68	53.28	31.596	-21.418
Z1.1.3.02	as above; Joint sizes: 150 x 50mm; Chipboard, butt joints; 18mm thick; pre-felted	m^2	0.24	28.15	1.55	10.95	40.65	13.754	-20.390
Z1.1.3.03	as above; Joint sizes: 150 x 50mm; Softwood, wrought, tongued and grooved jointed; 150mm wide, 25mm thick	m^2	0.40	46.92	2.58	14.36	63.86	7.889	-24.807
Z1.1.3.04	as above; Joint sizes: 200 x 50mm; Plywood, butt joints; 18mm thick	m^2	0.36	42.23	2.32	14.77	59.32	32.260	-23.254
Z1.1.3.05	as above; Joint sizes: 200 x 50mm; Chipboard, butt joints; 18mm thick; pre-felted	m^2	0.28	32.84	1.81	12.05	46.70	14.419	-22.226
Z1.1.3.06	as above; Joint sizes: 200 x 50mm; Softwood, wrought, tongued and grooved jointed; 150mm wide, 25mm thick	m^2	0.44	51.61	2.84	15.45	69.90	8.554	-26.643
Z1.1.3.07	as above; Joint sizes: 250 x 50mm; Plywood, butt joints; 18mm thick	m^2	0.40	46.92	2.58	15.88	65.38	32.925	-25.090
Z1.1.3.08	as above; Joint sizes: 250 x 50mm; Chipboard, butt joints; 18mm thick; pre-felted	m^2	0.32	37.54	2.06	13.15	52.75	15.083	-24.062
Z1.1.3.09	as above; Joint sizes: 250 x 50mm; Softwood, wrought, tongued and grooved jointed; 150mm wide, 25mm thick	m^2	0.48	56.30	3.10	16.56	75.96	9.219	-28.479

Z1 Carpentry And Joinery continued...

	Unit	Labour Hours	Labour Net £	Plant Net £	Materials Net £	Unit Net £	CO₂e Kg	

							CO$_2$e Kg	
Z1.1	**Structural and carcassing timber**							
Z1.1.4	Pitched roofs							
Z1.1.4.01	Timber pitched roof in traditional sawn softwood construction; including structural and carcassing timbers; roof with hipped ends; purlins; ridge; rafters and ceiling joists at 450mm centres, blinders; struts; hangers; ties; wall plates; no insulation no coverings; all timber treated; Roof pitch: 22.5 degrees; Rafters at 1800mm centres spanning; 8000mm	m²	0.25	29.32	1.61	12.97	43.90	5.542	-22.798
Z1.1.4.02	as above; Roof pitch: 22.5 degrees; Rafters at 1800mm centres spanning; 10000mm	m²	0.24	28.15	1.55	11.15	40.85	4.958	-19.530
Z1.1.4.03	as above; Roof pitch: 35 degrees; Rafters at 1800mm centres spanning; 8000mm	m²	0.28	32.96	1.81	15.34	50.11	6.626	-26.835
Z1.1.4.04	as above; Roof pitch: 35 degrees; Rafters at 1800mm centres spanning; 10000mm	m²	0.30	35.19	1.94	11.99	49.12	5.554	-20.973
Z1.1.4.05	as above; Roof pitch: 45 degrees; Rafters at 1800mm centres spanning; 8000mm	m²	0.31	36.71	2.02	16.28	55.01	7.100	-28.474
Z1.1.4.06	as above; Roof pitch: 45 degrees; Rafters at 1800mm centres spanning; 10000mm	m²	0.34	39.88	2.19	12.63	54.70	5.974	-22.101
Z1.1.4.07	as above; Roof pitch: 22.5 degrees; Trussed rafter spanning; 4600mm	m²	0.08	9.03	0.50	16.66	26.19	7.583	-21.860
Z1.1.4.08	as above; Roof pitch: 22.5 degrees; Trussed rafter spanning; 7000mm; Fink	m²	0.08	9.03	0.50	28.48	38.01	10.688	-29.823
Z1.1.4.09	as above; Roof pitch: 22.5 degrees; Trussed rafter spanning; 10000mm; Fink	m²	0.08	9.03	0.50	40.58	50.11	14.312	-41.074
Z1.1.4.10	Timber pitched roof; including structural and carcassing timbers with trussed rafters; sawn softwood trussed rafters at 600mm centres, roof with gable ends; sawn softwood wall plates, bracing and binders; galvanised steel straps to gable walls; no insulation; np coverings; all timber treated; Roof pitch: 35 degrees; Trussed rafter spanning; 4600mm; Fink	m²	0.09	9.97	0.55	17.98	28.50	7.306	-19.251
Z1.1.4.11	as above; Roof pitch: 35 degrees; Trussed rafter spanning; 7000mm; Fink	m²	0.09	9.97	0.55	31.70	42.22	10.585	-26.266
Z1.1.4.12	as above; Roof pitch: 35 degrees; Trussed rafter spanning; 10000mm; Fink	m²	0.10	11.96	0.66	45.66	58.28	14.402	-36.177
Z1.1.4.13	as above; Roof pitch: 45 degrees; Trussed rafter spanning; 4600mm; Fink	m²	0.12	14.19	0.78	19.06	34.03	7.240	-17.238

Z1 Carpentry And Joinery continued...

	Unit	Labour Hours	Labour Net £	Plant Net £	Materials Net £	Unit Net £	CO$_2$e Kg	
Z1.1	**Structural and carcassing timber**							
Z1.1.4	Pitched roofs							
Z1.1.4.14	as previous item; Roof pitch: 45 degrees; Trussed rafter spanning; 7000mm; Fink	m^2	0.12	14.19	0.78	34.31	49.28	10.657	-23.521
Z1.1.4.15	as above; Roof pitch: 45 degrees; Trussed rafter spanning; 10000mm; Fink	m^2	0.12	14.19	0.78	49.77	64.74	14.484	-32.394
Z1.1.5	Plates and bearers							
Z1.1.5.01	Untreated Softwood, sawn: 25 x 50mm	m	0.02	1.99	0.11	0.24	2.34	0.223	-0.989
Z1.1.5.02	as above; 25 x 75mm	m	0.02	1.99	0.11	0.37	2.47	0.293	-1.481
Z1.1.5.03	as above; 25 x 100mm	m	0.02	1.99	0.11	0.49	2.59	0.364	-1.976
Z1.1.5.04	as above; 38 x 50mm	m	0.02	1.99	0.11	0.37	2.47	0.296	-1.502
Z1.1.5.05	as above; 38 x 75mm	m	0.02	1.99	0.11	0.56	2.66	0.404	-2.253
Z1.1.5.06	as above; 38 x 100mm	m	0.02	2.46	0.14	0.74	3.34	0.531	-3.003
Z1.1.5.07	as above; 47 x 50mm	m	0.02	2.46	0.14	0.47	3.07	0.367	-1.858
Z1.1.5.08	as above; 47 x 75mm	m	0.02	2.46	0.14	0.69	3.29	0.500	-2.786
Z1.1.5.09	as above; 47 x 100mm	m	0.02	2.82	0.15	0.92	3.89	0.646	-3.714
Z1.1.5.10	as above; 75 x 75mm	m	0.03	3.28	0.18	1.10	4.56	0.770	-4.446
Z1.1.5.11	as above; 75 x 100mm	m	0.03	3.28	0.18	1.47	4.93	0.982	-5.927
Z1.1.5.12	as above; 75 x 150mm	m	0.04	4.93	0.27	2.21	7.41	1.473	-8.892
Z1.1.5.13	as above; 75 x 200mm	m	0.04	4.93	0.27	2.94	8.14	1.896	-11.855
Z1.1.5.14	as above; 100 x 100mm	m	0.04	4.93	0.27	1.96	7.16	1.332	-7.903
71.1.5.15	as above; 100 x 150mm	m	0.05	6.10	0.34	2.94	9.38	1.944	-11.855
Z1.1.5.16	as above; 100 x 200mm	m	0.07	8.09	0.45	3.92	12.46	2.591	-15.806
Z1.1.5.17	as above; 100 x 250 mm	m	0.09	10.21	0.56	4.91	15.68	3.242	-19.758
Z1.1.5.18	Treated Softwood, sawn: 25 x 50mm	m	0.02	1.99	0.11	0.36	2.46	0.186	-0.657
Z1.1.5.19	as above; 25 x 75mm	m	0.02	1.99	0.11	0.54	2.64	0.237	-0.984
Z1.1.5.20	as above; 25 x 100mm	m	0.02	1.99	0.11	0.73	2.83	0.289	-1.312
Z1.1.5.21	as above; 38 x 50mm	m	0.02	1.99	0.11	0.55	2.65	0.239	-0.997
Z1.1.5.22	as above; 38 x 75mm	m	0.02	1.99	0.11	0.83	2.93	0.318	-1.496
Z1.1.5.23	as above; 38 x 100mm	m	0.02	2.46	0.14	1.10	3.70	0.417	-1.994
Z1.1.5.24	as above; 47 x 50mm	m	0.02	2.46	0.14	0.68	3.28	0.297	-1.234
Z1.1.5.25	as above; 47 x 75mm	m	0.02	2.46	0.14	1.03	3.63	0.394	-1.849
Z1.1.5.26	as above; 47 x 100mm	m	0.02	2.82	0.15	1.36	4.33	0.506	-2.466
Z1.1.5.27	as above; 75 x 75mm	m	0.03	3.28	0.18	1.63	5.09	0.602	-2.951
Z1.1.5.28	as above; 75 x 100mm	m	0.04	4.93	0.27	2.18	7.38	0.826	-3.935
Z1.1.5.29	as above; 75 x 150mm	m	0.04	4.93	0.27	3.26	8.46	1.137	-5.903
Z1.1.5.30	as above; 75 x 200mm	m	0.04	4.93	0.27	4.35	9.55	1.447	-7.870
Z1.1.5.31	as above; 100 x 100mm	m	0.04	4.93	0.27	2.90	8.10	1.033	-5.247
Z1.1.5.32	as above; 100 x 150mm	m	0.05	6.10	0.34	4.35	10.79	1.495	-7.870
Z1.1.5.33	as above; 100 x 200mm	m	0.07	8.09	0.45	5.80	14.34	1.993	-10.493
Z1.1.5.34	as above; 100 x 250 mm	m	0.09	10.21	0.56	7.26	18.03	2.494	-13.116
Z1.1.6	Struts							
Z1.1.6.01	Untreated Softwood, sawn: 50 x 50mm	m	0.02	1.99	0.11	0.49	2.59	0.364	-1.976
Z1.1.6.02	as above; 50 x 100mm	m	0.04	4.11	0.23	0.98	5.32	0.734	-3.952
Z1.1.6.03	as above; 50 x 150mm	m	0.05	6.10	0.34	1.47	7.91	0.816	-3.952
Z1.1.6.04	as above; 50 x 200mm	m	0.07	8.09	0.45	1.96	10.50	1.576	-8.693
Z1.1.6.05	as above; 50 x 250 mm	m	0.09	10.21	0.56	2.45	13.22	1.832	-9.879
Z1.1.6.06	Treated Softwood, sawn: 50 x 50mm	m	0.02	1.99	0.11	0.73	2.83	0.289	-1.312
Z1.1.6.07	as above; 50 x 100mm	m	0.04	4.11	0.23	1.45	5.79	0.584	-2.623
Z1.1.6.08	as above; 50 x 150mm	m	0.05	6.10	0.34	2.18	8.62	0.874	-3.935
Z1.1.6.09	as above; 50 x 200mm	m	0.07	8.09	0.45	2.90	11.44	1.164	-5.247
Z1.1.6.10	as above; 50 x 250 mm	m	0.09	10.21	0.56	3.63	14.40	1.458	-6.558

Z1   Carpentry And Joinery continued...

	Unit	Labour Hours	Labour Net £	Plant Net £	Materials Net £	Unit Net £	CO$_2$e Kg	

Z1.1   Structural and carcassing timber

Z1.1.7   Cleats

Z1.1.7.01	Untreated Softwood, sawn: 50 × 100 × 200mm long	Nr	0.01	1.17	0.06	0.20	1.43	0.613	-3.952
Z1.1.7.02	as above; 75 × 100 × 225mm long	Nr	0.01	1.64	0.09	0.33	2.06	0.914	-5.927
Z1.1.7.03	as above; 75 × 150 × 300mm long	Nr	0.02	1.99	0.11	0.67	2.77	1.351	-8.892
Z1.1.7.04	as above; 100 × 200 × 400mm long	Nr	0.02	2.46	0.14	1.57	4.17	2.358	-15.806
Z1.1.7.05	Treated Softwood, sawn: 50 × 100 × 200mm long	Nr	0.01	1.17	0.06	0.29	1.52	0.463	-2.623
Z1.1.7.06	as above; 75 × 100 × 225mm long	Nr	0.01	1.64	0.09	0.49	2.22	0.690	-3.935
Z1.1.7.07	as above; 75 × 150 × 300mm long	Nr	0.02	1.99	0.11	0.98	3.08	1.015	-5.903
Z1.1.7.08	as above; 100 × 200 × 400mm long	Nr	0.02	2.46	0.14	2.32	4.92	1.760	-10.493

Z1.2   Strip boarding

Z1.2.1   Floors

Z1.2.1.01	Tongued and grooved joints; 150mm widths; Thickness: 22mm; Softwood, wrought	m²	0.10	9.18	0.32	15.46	24.96	2.355	-11.543
Z1.2.1.02	as above; 150mm widths; Thickness: 25mm; Softwood, wrought	m²	0.10	9.18	0.32	16.29	25.79	2.604	-13.116
Z1.2.1.03	as above; 75mm widths; Thickness: 22mm; Hardwood, wrought: Maple	m²	0.34	32.74	1.29	22.92	56.95	7.276	-14.760
Z1.2.1.04	as above; 75mm widths; Thickness: 22mm; Hardwood, wrought: Iroko	m²	0.29	26.88	0.97	21.87	49.72	7.033	-14.760
Z1.2.1.05	as above; 75mm widths; Thickness: 22mm; Hardwood, wrought: Oak	m²	0.34	32.74	1.29	22.92	56.95	7.276	-14.760

Z1.2.4   Soffits

Z1.2.4.01	At eaves; butt joints; including 38 × 50mm sawn softwood treated bearers; Widths: over 150mm; Masterboard; 6mm thick	m²	0.05	5.87	0.32	4.89	11.08	9.676	-7.230
Z1.2.4.02	as above; Widths: over 150mm; Softwood, wrought untreated; 16mm thick 100mm wide boards	m²	0.10	11.73	0.65	3.61	15.99	1.512	-6.505
Z1.2.4.03	as above; Widths: over 150mm; Plywood external quality; 18mm	m²	0.05	5.87	0.32	15.63	21.82	27.286	-13.715
Z1.2.4.04	as above; Widths: over 225mm; Masterboard; 6mm thick	m²	0.08	8.80	0.48	4.89	14.17	9.797	-7.230
Z1.2.4.05	as above; Widths: over 225mm; Softwood, wrought untreated; 16mm thick 100mm wide boards	m²	0.13	14.66	0.81	3.61	19.08	1.633	-6.505
Z1.2.4.06	as above; Widths: over 225mm; Plywood external quality; 18mm	m²	0.06	7.04	0.39	15.63	23.06	27.334	-13.715
Z1.2.4.07	as above; Widths: over 300mm; Masterboard; 6mm thick	m²	0.16	18.77	1.03	4.89	24.69	10.209	-7.230

Z1 Carpentry And Joinery continued...

	Unit	Labour Hours	Labour Net £	Plant Net £	Materials Net £	Unit Net £	CO₂e Kg	

Z1.2 Strip boarding

Z1.2.4 Soffits

Z1.2.4.08	as previous item; Widths: over 300mm; Softwood, wrought untreated; 16mm thick 100mm wide boards	m²	0.30	35.19	1.94	3.61	40.74	2.482	-6.505
Z1.2.4.09	as above; Widths: over 300mm; Plywood external quality; 18mm	m²	0.16	18.77	1.03	15.63	35.43	27.819	-13.715

Z1.3 Sheet boarding

Z1.3.1 Floors

Z1.3.1.01	Thickness: 15mm; Plywood, butt jointed	m²	0.11	12.90	0.71	5.05	18.66	18.144	-6.484
Z1.3.1.02	Thickness: 18mm; Plywood, butt jointed	m²	0.11	12.90	0.71	6.36	19.97	26.949	-9.727
Z1.3.1.03	Thickness: 18mm; Chipboard, tongued and grooved jointed	m²	0.06	7.04	0.39	3.82	11.25	9.646	-8.699
Z1.3.1.04	Thickness: 22mm; Plywood, butt jointed	m²	0.11	12.90	0.71	7.81	21.42	37.221	-13.509
Z1.3.1.05	Thickness: 22mm; Chipboard, tongued and grooved jointed	m²	0.06	7.04	0.39	5.63	13.06	13.283	-12.081

Z1.4 Stairs and walkways

Z1.4.1 Stairways and landings

Z1.4.1.01	Balustrade to one side, 900mm wide x 2600mm rise; stairs, 25mm treads, 19mm risers; 32mm strings each side; 75 x 75mm newel posts; balustrade, 38 x 38mm balusters at 150mm centres, 50 x 75mm handrail; fixing to masonry walls; no finishes; Wrought Softwood	Nr	4.00	469.20	25.80	457.98	952.98	116.040	-136.128
Z1.4.1.02	as above; Two flight with quarter landing	Nr	6.00	703.80	38.70	357.57	1,100.07	110.974	-160.786
Z1.4.1.03	as above; Wrought Softwood; Two flight with half landing	Nr	7.00	821.10	45.15	373.18	1,239.43	124.469	-185.444

Z1.4.3 Isolated balustrades

Z1.4.3.01	Balustrades 1100mm high; 32 x 32mm balusters; 50 x 75mm moulded handrails; jointed to newel posts; Wrought Softwood; Balustrade to landing, 3000mm long	Nr	1.00	117.30	6.45	127.83	251.58	42.989	-42.070
Z1.4.3.02	as above; Wrought Softwood; Balustrade to straight staircase, 2600mm rise	Nr	2.00	234.60	12.90	127.83	375.33	47.840	-42.070
Z1.4.3.03	as above; Wrought Oak; Balustrade to landing, 3000mm long	Nr	1.00	117.30	6.45	745.95	869.70	144.371	-42.070
Z1.4.3.04	as above; Wrought Oak; Balustrade to straight staircase, 2600mm rise	Nr	2.00	234.60	12.90	745.95	993.45	149.222	-42.070

Z1 Carpentry And Joinery continued...

	Unit	Labour Hours	Labour Net £	Plant Net £	Materials Net £	Unit Net £	CO_2e Kg	

Z1.5 Miscellaneous joinery

Z1.5.1 Skirtings

	Unit	Labour Hours	Labour Net £	Plant Net £	Materials Net £	Unit Net £	CO_2e Kg	
Z1.5.1.01 Skirtings including sawn softwood grounds plugged and screwed to blockwork; Rounded Softwood, wrought: 19 x 100mm	m	0.03	3.12	0.11	1.04	4.27	0.254	-0.997
Z1.5.1.02 as above; Rounded Softwood, wrought: 19 x 150mm	m	0.03	3.12	0.11	1.56	4.79	0.340	-1.496
Z1.5.1.03 as above; Rounded Softwood, wrought: 19 x 175mm	m	0.03	3.12	0.11	1.83	5.06	0.385	-1.744
Z1.5.1.04 as above; Rounded Softwood, wrought: 25 x 100mm	m	0.05	4.59	0.16	1.22	5.97	0.343	-1.312
Z1.5.1.05 as above; Rounded Softwood, wrought: 25 x 150mm	m	0.05	4.59	0.16	1.82	6.57	0.454	-1.968
Z1.5.1.06 as above; Rounded Softwood, wrought: 25 x 175mm	m	0.05	4.59	0.16	2.15	6.90	0.510	-2.296
Z1.5.1.07 as above; Rounded Oak, wrought: 19 x 100mm	m	0.09	7.37	0.16	1.63	9.16	2.924	-1.027
Z1.5.1.08 as above; Rounded Oak, wrought: 19 x 150mm	m	0.10	8.31	0.21	2.43	10.95	4.365	-1.540
Z1.5.1.09 as above; Rounded Oak, wrought: 19 x 175mm	m	0.10	8.31	0.21	2.87	11.39	2.974	-1.027
Z1.5.1.10 as above; Rounded Oak, wrought: 25 x 100mm	m	0.15	12.50	0.32	1.99	14.81	3.926	-1.351
Z1.5.1.11 as above; Rounded Oak, wrought: 25 x 150mm	m	0.15	12.50	0.32	2.98	15.80	5.768	-2.027
Z1.5.1.12 as above; Rounded Oak, wrought: 25 x 175mm	m	0.15	12.50	0.32	2.36	15.18	6.690	-2.364
Z1.5.1.13 as above; Moulded Softwood, wrought: 19 x 100mm	m	0.05	4.59	0.16	1.04	5.79	0.293	-0.997
Z1.5.1.14 as above; Moulded Softwood, wrought: 19 x 150mm	m	0.05	4.59	0.16	1.56	6.31	0.379	-1.496
Z1.5.1.15 as above; Moulded Softwood, wrought: 19 x 175mm	m	0.05	4.59	0.16	1.83	6.58	0.423	-1.744
Z1.5.1.16 as above; Moulded Softwood, wrought: 25 x 100mm	m	0.07	6.06	0.21	1.22	7.49	0.382	-1.312
Z1.5.1.17 as above; Moulded Softwood, wrought: 25 x 150mm	m	0.07	6.06	0.21	1.82	8.09	0.493	-1.968
Z1.5.1.18 as above; Moulded Softwood, wrought: 25 x 175mm	m	0.07	6.06	0.21	2.15	8.42	0.549	-2.296
Z1.5.1.19 as above; Moulded Oak, wrought: 19 x 100mm	m	0.10	8.31	0.21	1.63	10.15	2.963	-1.027
Z1.5.1.20 as above; Moulded Oak, wrought: 19 x 150mm	m	0.10	8.31	0.21	2.43	10.95	4.365	-1.540
Z1.5.1.21 as above; Moulded Oak, wrought: 19 x 175mm	m	0.10	8.31	0.21	2.87	11.39	5.065	-1.797
Z1.5.1.22 as above; Moulded Oak, wrought: 25 x 100mm	m	0.13	10.43	0.27	1.99	12.69	3.887	-1.351
Z1.5.1.23 as above; Moulded Oak, wrought: 25 x 150mm	m	0.13	10.43	0.27	2.98	13.68	5.729	-2.027
Z1.5.1.24 as above; Moulded Oak, wrought: 25 x 175mm	m	0.13	10.43	0.27	2.36	13.06	6.651	-2.364

Z1.5.2 Architraves

	Unit	Labour Hours	Labour Net £	Plant Net £	Materials Net £	Unit Net £	CO_2e Kg	
Z1.5.2.01 Rounded Softwood, wrought: 25 x 44mm	m	0.03	3.12	0.11	0.54	3.77	0.179	-0.577
Z1.5.2.02 as above 25 x 50mm	m	0.03	3.12	0.11	0.62	3.85	0.194	-0.657
Z1.5.2.03 as above 25 x 63mm	m	0.03	3.12	0.11	0.79	4.02	0.222	-0.827
Z1.5.2.04 as above: 25 x 75mm	m	0.03	3.12	0.11	0.97	4.20	0.249	-0.984

Z1 Carpentry And Joinery continued...

		Unit	Labour Hours	Labour Net £	Plant Net £	Materials Net £	Unit Net £	CO₂e	Kg

| | | | | | | | | CO_2e | |

Z1.5 Miscellaneous joinery

		Unit	Labour Hours	Labour Net £	Plant Net £	Materials Net £	Unit Net £	CO_2e Kg	
Z1.5.2	Architraves								
Z1.5.2.05	Rounded Oak, wrought: 25 x 44mm	m	0.07	6.06	0.21	0.86	7.13	1.780	-0.594
Z1.5.2.06	as above 25 x 50mm	m	0.07	6.06	0.21	0.99	7.26	2.002	-0.676
Z1.5.2.07	as above: 25 x 63mm	m	0.07	6.06	0.21	1.24	7.51	2.481	-0.851
Z1.5.2.08	as above: 25 x 75mm	m	0.07	6.06	0.21	1.51	7.78	2.924	-1.013
Z1.5.2.09	Moulded Softwood, wrought: 25 x 44mm	m	0.03	3.12	0.11	0.54	3.77	0.179	-0.577
Z1.5.2.10	as above: 25 x 50mm	m	0.03	3.12	0.11	0.62	3.85	0.194	-0.657
Z1.5.2.11	as above: 25 x 63mm	m	0.03	3.12	0.11	0.79	4.02	0.222	-0.827
Z1.5.2.12	as above: 25 x 75mm	m	0.03	3.12	0.11	0.97	4.20	0.249	-0.984
Z1.5.2.13	Moulded Oak, wrought: 25 x 44mm	m	0.07	6.06	0.21	0.86	7.13	1.780	-0.594
Z1.5.2.14	as above: 25 x 50mm	m	0.07	6.06	0.21	0.99	7.26	2.002	-0.676
Z1.5.2.15	as above: 25 x 63mm	m	0.07	6.06	0.21	1.24	7.51	2.481	-0.851
Z1.5.2.16	as above: 25 x 75mm	m	0.07	6.06	0.21	1.51	7.78	2.924	-1.013
Z1.5.3	Trims								
Z1.5.3.01	Rounded Softwood, wrought: 25 x 44mm	m	0.03	3.12	0.11	0.52	3.75	0.179	-0.577
Z1.5.3.02	as above: 25 x 50mm	m	0.03	3.12	0.11	0.60	3.83	0.194	-0.657
Z1.5.3.03	as above: 25 x 63mm	m	0.03	3.12	0.11	0.76	3.99	0.222	-0.827
Z1.5.3.04	as above: 25 x 75mm	m	0.03	3.12	0.11	0.93	4.16	0.249	-0.984
Z1.5.3.05	Rounded Oak, wrought: 25 x 44mm	m	0.07	6.06	0.21	0.86	7.13	0.467	-0.738
Z1.5.3.06	as above: 25 x 50mm	m	0.07	6.06	0.21	0.99	7.26	0.509	-0.839
Z1.5.3.07	as above: 25 x 63mm	m	0.07	6.06	0.21	1.24	7.51	0.600	-1.057
Z1.5.3.08	as above: 25 x 75mm	m	0.07	6.06	0.21	1.51	7.78	0.685	-1.258
Z1.5.3.09	Moulded Softwood, wrought: 25 x 44mm	m	0.03	3.12	0.11	0.52	3.75	0.179	-0.577
Z1.5.3.10	as above: 25 x 50mm	m	0.03	3.12	0.11	0.60	3.83	0.194	-0.657
Z1.5.3.11	as above: 25 x 63mm	m	0.03	3.12	0.11	0.76	3.99	0.222	-0.827
Z1.5.3.12	as above: 25 x 75mm	m	0.03	3.12	0.11	0.93	4.16	0.249	-0.984
Z1.5.3.13	Moulded Oak, wrought: 25 x 44mm	m	0.07	6.06	0.21	0.86	7.13	0.467	-0.738
Z1.5.3.14	as above: 25 x 50mm	m	0.07	6.06	0.21	0.99	7.26	0.509	-0.839
Z1.5.3.15	as above: 25 x 63mm	m	0.07	6.06	0.21	1.24	7.51	0.600	-1.057
Z1.5.3.16	as above: 25 x 75mm	m	0.07	6.06	0.21	1.51	7.78	0.685	-1.258
Z1.5.4	Shelves								
Z1.5.4.01	Shelving including bearers; Widths: 300mm; Softwood, wrought; 25mm thick	m	0.25	29.32	1.61	2.82	33.75	1.565	-2.422
Z1.5.4.02	as above; Widths: 300mm; Softwood, wrought; slatted 25 x 50mm spaced at 75mm; 25mm thick	m	0.20	23.46	1.29	1.94	26.69	1.676	-4.656
Z1.5.4.03	as above; Widths: 300mm; Blackboard, butt joints: 18mm thick	m	0.20	23.46	1.29	1.87	26.62	4.875	-4.531
Z1.5.4.04	as above; Widths: 300mm; Blackboard, butt joints: 25mm thick	m	0.20	23.46	1.29	2.41	27.16	6.281	-5.504
Z1.5.4.05	as above; Widths: 300mm; Chipboard, butt joints: 18mm thick	m	0.20	23.46	1.29	1.65	26.40	4.067	-4.637
Z1.5.4.06	as above; Widths: 300mm; Chipboard, butt joints: 25mm thick	m	0.20	23.46	1.29	2.19	26.94	5.158	-5.652

Z1   Carpentry And Joinery continued...

	Unit	Labour Hours	Labour Net £	Plant Net £	Materials Net £	Unit Net £	CO$_2$e Kg	
Z1.5	**Miscellaneous joinery**							
Z1.5.4	Shelves							
Z1.5.4.07	as previous item; Widths: 300mm; Plywood, butt joints: 18mm thick	m	0.20	23.46	1.29	2.41	27.16	9.185	-4.946
Z1.5.4.08	as above; Widths: 300mm; Plywood, butt joints: 25mm thick	m	0.20	23.46	1.29	2.85	27.60	12.266	-6.080
Z1.5.4.09	as above; Widths: 300mm; Chipboard, faced both sides with white melamine; 15mm thick	m	0.20	23.46	1.29	1.69	26.44	3.599	-4.202
Z1.5.4.10	as above; Widths: 450mm; Softwood, wrought; slatted 25 x 50mm spaced at 75mm; 25mm thick	m	0.25	29.32	1.61	2.92	33.85	2.271	-6.984
Z1.5.4.11	as above; Widths: 450mm; Blackboard, butt joints: 18mm thick	m	0.25	29.32	1.61	2.80	33.73	7.070	-6.796
Z1.5.4.12	as above; Widths: 450mm; Blackboard, butt joints: 25mm thick	m	0.25	29.32	1.61	3.62	34.55	9.178	-8.256
Z1.5.4.13	as above; Widths: 450mm; Chipboard, butt joints: 18mm thick	m	0.20	23.46	1.29	2.48	27.23	5.615	-6.956
Z1.5.4.14	as above; Widths: 450mm; Chipboard, butt joints: 25mm thick	m	0.20	23.46	1.29	3.29	28.04	7.251	-8.478
Z1.5.4.15	as above; Widths: 450mm; Plywood, butt joints: 18mm thick	m	0.25	29.32	1.61	3.62	34.55	13.534	-7.419
Z1.5.4.16	as above; Widths: 450mm; Plywood, butt joints: 25mm thick	m	0.25	29.32	1.61	4.27	35.20	18.157	-9.121
Z1.5.4.17	as above; Widths: 450mm; Chipboard, faced both sides with white melamine; 15mm thick	m	0.30	35.19	1.94	2.53	39.66	5.398	-6.304
Z1.5.4.18	as above; Widths: 600mm; Softwood, wrought; slatted 25 x 50mm spaced at 75mm; 25mm thick	m	0.30	35.19	1.94	3.89	41.02	2.867	-9.311
Z1.5.4.19	as above; Widths: 600mm; Blackboard, butt joints: 18mm thick	m	0.30	35.19	1.94	3.73	40.86	9.264	-9.061
Z1.5.4.20	as above; Widths: 600mm; Blackboard, butt joints: 25mm thick	m	0.30	35.19	1.94	4.82	41.95	12.076	-11.008
Z1.5.4.21	as above; Widths: 600mm; Chipboard, butt joints: 18mm thick	m	0.25	29.32	1.61	3.30	34.23	7.405	-9.275
Z1.5.4.22	as above; Widths: 600mm; Chipboard, butt joints: 25mm thick	m	0.25	29.32	1.61	4.39	35.32	9.588	-11.304
Z1.5.4.23	as above; Widths: 600mm; Plywood, butt joints: 18mm thick	m	0.30	35.19	1.94	4.82	41.95	17.884	-9.892
Z1.5.4.24	as above; Widths: 600mm; Plywood, butt joints: 25mm thick	m	0.30	35.19	1.94	5.69	42.82	24.047	-12.161
Z1.5.4.25	as above; Widths: 600mm; Chipboard, faced both sides with white melamine; 15mm thick	m	0.35	41.06	2.26	3.37	46.69	6.954	-8.405
Z1.5.4.26	as above; Widths: 900mm; Softwood, wrought; slatted 25 x 50mm spaced at 75mm; 25mm thick	m	0.35	41.06	2.26	5.11	48.43	3.607	-12.653

Z1   Carpentry And Joinery continued...

	Unit	Labour Hours	Labour Net £	Plant Net £	Materials Net £	Unit Net £	CO₂e Kg	

Z1.5   Miscellaneous joinery

Z1.5.4   Shelves

	Unit	Labour Hours	Labour Net £	Plant Net £	Materials Net £	Unit Net £	CO_2e Kg	
Z1.5.4.27 as previous item; Widths: 900mm; Blackboard, butt joints: 18mm thick	m	0.35	41.06	2.26	5.60	48.92	13.411	-13.592
Z1.5.4.28 as above; Widths: 900mm; Blackboard, butt joints: 25mm thick	m	0.35	41.06	2.26	7.24	50.56	17.629	-16.512
Z1.5.4.29 as above; Widths: 900mm; Chipboard, butt joints: 18mm thick	m	0.30	35.19	1.94	4.95	42.08	10.744	-13.912
Z1.5.4.30 as above; Widths: 900mm; Chipboard, butt joints: 25mm thick	m	0.30	35.19	1.94	6.58	43.71	14.018	-16.956
Z1.5.4.31 as above; Widths: 900mm; Plywood, butt joints: 18mm thick	m	0.35	41.06	2.26	7.24	50.56	26.341	-14.837
Z1.5.4.32 as above; Widths: 900mm; Plywood, butt joints: 25mm thick	m	0.35	41.06	2.26	8.54	51.86	35.586	-18.241
Z1.5.4.33 as above; Widths: 900mm; Chipboard, faced both sides with white melamine; 15mm thick	m	0.40	46.92	2.58	5.06	54.56	9.825	-12.607

Z1.6   Units and fittings

Z1.6.1   Base units

	Unit	Labour Hours	Labour Net £	Plant Net £	Materials Net £	Unit Net £	CO_2e Kg	
Z1.6.1.01 Framed construction fixed to blockwork with screws, plugging; including ironmongery; plinth to base; no worktop; Finish: White Matt Melamine; Wall unit; single door, single shelf; 500 x 500 x 882mm high	Nr	1.00	117.30	6.45	37.04	160.79	28.237	-21.746
Z1.6.1.02 as above; single door, single shelf; 1000 x 500 x 882mm high	Nr	1.10	129.03	7.10	51.20	187.33	43.274	-35.278
Z1.6.1.03 as above; single door, single shelf; 500 x 300 x 600mm high	Nr	0.75	87.97	4.84	34.86	127.67	16.630	-12.081
Z1.6.1.04 as above; double door, single shelf; 1000 x 300 x 600mm high	Nr	1.00	117.30	6.45	45.75	169.50	27.198	-20.780
Z1.6.1.05 as above; double door, double shelf; 1000 x 300 x 900mm high	Nr	1.20	140.76	7.74	51.20	199.70	36.483	-28.512
Z1.6.1.06 as above; single door, single shelf; 500 x 500 x 882mm high	Nr	1.50	175.95	9.68	71.89	257.52	73.314	-24.317
Z1.6.1.07 as above; single door, single shelf; 1000 x 500 x 882mm high	Nr	1.60	187.68	10.32	111.11	309.11	114.888	-39.447
Z1.6.1.08 as above; single door, single shelf; 500 x 300 x 600mm high	Nr	1.50	175.95	9.68	69.72	255.35	43.964	-13.509
Z1.6.1.09 as above; double door, single shelf; 1000 x 300 x 600mm high	Nr	2.00	234.60	12.90	102.39	349.89	72.804	-23.236
Z1.6.1.10 as above; double door, double shelf; 1000 x 300 x 900mm high	Nr	2.50	293.25	16.13	118.73	428.11	98.709	-31.882
Z1.6.1.11 Framed construction fixed to blockwork with screws, plugging; including ironmongery; plinth to base; no worktop; Finish: Oak; Wall unit; single door, single shelf; 500 x 500 x 882mm high	Nr	2.00	234.60	12.90	82.79	330.29	75.739	-24.317
Z1.6.1.12 as above; single door, single shelf; 1000 x 500 x 882mm high	Nr	2.20	258.06	14.19	143.79	416.04	117.798	-39.447
Z1.6.1.13 as above; single door, single shelf; 500 x 300 x 600mm high	Nr	1.50	175.95	9.68	86.05	271.68	43.964	-13.509

Z1 Carpentry And Joinery continued...

	Unit	Labour Hours	Labour Net £	Plant Net £	Materials Net £	Unit Net £	CO₂e Kg	

		Unit	Labour Hours	Labour Net £	Plant Net £	Materials Net £	Unit Net £	CO₂e Kg

Z1.6 Units and fittings

Z1.6.1 Base units

		Unit	Labour Hours	Labour Net £	Plant Net £	Materials Net £	Unit Net £	CO₂e Kg
Z1.6.1.14	as previous item; double door, single shelf; 1000 x 300 x 600mm high	Nr	2.00	234.60	12.90	135.07	382.57	72.804 / -23.236
Z1.6.1.15	as above; double door, double shelf; 1000 x 300 x 900mm high	Nr	2.50	293.25	16.13	159.04	468.42	98.709 / -31.882

Z1.6.3 Work tops

		Unit	Labour Hours	Labour Net £	Plant Net £	Materials Net £	Unit Net £	CO₂e Kg
Z1.6.3.01	High density chipboard faced and lipped with melamine laminates; post formed edge; fixing with screws; Thickness: 25mm; 3000 x 500mm	Nr	0.25	29.32	1.61	32.68	63.61	20.702 / -18.122
Z1.6.3.02	as above; Thickness: 25mm; 3000 x 600mm	Nr	0.25	29.32	1.61	40.85	71.78	24.599 / -21.746
Z1.6.3.03	as above; Thickness: 25mm; 4100 x 600mm	Nr	0.25	29.32	1.61	55.83	86.76	33.174 / -29.720
Z1.6.3.04	as above; Thickness: 50mm; 3000 x 500mm	Nr	0.25	29.32	1.61	59.91	90.84	40.190 / -36.244
Z1.6.3.05	as above; Thickness: 50mm; 3000 x 600mm	Nr	0.25	29.32	1.61	65.36	96.29	47.986 / -43.493
Z1.6.3.06	as above; Thickness: 50mm; 4100 x 600mm	Nr	0.25	29.32	1.61	89.32	120.25	65.135 / -59.440

Z2 Insulation

Z2.1 Sheets

Z2.1.1 Floors

		Unit	Labour Hours	Labour Net £	Plant Net £	Materials Net £	Unit Net £	CO₂e Kg
Z2.1.1.01	Expanded polystyrene butt jointed, laid loose; to floors; Thickness: 25mm; Standard grade	m²	0.02	1.99	-	4.08	6.07	11.816
Z2.1.1.02	as above; Thickness: 25mm; Non-inflammable	m²	0.02	1.99	-	4.08	6.07	11.816
Z2.1.1.03	as above; Thickness: 50mm; Standard grade	m²	0.02	1.99	-	5.67	7.66	23.632
Z2.1.1.04	as above; Thickness: 50mm; Non-inflammable	m²	0.02	1.99	-	5.67	7.66	23.632

Z2.1.3 Walls

		Unit	Labour Hours	Labour Net £	Plant Net £	Materials Net £	Unit Net £	CO₂e Kg
Z2.1.3.01	Sheets; expanded polystyrene butt jointed, fixing with adhesive; to walls; Thickness: 25mm; Standard grade	m²	0.02	1.94	-	4.08	6.02	11.816
Z2.1.3.02	as above; Thickness: 25mm; Non-inflammable	m²	0.02	1.94	-	4.08	6.02	11.816
Z2.1.3.03	as above; Thickness: 50mm; Standard grade	m²	0.02	1.94	-	9.91	11.85	47.171
Z2.1.3.04	as above; Thickness: 50mm; Non-inflammable	m²	0.02	1.94	-	9.91	11.85	47.171

Z2.2 Quilts

Z2.2.1 Floors

		Unit	Labour Hours	Labour Net £	Plant Net £	Materials Net £	Unit Net £	CO₂e Kg
Z2.2.1.01	Glass fibre insulation; laid loose; to floors; Thickness: 80mm; Pilkingtons "Crown wool"	m²	0.02	1.99	-	3.18	5.17	26.819
Z2.2.1.02	as above; Thickness: 100mm; Pilkingtons "Crown wool"	m²	0.02	1.99	-	3.18	5.17	26.819

Z2 Insulation continued...

	Unit	Labour Hours	Labour Net £	Plant Net £	Materials Net £	Unit Net £	CO_2e Kg
Z2.2 **Quilts**							
Z2.2.1 Floors							
Z2.2.1.03 as previous item; Thickness: 150mm; Pilkingtons "Crown wool"	m²	0.02	1.99	-	4.83	6.82	40.229
Z2.2.3 Walls							
Z2.2.3.01 Boards; "Styrofoam", Floormate 500, butt jointed, fixing with insulation retaining ties; to walls; Thickness: 50mm; Cavity wall insulation	m²	0.03	3.76	-	10.96	14.72	57.874
Z2.2.3.02 as above; Thickness: 80mm; Cavity wall insulation	m²	0.05	5.70	-	12.43	18.13	48.514

Z3 Windows, Doors And Glazing

Note(s): Timber windows; softwood side hung casement windows without glazing bars consisting of frame, mullions, transom and 140mm wide softwood cill; opening lights and casements on rustproof hinges with casement stays and fasteners; fully glazed; frame bedded in cement mortar pointed one side with mastic; knotting and priming by manufacturer prior to delivery, no decoration. Double glazing: 14mm overall; 4mm clear float, 6mm air gap + 4mm clear float.

	Unit	Labour Hours	Labour Net £	Plant Net £	Materials Net £	Unit Net £	CO_2e Kg
Z3.1 **Timber**							
Z3.1.1 Windows							
Z3.1.1.01 Softwood side hung reversible windows without glazing bars consisting of frame, mullions, transom and softwood cill; opening lights and casements on rustproof hinges with casement stays and fasteners; fully glazed; frame bedded in cement mortar pointed one side with mastic; knotting and priming by manufacturer prior to delivery, no decoration; 1200 x 1200mm; Jeld Wen reference number: LEC1212CFR	Nr	1.45	27.01	-	581.29	608.30	30.529
Z3.1.1.02 as above; 1800 x 1500mm; Jeld Wen reference number: LEC1815CFCR	Nr	2.05	38.19	-	1,007.10	1,045.29	56.756
Z3.1.1.03 Softwood top hung reversible windows without glazing bars; softwood cill; opening casements on rustproof hinges with casement stay and fastener; fully glazed in double glazing units; frame bedded in cement mortar pointed one side with mastic; knotting and priming by manufacturer prior to delivery, no decoration; 600 x 900 mm; Jeld Wen reference number: LEC0609AR	Nr	1.00	18.63	-	384.20	402.83	11.833
Z3.1.1.04 as above; 600 x 1200 mm; Jeld Wen reference number: LEC0612AR	Nr	1.00	18.63	-	423.56	442.19	15.572

Z3 Windows, Doors And Glazing continued...

	Unit	Labour Hours	Labour Net £	Plant Net £	Materials Net £	Unit Net £	CO$_2$e Kg	
Z3.1	**Timber**							
Z3.1.1	Windows							
Z3.1.1.05	as previous item; 900 x 750 mm; Jeld Wen reference number: LEC0907AR	Nr	1.00	18.63	-	405.96	424.59	14.637	
Z3.1.1.06	as above; 1200 x 1200 mm; Jeld Wen reference number: LEC1212AFR	Nr	1.00	18.63	-	663.49	682.12	30.529	
Z3.1.3	Doors							
Z3.1.3.01	Softwood interior flush doors; including frame and ironmongery; no decoration; Sizes: Single; 35mm thick: hardboard faced	Nr	1.50	27.95	-	95.38	123.33	28.152	-78.921
Z3.1.3.02	as above: Single; 35mm thick: sapele faced	Nr	1.50	27.95	-	117.53	145.48	28.152	-78.921
Z3.1.3.03	as above: Single; 40mm thick: hardboard faced	Nr	1.50	27.95	-	94.22	122.17	33.148	-91.238
Z3.1.3.04	as above: Single; 40mm thick: sapele faced	Nr	1.50	27.95	-	121.12	149.07	33.148	-91.238
Z3.1.3.05	as above: Double; 35mm thick: hardboard faced	Nr	1.65	30.74	-	190.18	220.92	56.144	-157.842
Z3.1.3.06	as above: Double; 35mm thick: sapele faced	Nr	1.65	30.74	-	234.48	265.22	56.144	-157.842
Z3.1.3.07	as above: Double; 40mm thick: hardboard faced	Nr	1.65	30.74	-	187.86	218.60	66.136	-182.476
Z3.1.3.08	as above: Double; 40mm thick: sapele faced	Nr	1.65	30.74	-	241.66	272.40	66.136	-182.476
Z3.1.3.09	Interior panelled doors; including frame and ironmongery; no decoration; Sizes: Single; Softwood	Nr	1.65	30.74	-	132.91	163.65	28.763	-82.343
Z3.1.3.10	as above: Double; Softwood	Nr	1.85	34.47	-	265.24	299.71	57.366	-164.686
Z3.2	**Metal**							
Z3.2.3	Doors							
Z3.2.3.01	Galvanised steel roller shutter doors; two hour rated fire resistance; including installation; Sizes: Single; Bolton Gate Co Fireroll E240 Rolling Shutters; opening sizes: 3050mm high, 3000mm wide	Nr	4.55	84.77	-	2,925.57	3,010.34	484.680	
Z3.2.3.02	as above; opening sizes: 4800mm high, 5000mm wide	Nr	5.50	102.46	-	6,851.46	6,953.92	1,292.479	
Z3.2.3.03	Galvanised steel roller shutter doors; two hour rated fire resistance; including installation; Sizes: Double; Bolton Gate Co Fireroll E240 Rolling Shutters; opening sizes: 3050mm high, 3000mm wide	Nr	5.55	103.40	-	3,134.54	3,237.94	514.054	
Z3.2.3.04	as above; opening sizes: 4800mm high, 5000mm wide	Nr	6.50	121.10	-	7,673.63	7,794.73	1,351.228	

Z3 Windows, Doors And Glazing continued...

		Unit	Labour Hours	Labour Net £	Plant Net £	Materials Net £	Unit Net £	CO$_2$e Kg	
Z3.3	**Plastics**								
Z3.3.1	Windows								
Z3.3.1.01	White UPVC tilt and turn windows; including ironmongery; 24mm low E factory double glazed units; clear glass; Frame fixed to masonry with screws, plugging; 620 x 1050mm	Nr	1.00	18.63	-	107.16	125.79	44.423	
Z3.3.1.02	as above; 1200 x 1200mm	Nr	1.55	28.88	-	190.42	219.30	98.386	
Z3.3.1.03	as above; 1200 x 1500mm	Nr	1.80	33.53	-	222.88	256.41	122.929	
Z3.3.6	Roof lights								
Z3.3.6.01	Cox Trade rooflights; GRP factory glazed double skin polycarbonate dome; fixed to concrete with screws, plugging; Rooflight sizes: 600 x 600mm; Non-ventilating; base frame	Nr	1.00	18.63	-	253.82	272.45	26.800	
Z3.3.6.02	as above; Rooflight sizes: 600 x 600mm; Ventilating; base frame	Nr	1.25	23.29	-	288.99	312.28	33.026	
Z3.3.6.03	as above; Rooflight sizes: 600 x 600mm; Non-ventilating; base frame and kerb	Nr	1.00	18.63	-	502.03	520.66	37.683	
Z3.3.6.04	as above; Rooflight sizes: 600 x 600mm; Ventilating; base frame and kerb	Nr	1.25	23.29	-	537.20	560.49	43.909	
Z3.3.6.05	as above; Rooflight sizes: 950 x 950mm; Non-ventilating; base frame	Nr	1.00	18.63	-	399.12	417.75	30.924	
Z3.3.6.06	as above; Rooflight sizes: 950 x 950mm; Ventilating; base frame	Nr	1.25	23.29	-	434.29	457.58	37.150	
Z3.3.6.07	as above; Rooflight sizes: 950 x 950mm; Non-ventilating; base frame and kerb	Nr	1.00	18.63	-	665.64	684.27	48.158	
Z3.3.6.08	as above; Rooflight sizes: 950 x 950mm; Ventilating; base frame and kerb	Nr	1.25	23.29	-	700.81	724.10	54.384	
Z3.3.6.09	as above; Rooflight sizes: 1200 x 1200mm; Non-ventilating; base frame	Nr	1.25	23.29	-	523.79	547.08	47.416	
Z3.3.6.10	as above; Rooflight sizes: 1200 x 1200mm; Ventilating; base frame	Nr	1.50	27.95	-	558.96	586.91	53.642	
Z3.3.6.11	as above; Rooflight sizes: 1200 x 1200mm; Non-ventilating; base frame and kerb	Nr	1.25	23.29	-	815.04	838.33	69.182	
Z3.3.6.12	as above; Rooflight sizes: 1200 x 1200mm; Ventilating; base frame and kerb	Nr	1.50	27.95	-	850.21	878.16	75.408	
Z3.5	**Glazing**								
Z3.5.1	Glass								
Z3.5.1.01	Glazing; clear sheet glass; in panes not exceeding 4m^2; Thickness: 4mm; To timber with bradded wood beads	m^2	0.65	11.82	-	46.97	58.79	12.170	-0.607
Z3.5.1.02	as above; Thickness: 4mm; To metal with screwed metal beads	m^2	0.65	11.82	-	48.13	59.95	15.026	-0.091
Z3.5.1.03	as above; Thickness: 6mm; To timber with bradded wood beads	m^2	0.65	11.82	-	67.67	79.49	17.556	-0.607

Z3 Windows, Doors And Glazing continued...

	Unit	Labour Hours	Labour Net £	Plant Net £	Materials Net £	Unit Net £	CO_2e Kg	
Z3.5	**Glazing**							
Z3.5.1	**Glass**							
Z3.5.1.04	as previous item; Thickness: 6mm; To metal with screwed metal beads	m²	0.65	11.82	-	68.83	80.65	20.412	-0.091
Z3.5.1.05	Glazing; rough cast patterned glass; in panes not exceeding 4m²; Thickness: 4mm; To timber with bradded wood beads	m²	0.65	11.82	-	50.12	61.94	13.541	-0.607
Z3.5.1.06	as above; Thickness: 4mm; To metal with screwed metal beads	m²	0.65	11.82	-	51.28	63.10	16.397	-0.091
Z3.5.1.07	as above; Thickness: 6mm; To timber with bradded wood beads	m²	0.65	11.82	-	76.28	88.10	19.613	-0.607
Z3.5.1.08	as above; Thickness: 6mm; To metal with screwed metal beads	m²	0.65	11.82	-	77.44	89.26	22.469	-0.091
Z3.5.1.09	Glazing; Georgian wired cast glass; in panes not exceeding 4m²; Thickness: 7mm; To timber with bradded wood beads	m²	0.65	11.82	-	52.74	64.56	22.648	-0.607
Z3.5.1.10	as above; To metal with screwed metal beads	m²	0.65	11.82	-	53.90	65.72	25.504	-0.091
Z3.5.3	**Special glass**							
Z3.5.3.01	Glazing; toughened safety glass; in panes not exceeding 4m²; Thickness: 4mm; To timber with bradded wood beads: pane size 1.00m²	Nr	0.65	11.82	-	71.36	83.18	13.541	-0.607
Z3.5.3.02	as above; Thickness: 4mm; To metal with screwed metal beads: pane size 1.00m²	Nr	0.65	11.82	-	72.52	84.34	16.397	-0.091
Z3.5.3.03	as above; Thickness: 6mm; To timber with bradded wood beads: pane size 1.00m²	Nr	0.80	14.55	-	105.93	120.48	19.613	-0.607
Z3.5.3.04	as above; Thickness: 6mm; To metal with screwed metal beads: pane size 1.00m²	Nr	0.80	14.55	-	107.09	121.64	22.469	-0.091
Z3.5.3.05	Glazing; clear laminated safety glass; in panes not exceeding 4m²; Thickness: 4.4mm; To timber with bradded wood beads; pane size	Nr	0.80	14.55	-	73.79	88.34	13.541	-0.607
Z3.5.3.06	as above; Thickness: 6.4mm; To timber with bradded wood beads; pane size 1.00m²	Nr	0.80	14.55	-	85.27	99.82	20.827	-0.607
Z3.5.3.07	Glazing; Lexan Exell "D" Clear-112 safety glass; extruded polycarbonate sheet, proprietary ultraviolet resistant surface on one side; to metal with screwed metal beads; Thickness: 5mm; Pane size 1.00m²	Nr	0.75	13.64	-	313.78	327.42	19.433	-0.091
Z3.5.3.08	as above; Thickness: 5mm; Pane size 2.00m²	Nr	1.05	19.10	-	627.04	646.14	37.547	-0.091
Z3.5.3.09	as above; Thickness: 6mm; Pane size 1.00m²	Nr	0.75	13.64	-	360.34	373.98	22.469	-0.091
Z3.5.3.10	as above; Thickness: 6mm; Pane size 2.00m²	Nr	1.05	19.10	-	720.16	739.26	43.619	-0.091

Z3   Windows, Doors And Glazing continued...

	Unit	Labour Hours	Labour Net £	Plant Net £	Materials Net £	Unit Net £	CO₂e Kg	

Z3.5   Glazing

Z3.5.4   Hermetically sealed units

	Unit	Labour Hours	Labour Net £	Plant Net £	Materials Net £	Unit Net £	CO_2e Kg	
Z3.5.4.01 Glazing; clear float double glazed units; to metal with screwed metal beads; nominal airspace 6mm; Thickness: 2 x 4mm; Unit size: 0.50m²	Nr	0.80	14.55	-	53.97	68.52	14.145	-0.091
Z3.5.4.02 as above; Thickness: 2 x 4mm; Unit size: 1.00m²	Nr	0.94	17.10	-	106.42	123.52	26.193	-0.118
Z3.5.4.03 as above; Thickness: 2 x 4mm; Unit size: 2.00m²	Nr	1.05	19.10	-	210.82	229.92	49.862	-0.182
Z3.5.4.04 as above; Thickness: 2 x 4mm; Unit size: 3.00m²	Nr	1.15	20.92	-	314.60	335.52	72.343	-0.182
Z3.5.4.05 as above; Thickness: 2 x 6mm; Unit size: 0.50m²	Nr	0.80	14.55	-	74.51	89.06	19.531	-0.091
Z3.5.4.06 as above; Thickness: 2 x 6mm; Unit size: 1.00m²	Nr	0.94	17.10	-	147.50	164.60	36.965	-0.118
Z3.5.4.07 as above; Thickness: 2 x 6mm; Unit size: 2.00m²	Nr	1.05	19.10	-	292.98	312.08	71.406	-0.182
Z3.5.4.08 as above; Thickness: 2 x 6mm; Unit size: 3.00m²	Nr	1.15	20.92	-	437.84	458.76	104.659	-0.182
Z3.5.4.09 as above; Thickness: 2 x 10mm; Unit size: 0.50m²	Nr	0.80	14.55	-	134.24	148.79	30.302	-0.091
Z3.5.4.10 as above; Thickness: 2 x 10mm; Unit size: 1.00m²	Nr	0.94	17.10	-	266.96	284.06	58.507	-0.118
Z3.5.4.11 as above; Thickness: 2 x 10mm; Unit size: 2.00m²	Nr	1.05	19.10	-	531.90	551.00	114.490	-0.182
Z3.5.4.12 as above; Thickness: 2 x 10mm; Unit size: 3.00m²	Nr	1.15	20.92	-	796.22	817.14	169.285	-0.182
Z3.5.4.13 Glazing; solar control double glazed units; to timber with bradded wood beads; bronze float glass to outside, clear glass to inside; nominal airspace 12mm; Thickness: 2 x 4mm; Unit size: 0.50m²	Nr	0.80	14.55	-	49.82	64.37	12.147	-0.457
Z3.5.4.14 as above; Thickness: 2 x 4mm; Unit size: 1.00m²	Nr	0.94	17.10	-	98.56	115.66	23.337	-0.634
Z3.5.4.15 as above; Thickness: 2 x 4mm; Unit size: 2.00m²	Nr	1.05	19.10	-	195.79	214.89	45.835	-0.910
Z3.5.4.16 as above; Thickness: 2 x 4mm; Unit size: 3.00m²	Nr	1.15	20.92	-	292.50	313.42	67.402	-1.075
Z3.5.4.17 as above; Thickness: 2 x 6mm; Unit size: 0.50m²	Nr	0.80	14.55	-	70.44	84.99	17.533	-0.457
Z3.5.4.18 as above; Thickness: 2 x 6mm; Unit size: 1.00m²	Nr	0.94	17.10	-	139.80	156.90	34.109	-0.634
Z3.5.4.19 as above; Thickness: 2 x 6mm; Unit size: 2.00m²	Nr	1.05	19.10	-	278.27	297.37	67.379	-0.910
Z3.5.4.20 as above; Thickness: 2 x 6mm; Unit size: 3.00m²	Nr	1.15	20.92	-	416.22	437.14	99.718	-1.075
Z3.5.4.21 as above; Thickness: 2 x 10mm; Unit size: 0.50m²	Nr	0.90	16.37	-	130.31	146.68	28.304	-0.457
Z3.5.4.22 as above; Thickness: 2 x 10mm; Unit size: 1.00m²	Nr	1.05	19.10	-	259.53	278.63	55.651	-0.634
Z3.5.4.23 as above; Thickness: 2 x 10mm; Unit size: 2.00m²	Nr	1.25	22.74	-	517.73	540.47	110.463	-0.910

Z3   Windows, Doors And Glazing continued...

	Unit	Labour Hours	Labour Net £	Plant Net £	Materials Net £	Unit Net £	CO_2e Kg	
Z3.5	**Glazing**								
Z3.5.4	Hermetically sealed units							
Z3.5.4.24	as previous item; Thickness: 2 x 10mm; Unit size: 3.00m^2	Nr	1.50	27.29	-	775.41	802.70	164.344	*-1.075*
Z3.5.5	Mirrors							
Z3.5.5.01	Silver backed 6mm clear glass; fixing to masonry with chromium headed screws, plugging; Edges polished; 450 x 300mm	Nr	0.75	13.97	-	18.42	32.39	2.461	
Z3.5.5.02	as above; Edges polished; 900 x 400mm	Nr	0.90	16.77	-	49.28	66.05	6.559	
Z3.5.5.03	as above; Edges polished; 1500 x 1500mm	Nr	2.00	37.26	-	203.50	240.76	40.983	
Z3.5.5.04	as above; Edges bevelled; 450 x 300mm	Nr	0.75	13.97	-	19.38	33.35	2.461	
Z3.5.5.05	as above; Edges bevelled; 900 x 400mm	Nr	0.90	16.77	-	51.87	68.64	6.559	
Z3.5.5.06	as above; Edges bevelled; 1500 x 1500mm	Nr	2.00	37.26	-	214.21	251.47	40.983	
Z3.6	**Patent glazing**								
Z3.6.1	Roofs							
Z3.6.1.01	2000mm long aluminium alloy bars at 600mm centres; Georgian wired polished glazing 6mm thick	m^2	1.93	114.36	-	220.28	334.64	46.710	
Z3.6.2	Opening lights							
Z3.6.2.01	2000mm long aluminium alloy bars at 600mm centres; Georgian wired polished glazing 6mm thick; 600 x 900mm; electric linear motor gear	Nr	2.01	119.12	-	715.95	835.07	20.129	
Z3.6.3	Vertical surfaces							
Z3.6.3.01	2000mm long aluminium alloy bars at 600mm centres; Georgian wired polished glazing 6mm thick	m^2	2.25	133.43	-	220.57	354.00	47.012	

Z4   Surface Finishes, Linings And Partitions

	Unit	Labour Hours	Labour Net £	Plant Net £	Materials Net £	Unit Net £	CO_2e Kg
Z4.1	**In situ finishes, beds and backings**							
Z4.1.1	Floors							
Z4.1.1.01	Trowelled finish, to concrete base; Thickness: 25mm; Cement and sand (1:3); surfaces of width: over 1m wide	m^2	0.04	1.21	0.73	6.08	8.02	7.424
Z4.1.1.02	as above; Thickness: 25mm; Cement and sand (1:3); surfaces of width: not exceeding 300mm	m	0.05	1.52	0.91	1.89	4.32	2.774
Z4.1.1.03	as above; Thickness: 25mm; Granolithic; surfaces of width: over 1m wide	m^2	0.04	1.21	0.73	12.20	14.14	13.525
Z4.1.1.04	as above; Thickness: 25mm; Granolithic; surfaces of width: not exceeding 300mm	m	0.05	1.52	0.91	3.80	6.23	4.672

Z4 Surface Finishes, Linings And Partitions continued...

	Unit	Labour Hours	Labour Net £	Plant Net £	Materials Net £	Unit Net £	CO$_2$e Kg
Z4.1	**In situ finishes, beds and backings**						
Z4.1.1	Floors						
Z4.1.1.05	as previous item; Thickness: 32mm; Cement and sand (1:3); surfaces of width: over 1m wide	m^2	0.04	1.21	0.73	7.84	9.78	9.425
Z4.1.1.06	as above; Thickness: 32mm; Cement and sand (1:3); surfaces of width: not exceeding 300mm	m	0.05	1.52	0.91	2.30	4.73	3.236
Z4.1.1.07	as above; Thickness: 32mm; Granolithic; surfaces of width: over 1m wide	m^2	0.04	1.21	0.73	15.73	17.67	17.289
Z4.1.1.08	as above; Thickness: 32mm; Granolithic; surfaces of width: not exceeding 300mm	m	0.05	1.52	0.91	4.61	7.04	5.541
Z4.1.1.09	as above; Thickness: 38mm; Cement and sand (1:3); surfaces of width: over 1m wide	m^2	0.04	1.21	0.73	9.19	11.13	10.965
Z4.1.1.10	as above; Thickness: 38mm; Cement and sand (1:3); surfaces of width: not exceeding 300mm	m	0.04	1.21	0.73	2.70	4.64	3.574
Z4.1.1.11	as above; Thickness: 38mm; Granolithic; surfaces of width: over 1m wide	m^2	0.04	1.21	0.73	15.73	17.67	17.289
Z4.1.1.12	as above; Thickness: 38mm; Granolithic; surfaces of width: not exceeding 300mm	m	0.05	1.52	0.91	5.42	7.85	6.410
Z4.1.1.13	as above; Thickness: 50mm; Cement and sand (1:3); surfaces of width: over 1m wide	m^2	0.05	1.52	0.91	12.17	14.60	14.476
Z4.1.1.14	as above; Thickness: 50mm; Cement and sand (1:3); surfaces of width: not exceeding 300mm	m	0.07	2.03	1.22	3.65	6.90	4.986
Z4.1.1.15	as above; Thickness: 50mm; Granolithic; surfaces of width: over 1m wide	m^2	0.05	1.52	0.91	24.40	26.83	26.679
Z4.1.1.16	as above; Thickness: 50mm; Granolithic; surfaces of width: not exceeding 300mm	m	0.07	2.03	1.22	7.32	10.57	8.647
Z4.1.3	Walls						
Z4.1.3.01	Rough case external render; cement and sand (1:3); trowelled finish with dry pebble dash coating, to brickwork base; surfaces of width; Thickness: Two coat 15mm; over 1m wide	m^2	0.33	10.10	0.93	4.43	15.46	4.965
Z4.1.3.02	as above; not exceeding 300mm	m	0.17	5.07	0.47	1.34	6.88	1.601
Z4.1.3.03	In situ finishes; plaster; pre-mixed Carlite; two coats 13mm thick; comprising 11mm floating coat of browning; 2mm finishing coat; steel trowelled; surfaces of width; Work to brickwork base; over 1m wide	m^2	0.17	5.07	0.47	6.19	11.73	2.877
Z4.1.3.04	as above; Work to brickwork base; not exceeding 300mm	m	0.13	3.79	0.35	1.71	5.85	0.949
Z4.1.3.05	as above; Work to metal lathing base; over 1m wide	m^2	0.20	6.07	0.56	6.19	12.82	2.936
Z4.1.3.06	as above; Work to metal lathing base; not exceeding 300mm	m	0.10	3.03	0.28	1.71	5.02	0.904

Z4 Surface Finishes, Linings And Partitions continued...

	Unit	Labour Hours	Labour Net £	Plant Net £	Materials Net £	Unit Net £	CO$_2$e Kg	
Z4.1	**In situ finishes, beds and backings**							
Z4.1.4	Soffits							
Z4.1.4.01	In situ finishes; plaster; Thistle Universal one coat 13mm thick; steel trowelled; Work to metal lathing base; surfaces of width: over 1m	m²	0.10	3.03	0.28	8.08	11.39	2.676	
Z4.1.4.02	as above; surfaces of width; surfaces of width: not exceeding 300mm	m	0.07	2.03	0.19	2.34	4.56	0.845	
Z4.2	**Tiles**							
Z4.2.1	Floors							
Z4.2.1.01	Butt joints straight both ways; fixing with adhesive to cement and sand base; Thickness: 2mm; Vinyl, Marleyflex Plus, 300 x 300mm units; surfaces of width: over 1m wide	m²	0.25	7.59	-	26.68	34.27	5.450	
Z4.2.1.02	as above; Thickness: 2mm; Vinyl, Marleyflex Plus, 300 x 300mm units; surfaces of width: not exceeding 300mm	m	0.13	3.79	-	8.88	12.67	1.815	
Z4.2.1.03	as above; Thickness: 2.5mm; Vinyl, Marleyflex Plus, 300 x 300mm units; surfaces of width: over 1m wide	m²	0.25	7.59	-	31.01	38.60	6.564	
Z4.2.1.04	as above; Thickness: 2.5mm; Vinyl, Marleyflex Plus, 300 x 300mm units; surfaces of width: not exceeding 300mm	m	0.13	3.79	-	10.33	14.12	2.186	
Z4.2.1.05	as above; Thickness: 2mm; Vinyl; cork 300 x 300mm units; Thickness: 3.2mm; surfaces of width: over 1m wide	m²	0.13	3.79	-	18.77	22.56	4.607	-2.974
Z4.2.1.06	as above; Thickness: 2mm; Vinyl; cork 300 x 300mm units; Thickness: 3.2mm; surfaces of width: not exceeding 300mm	m	0.13	3.79	-	6.25	10.04	1.534	-0.990
Z4.2.1.07	as above; Thickness: 2mm; Vinyl; cork 300 x 300mm units; Thickness: 4.5mm; surfaces of width: over 1m wide	m²	0.13	3.79	-	21.57	25.36	6.187	-4.275
Z4.2.1.08	as above; Thickness: 2mm; Vinyl; cork 300 x 300mm units; Thickness: 4.5mm; surfaces of width: not exceeding 300mm	m	0.13	3.79	-	7.18	10.97	2.060	-1.424
Z4.2.1.09	as above; Thickness: 2mm; Vinyl; cork 300 x 300mm units; Thickness: 6mm; surfaces of width: over 1m wide	m²	0.25	7.59	-	23.23	30.82	7.767	-5.576
Z4.2.1.10	as above; Thickness: 2mm; Vinyl; cork 300 x 300mm units; Thickness: 6mm; surfaces of width: not exceeding 300mm	m	0.13	3.79	-	7.74	11.53	2.586	-1.857
Z4.2.1.11	as above; Thickness: 2mm; Vinyl; rubber Polysafe stud tile 500 x 500mm units; Thickness: 4mm; surfaces of width: over 1m wide	m²	0.40	12.14	-	37.03	49.17	15.367	

Z4 Surface Finishes, Linings And Partitions continued...

	Unit	Labour Hours	Labour Net £	Plant Net £	Materials Net £	Unit Net £	CO_2e Kg
Z4.2 **Tiles**							
Z4.2.1 Floors							
Z4.2.1.12 as previous item; Thickness: 2mm; Vinyl; rubber Polysafe stud tile 500 x 500mm units; Thickness: 4mm; surfaces of width: not exceeding 300mm	m	0.20	6.07	-	12.33	18.40	5.117
Z4.2.3 Walls							
Z4.2.3.01 White glazed ceramic, butt joints straight both ways; fixing with adhesive on plaster base; pointing with white cement; Size: 108 x 108 x 4mm; surfaces of width: over 1m wide	m²	0.50	15.33	-	11.41	26.74	7.582
Z4.2.3.02 as above; Size: 108 x 108 x 4mm; surfaces of width: not exceeding 300mm	m	0.25	7.67	-	3.80	11.47	2.525
Z4.2.3.03 as above; Size: 152 x 152 x 5.5mm; surfaces of width: over 1m wide	m²	0.50	15.33	-	15.99	31.32	9.654
Z4.2.3.04 as above; Size: 152 x 152 x 5.5mm; surfaces of width: not exceeding 300mm	m	0.25	7.67	-	5.32	12.99	3.215
Z4.2.3.05 Flexible sheet coverings; to floors; linoleum sheet; 3.2mm thick; butt joints; fixing with adhesive to cement and sand base; Plain; surfaces of width: over 1m wide	m²	0.40	12.14	-	28.94	41.08	4.010
Z4.2.3.06 as above; Plain; surfaces of width: not exceeding 300mm	m	0.20	6.07	-	9.64	15.71	1.335
Z4.2.3.07 as above; Marbled patterns; surfaces of width: over 1m wide	m²	0.40	12.14	-	29.62	41.76	4.010
Z4.2.3.08 as above; surfaces of width: not exceeding 300mm	m	0.20	6.07	-	9.86	15.93	1.335
Z4.2.3.09 Flexible sheet coverings; to floors; fitted carpeting and underlay; heavy duty contract grade carpet; Plain; surfaces of width: over 1m wide	m²	0.40	12.14	-	32.26	44.40	12.402
Z4.2.3.10 as above; Plain; surfaces of width: not exceeding 300mm	m	0.20	6.07	-	10.74	16.81	4.130
Z4.2.3.11 as above; Patterned; surfaces of width: over 1m wide	m²	0.40	12.14	-	32.26	44.40	12.402
Z4.2.3.12 as above; Patterned; surfaces of width: not exceeding 300mm	m	0.20	6.07	-	10.74	16.81	4.130
Z4.2.3.13 Dry partitions and linings; walls; gypsum plasterboard; tapered edges fixed with galvanised nails to softwood studs; compound dabs, joints filled taped and finished flush; no decoration; Thickness: 9.5mm; surfaces of width: over 1m wide	m²	0.05	5.87	0.32	2.00	8.19	3.497
Z4.2.3.14 as above; Thickness: 9.5mm; surfaces of width: not exceeding 300mm	m	0.03	2.93	0.16	0.67	3.76	1.206
Z4.2.3.15 as above; Thickness: 12.5mm; surfaces of width: over 1m wide	m²	0.05	5.87	0.32	2.07	8.26	4.480

Z4   Surface Finishes, Linings And Partitions continued...

	Unit	Labour Hours	Labour Net £	Plant Net £	Materials Net £	Unit Net £	CO_2e Kg
Z4.2	**Tiles**						
Z4.2.3	**Walls**						
Z4.2.3.16	as previous item; Thickness: 12.5mm; surfaces of width: not exceeding 300mm	m	0.03	2.93	0.16	0.69	3.78	1.533
Z4.2.4	**Soffits**						
Z4.2.4.01	Dry partitions and linings; walls; gypsum plasterboard; tapered edges fixed with galvanised nails to softwood studs; compound dabs, joints filled taped and finished flush; no decoration; Thickness: 9.5mm; surfaces of width: over 1m wide	m²	0.07	7.86	0.43	2.00	10.29	3.579
Z4.2.4.02	as above; Thickness: 9.5mm; surfaces of width: not exceeding 300mm	m	0.03	3.87	0.21	0.67	4.75	1.244
Z4.2.4.03	as above; Thickness: 12.5mm; surfaces of width: over 1m wide	m²	0.07	7.86	0.43	2.07	10.36	4.562
Z4.2.4.04	as above; Thickness: 12.5mm; surfaces of width: not exceeding 300mm	m	0.03	3.87	0.21	0.69	4.77	1.572
Z4.5	**Suspended ceilings**						
Z4.5.1	**Depth of suspension system: not exceeding 150mm**						
Z4.5.1.01	Armstrong Microlook Suspended Ceiling system, galvanised steel comprising; runners and cross tees; runners at 1200mm centres fixed to concrete soffit on hangers at 1250 mm centres; including perimeter trim fixed to walls at 450mm intervals; Circus 600	m²	0.16	18.77	1.03	40.33	60.13	9.944
Z4.5.1.02	Gyproc M/F suspended ceiling system, comprising 12.7mm thick Gyproc wallboard on metal suspension grid; including perimeter trim fixed to walls, strap hangers fixed to concrete soffit, primary support channels and cleats; no decoration; 900 x 1800 x 12.7m	m²	0.12	14.08	0.77	11.60	26.45	15.776
Z4.5.2	**Depth of suspension system: 150 - 500mm**						
Z4.5.2.01	Armstrong Microlook Suspended Ceiling system, galvanised steel comprising; runners and cross tees; runners at 1200mm centres fixed to concrete soffit on hangers at 1250mm centres; including perimeter trim fixed to walls at 450mm intervals; Circus 600 x 60	m²	0.17	19.94	1.10	40.33	61.37	9.992

Z4 Surface Finishes, Linings And Partitions continued...

	Unit	Labour Hours	Labour Net £	Plant Net £	Materials Net £	Unit Net £	CO$_2$e Kg	

Z4.5 Suspended ceilings

Z4.5.2 Depth of suspension system: 150 - 500mm

Z4.5.2.02 Gyproc M/F suspended ceiling system, comprising 12.7mm thick Gyproc wallboard on metal suspension grid; including perimeter trim fixed to walls, strap hangers fixed to concrete soffit, primary support channels and cleats; no decoration; 900 x 1800 x 12.7m

	m^2	0.13	15.25	0.84	11.60	27.69	15.824

Z4.5.3 Depth of suspension system: 500 - 1000mm

Z4.5.3.01 Armstrong Microlook Suspended Ceiling system, galvanised steel comprising; runners and cross tees; runners at 1200mm centres fixed to concrete soffit on hangers at 1250mm centres; including perimeter trim fixed to walls at 450mm intervals; Circus 600 x 60

	m^2	0.18	21.11	1.16	40.33	62.60	10.041

Z4.5.3.02 Gyproc M/F suspended ceiling system, comprising 12.7mm thick Gyproc wallboard on metal suspension grid; including perimeter trim fixed to walls, strap hangers fixed to concrete soffit, primary support channels and cleats; no decoration; 900 x 1800 x 12.7m

	m^2	0.14	16.42	0.90	11.60	28.92	15.873

Z4.5.4 Bulkheads

Z4.5.4.01 Gyproc M/F suspended ceiling system, comprising 12.7mm thick Gyproc wallboard on metal suspension grid; including perimeter trim fixed to walls, strap hangers fixed to concrete soffit, primary support channels and cleats; no decoration; Vertical; 1000mm deep

	m	0.20	23.46	1.29	19.49	44.24	24.694

Z4.6 Raised access floors

Z4.6.0 Generally

		Unit	Labour Hours	Labour Net £	Plant Net £	Materials Net £	Unit Net £		CO$_2$e Kg
Z4.6.0.01	Pedestal height: 300mm; Diamond 600 medium grade panels	m^2	0.24	28.15	1.55	77.70	107.40	69.900	-17.430
Z4.6.0.02	as above heavy grade panels	m^2	0.24	28.15	1.55	84.46	114.16	77.095	-19.678
Z4.6.0.03	as above medium grade panels	m^2	0.24	28.15	1.55	81.06	110.76	74.835	-17.430
Z4.6.0.04	as above heavy grade panels	m^2	0.24	28.15	1.55	87.82	117.52	82.030	-19.678

Z4 Surface Finishes, Linings And Partitions continued...

	Unit	Labour Hours	Labour Net £	Plant Net £	Materials Net £	Unit Net £	CO$_2$e Kg	

Z4.7 Proprietary system partitions

Z4.7.1 Solid

		Unit	Labour Hours	Labour Net £	Plant Net £	Materials Net £	Unit Net £	CO$_2$e Kg	
Z4.7.1.01	Paramount dry partitions, with timber supports at vertical joints; tapered edges fixed with galvanised nails to softwood studs and floor and ceiling battens; joints filled, taped and finished flush; no decoration; Thickness: 57mm; Height of partition; 2100mm	m	0.50	58.65	3.23	13.46	75.34	36.489	-5.125
Z4.7.1.02	as above; Thickness: 57mm; Height of partition: 2400mm	m	0.75	87.97	4.84	15.19	108.00	42.480	-5.519
Z4.7.1.03	as above; Thickness: 57mm; Height of partition: 2700mm	m	1.00	117.30	6.45	16.91	140.66	48.473	-5.913
Z4.7.1.04	as above; Thickness: 63mm; Height of partition: 2100mm	m	0.50	58.65	3.23	21.26	83.14	44.156	-10.234
Z4.7.1.05	as above; Thickness: 63mm; Height of partition: 2400mm	m	1.00	117.30	6.45	23.42	147.17	52.170	-10.234
Z4.7.1.06	as above; Thickness: 63mm; Height of partition: 2700mm	m	1.00	117.30	6.45	26.05	149.80	57.958	-11.021

Z4.8 Framed panel cubicle sets

		Unit	Labour Hours	Labour Net £	Plant Net £	Materials Net £	Unit Net £	CO$_2$e Kg	
Z4.8.0	Komfort'; melamine faced high density chipboard doors, walls and dividers; lipped all round; thickness 18mm; including fibreboard lacquered pilaster, capping rails, aluminium wall fixings, channels, brackets and hinges; Blue / grey marble effect finish								
Z4.8.0.01	Sets made up from a combination of the following units: Dividers 1950 x 1500mm	Nr	0.50	58.65	3.23	202.76	264.64	53.804	-47.775
Z4.8.0.02	as above; Doors 1800 x 750mm	Nr	0.50	58.65	3.23	193.75	255.63	15.252	
Z4.8.0.03	as above; Doors 1800 x 950mm	Nr	0.50	58.65	3.23	168.07	229.95	12.506	-63.810
Z4.8.0.04	as above; Pilaster 1810 x 97mm	Nr	0.25	29.32	1.61	61.63	92.56	2.411	
Z4.8.0.05	as above; Pilaster 1810 x 110mm	Nr	0.25	29.32	1.61	79.09	110.02	2.571	
Z4.8.0.06	as above; Pilaster 1810 x 214mm	Nr	0.25	29.32	1.61	99.93	130.86	3.856	
Z4.8.0.07	as above; Head rails 1000mm	Nr	0.13	14.66	0.81	5.28	20.75	1.406	-0.743
Z4.8.0.08	as above; Head rails 2000mm	Nr	0.25	29.32	1.61	10.56	41.49	2.813	-1.486
Z4.8.0.09	as above; Head rails 3000mm	Nr	0.33	39.06	2.15	15.84	57.05	4.015	-2.229

Z5 Piped Building Services

Z5.1 Pipework

Z5.1.1 Pipes

		Unit	Labour Hours	Labour Net £	Plant Net £	Materials Net £	Unit Net £	CO$_2$e Kg
Z5.1.1.01	LTHW heating installation; copper pipes Table X to BS EN 2871; including fittings, valves, cocks and connections; fixed to backgrounds, plugging; nominal bore: 15mm; Without insulation	m	0.14	7.88	-	3.39	11.27	3.668
Z5.1.1.02	as above; nominal bore: 15mm; With mineral fibre foil faced insulation	m	0.18	10.31	-	5.09	15.40	3.769
Z5.1.1.03	as above; nominal bore: 20mm; Without insulation	m	0.14	7.88	-	5.33	13.21	6.783
Z5.1.1.04	as above; nominal bore: 20mm; With mineral fibre foil faced insulation	m	0.18	10.31	-	7.44	17.75	8.124
Z5.1.1.05	as above; nominal bore: 35mm; Without insulation	m	0.18	10.54	-	13.73	24.27	19.210

Z5   Piped Building Services continued...

	Unit	Labour Hours	Labour Net £	Plant Net £	Materials Net £	Unit Net £	CO_2e Kg
Z5.1	**Pipework**						
Z5.1.1 Pipes							
Z5.1.1.06 as previous item; nominal bore: 35mm; With mineral fibre foil faced insulation	m	0.27	15.35	-	16.24	31.59	21.221
Z5.1.1.07 as above; nominal bore: 54mm; Without insulation	m	0.23	13.15	-	21.96	35.11	32.467
Z5.1.1.08 as above; nominal bore: 54mm; With mineral fibre foil faced insulation	m	0.35	20.38	-	26.52	46.90	34.814
Z5.1.1.09 In sprinkler installations; black steel pipes to BS EN 10255; including fittings, valves, storage pump and sprinkler heads; Fixed to backgrounds, plugging; Rate per sprinkler head	Nr	0.80	46.33	-	262.09	308.42	52.541
Z5.2	**Equipment**						
Z5.2.2 Boiler plant and ancillaries							
Z5.2.2.01 Hot water heaters with thermostatic safety features; insulated; plugged and screwed to walls; connected to rising main including for fittings; not including electrical work; Loading: 3kW; Point of use; ABS moulded casing; vented taps and spout: capacity; 7 litres	Nr	3.00	173.73	-	73.97	247.70	42.463
Z5.2.2.02 as above; Multipoint Cistern type; copper casing; fitting for hot water outlet: Capacity: 25 litres	Nr	6.00	347.46	-	191.61	539.07	160.639
Z5.2.2.03 Domestic boiler plant and ancillaries; fully automatic, controlled by thermostat with electric controls; white stove enamelled finish; placing in position, balancing, assembly and connections; including electrical work; Gas fired: Floor standing; capacity; 11.80kW Rating	Nr	9.00	521.19	-	833.70	1,354.89	307.091
Z5.2.2.04 as above; Gas fired: Floor standing; capacity; 17.60 kW Rating	Nr	9.00	521.19	-	922.41	1,443.60	318.761
Z5.2.2.05 as above; Oil fired: Floor standing; capacity; 11.80 kW Rating	Nr	9.00	521.19	-	1,052.84	1,574.03	313.080
Z5.2.2.06 as above; Oil fired: Floor standing; capacity; 17.60 kW Rating	Nr	9.00	521.19	-	1,090.22	1,611.41	325.333
Z5.2.2.07 Heating coils; aluminium finned copper tubing; placing and fixing into 300mm deep slot; grille cover; connections for 120 degrees C flow temperature; L.P.H.W.; 75 x 50mm	m	-	-	-	-	187.00	-
Z5.2.2.08 as above; 108 x 108mm	m	-	-	-	-	218.00	-

Z5   Piped Building Services continued...

	Unit	Labour Hours	Labour Net £	Plant Net £	Materials Net £	Unit Net £	CO_2e Kg
Z5.2	**Equipment**						
Z5.2.2	Boiler plant and ancillaries						
Z5.2.2.09	Pressed steel complete with air cock and thermostatic radiator valves; fixing brackets to masonry with screws, plugging; Panels: Single; 450mm high; of length: 800mm	Nr	0.25	14.48	-	77.88	92.36	70.357
Z5.2.2.10	as above: Single; 450mm high; of length: 1120mm	Nr	0.30	17.37	-	86.13	103.50	80.718
Z5.2.2.11	as above: Single; 450mm high; of length: 1920mm	Nr	0.40	23.16	-	103.45	126.61	106.638
Z5.2.2.12	as above: Single; 450mm high; of length: 2720mm	Nr	0.50	28.95	-	118.91	147.86	132.559
Z5.2.2.13	as above: Single; 600mm high; of length: 800mm	Nr	0.30	17.37	-	84.61	101.98	75.479
Z5.2.2.14	as above: Single; 600mm high; of length: 1120mm	Nr	0.40	23.16	-	95.57	118.73	87.896
Z5.2.2.15	as above: Single; 600mm high; of length: 1920mm	Nr	0.50	28.95	-	120.05	149.00	118.939
Z5.2.2.16	as above: Single; 600mm high; of length: 2720mm	Nr	0.60	34.75	-	145.27	180.02	149.981
Z5.2.2.17	as above: Double; 450mm high; of length: 800mm	Nr	0.30	17.37	-	103.66	121.03	85.297
Z5.2.2.18	as above: Double; 450mm high; of length: 1120mm	Nr	0.35	20.27	-	114.67	134.94	101.652
Z5.2.2.19	as above: Double; 450mm high; of length: 1920mm	Nr	0.45	26.06	-	161.56	187.62	142.512
Z5.2.2.20	as above: Double; 450mm high; of length: 2720mm	Nr	0.55	31.85	-	214.44	246.29	183.391
Z5.2.2.21	as above: Double; 600mm high; of length: 800mm	Nr	0.40	23.16	-	116.82	139.98	93.329
Z5.2.2.22	as above: Double; 600mm high; of length: 1120mm	Nr	0.50	28.95	-	138.19	167.14	112.886
Z5.2.2.23	as above: Double; 600mm high; of length: 1920mm	Nr	0.60	34.75	-	195.98	230.73	161.777
Z5.2.2.24	as above: Double; 600mm high; of length: 2720mm	Nr	0.70	40.54	-	255.56	296.10	210.669
Z5.2.3	Pumps						
Z5.2.3.01	Centrifugal heating pump direct drive; 3 phase; 1450 rpm motor; max. temperature 110 degrees C; button starter; including electrical connections; Pump Size 40mm; Maximum Head 40kN per m^2; Maximum Delivery 4 litre/sec	Nr	3.00	173.73	-	697.32	871.05	215.467
Z5.2.3.02	as above; Pump Size 50mm; Maximum Head 35kN per m^2; Maximum Delivery 7 litre/sec	Nr	3.00	173.73	-	801.36	975.09	251.571
Z5.2.3.03	as above; Pump Size 80mm; Maximum Head 50kN per m^2; Maximum Delivery 16 litre/sec	Nr	3.00	173.73	-	1,113.48	1,287.21	498.283
Z5.2.3.04	as above; Pump Size 100mm; Maximum Head 80kN per m^2; Maximum Delivery 28 litre/sec	Nr	3.00	173.73	-	1,425.59	1,599.32	600.579
Z5.2.3.05	as above; Pump Size 150mm; Maximum Head 120kN per m^2; Maximum Delivery 70 litre/sec	Nr	3.00	173.73	-	2,153.87	2,327.60	702.874

Z5 Piped Building Services continued...

	Unit	Labour Hours	Labour Net £	Plant Net £	Materials Net £	Unit Net £	CO_2e Kg
Z5.3 **Sanitary appliances and fittings**							
Z5.3.0 Including waste pipes and fittings, valves, fittings, traps; placing in position, assembly and connections							
Z5.3.0.01 WC suites; white vitreous china with pan, cistern, flush pipe and seat	Nr	3.00	173.73	-	76.46	250.19	88.812
Z5.3.0.02 Bowl urinals; white vitreous china with concealed hangers, cistern and flush pipe	Nr	1.00	57.91	-	311.66	369.57	154.636
Z5.3.0.03 Slab urinals; white glazed fireclay with cistern, flush pipes and spreaders waste fitting: 2 persons	Nr	2.00	115.82	-	650.01	765.83	283.295
Z5.3.0.04 as above; 3 persons	Nr	3.00	173.73	-	919.87	1,093.60	385.523
Z5.3.0.05 as above; 4 persons	Nr	4.00	231.64	-	1,248.34	1,479.98	506.297
Z5.3.0.06 Hand rinse basins; vitreous china with brackets chromium plated waste, pillar taps, plug and chain	Nr	2.00	115.82	-	66.27	182.09	52.940

Z6 Ducted Building Services

	Unit	Labour Hours	Labour Net £	Plant Net £	Materials Net £	Unit Net £	CO_2e Kg
Z6.2 **Rectangular ductwork**							
Z6.2.1 Straight							
Z6.2.1.01 Air supply system; galvanised sheet steel; low pressure including supports; fittings; 1nr taper; 1nr 90 degree bend; 2nr grilles; 1nr fire damper; DW 142; overall length 20m; Without insulation; 0.6mm thick: sum of both sides 200mm	m	0.05	2.44	0.79	123.68	126.91	17.882
Z6.2.1.02 as above; Without insulation; 0.6mm thick: sum of both sides 400mm	m	0.05	2.64	0.85	152.60	156.09	33.987
Z6.2.1.03 as above; Without insulation; 0.6mm thick: sum of both sides 500mm	m	0.05	2.64	0.85	172.64	176.13	40.639
Z6.2.1.04 as above; Without insulation; 0.6mm thick: sum of both sides 600mm	m	0.05	2.64	0.85	196.03	199.52	50.647
Z6.2.1.05 as above; Without insulation; 0.8mm thick: sum of both sides 700mm	m	0.05	2.64	0.85	209.72	213.21	66.176
Z6.2.1.06 as above; Without insulation; 0.8mm thick: sum of both sides 900mm	m	0.06	2.83	0.92	312.22	315.97	95.419
Z6.2.1.07 as above; Without insulation; 0.8mm thick: sum of both sides 1100mm	m	0.06	2.83	0.92	383.06	386.81	107.379
Z6.2.1.08 as above; Without insulation; 0.8mm thick: sum of both sides 1500mm	m	0.07	3.27	1.06	451.96	456.29	148.784
Z6.2.1.09 as above; Without insulation; 1.0mm thick: sum of both sides 2000mm	m	0.08	3.66	1.18	513.56	518.40	156.119
Z6.2.1.10 as above; Without insulation; 1.0mm thick: sum of both sides 2400mm	m	0.08	4.05	1.31	604.89	610.25	214.830

Z6 Ducted Building Services continued...

	Unit	Labour Hours	Labour Net £	Plant Net £	Materials Net £	Unit Net £	CO$_2$e Kg
Z6.2	**Rectangular ductwork**						
Z6.2.1	Straight						
Z6.2.1.11 as previous item; Without insulation; 1.0mm thick: sum of both sides 2800mm	m	0.10	4.88	1.58	704.71	711.17	289.799
Z6.2.1.12 as above; Without insulation; 1.0mm thick:: sum of both sides 3200mm	m	0.12	5.71	1.85	770.54	778.10	364.581
Z6.2.1.13 as above; Rigid mineral fibre foil faced insulation; 0.6mm thick: sum of both sides 200mm	m	0.05	2.64	0.85	126.92	130.41	18.964
Z6.2.1.14 as above; Rigid mineral fibre foil faced insulation; 0.6mm thick: sum of both sides 400mm	m	0.06	2.83	0.92	159.08	162.83	36.141
Z6.2.1.15 as above; Rigid mineral fibre foil faced insulation; 0.6mm thick: sum of both sides 500mm	m	0.06	2.83	0.92	180.74	184.49	43.330
Z6.2.1.16 as above; Rigid mineral fibre foil faced insulation; 0.6mm thick: sum of both sides 600mm	m	0.06	2.83	0.92	205.75	209.50	53.874
Z6.2.1.17 as above; Rigid mineral fibre foil faced insulation; 0.8mm thick: sum of both sides 700mm	m	0.06	3.08	1.00	221.06	225.14	69.950
Z6.2.1.18 as above; Rigid mineral fibre foil faced insulation; 0.8mm thick: sum of both sides 900mm	m	0.07	3.27	1.06	326.80	331.13	100.265
Z6.2.1.19 as above; Rigid mineral fibre foil faced insulation; 0.8mm thick: sum of both sides 1100mm	m	0.07	3.27	1.06	400.88	405.21	113.298
Z6.2.1.20 as above; Rigid mineral fibre foil faced insulation; 0.8mm thick: sum of both sides 1500mm	m	0.08	3.66	1.19	476.26	481.11	156.846
Z6.2.1.21 as above; Rigid mineral fibre foil faced insulation; 1.0mm thick: sum of both sides 2000mm	m	0.08	4.05	1.31	545.96	551.32	166.864
Z6.2.1.22 as above; Rigid mineral fibre foil faced insulation; 1.0mm thick: sum of both sides 2400mm	m	0.09	4.49	1.45	643.77	649.71	227.723
Z6.2.1.23 as above; Rigid mineral fibre foil faced insulation; 1.0mm thick: sum of both sides 2800mm	m	0.12	5.71	1.85	750.07	757.63	304.854
Z6.2.1.24 as above; Rigid mineral fibre foil faced insulation; 1.0mm thick:: sum of both sides 3200mm	m	0.13	6.49	2.10	822.38	830.97	381.779
Z6.3	**Equipment**						
Z6.3.1	Conditioning and handling units						
Z6.3.1.01 Roof mounted air handling unit, air supply system; steel trays; insulation; mixing box filter; 240V centrifugal fan; attenuator; plenum box; heater battery; chiller; flexible connections; commissioning; Air volume: 0.3 m^3/sec; Heating Capacity 14.5 kW; Cooling Capacity: 7.1kW	Nr	15.00	777.85	180.45	1,867.69	2,825.99	298.906
Z6.3.1.02 as above; Air volume: 0.6 m^3/sec; Heating Capacity 25.8 kW; Cooling Capacity 12.4 kW	Nr	15.00	777.85	180.45	2,458.96	3,417.26	347.324

Z6 Ducted Building Services continued...

	Unit	Labour Hours	Labour Net £	Plant Net £	Materials Net £	Unit Net £	CO_2e Kg
Z6.3	**Equipment**						
Z6.3.1	Conditioning and handling units						
Z6.3.1.03	as previous item; Air volume: $1.4\,m^3/sec$; Heating Capacity 54.8 kW; Cooling Capacity 25.8 kW	Nr	18.00	933.42	216.54	3,446.31	4,596.27	573.821
Z6.3.3	Fans						
Z6.3.3.01	Ceiling mounted encased fan coil unit, air supply system; variable speed fan; filter; coils; grilles; condensate drip tray; fresh air capacity; including electrical connections; Heating capacity: 5.0kW; Cooling Capacity 2.3kW	Nr	2.00	106.74	24.06	383.63	514.43	146.293
Z6.3.3.02	as above; Heating capacity: 9.0kW; Cooling Capacity 4.7kW	Nr	2.00	106.74	24.06	437.09	567.89	189.524

Z7 Cabled Building Services

	Unit	Labour Hours	Labour Net £	Plant Net £	Materials Net £	Unit Net £	CO_2e Kg
Z7.1	**Cables**						
Z7.1.1	Laid or drawn into conduits; trunking or ducts						
Z7.1.1.01	Type: Two core; PVC insulated; S.W.A.; PVC sheathed copper: $1.5mm^2$	m	0.05	2.90	-	0.68	3.58	0.200
Z7.1.1.02	Type: Two core; PVC insulated; S.W.A.; PVC sheathed copper: $2.5mm^2$	m	0.05	2.90	-	0.84	3.74	0.223
Z7.1.1.03	Type: Two core; PVC insulated; S.W.A.; PVC sheathed copper: $4.0mm^2$	m	0.07	3.88	-	1.13	5.01	0.215
Z7.1.1.04	Type: Two core; PVC insulated; S.W.A.; PVC sheathed copper: $6.0mm^2$	m	0.10	5.79	-	1.52	7.31	0.368
Z7.1.1.05	Type: Two core; PVC insulated; S.W.A.; PVC sheathed copper: $10.0mm^2$	m	0.15	8.69	-	2.19	10.88	0.770
Z7.1.1.06	Type: Two core; PVC insulated; S.W.A.; PVC sheathed copper: $16.0mm^2$	m	0.17	9.67	-	3.02	12.69	1.193
Z7.1.1.07	Type: Two core; XLPE insulated; S.W.A.; PVC sheathed copper: $25mm^2$	m	0.25	14.48	-	3.96	18.44	2.209
Z7.1.1.08	Type: Two core; XLPE insulated; S.W.A.; PVC sheathed copper: $35mm^2$	m	0.33	19.28	-	5.13	24.41	2.945
Z7.1.1.09	Type: Two core; XLPE insulated; S.W.A.; PVC sheathed copper: $70mm^2$	m	0.42	24.15	-	10.26	34.41	4.628
Z7.1.1.10	Type: Three core; PVC insulated; S.W.A.; PVC sheathed copper: $1.5mm^2$	m	0.05	2.90	-	0.72	3.62	0.252
Z7.1.1.11	Type: Three core; PVC insulated; S.W.A.; PVC sheathed copper: $2.5mm^2$	m	0.05	2.90	-	0.96	3.86	0.265
Z7.1.1.12	Type: Three core; PVC insulated; S.W.A.; PVC sheathed copper: $4.0mm^2$	m	0.07	3.88	-	1.32	5.20	0.423

Z7 Cabled Building Services continued...

	Unit	Labour Hours	Labour Net £	Plant Net £	Materials Net £	Unit Net £	CO_2e Kg
Z7.1	**Cables**						
Z7.1.1	Laid or drawn into conduits; trunking or ducts						
Z7.1.1.13	Type: Three core; PVC insulated; S.W.A.; PVC sheathed copper: 6.0mm^2	m	0.10	5.79	-	1.75	7.54	0.574
Z7.1.1.14	Type: Three core; PVC insulated; S.W.A.; PVC sheathed copper: 10.0mm^2	m	0.15	8.69	-	2.86	11.55	0.980
Z7.1.1.15	Type: Three core; PVC insulated; S.W.A.; PVC sheathed copper: 16.0mm^2	m	0.17	9.67	-	4.08	13.75	1.466
Z7.1.1.16	Type: Three core; XLPE insulated; S.W.A.; PVC sheathed copper: 25mm^2	m	0.25	14.48	-	6.00	20.48	3.156
Z7.1.1.17	Type: Three core; XLPE insulated; S.W.A.; PVC sheathed copper: 35mm^2	m	0.33	19.28	-	7.43	26.71	3.787
Z7.1.1.18	Type: Three core; XLPE insulated; S.W.A.; PVC sheathed copper: 70mm^2	m	0.42	24.15	-	14.36	38.51	6.311
Z7.1.1.19	Type: Four core; PVC insulated; S.W.A.; PVC sheathed copper: 1.5mm^2	m	0.07	3.88	-	0.86	4.74	0.316
Z7.1.1.20	Type: Four core; PVC insulated; S.W.A.; PVC sheathed copper: 2.5mm^2	m	0.08	4.81	-	1.21	6.02	0.335
Z7.1.1.21	Type: Four core; PVC insulated; S.W.A.; PVC sheathed copper: 4.0mm^2	m	0.13	7.70	-	1.68	9.38	0.595
Z7.1.1.22	Type: Four core; PVC insulated; S.W.A.; PVC sheathed copper: 6.0mm^2	m	0.17	9.67	-	2.42	12.09	0.741
Z7.1.1.23	Type: Four core; PVC insulated; S.W.A.; PVC sheathed copper: 10.0mm^2	m	0.20	11.58	-	3.58	15.16	1.327
Z7.1.1.24	Type: Four core; PVC insulated; S.W.A.; PVC sheathed copper: 16.0mm^2	m	0.25	14.48	-	5.39	19.87	1.614
Z7.1.1.25	Type: Four core; XLPE insulated; S.W.A.; PVC sheathed copper: 25mm^2	m	0.25	14.48	-	7.49	21.97	3.787
Z7.1.1.26	Type: Four core; XLPE insulated; S.W.A.; PVC sheathed copper: 35mm^2	m	0.33	19.28	-	10.24	29.52	5.049
Z7.1.1.27	Type: Four core; XLPE insulated; S.W.A.; PVC sheathed copper: 70mm^2	m	0.33	19.28	-	19.13	38.41	8.626
Z7.1.1.28	Type: Single core; PVC insulated cable copper; Single Strand; 1.5mm^2	m	0.02	0.98	-	0.18	1.16	0.044
Z7.1.1.29	Type: Single core; PVC insulated cable copper; Single Strand; 2.5mm^2	m	0.02	0.98	-	0.25	1.23	0.074
Z7.1.1.30	Type: Single core; PVC insulated cable copper; Multi Strand; 1.5mm^2	m	0.02	0.98	-	0.16	1.14	0.044

Z7 Cabled Building Services continued...

		Unit	Labour Hours	Labour Net £	Plant Net £	Materials Net £	Unit Net £	CO_2e Kg
Z7.1	**Cables**							
Z7.1.1	Laid or drawn into conduits; trunking or ducts							
Z7.1.1.31	Type: Single core; PVC insulated cable copper; Multi Strand; 2.5mm^2	m	0.02	0.98	-	0.23	1.21	0.074
Z7.1.1.32	Type: Single core; PVC insulated cable copper; Multi Strand; 4.0mm^2	m	0.02	0.98	-	0.37	1.35	0.105
Z7.1.1.33	Type: Single core; PVC insulated cable copper; Multi Strand; 6.0mm^2	m	0.02	0.98	-	0.50	1.48	0.149
Z7.1.3	Fixed to surfaces							
Z7.1.3.01	Mineral Insulated; copper sheathed with copper conductors; Type: bare; Light duty 600 volt grade; 2L 1.0	m	0.03	1.45	-	3.67	5.12	0.250
Z7.1.3.02	as above; Type: bare; Light duty 600 volt grade; 2L 1.5	m	0.03	1.45	-	4.33	5.78	0.317
Z7.1.3.03	as above; Type: bare; Light duty 600 volt grade; 2L 2.5	m	0.03	1.45	-	5.62	7.07	0.425
Z7.1.3.04	as above; Type: bare; Light duty 600 volt grade; 2L 4.0	m	0.03	1.45	-	8.52	9.97	0.563
Z7.1.3.05	as above; Type: bare; Heavy duty 1000 volt grade; 1H 6.0	m	0.03	1.91	-	8.77	10.68	0.425
Z7.1.3.06	as above; Type: bare; Heavy duty 1000 volt grade; 1H 10.0	m	0.03	1.91	-	9.36	11.27	0.545
Z7.1.3.07	as above; Type: bare; Heavy duty 1000 volt grade; 2H 2.5	m	0.05	2.90	-	9.76	12.66	0.627
Z7.1.3.08	as above; Type: bare; Heavy duty 1000 volt grade; 2H 6.0	m	0.05	2.90	-	16.37	19.27	0.984
Z7.1.3.09	as above; Type: PV sheathed; Light duty 600 volt grade; 2L 1.0	m	0.03	1.45	-	4.28	5.73	0.250
Z7.1.3.10	as above; Type: PV sheathed; Light duty 600 volt grade; 2L 1.5	m	0.03	1.45	-	4.75	6.20	0.317
Z7.1.3.11	as above; Type: PV sheathed; Light duty 600 volt grade; 2L 2.5	m	0.03	1.45	-	6.00	7.45	0.425
Z7.1.3.12	as above; Type: PV sheathed; Light duty 600 volt grade; 2L 4.0	m	0.03	1.45	-	9.02	10.47	0.563
Z7.1.3.13	as above; Type: PV sheathed; Heavy duty 1000 volt grade; 1H 6.0	m	0.03	1.91	-	9.59	11.50	0.425
Z7.1.3.14	as above; Type: PV sheathed; Heavy duty 1000 volt grade; 1H 10.0	m	0.03	1.91	-	10.18	12.09	0.545
Z7.1.3.15	as above; Type: PV sheathed; Heavy duty 1000 volt grade; 2H 2.5	m	0.05	2.90	-	10.62	13.52	0.627
Z7.1.3.16	as above; Type: PV sheathed; Heavy duty 1000 volt grade; 2H 6.0	m	0.05	2.90	-	17.74	20.64	0.984
Z7.1.4	Laid in trenches							
Z7.1.4.01	PVC sheathed; S.W.A; PVC insulated copper; Type: Two core; 4mm^2	m	0.08	4.81	-	1.13	5.94	0.215
Z7.1.4.02	as above; S.W.A; PVC insulated copper; Type: Two core; 6mm^2	m	0.08	4.81	-	1.52	6.33	0.368
Z7.1.4.03	as above; S.W.A; PVC insulated copper; Type: Two core; 10mm^2	m	0.10	5.79	-	2.19	7.98	0.770

Z7   Cabled Building Services continued...

	Unit	Labour Hours	Labour Net £	Plant Net £	Materials Net £	Unit Net £	CO_2e Kg
Z7.1	**Cables**						
Z7.1.4	Laid in trenches						
Z7.1.4.04	as previous item; S.W.A; PVC insulated copper; Type: Two core; 16mm^2	m	0.17	9.67	-	3.02	12.69	1.193
Z7.1.4.05	as above; S.W.A; PVC insulated copper; Type: Three core; 4mm^2	m	0.08	4.81	-	1.32	6.13	0.423
Z7.1.4.06	as above; S.W.A; PVC insulated copper; Type: Three core; 6mm^2	m	0.12	6.78	-	1.75	8.53	0.574
Z7.1.4.07	as above; S.W.A; PVC insulated copper; Type: Three core; 10mm^2	m	0.17	9.67	-	2.86	12.53	0.980
Z7.1.4.08	as above; S.W.A; PVC insulated copper; Type: Three core; 16mm^2	m	0.20	11.58	-	4.08	15.66	1.466
Z7.1.4.09	as above; S.W.A; PVC insulated copper; Type: Four core; 4mm^2	m	0.15	8.69	-	1.68	10.37	0.595
Z7.1.4.10	as above; S.W.A; PVC insulated copper; Type: Four core; 6mm^2	m	0.20	11.58	-	2.42	14.00	0.741
Z7.1.4.11	as above; S.W.A; PVC insulated copper; Type: Four core; 10mm^2	m	0.20	11.58	-	3.58	15.16	1.327
Z7.1.4.12	as above; S.W.A; PVC insulated copper; Type: Four core; 16mm^2	m	0.25	14.48	-	5.39	19.87	1.614
Z7.1.4.13	XLPE insulated; S.W.A; PVC sheathed copper; Type: Two core; 25mm^2	m	0.17	9.67	-	3.96	13.63	2.209
Z7.1.4.14	as above; S.W.A; PVC sheathed copper; Type: Two core; 35mm^2	m	0.25	14.48	-	5.13	19.61	2.945
Z7.1.4.15	as above; S.W.A; PVC sheathed copper; Type: Two core; 70mm^2	m	0.25	14.48	-	10.26	24.74	4.628
Z7.1.4.16	as above; S.W.A; PVC sheathed copper; Type: Two core; 120mm^2	m	0.33	19.28	-	20.51	39.79	7.574
Z7.1.4.17	as above; S.W.A; PVC sheathed copper; Type: Two core; 240mm^2	m	0.33	19.28	-	35.89	55.17	14.516
Z7.1.4.18	as above; S.W.A; PVC sheathed copper; Type: Three core; 25mm^2	m	0.17	9.67	-	6.00	15.67	3.156
Z7.1.4.19	as above; S.W.A; PVC sheathed copper; Type: Three core; 35mm^2	m	0.25	14.48	-	7.43	21.91	3.787
Z7.1.4.20	as above; S.W.A; PVC sheathed copper; Type: Three core; 70mm^2	m	0.25	14.48	-	14.36	28.84	6.311
Z7.1.4.21	as above; S.W.A; PVC sheathed copper; Type: Three core; 120mm^2	m	0.25	14.48	-	21.54	36.02	10.414
Z7.1.4.22	as above; S.W.A; PVC sheathed copper; Type: Three core; 240mm^2	m	0.33	19.28	-	41.02	60.30	20.302
Z7.1.4.23	as above; S.W.A; PVC sheathed copper; Type: Four core; 25mm^2	m	0.08	4.81	-	7.49	12.30	3.787
Z7.1.4.24	as above; S.W.A; PVC sheathed copper; Type; 35mm^2	m	0.17	9.67	-	10.24	19.91	5.049
Z7.1.4.25	as above; S.W.A; PVC sheathed copper; Type; 70mm^2	m	0.17	9.67	-	19.13	28.80	8.626

Z7 Cabled Building Services continued...

	Unit	Labour Hours	Labour Net £	Plant Net £	Materials Net £	Unit Net £	CO_2e Kg
Z7.1 **Cables**							
Z7.1.4 Laid in trenches							
Z7.1.4.26 as previous item; S.W.A; PVC sheathed copper; Type; 120mm^2	m	0.17	9.67	-	27.69	37.36	14.095
Z7.1.4.27 as above; S.W.A; PVC sheathed copper; Type; 240mm^2	m	0.17	9.67	-	36.92	46.59	26.087
Z7.3 **Trunking**							
Z7.3.1 Plain							
Z7.3.1.01 Fixed to backgrounds; fittings included in the running length; Type: Single compartment; Galvanised steel; 3m lengths: 50 x 50mm	m	0.17	9.67	-	4.80	14.47	8.812
Z7.3.1.02 as above; Type: Single compartment; Galvanised steel; 3m lengths: 75 x 75mm	m	0.17	9.67	-	7.92	17.59	13.219
Z7.3.1.03 as above; Type: Single compartment; Galvanised steel; 3m lengths: 100 x 75mm	m	0.25	14.48	-	8.74	23.22	15.422
Z7.3.1.04 as above; Type: Single compartment; Galvanised steel; 3m lengths: 150 x 150mm	m	0.33	19.28	-	9.26	28.54	26.437
Z7.3.1.05 as above; Type: Single compartment; PVC; grey finish; clip on lid: 50 x 50mm	m	0.17	9.67	-	2.28	11.95	2.062
Z7.3.1.06 as above; Type: Single compartment; PVC; grey finish; clip on lid: 75 x 75mm	m	0.20	11.58	-	4.57	16.15	3.093
Z7.3.1.07 as above; Type: Single compartment; PVC; grey finish; clip on lid: 100 x 75mm	m	0.25	14.48	-	5.28	19.76	3.609
Z7.3.1.08 as above; Type: Single compartment; PVC; grey finish; clip on lid: 150 x 150mm	m	0.33	19.28	-	15.44	34.72	6.186
Z7.3.1.09 as above; Type: Double compartment; Galvanised steel; 3m lengths: 50 x 50mm	m	0.20	11.58	-	7.20	18.78	11.015
Z7.3.1.10 as above; Type: Double compartment; Galvanised steel; 3m lengths: 75 x 75mm	m	0.20	11.58	-	11.88	23.46	16.523
Z7.3.1.11 as above; Type: Double compartment; Galvanised steel; 3m lengths: 100 x 75mm	m	0.25	14.48	-	13.11	27.59	18.726
Z7.3.1.12 as above; Type: Double compartment; Galvanised steel; 3m lengths: 150 x 150mm	m	0.33	19.28	-	13.89	33.17	33.046
Z7.3.1.13 as above; Type: Double compartment; PVC; grey finish; clip on lid: 50 x 50mm	m	0.25	14.48	-	3.68	18.16	2.578
Z7.3.1.14 as above; Type: Double compartment; PVC; grey finish; clip on lid: 75 x 75mm	m	0.33	19.28	-	5.93	25.21	3.868
Z7.3.1.15 as above; Type: Double compartment; PVC; grey finish; clip on lid: 100 x 75mm	m	0.42	24.15	-	7.38	31.53	4.383
Z7.3.1.16 as above; Type: Double compartment; PVC; grey finish; clip on lid: 150 x 150mm	m	0.50	28.95	-	17.54	46.49	7.217

Z7 Cabled Building Services continued...

	Unit	Labour Hours	Labour Net £	Plant Net £	Materials Net £	Unit Net £	CO_2e Kg
Z7.3	**Trunking**						
Z7.3.1	Plain						
Z7.3.1.17	as previous item; Type: Triple compartment; Galvanised steel; 3m lengths: 75 x 75mm	m	0.17	9.67	-	17.82	27.49	19.828
Z7.3.1.18	as above; Type: Triple compartment; Galvanised steel; 3m lengths: 100 x 75mm	m	0.20	11.58	-	17.82	29.40	19.828
Z7.3.1.19	as above; Type: Triple compartment; Galvanised steel; 3m lengths: 150 x 150mm	m	0.33	19.28	-	17.82	37.10	19.828
Z7.3.1.20	as above; Type: Triple compartment; PVC; grey finish; clip on lid: 75 x 75mm	m	0.33	19.28	-	7.68	26.96	4.640
Z7.3.1.21	as above; Type: Triple compartment; PVC; grey finish; clip on lid: 100 x 75mm	m	0.42	24.15	-	9.48	33.63	5.155
Z7.3.1.22	as above; Type: Triple compartment; PVC; grey finish; clip on lid: 150 x 150mm	m	0.50	28.95	-	19.65	48.60	8.248
Z7.5	**Trays**						
Z7.5.1	Plain						
Z7.5.1.01	Cable tray; fittings included in the running length; Galvanised steel; 2.4m lengths: 50mm wide tray	m	0.17	9.67	0.72	1.67	12.06	2.507
Z7.5.1.02	as above: 100mm wide tray	m	0.25	14.48	1.09	3.46	19.03	4.861
Z7.5.1.03	as above; 2.4m lengths: 300mm wide tray	m	0.33	19.28	1.45	10.04	30.77	13.825
Z7.5.1.04	as above; 2.4m lengths: 600mm wide tray	m	0.75	43.43	3.25	14.87	61.55	27.803
Z7.5.1.05	Cable tray with return flange; fittings included in the running length; Galvanised steel; 2.4m lengths: 50mm wide tray	m	0.25	14.48	1.09	2.74	18.31	2.878
Z7.5.1.06	as above: 100mm wide tray	m	0.50	28.95	2.17	9.77	40.89	5.758
Z7.5.1.07	as above: 300mm wide tray	m	0.75	43.43	3.25	16.36	63.04	14.585
Z7.5.1.08	as above: 600mm wide tray	m	1.00	57.91	4.34	29.76	92.01	30.902
Z7.6	**Earthing and bonding**						
Z7.6.1	Tapes						
Z7.6.1.01	In accordance with the IEE Wiring Regulations; Copper tapes; PVC insulated single core cable; earth clamps; fixings; Rate per m	m	0.16	9.27	-	1.18	10.45	1.430
Z7.6.1.02	In lightning protection; Earth rods; strike points; 25 x 3mm copper tape; bonding links; test clamps; commissioning; Rate per m^2 of gross floor area	m^2	0.05	2.54	0.35	3.02	5.91	1.510

Z7 Cabled Building Services continued...

	Unit	Labour Hours	Labour Net £	Plant Net £	Materials Net £	Unit Net £	CO_2e Kg
Z7.7	**Final circuits**						
Z7.7.2	Cable and conduit						
Z7.7.2.01	With equipment and fittings to internal lighting; 2.5mm² PVC singles cable; galvanised heavy gauge mild steel conduit; luminaires; diffusers; control gear; switches; fixings; 1800mm 70W single batten	Point	2.00	115.82	8.68	49.01	173.51	30.920
Z7.7.2.02	as above; 1200mm 36W twin batten fluorescents	Point	2.00	115.82	8.68	54.66	179.16	31.450
Z7.7.2.03	as above; 1200mm 36W twin batten fluorescents, Zone 2 Hazard rated	Point	2.00	115.82	8.68	100.19	224.69	35.160
Z7.7.2.04	as above; 1200 x 600mm 36W recessed modular fluorescent	Point	3.00	173.73	13.02	74.91	261.66	35.391
Z7.7.2.05	as above; 1200 x 600mm 36W recessed modular fluorescent with emergency inverter / charger module and 3 hour battery pack	Point	4.00	231.64	17.36	127.32	376.32	39.862
Z7.7.2.06	With equipment and fittings to external lighting installations; 4 core 6mm² PVC/SWA/PVC cable in trenches; luminaires; control gear; brackets; columns; fixings; 250W SON XL floodlight and 7 metre column	Point	4.25	242.81	34.94	1,308.25	1,586.00	1,202.155
Z7.7.2.07	as above; 50W SON Aluminium bollards	Point	3.12	179.09	8.44	354.10	541.63	627.884
Z7.7.2.08	With equipment and fittings to suspended track lighting; Suspensions; brackets; 2.5mm² PVC single cable; galvanised conduit; control gear; switches; fixings; 58W single linear track (anodised finish)	Point	1.00	57.91	-	81.81	139.72	32.189
Z7.7.2.09	as above; 58W twin linear track (anodised finish)	Point	1.20	69.49	-	138.74	208.23	39.749
Z7.7.2.10	With equipment and fittings in emergency lighting installation; Luminaires; illuminated signs; MICC cable; galvanised cable tray; control gear; fixings; 11W Bulkhead	Point	1.20	69.49	-	92.19	161.68	29.173
Z7.7.2.11	as above; 11W Bulkhead Zone 2 Hazard rated	Point	1.50	86.86	-	178.70	265.56	29.931
Z7.7.2.12	as above; 24W "EXIT" Bulkhead sign	Point	1.20	69.49	-	191.00	260.49	32.537
Z7.7.2.13	as above; 24W "EXIT" Bulkhead sign Zone 2 Hazard rated	Point	1.50	86.86	-	332.52	419.38	33.525
Z7.7.2.14	With equipment and fittings in security alarm installation; Infra red detectors; card access system; perimeter intruder alarms; control alarms; control and mimic panels; PVC singles cable; galvanised conduit and cable trunking; fixings; Rate per point	Nr	0.33	19.28	-	158.55	177.83	44.528

Z7 Cabled Building Services continued...

	Unit	Labour Hours	Labour Net £	Plant Net £	Materials Net £	Unit Net £	CO_2e Kg
Z7.7	**Final circuits**						
Z7.7.2	Cable and conduit						
Z7.7.2.15	With equipment and fittings in fire alarm installation; Smoke detectors; heat detectors; alarm sounders; break glass units; control panel; MICC cable; galvanised steel cable tray; fixings; Rate per point	Nr	0.25	14.48	-	295.67	310.15	12.708
Z7.7.2.16	With equipment and fittings to trace heating to internal pipework; Parallel circuit heater cables; fibreglass wrap-a-round tape; connection glands; junction boxes; Rate per m	m	0.20	11.58	-	0.81	12.39	0.522
Z7.8	**Equipment and fittings**						
Z7.8.1	Equipment						
Z7.8.1.01	as above; Rate: SP & N; 500V 1nr 2-20 Amp HRC fuses: 4 way	Nr	2.00	115.82	-	222.12	337.94	48.026
Z7.8.1.02	as above; Rate: SP & N; 500V 1nr 2-20 Amp HRC fuses: 6 way	Nr	2.00	115.82	-	277.65	393.47	56.902
Z7.8.1.03	as above; Rate: SP & N; 500V 1nr 2-20 Amp HRC fuses: 8 way	Nr	2.00	115.82	-	333.17	448.99	63.991
Z7.8.1.04	as above; Rate: SP & N; 500V 6nr 63 Amp MCBs: 4 way	Nr	3.00	173.73	-	166.59	340.32	38.255
Z7.8.1.05	as above; Rate: SP & N; 500V 6nr 63 Amp MCBs: 8 way	Nr	3.00	173.73	-	194.36	368.09	42.693
Z7.8.1.06	as above; Rate: SP & N; 500V 6nr 63 Amp MCBs: 12 way	Nr	3.00	173.73	-	222.12	395.85	48.026
Z7.8.1.07	as above; Rate: TP & N; 500V 1nr 2-20 Amp HRC fuses: 4 way	Nr	3.00	173.73	-	333.17	506.90	63.991
Z7.8.1.08	as above; Rate: TP & N; 500V 1nr 2-20 Amp HRC fuses: 6 way	Nr	3.00	173.73	-	444.23	617.96	69.226
Z7.8.1.09	as above; Rate: TP & N; 500V 1nr 2-20 Amp HRC fuses: 8 way	Nr	3.00	173.73	-	499.76	673.49	76.314
Z7.8.1.10	as above; Rate: TP & N; 500V 6nr 63 Amp MCBs: 4 way	Nr	4.00	231.64	-	222.12	453.76	48.026
Z7.8.1.11	as above; Rate: TP & N; 500V 6nr 63 Amp MCBs: 8 way	Nr	4.00	231.64	-	388.70	620.34	65.714
Z7.8.1.12	as above; Rate: TP & N; 500V 6nr 63 Amp MCBs: 12 way	Nr	4.00	231.64	-	499.76	731.40	79.891
Z7.8.2	Switches						
Z7.8.2.01	Type: 1 gang; Fused Connection Unit 13 Amp; metal clad	Nr	0.25	14.48	-	8.72	23.20	3.088
Z7.8.2.02	Type: 1 gang; Fused Connection Unit 13 Amp; moulded plastic	Nr	0.25	14.48	-	4.67	19.15	2.059
Z7.8.2.03	Type: 1 gang; Switch 10 Amp; 1 way; weatherproof	Nr	0.33	19.28	-	29.02	48.30	3.088
Z7.8.2.04	Type: 1 gang; Switch 15 Amp; 2 way: metal clad	Nr	0.17	9.67	-	3.18	12.85	4.118
Z7.8.2.05	Type: 1 gang; Switch 15 Amp; 2 way: moulded plastic	Nr	0.17	9.67	-	0.97	10.64	4.118
Z7.8.2.06	Type: 1 gang; Switch 20 Amp: metal clad	Nr	0.25	14.48	-	5.59	20.07	5.147
Z7.8.2.07	Type: 1 gang; Switch 20 Amp: moulded plastic	Nr	0.25	14.48	-	3.38	17.86	5.147
Z7.8.2.08	Type: 2 gang; Fused Connection Unit 13 Amp; metal clad	Nr	0.25	14.48	-	13.08	27.56	4.118

Z7 Cabled Building Services continued...

	Unit	Labour Hours	Labour Net £	Plant Net £	Materials Net £	Unit Net £	CO$_2$e Kg
Z7.8	**Equipment and fittings**						
Z7.8.2	**Switches**						
Z7.8.2.09 Type: 2 gang; Switch 15 Amp; 2 way: moulded plastic	Nr	0.33	19.28	-	4.31	23.59	4.118
Z7.8.2.10 Type: 2 gang; Switch 20 Amp: metal clad	Nr	0.33	19.28	-	8.39	27.67	5.147
Z7.8.2.11 Type: 2 gang; Switch 20 Amp: moulded plastic	Nr	0.33	19.28	-	5.08	24.36	5.147
Z7.8.4	**Sockets**						
Z7.8.4.01 Type: 1 gang; Socket outlet 13 Amp metal clad; switched	Nr	0.33	19.28	-	6.47	25.75	1.029
Z7.8.4.02 Type: 1 gang; Socket outlet 13 Amp metal clad; switched with neon indicator	Nr	0.33	19.28	-	7.50	26.78	1.544
Z7.8.4.03 Type: 1 gang; Socket outlet 13 Amp; white moulded plastic: switched	Nr	0.33	19.28	-	5.73	25.01	1.029
Z7.8.4.04 Type: 1 gang; Socket outlet 13 Amp; white moulded plastic: switched with neon indicator	Nr	0.33	19.28	-	6.76	26.04	1.544
Z7.8.4.05 Type: 1 gang; Socket outlet 16 Amp industrial; 415V	Nr	1.00	57.91	-	6.67	64.58	2.059
Z7.8.4.06 Type: 1 gang; Socket outlet 16 Amp industrial; 240V	Nr	0.50	28.95	-	2.03	30.98	1.029
Z7.8.4.07 Type: 2 gang; Socket outlet 13 Amp metal clad; switched	Nr	0.42	24.15	-	8.23	32.38	2.059
Z7.8.4.08 Type: 2 gang; Socket outlet 13 Amp metal clad; switched with neon indicator	Nr	0.50	28.95	-	12.34	41.29	3.088
Z7.8.4.09 Type: 2 gang; Socket outlet 13 Amp; white moulded plastic: switched	Nr	0.50	28.95	-	6.89	35.84	2.059
Z7.8.4.10 Type: 2 gang; Socket outlet 13 Amp; white moulded plastic: switched with neon indicator	Nr	0.58	33.76	-	10.99	44.75	3.088
Z7.8.4.11 Type: 2 gang; Socket outlet 16 Amp industrial; 415V	Nr	1.50	86.86	-	10.00	96.86	4.118
Z7.8.4.12 Type: 2 gang; Socket outlet 16 Amp industrial; 240V	Nr	0.75	43.43	-	3.05	46.48	1.544

CLASS ZZ:
ALTERATIONS

Calculations used throughout Class ZZ - Alterations

Labour

			Qty		Rate		Total
L A0120ICE	**General Earthworks Labour Gang**						
	L A9016ICE	Ganger	1	x	17.64	=	£17.64
	L A9002ICE	Labourer (General Operative)	1	x	13.02	=	£13.02
	L A9015ICE	Banksman	1	x	14.03	=	£14.03
		Total hourly cost of gang				=	**£44.69**
L A0123ICE	**Landscape Labour Gang**						
	L A9003ICE	Labourer (Skill Rate 3)	1	x	14.86	=	£14.86
	L A9002ICE	Labourer (General Operative)	1	x	13.02	=	£13.02
		Total hourly cost of gang				=	**£27.88**
L A0130ICE	**Formwork (make, fix and strike) Labour Gang**						
	L A9000ICE	Craftsman WRA	1	x	17.33	=	£17.33
	L A9003ICE	Labourer (Skill Rate 3)	1	x	14.86	=	£14.86
	L A9001ICE	Carpenter (charge hand)	1	x	18.63	=	£18.63
		Total hourly cost of gang				=	**£50.82**
L A0140ICE	**Steel fixing Labour Gang**						
	L A9000ICE	Craftsman WRA	2	x	17.33	=	£34.66
	L A9002ICE	Labourer (General Operative)	1	x	13.02	=	£13.02
		Total hourly cost of gang				=	**£47.68**
L A0150ICE	**Provision of concrete Labour Gang**						
	L A9003ICE	Labourer (Skill Rate 3)	1	x	14.86	=	£14.86
	L A9002ICE	Labourer (General Operative)	1	x	13.02	=	£13.02
	L A9015ICE	Banksman	0.5	x	14.03	=	£7.01
		Total hourly cost of gang				=	**£34.89**
L A0155ICE	**Placing of concrete Labour Gang**						
	L A9003ICE	Labourer (Skill Rate 3)	1	x	14.86	=	£14.86
	L A9002ICE	Labourer (General Operative)	4	x	13.02	=	£52.06
	L A9015ICE	Banksman	0.25	x	14.03	=	£3.51
	L A9016ICE	Ganger	1	x	17.64	=	£17.64
	L A9000ICE	Craftsman WRA	0.5	x	17.33	=	£8.66
		Total hourly cost of gang				=	**£96.73**
L A0184ICE	**Small bore pipes in shallow trenches Labour Gang**						
	L A9015ICE	Banksman	1	x	14.03	=	£14.03
	L A9016ICE	Ganger	1	x	17.64	=	£17.64
	L A9017ICE	Pipelayer (standard rate)	1	x	14.86	=	£14.86
	L A9002ICE	Labourer (General Operative)	5	x	13.02	=	£65.08
		Total hourly cost of gang				=	**£111.61**
L A0300ICE	**Brickwork Labour Gang**						
	L A9033ICE	Bricklayer (chargehand)	1	x	18.63	=	£18.63
	L A9034ICE	Bricklayer	4	x	17.33	=	£69.31
	L A9002ICE	Labourer (General Operative)	2	x	13.02	=	£26.03
		Total hourly cost of gang				=	**£113.97**

L A0310ICE **Painting Labour Gang**

L A9041ICE	Painter (chargehand)	1	x	18.63	=	£18.63
L A9027ICE	Painter	2	x	17.33	=	£34.66
L A9028ICE	Brush hand (labourer)	1	x	13.02	=	£13.02
	Total hourly cost of gang				=	**£66.31**

L A0330ICE **Clearance Labour Gang**

L A9016ICE	Ganger	1	x	17.64	=	£17.64
L A9002ICE	Labourer (General Operative)	2	x	13.02	=	£26.03
	Total hourly cost of gang				=	**£43.67**

L K0200ICE **Manholes Labour Gang**

L A9016ICE	Ganger	1	x	17.64	=	£17.64
L A9002ICE	Labourer (General Operative)	3	x	13.02	=	£39.05
L A9000ICE	Craftsman WRA	1	x	17.33	=	£17.33
	Total hourly cost of gang				=	**£74.02**

L Z0100ICE **Concrete Drilling Labour Gang**

L A9002ICE	Labourer (General Operative)	1	x	13.02	=	£13.02
L A9003ICE	Labourer (Skill Rate 3)	2	x	14.86	=	£29.72
	Total hourly cost of gang				=	**£42.74**

L Z0101ICE **Skill Rate 3 Labour Gang**

L A9003ICE	Labourer (Skill Rate 3)	2	x	14.86	=	£29.72
	Total hourly cost of gang				=	**£29.72**

L S2070ICE **Brickwork cleaning Labour Gang**

L A9002ICE	Labourer (General Operative)	1	x	13.02	=	£13.02
L A9003ICE	Labourer (Skill Rate 3)	2	x	14.86	=	£29.72
	Total hourly cost of gang				=	**£42.74**

Plant

P A1040ICE **Formwork (make) Plant Gang**

P A0891ICE	Saw Bench - 24 inch Diesel / Electric	1	x	2.03	=	£2.03
P A0703ICE	Kango Type Tool - Electric Power Woodauger	1	x	0.61	=	£0.61
P A0704ICE	Kango Type Tool - Electric Nut Runner	1	x	0.52	=	£0.52
	Total hourly cost of gang				=	**£3.16**

P A1035ICE **Formwork (fix and strike) Plant Gang**

P A0703ICE	Kango Type Tool - Electric Power Woodauger	1	x	0.61	=	£0.61
P A0704ICE	Kango Type Tool - Electric Nut Runner	1	x	0.52	=	£0.52
P A0507ICE	Cranes Crawler - NCK 305B - 20t	0.25	x	48.55	=	£12.14
	Total hourly cost of gang				=	**£13.27**

P A1124ICE **General excavation Plant Gang**

P A0605ICE	Hydraulic Excavator - Cat 166kW	1	x	44.31	=	£44.31
P A0943ICE	Crawler Tractor / Dozer - Cat D6 LGP 160 Hp	0.5	x	52.02	=	£26.01
	Total hourly cost of gang				=	**£70.32**

P A1125ICE   **Trimming Plant Gang / preparation Plant Gang / spread and level Plant Gang**

P A0960ICE	Crawler Tractor / Dozer - Cat 16H Grader	0.34	x	87.86	=	£29.87
P A0592ICE	Hydraulic Excavator - Cat 320 96kW	0.34	x	36.36	=	£12.36
P A0943ICE	Crawler Tractor / Dozer - Cat D6 LGP 160 Hp	0.34	x	52.02	=	£17.69
	Total hourly cost of gang				=	**£59.92**

P A1126ICE   **General spoil haulage Plant Gang**

P A1011ICE	Dumper Truck - Volvo A25C 25t 6x6	3	x	60.30	=	£180.90
	Total hourly cost of gang				=	**£180.90**

P A1127ICE   **General compaction Plant Gang**

P A0868ICE	Roller - Bomag 90 900mm	1	x	3.50	=	£3.50
P A0945ICE	Crawler Tractor / Dozer - Dresser 1004 48kW	1	x	30.69	=	£30.69
	Total hourly cost of gang				=	**£34.19**

P A1140ICE   **Steel fixing Plant Gang**

P A0509ICE	Cranes Crawler - NCK 305C - 19t	0.25	x	46.29	=	£11.57
	Total hourly cost of gang				=	**£11.57**

P A1330ICE   **Clearance Plant Gang**

P A0945ICE	Crawler Tractor / Dozer - Dresser 1004 48kW	1	x	30.69	=	£30.69
P A1002ICE	Tipping Waggon - 16t 6-Wheel (24t Gr)	1	x	51.34	=	£51.34
P A0452ICE	Compressor - 2-Tool (Complete)	1	x	5.20	=	£5.20
P A0404ICE	Rock Drill	3	x	0.95	=	£2.85
	Total hourly cost of gang				=	**£90.08**

P A1331ICE   **Demolition Plant Gang**

P A0945ICE	Crawler Tractor / Dozer - Dresser 1004 48kW	1	x	30.69	=	£30.69
P A1002ICE	Tipping Waggon - 16t 6-Wheel (24t Gr)	1	x	51.34	=	£51.34
P A0452ICE	Compressor - 2-Tool (Complete)	1	x	5.20	=	£5.20
P A0404ICE	Rock Drill	3	x	0.95	=	£2.85
P A1031ICE	Cutting and Burning Gear	1	x	3.77	=	£3.77
P A0592ICE	Hydraulic Excavator - Cat 320 96kW	1	x	36.36	=	£36.36
	Total hourly cost of gang				=	**£130.21**

P K1200ICE   **Precast concrete manholes (shallow) Plant Gang**

P A0592ICE	Hydraulic Excavator - Cat 320 96kW	1	x	36.36	=	£36.36
P A0811ICE	Pump - Godwin ET50 23m3/h 4 inches	1	x	2.94	=	£2.94
P A0911ICE	Wheeled Tractor / Grader - Ford 3190H	1	x	25.17	=	£25.17
P A0961ICE	Trailer - Massey Tipping	1	x	1.81	=	£1.81
P A0544ICE	Dumper - 1.50t 2WD	1	x	2.94	=	£2.94
P A0803ICE	Trench Sheets	72	x	0.09	=	£6.48
P A0804ICE	Acrow Props	50	x	0.09	=	£4.50
P A0722ICE	Concrete Mixer - 4/3 Petrol	1	x	2.18	=	£2.18
	Total hourly cost of gang				=	**£82.38**

P Z0200ICE   **Concrete drilling Plant Gang**

P A0702ICE	Kango Type Tool - Electric Power Drill 19mm	1	x	1.12	=	£1.12
	Total hourly cost of gang				=	**£1.12**

P Z0202ICE   **Compressor Plant Gang**

P A0458ICE	Compressor - 480 cfm	1	x	20.17	=	£20.17
	Total hourly cost of gang				=	**£20.17**

Alterations

Note(s):   1) The rates in this section are for alteration works to an existing plant where the value of the alteration works is in the order of £250,000 to £500,000.
2) The rates in this section assume open and unrestricted access for the movement of labour and plant.
3) No costs have been included for the cost of disruption and loss of production that may occur to existing plant
4) No costs have been included in the disposal rates due to the wide variation in prices from tip to tip
5) Landfill Tax has not been included in the disposal rates as this is dependent on the nature of the material and statutory adjustment.

ZZ1   Alterations To Existing Works

	Unit	Labour Hours	Labour Net £	Plant Net £	Materials Net £	Unit Net £	CO_2e Kg
ZZ1.1 Buildings and other structures							
ZZ1.1.1 Diamond drilling to reinforced concrete walls							
ZZ1.1.1.01 125mm diameter holes 500mm deep	Nr	1.00	28.89	1.12	-	30.01	0.833
ZZ1.1.1.02 125mm diameter holes 750mm deep	Nr	1.25	53.43	1.40	-	54.83	1.041
ZZ1.1.1.03 200mm diameter holes 500mm deep	Nr	3.00	128.22	3.36	-	131.58	2.499
ZZ1.1.2 Forming holes in brickwork walls including making good to all edges and installation of lintel size 1.0 x 1.0m							
ZZ1.1.2.01 Thickness: 102.5mm	Nr	1.00	113.97	-	-	113.97	-
ZZ1.1.2.02 Thickness: 215mm	Nr	1.00	113.97	-	-	113.97	-
ZZ1.1.2.03 Thickness: 265mm cavity wall	Nr	2.00	227.94	-	-	227.94	-
ZZ1.2 Pipelines and other services							
ZZ1.2.1 Alterations to existing manholes							
ZZ1.2.1.01 Raising manhole cover 150mm high internal size of manhole 900 x 675mm with fair faced engineering bricks class B 215mm thick in cement mortar 1:3						773.55	176.955
ZZ1.2.1.02 Raising manhole cover 300mm high internal size of manhole 900 x 675mm with fair faced engineering bricks class B 215mm thick in cement mortar 1:3	Nr	7.20	532.87	588.82	51.80	1,173.49	288.314
ZZ1.2.1.03 Forming new 100mm diameter vitrified clay branch (including alteration of benching)	Nr	0.40	29.60	32.71	51.11	113.42	12.636
ZZ1.2.1.04 Take up and remove manhole cover and install new cover and frame 600 x 450mm grade B single seal cast iron	Nr	0.40	29.60	32.71	131.55	193.86	90.431
ZZ1.2.1.06 Filling in redundant manhole internal size 900 x 675mm 1m deep with granular sub base DTp type 1	Nr	1.00	74.01	81.78	15.45	171.24	34.171

ZZ1 Alterations To Existing Works continued...

	Unit	Labour Hours	Labour Net £	Plant Net £	Materials Net £	Unit Net £	CO_2e Kg
ZZ1.3	**Blocking up openings**						
ZZ1.3.1	Blocking up openings in brick walls with common bricks in cement mortar 1:3 fair face both sides including cut tooth and bond into existing brickwork						
ZZ1.3.1.01	102.5mm thick in stretcher bond	m²	0.85	96.87	-	15.54	112.41	38.457
ZZ1.3.1.02	215mm thick in English bond	m²	1.45	165.26	-	31.08	196.34	76.914
ZZ1.3.1.03	265mm thick in two 102.5mm stretcher bond including 50mm cavity	m²	1.80	205.15	-	31.08	236.23	76.914

ZZ2 Repairs And Renovation

	Unit	Labour Hours	Labour Net £	Plant Net £	Materials Net £	Unit Net £	CO_2e Kg
ZZ2.1	**Buildings and other structures**						
ZZ2.1.1	Pressure cleaning brickwork, masonry or concrete surfaces to receive treatment						
ZZ2.1.1.01	Vertical	m²	0.15	4.33	3.03	-	7.36	3.501
ZZ2.1.1.02	Horizontal (Bases)	m²	0.15	4.33	3.03	-	7.36	3.501
ZZ2.1.1.03	Horizontal (Soffits)	m²	0.20	5.78	4.03	-	9.81	4.669
ZZ2.1.2	Scrabble concrete surfaces 12mm deep to receive treatment, new works or repair (new works or repair measure separately)						
ZZ2.1.2.01	Vertical	m²	0.71	20.51	21.18	6.25	47.94	24.772
ZZ2.1.2.02	Horizontal (Bases)	m²	0.52	15.02	15.53	-	30.55	17.974
ZZ2.1.2.03	Horizontal (Soffits)	m²	1.00	28.89	28.24	-	57.13	32.680
ZZ2.1.3	Cut out cracks in concrete not exceeding 25 x 50mm and fill with approved polymer cement repair mortar						
ZZ2.1.3.01	Vertical	m	0.35	10.11	-	6.63	16.74	0.278
ZZ2.1.3.02	Horizontal (Bases)	m	0.31	8.96	-	6.63	15.59	0.278
ZZ2.1.3.03	Horizontal (Soffits)	m	0.45	13.00	-	6.63	19.63	0.278
ZZ2.1.4	Cut out half brick thick spalled brickwork and replace with common bricks in cement mortar 1:3						
ZZ2.1.4.01	Vertical	m²	1.80	205.15	-	15.54	220.69	38.457
ZZ2.1.4.02	Horizontal	m²	2.80	319.12	-	15.54	334.66	38.457
ZZ2.1.5	Rake out brickwork joints to a minimum depth of 25mm and point with cement mortar 1:3 flush joint						
ZZ2.1.5.01	Vertical	m²	1.50	170.95	-	0.51	171.46	2.033
ZZ2.1.5.02	Horizontal	m²	0.35	39.89	-	0.51	40.40	2.033
ZZ2.2	**Pipelines and other services**						
ZZ2.2.1	Dry grit blast (lead free) corroded pipework						
ZZ2.2.1.01	300mm diameter	m	0.75	22.29	15.13	-	37.42	17.507

ZZ2 Repairs And Renovation continued...

	Unit	Labour Hours	Labour Net £	Plant Net £	Materials Net £	Unit Net £	CO$_2$e Kg
ZZ2.2	**Pipelines and other services**						
ZZ2.2.1	Dry grit blast (lead free) corroded pipework						
ZZ2.2.1.02	450mm diameter	m	1.13	33.58	22.79	-	56.37	26.378
ZZ2.2.1.03	750mm diameter	m	1.88	55.87	37.92	-	93.79	43.885
ZZ2.2.1.04	1000mm diameter	m	2.44	72.52	49.21	-	121.73	56.957
ZZ2.2.2	Bituminous heavy duty paint in two coats to pipework						
ZZ2.2.2.01	300mm diameter	m	0.27	17.90	-	4.41	22.31	1.238
ZZ2.2.2.02	450mm diameter	m	0.41	27.18	-	6.62	33.80	1.857
ZZ2.2.2.03	750mm diameter	m	0.68	45.08	-	11.03	56.11	3.095
ZZ2.2.2.04	1000mm diameter	m	0.88	58.34	-	14.33	72.67	4.024

ZZ3 Miscellaneous Earthworks

	Unit	Labour Hours	Labour Net £	Plant Net £	Materials Net £	Unit Net £	CO$_2$e Kg
ZZ3.1	**Excavation**						
ZZ3.1.1	Excavation for foundations						
ZZ3.1.1.01	Topsoil maximum depth: not exceeding 250 mm	m^3	0.03	1.52	2.39	-	3.91	1.243
ZZ3.2	**Material other than topsoil, rock or artificial hard material**						
ZZ3.2.1	Maximum depth						
ZZ3.2.1.01	not exceeding 250 mm	m^3	0.04	1.97	3.09	-	5.06	1.609
ZZ3.2.1.02	250 - 500mm	m^3	0.05	2.23	3.52	-	5.75	1.828
ZZ3.2.1.03	500mm - 1m	m^3	0.05	2.37	3.73	-	6.10	1.938
ZZ3.3	**Disposal of excavated material**						
ZZ3.3.1	Generally						
ZZ3.3.1.01	Topsoil off site (transporting to skip distance not exceeding 5km)	m^3	0.06	2.68	10.85	-	13.53	6.794
ZZ3.3.1.02	Topsoil on site (adjacent to excavation)	m^3	0.02	0.67	2.71	-	3.38	1.698
ZZ3.3.1.03	Material other than topsoil, rock or artificial hard material off site (transporting to tip distance not exceeding 5km)	m^3	0.06	2.68	10.85	-	13.53	6.794
ZZ3.3.1.04	Material other than topsoil, rock or artificial hard material off site (adjacent to excavation)	m^3	0.03	1.34	5.43	-	6.77	3.397
ZZ3.4	**Filling**						
ZZ3.4.1	Generally						
ZZ3.4.1.01	Excavated topsoil	m^3	0.01	0.49	0.66	-	1.15	0.471
ZZ3.4.1.02	Excavated material other than topsoil or rock	m^3	0.03	1.47	1.79	-	3.26	0.824
ZZ3.4.1.03	Imported material other than topsoil of rock	m^3	0.02	1.07	3.47	15.24	19.78	2.582

ZZ3   Miscellaneous Earthworks continued...

	Unit	Labour Hours	Labour Net £	Plant Net £	Materials Net £	Unit Net £	CO$_2$e Kg	
ZZ3.4	**Filling**							
ZZ3.4.1	Generally							
ZZ3.4.1.04	Imported DTp type 1	m^2	0.01	0.49	1.79	24.72	27.00	11.421
ZZ3.4.1.05	Grass seeding n.e. 10 deg	m^2	0.00	0.11	-	2.50	2.61	-
ZZ3.4.1.06	Grass seeding 10 - 45 deg	m^2	0.00	0.11	-	2.50	2.61	-

ZZ4   Miscellaneous Concrete Works

	Unit	Labour Hours	Labour Net £	Plant Net £	Materials Net £	Unit Net £	CO$_2$e Kg	
ZZ4.1	**In situ concrete**							
ZZ4.1.1	Generally							
ZZ4.1.1.01	Provide Grade 20 (10mm aggregate)	m^3	-	-	-	93.42	93.42	261.473
ZZ4.1.1.02	Provide Grade 30 (20mm aggregate)	m^3	-	-	-	92.75	92.75	276.799
ZZ4.1.1.03	Place blinding concrete 80mm thick	m^3	0.55	36.46	-	-	36.46	-
ZZ4.1.1.04	Place reinforced ground slabs 150 - 300mm thick	m^3	0.67	43.96	-	-	43.96	-
ZZ4.1.1.05	Place reinforced ground slabs 300 - 500mm thick	m^3	0.62	40.54	-	-	40.54	-
ZZ4.1.1.06	Place reinforced walls 150 - 300mm thick	m^3	0.73	47.91	-	-	47.91	-
ZZ4.1.1.07	Place reinforced walls 300 - 500mm thick	m^3	0.61	40.01	-	-	40.01	-
ZZ4.2	**Formwork**							
ZZ4.2.1	Fair finish							
ZZ4.2.1.01	Fair finish: vertical 0.2 - 0.4m	m^2	0.76	38.62	4.99	6.34	49.95	5.721	-2.223
ZZ4.2.1.02	Fair finish: vertical 0.4 - 1.22m	m^2	1.16	58.95	7.67	7.58	74.20	7.115	-2.661
ZZ4.2.1.03	Fair finish: vertical exc. 1.22m	m^2	1.52	77.25	10.16	12.67	100.08	11.469	-4.446
ZZ4.2.1.04	Fair worked finish: vertical 0.2 - 0.4m	m^2	0.76	38.62	4.99	11.33	54.94	9.608	-3.975
ZZ4.2.1.05	Fair worked finish: vertical 0.4 - 1.22m	m^2	1.16	58.95	7.67	13.63	80.25	11.824	-4.783
ZZ4.2.1.06	Fair worked finish: vertical exc. 1.22m	m^2	1.52	77.25	10.16	22.85	110.26	19.391	-8.017
ZZ4.3	**Reinforcement**							
ZZ4.3.1	Fair finish							
ZZ4.3.1.01	High yield steel reinforcement 12mm diameter	t	16.00	762.72	185.12	686.80	1,634.64	620.004
ZZ4.3.1.02	High yield steel reinforcement 16mm diameter	t	14.00	667.38	161.98	649.06	1,478.42	616.068
ZZ4.3.1.03	Mesh reinforcement: A193	m^2	0.10	4.96	-	2.04	7.00	1.777
ZZ4.3.1.04	Mesh reinforcement: C385	m^2	0.15	7.15	-	2.30	9.45	2.007
ZZ4.3.1.05	Mesh reinforcement: A393	m^2	0.20	9.53	2.31	4.12	15.96	4.019

ZZ5   Site Clearance

	Unit	Labour Hours	Labour Net £	Plant Net £	Materials Net £	Unit Net £	CO_2e Kg
ZZ5.1 **Clearance of trees (including stumps) and backfilling void with suitable material**							
ZZ5.1.1 Girth							
ZZ5.1.1.01 100mm - 500mm	Nr	0.50	21.84	45.04	-	66.88	27.460
ZZ5.1.1.02 1m - 2m	Nr	1.00	43.67	90.08	-	133.75	54.920
ZZ5.1.1.03 500mm - 1m	Nr	1.25	54.59	112.60	-	167.19	68.650
ZZ5.2 **Demolition of buildings and other structures**							
ZZ5.2.1 Demolition brick building to ground slab level and dispose of material off site							
ZZ5.2.1.01 Volume: not exceeding 100m³	Nr	6.60	288.22	859.39	-	1,147.61	469.405
ZZ5.2.1.02 100 - 250 m³	Nr	10.50	458.54	1,367.21	-	1,825.75	746.781
ZZ5.2.1.03 250 - 500m³	Nr	12.25	534.96	1,595.07	-	2,130.03	871.245
ZZ5.2.1.04 500 - 1000m³	Nr	18.50	807.89	2,408.89	-	3,216.78	1,315.757
ZZ5.2.2 Demolition steel framed building with sheet cladding and dispose of material off site (including credit for scrap steel)							
ZZ5.2.2.01 Volume: not exceeding 100m³	Nr	7.75	338.44	1,009.13	-	1,347.57	551.196
ZZ5.2.2.02 100 - 250 m³	Nr	12.25	534.96	1,595.07	-	2,130.03	871.245
ZZ5.2.2.03 250 - 500m³	Nr	15.00	655.05	1,953.15	-	2,608.20	1,066.830
ZZ5.2.2.04 500 - 1000m³	Nr	23.00	1,004.41	2,994.83	-	3,999.24	1,635.806
ZZ5.2.3 Demolish walls to foundation level and dispose off site							
ZZ5.2.3.01 Reinforced concrete wall, thickness: 150 - 300mm	m²	2.00	87.34	180.16	-	267.50	109.840
ZZ5.2.3.02 Brick wall, thickness 225mm	m²	0.50	21.84	45.04	-	66.88	27.460
ZZ5.2.4 Take down fencing (including grubbing up foundations) and dispose off site							
ZZ5.2.4.01 Chestnut pale fencing 1200mm high (with posts)	m	0.25	10.92	-	-	10.92	-
ZZ5.2.4.02 Chain link fencing 1800mm high (with posts)	m	0.50	21.84	-	-	21.84	-
ZZ5.3 **Removal and grouting of services**							
ZZ5.3.1 Removal of redundant services							
ZZ5.3.1.01 Clay sewer or drain, diameter 100mm	m	1.00	29.72	-	-	29.72	-
ZZ5.3.1.02 Ductile iron sewer or drain, diameter: 300mm	m	1.50	44.58	-	-	44.58	-
ZZ5.3.2 Grouting redundant drains or sewers							
ZZ5.3.2.01 Clay sewer or drain, diameter 100mm	m	1.00	111.61	-	0.92	112.53	2.287
ZZ5.3.2.02 Ductile iron sewer or drain, diameter: 300mm	m	1.75	195.32	-	7.96	203.28	19.817

SECTION 2:
APPROXIMATE ESTIMATING

SECTION 2: APPROXIMATE ESTIMATING

The prices contained in this section are intended to assist in the preparation of estimates for work where an outline design has been formulated and for comparative pricing of various alternative design proposals.

Prices are based on the same sources of information used in the 'Unit Pricing' section. The prices do not include for any Preliminary type items nor for any incidental items of mechanical plant or equipment but do include normal services. Suitable percentage additions should be made to cover these and General Items. The latter is commonly covered by 15% of the measured work.

Costing by individual unit can be found in the 'Unit Pricing' section.

	EARTHWORKS	Amount £	Unit
	Excavation		
	General excavation in a large pit or into a large embankment with no restriction of movement of the machine or wagons; in moderately hard illy sandy clay ground conditions.		
a)	Plant; 2m³ hydraulic face shovel excavator; rear dump wagons; D6 tractor and blade	3.68	m³
b)	add for disposal on site not exceeding 100m	2.89	m³
c)	add for disposal off-site distance not exceeding 15km	14.53	m³
	Excavating foundations; not exceeding 500m³; in moderately hard dry sandy clay ground conditions		
d)	Plant; 2m³ hydraulic face shovel excavator; rear dump wagons; D6 tractor and blade	4.99	m³
e)	add for disposal on site not exceeding 100m	2.89	m³
f)	add for disposal off-site distance not exceeding 15km	14.53	m³
	Trunk road cutting; in moderately hard dry sandy clay ground conditions; material deposited on site within 1km		
g)	Plant; 2m motorised scrapers; pusher tractor; blade tractor and grader	3.21	m³
	Filling and compacting excavated material on site		
	General excavation; deposited in layers and compacted in open areas		
h)	Plant; 0.5m³ hydraulic excavator; tractor with blade; compactor; grader	2.88	m³
	Trunk road embankment; deposited in layers and compacted in open areas		
i)	Plant; tractor with blade; compactor; grader	1.05	m³

CONCRETE STRUCTURES

Concrete including blinding, mass and reinforced concrete used in a typical civil engineering structure such as tanks and biological filters including miscellaneous metal work and related pipework but not mechanical and electrical plant and equipment.

		Amount £	Unit
a)	Biological filters; per cubic metre of filter media		
	20,000 - 30,000 m³	80.01	m³
	10,000 -20,000 m³	106.33	m³
	2,000 -10,000 m³	125.43	m³
	Less than 2,000 m³	146.34	m³
b)	Humus tanks; per cubic metre of capacity		
	1,500 m³ and above	180.66	m³
	500 -1,500 m³	216.79	m³
	Less than 500 m³	485.20	m³

		Amount £	Unit

Bridges

c)	Reinforced in situ concrete; per square metre of deck area	1,393.66	m²
d)	Reinforced in situ deck with precast prestressed beams; per square metre of deck area	1,316.24	m²
e)	Reinforced in situ deck; with steel beams; per square metre of deck area	1,223.33	m²

OTHER STRUCTURES

Conventional single storey brick and block structure including earthworks, foundations, normal services, fixtures and fittings but excluding mechanical and electrical work.

Expressed as price per square metre of floor area measured between external walls and over internal walls.

| f) | Pumping Station including sump pump | 1,675 - 2,565 | m² |
| g) | Administration building | 1,485 - 1,795 | m² |

Typical water or sewage treatment works portal frame with cladding superstructure including earthworks, concrete and associated pipework, excluding mechanical and electrical work

| h) | Inlet building | 1,325 - 1,975 | m² |

DAMS

Typical mass concrete gravity dam; overall length 200x40m high; consisting of abutment blocks, main embankment and central intake block, single carriage. Crest roadway comprising precast concrete beams, parapets, insitu deck and piers, excluding external works, approach roads and Hydro electric works.

Ground stabilization	1,344,108
Concrete	5,120,416
Formwork	1,217,921
Reinforcement	413,280
Precast concrete	474,003
Metalwork	168,067
Brickwork	210,074
Miscellaneous	235,219
	9,183,088
Contingencies @ 10%	918,309
	10,101,397

Volume of concrete 54,000m³: cost per cubic metre	£187	per m³

ROADWORKS

Road Construction Excluding Earthworks and Structures

Trunk road pavement construction; alternative combinations in accordance with DTp specification for Road and Bridge Works assumes three lane dual carriageway road 4km in length with hard shoulders total width 33m. Prices derived from information supplied by specialist subcontractors.

a) Hot rolled asphalt
Rolled asphalt - 911 40 Wearing course

Rolled asphalt - 905 60 Base course	9.67	m²
Rolled asphalt - 904 205 Road base	30.18	m²

Granular subbase

material 803 150 Sub base +804 +805	6.66	m²
TOTAL	£46.51	m²

Road Construction Excluding Earthworks and Structures continued/...

		Amount £	Unit
b)	**Dense Bitumen macadam**		
	Rolled asphalt - 911 40 Wearing course		
	Rolled asphalt - 906 60 Base course	8.87	m²
	Rolled asphalt - 903 205 Road base	33.77	m²
	Granular subbase		
	material 803 150 Sub base	6.66	m²
	804		
	805		
	TOTAL	£49.30	m²
c)	**Lean mix concrete**		
	Rolled asphalt - 911 40 Wearing course	9.48	m²
	Dense Bitumen - 906 60 Base course		
	Dense Bitumen - 906 100 Base course	14.07	m²
	Lean Concrete - 1030 210 Road base	17.83	m²
	Granular subbase material 803 150 Sub base	6.66	m²
	TOTAL	£48.04	m²
d)	**Rigid pavement**		
	Concrete pavement - 1000 280	41.25	m²
	Granular subbase material - 803 170 Sub base	7.54	m²
	TOTAL	£48.79	m²

e) **Typical Trunk Road costs; 4km in length and 33m wide**
Average cost of pavement: £51.00 per m²

	Amount £	Unit
Total width of pavement: 33m, therefore cost per metre run of pavement	1,683.00	
Site clearance, soiling, seeding and hedging	28.00	
Lighting and telephones	51.52	
Fencing including safety barriers	117.93	
Signs and marking	31.36	
Associated drainage	365.15	
PC kerbs and channels	23.96	
	2,300.93	
Earthworks (assume value of £2,800,000)	700.00	
	3,000.93	
Contingencies @10%	300.09	
Cost of Trunk Road per m run (excluding structures and profits and overheads)	**£3,301.02**	**per m**

BRIDGES

a) Typical single span road bridge with two abutments carrying dual carriageway trunk road over two track rail line comprising precast concrete beams and in situ deck; excluding excavation and piling

	Amount £	Unit
Formwork	167,876	
Steel reinforcement	167,034	
In situ concrete	271,506	
Precast concrete	235,439	
Bearings	37,371	
Joints	20,162	
Parapets	37,504	
Finishings	3,187	
Waterproofing	67,743	
	1,007,820	
Contingencies @10%	100,782	
	1,108,602	
Deck area of bridge 928 m²: cost per square metre	**£1,195**	**per m²**

Bridges continued/...

b) Single span railway bridge with two abutments carrying two track main line trains and sidings over a proposed new road including partially constructing abutments beneath existing bridge with demolition of existing bridge and slide/roll of superstructure deck during weekend possession; excluding excavation and piling

Demolition	29,651
Formwork	59,250
Steel reinforcement	74,907
In situ concrete	125,946
Structural steel deck	249,890
Brickwork	73,256
Waterproofing	21,024
Miscellaneous	70,566
Slide/roll operation	108,649
	813,139
Contingencies @ 10%	81,314
	894,452

Deck area of bridge = 332 m²: cost per square metre	**£2,694**	per m²

PIPEWORK

The tables shown on the following pages indicate approximate guide prices for the constituent elements for pipework.

Access for delivery of large quantities of materials will probably warrant the construction of a temporary access road at an approximate cost of £33 per metre run. Smaller quantities will warrant an appropriate gang to deliver the materials along the line.

UNDERGROUND CHAMBERS

1 Precast Concrete Manholes
Excavation and backfilling in natural material; in situ concrete grade 20 in 300mm base;
300mm diameter half round channel: 300/150mm branch junction channel; 150mm half round bend channel; cement mortar benching (1:3); precast concrete manhole rings, taper section and cover slab to BS 5911 150mm in situ concrete grade 20/20 surround to rings and taper section, 600mm diameter heavy duty cast iron access cover and frame; galvanised malleable step irons to BS EN13101 at 305mm centres cast into side of chamber rings.

		Nominal diameter:			
		1200mm	1350mm	1500mm	Unit
		£	£	£	
a)	Depth; not exceeding 2m	2,059	2,303	2,818	nr
b)	2-3m	2,417	2,745	3,330	nr
c)	3-4m	2,916	3,386	4,017	nr
d)	4-5m	3,469	4,033	4,679	nr

2 Shafts
Vertical shafts formed out of precast concrete segmental bolted rings, including excavating in sand, silts and clays, disposal off site, reinstatement of ground, benching, landings, reducing slabs, ladders, handrails, flooring, chains and cover.

		Amount	Unit
		£	
	Depth 4 -8m		
a)	Internal diameter: 2.74m	6,441	m
b)	Internal diameter: 3.97m	9,588	m

PIPEWORK

Supply and installation (including earthworks and 10% allowance for fittings)

	300mm £	375mm £	450mm £	600mm £	750mm £	900mm £	1200mm £	UNIT
a) Vitrified clay pipes, BS 65, spigot and socket flexible joints								
Depth: not exceeding 1.5m	90.51	161.60	246.65	-	-	-	-	m
1.5-2m	92.37	163.58	254.61	-	-	-	-	m
2-3m	101.32	177.72	264.16	-	-	-	-	m
3-4m	148.13	230.50	326.26	-	-	-	-	m
4-5m	238.58	326.49	421.24	-	-	-	-	m
b) Ductile spun iron pipes; concrete lined, BS EN598, spigot and socket Tyton joints								
Depth: not exceeding 1.5m	127.22	-	204.77	-	-	-	-	m
1.5-2m	129.40	-	209.80	306.20	-	-	-	m
2-3m	137.06	-	216.88	315.25	-	-	-	m
3-4m	162.82	-	246.55	351.05	-	-	-	m
4-5m	202.30	-	288.27	387.35	-	-	-	m
c) UPVC pipes; BS EN1452 and BS 3506; compression joints with rubber rings								
Depth: not exceeding 1.5m	187.85	-	-	-	-	-	-	m
1.5-2m	192.22	-	-	-	-	-	-	m
2-3m	199.86	-	-	-	-	-	-	m
3-4m	227.85	-	-	-	-	-	-	m
4-5m	282.27	-	-	-	-	-	-	m

Supply and installation (including earthworks and 10% allowance for fittings)

	74mm £	102mm £	131mm £	204mm £	258mm £	290mm £	327mm £	UNIT
d) Blue MDPE (SDR 11) pipe to WIS 4-32-03 butt welded joints								
Depth: not exceeding 1.5m	22.59	30.75	40.62	48.29	72.43	101.33	123.26	m
1.5-2m	25.37	36.33	48.99	53.64	78.01	106.92	126.04	m
2-3m	36.54	47.49	57.36	62.25	87.79	116.68	138.60	m
3-4m	53.25	65.92	78.48	85.19	113.83	145.40	168.19	m
4-5m	75.15	95.11	116.78	122.53	164.91	209.25	239.33	m

Supply and installation (including earthworks and 10% allowance for fittings)

	300mm £	375mm £	450mm £	600mm £	750mm £	900mm £	1200mm £	UNIT
Concrete pipes; BS 5911; rebated flexible joints with mastic sealant to internal faces								
e) Class H								
Depth: not exceeding 1.5m	-	52.55	59.68	-	-	-	-	m
1.5-2m	-	57.12	65.26	83.37	122.07	155.05	203.38	m
2-3m	-	63.60	73.51	92.12	129.34	163.80	218.37	m
3-4m	-	90.65	106.18	126.53	164.36	202.82	269.16	m
4-5m	-	128.74	139.24	168.89	201.17	245.20	321.70	m
f) Class L								
Depth: not exceeding 1.5m	46.45	52.55	59.68	-	-	-	-	m
1.5-2m	50.55	57.12	65.26	83.37	122.07	155.05	203.38	m
2-3m	56.39	63.60	73.51	92.12	129.34	163.80	218.37	m
3-4m	81.36	90.65	106.18	126.53	164.36	202.82	269.16	m
4-5m	113.17	128.74	139.24	168.89	201.17	245.20	321.70	m

PIPEWORK continued/...

	300mm £	375mm £	450mm £	600mm £	750mm £	900mm £	1200mm £	UNIT
g) Class M								
Depth: not exceeding 1.5m	46.45	52.55	59.68	-	-	-	-	m
1.5-2m	50.55	57.12	65.26	83.37	122.07	155.05	203.38	m
2-3m	56.39	63.60	73.51	92.12	129.34	163.80	218.37	m
3-4m	81.36	90.65	106.18	126.53	164.36	202.82	269.16	m
4-5m	113.17	128.74	139.24	168.89	201.17	245.20	321.70	m

Trench Support

	300mm £	375mm £	450mm £	600mm £	750mm £	900mm £	1200mm £	UNIT
h) Depth: 1.5-2m	67.57	67.57	67.57	67.57	67.57	67.57	67.57	m
i) 2-3m	100.13	100.13	100.13	100.13	100.13	100.13	100.13	m
j) 3-4m	128.09	128.09	128.09	128.09	128.09	128.09	128.09	m
k) 4-5m	165.25	165.25	165.25	165.25	165.25	165.25	165.25	m

Excavation Through Obstructions

	300mm £	375mm £	450mm £	600mm £	750mm £	900mm £	1200mm £	UNIT
l) Extra for breaking through rock (say 0.5m depth of rock in trench)	28.18	30.83	33.46	38.75	44.03	49.32	68.67	m

Beds, Haunches and Surrounds

	300mm £	375mm £	450mm £	600mm £	750mm £	900mm £	1200mm £	UNIT
m) Sand; thickness 150mm								
Bed	10.79	12.13	13.13	15.86	18.17	20.00	25.35	m
Surround (incl bed)	27.92	31.92	35.36	43.09	50.90	59.47	87.67	m
n) Selected excavated granular material; thickness 150mm								
Bed	10.04	11.64	12.70	15.88	18.52	20.12	23.81	m
Surround (incl bed)	22.19	24.85	26.45	30.15	34.38	39.15	44.44	m
o) Imported granular material; thickness 150mm								
Bed	10.18	11.50	12.40	14.60	17.25	18.96	23.90	m
Surround (incl bed)	25.92	29.65	32.61	39.56	46.71	54.34	78.78	m
p) Mass concrete; grade 15, 20mm aggregate; thickness 150mm								
Bed	20.46	27.83	30.75	36.57	41.25	46.31	60.84	m
Haunches (incl bed)	32.76	45.70	53.82	73.99	105.44	137.99	185.88	m
Surround (incl bed)	55.50	74.24	85.59	108.00	141.99	175.56	274.50	m
q) Reinforced concrete; grade 25, 20mm aggregate; 1 layer A252 mesh reinforcement; thickness 150mm								
Bed	25.33	27.83	30.75	36.57	41.25	46.31	60.84	m
Haunches (incl bed)	38.14	45.70	53.82	73.99	105.44	137.99	185.88	m
Surround (incl bed)	62.79	74.24	85.59	108.00	141.99	175.56	274.50	m

PIPEWORK continued/...

Reinstatement

Breaking up and temporary reinstatement of roads

	300mm £	375mm £	450mm £	600mm £	750mm £	900mm £	1200mm £	UNIT
r) Break up 150mm flexible pavement and reinstate with 150mm lean mix, 40mm base course and 10mm wearing course; in bituminous macadam	60.42	-	-	72.60	-	86.81	101.93	m
s) Break up 325mm flexible pavement and reinstate with 200mm lean mix, 85mm base course and 15mm wearing course; in bituminous macadam	90.87	-	-	111.12	-	135.58	161.68	m
t) Break up 250mm rigid pavement and reinstate with 150mm. lean mix and 150mm concrete grade 20	73.18	-	-	89.62	-	110.72	133.09	m
u) Break up 300mm reinforced rigid pavement and reinstate with 200mm lean mix and 150mm concrete grade 20, reinforced (1 layer A252)	87.76	-	-	98.85	-	133.67	160.31	m

SECTION 3:
PLANT HIRE RATES AND OUTPUTS

SECTION 3: PLANT HIRE RATES AND OUTPUTS

PLANT HIRE RATES

The first part of this section details a comprehensive list of hire rates of plant, likely to be used in Civil Engineering and has been calculated on the basis of a 39 hour week.

Rates for plant hire vary considerably between hire companies depending upon many factors and readers are advised to enquire widely to obtain the most competitive rates. The rates quoted in this section have been obtained from several companies and show significant differences and the rates given will be strongly influenced by the availability of plant and by the duration of hire – longer hire periods giving rise to noticeably lower charges. Readers must allow for transporting plant to and from the site and for movement around the site as required.

Readers may wish to substitute different plant items in lieu of those detailed in the plant gang build-ups at the front of the Unit Pricing Sections. It is important to note, that the cost of plant operators has been based on the following, where applicable, which are 'average' plus rates under the W.R.A classified into 4 categories in order to simplify the wide range of plus rates for estimating purposes.

Plant Operator	Class 1	£18.01 per hour
	Class 2	£17.33 per hour
	Class 3	£16.21 per hour
	Class 4	£15.30 per hour

Fuel has been included in the total rate at the following Prime Cost sums:

Gas Oil (Red)	53.0p per litre
Diesel	145.0p per litre
Unleaded petrol	140.0p per litre

The reader can substitute current fuel costs when fine tuning the rates, to suit his own circumstances.

Fuel consumption has been based on the assumption of 75% utilization over 39 hours per week, other than the following:

Crawler cranes	= 25%
Cable excavators/bulldozer	= 85%
Mobile cranes	= 40%
Vans	= 20%
Dumpers	= 90%
Mixers, generators, pumps	= 100%
Waggons	= 112%

The second part of this section provides details of likely outputs which may be expected from commonly used Civil Engineering Plant.

Price Code	Description	Weekly Hire Charges	Consumables percent	Operator cost per hour	Fuel cost per hour	TOTAL COST PER HOUR
		£	%	£	£	£
Air Tools and Fittings						
PA0401ICE	Thor 16D/Maco SK8 Medium Duty Breaker	32.11	-	-	-	0.82
PA0402ICE	CP9 Air Breaker drill	17.66	-	-	-	0.45
PA0403ICE	CP222 Clay Digger	14.13	-	-	-	0.36
PA0404ICE	Rock Drill	37.03	-	-	-	0.95
PA0405ICE	Air Vibrating Poker up to 75mm	52.01	-	-	-	1.33
PA0406ICE	Scabbler - Single Headed Hand	38.10	-	-	-	0.98
PA0407ICE	Scabbler - Three Headed Hand	44.09	-	-	-	1.13
PA0408ICE	Scabbler - Floor - 3 Headed	102.95	-	-	-	2.64
PA0409ICE	Scabbler - Floor - 5 Headed	147.04	-	-	-	3.77
PA0410ICE	Air Pump	43.02	-	-	-	1.10
PA0411ICE	Air Hose - 3/4 inch - 15m length	5.99	-	-	-	0.15
PA0412ICE	Air Hose - 1 inch - 15m length	14.66	-	-	-	0.38
PA0413ICE	Air Hose - 1.5 inch - 15m length	20.65	-	-	-	0.53
PA0414ICE	Points and Chisels	2.14	-	-	-	0.05
PA0415ICE	Tarmac Cutters	7.38	-	-	-	0.19
PA0416ICE	Clay Spades	7.38	-	-	-	0.19
PA0417ICE	Piling Attachment	19.05	-	-	-	0.49
PA0418ICE	Vertical Rammer - 2 Stroke	64.21	-	-	-	1.65
PA0420ICE	Waggon Drill with Steel & Bits	191.13	-	-	-	4.90
PA0421ICE	Exploder with curcuit Tester	102.95	-	-	-	2.64
Bar Benders and Croppers						
PA0426ICE	Bar Bender - Manual	22.37	-	-	-	0.57
PA0427ICE	Bar Bender - Electric (to 40mm capacity)	147.04	-	-	-	3.77
PA0431ICE	Bar Croppers - Manual	22.37	-	-	-	0.57
PA0432ICE	Bar Croppers - Electric (to 40mm capacity)	147.04	-	-	-	3.77
Bowsers						
PA0441ICE	Bowsers (200 gallon water/fuel)	29.43	-	-	-	0.75
PA0442ICE	Bowsers (250 gallon water/fuel)	36.71	-	-	-	0.94
PA0443ICE	Bowsers (400 gallon water/fuel)	51.48	-	-	-	1.32
PA0444ICE	Bowsers (500 gallon water/fuel)	58.86	-	-	-	1.51
Cars and Vans						
PA1021ICE	Ford Transit 18v	137.67	-	-	5.47	9.00
PA1022ICE	Ford Escort Van 9v	115.11	-	-	3.91	6.86
PA1023ICE	Landrover 4WD	416.99	-	-	5.47	16.16
PA1024ICE	Minibus 12-seater	262.67	-	-	5.47	12.21
PA1026ICE	1300 Saloon	109.60	-	-	1.82	4.63
PA1027ICE	1500 Saloon	210.26	-	-	2.34	7.74
PA1028ICE	2000 Saloon	222.64	-	-	2.87	8.57
PA1029ICE	2000 Ghia	237.30	-	-	2.87	8.95

Price Code	Description	Weekly Hire Charges	Consumables percent	Operator cost per hour	Fuel cost per hour	TOTAL COST PER HOUR
		£	%	£	£	£
Compressors						
PA0451ICE	Compressor - Single Tool (Complete)	58.77	7.50%	-	1.33	2.95
PA0452ICE	Compressor - 2 Tool (Complete)	59.51	7.50%	-	3.56	5.20
PA0453ICE	Compressor - 100 cfm	68.14	7.30%	-	2.78	4.65
PA0454ICE	Compressor - 180 cfm	110.79	7.50%	-	4.60	7.66
PA0455ICE	Compressor - 200 cfm	172.58	7.50%	-	5.20	9.95
PA0456ICE	Compressor - 250 cfm	175.03	7.50%	-	6.74	11.57
PA0457ICE	Compressor - 375 cfm	228.18	7.50%	-	9.71	16.00
PA0458ICE	Compressor - 480 cfm	273.98	7.50%	-	12.62	20.17
PA0459ICE	Compressor - 600 to 630 cfm	342.73	7.50%	-	15.80	25.25
Concrete Equipment						
PA0461ICE	Hose & Breaker 7 3 Steels	14.66	-	-	-	0.38
PA0471ICE	Power Float 900mm etrol (Blades extra)	57.97	-	-	0.62	2.11
PA0472ICE	Cement Silo 30t	73.52	-	-	-	1.89
PA0473ICE	Cement Silo 50t	102.95	-	-	-	2.64
PA0476ICE	Concrete Skip	29.43	-	-	-	0.75
Cranes Transit						
PA0481ICE	Cranes Transit - 25t	755.34	6.00%	16.21	23.82	60.56
PA0482ICE	Cranes Transit - 12/15t	596.63	6.00%	16.21	17.73	50.16
PA0483ICE	Cranes Transit - 10t	557.08	6.00%	16.21	15.70	47.06
PA0484ICE	Cranes Transit - 8t	374.96	6.00%	16.21	14.33	40.73
Cranes Wheeled						
PA0491ICE	Cranes Wheeled - Jones Iron Fairy - 8t	398.28	6.00%	15.30	1.56	27.69
PA0492ICE	Cranes Wheeled - Jones Iron Fairy - 10t	444.13	6.00%	15.30	2.30	29.68
PA0493ICE	Cranes Wheeled - Jones Iron Fairy - 12t	535.63	6.00%	15.30	2.30	32.16
PA0494ICE	Cranes Wheeled - Jones Iron Fairy - 15t	629.27	6.00%	15.30	2.30	34.71
Cranes Crawler						
PA0502ICE	Cranes Crawler - NCK 605C - 42t	1580.87	6.00%	16.21	3.57	62.75
PA0503ICE	Cranes Crawler - NCK 407C - 32t	1501.36	6.00%	16.21	3.57	60.59
PA0505ICE	Cranes Crawler - NCK 406C - 30t	1250.53	6.00%	16.21	2.10	52.30
PA0506ICE	Cranes Crawler - NCK - 26t	1116.16	6.00%	16.21	2.71	49.25
PA0507ICE	Cranes Crawler - NCK 305B - 20t	1135.12	6.00%	16.21	1.82	48.89
PA0508ICE	Cranes Crawler - NCK - 20t Pentland	1118.62	6.00%	16.21	2.11	48.73
PA0509ICE	Cranes Crawler - NCK 305C - 19t	962.81	6.00%	16.21	4.25	46.64
PA0510ICE	Cranes Crawler - 30RB - 35t	1320.49	6.00%	16.21	2.03	54.13
PA0511ICE	Cranes Crawler - 22RB - 15t	852.29	6.00%	16.21	2.03	41.40

Price Code	Description	Weekly Hire Charges	Consumables percent	Operator cost per hour	Fuel cost per hour	TOTAL COST PER HOUR
		£	%	£	£	£
Dumpers						
PA0541ICE	Dumper - 0.60t 2WD	41.16	6.00%	-	0.58	1.70
PA0542ICE	Dumper - 0.75t 2WD	46.20	6.00%	-	0.68	1.94
PA0543ICE	Dumper - 1.25t 2WD	62.57	6.00%	-	0.68	2.38
PA0544ICE	Dumper -1.50t 2WD	59.70	4.00%	-	1.35	2.94
PA0545ICE	Dumper - 1.75t 2WD	64.14	4.00%	-	1.56	3.27
PA0546ICE	Dumper - 2.0t 2WD	79.72	4.00%	-	1.75	3.88
PA0547ICE	Dumper - 2.0t 4WD	149.52	4.00%	15.30	2.17	21.46
PA0548ICE	Dumper - 2.25t 4WD	157.38	4.00%	15.30	2.89	22.39
PA0549ICE	Dumper - 2.5t 4WD	165.13	4.00%	15.30	3.62	23.33
PA0550ICE	Dumper - 3.0t 4WD	202.58	4.00%	15.30	5.07	25.77
PA0551ICE	Dumper - 4.0t 4WD	250.87	4.00%	15.30	6.53	28.52
PA0552ICE	Dumper - 5.0t 4WD	159.91	4.00%	15.30	7.29	26.86
PA0553ICE	Dumper - 1.25t Swivel Skip	224.45	4.00%	15.30	3.62	24.91
PA0554ICE	Dumper - 2.0t 4WD Swivel Skip	203.38	4.00%	15.30	2.17	22.90
Dump Trucks						
PA1011ICE	Dumper Truck - Volvo A25C 25t 6x6	972.03	-	15.30	20.41	60.63
PA1012ICE	Dumper Truck - Volvo A35	1243.59	-	15.30	24.02	71.21
Excavators Cable						
PA0522ICE	Excavators Cable - NCK 6052C 1.34m³	1430.27	6.00%	16.21	8.90	63.99
PA0523ICE	Excavators Cable - NCK 407 0.60m³	1377.92	6.00%	16.21	9.19	62.85
PA0526ICE	Excavators Cable - NCK 305A 0.67m³	798.32	6.00%	16.21	7.15	45.06
PA0527ICE	Excavators Cable - 30RB 1.34m³	1180.16	6.00%	16.21	14.33	62.62
PA0528ICE	Excavators Cable - 22RB ICD 0.72m³	770.07	6.00%	16.21	6.89	44.03
PA0529ICE	Excavators Cable - 22RB HD 1.00m³	770.07	6.00%	16.21	6.89	44.03
Fork Lift Trucks						
PA0561ICE	Fork Lift Truck - 1.0t 2WD	196.94	3.00%	15.30	2.66	23.17
PA0562ICE	Fork Lift Truck - 2.5t 2WD	262.80	3.00%	15.30	4.46	26.70
PA0563ICE	Fork Lift Truck - 2.0t 4WD	284.85	3.00%	15.30	4.46	27.28
PA0564ICE	Fork Lift Truck - 2.5t 4WD	350.25	3.00%	15.30	4.46	29.01
Generators						
PA0571ICE	Generator - 1.5 kvA Petrol	28.38	3.00%	-	0.34	1.09
PA0572ICE	Generator - 2.5 kvA Petrol	35.66	3.00%	-	0.34	1.28
PA0573ICE	Generator - 3 kvA Petrol	50.43	3.00%	-	0.34	1.67
PA0574ICE	Generator - 4kvA Diesel	57.01	3.00%	-	0.43	1.94
PA0575ICE	Generator - 5kvA Diesel	70.99	3.00%	-	0.59	2.47
PA0576ICE	Generator - 6kvA Diesel	96.20	3.00%	-	0.68	3.22
PA0577ICE	Generator - 7kvA Diesel	92.42	3.00%	-	0.74	3.18
PA0578ICE	Generator - 10 kvA Diesel	97.77	3.00%	-	1.22	3.80
PA0579ICE	Generator - 57 kvA Diesel	487.67	3.00%	-	2.84	15.72
PA0580ICE	Generator - 150 kvA 3-Phase Skid Mounted Silenced	1223.53	3.00%	-	14.70	47.02
PA0581ICE	Generator - 250 kvA 3-Phase Skid Mounted Silenced	1690.53	3.00%	-	23.62	68.27

Price Code	Description	Weekly Hire Charges	Consumables percent	Operator cost per hour	Fuel cost per hour	TOTAL COST PER HOUR
		£	%	£	£	£
Hydraulic Excavators						
PA0592ICE	Hydraulic Excavator - Cat 320 96kW	386.70	3.0	17.19	6.58	33.98
PA0594ICE	Hydraulic Excavator - JCB 3CX Sitemaster	249.30	3.0	17.19	3.24	27.01
PA0595ICE	Hydraulic Excavator - JCB 803 R/T	305.30	3.0	17.19	1.82	27.08
PA0596ICE	Hydraulic Excavator - JCB 803 S/T	330.70	3.0	17.19	1.82	27.75
PA0597ICE	Hydraulic Excavator - JCB 802 R/T	239.10	3.0	17.19	1.82	25.33
PA0598ICE	Hydraulic Excavator - JCB 801 R/T	244.20	3.0	17.19	6.48	30.12
PA0605ICE	Hydraulic Excavator - Cat 166kW	544.40	3.0	17.19	9.84	41.41
PA0606ICE	Hydraulic Excavator - Case 588A	559.60	3.0	17.19	7.54	39.51
PA0607ICE	Hydraulic Excavator - Samsung SE130W-4	636.00	3.0	17.19	7.54	41.52
PA0608ICE	Hydraulic Excavator - Komatsu PC130	686.80	3.0	17.19	7.54	42.86
PA0609ICE	Hydraulic Excavator - Komatsu PW130	686.80	3.0	17.19	7.54	42.86
PA0610ICE	Hydraulic Excavator - Hyundai 210	747.90	3.0	17.19	5.42	42.37
PA0611ICE	Hydraulic Excavator - Cat 215	381.60	3.0	17.19	5.42	32.69
Hoists						
PA0641ICE	Scaffold Hoist 5 cut	40.70	-	-	-	1.04
PA0642ICE	Platform Hoist 10 Cut Friction	96.70	-	-	-	2.48
PA0643ICE	Platform Hoist 10 Cut Geared	96.70	-	-	-	2.48
PA0644ICE	Wyselift Hoist GL75	96.70	-	-	-	2.48
PA0645ICE	Elephante hoist 6 East	99.00	-	-	-	2.54
PA0646ICE	5 Omers Bocker Hoist	117.00	-	-	-	3.00
PA0647ICE	Elevator CEDA Hoist	101.80	-	-	-	2.61
Hydraulic Breakers - as attachments						
PA0661ICE	Hydraulic Breaker - BRH40	226.70	-	-	-	5.81
PA0662ICE	Hydraulic Breaker - BRH125	302.20	-	-	-	7.75
PA0663ICE	Hydraulic Breaker - BRH250	384.60	-	-	-	9.86
PA0664ICE	Hydraulic Breaker - BRH501	439.60	-	-	-	11.27
Hydraulic Breakers - Independent						
PA0671ICE	Hydraulic Breaker - Single Tool Static Breaker	56.00	-	-	1.22	2.66
PA0672ICE	Hydraulic Breaker - Single Tool Towable	61.10	-	-	0.67	2.24
PA0673ICE	Hydraulic Breaker - Two Tool Towable	68.70	-	-	0.98	2.75
PA0674ICE	Hydraulic Breaker - Single Tool-metreater	86.50	-	-	1.97	4.19
Kango Type Tools						
PA0701ICE	Kango Type Tool - 950 Kango	48.30	-	-	-	1.24
PA0702ICE	Kango Type Tool - Electric Power Drill 19mm	40.70	-	-	-	1.04
PA0703ICE	Kango Type Tool - Electric Power Woodauger	22.40	-	-	-	0.57
PA0704ICE	Kango Type Tool - Electric Nut Runner	18.80	-	-	-	0.48
PA0705ICE	Kango Type Tool - Electric Screwdriver	18.80	-	-	-	0.48
Lighting Towers						
PA0711ICE	Towable Lighting Tower 10m Diesel	96.20	-	-	-	2.47
PA0712ICE	Towable Lighting Tower 20m Diesel	109.90	-	-	-	2.82
PA0713ICE	Lighting Tower - Lighting the Works	193.30	-	-	-	4.96

Price Code	Description	Weekly Hire Charges	Consumables percent	Operator cost per hour	Fuel cost per hour	TOTAL COST PER HOUR
		£	%	£	£	£
Concrete Mixers						
PA0721ICE	Cement Mixer - 3/2 Petrol	36.79	2.00%	-	1.15	2.11
PA0722ICE	Cement Mixer - 4/3 Petrol	39.47	2.00%	-	1.15	2.18
PA0723ICE	Cement Mixer - 5/3 Petrol	69.88	2.00%	-	0.97	2.80
PA0724ICE	Cement Mixer - 5/3.5 Diesel Road Tow	80.80	2.00%	-	0.97	3.08
PA0725ICE	Cement Mixer - 7/5 Diesel	85.08	2.00%	-	0.97	3.19
PA0729ICE	Cement Mixer - 10/7 Diesel	84.46	2.00%	-	3.65	5.86
PA0730ICE	Cement Mixer - Mortar Pan Mixer	146.61	2.00%	-	0.97	4.80
PA0735ICE	Cement Mixer - Liner Rolpanit	820.12	2.00%	16.21	2.66	40.32
PA0737ICE	Cement Mixer - Schwing BP1000R Concrete Pumps	1059.36	2.00%	16.21	2.75	46.67
PA0738ICE	Cement Mixer - 18/12 Diesel	234.03	2.00%	16.21	3.99	26.33
Tunneling Equipment						
PA1049ICE	Micro Tunneling Equipment (rock) 1.2m dia	11676.01	20.00%	-	-	359.26
PA1050ICE	Micro Tunneling Equipment (rock) 1.5m dia	12028.31	20.00%	-	-	370.10
PA1051ICE	Micro Tunneling Equipment (rock) 1.8m dia	12783.10	20.00%	-	-	393.33
PA1052ICE	Micro Tunneling Equipment (rock) 3.0m dia	16859.49	20.00%	-	-	518.75
PA1053ICE	Micro Tunneling Equipment (soil) 1.2m dia	10417.70	20.00%	-	-	320.54
PA1054ICE	Micro Tunneling Equipment (soil) 1.5m dia	10518.29	20.00%	-	-	323.64
PA1055ICE	Micro Tunneling Equipment (soil) 1.8m dia	10770.00	20.00%	-	-	331.38
PA1056ICE	Micro Tunneling Equipment (soil) 3.0m dia	13839.79	20.00%	-	-	425.84
Piling Equipment						
PA1046ICE	Pile Jetting Pump and Accessories	436.52	30.00%	-	21.86	36.41
PA1047ICE	Pile Testing Rig and Equipment	4862.45	45.00%	18.01	36.32	235.12
PA1066ICE	Driven Piling Rig Large	4459.64	45.00%	18.01	36.32	220.14
PA1067ICE	Driven Piling Rig Medium	2990.29	45.00%	18.01	36.32	165.51
PA1062ICE	Bored Piling Rig Large	4459.64	45.00%	18.01	36.32	220.14
PA1063ICE	Bored Piling rig Medium	2990.29	45.00%	18.01	36.32	165.51
PA1079ICE	Sheet Piling Rig Large	4459.64	45.00%	18.01	36.32	220.14
PA1080ICE	Sheet Piling rig Medium	2990.29	45.00%	18.01	36.32	165.51
Drilling Equipment						
PA1060ICE	Anchor Drilling Rig (rock)	643.71	45.00%	18.01	43.72	85.67
PA1061ICE	Anchor drilling Rog (soil)	473.02	45.00%	18.01	43.72	79.32
Pipework and Plumbing Equipment						
PA1083ICE	Vactor Unit for Sewer Cleaning	145.20	10.00%	-	4.40	8.50
PA1084ICE	Pipe Thrust Boring Equipment	6846.96	25.00%	-	-	219.45
PA1085ICE	Pipe Jacking Equipment	13630.36	20.00%	-	-	419.40
PA1041ICE	Pressure Testing Equipment	436.52	30.00%	-	21.86	36.41
PA1065ICE	CCTV unit for Sewer Inspection	2938.67	10.00%	15.30	4.40	102.59
Fencing Equipment						
PA1335ICE	Fence Post Rammer	31.89	0.00%	-	-	0.82
PA1336ICE	Post Hole Borer: Two Man	108.85	20.00%	0.00	1.26	4.61
PA1059ICE	Agricultural Tractor: Fencing Auger	202.56	10.00%	15.30	4.40	25.42

Price Code	Description	Weekly Hire Charges	Consumables percent	Operator cost per hour	Fuel cost per hour	TOTAL COST PER HOUR
		£	%	£	£	£
Sheet Piling Equipment (Excluding Compressors)						
PA0752ICE	160 cfm - Air Hammer	177.01	-	-	-	4.54
PA0753ICE	250 cfm - Air Hammer	240.68	-	-	-	6.17
PA0754ICE	480 cfm - Air Hammer	290.23	-	-	-	7.44
PA0755ICE	600 cfm - Air Hammer	325.55	-	-	-	8.35
PA0761ICE	BSP 300 - Air Hammer	240.68	-	-	-	6.17
PA0762ICE	BSP 500 - Air Hammer	268.93	-	-	-	6.90
PA0763ICE	BSP 600 - Air Hammer	297.29	-	-	-	7.62
PA0764ICE	Zeneth 20 Sheet Pile Extractor	254.81	-	-	-	6.53
PA0765ICE	Zeneth 80 Sheet Pile Extractor	283.17	-	-	-	7.26
PA0766ICE	Zeneth 120 Sheet Pile Extractor	594.59	-	-	-	15.25
PA0767ICE	HD7 Sheet Pile Extractor	530.81	-	-	-	13.61
PA0768ICE	HD10 Sheet Pile Extractor	637.07	-	-	-	16.34
PA0769ICE	HD15 Sheet Pile Extractor	778.55	-	-	-	19.96
Air Hoses						
PA0781ICE	Air Hose - 3/4 inch x 15m length	5.89	-	-	-	0.15
PA0782ICE	Air Hose - 1.5 inch x 15m length	20.55	-	-	-	0.53
PA0783ICE	Air Hose - 2 inch x 15m length	22.05	-	-	-	0.57
PA0784ICE	Air Hose - 2.5 inch x 15m length	29.43	-	-	-	0.75
Diesel Hammers						
PA0792ICE	Diesel Hammer up to 1500kg	479.79	-	-	3.57	15.88
PA0793ICE	Diesel Hammer up to 2500kg	566.91	-	-	5.79	20.33
PA0794ICE	Diesel Hammer up to 5000kg	964.96	-	-	7.41	32.15
PA0795ICE	Diesel Hammer up to 7500kg	1619.37	-	-	7.41	48.93
Quick Release Shackles						
PA0801ICE	Quick Release Shackles - Larssen Single	26.11	-	-	-	0.67
PA0802ICE	Quick Release Shackles - Frodingham Double	46.87	-	-	-	1.20
Trench Supports						
PA0803ICE	Trench Sheets	3.32	-	-	-	0.09
PA0804ICE	Acrow Props	3.32	-	-	-	0.09
PA0805ICE	Timber Baulks	25.47	-	-	-	0.65
Pumps						
PA0811ICE	Godwin ET50 23m³/h 4 inches	79.91	20.00%	-	0.48	2.94
PA0812ICE	Godwin ET75 74m³/h 4 inches	116.79	20.00%	-	0.81	4.40
PA0813ICE	Godwin Univac 4 inches 500 gpm	173.36	20.00%	-	2.60	7.94
PA0814ICE	Godwin Univac 6 inches 1000 gpm	229.39	20.00%	-	4.25	11.31
PA0815ICE	Godwin HL4 200m³/h	335.36	20.00%	-	3.57	13.89
PA0816ICE	Godwin HL5 150/100mm 1200m³/h	392.80	20.00%	-	4.25	16.34
PA0817ICE	Godwin HL8 200mm 2000 gpm	313.91	20.00%	-	9.44	19.09
PA0818ICE	Godwin HL6 200/150mm 1250 gpm	479.23	20.00%	-	7.42	22.17
PA0819ICE	Flygt B2050 2 inch Discharge	70.27	20.00%	-	1.91	4.07
PA0820ICE	Flygt B2066 3 inch Discharge	83.85	20.00%	-	2.97	5.55
PA0821ICE	Flygt B2102 4 inch Discharge	101.91	20.00%	-	3.70	6.84
PA0822ICE	Flygt B2125 6 inch Discharge	144.73	47.00%	-	9.87	15.33
PA0823ICE	Flygt B2151 6 inch Discharge	160.70	41.00%	-	14.82	20.63

Price Code	Description	Weekly Hire Charges	Consumables percent	Operator cost per hour	Fuel cost per hour	TOTAL COST PER HOUR
		£	%	£	£	£
Pumps continued/...						
PA0824ICE	Flygt C53085MT 3 inch Discharge	80.72	10.00%	-	2.97	5.24
PA0825ICE	Flygt CS3102 4 inch Discharge	150.20	30.00%	-	3.88	8.89
PA0826ICE	Flygt CS3126MT 6 inch Discharge	163.73	50.00%	-	9.87	16.17
PA0827ICE	Weda 2 inches L154 30m³/h	65.30	10.00%	-	1.92	3.77
PA0828ICE	Weda 3 inches L262/4 87m³/h	75.58	50.00%	-	3.94	6.84
PA0829ICE	Weda 4 inches L502 107m³/h	131.80	45.00%	-	4.89	9.79
PA0830ICE	Weda 4 inches L554 180m³/h	123.73	50.00%	-	7.81	12.57
PA0832ICE	Godwin Heidra 100mm Pump (Hydrosubmersible) Complete with 10m Hoses and Power Pack on Site Trolley	209.39	40.00%	-	4.89	12.41
PA0833ICE	Pump - Godwin Heidra 100mm Pump Sewer Cleaner (Hydrosubmersible) Complete with 10m Hoses and Power Pack on Site Trolley	239.40	40.00%	-	5.56	14.15
PA0834ICE	Pump - Godwin Hycon 100mm Pump (Hydrosubmersible) Complete with 10m Hoses, Strainer and Power Pack on Site Trolley	266.16	40.00%	-	5.56	15.11
PA1081ICE	Shotcrete Pump	381.98	20.00%	-	18.98	30.74
PA1334ICE	Pump 100mm 1500 l/min	105.95	10.00%	-	1.45	4.44
PA1077ICE	Screed Pump	503.09	20.00%	-	-	15.48
Diaphragm Pumps						
PA0836ICE	Petrol Pump c/w 6m 50mm Suction Hose and 6m 75mm Delivery Hose	70.17	20.00%	-	0.22	2.38
PA0837ICE	Diesel Pump 75mm c/w 6m Suction Hose and 6m Delivery Hose	101.79	20.00%	-	0.46	3.60
PA0838ICE	Extra for Additonal 50mm x 6m Suction Hose	10.92	-	-	-	0.28
PA0839ICE	Extra for Additonal 75mm x 6m Delivery Hose	6.53	-	-	-	0.17
Rammers and Compacters						
PA0851ICE	Vibrating Plate Light	42.09	-	-	0.33	1.41
PA0852ICE	Vibrating Plate Light Medium	47.55	-	-	0.33	1.55
PA0853ICE	Vibrating Plate Diesel 24kN	85.65	-	-	0.33	2.53
PA0854ICE	Vibrating Plate Diesel 33.5kN	102.02	-	-	0.33	2.95
Rollers						
PA0861ICE	Single Drum - 28 inch Pedestrian	53.24	2.00%	-	0.54	1.93
PA0862ICE	Single drum - 28 inch Roller Breaker c/w Trailer	92.40	2.00%	-	0.54	2.95
PA0863ICE	28 inch Vibratory Roller 0.37t	63.08	2.00%	-	0.54	2.19
PA0864ICE	Bomag (Trench) 350mm	77.77	2.00%	-	0.41	2.44
PA0865ICE	Bomag 55 550mm	57.62	2.00%	-	0.54	2.04
PA0866ICE	Bomag 60 600mm	59.74	2.00%	-	0.67	2.23
PA0867ICE	Bomag 75 750mm	72.08	2.00%	-	0.96	2.84
PA0868ICE	Bomag 90 900mm	87.33	2.00%	-	1.22	3.50
PA0869ICE	800mm Ride on Roller	197.38	2.00%	-	1.33	6.49
PA0870ICE	1000mm Ride on Roller	222.58	3.00%	-	3.99	9.87
PA0871ICE	1200mm Ride on Roller	259.61	3.00%	-	3.99	10.85

Price Code	Description	Weekly Hire Charges	Consumables percent	Operator cost per hour	Fuel cost per hour	TOTAL COST PER HOUR
		£	%	£	£	£
Rollers continued/...						
PA0872ICE	1300mm Ride on Roller	271.48	3.00%	-	3.99	11.16
PA0874ICE	Aveling Barford DCI2 8.5t	222.58	3.00%	-	3.99	9.87
PA0876ICE	Two Drum Ride On	148.52	3.00%	-	3.99	7.92
PA0878ICE	Case Vibromax W651	154.78	3.00%	-	3.53	7.62
Sawing						
PA0891ICE	Saw Bench - 24 inch Diesel/Electric	65.76	3.00%	-	0.29	2.03
PA1058ICE	9" Circular Saw	37.03	10.00%	-	-	1.04
Road Plant						
PA0900ICE	Tar Sprayer	51.69	-	-	-	1.33
PA0901ICE	Barber Greene BGP 255 Paver	3563.87	3.00%	16.21	-	110.33
PA0902ICE	Barber Greene BGP 200 Paver	1483.99	3.00%	16.21	-	55.40
PA1048ICE	Road Marking Vehicle	455.95	10.00%	-	34.45	47.31
Wheeled Tractors/Graders						
PA0911ICE	Ford 3190H	228.44	4.00%	15.30	4.11	25.51
PA0912ICE	Ford 4610H	255.23	2.00%	15.30	4.40	26.38
PA0913ICE	Ford 5610H	266.11	2.00%	15.30	5.15	27.42
PA0914ICE	Ford 6610H	290.65	2.00%	15.30	4.40	27.30
PA0916ICE	Massey Ferguson MF20B 35kW	189.95	2.00%	15.30	4.40	24.67
PA0917ICE	Massey Ferguson MF50E 52kW	204.85	4.00%	15.30	4.40	25.17
Crawler Tractors/Dozers						
PA0931ICE	Cat D3C LGP 52kW	582.77	4.00%	16.21	5.62	37.37
PA0932ICE	Cat D4C LGP 60kW	580.42	4.00%	16.21	6.83	38.52
PA0933ICE	Cat D5M LGP 82kW	669.39	4.00%	16.21	9.61	43.67
PA0934ICE	Cat D6R LGP 138kW	759.18	4.00%	16.21	12.85	49.31
PA0935ICE	Cat D8R LGP	1351.24	4.00%	16.21	24.78	77.02
PA0937ICE	Cat 302kW Pusher	1716.50	4.00%	16.21	28.52	90.51
PA0938ICE	Cat 931	528.41	4.00%	16.21	5.62	35.92
PA0939ICE	Cat 941	548.26	4.00%	16.21	7.35	38.18
PA0940ICE	Cat 951	597.27	4.00%	16.21	7.35	39.49
PA0941ICE	Cat 621F 246kW 153m³	1245.82	4.00%	16.21	30.28	79.71
PA0942ICE	Cat 631F 365kW 23.7m³	1390.89	4.00%	16.21	37.69	90.99
PA0943ICE	Cat D6 LGP 160Hp	762.98	4.00%	16.21	15.80	52.36
PA0944ICE	Cat D456kW	577.64	4.00%	16.21	6.83	38.45
PA0945ICE	Dresser 1004 48kW	384.16	4.00%	16.21	4.57	31.03
PA0946ICE	Cat 561 Sideboom	707.49	4.00%	16.21	9.61	44.69
PA0953ICE	Fiatallis FD145 93kW	661.56	-	16.21	22.29	55.46
PA0956ICE	Terex TS14 D Scraper (Wheeled) 10.7m³	1334.82	3.00%	16.21	22.23	73.69
PA0957ICE	Terex TS24 C Scraper (Wheeled) 18.4m³	1681.57	2.50%	16.21	42.95	103.36
PA0958ICE	Cat 633E Scraper	1334.41	2.50%	16.21	45.50	96.78
PA0959ICE	Tractor D8 Blade	1148.43	2.50%	16.21	22.00	68.39
PA0960ICE	Cat 16H Grader	1052.14	2.50%	16.21	44.34	88.21
Tipping Wagon						
PA1001ICE	10t 4-Wheel (16t Gr)	591.89	-	15.30	14.73	45.21

Price Code	Description	Weekly Hire Charges	Consumables percent	Operator cost per hour	Fuel cost per hour	TOTAL COST PER HOUR
		£	%	£	£	£
Tipping Wagon continued/...						
PA1002ICE	16t 6-Wheel (24t Gr)	723.08	-	15.30	17.83	51.67
PA1003ICE	20t 8-Wheel (30t Gr)	929.35	-	15.30	19.45	58.58
Trailers						
PA0961ICE	Massey Tipping	70.74	-	-	-	1.81
PA0962ICE	28 inch Vib Roller Trailer	27.18	-	-	-	0.70
PA0963ICE	D/D Vib Roller Trailer	43.56	-	-	-	1.12
PA0964ICE	Road Sweeper	136.13	-	-	-	3.49
Traffic Lights						
PA0971ICE	Main Generator 2-Way with 100m Cables	125.21	-	-	-	3.21
PA0981ICE	Petrol up to 75mm	39.02	-	-	0.42	1.42
PA0982ICE	Diesel up to 75mm	42.79	-	-	0.17	1.27
Welding Sets						
PA0991ICE	300 amp Diesel Electric Start Sil	122.68	-	-	1.33	4.48
PA0992ICE	250 amp Diesel Electric Start Sil	101.61	-	-	1.16	3.76
PA0993ICE	200 amp Petrol Recoil Start	60.68	-	-	1.12	2.67
PA0994ICE	Plastic Pipe Welding (Small)	266.79	-	-	-	6.84
PA0995ICE	Plastic Pipe Welding (Large)	353.91	-	-	-	9.07
Miscellaneous						
PA0740ICE	12m Hydraulic Platform	403.99	-	-	-	10.36
PA0741ICE	Asphalt Boiler	81.12	-	-	-	2.08
PA0742ICE	Towered Brush - Canline	75.13	-	-	-	1.93
PA0744ICE	Shotblast Equipment - 160 (1.5 Bag Pot)	80.91	-	-	-	2.07
PA0745ICE	Shotblast Equipment - 350 (3.0 Bag Pot)	95.57	-	-	-	2.45
PA0746ICE	Shotblast Equipment (3.0 Bag Pot) with 250 cfm compressor	139.98	-	-	-	3.59
PA0747ICE	Paint Spray Equipment - HD 1 Gun	81.78	-	-	0.97	3.07
PA0730ICE	Rammer Benjo	62.07	-	-	-	1.59
PA0731ICE	Cutting and Burning Gear	147.04	-	-	-	3.77
PA0739ICE	Bentonite Plant	1683.54	30.00%	-	59.40	115.52
PA1044ICE	Articulated Lorry	951.71	25.00%	18.01	20.41	68.92
PA1045ICE	Low Loader Trailer	544.50	-	-	-	13.96
PA1064ICE	Cable Percussion Rig	188.59	45.00%	-	1.62	8.63
PA1068ICE	Ground Anchor Drilling Rig	1804.51	25.00%	18.01	43.72	119.57
PA1069ICE	Grouting Rig (Anchorage Items)	283.17	45.00%	-	-	10.53
PA1073ICE	Robotic Sewer Repair Unit	1282.99	20.00%	-	34.45	73.93
PA1074ICE	Rotary Borehole Rig	300.28	45.00%	-	2.01	13.17
PA1076ICE	Scissor Lift - 2032	140.94	20.00%	-	-	4.34
PA1078ICE	Self Propelled Chip Spreader	406.96	10.00%	15.30	4.40	31.18

TYPICAL PLANT OUTPUTS

The information used in this section has been derived from generally available information published by certain manufacturers but in assessing capacities and outputs readers are advised to consult the makers of the equipment being assessed to confirm capacities and outputs.

Civil engineering activities often require large quantities of material to be excavated and removed. The types of excavation determine the type of machine used assuming it to be available. The most common types are:

Trenches
Pits
Cuttings/Banks
Foundations
each can vary considerably in size.

To illustrate the calculation of machine outputs a reasonable sized task has been chosen for each operation.

Excavation to:

Trenches
A trench 7 metres deep and 2 metres wide at the bottom in stiff sandy clay with battered sides. The material is dumped by the side of the trench ready for backfilling.

Use: Case Poclain 170B.

Output which can be expected depends on the following factors:
1)   Activity   (A)   The type of activity can vary from very difficult restricted work to straight forward unrestrained work. Factors vary from 0.5 to 1.0.
2)   Cycle time   (C)   Comprises four movements, dig, slew loaded, dump and slew empty.

The movements are affected by:
a) the nature of the material being dug; see table I.
b) the position of the point of dig in relation to the point of dump; see table 2.
3)   Bucket capacity   (B)   The nature of the material affects the makers rated capacity. This is normally expressed as a CECE* rating.
4)   Job Efficiency   (E)   This takes into account the actual time in a shift that its machine is usefully engaged. A normal rating is 0.83%.

Applying these factors to this task.

1)   Activity   (A)   The machine has to move frequently otherwise straight forward. Use factor 0.9 (A2).

2)   Cycle time   apply factors to theoretical timing.
dig 4.0 seconds from table 1 apply factor of 1.42 (D3)
4.0 x 1.42   = 5.68 sec.
slew loaded 5.0 sec from table 2 use factor of 1.5 (S3)
5.0 x 1.50   = 7.50 sec.
dump 1.5 seconds   = 1.50 sec.
slew empty   = 4.00 x 1.42 (S3)
                           = 5.68 sec.
TOTAL   20.36 secs.
Equals   176 cycles per hour.

3) Bucket capacity   CECE rating 1.6m^3 apply factor from table 3 for stiff clay 0.85 (B3).
Capacity = 1.60 x 0.85 = 1.36m^3.

NOTE: *CECE - Committee on European Construction Equipment

TYPICAL PLANT OUTPUTS continued/...

Trenches continued/...

4) Job Efficiency use 0.83 (E2).

Thus:		
	Activity	= 0.9
	Cycles	= 176 per hour
	Bucket Capacity	= 1.36 m³
	Efficiency	= 0.83
	Output equals	= 0.9 x 176 x 1.36 x 0.83
		= 178m³ per hour

Output equals 178m³ per hour. This is loose material. Apply bulking factor of 0.85 from table 4. Output measured in the solid equals 178 x 0.85 = 152m³ per hour.

This is excavation only, no allowance has been made for fairing up the bottom of the trench or backfilling or double handling.

Excavation to:

Pits

Assume a large cut excavation on an open cast coal site. Typical coal measure overburden – loose shale and sandy clay.

Use Caterpillar 245 Excavator with face shovel equipment, bottom dump bucket rated capacity 3.1m³ loading into dump-trucks. Unrestricted space for trucks to manoeuvre.

Activity	Straight forward assuming adequate number of trucks.	
	Use factor 1.0 (A1).	
Cycle	Makers theoretical rated cycle times.	
	Dig 9 seconds x 1.3 (D2)	11.7
	Slew loaded 4.0 x 1.3 (S2)	5.2
	This allows for digging from track level and slewing 45 degrees to load into dump-truck 4m high.	
	Dump 2.5 sec	2.5
	Slew empty 4.0 x 1.3 (S2)	5.2
		24.6 secs.

24.6 secs. equals 146 cycles per hour.

Bucket Capacity	3.1m³. Bucket fill factor 0.85 (B2) = 2.65
Job Efficiency	= 0.83
Output	= 1 x 146 x 2.65 x 0.83
	= 321m3 per hour

This is loose cubic metres per hour allowing a bulking factor of 0.75 (from table 4) the solid measure equals 321 x 0.75 = 240m³ per hour.

Excavation to:

Cuttings

Assume a motorway cut and fill operation. The average length of haul is 500 metres. Material stiff, sandy, clay, dry. Haul grade no more than 4 per cent.

A typical plant spread for this operation would consist of:

4/6 motorized scrapers
Tractor D9 pusher
Tractor D8 blade
Grader Caterpillar

TYPICAL PLANT OUTPUTS continued/...

Cuttings continued/...

The production units being the scrapers; the output of one scraper is examined here.

Use a Cat 63lE.

Activity	The task is a normal one and a factor of 1.0 (A1) is used.	
Cycle	Load 36 sees. Use factor of 0.9 (DI)	
	36 x 0.9	= 32 secs.
	Haul loaded	= 84 secs.
	Use factor of 1.0 (S1)	= 84 secs.
	Spread and manoeuvre 42 secs.	= 42 secs.
	Haul empty 60 secs.	
	Use factor of 1.0 (S1)	= <u>60</u> secs.
	TOTAL	= <u>218</u> secs.
Cycle	equals 218 secs. equals 16.5 per hour.	
Capacity	rated capacity average 20m³ 20m³ loose. Use factor 0.85 (B2). Capacity equals 20 x 0.85 = 17m³.	
Efficiency	use 0.83.	
Output	Activity	= 1.0
	Cycle	= 16.5 per hour
	Capacity	= 17.0 m³
	Efficiency	= 0.83
	output equals 1 x 16.5 x 17 x 0.83	= 232m³

To convert to solid cubic metres refer to table 4 use a bulking factor of 0.79.
Output in solid m³ equals 232 x 0.79 = 183m³ per hour

Excavation to:

Pile Caps

1)	Type of activity	Use a JCB 3CX Sitemaster. This means a lot of travel is required by the machine and a set up at each position. Use an activity factor of 0.5 (A4)
2) Cycle Time		Dig will be slower because of much higher proportion of squaring off and cleaning bottom.
		Use factor (D5) 2.5.
		Use factor (S3) 1.67.

Dig 2.6 x 2.50	= 6.5
Slew 2.1 x 1.67	= 3.5
Dump 1.5	= 1.5
Slew 1.7 x 1.67	= <u>2.8</u>
Total cycle time	<u>14.3</u> seconds
= 251 cycles per hour	

3) Bucket		Capacity slightly less than before because of obstructions of pile heads and steel. Use factor (B4) 0.6.
4) Job Efficiency		This will be low because of constant moving. Use factor 0.75.

Output is calculated as follows:
activity x cycles x bucket x job efficiency
= 0.5 (A4) x 251 x 0.24 x 0.6 (B4) x 0.75 = 13.55m³ per hour

TYPICAL PLANT OUTPUTS continued/...

Loading excavated materials

Load, excavated, medium hard clay spoil into 3 tipper wagons. Use Drott International Crawler Loader.

1)	Activity	No impediment to movement. Use activity factor 1.0 of (A1).

2)   Cycle time   Use factors shown:

Load	3 x (D2) 1.3	= 3.9	
Manoeuvre	4 x (S2) 1.3	= 5.2	
Dump	1.5	= 1.5	
Manoeuvre	3 x (S2) 1.3	= 3.9	
Total cycle time		14.5 seconds	

= 248 cycles per hour

3)   Bucket   CECE rated capacity 1m3. Use factor 0.85 (B2).

4)   Job Efficiency   = 0.83

Output

1.0 (A1) x 248 cycles x bucket

0.85 (B2) x efficiency 0.83

1.0 x 248 x 0.85 x 0.83

175m^3 per hour

Weight of material bulked; see table 4

Medium hard clay dry. 1.480 tonne m^3

= 175 x 1.480 = 259 tonne per hour.

Tables to be used for excavation output calculations

Table 1 - Dig Factors

(DI)	0.9 to 1.0	Soft soil or sand
(D2)	1.0 to 1.3	Medium hard clay or gravel with cobbles
(D3)	1.3 to 2.0	Hard clay with some boulders
(D4)	2.0 to 2.5	Well blasted rock
(D5)	2.5 to 4.0	Frequent obstructions, working with trench shield

Table 2 -Slew Factors

(S1)	0.9 to 1.0	Trench not less than 2m deep. Spoil can be dumped on the ground by the trench
(S2)	1.0 to 1.3	Trench up to 3m deep. Spoil dump on ground or into wagons not more than 3m above track level. Slew not more than 900.
(S3)	1.3 to 2.0	Trench up to 4m deep. Spoil to be dumped into wagons at rear of m/c.
(S4)	2.0 to 2.5	Trench over 4m deep. Spoil to be dumped into wagon.

Table 3 - Bucket Fill Factors

The Committee on European Construction Equipment (CECE) rates heaped bucket pay loads on a 2:1 angle of repose.

CECE capacity times

(BI)	1.0	Soft soil
(B2)	0.85 to 1.0	Medium hard clay or gravel
(B3)	0.7 to 0.85	Hard clay
(B4)	0.5 to 0.75	Well blasted rock
(B5)	0.4 to 0.5	Poorly blasted rock

TYPICAL PLANT OUTPUTS continued/...

Table 4 - Bulking Factors

WEIGHT OF MATERIALS*	LOOSE		SOLID		BULKING FACTOR
	kg/m^3	lb/yd^3	kg/m^3	lb/yd^3	
Basalt	1960	3300	2970	5000	0.67
Bauxite, Kaolin	1420	2400	1900	3200	0.75
Caliche	1250	2100	2260	3800	0.55
Carnotite, uranium ore	1630	2750	2200	3700	0.74
Cinders	560	950	860	1450	0.66
Clay - Natural bed	1660	2800	2020	3400	0.82
Dry	1480	2500	1840	3100	0.81
Wet	1660	2800	2080	3500	0.80
Clay & Gravel - Dry	1420	2400	1660	2800	0.85
Wet	1540	2600	1840	3100	0.85
Coal - Anthracite Raw	1190	2000	1600	2700	0.74
Washed	1100	1850			0.74
Ash, Bituminous Coal	540-650	900-1100	590-890	1000-1500	0.93
Bituminous, Raw	950	1600	1280	2150	0.74
Washed	830	1400			0.74
Decomposed rock -					
75% Rock, 25% Earth	1960	3300	2790	4700	0.70
50% Rock, 50% Earth	1720	2900	2280	3850	0.75
25% Rock, 75% Earth	1570	2650	1960	3300	0.80
Earth - Dry packed	1510	2550	1900	3200	0.80
Wet excavated	1600	2700	2020	3400	0.79
Loam	1250	2100	1540	2600	0.81
Granite - Broken	1660	2800	2730	4600	0.61
Gravel - Pitrun	1930	3250	2170	3650	0.89
Dry	1510	2550	1690	2850	0.89
Dry 6-50mm (0.25" - 2")	1690	2850	1900	3200	0.89
Wet 6-50mm (0.25" - 2")	2020	3400	2260	3800	0.89
Gypsum - Broken	1810	3050	3170	5350	0.57
Crushed	1600	2700	2790	4700	0.57
Hematite, iron ore, high grade	1810-2450	4000-5400	2130-2900	4700-6400	0.85
Limestone - Broken	1540	2600	2610	4400	0.59
Crushed	1540	2600	-	-	-
Magnetite, iron ore	2790	4700	3260	5500	0.85
Pyrite, iron ore	2580	4350	3030	5100	0.85
Sand - Dry loose	1420	2400	1600	2700	0.89
Damp	1690	2850	1900	3200	0.89
Wet	1840	3100	2080	3500	0.89

* Varies with moisture content, grain size, degree of compaction, etc. Tests must be made to determine exact material characteristics.

TYPICAL PLANT OUTPUTS continued/...

Table 4 - Bulking Factors continued/...

WEIGHT OF MATERIALS*	LOOSE		SOLID		BULKING FACTOR
	kg/m³	lb/yd³	kg/m³	lb/yd³	
Sand and clay - Loose	1600	2700	2020	3400	0.79
Compacted	2400	4050	-	-	-
Sand & Gravel - Dry	1720	2900	1930	3250	0.89
Wet	2020	3400	2230	3750	0.91
Sandstone	1510	2550	2520	4250	0.60
Shale	1250	2100	1660	2800	0.75
Slag - broken	1750	2950	2940	4950	0.60
Snow - Dry	130	220	-	-	-
Wet	520	860	-	-	-
Stone - Crushed	1600	2700	2670	4500	0.60
Taconite	1630-1900	3600-4200	2360-2700	5200-6100	0.58
Top Soil	950	1600	1370	2300	0.70
Traprock - broken	1750	2950	2610	4400	0.67

* Varies with moisture content, grain size, degree of compaction, etc. Tests must be made to determine exact material characteristics.

Summary of Tables 1, 2 and 3

	Factors	1	2	3	4	5
(A)	Activity	1.0	0.85	0.75	0.6	0.5
(D)	Dig	0.9 - 1.0	1.0 - 1.3	1.3 - 2.0	2.0 - 2.5	2.5 - 4.0
(S)	Slew	0.9 - 1.0	1.0 - 1.3	1.3 - 2.0	2.0 - 2.5	-
(B)	Bucket	1.1 - 1.0	1.0 - 0.85	0.85 - 0.70	0.7 - 0.5	-
(E)	Efficiency	0.9	0.83	0.75	0.6	-

Output in cubic metres per hour equals:
Activity factor (A) x Cycles per hour x Bucket CECE rating x factor (B) x Efficiency factor (E)

TYPICAL PLANT OUTPUTS continued/...

Compaction by vibrating roller – light

Bomag BW 655 width 650rnm double vibrating roller pedestrian controlled Mass 600kg.
Speed 1st gear 0.875mph = 1.4km/h

Compaction is specified as: mass per metre width of roll. Number of passes depends on the thickness of material.

Compact granular material on footpath formation 75rnm thick 2m wide 100m long.
Use Bomag BW 655. DTp clause 802 requires 6 passes of this mass of double roller.

Factors affecting this output. Uneven surface – 0.75 to 1.0.
Job efficiency normal 0.83.

Area to be consolidated 2 x 100 = 200m² width of roll .65m length to be rolled
200 divided by 0.65m = 308m @ 1.4km per hr. x .75 surface factor x 0.83
efficiency factor = 0.87km hour.

308m @0.870kmlh = 0.35 hr. 6 passes 2.12 hr

100m² 75mm granular material consolidated in 0.69 hr

Compaction by vibrating roller – heavy

Bomag BW 90S width .90m

Double vibrating roller
Pedestrian controlled.
Mass	= 1300kg
Speed 1st gear .67mph	= 1.08km/hr
2nd gear 1.75	= 2.82

Compaction is specified as:

Mass per metre width of roll
Number of passes for varying thickness of material

BW90S Bomag Example 2.

Compact granular material in road formation.
li0rnm thick 5m wide 100m long.
Mass	= 1300kg per .90m
	= 1440kg per m

DTp clause 802 requires 3 passes of this mass of a double roller. Factors affecting this output.

Uneven surface, factors vary .75 to 1.0
Job efficiency normal	= 0.83
Area to be consolidated	= 100 x 5 = 500m²
Width of roll = .9m	
Length to be rolled 500 divided by .9	= 555m

555m @ 1.08km/h x .75 surface factor
x 0.83 efficiency factor
	= 0.67km/h
555 @ 670m/hr	= 0.828 hr
3 passes	= 2.484 hr

100m², 110mm granular material consolidated in **0.45 hours**

SECTION 4:
ECONOMIC FORECAST

SECTION 4 - ECONOMIC FORECAST

The UK economy is struggling to recover from the worst recession for a generation. Whilst the long predicted 'double-dip' finally appeared in 1Q12, the economy grew 0.3% during the first quarter of 2013. There is some light at the end of the tunnel with the latest forecasts from HM Treasury (February 2013) showing positive growth in the UK for the next five years.

Whilst GDP figures continue to move between losses and gains, one constant since the recession hit has been the Bank of England's base rate of inflation. Staying at 0.5% since March 2009, the lowest rate in the Bank's three hundred year old history, it is largely seen as a measure to stimulate growth within the economy. Whilst this has had mixed results another of the Bank's targets, to keep inflation at 2%, has begun to see some success.

The consumer price index (CPI), had been inflating at over 4% for the whole of 2011, more than double the Bank of England's target. It is currently (March 2013) showing annual growth of 2.8%. Whilst much of the blame for the high inflation in 2011 belonged to the increase in VAT to 20%, it was still a major cause of concern within the economy. It is hoped that the current trend toward the Bank's target will encourage more confidence in the economy as a whole.

At the time of writing the average weekly earnings index for the whole economy (excluding bonuses) recorded an annual inflation rate of 1.4%. Although this figure is below the CPI rate of 2.8% quoted above, consumer spending rose by 0.4% in the 3Q12, largely as a result of Olympic ticket sales and purchases of cars. Growth in consumer spending will need to continue if the GDP forecasts quoted above are to come to fruition.

House prices, an important indicator of health within the UK economy, rose by 3.3% from December 2011 to December 2012. This follows on from average growth over the whole of 2012 reaching 1.6% whist 2011's figure was -1%. However, much of the growth continues to be traced back to London with the figure for 2012 for the UK excluding London reaching only 0.8%.

The unemployment rate for October to December 2012 was 7.8%. This represents an increase of 0.1% from the figure for July to September. Whilst the rate had improved over much of 2012 many analysts questioned the apparent strength of the market against the backdrop of an economy which is struggling to grow. This raises concerns about the underlying productivity rate that exists within the economy as there are more jobs being filled but this is not translating into growth in the economy.

Moving onto the construction sector; one of the major components of GDP, which has been hit hard by the recession and continues to struggle. However, the latest figures show an increase in total output of 0.9% for 4Q12 in comparison to 3Q12. This represents the first increase in output in over a year. One of the key contributors to this success was the infrastructure sector which recorded a quarter-on-quarter growth figure of 4.2% after a 10.5% rise the previous quarter. Despite this the Construction Skills Network 'Blueprint for Construction 2013-2017' predicts the construction sector's output to contract again in 2013 before starting to grow in 2014.

Infrastructure spending, which amounts to 12% of total construction output, is expected to grow by 3% over 2013 before growing a further 3% in 2014. New order figures for infrastructure showed an increase of 1.5% from 3Q12 to 4Q12, largely due to an increase of 8.2% in the electricity sector.

It is expected that energy will provide a good source of growth over the next few years with the new Hinkley Point nuclear plant sitting at the forefront of this. Work on Crossrail is still continuing and is likely to peak in the coming years, meaning that the rail sector is also likely to contribute to the predicted success in infrastructure.

In the latest budget announcement the Chancellor of the Exchequer allowed for an extra £3bn per year to be spent on infrastructure projects from 2015. Coming after dramatic cuts in public sector expenditure across the board since the Coalition Government came to power it is welcome news and should provide further reason for optimism in the sector.

Building tender prices in the UK were recorded at 1.4% less than the previous quarter in 4Q12. However, forecasts from the BCIS predict positive growth will return in the mid 2013 after stagnating at the start of the year. Tender prices in the civil engineering sector have shown positive annual growth since 2Q09. This growth is, in part, due to strong growth in input costs. Civil engineering input costs have inflated at an average annual rate of over 5% since 2010.

Contractor's input costs in the building sector have risen at an average annual rate of 2.75% since January 2010 and are forecast to remain close to this level, on average, for the next four years. Plant costs have been one of the biggest contributors to inflation within the building sector with average annual inflation of 4.9% since February 2010.

Two of the most important commodities in the construction industry; oil and steel, showed differing price trends over 2012. Oil (Brent Crude) showed a small increase of just under 1% at $112 per barrel whilst steel saw a fall of 11.5% from 2011 to 2012. The decline in steel prices is seen by many as a reaction to the fall in iron ore prices, a key component of the steel industry, as much as a cooling interest in some of the developing countries such as China and India.

With the economy still struggling it is likely that 2013 will be another tough year in the construction sector. However, with output forecast to increase from 2014, along with tender prices, there is some room for optimism in the years ahead.

SECTION 5:
WORKING RULE AGREEMENT

SECTION 5 - WORKING RULE AGREEMENT

The employment of labour within the civil engineering industry in Great Britain is governed by the Working Rule Agreement - often abbreviated to WRA – produced by the Construction Industry Joint Council. This Council is made up of representatives from Unions, Trade bodies and Employers and meets regularly to modify and amend the WRA from time to time.

This section summarises some of the WRA that commonly has an impact upon the rates of wages paid to operatives.

The WRA is available from the Construction Industry Joint Council (at the address shown in Section 6: Professional, Government and Trade Bodies).

The current version of the WRA is effective from 1st March 2011. The following selected extracts are reproduced with kind permission of the Construction Industry Joint Council who holds the copyright.

Scope of these Extracts

The total scope and coverage of the WRA is best derived from an appraisal of the Index of the Agreement which is reproduced in full below.

Those sections which are reproduced are marked with an asterisk.

INDEX

WORKING RULE

WR.1 Entitlement to Basic and Additional Rates of Pay*
1.1.1   General Operatives
1.2.1   Skilled Operatives
1.3   Craft Operatives
1.4   Conditions of Employment of Apprentices

WR.2 Bonus*

WR.3 Working Hours
3.1   Rest/Meal Breaks
3.2   Average Weekly Working Hours

WR.4 Overtime Rates

WR.5 Daily Fare and Travel Allowances
5.1   Extent of Payment
5.2   Measurement of Distance
5.3   Transport provided free by the Employer
5.4   Transfer during working day
5.5   Emergency Work

WR.6 Shift Working

WR.7 Night Work
7.1   Night Work Allowance and Overtime
7.2   Average Weekly Working Hours
7.3   Health and Risk Assessments

WR.8 Continuous Working

WR.9 Tide Work

WR.10 Tunnel Work

WR.11 Refuelling, Servicing, Maintenance and Repairs

WR.12 Storage of Tools

WR.13 Loss of Clothing (deleted)

WR.14 Transfer Arrangements

WR.15 Subsistence Allowance

SCHEDULE 1*

Specified Work establishing entitlement to a Skilled Rate or the Craft Rate
Bar Benders and Reinforcement Fixers
Concrete
Drilling and Blasting
Dryliners
Formwork Carpenters
Gangers and Trade Chargehands
Gas Network Operations
Linesmen – Erectors
Mason Paviors
Mechanics

Mechanical Plant Drivers and Operators
Backhoe Loaders
Banksman
Compressors and Generators
Concrete Mixers
Concrete Placing Equipment
Cranes Dozers
Dumpers and Dump Trucks
Excavators (360 degree slewing)
Fork Lift Trucks and Telehandlers
Locos
Motor Graders
Motorised Scrapers
Motor Vehicles (Road Licensed Vehicles)
Power Driven Tools
Power Rollers
Pumps
Shovel Loaders, wheeled or tracked including skid steer
Slinger/Signaller
Tractors (Wheeled or Tracked)
Trenchers (Type wheel, chain or saw)
Winches

Piling

Pipe Jointers

Pipelayers

Pre-Stressing Concrete

Road Surfacing Work

Scaffolders

Steelwork Construction

Timberman

Tunnels

Welders

Young Workers

WR.1 ENTITLEMENT TO BASIC RATES OF PAY

Operatives employed to carry out work in the Building and Civil Engineering Industry are entitled to basic pay in accordance with this Working Rule (WR.1). Rates of pay are set out in a separate Schedule, published periodically by the Council.

Classification of basic rates of pay for operatives:

General Operative
Skilled Operative Rate   4
Skilled Operative Rate   3
Skilled Operative Rate   2
Skilled Operative Rate   1
Craft Operative

1.1 General Operatives

1.1.1   General Operatives employed to carry out general building and/or civil engineering work are entitled to receive the General Operatives Basic Rate of Pay.

Payment for Occasional Skilled Work

1.1.2   General Operatives, employed as such, who are required to carry out building and/or civil engineering work defined in Schedule 1, on an occasional basis, are entitled to receive the General Operative Basic Rate of Pay increased to the rate of pay specified in Schedule 1 for the hours they are engaged to carry out the defined work.

1.2 Skilled Operatives

1.2.1   Skilled Operatives engaged and employed whole time as such, who are required to carry out skilled building and/or civil engineering work defined in Schedule 1 on a continuous basis, are entitled to the Basic Rate of Pay specified in Schedule 1.

1.3 Craft Operatives

Craft Operatives employed to carry out craft building and/or civil engineering work are entitled to receive the Craft Operative Basic Rate of Pay.

1.4 Conditions of Employment of Apprentices

1.4.1   Conditions

An apprentice who has entered into a training service agreement is subject to the same conditions of employment as other operatives employed under the Working Rule Agreement except as provided in WR.1.4.2 to 1.4.6.

1.4.2   Wages

Rates of pay are set out in a separate schedule, published periodically by the Council. Payment under the scale is due from the date of entry into employment as an apprentice, whether the apprentice is working on site or undergoing full-time training on an approved course, subject to the provisions of WR.1.4.3. Payment under the scale is due from the beginning of the pay week during which the specified period starts.

1.4.3   Payment During Off-the-Job Training

Apprentices are entitled to be paid during normal working hours to attend approved courses off-the-job training in accordance with the requirement of their apprenticeship. Payment during such attendance shall be at their normal rate of pay, but the employer may withhold payment for hours during which an apprentice, without authorisation fails to attend the course.

1.4.4 Overtime

The working of overtime by apprentices under 18 years of age shall not be permitted. Where an apprentice age 18 or over is required to work overtime payment shall be in accordance with the provisions of WR.4.

1.4.5 Daily Fare and Travel Allowances

The apprentice shall be entitled to fare and travel allowances in accordance with WR.5.

1.4.6   Absence and Sick Pay

The employer must be notified at the earliest practical time during the first day of any absence and no later than mid-day. The first seven days may be covered by self certification. Thereafter absence must be covered by a certificate or certificates given by a registered medical practitioner. The apprentice shall be entitled to Statutory Sick Pay (SSP) plus Industry sick pay in accordance with WR.20 save the aggregate amount of SSP plus Industry sick pay shall not exceed a normal week's pay in accordance with WR.1.4.2.

1.4.7   Other Terms and Conditions of Engagement

The apprentice shall be subject to all other provisions and entitlements contained within the Working Rule Agreement.

Note: Normal hourly rate

The expression 'normal hourly rate' in this Agreement means the craft, skilled operative, general operative or apprentice weekly basic rate of pay as above, divided by the hours defined in WR.3 "Working Hours". Additional payments for occasional skilled work or bonus payments are not taken into account for calculating the "normal hourly rate".

WR.1 ENTITLEMENT TO BASIC AND ADDITIONAL RATES OF PAY

Classification	Basic pay (pence per hour)	Weekly Rates based on 39 hours £
General Operative	(803)	313.17
Skill Rate 4	(865)	337.35
3	(916)	357.24
2	(979)	381.81
1	(1017)	396.63
Craft Rate	(1067)	416.13

Apprentice Rates

Stage of Training	Basic Pay (pence per hour)	Weekly Rates based on 39 hours £
Year 1	(444)	173.16
Year 2	(573)	223.47
Year 3 without NVQ 2	(671)	261.69
Year 3 with NVQ 2	(853)	332.67
Year 3 with NVQ 3	(1067)	416.13
On completion with NVQ 2	(1067)	416.13

WR.2 BONUS

It shall be open to employers and operatives on any job to agree a bonus scheme based on measured output and productivity for any operation or operations on that particular job.

SCHEDULE 1

Specified Work Establishing Entitlement to the Skilled Operative Pay Rate 4, 3, 2, 1 or Craft Rate

Basic Rate of Pay

BAR BENDERS AND REINFORCEMENT FIXERS
Bender and fixer of Concrete Reinforcement capable of reading and
understanding drawings and bending schedules and able to set out work ... Craft Rate

CONCRETE
Concrete Leveller or Vibrator Operator ... 4

Screeder and Concrete Surface Finisher working from datum
such as road-form, edge beam or wire ... 4

Operative required to use trowel or float (hand or powered) to produce
high quality finished concrete ... 4

DRILLING AND BLASTING
Drills, rotary or percussive: mobile rigs, operator of ... 3

Operative attending drill rig ... 4

Shotfirer, operative in control of and responsible for explosives
including placing, connecting and detonating charges ... 3

Operatives attending on shotfirer, including stemming ... 4

DRYLINERS
Operatives undergoing approved training in drylining ... 4

Operatives who can produce a certificate of training achievement
indicating satisfactory completion of at least one unit
of approved drylining training ... 3

Dryliners who have successfully completed their training
In drylining fixing and finishing ... Craft Rate

FORMWORK CARPENTERS
1st year trainee ... 4

2nd year trainee ... 3

Formwork Carpenters ... Craft Rate

GANGERS AND TRADE CHARGEHANDS
(Higher grade payments may be made at the employer's discretion) ... 4

GAS NETWORK OPERATIONS
Operatives who have successfully completed approved training to the standard of:
GNO Trainee ... 4

GNO Assistant ... 3

Team Leader – Services ... 2

Team Leader – Mains ... 1

Team Leader – Mains and Services ... 1

LINESMEN – ERECTORS

1st grade…………..

(Skilled in all works associated with the assembly, erection, maintenance
and dismantling of Overhead Lines Transmission Lines on steel towers,
concrete or wood poles, including all overhead lines construction elements.)   2

2nd Grade

(As above but lesser degree of skill – or competent and
fully skilled to carry out some of the elements of construction listed above.)   3

Linesmen-erector's mate

(Semi-skilled in works specified above and a general helper)   4

MASON PAVIORS

Operative assisting a Mason Pavior undertaking kerb laying, block and
sett paving, flag laying, in natural stone and precast products   4

Operative engaged in stone pitching or dry stone walling   3

MECHANICS

Maintenance Mechanic capable of carrying out field service duties,
maintenance activities and minor repairs   2

Plant Mechanic capable of carrying out major repairs and overhauls including
welding work, operating metal turning lathe or similar machine and using
electronic diagnostic equipment   1

Maintenance/Plant Mechanics' Mate on site or in depot   4

Tyre Fitter, heavy equipment tyres   2

MECHANICAL PLANT DRIVERS AND OPERATORS

Backhoe Loaders (with rear excavator bucket and front shovel and additional
equipment such as blades, hydraulic hammers, and patch planers)

Backhoe, up to and including 50kW net engine power; driver of   4

Backhoe, over 50kW up to and including 100kW net engine power; driver of   3

Backhoe, over 100kW net engine power; driver of   2

Concrete Mixers

Operative responsible for operating a concrete mixer or mortar pan up to
and including 400 litres drum capacity   4

Operative responsible for operating a concrete mixer over
400 litres and up to and including 1,500 litres drum capacity   3

Operative responsible for operating a concrete mixer over 1,500 litres
drum capacity   2

Operative responsible for operating a mobile self-loading and batching
concrete mixer up to 2,500 litres drum capacity   2

Operative responsible for a operating a mechanical drag-shovel   4

Concrete Placing Equipment

Trailer mounted or static concrete pumps: self-propelled concrete
Placers: concrete placing booms; operator of   3

Self-propelled Mobile Concrete Pump, with or without boom,
mounted on lorry or lorry chassis; driver/operator of   2

Cranes

Mobile Cranes
Self-propelled mobile crane on road wheels, rough terrain wheels or caterpillar tracks including lorry mounted:

Max. lifting capacity at min. radius, up to and including 5 Tonne; driver of	4
Max. lifting capacity at min. radius, over 5 Tonne and up to and including 10 Tonne; driver of	3
Max. lifting capacity at min. radius, over 10 Tonne	Craft Rate

Where grabs are attached to cranes the next higher skill rate of pay applies except over 10 Tonne where the rate is at the employer's discretion.

Tower Cranes (including static or travelling: standard trolley or luffing jib)

Up to and including 2 Tonne max. lifting capacity at min. radius; driver of	4
Over 2 Tonne up to and including 10 Tonne max. lifting capacity at min. radius; driver of	3
Over 10 Tonne up to and including 20 Tonne max. lifting capacity at min. radius; driver of	2
Over 20 Tonne max. lifting capacity at min. radius; driver of	1

Miscellaneous Cranes and Hoists

Overhead bridge crane or gantry crane up to and including 10 Tonne capacity; driver of	3
Overhead bridge crane or gantry crane over 10 Tonne up to and including 20 Tonne capacity; driver of	2
Power driven hoist or jib crane; operator of	4

Slinger / Signaller appointed to attend Crane or hoist to be responsible for fastening or slinging loads and generally to direct lifting operations   4

Dozers

Crawler dozer with standard operating weight up to and including 10 Tonne; driver of	3
Crawler dozer with standard operating weight over 10 Tonne and up to and including 50 tonne; driver of	2
Crawler dozer with standard operating weight over 50 Tonne; driver of	1

Dumpers and Dump Trucks

Up to and including 10 Tonne rated payload; driver of	4
Over 10 Tonne and up to and including 20 Tonne rated payload; driver of	3
Over 20 Tonne and up to and including 50 Tonne rated payload; driver of	2
Over 50 Tonne and up to and including 100 Tonne rated payload; driver of	1
Over 100 Tonne rated payload; driver of	Craft Rate

Excavators (360 degree slewing)

Excavators with standard operating weight up to and including 10 Tonne; driver of — 3

Excavator with standard operating weight over 10 Tonne and up to and including 50 Tonne; driver of — 2

Excavator with standard operating weight over 50 Tonne; driver of — 1

Banksman appointed to attend excavator or responsible for positioning vehicles during loading or tipping — 4

Fork-Lifts Trucks and Telehandlers

Smooth or rough terrain fork lift trucks (including side loaders) and telehandlers up to and including 3 Tonne lift capacity; driver of — 4

Over 3 Tonne lift capacity; driver of — 3

Motor Graders: driver of — 2

Motorised Scrapers: driver of — 2

Motor Vehicles (Road Licensed Vehicles)
Driver and Vehicle Licensing Agency (DVLA)

Vehicles requiring a driving licence of category C1; driver of — 4
(Goods vehicle with maximum authorised mass (mam) exceeding 3.5 Tonne but not exceeding 7.5 Tonne and including such a vehicle drawing a trailer with a mam not over 750kg)

Vehicles requiring a driving licence of category C; driver of — 2
(Goods vehicle with a maximum authorised mass (mam) exceeding 3.5 Tonne and including such a vehicle drawing a trailer with mam not over 750kg)

Vehicles requiring a driving licence of category C plus E; driver of — 1
(Combination of a vehicle in category C and a trailer with maximum authorised mass over 750kg)

Power Driven Tools

Operatives using power-driven tools such as breakers, percussive drills, picks and spades, rammers and tamping machines — 4

Power Rollers

Roller, up to and including 4 Tonne operating weight; driver of — 4

Roller, over 4 Tonne operating weight and upwards; driver of — 3

Pumps, Power-driven pump(s); attendant of — 4

Shovel Loaders, (Wheeled or tracked, including skid steer)
Up to and including 2 cubic metre shovel capacity; driver of — 4

Over 2 cubic metre and up to and including 5 cubic metre shovel capacity; driver of — 3

Over 5 cubic metre shovel capacity; driver of — 2

Tractors (Wheeled or Tracked)

Tractor, when used to tow trailer and/or with mounted compressor, up to and including 100kW rated engine power; driver of — 4

Tractor, ditto, over 100KW up to and including 250kW rated engine power; driver of — 3

Tractor, ditto, over 250kW rated engine power; driver of — 2

Trenchers (Type wheel, chain or saw)

Trenching Machine, up to and including 50kW gross engine power; driver of	4
Trenching Machine, over 50kW and up to and including 100kW gross engine power; driver of	3
Trenching Machine, over 100kW gross engine power; driver of	2
Winches, Power driven winch; driver of	4

PILING

General Skilled Piling Operative	4
Piling Chargehand / Ganger	3
Pile Tripod Frame Winch Driver	3
CFA or Rotary or Driven Mobile Piling Rig Driver	2
Concrete Pump Operator	3

PIPE JOINTERS

Jointers, pipes up to and including 300mm diameter	4
Jointers, pipes over 300mm diameter and up to 900mm diameter	3
Jointers, pipes over 900mm diameter	2
except in HDPE mains when experienced in butt fusion and/or electrofusion jointing operations	2

PIPELAYERS

Operative preparing the bed and laying pipes up to and including 300 mm diameter	4
Operative preparing the bed and laying pipes over 300 mm diameter and up to and including 900mm diameter	3
Operative preparing the bed and laying pipes over 900mm diameter	2

PRE-STRESSING CONCRETE

Operative in control of and responsible for hydraulic jacks and other tensioning devices engaged in post-tensioning and/or pre-tensioning concrete elements	3

ROAD SURFACING WORK (includes rolled asphalt, dense bitumen macadam and surface dressings)

Operatives employed in this category of work to be paid as follows: Chipper	4
Gritter Operator	4
Raker	3
Paver Operator	3
Leveller on Paver	3
Road Planer Operator	3
Road Roller Driver, 4 Tonne and upwards	3
Spray Bar Operator	4

SCAFFOLDERS
See WR.26 above.

STEELWORK CONSTRUCTION
A skilled steel erector engaged in the assembly, erection and fixing
into position of steel-framed buildings and structures — 1

Operative capable of and engaged in fixing simple steelwork components
such as beams, girders and metal decking — 3

TIMBERMAN
Timberman, installing timber supports — 3

Highly skilled timberman working on complex supports using
timbers of size 250mm by 125mm and above — 2

Operative attending — 4

TUNNELS
Operative working below ground on the construction of tunnels and
underground spaces or sinking shafts:

Tunnel Boring Machine operator — 2

Tunnel Miner (skilled operative working at the face) — 3

Tunnel Miner's assistants (operative who assists the tunnel miner) — 4

Other operatives engaged in driving headings in connection with cable and pipe laying — 4

Operative driving loco — 4

WELDERS
Grade 4 (Fabrication Assistant)
Welder able to tack weld using SMAW or MIG welding processes in
accordance with verbal instructions and including mechanical preparation
such as cutting and grinding — 3

Grade 3 (Basic Skill Level)
Welder able to weld carbon and stainless steel using at least one of the
following processes SMAW, GTAW, GMAW for plate-plate fillet welding
in all major welding positions, including mechanical preparation and
complying with fabrication drawings. — 2

Grade 2 (Intermediate Skill Level)
Welder able to weld carbon and stainless steel using manual
SMAW, GTAW, semi-automatic MIG or MAG, and FCAW welding
processes including mechanical preparation, and complying with
welding procedures, specifications and fabrication drawings. — 1

Grade 1 (Highest Skill Level)
Welder able to weld carbon and stainless steel using manual SMAW, GTAW,
semi-automatic GMAW or MIG or MAG, and FCAW welding processes in all
modes and directions in accordance with BSEN 287-1 and/or 287-2 Aluminium
Fabrications including mechanical preparation and complying with welding
procedures, specifications and fabrication drawings. — Craft Rate

YOUNG WORKERS
Operatives below 18 years of age will receive payment 60% of the General Operative Basic Rate.

At 18 years of age or over the payment is 100% of the relevant rate.

SECTION 6:
PROFESSIONAL, GOVERNMENT AND TRADE BODIES

SECTION 6 - PROFESSIONAL, GOVERNMENT AND TRADE BODIES

List below are some of the organisations in central and local government, public utilities, trade and professional bodies

Advisory, Conciliation and Arbitration Service

(National, Head Office)
Euston Tower
286 Euston Road
London
NW1 3JJ
Telephone: Helpline: 08457 47 47 47
Customer Service: 08457 38 37 36

The Architect and Surveyors Institute

St Mary House
15 St. Mary Street
Chippenham
Wiltshire
SN15 3WD
Telephone: 0124 944 4505
Fax: 0124 944 3602

Association of Consulting Engineers

Alliance House
12 Caxton Street
London
SW1H 0QL
Telephone:0207 222 6557
Fax: 020 7990 9202

Association of Cost Engineers

Lea House
5 Middlewich Road
Sandbach
Cheshire
CW11 1XL
Telephone: 0127 076 4798
Fax: 0127 076 6180

British Board of Agreement

PO Box No 195
Bucknalls Lane
Garston
Watford
Hertfordshire
WD2 7NG
Telephone: 0192 366 5300
Fax: 0192 366 5301

British Cement Association

Century House
Telford Avenue
Crowthorne
Berks
RG45 6YS
Telephone: 0134 476 2676
Fax: 0134 476 1214

British Constructional Steelwork Association

4 Whitehall Court
Westminster
London
SW1A 2ES
Telephone: 0207 839 8566
Fax: 0207 976 1634

British Geotechnical Society

Administered by the BGS Secretary at the Institution of Civil Engineers

British Precast Concrete Federation

60 Charles Street
4th Floor
Leicester
Leicestershire
LE1 1FB
Telephone: 0116 253 6161
Fax: 0116 251 4568

British Railway Board

Whittles House
14 Pentonville Road
LONDON
N1 9HF
Telephone: 0207 904 5079

British Road Federation

Pillar House 194-202 Old Kent Road
London
SE1 5TG
Telephone: 0207 703 9769
Fax: 0207 701 0029

British Standards Institute (BSI)

389 Chiswick High Road
London
W4 4AL
Telephone: 0208 996 9000
Fax: 0208 996 7001

British Waterways Board

64 Clarendon Road
Watford
Hertfordshire
WD17 1DA
Telephone: 0192 320 1120
Fax: 0192 320 1304

Building Research Establishment

Bucknalls Lane
Garston
Watford
Herts
WD25 9XX
Telephone: 0192 366 4000
Fax: 0192 366 4010

Building Services Research and Information Association

Old Bracknell Lane West
Bracknell
Berkshire
RG12 7AH
Telephone: 0134 446 5600
Fax: 0134 446 5626

Chartered Institute of Building

Englemere
King's Ride
Ascot
Berkshire
SL5 7TB
Telephone: 0134 463 0700
Fax: 0134 463 0777

The Chartered Institution of Building Services Engineers

222 Balham High Road
Balham
London
SW12 9BS
Telephone: 0208 675 5211
Fax: 0208 675 5449

Clay Pipe Development Association Limited

Treetops
Bellingdon
Chesham
Bucks
HP5 2XL
Telephone: 0149 479 1456
Fax: 0149 479 2378

The Concrete Society

Riverside House
4 Meadows Business Park
Station Approach
Blackwater, Camberley
Surrey
GU17 9AB
Telephone: 0127 660 7140
Fax: 0127 660 7141

Construction Confederation

55 Tufton Street
Westminster
London
SW1P 3QL
Telephone: 0870 898 9090
Fax: 0870 898 9095

Construction Industry Computing Association

National Computing Centre
Oxford House
Oxford Road
Manchester
M1 7ED
Telephone: 0161 242 2262

Construction Federation

55 Tufton Street
London
SW1P 3QL
Telephone: 0207 227 4500
Fax: 0207 227 4501

Construction Industry Research and Information Association (CIRIA)

Classic House
174 - 180 Old Street
London
EC1V 9BP
Telephone: 0207 549 3300
Fax: 0207 253 0523

Construction Industry Training Board

Bircham Newton
Kings Lynn
Norfolk
PE31 6RH
Telephone: 0344 994 4400
Fax: 0148 557 7793

Construction Plant Hire Association (CPA)

27/28 Newbury Street
Barbican
London
EC1A 7HU
Telephone: 0207 796 3366
Fax: 0207 796 3399

Department for Environment, Food & Rural Affairs (Defra)

Eastbury House
30 - 34 Albert Embankment
London
SE1 7TL
Telephone: 0207 238 2188
Fax: 0207 238 2188

Department for Business, Innovation & Skills

1 Victoria Street
London
SW1H 0ET
Telephone: 0207 215 5000
Fax: 0207 215 0105

Ductile Iron Pipe Association (DIPA)

The National Metalforming Centre
47 Birmingham Road
West Bromwich
West Midlands
B70 6PY
Telephone: 0121 601 6390
Fax: 0121 601 6391

Health and Safety Executive

Rose Court
2 Southwark Bridge
London
SE1 9HS
Telephone: 0207 556 2100
Fax: 0207 556 2102

Heating and Ventilating Contractors' Association

HVCA Head Office
Esca House
34 Palace Court
London
W2 4JG
Telephone: 0207 313 4900
Fax: 0207 727 9268

Centre for Ecology and Hydrology (CEH)

CEH Wallingford
Maclean Building
Benson Lane
Crowmarsh Gifford
Wallingford
Oxfordshire
OX10 8BB
Telephone: 0149 183 8800
Fax: 0149 169 2424

Chartered Management Institute

3rd Floor
2 Savoy Court
Strand
London
WC2R 0EZ
Telephone: 0207 497 0580
Fax: 0207 497 0463

Institute of Measurement and Control

87 Gower Street
London
WC1E 6AF
Telephone: 0207 387 4949
Fax: 0207 388 8431

The Chartered Quality Institute

2nd Floor North
Chancery Exchange
10 Furnival Street
London
EC4A 1AB
Telephone: 0207 245 6722
Fax: 0207 245 6788

Institution of Civil Engineering Surveyors (ICES)

Dominion House
Sibson Road
Sale
Cheshire
M33 7PP
Telephone: 0161 972 3100
Fax: 0161 972 3118

Institution of Civil Engineers (ICE)

One Great George Street
Westminster
London
SW1P 3AA
Telephone: 0207 222 7722

The Institution of Engineering Designers

Courtleigh
Westbury Leigh
Westbury
Wiltshire
BA13 3TA
Telephone: 0137 382 2801
Fax: 0137 385 8085

Institution of Gas Engineers and Managers

IGEM House
28 High Street
Kegworth
Derbyshire
DE74 2DA
Telephone: 0844 375 4436
Fax: 01509 678198

Institution of Mechanical Engineers

1 Birdcage Walk
Westminster
London
SW1H 9JJ
Telephone: 0207 222 7899
Fax: 0207 222 4557

Institute of Materials, Minerals and Mining

1 Carlton House Terrace
London
SW1Y 5DB
Telephone: 0130 232 0486
Fax: 0130 238 0900

The Institution of Nuclear Engineers

Allan House
1 Penerley Road
London
SE6 2LQ
Telephone: 0208 698 1500
Fax: 0208 695 6409

Institution of Plant Engineers (IPlantE)

22 Greencoat Place
Westminster London
SW1P 1PR
Telephone: 0207 630 1111
Fax: 0207 630 6677

The Institution of Structural Engineers

11 Upper Belgrave Street
London
SW1X 8BH
Telephone: 0207 235 4535
Fax: 0207 235 4294

Chartered Institution of Water and Environmental Management (CIWEM)

15 John Street
London
WC1N 2EB
Telephone: 0207 831 3110
Fax: 0207 405 4967

(ENDR Office) International Arbitration Centre
24 Angel Gate
London
EC IV 2RS
Telephone: 0207 837 4483
Fax: 0207 837 4185

London Court of International Arbitration
70 Fleet Street
London
EC4Y 1EU
United Kingdom
Telephone: 020 7936 6200
Fax: 020 7936 6211

Laboratory of the Government Chemist (LGC)
Queens Road
Teddington
Middlesex
TW11 0LY
Telephone: 0208 943 7000
Fax: 0208 943 2767

Meteorological Office (Met Office)
FitzRoy Road
Exeter
Devon
EX1 3PB
Telephone: 0139 288 5680
Fax: 0139 288 5681

National Assembly for Wales
Cardiff Bay
Cardiff
CF99 1NA
Telephone: 0845 010 5500

National Association of Scaffolding Contractors
18 Mansfield Street
London
W1M 9FG
Telephone: 0207 580 5404

The National Access and Scaffolding Confederation (NASC)
4th Floor
12 Bridewell Place
London
EC4V 6AP

National Federation of Painting & Decorating Contractors
18 Mansfield Street
London
W1M 9FG
Telephone: 0207 580 5404
Fax: 0207 636 5984

OFGEM (Office of Gas and Electricity Markets)
9 Millbank
London
SW1P 3GE
Telephone: 0207 901 7000
Fax: 0207 901 7066

Office of Rail Regulation (OFRR)
One Kemble Street
London
WC2B 4AN
Telephone: 0207 282 2000
Fax: 0207 282 2040

OFCOM (Office of Communications)
Riverside House
2a Southwark Bridge Road
London
SE1 9HA
Telephone: 0300 123 3333 or 020 7981 3040
Fax: 0207 981 3333

OFWAT (Office of Water Services)
Centre City Tower
7 Hill Street
Birmingham
B5 4UA
Telephone: 0121 644 7500
Fax: 0121 644 7559

Ordnance Survey
Customer Service Centre
Ordnance Survey
Adanac Drive
Southampton
SO16 0AS
Telephone: 0845 605 0505
Fax: 0845 0990 494

Pipe Jacking Association
10 Greycoat Place
London
SW1P 1SB
Telephone: 0845 070 5201
Fax: 0845 070 5202

The Pipeline Industries Guild (Head Office)
F150 First Floor
Cherwell Business Village
Southam Road
Banbury
OX16 2SP
Telephone: 0207 235 7938
Fax: 0207 235 0074

Property Services Agency
Carillion Head Office
24 Birch Street
Wolverhampton
West Midlands
WV1 4HY
Telephone: 0190 242 2431

Quarry Products Association
Gillingham House
38-44 Gillingham Street
London
SW1V 1HU
Telephone: 0207 963 8000
Fax: 0207 963 8001

Royal Institute of British Architects
66 Portland Place
London
W1B 1AD
Telephone: 0207 580 5533
Fax: 0207 255 1541

Royal Institution of Chartered Surveyors
12 Great George Street
Parliament Square
London
SW1P 3AD
Telephone: 024 7686 8555
Fax: 0207 334 3811

The Royal Town Planning Institute
41 Botolph Lane
London
EC3R 8DL
Telephone: 0207 929 9494
Fax: 0207 929 9490

Scottish Executive Development Department
Victoria Quay
Edinburgh
EH6 6QQ
Telephone: 0131 556 8400
Fax: 0139 779 5001

Society of Construction Law (UK)
The Cottage
Bullfurlong Lane
Burbage, Hinckley
LE10 2HQ
Telephone: 07730 474074

Society of Surveying Technicians
Surveyor Court
Westwood Way
Coventry
CV4 8JE
Telephone: 0207 222 7000

Water UK
1 Queen Anne's Gate
London
SW1H 9BT
Telephone: 0207 344 1844
Fax: 020 7344 1853

WRc Plc (Water Research Centre plc)
Frankland Road
Blagrove
Swindon
Wiltshire
SN5 8YF
Telephone: 0179 386 5000
Fax: 0179 386 5001

SECTION 7:
TECHNICAL INFORMATION

SECTION 7 - TECHNICAL INFORMATION

Generally

Metric Conversion Factors

	Metric	Imperial		Imperial	Metric
LENGTH	1 mm	= 0.0394 inches		1 inch	= 25.4 (exact) mm
	1 m	= 3.2808 feet		1 foot	= 0.3048 m
	1 m	= 1.0936 yards		1 yard	= 0.9144 m
	1 km	= 0.6214 miles		1 mile	= 1.6093 km
AREA	1 mm^2	= 0.0016 sq inches		1 sq inch	= 645.1600 mm^2
	1 m^2	= 10.7639 sq feet		1 sq foot	= 0.0929 m^2
	1 m^2	= 1.1960 sq yards		1 sq yard	= 0.8361 m^2
	1 km^2	= 0.3861 sq mile		1 sq mile	= 2.5900 km^2
	1 ha	= 2.4711 acres		1 acre	= 0.4047 ha
VOLUME	1 m^3	= 35.3147 cubic feet		1 cubic foot	= 0.0283 m^3
	1 m^3	= 1.3080 cubic yards		1 cubic yard	= 0.7646 m^3
	1 litre	= 1.7598 pints		1 pint	= 0.5683 litres
	1 litre (UK)	= 0.2200 gallons (UK)		1 gallon	= 4.5461 litres
	NOTE: 1 litre	= 1000 cm^3	1 m^3	= 1000 litres	
WEIGHT	1 kg	= 2.2046 pounds		1 pound	= 0.4536 kg
	1 kg	= 0.0197 hundredweight		1 hundredweight	= 50.8024 kg
	1 tonne	= 0.9842 ton		1 ton	= 1.0161 tonnes
FORCE	1 N	= 0.2248 lbf		1 pdf	= 4.4482 N
	1 kN	= 225 lbf		1 lbf	= 0.0044 kN
	1 MN	= 9.3197 tonf		1 tonf	= 0.1073 MN

NOTE: Standard gravity (Gn) = 9.80665 m/s^2 (exactly)

	Metric	Imperial		Imperial	Metric
PRESSURE	1 kpa	= 0.1450 lbf/sq inch		1 lbf/sq inch	= 6.8948 kpa

NOTE: 1 kpa = 1 kN/m^2 = 1 N/mm^2

	Metric	Imperial		Imperial	Metric
ENERGY (work and heat)	1 kJ	= 0.9478 Btu			
	1 Btu	= 1.055 kJ			
POWER	1 W	= 3.4121 Btu		1 Btu	= 0.2931 W
	1 kW	= 1.3410 hp		1 hp	= 0.7457 kW

Conversion to Metric

On the following page is a set of Conversion Tables – metres to yards (Linear, square and cubic). Readers who either by choice or compulsion are estimating in Imperial measurements may use the Metric analysis detail and convert to Imperial by using these tables, or if more convenient, the following 'rule of thumb' formula may be used for quick calculation where exact accuracy is not essential.

To convert	Deduct from metric cost
Linear metre cost to linear yard cost	one eleventh
Square metre cost to square yard cost	one sixth
Cubic metre cost to cubic yard cost	one quarter

Worked examples:

1) If the Unit Price of a cubic metre item = £9.00 per cubic metre, then the cubic yard rate = £9.00 less one quarter = £6.75. Referring to the tables, the equivalent is £6.881.

2) If the Unit Price of a square metre item = £3.40 per square metre then the square yard rate = £3.40 less one sixth = £2.83. Referring to the tables, the equivalent is £2.842.

3) If the Unit Price of a linear metre item = £1.26 per linear metre, then the linear yard rate = £1.26 less one eleventh = £1.145. Referring to the tables, the equivalent is £1.152.

Cost Conversion Tables Metres to Yards

	LINEAR		SQUARE		CUBIC
£ Metre	£ Yard	£ Metre	£ Yard	£ Metre	£ Yard
0.01	0.01	0.01	0.01	0.01	0.01
0.02	0.02	0.02	0.02	0.02	0.015
0.03	0.025	0.03	0.025	0.03	0.025
0.04	0.035	0.04	0.03	0.04	0.03
0.05	0.045	0.05	0.04	0.05	0.04
0.06	0.055	0.06	0.05	0.06	0.045
0.07	0.065	0.07	0.06	0.07	0.055
0.08	0.075	0.08	0.065	0.08	0.06
0.09	0.085	0.09	0.075	0.09	0.07
0.10	0.09	0.10	0.08	0.10	0.075
0.11	0.10	0.11	0.09	0.11	0.085
0.12	0.11	0.12	0.10	0.12	0.09
0.13	0.12	0.13	0.11	0.13	0.10
0.14	0.13	0.14	0.12	0.14	0.11
0.15	0.135	0.15	0.125	0.15	0.115
0.16	0.145	0.16	0.135	0.16	0.12
0.17	0.155	0.17	0.14	0.17	0.13
0.18	0.165	0.18	0.15	0.18	0.14
0.19	0.175	0.19	0.16	0.19	0.145
0.20	0.185	0.20	0.17	0.20	0.155
0.25	0.23	0.25	0.21	0.25	0.19
0.30	0.275	0.30	0.25	0.30	0.23
0.35	0.32	0.35	0.29	0.35	0.27
0.40	0.365	0.40	0.33	0.40	0.305
0.45	0.41	0.45	0.375	0.45	0.345
0.50	0.46	0.50	0.42	0.50	0.38
0.60	0.55	0.60	0.50	0.60	0.46
0.70	0.64	0.70	0.585	0.70	0.53
0.80	0.73	0.80	0.67	0.80	0.61
0.90	0.825	0.90	0.75	0.90	0.69
1.00	0.915	1.00	0.835	1.00	0.765
1.50	1.37	1.50	1.255	1.50	1.145
2.00	1.83	2.00	1.67	2.00	1.53
2.50	2.285	2.50	2.09	2.50	1.91
3.00	2.745	3.00	2.505	3.00	2.295
3.50	3.20	3.50	2.925	3.50	2.675
4.00	3.66	4.00	3.34	4.00	3.06
4.50	4.115	4.50	3.765	4.50	3.44
5.00	4.575	5.00	4.185	5.00	3.825
6.00	5.49	6.00	5.02	6.00	4.59
7.00	6.405	7.00	5.855	7.00	5.355
8.00	7.32	8.00	6.69	8.00	6.13
9.00	8.235	9.00	7.525	9.00	6.895
10.00	9.15	10.00	8.36	10.00	7.65

MEASUREMENT FORMULAE

Perimeters or circumferences of planes

Circle: 3.14159 x Diameter

Ellipse: 3.14159 x *(major axis; minor axis)* / 2

$$3.14159 \times \frac{\textit{(major axis; minor axis)}}{2}$$

Sector: $\dfrac{\textit{Radius x Degrees in Arc}}{57.3}$

Surface areas of planes and solids

Circle: 3.14159 x Radius Sq

Sphere: 3.14159 x Diameter Sq

Ellipse: 0.7854 *(major axis x minor axis)*

Cylinder: *(circumference x length)* + *(2 x area of end)*

Cone: Area of base + $\dfrac{\textit{(circumference x slant height)}}{2}$

Frustum of cone: 3.14159 x *slant height (radius at top +radius at bottom) + area of top + area of bottom*

Pyramid: $\dfrac{\textit{(sum of base perimeters)}}{2}$ *slant height + area of base*

Sector of circle: $\dfrac{\textit{3.14159 x Degrees in Arc x Radius Sq}}{360}$

Segment of circle: Area of sector less area of triangle

Segment of arc: 2 (chord x rise) + $\dfrac{\textit{rise}^3}{\textit{2 x chord}}$

Bellmouth at road junction: *Area 'A'* = $\dfrac{3}{14}$ x *Radius²*

Volume of solids

Sphere: 4.1888 x *Radius³*

Cone: $\dfrac{\textit{height (area of base)}}{3}$

Frustum of cone: $\dfrac{\textit{height}}{3}$ *(3.14159 x R² + r² + Rr) where R and r are radius of base and top*

Pyramid: $\dfrac{\textit{height (area of base)}}{3}$

Frustum of pyramid: $\dfrac{\textit{height}}{3}$ *(A + B + √AB) where A and B are areas of base and top*

Volumes of earthworks

a) Simpsons Rule for Volumes (Prismoidal formula)

The volume must be divided into an even number of prisms by an odd number of cross-sectional areas.
(*V* = Volume, *d* = constant distance between sections, *A* = Area)

$$V = \frac{d}{3} \left[(A1 + An) + 2(A3 + A5 + \ldots + A(n\text{-}2)) + 4(A2 + A4 + \ldots + A(n\text{-}l)) \right]$$

i.e.

$$V = \frac{d}{3} \left[1st + last\ Area + 2(Sum\ of\ odd\ Areas) + 4(Sum\ of\ even\ Areas) \right]$$

b) End Areas Formula (Trapezoidal rule)

The total volume is divided into any number of equal portions by cross-sectional planes

$$V = \frac{d}{2} \left[(A1 + An) + A2 + A3 + \ldots + A(n\text{-}1) \right]$$

i.e.

$$V = \frac{d}{2} \left[(1st + last\ Area) + Sum\ of\ other\ Areas \right]$$

GENERALLY

Density and Thermal Conductivity of Building Materials

Material	Weight (kg m⁻³)	Thermal conductivity (W m⁻¹ K⁻¹)
Aggregates:		
Undried	2240	1.80
Oven dried	2240	1.30
Asphalt:		
Poured	2100	1.20
Reflective coat	2300	1.20
Roofing, mastic	2330	1.50
Asbestos-related materials:		
Asbestos cement	1750	1.02
Asbestos cement building board	1920	0.60
Asbestos cement decking	1500	0.36
Asbestos cement sheet	700	0.36
Asbestos fibre	640	0.06
Asbestos mill board	1400	0.25
Bitumen:		
Composite, flooring	2400	0.85
Insulation, all types	1000	0.20
Brick:		
Aerated	1000	0.30
Brickwork, inner leaf	1700	0.62
Brickwork, outer leaf	1700	0.84
Paviour	2000	0.96
Reinforced	1920	1.10
Brass:	8500	110
Bronze:	8150	64
Carpet/underlay:		
with cellular rubber underlay	400	0.10
synthetic	160	0.06
Cement/plaster/ mortar:		
Cement	1860	0.72
Cement mortar	1650	0.72
Cement/lime plaster	1600	0.80
Cement plaster	1760	0.72
Cement screed	2100	1.40
Gypsum	1200	0.42
Gypsum plaster	1120	0.51
Gypsum plasterboard	800	0.16
Gypsum plastering	1300	0.80
Limestone mortar	1600	0.70
Plaster	800	0.22
Plasterboard	950	0.16
Ceramic, glazed:	2500	1.40
Ceramic/clay tiles:		
Ceramic tiles	2000	1.20
Ceramic floor tiles	1700	0.80
Clay tiles	1900	0.85
Clay tile, pavior	1920	1.803
Concrete blocks/tiles:		
block, aerated	750	0.24
block, heavyweight, 300 mm	2240	1.31
block, lightweight, 150 mm	1760	0.66
block, lightweight, 300 mm	1800	0.73
block, mediumweight, 150 mm	1900	0.77
block, mediumweight, 300 mm	1940	0.83
block, hollow, heavyweight, 300 mm	1220	1.35
block, hollow, lightweight, 150 mm	880	0.48
block, hollow, lightweight, 300 mm	780	0.76
block, hollow, mediumweight, 150 mm	1040	0.62
block, hollow, mediumweight, 300 mm	930	0.86
Concrete, cast:		
dense	2200	1.70
compacted	2400	2.20
dense, reinforced	2300	1.90
glass reinforced	1950	0.90

Material	Weight (kg m^{-3})	Thermal conductivity (W m^{-1} K^{-1})
Concrete cast:		
heavyweight, dry	2000	1.30
heavyweight, moist	2000	1.70
lightweight, dry	770	0.23
lightweight, moist	700	0.23
mediumweight, dry	1350	0.59
mediumweight, moist	1050	0.59
roofing slab, areated	500	0.16
Copper:	8600	384
Iron:		
iron	7900	72
iron, cast	7500	56
Lead:	11340	35
Linoleum:	1200	0.19
Masonry:		
block, lightweight	600	0.22
block, mediumweight	1650	0.85
heavyweight	1850	0.9
lightweight	750	0.22
mediumweight	1050	0.32
Paper:		
bitumen impregnated paper	1090	0.06
laminated paper	480	0.072
Plastics:		
Polyvinylchloride (PVC)	1380	0.16
Tiles	1200	0.19
Rubber:		
rubber	1500	0.17
expanded board, rigid	70	0.032
hard	1200	0.15
tiles	1600	0.30
Sand:		
sand	2240	1.74
Soil:		
earth, common	1460	1.28
earth, gravel-based	2050	0.52
Steel:		
stainless steel, 5% Ni	7850	29
stainless steel, 20% Ni	8000	16
steel	7800	45
Stone:		
stone chippings for roofs	1800	0.96
granite	2880	3.49
granite, red	2650	2.90
hard stone (unspecified)	2880	3.49
limestone	2180	1.50
Timber:		
fir, pine	510	0.12
hardwood (unspecified)	90	0.05
maple, oak and similar hardwoods	720	0.16
oak, radial	700	0.19
oak, beech, ash, walnut	650	0.23
pine, pitch pine	650	0.17
softwood	510	0.12
timber	480	0.072
timber flooring	650	0.14
chipboard	430	0.067
melamine	630	0.25
hardboard	600	0.08
particle board	750	0.098
plywood	540	0.12
Tin:	7300	65
Vinyl floor covering	1200	0.19
Zinc:	7000	113

EARTHWORKS

Bearing Capacities of Soils

Nature of soil:	Approximate bearing capacity; kN/m²
Peat and bog	0 - 20
Clay, marl, loam	330 - 750
Solid chalk	110 - 450
Solid rock (unweathered)	220 - 2000
Gravel, coarse	660 - 900
Gravel, fine	450 - 660
Sand	220 - 550

Bulkage of Soils after Excavation

Nature of soil:	Approximate bulkage of 1m³ after excavation
Vegetable soil and loam	1.25 - 1.30 m³
Soft clay, marl	1.30 - 1.40 m³
Sand	1.10 - 1.15 m³
Gravel	1.20 - 1.25.m³
Chalk	1.40 - 1.50 m³
Stiff clay	1.40 - 1.50 m³
Rock, weathered	1.30 - 1.40 m³
Rock, unweathered	1.50 - 1.60 m³

Shrinkage of Materials

NOTE: Materials increase in bulk when excavated and, on being deposited, shrink. Wet soils shrink more than dry. Rock increases in bulk when broken up and does not settle to less than its original volume. The increase in bulk is dependant on the size of the broken pieces, and varies between 40 and 60 per cent.

Nature of material	Percentage shrinkage
Clay	10%
Gravel	8%
Gravel and sand	9%
Loam and light sandy soils	12%
Loose vegetable soils	15%

Cubic metres of solid material hauled per load by various types of transporting plant

NOTES:

1) In transporting by dumpers or wagons the bulkage of the material should be taken into account. Such vehicles are rated on a volume basis, their struck measured capacity being stated. The vehicles, however, have a Heaped Capacity, the load being heaped in the vehicle. In the table the cubic metres hauled per cubic metre of Struck Measured Capacity allow for normal heaping of the loads and the bulkage of the various materials. Thus a 2 cubic metre dumper hauls 2 x 0.80 = 1.60 cubic metre of stiff clay (solid) per load.

2) In hauling excavated materials by lorry, the weight of load carried should not exceed that which the vehicle is designed to carry.

Nature of material hauled	Weight of the material in the solid. m³ per tonne	Haulage by lorries. Cubic metres of material (solid) matrial hauled per load						The cubic metres of solid material hauled by dumpers or in wagons per m³ of struck measured capacity
		1 tonne lorry	2 tonne lorry	3 tonne lorry	4 tonne lorry	5 tonne lorry	6 tonne lorry	
Chalk	0.44	0.44	0.88	1.32	1.76	2.20	2.64	0.70
Soft or Sandy Clay	0.57	0.57	0.14	1.71	2.28	2.85	3.42	0.85
Stiff Clay	0.53	0.53	1.06	1.59	2.12	2.65	3.18	0.80
Gravel	0.57	0.57	1.14	1.71	2.28	2.85	3.42	1.14
Loam	0.67	0.67	1.34	2.01	2.68	3.35	4.02	0.92
Marl	0.57	0.57	1.14	1.71	2.28	2.85	3.42	0.98
Sand	0.67	0.67	1.34	2.01	2.68	3.35	4.02	1.09
Soil	0.63	0.63	1.26	1.89	2.52	3.15	3.78	0.90

CONCRETE WORK

Quantities of Materials per 1m³ **Hardened Concrete**

NOTE: The following quantities allow for the increase in bulk of moist sand and moist all-in aggregate.

Nominal mix by volume	Cement tonnes	Moist sand m³	Gravel m³
1:3:6	0.215	0.55	0.88
1:2:4	0.304	0.53	0.84
1:1.5:3	0.389	0.50	0.80
		Moist all-in aggregate (m³)	
1:6	0.304	0.45	
1:9	0.214	0.52	
1:12	0.167	0.55	

Nominal mix by weight	Cement tonnes	Moist sand tonnes	Gravel tonnes
1:3:6	0.216	0.81	1.31
1:2:4	0.312	0.81	1.24
1:1.5:3	0.391	0.74	1.17
	Cement tonnes	Moist all-in aggregate (tonnes)	
1:6	0.312	2.15	
1:9	0.217	2.26	
1:12	0.172	2.38	

Quantities of Material per 1m³ of Concrete

Grade	Quantity/m3 including waste (in tonnes)	Material
20/20	0.320	Cement
	0.630	Sand
	1.170	Gravel
25/20	0.360	Cement
	0.612	Sand
	1.137	Gravel
30/20	0.400	Cement
	0.595	Sand
	1.105	Gravel
7/40 All-in	0.180	Cement
	1.950	Aggregate
20/20 All-in	0.320	Cement
	0.180	Aggregate
25/20 All-in	0.360	Cement
	1.750	Aggregate

Quantities of Materials for Prescribed Mix Concrete

Prescribed Mix	Workability medium/high	Aggregate	Cement	(kg per m³) Aggregate	Sand
C7.5P	Medium	20mm	235	1272	848
C7.5P	High	20mm	268	1252	835
C7.5P	Medium	40mm	240	1400	754
C7.5P	High	40mm	231	1381	743
C10P	Medium	20mm	271	1251	834
C10P	High	20mm	302	1250	833
C10P	Medium	40mm	235	1378	742
C10P	High	40mm	262	1361	733
C15P	Medium	20mm	302	1232	821
C15P	High	20mm	346	1205	804
C15P	Medium	40mm	268	1370	741
C15P	High	40mm	298	1338	718
C20P	Medium	10mm	386	1081	888
C20P	High	10mm	421	967	967
C20P	Medium	14mm	357	1070	714
C20P	High	14mm	413	1066	876
C20P	Medium	20mm	336	1312	706
C20P	High	20mm	374	1189	793

Quantities of Material for Prescribed Mix Concrete

Prescribed Mix	Workability medium/high	Aggregate	Cement	(kg per m³) Aggregate	Sand
C20P	Medium	40mm	310	1432	614
C20P	High	40mm	336	1312	706
C25P	Medium	10mm	428	1177	749
C25P	High	10mm	501	927	927
C25P	Medium	14mm	399	1134	822
C25P	High	14mm	462	1039	855
C25P	Medium	20mm	386	1278	691
C25P	High	20mm	421	1161	775
C25P	Medium	40mm	357	1399	600
C25P	High	40mm	386	1278	691
C30P	Medium	10mm	491	1026	840
C30P	High	10mm	512	1014	829
C30P	Medium	14mm	471	1159	725
C30P	High	14mm	512	1014	829
C30P	Medium	20mm	421	1257	678
C30P	High	20mm	471	1130	754
C30P	Medium	40mm	380	1356	619
C30P	High	40mm	421	1257	678

Cubic Metres of Concrete Hauled per Load Using Various Sized Lorries and Dumpers, etc.

NOTE: The data shown are based on concrete weighing 2.242 tonnes per cubic metre

Type of plant used to transport concrete	Cubic metres of concrete hauled per load or per wagon
1 tonne lorry	0.46
2 tonne lorry	0.92
3 tonne lorry	1.38
4 tonne lorry	1.83
5 tonne lorry	2.29
6 tonne lorry	2.75
1.5 m³ dumper	1.22
2m³ dumper	1.83
0.5 m³ Decauville wagon	0.46
0.7 m³ Decauville wagon	0.61

Aggregates, Cement, etc. - Approximate Average Properties

Material	Loose weight Per m3 kg	Per cu. ft. lb	Per cu. yd. cwt	tons	Voids %	Absorption by Weight %
Sand	1862					
Bone dry		116	28	1.40	30	0.4
Average	1595	100	24	1.20	39	-
Moist	1277	80	19	0.96	52	-
Gravel						
10mm to sand	1408	88	21	1.06	47	1.0
20mm to sand	1461	91	22	1.10	45	1.0
25mm to sand	1488	93	22	1.12	44	1.0
38mm to sand	1530	95	23	1.15	42	1.0
50mm to sand	1556	97	23	1.17	41	1.0
75mm to sand	1595	100	24	1.20	39	1.0
150mm to sand	1635	102	25	1.23	38	1.0
38 to 20mm	1635	102	25	1.23	38	1.0
75 to 38mm	1556	97	23	1.17	41	1.0
150 to 75mm	1461	91	22	1.10	45	1.0
225 to 150mm	1382	86	21	1.04	48	1.0
Broken stone						
10mm to sand	1250	78	19	0.94	53	2.0
20mm to sand	1303	81	20	0.98	51	2.0
25mm to sand	1330	83	20	1.00	50	2.0
38mm to sand	1370	85	20	1.03	49	2.0
50mm to sand	1382	86	21	1.04	48	2.0
Granite						
6mm to sand	1330	83	20	1.00	51	0.5
10mm to sand	1370	85	20	1.08	50	0.5

20mm to sand	1408	88	21	1.06	48	0.5
25mm to sand	1435	90	22	1.08	47	0.5
38mm to sand	1435	90	22	1.08	47	0.5
50mm to sand	1461	91	22	1.10	46	0.5
'All-in' Ballast	1795	112	27	1.35	32	0.6
Brick Hardcore	1197	75	18	0.90	35	25
Brick Aggregate						
Coarse	878	55	13	0.66	52	25
Fine	1118	70	17	0.84	39	25
Breeze Clinker						
Coarse	18	45	11	0.54	50	-
Fine	878	55	13	0.66	45	-
Foamed Slag						
Coarse	557	35	8	0.42	-	30
Fine	718	45	11	0.54	-	30
Pumice	557	35	8	0.42	-	-
Vermiculite	66	4	1	0.05	80	-
Portland Cement	1436	90	22	1.08	54	-
Chalk Lime						
Lump	718	54	11	0.45	73	-
Ground	798	50	12	0.45	70	-
Slaked	557	35	8	0.42	60	-
Gypsum Plaster						
Heavy	1000	62	15	0.75	57	-
Light	798	50	12	0.60	65	-
Plaster of Paris	918	56	14	0.69	66	-

Curing Times (at 10C)

Type of Cement	Wet curing time after completion of placing concrete (not less than)
Ordinary Portland	4 days
Sulphate-resisting Portland	4 days
Portland blast-furnace	4 days
Super-Sulphated	4 days
Rapid hardening Portland	3 days

Formwork Stripping Times

Minimum time in days before stripping	Normal weather (about 60F/16C)		*Cold weather (about 35F/2C)	
	Portland cement	Rapid hardening	Portland cement	Rapid hardening
Sides of beams, columns or walls	1	1	6	5
Soffits (props left under)				
of slabs	3	2	10	7
of beams	7	4	12	10
Removal of props				
from slabs	7	4	14	14
from beams	16	8	28	21

*These curing periods may be reduced when the materials are heated and the concrete is insulated.

Weight of Steel Bar Reinforcement

Diameter	kg/m	m/tonne	Cross-sectional area mm²
6mm	0.222	4505	28.3
8mm	0.395	2532	50.3
10mm	0.616	1624	78.5
12mm	0.888	1126	113.1
16mm	1.579	634	201.1
20mm	2.466	406	314.2
25mm	3.854	260	490.9
32mm	6.313	158	804.2
40mm	9.864	101	1256.6
50mm	15.413	65	1963.3

Weights of steel bar reinforcement in various percentages of 1m³ of concrete

Percentage of reinforcement

Weights	0.50	0.75	1.00	1.25	1.50	1.75	2.00	2.50	3.00	3.50	4.00	4.50	5.00
kg/m³	39.00	59.00	79.00	98.00	118.00	137.00	157.00	196.00	236.00	275.00	314.00	353.00	393.00

Weights of Stainless Steel Bar Reinforcement

Diameter	kg/m	m/tonne	Cross-sectional area mm²
10mm	0.667	1499	78.5
12mm	0.938	1066	113.1
16mm	1.628	614	201.1
20mm	2.530	395	314.2
25mm	4.000	250	490.9
32mm	6.470	155	804.2

Weights of steel fabric reinforcement

Fabric reinforcement to BS 4483

Square Mesh Fabric BS reference	Mesh Size Main mm	Cross mm	Wire Size Main mm	Cross mm	Weight/m² kg
A393	200	200	10	10	6.16
A252	200	200	8	8	3.95
AI93	200	200	7	7	3.02
AI42	200	200	6	6	2.22
A98	200	200	5	5	1.54

Structural Mesh Fabric BS reference	Mesh Size Main mm	Cross mm	Wire Size Main mm	Cross mm	Weight/m² kg
B1131	100	200	12	8	10.90
B785	100	200	10	8	8.14
B503	100	200	8	8	5.93
B385	100	200	7	7	4.53
B283	100	200	6	7	3.73
BI96	100	200	5	7	3.05

Long Mesh Fabric BS reference	Mesh Size Main mm	Cross mm	Wire Size Main mm	Cross mm	Weight/m² kg
C785	100	400	10	6	6.72
C503	100	400	8	5	4.34
C385	100	400	7	5	3.41
C283	100	400	6	5	2.61

Wrapping Fabric BS reference	Mesh Size Main mm	Cross mm	Wire Size Main mm	Cross mm	Weight/m² kg
D98	200	200	5	5	1.54
D49	100	100	2.5	2.5	0.77

Carriageway Fabric BS reference	Mesh Size		Wire Size		Weight/m² kg
	Main mm	Cross mm	Main mm	Cross mm	
C636	80-130	400	8-10	6	5.55

Floor Finishes

Description	Thickness (mm)	kg/m² (including waste)	
		Cement	Grano-Chippings
Granolithic paving (1:2:5) to floors	12	6.6	24.8
	25	13.8	51.7
	32	17.7	66.2
	38	21.0	78.6
	50	27.7	103.4

Description	Thickness (mm)	kg/m² (including waste)	
		Cement	Sand
Cement and sand (1:3) to floors	10	4.5	20.1
	25	11.3	50.3
	32	14.4	64.4
	38	17.1	76.5
	50	22.6	100.7

PIPEWORK

Beds, Haunches and Surrounds

100mm Concrete to Pipes – m³ per linear metre

Pipe diameter (mm)	Bed only (m³)	Bed and haunched (m³)	Bed and surround (m³)
100	0.036	0.061	0.108
150	0.041	0.074	0.133
225	0.049	0.095	0.172
300	0.057	0.118	0.213
450	0.075	0.175	0.321
600	0.093	0.237	0.442
750	0.110	0.301	0.563
900	0.127	0.371	0.699
1200	0.163	0.538	1.026
1500	0.197	0.715	1.375

150mm Concrete to Pipes – m³ per linear metre

Pipe diameter (mm)	Bed only (m³)	Bed and haunched (m³)	Bed and surround (m³)
100	0.070	0.105	0.191
150	0.077	0.125	0.226
225	0.090	0.156	0.282
300	0.100	0.188	0.340
450	0.128	0.264	0.483
600	0.155	0.346	0.635
750	0.180	0.427	0.791
900	0.206	0.528	0.960
1200	0.259	0.729	1.361
1500	0.311	0.953	1.798

225mm Concrete to Pipes – m³ per linear metre

Pipe diameter (mm)	Bed only (m³)	Bed and haunched (m³)	Bed and surround (m³)
100	0.139	0.192	0.353
150	0.152	0.225	0.402
225	0.169	0.271	0.483
300	0.187	0.318	0.562
450	0.227	0.432	0.760
600	0.268	0.547	0.970
750	0.305	0.661	1.170
900	0.343	0.782	1.395
1200	0.423	1.052	1.900
1500	0.500	1.338	2.435

Pipe Trench Excavation - Battered Sides

Based on OG Pipes (m³ per linear metre)

Invert depth metres	Pipe Diameter (mm) Not exceeding							
	225	300	450	600	750	900	1200	1500
0.60	0.57	0.62	0.73	0.87	0.99	1.11	1.37	1.64
0.80	0.84	0.90	1.04	1.22	1.39	1.53	1.86	2.20
1.00	1.14	1.22	1.39	1.62	1.82	1.99	2.39	2.81
1.25	1.58	1.68	1.90	2.17	2.42	2.63	3.08	3.61
1.50	2.09	2.20	2.47	2.78	3.08	3.33	3.90	4.48
1.75	2.65	2.78	3.09	3.45	3.79	4.09	4.73	5.40
2.00	3.27	3.42	3.78	4.17	4.57	4.90	5.63	6.39
2.50	4.83	4.90	5.37	5.84	6.32	6.76	7.64	8.57
3.00	6.39	6.62	7.15	7.74	8.31	8.80	9.88	10.98
3.50	8.34	8.56	9.22	9.91	10.57	11.13	12.38	13.65
4.00	10.52	10.82	11.53	12.30	13.06	13.70	15.12	16.56
5.00	15.64	16.02	16.90	17.87	18.80	19.60	21.36	23.15
6.00	21.77	22.22	23.28	24.43	25.60	26.56	28.60	30.73
7.00	28.83	29.42	30.65	31.98	33.34	34.46	36.85	39.32

Pipe Trench Excavation – Vertical Sides

OG Pipes with Light or Medium Timber (m³ per linear metre)

Invert depth metres	Pipe Diameter (mm) Not exceeding							
	225	300	450	600	750	900	1200	1500
0.60	0.47	0.52	0.62	0.75	0.87	1.00	1.24	1.50
0.80	0.62	0.68	0.82	0.98	1.14	1.30	1.60	1.93
1.00	0.77	0.85	1.01	1.21	1.40	1.60	1.96	2.36
1.25	0.96	1.05	1.26	1.50	1.74	1.98	2.41	2.90
1.50	1.15	1.26	1.50	1.79	2.07	2.35	2.86	3.44
1.75	1.33	1.47	1.74	2.07	2.40	2.73	3.31	3.97
2.00	1.52	1.67	1.99	2.36	2.73	3.10	3.76	4.51
2.50	1.90	2.09	2.48	2.94	3.39	4.85	4.56	5.59
3.00	2.27	2.50	2.96	3.51	4.05	4.60	5.56	6.66
3.50	2.85	3.11	3.65	4.29	4.92	5.55	6.66	7.94
4.00	3.42	3.72	4.33	5.06	5.78	6.50	7.76	9.21
5.00	4.57	4.95	5.70	6.61	7.50	8.40	9.95	11.76
6.00	5.72	6.17	7.08	8.16	9.22	10.30	12.15	14.31
7.00	7.17	7.70	8.75	10.01	11.25	12.50	14.65	17.16

OG Pipes with Medium-heavy/Heavy Timber (m³ per linear metre)

Invert depth metres	Pipe Diameter (mm) Not exceeding							
	225	300	450	600	750	900	1200	1500
0.60	0.52	0.56	0.67	0.79	0.92	1.04	1.28	1.55
0.80	0.68	0.74	0.88	1.04	1.20	1.36	1.66	1.99
1.00	0.84	0.92	1.09	1.28	1.48	1.69	2.03	2.44
1.25	1.05	1.15	1.35	1.59	1.83	2.07	2.50	2.99
1.50	1.26	1.37	1.62	1.90	2.18	2.46	2.92	3.55
1.75	1.46	1.60	1.88	2.20	2.53	2.85	3.45	4.11
2.00	1.67	1.82	2.14	2.51	2.88	3.25	3.91	4.66
2.50	2.08	2.27	2.67	3.12	3.58	4.04	4.85	5.78
3.00	2.49	2.72	3.19	3.73	4.28	4.82	5.78	6.89
3.50	3.10	3.37	3.88	4.54	5.17	5.81	6.92	8.20
4.00	3.71	4.01	4.52	5.35	6.07	6.79	8.05	9.51
5.00	4.94	5.32	6.00	6.98	7.87	8.77	10.32	12.13
6.00	6.16	6.62	7.52	8.60	9.67	10.74	12.60	14.76
7.00	7.69	8.22	9.27	10.53	11.77	13.02	15.17	17.68

Lay and Joint

Concrete Pipes

Pipe Diameter (mm)	Lay only per linear metre (hrs)	Per Joint Ogee Pipes (hrs)
150	0.27	0.20
225	0.33	0.24
300	0.49	0.32
450	0.88	0.64
600	1.10	0.87
750	1.65	1.52
900	2.75	1.96
1200	5.09	3.85
1500	7.53	5.63

Where pipes are laid in battered trenches take 0.66 of the above allowances for lay and joint. For porous concrete pipes take the lay only labour allowances.

Spun Iron Pipes

Pipe Diameter (mm)	Lay only per linear metre (hrs)	Tyton Joints per Joint (hrs)	Mechanical Joints per Joint (hrs)
102	0.22	0.12	0.36
150	0.36	0.17	0.50
225	0.55	0.28	0.83
300	0.90	0.37	1.10
450	1.49	0.62	1.87
600	2.09	0.83	2.50
750	3.08	1.05	3.15
900	3.85	1.25	3.75
1200	5.67	1.67	5.00

The above figures are for short lengths of sewers. For long runs of sewers or rising mains (more akin to gas and water works) take 0.66 of the allowances. For awkward lengths such as in sewage works take 1.5 times the allowances.

Trench Reinstatement

OG Pipes (super metres per linear metre)

Pipe diameter (mm)	Depth 0-3m	Depth 3-6m	Depth 6-8m
225	0.750	1.150	1.450
300	0.825	1.225	1.525
450	0.975	1.375	1.675
600	1.150	1.550	1.850
750	1.325	1.725	2.025
900	1.500	1.900	2.200
1200	1.800	2.200	2.500
1500	2.150	2.550	2.850

Above is for trenches with light or medium timbering

a) For medium-heavy or heavy timber add 0.08 m²
b) In road with heavy traffic over 3 m deep add 0.37 m²
c) For final coat or permanent reinstatement add 0.15 m²

Weights of Pipes

Weight of concrete pipes

NOTE: The data show average weights, as there is slight variation in weight between the pipes supplied by various manufacturers. The data shown are for unreinforced pipes. If pipes are reinforced add 2.5 percent to weights shown, and reduce the metres of pipes per tonne accordingly.

Ogee joints

Internal Diameter of Pipe (mm)	Metres per tonne	Tonnes per metre
300	11.89	0.083
375	8.74	0.115
450	6.93	0.144
525	5.07	0.197
600	4.15	0.240
675	3.52	0.283
750	2.82	0.355
825	2.48	0.402
900	2.12	0.473
975	1.89	0.529
1050	1.64	0.610
1125	1.36	0.732
1200	1.26	0.796
1350	0.99	1.100
1500	0.82	1.222

Nominal Internal Diameter (mm)	Nominal Wall thickness (mm)
150	25
225	29
300	32
375	35
450	38
600	48

Nominal Internal Diameter (mm)	Nominal Wall thickness (mm)
750	54
900	60
1200	76
1500	89
1800	102
2100	127

Wall thickness will vary, depending on type of pipe required.

The particulars shown above represent a selection of available diameters and are applicable to strength class 1 pipes with flexible rubber ring joints.

Tubes with Ogee joints are also available.

Weight of clayware pipes

Internal Diameter of Pipe (mm)	Metres per tonne	Tonnes per metre
100	51.0	0.012
150	49.0	0.020
225	27.7	0.036
300	16.5	0.060
375	11.2	0.091
450	7.3	0.137
525	5.4	0.183
600	4.1	0.244
675	3.2	0.310
750	2.7	0.366
900	1.8	0.515

Clayware gully pots

Size of gully pots in mm

Diameter	Depth	Number of gully pots per tonne
300	750	17.0
300	900	16.0
375	750	12.0
375	900	10.0
375	1050	9.5
450	750	9.0
450	900	8.0
450	1050	6.5
450	1200	6.0

Weight of Stanton Socket and Spigot Spun Iron Pipes

Weights of Standard Lengths

PIPES 5.4m LONG

Test pressure nominal bore mm	Class B 120m head kg	Class C 180m head kg	Class D 240m head kg
100	116	119	132
150	186	206	235
225	332	369	420
300	494	572	651
375	630	794	904
450	832	1043	1190
525	1072	1275	1476
600	1277	1598	1795
675	1629	2005	2296

Weights of Standard Lengths

PIPES 3.6m LONG

Test pressure nominal bore mm	Class B 120m head kg	Class C 180m head kg	Class D 240m head kg
75	60	60	61
100	80	83	91
150	128	140	161
225	214	254	288
250	249	293	336
300	318	394	446
375	435	547	620
450	574	718	816

Weight of Steel Pipes

Pipe Inside diameter (mm)	Outside diameter (mm)	Class A Weight kg/m	Gauge and thickness (mm)	Class B Weight kg/m	Gauge and thickness (mm)	Class C Weight kg/m	Gauge and thickness (mm)	Class D Weight kg/m	Gauge and thickness (mm)
50	59	3.81	12g	4.20	11g	4.60	109	5.14	9g
75	87	6.29	11g	7.84	9g	9.37	7g	10.17	6g
100	112	8.99	10g	10.05	9g	12.19	7g	13.26	6g
150	162	14.67	9g	17.83	7g	19.41	6g	21.36	5g
225	237	28.63	6g	31.05	5g	37.05	6.3	41.50	7.1
300	312	37.86	6g	41.75	5g	49.05	6.3	54.06	7.1
450	462	56.32	6g	64.06	5.5	73.11	6.3	91.06	7.9
600	612	97.14	6.4	121.11	7.9	133.05	8.7	144.95	9.5
750	768	153.00	7.9	168.00	8.7	198.00	9.5	195.00	10.3
900	918	183.00	7.9	219.00	9.5	237.00	10.3	255.00	11.1
1200	1225	292.00	9.5	340.00	11.1	389.00	12.7	436.00	14.3
1500	1525	485.00	12.7	544.00	14.3	605.00	15.9	665.00	17.5

Circular Sewers

Discharge through 225mm diameter pipe in litres per minute

Gradient	One-eighth	Depth of flow in proportion to diameter of pipe		
		One-quarter	One-half	Seven-eighths
1 in 20	264	1012	3432	6978
1 in 30	218	887	2796	5660
1 in 40	182	718	2409	4932
1 in 50	168	650	2160	4432
1 in 80	136	509	1714	3491
1 in 100	118	455	1500	3100

Discharge through 300mm diameter pipe in litres per minute

Gradient	One-eighth	Depth of flow in proportion to diameter of pipe		
		One-quarter	One-half	Seven-eighths
1 in 30	446	1727	5796	11730
1 in 40	391	1500	5000	10164
1 in 50	345	1327	4455	9092
1 in 80	273	1068	3296	7182
1 in 100	241	963	3136	6418
1 in 200	173	659	2227	4546

Discharge through 375mm diameter pipe in litres per minute

Gradient	One-eighth	Depth of flow in proportion to diameter of pipe		
		One-quarter	One-half	Seven-eighths
1 in 40	682	2682	8637	17729
1 in 50	614	2387	7728	15819
1 in 80	478	1887	6090	12500
1 in 100	432	1690	5478	11183
1 in 200	305	1205	4036	7887
1 in 440	205	796	2591	5295
1 in 660	164	660	2114	4523
1 in 880	145	568	1818	3751

Discharge through 450mm diameter pipe in litres per minute

Gradient	One-eighth	Depth of flow in proportion to diameter of pipe		
		One-quarter	One-half	Seven-eighths
1 in 50	955	3773	12201	25003
1 in 66	827	3114	10820	21707
1 in 80	745	2841	9637	19710
1 in 100	668	2614	8651	17761
1 in 200	478	1882	6100	12480
1 in 440	318	1236	4123	8388
1 in 880	227	873	2909	5919

Manholes

Concrete to manhole chamber rings

150mm concrete to chamber rings

Internal diameter (mm)	Concrete m³/linear metre	Shutter m²/linear metre	m²/ m³
675	0.474	3.59	7.57
750	0.511	3.84	7.51
900	0.584	4.34	7.43
975	0.619	4.55	7.35
1050	0.656	4.80	7.32
1200	0.730	5.26	7.21
1350	0.802	5.77	7.19
1500	0.875	6.23	7.12
1650	0.947	6.72	7.10
1800	1.020	7.19	7.05
2400	1.486	10.38	7.02

Weight of precast concrete manhole parts

Diameter of manhole (mm)	Precast concrete bases 600mm high Weight per base (kg)	Tapers Height of taper (mm)	Weight per taper (kg)	Cover slabs 150mm thick Diameter of cover slab	Weight per cover slab (kg)
900	600	600	265	1050	225
1050	900	600	350	1200	300
1200	1300	600	500	1350	425
1350	1750	600	615	1500	625
1500	2000	900	775	1800	915
1800	2500	900	975	2000	1100

Weight of precast concrete manhole rings

Diameter of ring (mm)	Weight per metre (kg)
675	281
900	467
1050	624
1200	760
1350	869
1500	1036
1800	1449

METALWORK

Weight of metals in kg/m²

Type of metal	Thickness in mm 1.6	3.2	6.4	9.5	12.7	15.9	19.1	22.2	25.0
Brass	13.9	27.7	55.4	83.2	108.9	139.0	168.0	190.4	222.0
Cast iron	11.5	22.1	45.7	68.7	91.8	114.4	137.0	161.0	191.2
Copper	14.1	28.2	56.4	84.6	112.8	141.0	169.2	197.4	225.6
Lead, cast	18.1	36.2	72.4	108.6	144.8	181.0	217.2	253.4	289.6
Steel	12.5	25.0	50.0	75.0	100.0	125.0	150.0	175.0	200.0
Wrought iron	12.2	24.4	48.8	73.2	97.4	121.9	146.4	170.8	193.8

Type of material	Weight in tonnes/m³
Cast brass	8.474
Cast copper	8.769
Cast iron	7.348
Cast lead	11.566
Gun-metal	8.618
Milled lead	11.643
Sheet copper	8.981
Steel	7.894
Tin	7.304
Wrought iron	7.668
Zinc	7.048

TIMBER

Lengths of timber per m³

Dimensions (mm)	m/m³	Dimensions (mm)	m/m³
25 x 25	1600	50 x 50	400
50	800	75	267
75	533	100	200
100	400	125	160
125	320	150	133
150	267	175	114
		200	100
		225	89
		250	80
		300	67
75 x 75	178	100 x 100	100
100	133	150	67
125	107	200	50
150	89	250	40
175	76	300	33
200	67		
225	59		
250	53		
300	44		
150 x 200	33	200 x200	25
250	27	250 x 250	16
300	22	300 x 300	11

m³ per tonne of timber

Wood	m³ per tonne Air dry (15%)	Shipping dry (20%)	Green
Ash, European	1.47	1.41	1.24
Beech	1.44	1.38	1.06
Redwood	2.01	1.96	1.7
Elm, Common	1.84	1.78	0.98
Mahogany, Honduras	1.9	1.84	1.44
Oak, European	1.44	1.38	0.95

Drying times

Species	Time (weeks)	Species	Time (weeks)
Oak	10-12	Ash	4-5
Elm	4-5	Pitch pine	3-4
Beech	5-6	Scots pine	1-2

Drying times are for 50mm timber from freshly felled to 12% moisture content in a moderately efficient fan kiln.

m² per m³ and m³ per 100 m² for timber of various thicknesses

Thickness of timber (mm)	m² per m³	m³ per 100 m²	Thickness of timber (mm)	m² per m³	m³ per 100 m²
12	83.3	120	50	20.0	500
19	52.6	190	57	17.6	570
25	40.0	250	63	15.9	630
32	31.2	320	75	13.5	750
40	25.0	400	89	11.3	890
45	22.2	450	100	10.0	1000

Number of nails per kilogramme

Oval Wire Nails

Length	18mm	25mm	31mm	38mm	44mm	50mm	63mm	75mm	88mm	100mm
No. nails per kg	4500	2425	1365	880	620	375	240	110	100	70

Cut Clasp Nails

Length	25mm	38mm	50mm	63mm	75mm	88mm	100mm	112mm	125mm	150mm
No. nails per kg	1455	550	310	175	110	90	55	45	33	27

Round Wire Nails

Length	25mm	31mm	38mm	50mm	63mm	75mm	88mm	100mm	112mm	125mm	175mm
Gauge	14	13	12	10	10	8	8	8	8	6	5
No. nails per kg	1322	1145	620	265	200	130	110	100	90	65	34

ROADS AND PAVINGS

Weight of road foundation materials

NOTE: The data shown are based on 45 percent voids

Nature of the material	Weight in the solid kg per m³	Weight in the loose as delivered on the site	
		Tonnes per m³	m³ per tonne
Ashes	-	0.96	1.04
Clinker	-	0.8	1.24
Gravel	-	1.52	0.66
Hardcore, brick	2123	1.16	0.86
Hardcore, chalk	2286	1.24	0.81
Hardcore, concrete	2286	1.24	0.81
Pitching, granite	2711	1.48	0.68
Pitching, limestone	2367	1.29	0.78
Pitching, sandstone	2449	1.33	0.75
Pitching, whinstone	2804	1.53	0.65

Weight of road surfacing materials

Nature of the material	Weight in the solid kg per m³	Weight in the loose as delivered on the site	
		Tonnes per m³	m³ per tonne
Asphalt, bottom course	-	2.12	0.51
Asphalt, top course	-	2.24	0.45
Asphalt, mastic	2711	2.33	0.43
Tarred granite	-	1.53	0.62
Tarred gravel	2449	1.53	0.65
Tarred limestone	2531	1.57	0.64
Tarred slag	2804	1.4	0.70
Tarred whinstone	-	1.56	0.61

Stone pitching (m² covered by one tonne of material)

Weight of stone in the solid tonnes per m³	Consolidated thickness laid in mm and m² covered by 1 tonne of material				
	150mm	175mm	200mm	225mm	300mm
2.082	4.3	3.69	3.22	2.86	2.15
2.162	4.18	3.59	3.14	2.78	2.08
2.242	4.06	3.49	3.06	2.71	2.03
2.322	3.95	3.39	2.97	2.63	1.98
2.402	2.84	3.29	2.89	2.56	1.92
2.482	3.73	3.2	2.81	2.48	1.87
2.562	3.61	3.1	2.72	2.41	1.81
2.643	3.5	3.01	2.65	2.33	1.76
2.724	3.39	2.91	2.56	2.26	1.7
2.803	3.28	2.82	2.47	2.18	1.65
2.883	3.16	2.72	2.39	2.11	1.59

Surface dressing roads (m² covered by various sizes of chippings, gravel or sand per m³ and per tonne)

| Size in mm | per m³ | Sand | Square metres covered per tonne | | |
			Granite chips	Gravel	Limestone chips
Sand	242	168	-	-	-
3	198	-	148	152	165
6	176	-	130	133	144
9	154	-	111	114	123
13	121	-	85	87	95
19	99	-	68	71	78

Weight of kerb per metre

| Size of kerb in mm | Type of kerb and weight in tonnes per metre | |
	Concrete	Granite
125 x 50	0.011	0.013
175 x 50	0.022	0.025
250 x 100	0.066	0.077
250 x 125	0.077	0.088
250 x 150	0.088	0.099
250 x 200	0.11	0.132
300 x 150	0.099	0.121
300 x200	0.132	0.165

Weight of artificial stone paving

Thickness of stone	Tonnes per m²
50mm	0.12
63mm	0.15

Weight of York and Purbeck stone paving

| Thickness in mm | m² per tonne | |
	York	Purbeck
50	7.8	7.5
63	6.4	6
75	5.3	5
100	3.9	3.7

Number of paving flags per m²

Paving Flags	Number/m² (incl. waste)
600 x 450mm	3.527
600 x 600mm	2.646
600 x 750mm	2.116
600 x 900mm	1.764

Asphalt Work

Weight of asphalt per m²

Description	Thickness (mm)	Tonnes of asphalt/m²
Single coat cold asphalt	13	0.029
	25	0.058
	50	0.115
	75	0.173
Two coats cold asphalt	88	0.194
	100	0.222
Single coat hot asphalt	25	0.062
	50	0.125
	75	0.187
Two coats hot asphalt	88	0.206
	100	0.238
Mastic asphalt	25	0.060
	50	0.120

BRICKWORK, BLOCKWORK AND MASONRY

Bricks and mortar required per m²

Bricks 215 x 102.5 x 65mm with 10mm joints:

Wall thickness	Number of bricks (no waste allowance)	Mortar (m³) no frogs	one frog	two frogs
102.5mm (half brick)	59.25	0.017	0.024	0.031
215mm (one brick)	118.50	0.045	0.059	0.073
327.5mm (one and a half brick)	177.75	0.072	0.093	0.114
440mm (two brick)	237.00	0.101	0.128	0.155

Extra over common brickwork for facing and pointing one side in:

		Length of pointing to one face per m²		
	Number of bricks (no waste allowance)	Horizontal joints	Vertical joints	Combined
English bond	89	13.3m	5.8m	19.1m
Flemish bond	79	13.3m	5.1m	18.4m
English garden wall bond	74	13.3m	4.8m	18.1m
Flemish garden wall bond	68	13.3m	4.4m	17.7m

Blocks and mortar required per m²

Blocks 440 x 215mm on face with 10mm joints require 9.88 blocks per m² (exclusive of waste)

Wall thickness (mm)	50	60	70	75	90	100	115	125
Mortar (m³/m²)	0.003	0.004	0.005	0.005	0.006	0.007	0.008	0.008

Wall thickness (mm)	140	150	190	215	220	250	275	305
Mortar (m³/m²)	0.009	0.010	0.013	0.014	0.015	0.017	0.019	0.020

Approximate weight of brickwork

Type of brick	Approximate weight of brickwork in kg/m³
Facing bricks	2025
Flettons	1602
Staffordshire blue wirecuts	2123
Staffordshire blue pressed facing bricks	2188
Stocks	1945
Wirecuts	2091
Heavy engineering bricks	2286

Approximate weight of blockwork

Thickness of blocks in mm	Weight per m² in kg	m² of blocks per tonne
50	43	23
63	54	18
75	65	16
100	87	12

Quantity of brickwork (laid in mortar) and number of bricks per metre of sewer or culvert

Diameter in metres	Half a brick thick		One brick thick	
	No. of bricks	m³ of brickwork	No. of bricks	m³ of brickwork
0.3	72	0.15	181	0.39
0.6	127	0.26	303	0.61
0.9	177	0.37	415	0.83
1.2	240	0.48	524	1.07
1.5	279	0.59	633	1.26
1.8	353	0.7	746	1.46
2.1	406	0.81	858	1.70
2.4	462	0.92	963	1.93
2.7	514	1.03	1075	2.15
3.0	574	1.14	1188	2.37

Basic compressive strengths of bricks

Type of brick	N/mm² compressive strength

Clay bricks
Engineering types
BS 3921: 1985 engineering bricks:

Class A	100 upwards
Class B	70 upwards

Sand-lime bricks
Range
BS 187: 1978 classes:
Special purposes

Class 3	20.5
Class 4	27.5
Class 5	34.5
Class 6	41.5
Class 7	48.5

Concrete bricks
Range with natural aggregates
BS 6073, Pt 1: 1981   7-40

Random rubble walling: weight per m²

Thickness of wall (mm)	Average stone thickness (mm)	Tonnes of stone/m²	m³ of mortar/m²
300	75	0.615	0.143
300	150	0.64575	0.121
300	225	0.661	0.099
300	300	0.6765	0.077
600	75	1.23	0.297
600	150	1.2915	0.231
600	225	1.322	0.187
600	300	1.353	0.165
900	75	1.845	0.44
900	150	1.93725	0.352
900	225	1.98338	0.297
900	300	2.0295	0.231

PAINTING

Average coverages of paint per coat

Coating (m² per litre)	Type of Surface						
	Smooth concrete/ Cement	Fair-faced Brickwork	Blockwork	Roughcast/ pebble-dash	Structural Steelwork	Metal Sheeting	Joinery
Woodprimer (oil based)	-	-	-	-	-	-	8-11
Metal primer: conventional	-	-	-	-	7-10	10-13	-
Alkali resistant primer	7-11	6-8	4-6	2-4	-	-	-
External wall primer sealer	6-8	6-7	4-6	2-4	-	-	-
Undercoat	7-9	6-8	6-8	3-4	10-12	10-12	10-12
Gloss finish	8-10	7-9	6-8	-	10-12	10-12	10-12
Emulsion Paint:							
standard	11-14	8-12	6-10	2-4	-	-	10-12
contract	10-12	7-10	5-9	2-4	-	-	10-12
Masonry paint	5-7	4-6	3-5	2-4	-	-	-
Cement based paint	6-7	3-6	3-6	2-3	-	-	-

WATERPROOFING

Quantities of material per m²

Material	Quantity per m²
1 layer 1200 gauge polythene sheet	0.01237 Number (rolls 25 x 4m)
2 coats RIW solution	0.2120 litres
2 coats synthaprufe horizontally	1.6545 litres
3 coats synthaprufe vertically	1.4320 litres

			kg per m² (including waste)	
Wall Renders	Thickness (mm)	Sand	Cement	Lime
Cement and sand	12	24.2	5.4	-
(1:3) to brickwork,	18	36.0	8.1	-
blockwork or concrete				
Cement, lime and sand	12	25.5	2.8	1.5
1: 1:6) external rendering to walls	18	38.2	4.2	2.2

INDEX

How to activate your 12-month period of access to CapIT: Online Carbon & Cost Estimator

As a purchaser of the CESMM4 Carbon & Price Book you are entitled to a free 12-month period of access to the CapIT online carbon and price estimator (see introduction for more details).

To activate your account, please follow the three simple steps below. The account can be activated any time between May 2013 and May 2014.

1. Go to www.capit-online.com/cesmm

2. Enter the account activation code at the bottom of this page.

3. Click on "activate".

For technical support please contact eru@mottmac.com
For information on CapIT: Online Carbon & Cost Estimator visit www.capit-online.com